Lecture Notes in Networks and Systems

Volume 476

The series "Lecture Notes in Networks and Systems" publishes the latest developments in Networks and Systems—quickly, informally and with high quality. Original research reported in proceedings and post-proceedings represents the core of LNNS.

Volumes published in LNNS embrace all aspects and subfields of, as well as new challenges in, Networks and Systems.

The series contains proceedings and edited volumes in systems and networks, spanning the areas of Cyber-Physical Systems, Autonomous Systems, Sensor Networks, Control Systems, Energy Systems, Automotive Systems, Biological Systems, Vehicular Networking and Connected Vehicles, Aerospace Systems, Automation, Manufacturing, Smart Grids, Nonlinear Systems, Power Systems, Robotics, Social Systems, Economic Systems and other. Of particular value to both the contributors and the readership are the short publication timeframe and the world-wide distribution and exposure which enable both a wide and rapid dissemination of research output.

The series covers the theory, applications, and perspectives on the state of the art and future developments relevant to systems and networks, decision making, control, complex processes and related areas, as embedded in the fields of interdisciplinary and applied sciences, engineering, computer science, physics, economics, social, and life sciences, as well as the paradigms and methodologies behind them.

Indexed by SCOPUS, INSPEC, WTI Frankfurt eG, zbMATH, SCImago.

All books published in the series are submitted for consideration in Web of Science.

For proposals from Asia please contact Aninda Bose (aninda.bose@springer.com).

More information about this series at https://link.springer.com/bookseries/15179

Said Melliani · Oscar Castillo
Editors

Recent Advances in Fuzzy Sets Theory, Fractional Calculus, Dynamic Systems and Optimization

Springer

Editors
Said Melliani
Department of Mathematics
Sultan Moulay Slimane University
Beni Mellal, Morocco

Oscar Castillo
Division of Graduate Studies and Research
Tijuana Institute of Technology
Tijuana, Baja California, Mexico

ISSN 2367-3370 ISSN 2367-3389 (electronic)
Lecture Notes in Networks and Systems
ISBN 978-3-031-12415-0 ISBN 978-3-031-12416-7 (eBook)
https://doi.org/10.1007/978-3-031-12416-7

This Springer imprint is published by the registered company Springer Nature Switzerland AG
The registered company address is: Gewerbestrasse 11, 6330 Cham, Switzerland

Preface

This book contains the written versions of most of the contributions presented during the International Conference on Partial Differential Equations and Applications, Modeling and Simulation organized by "Applied Mathematics and Scientific Computing Laboratory" and "Systems Engineering Laboratory", at Sultan Moulay Slimane University Beni Mellal, Morocco, from June 1 to 2, 2021.

The workshop has provided a setting for discussing recent developments in a wide variety of topics including the partial differential equations, dynamic systems, optimization, numerical analysis, fuzzy sets theory, fractional calculus and its applications, to name the main topics. The workshop has brought together leading academic scientists, researchers, and research scholars to exchange and share their experiences and research results on all aspects of applied mathematics, modeling, algebra, economics, finance, and applications in various domains. It has also provided an interdisciplinary platform for researchers, practitioners, and educators to present and discuss the most recent innovations, trends, and concerns as well as practical challenges encountered and solutions adopted in the fields of the broadly perceived applied mathematics.

The chapters address various aspects that are of interest and concern for academic scientists, researchers, and research scholars in many fields, both in the sense of theoretical and applied challenges.

We hope that the various contributions of this book will be useful for researchers and graduate students in applied mathematics, engineering, physics, and even for other, notably practitioners in these and related fields. Some basis knowledge of analysis and control in systems theory and dynamical systems is just needed.

We wish to thank the authors for their interesting and novel contributions and the peer reviewers for their deep and insightful reviews which have greatly helped improve the papers included.

And last but not least, we wish to thank Dr. Tom Ditzinger and Mr. Holger Schaepe for their dedication and help to implement and finish this important publication project on time, while maintaining the highest publication standards.

Said Melliani
Oscar Castillo

Contents

A Generalized Coupled System of Impulsive Integro-Differential Evolution Equations with Mutual Boundary Values

Abdelati El Allaoui[1], Said Melliani[2(✉)], JinRong Wang[3,4], Youssef Allaoui[2], and Lalla Saadia Chadli[2]

[1] Laboratoire Maths Info et Systèmes de Communication (MISCOM), ENSAS, Cadi Ayyad University, Route Sidi Bouzid, Safi, Morocco
[2] LMACS, Sultan Moulay Slimane University, BP 523, 23000 Beni Mellal, Morocco
s.melliani@usms.ma, sa.chadli@yahoo.fr
[3] Mathematics, Guizhou University, Guiyang 550025, Guizhou, China
jrwang@gzu.edu.cn
[4] School of Mathematical Sciences, Qufu Normal University, Qufu 273165, Shandong, China

Abstract. The main objective of this paper is to study a coupled system of general class of impulsive integro-differential evolution equations with alternatives boundary conditions. The existence and uniqueness result of the solution is established via Banach fixed point theorem, and by introducing a measure of noncompactness we show the existence result. Some examples are given to illustrate obtained results at the end of this paper.

1 Introduction

Recently, impulsive evolution equations have attracted much attention of several researchers in different fields due to its applications to different real world problems including, physics, medicine, engineering, electrochemistry, etc. (see [1, 10, 12]). It is a very interesting field which is still in development by mathematicians [2, 4, 5, 8]. The study of impulsive integrro-differential evolution equations which is a sub-field of evolution equations also has enough importance. Lately, several researchers have studied various types of these problems [3, 7, 13]. In [14] the authors discussed with more details the following integro-differential equation:

$$\begin{cases} x'(t) = Ax(t) + \int_0^t B(t-s)x(s)ds + \varphi(t,x(t)) \text{ for } t \in [0,a] \text{ and } t \neq t_1 \\ \Delta x(t_i) = I_i(x(t_i)) \text{ for } i = 1,\ldots,p \text{ and } 0 < t_1 < t_2 < \cdots < t_p < t_{p+1} = a \\ x(0) = g(x) \end{cases}$$

where A and B are two closed linear operators. To show the existence of solution for this problem, they used Darbo's fixed point theorem as a tool.

The following periodic boundary value problem:

$$\begin{cases} x'(t) = Ax(t) + f(t,x(t),x(\rho(t))) + B(t)u(t),\ t \in (s_i,t_{i+1}],\ i = 0,1,2,\ldots,m,\ u \in \mathcal{U}_{\text{ad}} \\ x(t) = T(t-t_i)g_i(t,x(t)), \quad t \in (t_i,s_i],\ i = 1,2,\ldots,m \\ x(0) = x(a) \in X \end{cases}$$

S. Melliani and O. Castillo (Eds.): ICPAMS 2021, LNNS 476, pp. 1–16, 2023.
https://doi.org/10.1007/978-3-031-12416-7_1

was the object of study by the authors of [9]. They showed the existence and the uniqueness of its solution using the Banach fixed point theorem, also Krasnoselskii's fixed point theorem was their second way to show the existence of a solution.

Suppose we have two-balance scale, the first contains a specific mass while the second is empty, for the moment the balance is stable. We start adding segmented masses to the empty scale successively, at some point we will have a fall of this scale versus rise of the other with a certain speed. Then we can model these dependent speeds by a coupled system and in this case the initial state (initial position) of a scale will be the final state of the other and vice versa. Similarly, we can take the example of two athletes who run a 400 m course alternately. Motivated by the two problems above, the objective of this paper is the study of a coupled system of more general impulsive intgro-differential evolution equations, and according to two real examples cited above, we have thought to link this proposed system with alternatives(mutual) boundary conditions, then the proposed model is in the following form:

$$\begin{cases} x'(t) = Ax(t) + \int_0^t B_1(t-\tau)x(\tau)d\tau + \varphi_1(t,x(t),y(t)), & t \in (s_i,t_{i+1}], \quad i = 0,1,2,\cdots,m, \ (1) \\ y'(t) = By(t) + \int_0^t B_2(t-\tau)y(\tau)d\tau + \varphi_2(t,x(t),y(t)), & t \in (s_i,t_{i+1}], \quad i = 0,1,2,\cdots,m, \ (2) \\ x(t) = T(t-t_i)\psi_{1i}(t,x(t),y(t)), & t \in (t_i,s_i], \quad i = 1,2,\cdots,m, \\ y(t) = S(t-t_i)\psi_{2i}(t,x(t),y(t)), & t \in (t_i,s_i], \quad i = 1,2,\cdots,m, \\ x(s_i) + g_1(x,y) = x_i \in X, & i = 1,\cdots,m, \\ y(s_i) + g_2(x,y) = y_i \in X, & i = 1,\cdots,m, \\ x(0) = y(a), \\ y(0) = x(a). \end{cases} \tag{1}$$

Here the operators $A : D(A) \subset X \longrightarrow X$ and $B : D(B) \subset X \longrightarrow X$ are the infinitesimal generators of a uniformly continuous semigroup $\{T(t), t \geq 0\}$ and $\{S(t), t \geq 0\}$ respectively on a Banach space X provided with a norm $\|.\|$, where they satisfy $\|T(t)\| \leq M_T e^{\omega_T a}$ and $\|S(t)\| \leq M_S e^{\omega_S a}$, B_1 and B_2 are two closed linear operators on X which satisfy $D(A) \subset D(B_1)$ and $D(B) \subset D(B_2)$, and for each $x \in X$ the maps $t \longmapsto B_1(t)x$ and $t \longmapsto B_2(t)x$ are bounded differentiable and the maps $t \longmapsto B_1'(t)x$ and $t \longmapsto B_2'(t)x$ are bounded uniformly continuous on $[0, +\infty)$.

The fixed points s_i and t_i satisfy

$$0 = s_0 < t_1 \leq s_1 \leq t_2 < \cdots < t_m \leq s_m \leq t_{m+1} = a$$

are pre-fixed numbers, $\varphi_1, \varphi_2 : , (s_i, t_{i+1}] \times X \times X \longrightarrow X$, ψ_{1i}, $\psi_{2i} :, (t_i, s_i] \times X \times X \longrightarrow X$ and $g_1, g_2 : \mathscr{PC}([0,a], X) \times \mathscr{PC}([0,a], X) \longrightarrow X$ are given functions, such that $T(t - t_i)\psi_{1i}(t, x(t), y(t))|_{t=s_i} = x_i - g_1(x,y)$ and $S(t-t_i)\psi_{2i}(t,x(t),y(t))|_{t=s_i} = y_i - g_2(x,y)$; $i = 1, \cdots, m$.

For our knowledge, mutual boundary conditions in a coupled system of differentiated equations it is a new property. Also, the two Eqs. (1)–(2) of problem (1) have never been treated in combination with the third and fourth equations. In addition to that, the study of existence result of solution for this system is done by the Monch's fixed point theorem including measure of noncompactness to get out of frequently theorems style used to study coupled systems. And Banach's fixed point theorem is used to show the uniqueness of the solution.

2 Preliminaries

In this section we recall same basic notions used to build our results.

Denote by $\mathscr{B}(Y)$ the set of all bounded subsets of a Banach space Y.

Definition 1. We say that $m : \mathscr{B}(Y) \longrightarrow \mathbb{R}^+$ is a measure of noncompactness on Y if the following proprieties are satisfied:

1. $m(A) = 0$ if and only if A is precompact.
2. $m(A) = m(\overline{A})$, for all $A \in \mathscr{B}(Y)$.
3. $m(A \cup B) = \max\{m(A), m(B)\}$, for all $A, B \in \mathscr{B}(Y)$.

We recall the Kuratowski measure of noncompactness defined by

$$m(A) = \inf\left\{\rho > 0 : A \subset \cup_{j=1}^{m} A_j, \operatorname{diam}(A_j) \leq \rho\right\}, \text{ for } A \in \mathscr{B}(Y).$$

Now, we present the following Theorem called Monch's fixed point Theorem on which we will be based to show the existence of our solution.

Theorem 1. *[17] Let Ω be a bounded, closed, and convex subset of Y such that $0 \in \Omega$, $\Upsilon : \Omega \longrightarrow \Omega$ is a continuous mapping. Then, Υ has at least a fixed point if $C = \overline{co}(\Upsilon(C))$ or $C = \Upsilon(C) \cup \{0\} \Rightarrow \overline{C}$ is compact for each $C \subset \Omega$. Where $\overline{co}(\Upsilon(C))$ is the closed convex hull of $\Upsilon(C)$.*

Let

$$L^{\infty}([0,a]) = \{l : [0,a] \longrightarrow \mathbb{R} : l \text{ is measurable and essentially bounded}\}.$$

With the following norm

$$\|l\|_{L^{\infty}} = \inf\{\beta > 0 : |l(t)| \leq \beta, \text{ a.e.} t \in [0,a]\}$$

$L^{\infty}([0,a])$ is Banach space.

Definition 2. [15] A resolvent operator for the problem

$$\begin{cases} x'(t) = Ax(t) + \int_0^t B(t-\tau)x(\tau)d\tau, & t \in [0,+\infty) \\ x(0) = x_0 \in Y. \end{cases}$$

is a bounded linear operator-valued function $\Gamma(t)$ satisfying the following proprieties:

1. $\Gamma(0) = I_Y$. (I_Y the identity of Y) and there exist two constants $N > 0$, and $b \in \mathbb{R}$, such that $\|\Gamma(t)\| \leq Ne^{bt}$.
2. $t \longrightarrow \Gamma(t)y$ is strongly continuous for each $y \in Y$.
3. $\Gamma(t)$ is bounded for $t \geq 0$. And for $x \in D(A)$, $\Gamma(\cdot)x \in \mathscr{C}(\mathbb{R}_+, D(A)) \cap \mathscr{C}^1(\mathbb{R}_+, Y)$ and satisfying the following propriety

$$\Gamma'(t)x = A\Gamma(t)x + \int_0^t B(t-\tau)\Gamma(\tau)xd\tau = \Gamma(t)Ax + \int_0^t \Gamma(t-\tau)B(\tau)xd\tau; t \in [0,\infty).$$

We refer [6, 11, 16] for more information on the mathematical notions used.

3 Main Results

Firstly, we provide the following result we need:
We define on $\mathscr{B}(X\times X)$ the map $\widehat{m}$ by

$$\widehat{m}(D\times E)=\max\{m(D),m(E)\},\ \text{for,}\ C\times D\in\mathscr{B}(X\times X)\subset\mathscr{B}(X)\times\mathscr{B}(X).$$

For $D\times E$, $F\times G$, $\in\mathscr{B}(X\times X)$, we have

$$\begin{aligned}&\widehat{m}(D\times E)=0\Leftrightarrow m(D)=0\,\text{and}\,m(E)=0\Leftrightarrow D\times E\,\text{is precompact},\\&\widehat{m}(\overline{D\times E})=\widehat{m}(\overline{D}\times\overline{E})=\max\{m(\overline{D}),m(\overline{E})\}=\max\{m(D),m(E)\}=\widehat{m}(D\times E),\end{aligned}$$

and

$$\begin{aligned}\widehat{m}\big((D\times E)\cup(F\times G)\big)=\widehat{m}\big((D\cup F)\times(E\cup G)\big)&=\max\{m(D\cup F),m(E\cup G)\}\\&=\max\{m(D),m(F),m(E),m(F)\}\\&=\max\{\widehat{m}\big(D\times E\big),\widehat{m}\big(F\times G\big)\}.\end{aligned}$$

So, $\widehat{m}$ is a measure of noncompactness on $X\times X$.

Now, we define the following spaces

$$\begin{aligned}\mathscr{PC}([0,a],X)=\{x:[0,a]\longrightarrow X: x\in\mathscr{C}([0,t_1]\cup(t_i,s_i]\cup(s_i,t_{i+1}],X); i=1,\cdots m,\\ x(t_i^-),x(t_i^+),x(s_i^-)\,\text{and}\,x(s_i^+)\,\text{exist, with}\,x(t_i^-)=x(t_i)\,\text{and}\,x(s_i^-)=x(s_i)\}\end{aligned}$$

endowed with the norm $\|x\|_{\mathscr{PC}}=\sup\limits_{t\in[0,a]}\|x(t)\|$. And

$$\mathscr{PC}^2:=\mathscr{PC}([0,a],X)\times\mathscr{PC}([0,a],X),$$

which is a Banach space with the following norm

$$\|(x,y)\|_2=\|x\|_{\mathscr{PC}}+\|y\|_{\mathscr{PC}},\ \text{for}\ (x,y)\in\mathscr{PC}^2.$$

Firstly, we give the expression of mild solution for the following impulsive integro-differential evolution equation:

$$\begin{cases}x'(t)=Ax(t)+\int_0^t B(t-\tau)x(\tau)d\tau+\varphi(t,x(t)), & t\in(s_i,t_{i+1}], \quad i=0,1,2,\cdots,m,\\ x(t)=T(t-t_i)\psi_i(t,x(t)), \quad t\in(t_i,s_i], & i=1,2,\cdots,m,\\ x(s_i)+g(x)=x_i\in X, \quad i=1,\cdots,m,\\ x(0)=x(a).\end{cases}$$

For $t\in[0,t_1]$, we have

$$\begin{aligned}x(t)&=\Gamma(t)x(0)+\int_0^t\Gamma(t-\tau)\varphi(\tau,x(\tau))d\tau\\&=\Gamma(t)x(a)+\int_0^t\Gamma(t-\tau)\varphi(\tau,x(\tau))d\tau\\&=\Gamma(t)\left[\Gamma(a)\big(x_m-g(x)\big)+\int_{s_m}^a\Gamma(a-\tau)\varphi(\tau,x(\tau))d\tau\right]+\int_0^t\Gamma(t-\tau)\varphi(\tau,x(\tau))d\tau\\&=\Gamma(t)\Gamma(a)\big(x_m-g(x)\big)+\Gamma(t)\int_{s_m}^a\Gamma(a-\tau)\varphi(\tau,x(\tau))d\tau+\int_0^t\Gamma(t-\tau)\varphi(\tau,x(\tau))d\tau\end{aligned}$$

Let Γ_1, Γ_2 the resolvents associated with Eqs. (1) and (2) respectively.

From above and problem (1), we can verify that:

For $t \in [0,t_1]$, we have

$$\begin{aligned} x(t) &= \Gamma_1(t)x(0) + \int_0^t \Gamma_1(t-\tau)\varphi_1(\tau,x(\tau),y(\tau))d\tau \\ &= \Gamma_1(t)y(a) + \int_0^t \Gamma_1(t-\tau)\varphi_1(\tau,x(\tau),y(\tau))d\tau \\ &= \Gamma_1(t)\left[\Gamma_2(a)(y_m - g_2(x,y)) + \int_{s_m}^a \Gamma_2(a-\tau)\varphi_2(\tau,x(\tau),y(\tau))d\tau\right] + \int_0^t \Gamma_1(t-\tau)\varphi_1(\tau,x(\tau),y(\tau))d\tau \\ &= \Gamma_1(t)\Gamma_2(a)(y_m - g_2(x,y)) + \Gamma_1(t)\int_{s_m}^a \Gamma_2(a-\tau)\varphi_2(\tau,x(\tau),y(\tau))d\tau + \int_0^t \Gamma_1(t-\tau)\varphi_1(\tau,x(\tau),y(\tau))d\tau \end{aligned}$$

Similarly, we have

$$y(t) = \Gamma_2(t)\Gamma_1(a)(x_m - g_1(x,y)) + \Gamma_2(t)\int_{s_m}^a \Gamma_1(a-\tau)\varphi_1(\tau,x(\tau),y(\tau))d\tau + \int_0^t \Gamma_2(t-\tau)\varphi_2(\tau,x(\tau),y(\tau))d\tau$$

Now, we can define the form of our solution, it's given in the following definition

Definition 3. We say that (x,y) is a mild solution of the problem (1) if $(x,y) \in \mathscr{PC}^2$ and satisfies the following system

$$\begin{pmatrix} x(t) \\ y(t) \end{pmatrix} = \begin{cases} \begin{pmatrix} \Gamma_1(t)\Gamma_2(a)(y_m - g_2(x,y)) + \Gamma_1(t)\int_{s_m}^a \Gamma_2(a-\tau)\varphi_2(\tau,x(\tau),y(\tau))d\tau + \int_0^t \Gamma_1(t-\tau)\varphi_1(\tau,x(\tau),y(\tau))d\tau \\ \Gamma_2(t)\Gamma_1(a)(x_m - g_1(x,y)) + \Gamma_2(t)\int_{s_m}^a \Gamma_1(a-\tau)\varphi_1(\tau,x(\tau),y(\tau))d\tau + \int_0^t \Gamma_2(t-\tau)\varphi_2(\tau,x(\tau),y(\tau))d\tau \end{pmatrix} & \text{for } t \in [0,t_1] \\ \begin{pmatrix} \Gamma_1(t)(x_i - g_1(x,y)) + \int_{s_i}^t \Gamma_1(t-\tau)\varphi_1(\tau,x(\tau),y(\tau))d\tau \\ \Gamma_2(t)(y_i - g_2(x,y)) + \int_{s_i}^t \Gamma_2(t-\tau)\varphi_2(\tau,x(\tau),y(\tau))d\tau \end{pmatrix} & \text{for } t \in (s_i,t_{i+1}], \quad i = 1,2,\cdots,m \\ \begin{pmatrix} T(t-t_i)\psi_{1i}(t,x(t),y(t)) \\ S(t-t_i)\psi_{2i}(t,x(t),y(t)) \end{pmatrix} & \text{for } t \in (t_i,s_i], \quad i = 1,2,\cdots,m \end{cases}$$

Now, we pose the following hypotheses on which our existence result is based.

A_1 The functions $t \longmapsto \varphi_j(t,x,y)$ and $t \longmapsto \psi_{ji}(t,x,y)$; ; $j = 1,2$, are measurable on $[0,a]$ for all $(x,y) \in X \times X$, and continuous on $X \times X$ for a.e. $t \in (s_i,t_{i+1}]$ and $(t_i,s_i]$, respectively.

A_2 There exit μ_1, μ_2, ν_{1i}, $\nu_{2i} \in \mathscr{L}^\infty([0,a])$; $i = 1,\cdots,m$, which satisfy

$$\|\varphi_j(t,x,y)\| \le \mu_j(t)(1 + \|x\| + \|y\|);\ a.e\ t \in (s_i,t_{i+1}], \text{ and for all } x,\ y \in X;\ j = 1,2,$$

and

$$\|\psi_{ji}(t,x,y)\| \le \nu_{ji}(t)(1 + \|x\| + \|y\|);\ i = 1\cdots,m,\ a.e\ t \in (t_i,s_i], \text{ and for all } x,\ y \in X;\ j = 1,2,$$

A_3 There exists a constant α_1, $\alpha_2 > 0$, such that

$$\|g_j(x,y)\| \le \alpha_j(1 + \|x\|_{\mathscr{PC}} + \|y\|_{\mathscr{PC}})\ a.e\ t \in [0,a], \text{ and for all } x,\ y \in \mathscr{PC}([0,a],X);\ j = 1,2.$$

A_4 For all bounded set $\Theta \subset X \times X$, and $t \in [0,a]$, we have

$$\widehat{m}(\varphi_j(t,\Theta)) \le \mu_j(t)\widehat{m}(\Theta), \text{ and } \widehat{m}(\psi_{ji}(t,\Theta)) \le \nu_{ji}(t)\widehat{m}(\Theta);\ i = 1,\cdots,m;\ j = 1;2,$$

and for all bounded set $\tilde{\Theta} \subset \mathscr{PC}^2$, we have

$$\widehat{m}(g_j(\tilde{\Theta})) \le \alpha_j \sup_{t\in[0,a]} \widehat{m}(\tilde{\Theta}(t)),\ j = 1,2,$$

where $\tilde{\Theta}(t) = \{(x(t),y(t)) : (x,y) \in \mathscr{PC}^2\}$, for all $t \in [0,a]$.

H$_1$ The functions $\varphi_j \in \mathscr{C}([0,a]\times X\times X, X)$, $\psi_{ji} \in \mathscr{C}([s_i,t_i]\times X\times X, X)$; $i=1,\cdots,m$; $j=1,2$, and g_1, g_2 are continuous.

H$_2$ There exist constants $L_{\varphi_j}, L_{\psi_{ji}}, L_{g_j} > 0$; $j=1,2$, $i=1,\cdots,m$, such that

$$\|\varphi_j(t,x_1,y_1)-\varphi_j(t,x_2,y_2)\| \leq L_{\varphi_j}\left(\|x_1-x_2\|+\|y_1-y_2\|\right), \text{ for each } t\in[s_i,t_{i+1}];\, i=0,\cdots,m;,\, x_j,y_j\in X,\, j=1;2$$
$$\|\psi_{ji}(t,x_1,y_1)-\psi_{ji}(t,x_2,y_2)\| \leq L_{\psi_{ji}}\left(\|x_1-x_2\|+\|y_1-y_2\|\right), \text{ for each } t\in[t_i,s_i], i=1,\cdots,m,\, x_j,y_j\in X,\, j=1,2$$
$$\|g_j(x_1,y_1)-g_j(x_2,y_2)\| \leq L_{g_j}\left(\|x_1-x_2\|_{\mathscr{PC}}+\|y_1-y_2\|_{\mathscr{PC}}\right), \text{ for each } x_j,y_j\in \mathscr{PC}([0,a],X),\, j=1,2.$$

To reduce the form of mathematical expressions, we use the following notations:

$$\rho_j = \|\mu_j\|_{L^\infty},\ \sigma_j = \max_{i=1,\cdots,m}\|\nu_{ji}\|_{L^\infty},\ N_j = \sup_{t\in[0,a]}\|\Gamma_j(t)\|_{\mathscr{B}},\ j=1,2,$$
$$r_1 = \frac{N_1\left(N_2\|y_m\|+N_2\alpha_2+aN_2\rho_2+a\rho_1\right)+N_2\left(N_1\|x_m\|+N_1\alpha_1+aN_1\rho_1+a\rho_2\right)}{1-\eta_1},$$
$$r_2 = \frac{N_1\left(\max_i\|x_i\|+\alpha_1+a\rho_1\right)+N_2\left(\max_i\|y_i\|+\alpha_2+a\rho_2\right)}{1-\left(N_1\left(\alpha_1+a\rho_1\right)+N_2\left(\alpha_2+a\rho_2\right)\right)},$$
$$r_3 = \frac{\eta_2}{1-\eta_2},$$
$$\eta_1 = N_1\left(N_2\alpha_2+aN_2\rho_2+a\rho_1\right)+N_2\left(N_1\alpha_1+aN_1\rho_1+a\rho_2\right),$$
$$\eta_2 = \sigma_1 M_T e^{\omega_T a}+\sigma_2 M_S e^{\omega_S a},\ \eta=\max\{\eta_1,\eta_2\},$$
$$\xi_{11} = N_1\left(L_{g_2}N_2+L_{\varphi_2}N_2\max_{i=1,\cdots,m}(t_{i+1}-s_i)+L_{\varphi_1}t_1\right),$$
$$\xi_{12} = N_2\left(L_{g_1}N_1+L_{\varphi_1}N_1\max_{i=1,\cdots,m}(t_{i+1}-s_i)+L_{\varphi_2}t_1\right),$$
$$\xi_{21} = \max_{i=1,\cdots,m} L_{\psi_{1i}}M_T e^{\omega_T a},\ \text{and}\ \xi_{22} = \max_{i=1,\cdots,m} L_{\psi_{2i}}M_S e^{\omega_S a}.$$

After provided assumptions, now we are in a position to present our first existence result based on Banach's fixed point theorem.

Theorem 2. *Let assumptions* $\boldsymbol{H}_1$ *and* $\boldsymbol{H}_2$ *be satisfied. Suppose also that*

$$\xi := \max\{\xi_{11}+\xi_{12}, \xi_{21}+\xi_{22}\} < 1.$$

Then, the problem (1) has a unique mild solution on $[0,a]$.

Proof. We define on $\mathscr{PC}^2$ the following operator

$$(\Upsilon(x,y))(t) = \left(\Upsilon_1(x,y)(t), \Upsilon_2(x,y)(t)\right), \tag{2}$$

where

$$\Upsilon_1(x,y)(t) = \begin{cases} \Gamma_1(t)\Gamma_2(a)\left(y_m-g_2(x,y)\right)+\Gamma_1(t)\int_{sm}^a\Gamma_2(a-\tau)\varphi_2(\tau,x(\tau),y(\tau))d\tau+\int_0^t\Gamma_1(t-\tau)\varphi_1(\tau,x(\tau),y(\tau))d\tau, & \text{for } t\in[0,t_1] \\ \Gamma_1(t)\left(x_i-g_1(x,y)\right)+\int_{s_i}^t\Gamma_1(t-\tau)\varphi_1(\tau,x(\tau),y(\tau))d\tau,\ t\in(s_i,t_{i+1}], & i=1,2,\cdots,m \\ T(t-t_i)\psi_{1i}(t,x(t),y(t)), \quad \text{for } t\in(t_i,s_i], & i=1,2,\cdots,m \end{cases}$$

and

$$\Upsilon_2(x,y)(t) = \begin{cases} \Gamma_2(t)\Gamma_1(a)\left(x_m-g_1(x,y)\right)+\Gamma_2(t)\int_{sm}^a\Gamma_1(a-\tau)\varphi_1(\tau,x(\tau),y(\tau))d\tau+\int_0^t\Gamma_2(t-\tau)\varphi_2(\tau,x(\tau),y(\tau))d\tau, & \text{for } t\in[0,t_1] \\ \Gamma_2(t)\left(y_i-g_2(x,y)\right)+\int_{s_i}^t\Gamma_2(t-\tau)\varphi_2(\tau,x(\tau),y(\tau))d\tau, \quad \text{for } t\in(s_i,t_{i+1}], & i=1,2,\cdots,m \\ S(t-t_i)\psi_{2i}(t,x(t),y(t)), \quad \text{for } t\in(t_i,s_i], & i=1,2,\cdots,m \end{cases}$$

Let (x_1,y_1), $(x_2,y_2)\in\mathscr{PC}^2$, we discuss all possible cases.

Case 1: For $t \in [0, t_1]$, we have

$$\begin{aligned}
\|\Upsilon_1(x_1,y_1)(t) - \Upsilon_1(x_2,y_2)(t)\| &\leq \|\Gamma_1(t)\Gamma_2(a)g_2(x_1,y_1) - \Gamma_1(t)\Gamma_2(a)g_2(x_2,y_2)\| \\
&\quad + \|\Gamma_1(t)\|_{\mathscr{B}} \int_{s_m}^{a} \|\Gamma_2(a-\tau)\|_{\mathscr{B}} \|\varphi_2(\tau,x_1(\tau),y_1(\tau)) - \varphi_2(\tau,x_2(\tau),y_2(\tau))\| d\tau \\
&\quad + \int_0^t \|\Gamma_1(t-\tau)\|_{\mathscr{B}} \|\varphi_1(\tau,x_1(\tau),y_1(\tau)) - \varphi_1(\tau,x_2(\tau),y_2(\tau))\| d\tau \\
&\leq L_{g_2} N_1 N_2 (\|x_1 - x_2\|_{\mathscr{PC}} + \|y_1 - y_2\|_{\mathscr{PC}}) \\
&\quad + L_{\varphi_2} N_1 N_2 \int_{s_m}^{a} (\|x_1(\tau) - x_2(\tau)\| + \|y_1(\tau) - y_2(\tau)\|) d\tau \\
&\quad + L_{\varphi_1} N_1 \int_0^t (\|x_1(\tau) - x_2(\tau)\| + \|y_1(\tau) - y_2(\tau)\|) d\tau \\
&\leq N_1 \left(L_{g_2} N_2 + L_{\varphi_2} N_2 (a - s_m) + L_{\varphi_1} t_1\right) (\|x_1 - x_2\|_{\mathscr{PC}} + \|y_1 - y_2\|_{\mathscr{PC}}) \\
&\leq N_1 \left(L_{g_2} N_2 + L_{\varphi_2} N_2 (a - s_m) + L_{\varphi_1} t_1\right) \|(x_1,y_1) - (x_2,y_2)\|_2 \\
&\leq \xi_{11} \|(x_1,y_1) - (x_2,y_2)\|_2
\end{aligned}$$

Similarly, we have

$$\begin{aligned}
\|\Upsilon_2(x_1,y_1)(t) - \Upsilon_2(x_2,y_2)(t)\| &\leq N_2 \left(L_{g_1} N_1 + L_{\varphi_1} N_1 (a - s_m) + L_{\varphi_2} t_1\right) \|(x_1,y_1) - (x_2,y_2)\|_2 \\
&\leq \xi_{12} \|(x_1,y_1) - (x_2,y_2)\|_2
\end{aligned}$$

Then, we obtain

$$\begin{aligned}
\|\Upsilon(x_1,y_1)(t) - \Upsilon(x_2,y_2)(t)\| &= \|\Upsilon_1(x_1,y_1)(t) - \Upsilon_1(x_2,y_2)(t)\| + \|\Upsilon_2(x_1,y_1)(t) - \Upsilon_2(x_2,y_2)(t)\| \\
&\leq (\xi_{11} + \xi_{12}) \|(x_1,y_1) - (x_2,y_2)\|_2
\end{aligned}$$

Case 2: For $t \in (s_i, t_{i+1}]$; $i = 1, \cdots, m$, we have

$$\begin{aligned}
\|\Upsilon_1(x_1,y_1)(t) - \Upsilon_1(x_2,y_2)(t)\| &\leq \|\Gamma_1(t)\|_{\mathscr{B}} \|g_1(x_1,y_1) - g_1(x_2,y_2)\| \\
&\quad + \int_{s_i}^{t} \|\Gamma_1(t-\tau)\|_{\mathscr{B}} \|\varphi_1(\tau,x_1(\tau),y_1(\tau)) - \varphi_1(\tau,x_2(\tau),y_2(\tau))\| d\tau \\
&\leq L_{g_1} N_1 (\|x_1 - x_2\|_{\mathscr{PC}} + \|y_1 - y_2\|_{\mathscr{PC}}) \\
&\quad + L_{\varphi_1} N_1 \int_{s_i}^{t} (\|x_1(\tau) - x_2(\tau)\| + \|y_1(\tau) - y_2(\tau)\|) d\tau \\
&\leq N_1 \left(L_{g_1} + L_{\varphi_1} \max_{i=1,\cdots,m} (t_{i+1} - s_i)\right) \|(x_1,y_1) - (x_2,y_2)\|_2 \\
&\leq \xi_{11} \|(x_1,y_1) - (x_2,y_2)\|_2
\end{aligned}$$

Likewise, we get

$$\begin{aligned}
\|\Upsilon_2(x_1,y_1)(t) - \Upsilon_2(x_2,y_2)(t)\| &\leq N_2 \left(L_{g_2} + L_{\varphi_2} \max_{i=1,\cdots,m} (t_{i+1} - s_i)\right) \|(x_1,y_1) - (x_2,y_2)\|_2 \\
&\leq \xi_{12} \|(x_1,y_1) - (x_2,y_2)\|_2
\end{aligned}$$

Hence,

$$\|\Upsilon(x_1,y_1)(t) - \Upsilon(x_2,y_2)(t)\| \leq (\xi_{11} + \xi_{12}) \|(x_1,y_1) - (x_2,y_2)\|_2$$

Case 3: For $t \in (t_i, s_i]$; $i = 1, \cdots, m$, we have

$$\begin{aligned}
\|\Upsilon_1(x_1,y_1)(t) - \Upsilon_1(x_2,y_2)(t)\| &\leq \|T(t - t_i)\| \|\psi_{1i}(t,x_1(t),y_1(t)) - \psi_{1i}(t,x_2(t),y_2(t))\| \\
&\leq L_{\psi_{1i}} M_T e^{\omega_T a} \|(x_1,y_1) - (x_2,y_2)\|_2 \\
&\leq \xi_{21} \|(x_1,y_1) - (x_2,y_2)\|_2
\end{aligned}$$

Similarly, we get

$$\begin{aligned}\|\Upsilon_2(x_1,y_1)(t)-\Upsilon_2(x_2,y_2)(t)\| &\leq L_{\psi_{2i}} M_S e^{\omega_S a}\|(x_1,y_1)-(x_2,y_2)\|_2\\ &\leq \xi_{22}\|(x_1,y_1)-(x_2,y_2)\|_2\end{aligned}$$

Then, we have

$$\|\Upsilon(x_1,y_1)(t)-\Upsilon(x_2,y_2)(t)\| \leq (\xi_{21}+\xi_{22})\,\|(x_1,y_1)-(x_2,y_2)\|_2$$

Finally, we get the following inequality

$$\|\Upsilon(x_1,y_1)-\Upsilon(x_2,y_2)\|_2 \leq \max\{\xi_{11}+\xi_{12},\xi_{21}+\xi_{22}\}\,\|(x_1,y_1)-(x_2,y_2)\|_2$$

Therefore, Υ is a contraction. So, according to Banach fixed point theorem, problem (1) has a unique mild solution.

Using Monch's fixed point theorem, we present the second result of existence as follows:

Theorem 3. *Suppose that assumptions A_1–A_4 are satisfied, in addition*

$$\eta < 1. \tag{3}$$

Then, problem (1) has at least one mild solution on $[0,a]$.

Proof. To proof this result we transform our problem into fixed point, for this, we consider the operator $\Upsilon : \mathscr{PC}^2 \longrightarrow \mathscr{PC}^2$ defined in (2), and we define the ball $B_r := \{(x,y)\in\mathscr{PC}^2 : \|(x,y)\|_2 \leq r\}$, where

$$r \geq \max\{r_1,r_2,r_3\}.$$

Firstly, we prove that Υ is defined from B_r into itself. Indeed:

Case 1: For $(x,y)\in B_r$, and $t\in[0,t_1]$, we have

$$\begin{aligned}\|\Upsilon_1(x,y)(t)\| \leq{}& \|\Gamma_1(t)\|_{\mathscr{B}}\,\|\Gamma_2(a)\|_{\mathscr{B}}\,(\|y_m\|+\|g_2(x,y)\|)+\|\Gamma_1(t)\|_{\mathscr{B}}\int_{s_m}^{a}\|\Gamma_2(a-\tau)\|_{\mathscr{B}}\,\|\varphi_2(\tau,x(\tau),y(\tau))\|d\tau\\ &+\int_0^t\|\Gamma_1(t-\tau)\|_{\mathscr{B}}\,\|\varphi_1(\tau,x(\tau),y(\tau))\|d\tau\\ \leq{}& N_1N_2(\|y_m\|+\alpha_2(1+\|x\|_{\mathscr{PC}}+\|y\|_{\mathscr{PC}}))+aN_1N_2\rho_2(1+\|x\|_{\mathscr{PC}}+\|y\|_{\mathscr{PC}})+aN_1\rho_1(1+\|x\|_{\mathscr{PC}}+\|y\|_{\mathscr{PC}})\\ \leq{}& N_1\,(N_2\|y_m\|+N_2\alpha_2+aN_2\rho_2+a\rho_1)+N_1\,(N_2\alpha_2+aN_2\rho_2+a\rho_1)\,r\end{aligned}$$

Similarly, we get

$$\|\Upsilon_2(x,y)(t)\| \leq N_2\,(N_1\|x_m\|+N_1\alpha_1+aN_1\rho_1+a\rho_2)+N_2\,(N_1\alpha_1+aN_1\rho_1+a\rho_2)\,r$$

Then,

$$\begin{aligned}\|\Upsilon(x,y)\|_2 ={}& \|\Upsilon_1(x,y)\|_2+\|\Upsilon_2(x,y)\|_2\\ \leq{}& N_1\,(N_2\|y_m\|+N_2\alpha_2+aN_2\rho_2+a\rho_1)\\ &+N_2\,(N_1\|x_m\|+N_1\alpha_1+aN_1\rho_1+a\rho_2)\\ &+[N_1\,(N_2\alpha_2+aN_2\rho_2+a\rho_1)+N_2\,(N_1\alpha_1+aN_1\rho_1+a\rho_2)]\,r\\ \leq{}& (1-\eta_1)r_1+\eta_1 r\\ \leq{}& r\end{aligned}$$

Case 2: For $(x,y) \in B_r$, and $t \in (s_i, t_{i+1}]$; $i = 1, \cdots, m$, we have

$$\begin{aligned}
\|\Upsilon_1(x,y)(t)\| &\leq \|\Gamma_1(t)\|_{\mathscr{B}} \left(\|x_i\| + \|g_1(x,y)\|\right) + \int_{s_i}^{t} \|\Gamma_1(t-\tau)\|_{\mathscr{B}} \|\varphi_1(\tau, x(\tau), y(\tau))\| d\tau \\
&\leq N_1 \left(\|x_i\| + \alpha_1 (1 + \|x\|_{\mathscr{PC}} + \|y\|_{\mathscr{PC}})\right) + aN_1\rho_1 (1 + \|x\|_{\mathscr{PC}} + \|y\|_{\mathscr{PC}}) \\
&\leq N_1 \left(\|x_i\| + \alpha_1 + a\rho_1\right) + N_1 \left(\alpha_1 + a\rho_1\right) r
\end{aligned}$$

In the same way, we get

$$\|\Upsilon_2(x,y)(t)\| \leq N_2 \left(\|y_i\| + \alpha_2 + a\rho_2\right) + N_2 \left(\alpha_2 + a\rho_2\right) r.$$

Therefore,

$$\begin{aligned}
\|\Upsilon(x,y)\| &\leq N_1 \left(\|x_i\| + \alpha_1 + a\rho_1\right) + N_2 \left(\|y_i\| + \alpha_2 + a\rho_2\right) + \left[N_1 \left(\alpha_1 + a\rho_1\right) + N_2 \left(\alpha_2 + a\rho_2\right)\right] r \\
&\leq \left(1 - \left[N_1 \left(\alpha_1 + a\rho_1\right) + N_2 \left(\alpha_2 + a\rho_2\right)\right]\right) r_2 + \left[N_1 \left(\alpha_1 + a\rho_1\right) + N_2 \left(\alpha_2 + a\rho_2\right)\right] r \\
&\leq r.
\end{aligned}$$

Case 3: For $(x,y) \in B_r$, and $t \in (t_i, s_i]$; $i = 1, \cdots, m$, we have

$$\begin{aligned}
\|\Upsilon_1(x,y)(t)\| &\leq \|T(t-t_i)\| \|\psi_{1i}(t, x(t), y(t))\| \\
&\leq \sigma_1 M_T e^{\omega_T a} (1 + r)
\end{aligned}$$

Similarly, we obtain

$$\|\Upsilon_2(x,y)(t)\| \leq \sigma_2 M_S e^{\omega_S a} (1 + r)$$

Then,

$$\begin{aligned}
\|\Upsilon(x,y)\|_2 &\leq \left(\sigma_1 M_T e^{\omega_T a} + \sigma_2 M_S e^{\omega_S a}\right)(1 + r) \\
&\leq (1 - \eta_2) r_3 + \eta_2 r \\
&\leq r.
\end{aligned}$$

which shows that Υ is defined from B_r into itself.

The rest of proof will be done in four steps by discussing all cases in each step.

Step 1: Υ is continuous:

Let $(x_n, y_n)_{n \geq 0} \subset B_r$ be a sequence, such that $\lim_{n \to +\infty} (x_n, y_n) = (x,y)$ in B_r. Clearly, we have $\|(x_n, y_n) - (x,y)\|_2 = \|x_n - x\|_{\mathscr{PC}} + \|y_n - y\|_{\mathscr{PC}}$ which implies that

$$\lim_{n \to +\infty} (x_n, y_n) = (x,y) \text{ in } B_r \text{ if and only if } \lim_{n \to +\infty} x_n = x \text{ and } \lim_{n \to +\infty} y_n = y \text{ in } \{x \in \mathscr{PC}([0,a], X) : \|x\|_{\mathscr{PC}} \leq r\}.$$

Case 1: For $t \in [0, t_1]$, we have

$$\begin{aligned}
\|\Upsilon_1(x_n, y_n)(t) - \Upsilon_1(x,y)(t)\| &\leq \|\Gamma_1(t)\|_{\mathscr{B}} \|\Gamma_2(a)\|_{\mathscr{B}} \|g_2(x_n, y_n) - g_2(x,y)\| \\
&\quad + \|\Gamma_1(t)\|_{\mathscr{B}} \int_{s_m}^{a} \|\Gamma_2(a-\tau)\|_{\mathscr{B}} \|\varphi_2(\tau, x_n(\tau), y_n(\tau)) - \varphi_2(\tau, x(\tau), y(\tau))\| d\tau \\
&\quad + \int_0^t \|\Gamma_1(t-\tau)\|_{\mathscr{B}} \|\varphi_1(\tau, x_n(\tau), y_n(\tau)) - \varphi_1(\tau, x(\tau), y(\tau))\| d\tau \\
&\leq N_1 N_2 \|g_2(x_n, y_n) - g_2(x,y)\| + N_1 N_2 \int_{s_m}^{a} \|\varphi_2(\tau, x_n(\tau), y_n(\tau)) - \varphi_2(\tau, x(\tau), y(\tau))\| d\tau \\
&\quad + N_1 \int_0^t \|\varphi_1(\tau, x_n(\tau), y_n(\tau)) - \varphi_1(\tau, x(\tau), y(\tau))\| d\tau
\end{aligned}$$

And

$$\|\Upsilon_2(x_n,y_n)(t)-\Upsilon_2(x,y)(t)\| \leq N_1N_2\|g_1(x_n,y_n)-g_1(x,y)\|+N_1N_2\int_{s_m}^{a}\|\varphi_1(\tau,x_n(\tau),y_n(\tau))-\varphi_1(\tau,x(\tau),y(\tau))\|d\tau$$
$$+N_2\int_0^t\|\varphi_2(\tau,x_n(\tau),y_n(\tau))-\varphi_2(\tau,x(\tau),y(\tau))\|d\tau$$

Case 2: For $t \in (s_i,t_{i+1}]$; $i=1,\cdots,m$, we have

$$\begin{aligned}\|\Upsilon_1(x_n,y_n)(t)-\Upsilon_1(x,y)(t)\| &\leq \|\Gamma_1(t)\|_{\mathscr{B}}\|g_1(x_n,y_n)-g_1(x,y)\|\\ &+\int_{s_i}^{t}\|\Gamma_1(t-\tau)\|_{\mathscr{B}}\|\varphi_1(\tau,x_n(\tau),y_n(\tau))-\varphi_1(\tau,x(\tau),y(\tau))\|d\tau\\ &\leq N_1\|g_1(x_n,y_n)-g_1(x,y)\|+N_1\int_{s_i}^{t}\|\varphi_1(\tau,x_n(\tau),y_n(\tau))-\varphi_1(\tau,x(\tau),y(\tau))\|d\tau\end{aligned}$$

Similarly, we get

$$\|\Upsilon_2(x_n,y_n)(t)-\Upsilon_2(x,y)(t)\| \leq N_2\|g_2(x_n,y_n)-g_2(x,y)\|+N_2\int_{s_i}^{t}\|\varphi_2(\tau,x_n(\tau),y_n(\tau))-\varphi_2(\tau,x(\tau),y(\tau))\|d\tau$$

Case 3: For $t \in (t_i,s_i]$; $i=1,\cdots,m$, we have

$$\begin{aligned}\|\Upsilon_1(x_n,y_n)(t)-\Upsilon_1(x,y)(t)\| &\leq \|T(t-t_i)\|\|\psi_{1i}(t,x_n(t),y_n(t))-\psi_{1i}(t,x(t),y(t))\|\\ &\leq M_Te^{\omega_T a}\|\psi_{1i}(t,x_n(t),y_n(t))-\psi_{1i}(t,x(t),y(t))\|\end{aligned}$$

And

$$\|\Upsilon_2(x_n,y_n)(t)-\Upsilon_2(x,y)(t)\| \leq M_Se^{\omega_S a}\|\psi_{2i}(t,x_n(t),y_n(t))-\psi_{2i}(t,x(t),y(t))\|$$

We know that, φ_j, ψ_{ji} and g_j; $j=1,2$; $i=1,\cdots,m$ are continuous, then according to Lebesgue-dominated convergence theorem, we get from each previous step

$$\lim_{n\to+\infty}\|\Upsilon_1(x_n,y_n)-\Upsilon_1(x,y)\|_2=0 \text{ and } \lim_{n\to+\infty}\|\Upsilon_2(x_n,y_n)-\Upsilon_2(x,y)\|_2=0,$$

and since from (2) we have

$$\lim_{n\to+\infty}\|\Upsilon(x_n,y_n)-\Upsilon(x,y)\|_2=0 \Leftrightarrow \left(\lim_{n\to+\infty}\|\Upsilon_1(x_n,y_n)-\Upsilon_1(x,y)\|=0 \text{ and } \lim_{n\to+\infty}\|\Upsilon_2(x_n,y_n)-\Upsilon_2(x,y)\|=0\right).$$

So, we deduce that $\lim_{n\to+\infty}\|\Upsilon(x_n,y_n)-\Upsilon(x,y)\|_2=0$.

Step 2: $\Upsilon(B_r)$ is bounded. Indeed:
We have Υ is defined on B_r into itself. So, $\Upsilon(B_r)\subset B_r$ which prove that $\Upsilon(B_r)$ is bounded.

Step 3: Υ is equicontinuous.

Case 1: For $(x,y)\in B_r$ and $0\leq\tau_1<\tau_2\leq t_1$, we have

$$\begin{aligned}\|\Upsilon_1(x,y)(\tau_2)-\Upsilon_1(x,y)(\tau_1)\| &\leq \|\Gamma_2(a)\|_{\mathscr{B}}\left(\|y_m\|+\|g_2(x,y)\|\right)\|\Gamma_1(\tau_2)-\Gamma_1(\tau_1)\|_{\mathscr{B}}\\ &+\|\Gamma_1(\tau_2)-\Gamma_1(\tau_1)\|_{\mathscr{B}}\int_{s_m}^{a}\|\Gamma_2(a-\tau)\|_{\mathscr{B}}\|\varphi_2(\tau,x(\tau),y(\tau))\|d\tau\\ &+\int_0^{\tau_1}\|\Gamma_1(\tau_2-\tau)-\Gamma_1(\tau_1-\tau)\|_{\mathscr{B}}\|\varphi_1(\tau,x(\tau),y(\tau))\|d\tau\\ &+\int_{\tau_1}^{\tau_2}\|\Gamma_1(\tau_2-\tau)\|_{\mathscr{B}}\|\varphi_1(\tau,x(\tau),y(\tau))\|d\tau\\ &\leq N_2\left[\left(\|y_m\|+\alpha_2(1+r)\right)+(a-s_m)\rho_2(1+r)\right]\|\Gamma_1(\tau_2)-\Gamma_1(\tau_1)\|_{\mathscr{B}}\\ &+\rho_1(1+r)\int_0^{\tau_1}\|\Gamma_1(\tau_2-\tau)-\Gamma_1(\tau_1-\tau)\|_{\mathscr{B}}d\tau+N_1\rho_1(1+r)(\tau_2-\tau_1)\end{aligned}$$

In the same manner, we get

$$\begin{aligned}\|\Upsilon_2(x,y)(\tau_2)-\Upsilon_2(x,y)(\tau_1)\| &\leq N_1\left[\left(\|x_m\|+\alpha_1(1+r)\right)+(a-s_m)\rho_1(1+r)\right]\|\Gamma_2(\tau_2)-\Gamma_2(\tau_1)\|_{\mathscr{B}}\\ &+\rho_2(1+r)\int_0^{\tau_1}\|\Gamma_2(\tau_2-\tau)-\Gamma_2(\tau_1-\tau)\|_{\mathscr{B}}\,d\tau+N_2\rho_2(1+r)(\tau_2-\tau_1)\end{aligned}$$

Case 2: For $(x,y)\in B_r$ and $s_i<\tau_1<\tau_2\leq t_{i+1}$; $i=1,\cdots,m$, we have

$$\begin{aligned}\|\Upsilon_1(x,y)(\tau_2)-\Upsilon_1(x,y)(\tau_1)\| &\leq \|\Gamma_1(\tau_2)-\Gamma_1(\tau_1)\|_{\mathscr{B}}\left(\|x_i\|+\|g_1(x,y)\|\right)\\ &+\int_{s_i}^{\tau_1}\|\Gamma_1(\tau_2-\tau)-\Gamma_1(\tau_1-\tau)\|_{\mathscr{B}}\,\|\varphi_1(\tau,x(\tau),y(\tau))\|d\tau\\ &+\int_{\tau_1}^{\tau_2}\|\Gamma_1(\tau_2-\tau)\|_{\mathscr{B}}\,\|\varphi_1(\tau,x(\tau),y(\tau))\|d\tau\\ &\leq \left(\|x_i\|+\alpha_1(1+r)\right)\|\Gamma_1(\tau_2)-\Gamma_1(\tau_1)\|_{\mathscr{B}}+\rho_1(1+r)\int_{s_i}^{\tau_1}\|\Gamma_1(\tau_2-\tau)-\Gamma_1(\tau_1-\tau)\|_{\mathscr{B}}\,d\tau\\ &+N_1\rho_1(1+r)(\tau_2-\tau_1)\end{aligned}$$

And

$$\begin{aligned}\|\Upsilon_2(x,y)(\tau_2)-\Upsilon_2(x,y)(\tau_1)\| &\leq \left(\|y_i\|+\alpha_2(1+r)\right)\|\Gamma_2(\tau_2)-\Gamma_2(\tau_1)\|_{\mathscr{B}}+\rho_2(1+r)\int_{s_i}^{\tau_1}\|\Gamma_2(\tau_2-\tau)-\Gamma_2(\tau_1-\tau)\|_{\mathscr{B}}\,d\tau\\ &+N_2\rho_2(1+r)(\tau_2-\tau_1)\end{aligned}$$

Case 3: For $(x,y)\in B_r$ and $t_i<\tau_1<\tau_2\leq s_i$; $i=1,\cdots,m$, we have

$$\begin{aligned}\|\Upsilon_1(x,y)(\tau_2)-\Upsilon_1(x,y)(\tau_1)\| &\leq \|T(\tau_1-t_i)\|\|T(\tau_2-\tau_1)\psi_{1i}(\tau_2,x(\tau_2),y(\tau_2))-\psi_{1i}(\tau_1,x(\tau_1),y(\tau_1))\|\\ &\leq M_T e^{\omega_T a}\|T(\tau_2-\tau_1)\psi_{1i}(\tau_2,x(\tau_2),y(\tau_2))-\psi_{1i}(\tau_1,x(\tau_1),y(\tau_1))\|\end{aligned}$$

Similarly, we obtain

$$\|\Upsilon_2(x,y)(\tau_2)-\Upsilon_2(x,y)(\tau_1)\|\leq M_S e^{\omega_S a}\|S(\tau_2-\tau_1)\psi_{2i}(\tau_2,x(\tau_2),y(\tau_2))-\psi_{2i}(\tau_1,x(\tau_1),y(\tau_1))\|$$

In all previous cases, we have

$$\|\Upsilon(x,y)(\tau_2)-\Upsilon(x,y)(\tau_1)\|=\|\Upsilon_1(x,y)(\tau_2)-\Upsilon_1(x,y)(\tau_1)\|+\|\Upsilon_2(x,y)(\tau_2)-\Upsilon_2(x,y)(\tau_1)\|\to 0 \text{ as } \tau_1\to\tau_2.$$

This allows us to conclude that Υ is equicontinuous.

Step 4: Let $C\subset B_r$ be a non empty subset, such that

$$C\subset\overline{\Upsilon(C)}\cup\{(0,0)\}=(\overline{\Upsilon_1(C)}\times\overline{\Upsilon_2(C)})\cup\{(0,0)\}=(\overline{\Upsilon_1(C)}\cup\{0\})\times\overline{\Upsilon_2(C)}\cup\{0\}).$$

Clearly, it is bounded and equicontinuous.
Consider the function l defined by

$$l(t)=\widehat{m}(C(t)),\ t\in[0,a],$$

which is continuous.

Case 1: For $t\in[0,t_1]$, we have

$$l(t)=\widehat{m}(C(t))\leq\widehat{m}\left(\overline{\Upsilon(C)}(t)\cup\{(0,0)\}\right)\leq\widehat{m}\left(\overline{\Upsilon(C)}(t)\right)=\widehat{m}(\Upsilon(C)(t))=\max\{m(\Upsilon_1(C)(t)),m(\Upsilon_2(C)(t))\}.$$

Since, we have

$$\begin{aligned} m(\Upsilon_1(C)(t)) &\leq \|\Gamma_1(t)\|_{\mathscr{B}} \|\Gamma_2(a)\|_{\mathscr{B}}\, \alpha_2 \sup_{t\in[0,a]} \widehat{m}(C(t)) + \|\Gamma_1(t)\|_{\mathscr{B}} \int_{s_m}^{a} \|\Gamma_2(a-\tau)\|_{\mathscr{B}}\, \rho_2 \widehat{m}(C(\tau))d\tau \\ &+ \int_0^t \|\Gamma_1(t-\tau)\|_{\mathscr{B}}\, \rho_1 \widehat{m}(C(\tau))d\tau \\ &\leq N_1\left(N_2\alpha_2 + aN_2\rho_2 + a\rho_1\right)\|l\|_\infty \\ &\leq \eta_1\|l\|_\infty \end{aligned}$$

and

$$\begin{aligned} m(\Upsilon_2(C)(t)) &\leq N_2\left(N_1\alpha_1 + aN_1\rho_1 + a\rho_2\right)\|l\|_\infty \\ &\leq \eta_1\|l\|_\infty \end{aligned}$$

Therefore,

$$l(t) \leq \eta\|l\|_\infty.$$

Case 2: For $t \in (s_i, t_{i+1}]$; $i = 1, \cdots, m$, we have

$$\begin{aligned} m(\Upsilon_1(C)(t)) &\leq \|\Gamma_1(t)\|_{\mathscr{B}}\alpha_1 \sup_{t\in[0,a]} \widehat{m}\left((C(t))\right) + \int_{s_i}^{t} \|\Gamma_1(t-\tau)\|_{\mathscr{B}}\rho_1\widehat{m}(C(\tau))d\tau \\ &\leq N_1\left(\alpha_1 + a\rho_1\right)\|l\|_\infty \\ &\leq \eta_1\|l\|_\infty \end{aligned}$$

and

$$\begin{aligned} m(\Upsilon_2(C)(t)) &\leq N_2\left(\alpha_2 + a\rho_2\right)\|l\|_\infty \\ &\leq \eta_1\|l\|_\infty \end{aligned}$$

Then,

$$l(t) \leq \eta\|l\|_\infty.$$

Case 3: For $t \in (t_i, s_i]$; $i = 1, \cdots, m$, we have

$$\begin{aligned} m(\Upsilon_1(C)(t)) &\leq \|T(t-t_i)\|\sigma_1 \sup_{t\in[0,a]} \widehat{m}(C(t)) \\ &\leq \sigma_1 M_T e^{\omega_T a}\|l\|_\infty \\ &\leq \eta_2\|l\|_\infty \end{aligned}$$

and

$$\begin{aligned} m(\Upsilon_2(C)(t)) &\leq \sigma_2 M_S e^{\omega_S a}\|l\|_\infty \\ &\leq \eta_2\|l\|_\infty \end{aligned}$$

Then,

$$l(t) \leq \eta\|l\|_\infty.$$

Hence, from above cases we can deduce that

$$\|l\|_\infty \leq \eta \|l\|_\infty.$$

Since $\eta < 1$, so obviously we have $\|l\|_\infty = 0$ which is equivalent to saying that $\widehat{m}(C(t)) = 0$. So, according to the first property of Definition 3, $C(t)$ is relatively compact in $X \times X$. Then, by the Ascoli-Arzel theorem, it is relatively compact in B_r.

Thus, all conditions of Theorem 3 are satisfied, and consequently our problem has a solution.

4 Examples

In this section we present two examples to illustrate our existence results.

Example 1. We consider the following problem:

$$\begin{cases} \frac{\partial}{\partial t}u(t,x) = \frac{\partial^2}{\partial x^2}u(t,x) + \int_0^t L_1(t-\tau)\frac{\partial^2}{\partial x^2}u(\tau,x)d\tau + \frac{1}{18N_1N_2}\big(\cos(u(t,x)) + \sin(v(t,x))\big),\ t \in (0,1] \cup (2,3], x \in [0,1] \\ \frac{\partial}{\partial t}v(t,x) = \frac{\partial^2}{\partial x^2}v(t,x) + \int_0^t L_2(t-\tau)\frac{\partial^2}{\partial x^2}v(\tau,x)d\tau + \frac{1}{18N_1N_2}\big(\cos(u(t,x)) + v(t,x)\big),\ t \in (0,1] \cup (2,3], x \in [0,1] \\ u(t,x) = T(t-1)\frac{1}{14}\big(\sin(u(t,x)) + \sin(v(t,x))\big),\ t \in (1,2], x \in [0,1] \\ v(t,x) = T(t-1)\frac{1}{14}\big(\cos(u(t,x)) + \sin(u(t,x))\big),\ t \in (1,2], x \in [0,1] \\ u(t,0) = v(t,0) = u(t,1) = v(t,1) = 0,\ t \in (0,1] \cup (2,3] \\ u(0,x) = v(3,x),\ x \in [0,1] \\ v(0,x) = u(3,x),\ x \in [0,1] \\ u(0,x) + \frac{1}{8N_1N_2}(1 + \sin(u) + v) = 1 + e^x,\ x \in [0,1] \\ u(2,x) + \frac{1}{8N_1N_2}(1 + \sin(u) + v) = 2 + e^x,\ x \in [0,1] \\ v(0,x) + \frac{1}{8N_1N_2}(1 + \cos(u) + v) = 1 + e^x,\ x \in [0,1] \\ v(2,x) + \frac{1}{8N_1N_2}(1 + \cos(u) + v) = 2 + e^x,\ x \in [0,1] \end{cases} \tag{4}$$

where $L_1, L_2 \in \mathscr{C}^1([0,3],\mathbb{R})$.

The previous problem can be abstracted into problem (1), where $X = L^2([0,1])$ endowed with the norm $\|u\| = \left(\int_0^1 |u(x)|^2 dx\right)^{\frac{1}{2}}$ which is a Banach space, and $Au = Bu = \frac{\partial^2}{\partial x^2}u$, for $u \in D(A) = \left\{u \in X : \frac{\partial}{\partial x}u,\ \frac{\partial^2}{\partial x^2}u \in X,\ u(0) = u(1) = 0\right\}$.

A is the generator of a strongly continuous and compact semigroup $\{T(t), t \geq 0\}$ on X and $\|T(t)\| \leq 1$, for all $t \geq 0$.

$$B_1(t) = L_1(t)A,\ B_2(t) = L_2(t)A,\ \varphi_1(t,u,v) = \frac{1}{18N_1N_2}\big(\cos(u(t,x)) + \sin(v(t,x))\big),$$

$$\varphi_2(t,u,v) = \frac{1}{18N_1N_2}\big(\cos(u(t,x)) + v(t,x)\big),\ \psi_{11}(t,u,v) = \frac{1}{14}\big(\sin(u(t,x)) + \sin(v(t,x))\big),$$

$$\psi_{21}(t,u,v) = \frac{1}{14}\left(\cos(u(t,x)) + \sin(v(t,x))\right),$$
$$g_1(u,v) = \frac{1}{8N_1N_2}(1+\sin(u)+v),\ g_2(u,v) = \frac{1}{8N_1N_2}(1+\cos(u)+v).$$

Clearly, we have $L_{\varphi_1} = L_{\varphi_2} = \frac{1}{18N_1N_2}$, $L_{\psi_{11}} = L_{\psi_{21}} = \frac{1}{14N}$, $L_{g_1} = L_{g_2} = \frac{1}{8N_1N_2}$, $M_T = M_S = 1$, $a = 3$ and $\omega_T = \omega_S = 0$. Then $\xi_{11} = \xi_{12} \leq \frac{17}{72}$, $\xi_{21} = \xi_{22} = \frac{1}{14}$, therefore $\xi \leq \frac{17}{36} < 1$.

Hence, according to Theorem 2, problem (4) has a unique mild solution.

Example 2. To illustrate our second existence result, we present the following problem:

$$\begin{cases} \frac{\partial}{\partial t}u(t,x) = \frac{\partial^2}{\partial x^2}u(t,x) + \int_0^t L_1(t-\tau)\frac{\partial^2}{\partial x^2}u(\tau,x)d\tau + \left(\frac{1}{e^9} + \frac{1}{e^{t+x+9}}\right)\frac{t^2(1+u(t,x)+v(t,x))}{36N_1N_2(1+\|u\|+\|v\|)},\ t\in(0,1]\cup(2,3],\ x\in[0,1] \\ \frac{\partial}{\partial t}v(t,x) = \frac{\partial^2}{\partial x^2}v(t,x) + \int_0^t L_2(t-\tau)\frac{\partial^2}{\partial x^2}u(\tau,x)d\tau + \left(\frac{1}{e^9} + \frac{1}{e^{t+x+9}}\right)\frac{t^2(1+u(t,x))}{36N_1N_2(1+\|u\|+\|v\|)},\ t\in(0,1]\cup(2,3],\ x\in[0,1] \\ u(t,x) = T(t-1)\frac{u(t,x)}{24(1+\|u\|+\|v\|)},\ t\in(1,2],\ x\in[0,1] \\ v(t,x) = T(t-1)\frac{v(t,x)}{24(1+\|u\|+\|v\|)},\ t\in(1,2],\ x\in[0,1] \\ u(t,0) = u(t,1) = v(t,0) = v(t,1) = 0,\ t\in(0,1]\cup(2,3] \\ u(0,x) = v(3,x),\ x\in[0,1] \\ v(0,x) = u(3,x),\ x\in[0,1] \\ u(0,x) + \frac{1}{8N_1N_2}(1+\sin(u)+cos(v)) = 1+e^x,\ x\in[0,1] \\ u(2,x) + \frac{1}{8N_1N_2}(1+\sin(u)+cos(v)) = 2+e^x,\ x\in[0,1] \\ v(0,x) + \frac{1}{8N_1N_2}(1+\sin(u)+v) = 1+e^x,\ x\in[0,1] \\ v(2,x) + \frac{1}{8N_1N_2}(1+\sin(u)+v) = 2+e^x,\ x\in[0,1] \end{cases} \tag{5}$$

The previous problem can be written as problem (1), where

$$\varphi_1(t,u,v) = \left(\frac{1}{e^9} + \frac{1}{e^{t+x+9}}\right)\frac{t^2(1+u(t,x)+v(t,x))}{36N_1N_2(1+\|u\|+\|v\|)},$$
$$\varphi_2(t,u,v) = \left(\frac{1}{e^9} + \frac{1}{e^{t+x+9}}\right)\frac{t^2(1+u(t,x))}{36N_1N_2(1+\|u\|+\|v\|)},$$
$$\psi_{11}(t,u,v) = \frac{u(t,x)}{24(1+\|u\|+\|v\|)},$$
$$\psi_{21}(t,u,v) = \frac{v(t,x)}{24(1+\|u\|+\|v\|)},$$
$$g_1(u,v) = \frac{1}{8N_1N_2}(1+\sin(u)+cos(v)),$$
$$g_2(u,v) = \frac{1}{8N_1N_2}(1+\sin(u)+v).$$

It's easy to verify that $\rho_1 = \rho_2 = \frac{1}{2N_1N_2e^9}$, $\sigma_1 = \sigma_2 = \frac{1}{24}$, and $\alpha_1 = \alpha_2 = \frac{1}{8N_1N_2}$. Thus, $\eta_1 \leq \frac{24+e^9}{4e^9} < 1$ and $\eta_2 = \frac{1}{12} < 1$. Therefore, by using Theorem 3, our problem (5) has a solution.

Acknowledgements. The authors express their sincere thanks to the anonymous referees for numerous helpful and constructive suggestions which have improved the manuscript.

Author contributions. All authors contributed equally to the writing of this paper. All authors read and approved the final manuscript.

Data Availibility Statement. The data used to support the findings of this study are available from the corresponding author upon request.

Competing Interests. The authors declare that they have no competing interests.

Conflict of Interests. The authors declare that there is no conflict of interests regarding the publication of this paper.

References

1. Luo, A.C.J., Merdan, H. (eds.): Mathematical Modeling and Applications in Nonlinear Dynamics. NSC, vol. 14. Springer, Cham (2016). https://doi.org/10.1007/978-3-319-26630-5
2. Bainov, D.D., Simeonov, P.S.: Impulsive Differential Equations: Periodic Solutions and Applications. Longman Scientific and Technical, New York (1993)
3. Balachandran, K., Ravikumar, R.: Existence of solutions of integrodifferential evolution equations with time varying delays. Appl. Math. E-Notes **7**, 1–8 (2007)
4. El Allaoui, A., Allaoui, Y., Melliani, S., Chadli, L.S.: Coupled system of mixed hybrid differential equations: linear perturbation of first and second type. J. Univ. Math. **1**(1), 24–31 (2018)
5. Feckan, M., Wang, J.R.: A general class of impulsive evolution equations. Topol. Methods Nonlinear Anal. **46**, 915–934 (2015)
6. Lakshmikantham, V., Bainov, D.D., Simeonov, P.S.: Theory of Impulsive Differential Equations. World Scientific, Singapore (1989)
7. Liu, J.H.: Integrodifferential equations with non-autonomous operators. Dynam. Syst. Appl. **7**, 427–440 (1998)
8. Melliani, S., Chadli, L.S., El Allaoui, A.: Periodic boundary value problems for controlled nonlinear impulsive evolution equations on Banach spaces. Int. J. Nonlinear Anal. Appl. **8**(1), 301–314 (2017)
9. Melliani, S., El Allaoui, A., Chadli, L.S.: A general class of periodic boundary value problems for controlled nonlinear impulsive evolution equations on Banach spaces. Adv. Differ. Equ. **2016**(1), 1–13 (2016). https://doi.org/10.1186/s13662-016-1004-2
10. Wang, J.R., Feckan, M., Zhou, Y.: Noninstantaneous impulsive models for studying periodic evolution processes in pharmacotherapy. In: Luo, A., Merdan, H. (eds.) Mathematical Modeling and Applications in Nonlinear Dynamics. Nonlinear Systems and Complexity, vol. 14, pp. 87–107. Springer, Cham (2016)

11. Yu, X., Wang, J.: Periodic boundary value problems for nonlinear impulsive evolution equations on Banach spaces. Commun. Nonlinear Sci. Numer. Simul. **22**, 980–989 (2015)
12. Rogovchenko, Y.V.: Nonlinear impulse evolution systems and applications to population models. J. Math. Anal. Appl. **207**, 300–315 (1997)
13. Wei, W., Xiang, X., Peng, Y.: Nonlinear impulsive integro-differential equations of mixed type and optimal controls. Optimization **55**, 141–156 (2006)
14. Diallo, M.A., Ezzinbi, K., Sene, A.: Impulsive integro-differential equations with nonlocal conditions in Banach spaces. Trans. A. Razmadze Math. Institule **171**, 304–315 (2017)
15. Dieye, M., Diop, M.A., Ezzinbi, K., Hmoyed, H.: On the existence of mild solutions for nonlocal impulsive integrodifferential equations in Banach spaces. Le Matematiche **LXXIV**(1), 13–34 (2019)
16. Pazy, A.: Semigroups of Linear Operators and Applications to Partial Differential Equations. Springer, Berlin (1983). https://doi.org/10.1007/978-1-4612-5561-1
17. Monch, H.: Boundary value problems for nonlinear ordinary differential equations of second order in Banach spaces. Nonlinear Anal. **4**(5), 985–999 (1980)

Density Functional Theory Study on the Electronic and Optical Properties of Graphene, Single-Walled Carbon Nanotube and C60

Abdelhafid Najim[1(✉)], Omar Bajjou[2], Mustapha Boulghallat[1], Khalid Rahmani[4], Mohammed Khenfouch[3], Lhouceine Moulaoui[4], Younes Chrafih[5], and Mohamed Al-Hattab[4]

[1] LDD, FST, Sultan Moulay Slimane University, Beni-Mellal, Morocco
abdelhafid.najim@usms.ma
[2] Material Physics Laboratory, FST, Sultan Moulay Slimane University, Beni-Mellal, Morocco
[3] Faculty of Applied Sciences, Ibn Zohr University, Ait-Melloul, Morocco
[4] ERPTM, Polydisciplinary Faculty, Sultan Moulay Slimane University, Beni-Mellal, Morocco
[5] LPHE-Modeling and Simulation Faculty of Science, Mohammed V University in Rabat, Rabat, Morocco

Abstract. In this study, the physical properties of buckyball (zero-dimensional), Single-Walled Carbon Nanotube (one-dimensional) and graphene (two-dimensional) nanostructures, were investigated using the density functional theory (DFT) calculations. The Visualizer module of Material Studio software is used to construct the structures of these materials. Then, the CASTEP code is used to optimize and calculate the band structures, total density of states (TDOS) and optical properties. We focus on such these three carbon materials, by reason of the appealing interest in the next generation in optoelectronic devices. Change the form of graphene to Single-Walled Carbon Nanotube (SWCNT) and buckyball (C60) leads to change its bandgap, TDOS, absorption coefficient, dielectric function and refractive index. The peaks of TDOS of graphene around the fermi level are very weak. The bandgap energy of graphene, SWCNT and C60 materials are 0, 0.198 and 0.102 eV, respectively. The peaks of absorption coefficient of graphene, SWCNT and C60 structures are at 268.26, 251.87 and 296.13 nm, respectively. It is found that the bandgap energy, TDOS and the absorption coefficient could be affected by the change of the form of graphene. These results, give the fundamental information's about understanding of the electronic and optical properties of various dimensional crystals (0D, 1D and 2D). This study can provide certain theoretical support for our future experimental research of graphene, SWCNT and C60 properties.

1 Introduction

Electronic states of carbon allotropes (graphene, SWCNT and C60), are well described by the tight binding Hamiltonian of π electrons of carbon atoms. In the carbon atoms form these allotropes, three atomic orbitals, $2s$, $2p_x$ and $2p_y$, are hybridized to form

S. Melliani and O. Castillo (Eds.): ICPAMS 2021, LNNS 476, pp. 17–25, 2023.
https://doi.org/10.1007/978-3-031-12416-7_2

three sp^2 hybrid orbitals, and $2p_z$ orbital remains perpendicular to other orbitals [1]. The hybridized orbitals are responsible of three σ bonds between the adjacent carbon atoms, and $2p_z$ orbital results in π bonds. Graphene, is a semimetal with zero bandgap, and it is the representative member of 2D-family that have been receiving much attention in many fields, by reason of its remarkable physical properties. Its band structures show zero gap at the Dirac point [2].

In recent years, the carbon nanotube material has been intensively studied, by reason of its importance as building block in nanotechnology. Three different types of carbon nanotubes are experimentally observed: armchair, zigzag and chiral [3]. The electronic properties of CNTs are result by reason of the sp^2-hybridized carbon atoms, and the delocalized π network perpendicular to nanotube surface.

Wrapping graphene into a sphere produces the fullerene family. Among different types of fullerene, the largest known and better stable species is C60. The C60 form the ball type structure, contains 12 pentagons and 20 hexagons [4]. The importance of C60 material has encouraged major studies in various fields. C60 has been applied in several applications, such as the nanoelectronics disciplines by reason of its exceptional physical properties. In addition, the nearly sp^2-hybridized carbon in C60 material backing the great electronic conductivity [5]. Therefore, C60 material have the diver's potential uses in optoelectronics devices and the big position in industry, by reason of its electron-taking characteristics and geometry.

Our paper is outlined as follows: In Sect. 2, we briefly present the Computational Methods. Section 3 is devoted to discuss the numerical results and give our interpretations, and finally, the conclusions of our study is included in Sect. 4.

2 Computational Methods

The electronic structures simulations and optical properties of graphene, SWCNT and C60 structures, were calculated based on the density functional theory (DFT), all calculations were performed using the CASTEP code by OTFG ultrasoft pseudopotentials [6, 7]. In this study, the exchange-correlation energy was employed using the Perdew-Burke-Ernzerhof (PBE) functional, within the generalized gradient approximation (GGA) [8]. A plane-wave energy cut-off was set to 600 eV for all calculations. The K-point of the Brillouin zone was sampled using $6 \times 6 \times 1$ gamma-centered Monkhorst-Pack grid during the geometry optimizations of graphene, SWCNT and C60 structures [9]. However, during all structural relaxations, the convergence tolerance criteria for the geometry optimization, were set to 2.10^{-6} eV/atom for the energy. During the atomic relaxations, the positions of atoms were optimized until the convergence of the force on each atom, was less than 0.05 eV/Å, and 0.03 Å for the displacement. The self-consistent field (SCF) convergence tolerance was set to 2.10^{-6} eV/atom. The maximum stress is set to 0.1 GPa. In the present simulations, the graphene, SWCNT and C60 structures are investigated in Fig. 1.

Fig. 1. Crystal structures of graphene, SWCNT and C60. The gray spheres represent the carbon atoms.

3 Results and Discussions

3.1 Electronic Structure

3.1.1 Optical Gap

The band structures of graphene, SWCNT and C60 materials, calculated along high symmetry directions in the Brillouin zone, are plotted in Fig. 2. The band structures show that the conduction band minimum and the valence band maximum are located at various points of the Brillouin zone. Hence, the SWCNT and C60 structures has a direct bandgap. The bandgap energy of graphene, SWCNT and C60 materials are 0 eV, 0.198 eV and 0.102 eV respectively. Thus, SWCNT and C60 structures have the semi-conducting properties. The wrapping graphene into C60, and folding into SWCNT, can lead to change its electronic states by reason of the symmetry breaking. These results, indicates that change the form of graphene sheet leads to modify its electronic properties and open its bandgap energy. This makes graphene and its derivatives (SWCNT and C60) are materials with bandgap energy used for developing the optoelectronics devices.

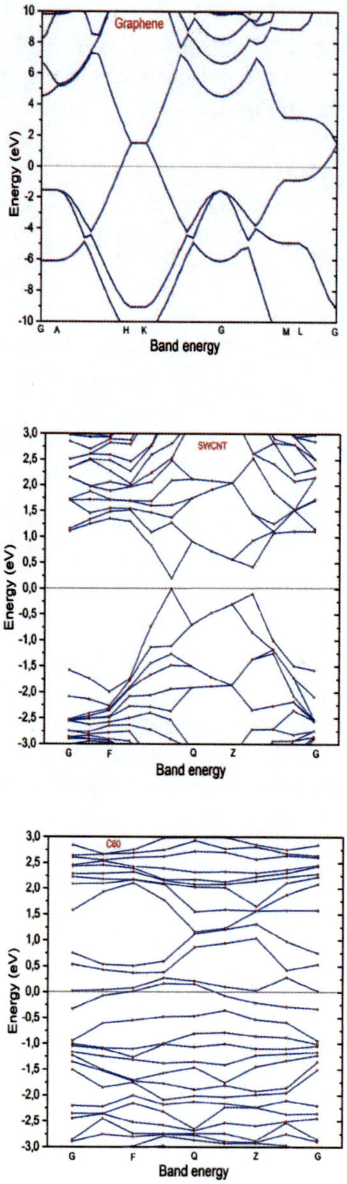

Fig. 2. Bandgap energy of graphene, SWCNT and C60 structures.

3.1.2 Density of States

The concept of the density of electronic states corresponds to the number of allowed electron energy states per unit energy interval around an energy E. The total density of states (TDOS) of graphene, SWCNT and C60 structures, were plotted as a function of the energy, as shown in Fig. 3. The TDOS near the Fermi level of graphene exhibits low population compared to the SWCNT and C60. Hence, the probability of occupation of the electronic states by the electrons in C60 around the fermi level is more important. These results, clarify the strong effect of change the form of graphene sheet to the SWCNT and C60 on its electronic properties.

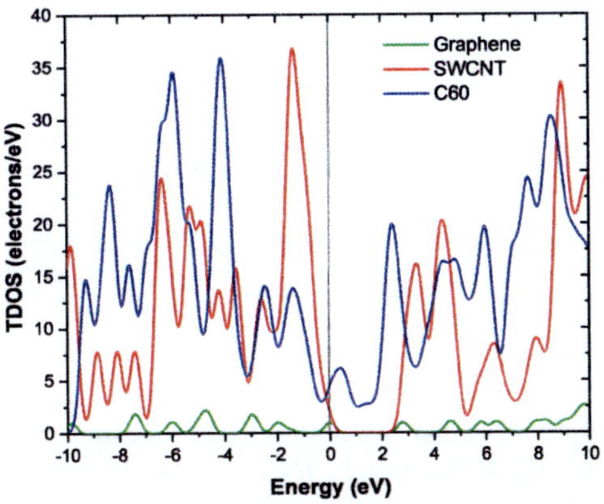

Fig. 3. The TDOS of graphene, SWCNT and C60 structures.

3.2 Optical Properties

3.2.1 Absorption

In the present study, we have presented the variation of absorption coefficient (α) as a function of wavelength of graphene, SWCNT and C60 materials, in 200–800 nm range, as shown in Fig. 4. The absorption spectrum indicates that the first peaks of absorption by graphene, SWCNT and C60 structures are at 268.26, 251.87 and 296.13 nm, respectively. The origin of these peaks arises from the important transitions that occur between the electronic states. These peaks correspond to the $\pi - \pi^*$ transition of C-C in sp^2 hybrid regions, which are close to the Fermi level. According to these peaks, graphene and C60 structures exhibits the absorption of light in both UV and visible ranges. But, the SWCNT structure absorbs only the UV light. The absorption of visible light by graphene and C60 materials, is essential for the solar cells applications.

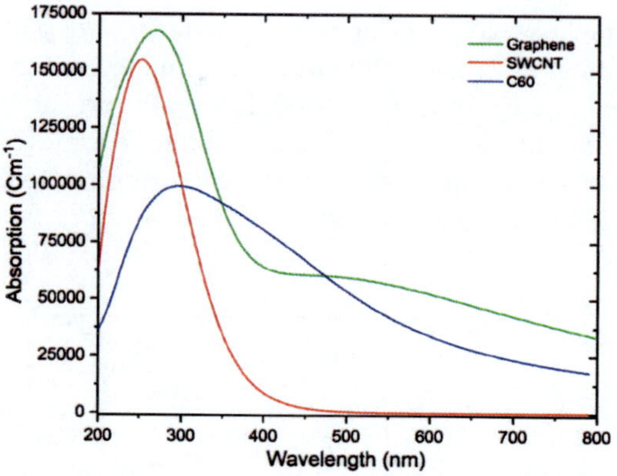

Fig. 4. Calculated α as a function of wavelength of graphene, SWCNT and C60 structures.

3.2.2 Dielectric Function

The complex frequency dependent dielectric function, $\varepsilon(\omega)$, can be used to descript the optical properties of (0D, 1D and 2D) materials, and describes how light interacts when propagating through matter. It determines the dispersion effects by its real part, $\varepsilon_1(\omega)$, and the absorption by the imaginary part, $\varepsilon_2(\omega)$. The complex dielectric function, $\varepsilon(\omega)$, is the sum of real and imaginary parts:

$$\varepsilon(\omega) = \varepsilon_1(\omega) + i\varepsilon_2(\omega) \tag{1}$$

In the present study, the $\varepsilon_1(\omega)$ and $\varepsilon_2(\omega)$ parts of graphene, SWCNT and C60 structures, are calculated in 200–800 nm wavelength range, as presented in Fig. 5. At low frequency, the $\varepsilon(\omega)$ is associated with the electronic intraband transitions, within the conduction band. In this spectral range the optical response is dominated by the free electron behavior. At higher frequency, the $\varepsilon(\omega)$ function reflects the electronic interband transitions. The peaks of $\varepsilon_1(\omega)$ part of graphene and SWCNT are at 368.77 and 347.29, respectively. The first peak of $\varepsilon_2(\omega)$ part of graphene, SWCNT and C60 are at 310.59, 285.30 and 454.50 nm, respectively. On the other hand, for these three materials, the peaks of $\varepsilon_2(\omega)$ are always related to the electron excitations. So, we can introduce the graphene and its derivative in transparent conducting films and photovoltaics devices. In addition, graphene, SWCNT and C60 are the absorbent materials of the visible light, which indicates that these three structures can be used as an important element in a number of optoelectronic devices.

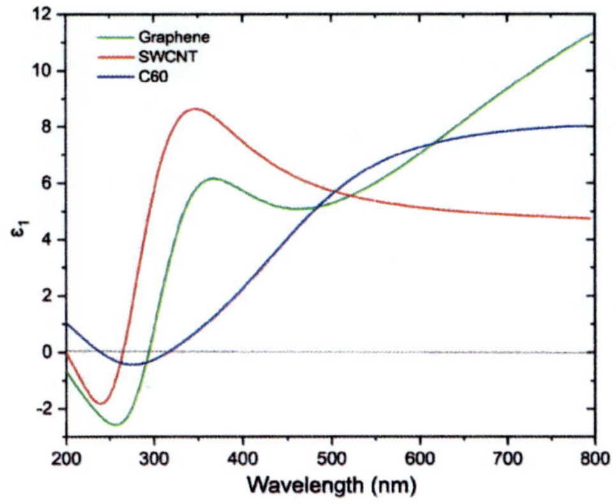

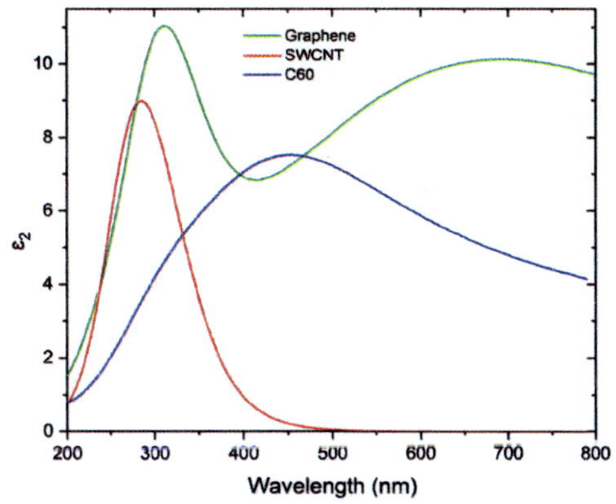

Fig. 5. Calculated ε of graphene, SWCNT and C60 structures as a function of wavelength

3.2.3 Refractive Index

Propagation in absorbing materials, can be described using a complex-valued of the refractive index, $n^*(\omega)$. The imaginary part, $k(\omega)$ then handles the attenuation, while the real part, $n(\omega)$, accounts of the refraction by [10]:

$$n^*(\omega) = n(\omega) + ik(\omega) \tag{2}$$

with,

$$n(\omega) = \sqrt{\frac{|\varepsilon(\omega)| + \varepsilon_1(\omega)}{2}} \tag{3}$$

$$k(\omega) = \sqrt{\frac{|\varepsilon(\omega)| - \varepsilon_1(\omega)}{2}} \tag{4}$$

The variation of $n(\omega)$ and $k(\omega)$ parts of graphene, SWCNT and C60 structures in terms of wavelength, found using the CASTEP code are depicted in Fig. 6. The above values of the static dielectric constant, $\varepsilon_1(0)$, and the static refractive index, $n(0)$, are validate the relation $n = \sqrt{\varepsilon_1}$ (see Table 1). The $n(\omega)$ part of these three structures are varie with the wavelength of photons incident. So, these materials are dispersive mediums. The $n(\omega)$ part of graphene, SWCNT and C60 is greater than 0.1662, 0.4514 and 0.4238, respectively, because the incident photons are slowed down by the interaction with electrons, inside these materials. On the other hand, the $k(\omega)$ part of graphene and C60 is more important in both ultraviolet and the visible ranges (see Fig. 6). When we analyze the graphs of the $\varepsilon_2(\omega)$ and $k(\omega)$, a similar physical behavior is observed in Fig. 5 and Fig. 6. Its two physical quantities give the information of the absorption of light by the graphene, SWCNT and C60 materials.

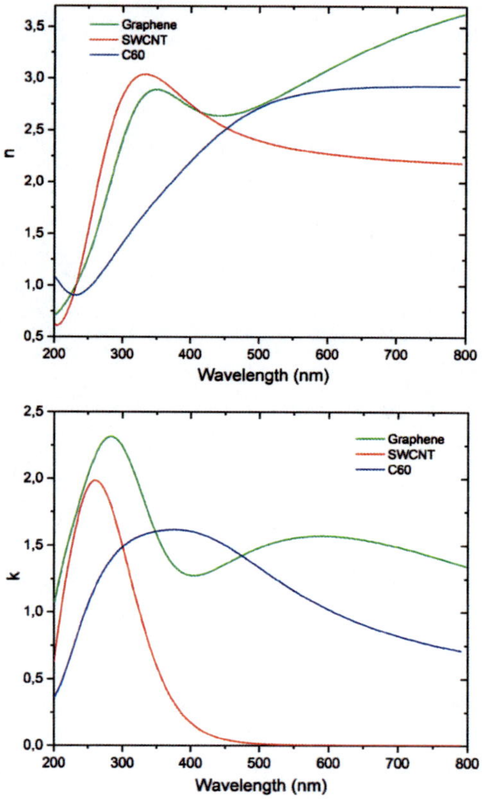

Fig. 6. Calculated n and k of graphene, SWCNT and C60 structures as a function of wavelength.

Table 1. The static dielectric constant, $\varepsilon_1(0)$, and the static refractive index, $n(0)$, of graphene, SWCNT and C60 structures

Parameters	Graphene	SWCNT	C60
$n(0)$	0.81547	0.88326	0.88287
$\varepsilon_1(0)$	0.66499	210.78015	0.77946

4 Conclusion

In conclusion, we have studied the electronic and optical properties of graphene, SWCNT and C60 structures, using the DFT calculations. We have shown that the SWCNT and C60 structures are semiconductors materials, and the graphene material chow zero bandgap. The bandgap energy of graphene can be significantly changed by changes its form to SWCNT and C60. We observed the changes in the dielectric function by reason of the change of the bandgap energy. The intensity of the absorption peaks is more important of graphene and C60 in both UV and visible ranges. But, the SWCNT structure have a maximum value of α peak in UV range. These insights provide a basis of the applications of graphene, Single-Walled Carbon Nanotube and C60 materials in the optoelectronic devices.

Acknowledgements. In conclusion, the authors sincerely thank all that contributed to this scientific work and particularly Sultan Moulay Slimane University and Africa Graphene Center.

References

1. Lehtinen, P.O., Foster, A.S., Ayuela, A., Krasheninnikov, A., Nordlund, K., Nieminen, R.M.: Magnetic properties and diffusion of adatoms on a graphene sheet. Phy. Rev. Lett. https://doi.org/10.1103/PhysRevLett.91.017202
2. Trevisanutto, P.E., Giorgetti, C., Reining, L., Ladisa, M., Olevano, V.: Ab initio GW many-body effects in graphene. Cond-mat (2008)
3. Prabhu, S., Bhaumik, S., Vinayagam, B.K.: Finite element modeling and analysis of zigzag and armchair type single wall carbon nanotube. J. Mech. Eng. Res. **4**(8), 260–266 (2012). https://doi.org/10.5897/JMER12.025
4. Inani, H., Singhal, R., Sharma, P., Vishnoi, R., Aggarwal, S., Sharma, G.D.: Effect of low fluence radiation on nanocomposite thin films of Cu nanoparticles embedded in fullerene C60. Vacuum **142**, 5–12 (2017)
5. Saraswati, T.E., Setiawan, U.H., Ihsan, M.R., Isnaeni, I., Herbani, Y.: The study of the optical properties of C60 fullerene in different organic solvents. Open Chem. **17**, 1198–1212 (2019). https://doi.org/10.1515/chem-2019-0117
6. Delley, B.: An all-electron numerical method for solving the local density functional for polyatomic molecules. J. Chem. Phys. **92**, 508–517 (1990)
7. McNellis, E.R., Meyer, J., Reuter, K.: Azobenzene at coinage metal surfaces: role of dispersive van der Waals interactions. Phys. Rev. B **80**(20), 205414 (2009)
8. Perdew, J.P., Burke, K., Ernzerhof, M.: Generalized gradient approximation made simple. Phys. Rev. Lett. **77**(18), 3865–3868 (1996)
9. Monkhorst, H.J., Pack, J.D.: Special points for Brillouin-zone integrations. Phys. Rev. B **13**(12), 5188 (1976)
10. Lashgari, H., Boochani, A., Shekaari, A., Solaymani, S., Sartipi, E., Mendi, R.T.: Electronic and optical properties of 2D graphene-like ZnS: DFT calculations. Appl. Surf. Sci. **369**, 76–81 (2016)

A SIR Epedimic Model Involving Fractional Atangana Derivative

H. El-Houari, A. Kraita, L. S. Chadli(✉), and H. Moussa

Faculté des Sciences et Techniques Béni Mellal, MAROC, Laboratoire de recherche "Mathématiques Appliquées et Calcul Scientifique", Beni Mellal, Morocco
sa.chadli@yahoo.fr

Abstract. This paper aime to analyzes a fractional order SIR epidemic model with a nonlinear incidence rate. First, we prove the global existence, positivity and boundedness of the solutions. The stability is studied. Finally, numerical simulations are presented to illustrate our theoretical results.

1 Introduction

Fractional calculus could be a generalization of integral and by-product to non-integer order that was 1st applied by Abel in his study of the tautochrone drawback [1]. Therefore, it's been largely applied in several fields like mechanics, viscoelasticity, technology, finance, and management theory [2–6]. As opposition the normal by-product, that could be a native operator, the fragmental order derivative has the most property referred to as memory result. A lot of exactly, ensuing state of fractional by-product for any given operate f depends not solely on their current state, but also upon all of their historical states. Because of this property, the fragmental order by-product is more suited to modeling issues involving memory, that is that the case in most biological systems [7,8]. Also, another advantage for exploitation fragmental order by-product is enlarging the stability region of the propellent systems. In medical specialty, several works involving fragmental order by-product are done, and most of them ar principally involved with SIR-type models with linear incidence rate [9–12]. In [13], Saeedian et al. studied the memory result of associate degree SIR epidemic model exploitation the Caputo fragmental by-product. They well-tried that this result plays an important role within the spreading of diseases. During this work, we have a tendency to any propose a fragmental order SIR model with nonlinear incidence rate given by

$$\begin{cases} {}^{AC}D^{\alpha}S(t) = \Lambda - \nu S - \dfrac{\beta SI}{1+\alpha_1 S+\alpha_2 I+\alpha_3 SI}, \\ {}^{AC}D^{\alpha}I(t) = \frac{\beta SI}{1+\alpha_1 S+\alpha_2 I+\alpha_3 SI} - (\nu+d+r)I, \\ {}^{AC}D^{\alpha}R(t) = rI - \nu R, \end{cases} \tag{1}$$

where ${}^{AC}D^{\alpha}$ denotes the Atangana-Caputo fractional derivative of order $0 < \alpha \leq 1$ defined for an arbitrary function $f(t)$ as follows:

$${}^{AC}_{a}D^{\alpha}_{t}f(t) = \frac{B(\alpha)}{1-\alpha}\int_a^t f'(x)E_{\alpha}\big[-\alpha\frac{(t-x)^{\alpha}}{1-\alpha}\big]dx, \tag{2}$$

S. Melliani and O. Castillo (Eds.): ICPAMS 2021, LNNS 476, pp. 26–36, 2023.
https://doi.org/10.1007/978-3-031-12416-7_3

where the Mittag-Leffler function $E_\alpha(z)$ defined by:

$$E_\alpha(z) = \sum_{k=0}^{\infty} \frac{z^k}{\Gamma(\alpha k+1)}, \quad (z \in \mathbb{C}, \alpha > 0),$$

and his Laplace Transformation is given by

$$\mathfrak{L}[x^{\beta-1}E_{\alpha,\beta}(ax^\alpha)](s) = \frac{s^{\alpha-\beta}}{s^\alpha - a} \qquad Re(\alpha > 0), Re(\beta) > 0. \tag{3}$$

Also the Laplace Transform of the fractional operator ${}_0^{ABC}D_t^\alpha$ is given by

$$\mathfrak{L}\left[{}_0^{ABC}D_t^\alpha f(t)\right](p) = \frac{B(\alpha)}{1-\alpha} \frac{p^\alpha \mathfrak{L}[f(t)](p) - p^{\alpha-1}f(0)}{p^\alpha + \frac{\alpha}{1-\alpha}}. \tag{4}$$

In system (1), $S(t)$, $I(t)$, and $R(t)$ represent the numbers of susceptible, infective, and recovered individuals at time t, respectively. Λ is the recruitment rate of the population, μ is the natural death rate, while d is the death rate due to disease and r is the recovery rate of the infective individuals. The incidence rate of disease is modeled by the specific functional response $\frac{\beta SI}{1+\alpha_1 S+\alpha_2 I+\alpha_3 SI}$ presented by Hattaf et al. [15], where $\beta > 0$ is the infection rate and α_1,α_2, α_3 are non-negative constants. Since the two first equations in system (1) are independent of the third equation, system (1) can be reduced to the following equivalent system:

$$\begin{cases} {}^{AC}D^\alpha S(t) = \Lambda - \nu S - \frac{\beta SI}{1+\alpha_1 S+\alpha_2 I+\alpha_3 SI}, \\ {}^{AC}D^\alpha I(t) = \frac{\beta SI}{1+\alpha_1 S+\alpha_2 I+\alpha_3 SI} - (\nu+d+r)I, \end{cases} \tag{5}$$

The rest of our paper is organized as follows. In Sect. 2, we show that our model is biologically and mathematically well posed. In Sect. 3, the existence of equilibria and their local stability are investigated. The global stability is studied in Sect. 4. Numerical simulations given in Sect. 5. We end up our paper with a conclusion.

2 Properties of Solutions

Let $Y(t) = (S(t), I(t))T$, then system (5) can be reformulated as follows

$${}^{AC}D^\alpha Y(t) = G(Y(t)) \tag{6}$$

where $G(Y(t)) = \begin{pmatrix} \Lambda - \nu S - \frac{\beta SI}{1+\alpha_1 S+\alpha_2 I+\alpha_3 SI} \\ \frac{\beta SI}{1+\alpha_1 S+\alpha_2 I+\alpha_3 SI} - (\nu+d+r)I \end{pmatrix}$.
In order to prove the global existence of solutions for system (5), we need the following lemma which is a direct corollary from [2, Lemma 3.1].

Lemma 2.1. *Assume that the vector function $F : \mathbb{R}^2 \to \mathbb{R}^2$ satisfies the following conditions:*

i) $G(Y)$ *and* $\frac{\partial G}{\partial Y}$ *are continuous.*
ii) $||G(Y)|| \leq \rho + \lambda ||Y||$ $\forall Y \in \mathbb{R}^2$ *where* ρ *and* λ *are two positive constants. Then system (12) has a unique solution.*

For biological reasons, we consider system (5) with the following initial conditions:

$$S(0) \geq 0 \qquad I(0) \geq 0. \tag{7}$$

Theorem 2.2. *For any given initial conditions satisfying (13), there exists a unique solution of system (5) defined on* $[0, +\infty)$*, and this solution remains non-negative and bounded for all* $t \geq 0$*. Moreover, we have*

$$P(t) \leq P(0) + \frac{\Lambda}{\nu},$$

where $P(t) = S(t) + I(t)$

Proof. Since the vector function G satisfies the first condition of Lemma 2.1, we only need to prove the last one. Denote

$$\varepsilon = \begin{pmatrix} \Lambda \\ 0 \end{pmatrix}, A_1 = \begin{pmatrix} -\nu & 0 \\ 0 & -(\nu + d + r) \end{pmatrix}, A_2 = \begin{pmatrix} 0 & \frac{-\beta}{\alpha_1} \\ 0 & \frac{\beta}{\alpha_1} \end{pmatrix}, A_3 = \begin{pmatrix} \frac{-\beta}{\alpha_2} & 0 \\ \frac{\beta}{\alpha_2} & 0 \end{pmatrix},$$

$$A_4 = \begin{pmatrix} \frac{-\beta}{\alpha_3} \\ \frac{\beta}{\alpha_3} \end{pmatrix}, \qquad A_5 = \begin{pmatrix} -\beta & 0 \\ \beta & 0 \end{pmatrix}$$

Hence, we discuss four cases as follows.
Case 1: If $\alpha_1 \neq 0$, we have

$$G(Y) = \varepsilon + A_1 Y + \frac{\alpha_1 S}{1 + \alpha_1 S + \alpha_2 I + \alpha_3 SI} A_2 Y$$

Then

$$||G(Y)|| \leq ||\varepsilon|| + ||A_1 Y|| + ||A_2 Y|| = ||\varepsilon|| + (||A_1|| + ||A_2||)||Y||.$$

Case 2: If $\alpha \neq 0$ we have

$$G(Y) = \varepsilon + A_1 Y + \frac{\alpha_2 S}{1 + \alpha_1 S + \alpha_2 I + \alpha_3 SI} A_3 Y$$

Then

$$||G(Y)|| \leq ||\varepsilon|| + (||A_1|| + ||A_3||)||Y||$$

Case 3: If $\alpha_3 \neq 0$ we have

$$G(Y) = \varepsilon + A_1 Y + \frac{\alpha_3 SI}{1 + \alpha_1 S + \alpha_2 I + \alpha_3 SI} A_4.$$

Then

$$||G(Y)|| \leq ||\varepsilon|| + (||A_1|| + ||I||A_5)||Y||$$

By Lemma 2.1, system (5) has a unique solution. Next, we prove the non-negativity of this solution. Since ${}^{AC}D^{\alpha}S|_{s=0} = \Lambda \geq 0$, ${}^{AC}D^{\alpha}I|_{I=0} = 0$. Based on Lemmas 2.5 and 2.6 in [16], it is not hard to deduce that the solution of (2) remains non-negative for all $t \geq 0$

Finally, we establish the boundedness of solution. By summing all the equations of system (5), we obtain

$$\begin{aligned} {}^{AC}DP(t) &= \Lambda - \nu S(t) - (\nu + d + r)I(r) \\ &\leq \Lambda - \nu P(t). \end{aligned} \tag{8}$$

By using (4), we get

$$\frac{B(\alpha)}{1-\alpha}\frac{\mathfrak{L}(P)(s)s^{\alpha} - s^{\alpha-1}P(0)}{s^{\alpha} + \frac{\alpha}{1-\alpha}} \leq \frac{\Lambda}{s} - \nu\mathfrak{L}(P)(s)$$

Applying Laplace inverse to last inequality we obtain

$$\begin{aligned} P(t) &\leq \frac{\alpha\Lambda}{1-\alpha}\mathfrak{L}^{-1}\Big(\frac{s^{\alpha-(1+\alpha)}}{s^{\alpha} + \frac{B(\alpha)+\alpha\nu}{1-\alpha}}\Big) + (\Lambda + B(\alpha)P(0))\mathfrak{L}^{-1}\Big(\frac{s^{\alpha-1}}{s^{\alpha} + \frac{B(\alpha)+\alpha\nu}{1-\alpha}}\Big) \\ &\leq \frac{\alpha\Lambda}{1-\alpha}t^{\alpha}E_{\alpha,\alpha+1}\Big(\frac{B(\alpha)+\alpha\nu}{\alpha-1}t^{\alpha}\Big) + (\Lambda + B(\alpha)P(0))E_{\alpha,1}\Big(\frac{B(\alpha)+\alpha\nu}{\alpha-1}t^{\alpha}\Big) \\ &\leq E_{\alpha,1}\Big(-\frac{B(\alpha)+\alpha\nu}{1-\alpha}t^{\alpha}\Big)\Big[\frac{-\alpha\Lambda}{B(\alpha)+\alpha\nu} + \Lambda + B(\alpha)P(0)\Big] + \frac{\alpha\Lambda}{(B(\alpha)+\alpha\nu)\Gamma(1)} \end{aligned}$$

Since $0 < E_{\alpha,1}\Big(-\frac{B(\alpha)+\alpha\nu}{1-\alpha}t^{\alpha}\Big) < 1$ then we get the result.

Lemma 2.3 *Let $x(t) \subset \mathbb{R}^{+}$ be a continuous and derivable function, then for any time instant $t \geq t_0$*

$$ {}^{AC}_{t_0}D^{\alpha}_{t}\Big[x(t) - x^* - x^*Ln(\frac{x(t)}{x^*})\Big] \leq \Big(1 - \frac{x^*}{x(t)}\Big){}^{AC}_{t_0}D^{\alpha}_{t}x(t) \quad x^* \in \mathbb{R}^{+}\forall\alpha \in (0,1). \tag{9}$$

Proof. We have that

$${}^{AC}_{t_0}D^{\alpha}_{t}\Big[x(t) - x^* - x^*Ln(\frac{x(t)}{x^*})\Big] \leq \Big(1 - \frac{x^*}{x(t)}\Big){}^{AC}_{t_0}D^{\alpha}_{t}x(t)$$

then

$${}^{AC}_{t_0}D^{\alpha}_{t}x(t) - {}^{AC}_{t_0}D^{\alpha}_{t}x^* - x^*{}^{AC}_{t_0}D^{\alpha}_{t}Ln(\frac{x(t)}{x^*}) \leq \Big(x(t) - \frac{x^*}{x(t)}\Big){}^{AC}_{t_0}D^{\alpha}_{t}x(t)$$

Since

$${}^{AC}_{t_0}D^{\alpha}_{t}x(t) - x(t){}^{AC}_{t_0}D^{\alpha}_{t}Ln(\frac{x(t)}{x^*}) \leq 0$$

Last inequality implies that

$$\frac{B(\alpha)}{1-\alpha}\int_{t_0}^{t}\dot{x}(u)E_\alpha[\frac{-\alpha(t-u)^\alpha}{1-\alpha}]du - x(t)\frac{B(\alpha)}{1-\alpha}\int_{t_0}^{t}\frac{\dot{x}(u)}{x(t)}E_\alpha[\frac{-\alpha(t-u)^\alpha}{1-\alpha}]du \le 0$$

then

$$\frac{B(\alpha)}{1-\alpha}\int_{t_0}^{t}\dot{x}(u)E_\alpha[\frac{-\alpha(t-u)^\alpha}{1-\alpha}](1-\frac{x(t)}{x(u)})du \le 0.$$

We set $w(u)=\frac{x(u)-x(t)}{x(t)}$ then $\dot{w}(u)=\frac{\dot{x}(u)}{x(t)}$ and

$$\frac{B(\alpha)}{1-\alpha}\int_{t_0}^{t}\dot{w}(u)E_\alpha[\frac{-\alpha(t-u)^\alpha}{1-\alpha}]x(t)(1-\frac{1}{1+w(u)})du \le 0.$$

Using part integration with setting

$$v=\frac{B(\alpha)}{1-\alpha}E_\alpha[\frac{-\alpha(t-u)^\alpha}{1-\alpha}] \Rightarrow dv=\frac{B(\alpha)\alpha}{(1-\alpha)^2}(t-u)^{\alpha-1}E_\alpha\left[\alpha\frac{(t-u)^\alpha}{\alpha-1}\right]$$

and

$$dy=x(t)(1-\frac{1}{1+w(u)})\dot{w}(u)du$$

then we have

$$\begin{aligned}
&\left[x(t)(w(u)\ -\ ln(w(u)+1))\frac{B(\alpha)}{1-\alpha}E_\alpha[\frac{-\alpha(t-u)^\alpha}{1-\alpha}]\right]_{t_0}^{t}\\
&\quad-\frac{B(\alpha)}{1-\alpha}\int_{t_0}^{t}\frac{\alpha(t-u)^{\alpha-1}}{1-\alpha}E_\alpha[\frac{-\alpha(t-u)^\alpha}{1-\alpha}]x(t)(w(u)-ln1+w(u))du\\
&\quad=-x(t)\left[(w(t_0)-\ln(w(t_0)+1))\frac{B(\alpha)}{1-\alpha}E_\alpha[\frac{-\alpha(t-u)^\alpha}{1-\alpha}(t-t_0)]\right.\\
&\quad-\frac{B(\alpha)}{1-\alpha}\int_{t_0}^{t}\frac{\alpha(t-u)^{\alpha-1}}{1-\alpha}E_\alpha[\frac{-\alpha(t-u)^\alpha}{1-\alpha}]x(t)(w(u)-ln1+w(u))du\\
&\quad\le 0
\end{aligned}$$

the proof is complete.

3 Equilibria and Their Local Stability

Now, we discuss the existence and the local stability of equilibria for system (5). For this, we define the basic reproduction number R_0 of our model by

$$R_0=\frac{\beta\Lambda}{(\nu+\alpha_1\lambda)(\nu+d+r)}$$

It is not hard to see that system (5) has always a disease-free equilibrium of the form $E_0=(\frac{\Lambda}{\nu},0)$. In the following result, we prove that there exists another equilibrium point when $R_0>1$.

Theorem 3.1 *i) If $R_0 \leq 1$, then system (5) has a unique disease-free equilibrium of the form $E_0(S_0, 0)$, where $S_0 = \frac{\Lambda}{\nu}$.*
ii) If $R_0 > 1$, the disease-free equilibrium is still present and system (5) has a unique endemic equilibrium of the form $E^*(S^*, \frac{\Lambda - \nu S^*}{a})$, where $S^* = \frac{2(a+\alpha_2\Lambda)}{\beta - \alpha_1 a + \alpha_2 \nu - \alpha_3 \Lambda + \sqrt{\sigma}}$ with $a = \nu + d + r$ and $\sigma = (\beta - \alpha_1 a + \alpha_2 \nu - \alpha_3 \Lambda)^2 + 4\alpha_3 \nu (a + \alpha_2 \Lambda)$.

Next, we study the local stability of the disease-free equilibrium E_0 and the endemic equilibrium E^*. We define the Jacobian matrix of system (5) at any equilibrium $\overline{E}(\overline{S}, \overline{I})$ by

$$J_E = \begin{pmatrix} -\nu - \frac{\beta \overline{I}(1+\alpha_2 \overline{I})}{(1+\alpha_1 \overline{S}+\alpha_2 \overline{I}+\alpha_3 \overline{S}\overline{I})^2} & \frac{-\beta \overline{S}(1+\alpha_1 \overline{S})}{(1+\alpha_1 \overline{S}+\alpha_2 \overline{I}+\alpha_3 \overline{S}\overline{I})^2} \\ \frac{\beta \overline{I}(1+\alpha_2 \overline{I})}{(1+\alpha_1 \overline{S}+\alpha_2 \overline{I}+\alpha_3 \overline{S}\overline{I})^2} & \frac{\beta \overline{S}(1+\alpha_1 \overline{S})}{(1+\alpha_1 \overline{S}+\alpha_2 \overline{I}+\alpha_3 \overline{S}\overline{I})^2} - a \end{pmatrix} \tag{10}$$

We recall that a sufficient condition for the local stability of $\overline{E}$ is

$$|arg(\xi_i)| > \frac{\alpha \pi}{2}, \quad i = 1, 2 \tag{11}$$

where ξ_i are the eigenvalues of J_E [17]. First, we establish the local stability of E_0.

Theorem 3.2 *The disease-free equilibrium E_0 is locally asymptotically stable if $R_0 < 1$ and unstable whenever $R_0 > 1$*

Proof. At E_0, (10) becomes $J_{E_0} \begin{pmatrix} -\nu & \frac{-\beta\Lambda}{\nu+\alpha_1\Lambda} \\ 0 & \frac{\beta\Lambda}{\nu+\alpha_1\Lambda} - a \end{pmatrix}$

Hence, the eigenvalues of J_{E_0} are $\xi_1 = -\nu$ and $\xi_2 = a(R_0 - 1)$. Clearly, ξ_2 satisfies condition (11) if $R_0 < 1$, and since ξ_1 is negative, the proof is complete.

Now, to investigate the local stability of E^*, we assume that $R_0 > 1$. After evaluating (10) at E^* and calculating its characteristic equation, we get $\lambda^2 + a_1 \lambda + a_2 = 0$ where

$$a_1 = \nu + \frac{\beta I^*(1+\alpha_2 I^*) + \beta S^* I^*(\alpha_2 + \alpha_3 S^*)}{(1+\alpha_1 S^* + \alpha_2 I^* + \alpha_3 S^* I^*)^2},$$

$$a_2 = \frac{a\beta I^*(1+\alpha_2 I^*) + \nu\beta S^* I^*(\alpha_2 + \alpha_3 S^*)}{(1+\alpha_1 S^* + \alpha_2 I^* + \alpha_3 S^* I^*)^2},$$

It is clear that $a_1 > 0$ and $a_2 > 0$. Hence, the Routh-Hurwitz conditions are satisfied. From [22, Lemma 1], we get the following result.

Theorem 3.3 *If $R_0 > 1$, then the endemic equilibrium E^* is locally asymptotically stable.*

4 Global Stability

In this section, we investigate the global stability of both equilibria

Theorem 4.1 *The disease-free equilibrium E_0 is globally asymptotically stable whenever $R_0 \leq 1$*

Proof. Consider the following Lyapunov functional:

$$L_0(t) = \frac{S_0}{1+\alpha_1 S_0}\Psi(\frac{S}{S_0}) + I,$$

where $\Psi(x) = x - 1 - ln(x)$, $x > 0$. It is obvious that $\Psi(x) \geq 0$. Then we calculate the fractional time derivation of L_0 along the solution of system. By using Lemma 2.3, we have

$$^{AC}D^{\alpha}L_0(t) \leq \frac{1}{1+\alpha_1 S_0}(1-\frac{S_0}{S})^{AC}D^{\alpha}S + {}^{AC}D^{\alpha}I.$$

Using $\Lambda = \nu S_0$, we get

$$\begin{aligned} ^{AC}D^{\alpha}L_0(t) \leq & \frac{1}{1+\alpha_1 S_0}(1-\frac{S_0}{S})\nu(S_0 - S) \\ & - \frac{1}{1+\alpha_1 S_0}(1-\frac{S_0}{S})\frac{\beta SI}{1+\alpha_1 S + \alpha_2 I + \alpha_3 SI} \\ & + \frac{\beta SI}{1+\alpha_1 S + \alpha_2 I + \alpha_3 SI} - aI \\ \leq & \frac{-\nu}{(1+\alpha_1 S_0)S}(S - S_0)^2 + a(R_0 - 1)I. \end{aligned}$$

Since $R_0 \leq 1$, then $^{AC}D^{\alpha}L_0(t) \leq 0$. Furthermore $^{AC}D^{\alpha}L_0(t) = 0$ if and only if $S = S_0$ and $(R_0 - 1)I = 0$. We discuss two cases:

- If $R_0 < 1$, then $I = 0$.
- If $R_0 = 1$, from the first equation in (5) and $S = S_0$, we have

$$0 = \Lambda - \nu S_0 - \frac{\beta S_0 I}{1+\alpha_1 S_0 + \alpha_2 I + \alpha_3 S_0 I},$$

which implies that $\frac{\beta S_0 I}{1+\alpha_1 S_0+\alpha_2 I+\alpha_3 S_0 I} = 0$. Consequently, we get $I = 0$. From the above discussions, we conclude that the largest invariant set of $\{(S,I) \in \mathbb{R}^2_+ : {}^{AC}D^{\alpha}L_0(t) = 0\}$ is the singleton $\{E_0\}$. Consequently, from [24, Lemma 4.6], E_0 is globally asymptotically stable.

Theorem 4.2 *Assume that $R_0 > 1$. Then the endemic equilibrium E^* is globally asymptotically stable*

Proof. Consider the following Lyapunov functional:

$$L_1(t) = \frac{1+\alpha_2 S^*}{1+\alpha_1 S^* + \alpha_2 I^* + \alpha_3 S^* I^*} S^*\Psi(\frac{S}{S^*}) + I^*\Psi(\frac{I}{I^*}).$$

Hence, we have

$$^{AC}D^{\alpha}L_1(t) \leq \frac{1+\alpha_2 S^*}{1+\alpha_1 S^* + \alpha_2 I^* + \alpha_3 S^* I^*}(1-\frac{S}{S^*})^{AC}D^{\alpha}S + (1-\frac{I}{I^*})^{AC}D^{\alpha}I.$$

Note that $\Lambda = \nu S^* + aI^*$ and $\frac{\beta S^*}{1+\alpha_1 S^*+\alpha_2 I^*+\alpha_3 S^* I^*} = a$. Thus

$$
\begin{aligned}
{}^{AC}D^{\alpha}L_1(t) \leq & \frac{1+\alpha_2 S^*}{1+\alpha_1 S^*+\alpha_2 I^*+\alpha_3 S^* I^*}(1-\frac{S}{S^*})\nu(S^*-S) \\
& + aI^*\Big(\frac{(1+\alpha_2 S^*)(S-S^*)}{(1+\alpha_1 S^*+\alpha_2 I^*+\alpha_3 S^* I^*)S} - \frac{(1+\alpha_1 S^*+\alpha_2 I^*+\alpha_3 S^* I^*)I}{(1+\alpha_1 S^*+\alpha_2 I^*+\alpha_3 S^* I^*)I^*}\Big) \\
& + aI^*\Big(1-\frac{I}{I^*}-\frac{(1+\alpha_1 S^*+\alpha_2 I^*+\alpha_3 S^* I^*)S}{(1+\alpha_1 S+\alpha_2 I+\alpha_3 SI)S^*}\Big) \\
\leq & \frac{-\nu(1+\alpha_2 S^*)(S-S^*)^2}{(1+\alpha_1 S^*+\alpha_2 I^*+\alpha_3 S^* I^*)S} \\
& + aI^*\Big(3-\frac{(1+\alpha_1 S+\alpha_2 I^*+\alpha_3 SI^*)S^*}{(1+\alpha_1 S^*+\alpha_2 I^*+\alpha_3 S^* I^*)S} - \frac{1+\alpha_1 S+\alpha_2 I+\alpha_3 SI}{1+\alpha_1 S+\alpha_2 I^*+\alpha_3 SI^*} \\
& - \frac{(1+\alpha_1 S^*+\alpha_2 I^*+\alpha_3 S^* I^*)S}{(1+\alpha_1 S+\alpha_2 I+\alpha_3 SI)S^*}\Big) + aI^*\Big(-1-\frac{I}{I^*} \\
& + \frac{1+\alpha_1 S+\alpha_2 I+\alpha_3 SI}{1+\alpha_1 S+\alpha_2 I^*+\alpha_3 SI^*} + \frac{(1+\alpha_1 S+\alpha_2 I^*+\alpha_3 SI^*)I}{(1+\alpha_1 S+\alpha_2 I+\alpha_3 SI)I}\Big) \\
\leq & \frac{-\nu(1+\alpha_2 S^*)(S-S^*)^2}{(1+\alpha_1 S^*+\alpha_2 I^*+\alpha_3 S^* I^*)S} \\
& - aI^*\Big(\Psi(\frac{(1+\alpha_1 S+\alpha_2 I^*+\alpha_3 SI^*)S^*}{(1+\alpha_1 S^*+\alpha_2 I^*+\alpha_3 S^* I^*)S}) + \Psi(\frac{1+\alpha_1 S+\alpha_2 I+\alpha_3 SI}{1+\alpha_1 S+\alpha_2 I+\alpha_3 SI^*}) \\
& + \Psi(\frac{(1+\alpha_1 S^*+\alpha_2 I^*+\alpha_3 S^* I^*)S}{(1+\alpha_1 S+\alpha_2 I+\alpha_3 SI)S^*})\Big) \\
& - \frac{(a(\alpha_2+\alpha_3 S)(1+\alpha_1 S)(I-I^*)^2)}{(1+\alpha_1 S+\alpha_2 I^*+\alpha_3 SI^*)((1+\alpha_1 S+\alpha_2 I+\alpha_3 SI)}.
\end{aligned}
$$

5 Numerical Result

We use the conformable Euler's method. The the fraction power series expansions to the initial value problem

$$
\begin{cases} {}^{AC}D^{\alpha}x(t) = f(x(t)), & a \leq t \leq b \\ x(0) = c \end{cases} \tag{12}
$$

where $x(t) = (S(t), I(t), R(t)), f(x(t)) = \begin{pmatrix} \Lambda - \nu S - \frac{\beta SI}{1+\alpha_1 S+\alpha_2 I+\alpha_3 SI} \\ \frac{\beta SI}{1+\alpha_1 S+\alpha_2 I+\alpha_3 SI} - (\nu+d+r)I \\ rI - \nu R \end{pmatrix}, (a,b) =$ $(0,90)$, $\alpha = 0.8$ and $c = (5,2,-2.9)$.

Can be written as,

$$x(t_i+h)=x(t_{i+1})=x(t_i)+\frac{h^\alpha}{\alpha}{}^{AC}D^\alpha x(t_i)+\frac{h^{2\alpha}}{2\alpha^2}{}^{AC}D^{2\alpha}x(t_i+\theta_i h) \quad (13)$$

with $h=\frac{b-a}{N}$, $t_i=a+(i-1)h$, for $i=1,...,N+1$

We get the numerical schema

$$\begin{cases} S_{i+1}=S_i+\frac{h^\alpha}{\alpha}\left(\Lambda-\nu S_i-\frac{\beta S_i I_i}{1+\alpha_1 S_i+\alpha_2 I_i+\alpha_3 S_i I_i}\right), \\ I_{i+1}=I_i+\frac{h^\alpha}{\alpha}\left(\frac{\beta S_i I_i}{1+\alpha_1 S_i+\alpha_2 I_i+\alpha_3 S_i I_i}-(\nu+d+r)I_i\right), \\ R_{i+1}=R_i+\frac{h^\alpha}{\alpha}(rI_i-\nu R_i) \\ (S_1,I_1,R_1)=(5,2,-2.9) \end{cases} \quad (14)$$

Now we analyze the stability of the system for different values of (r,d,ν), $(\alpha_1,\alpha_2,\alpha_3)$ and (Λ,β).

Table 1. Stability analysis of the system (1) for different values.

(r,d,ν)	$(\alpha_1,\alpha_2,\alpha_3)$	(Λ,β)	Stability
$(3,-4.04,1)$	$(9,8,7)$	$(0.07,-6)$	Yes
$(3,-2,1)$	$(5,11,9)$	$(0.9,8)$	No
$(3,-4.18,1.15)$	$(5,6,7)$	$(5,\frac{5}{2})$	No

The numerical solutions $S(t), I(t)$ and $R(t)$ are shown in Figs. 1, 2 and 3 with fixed α and different parameter values.

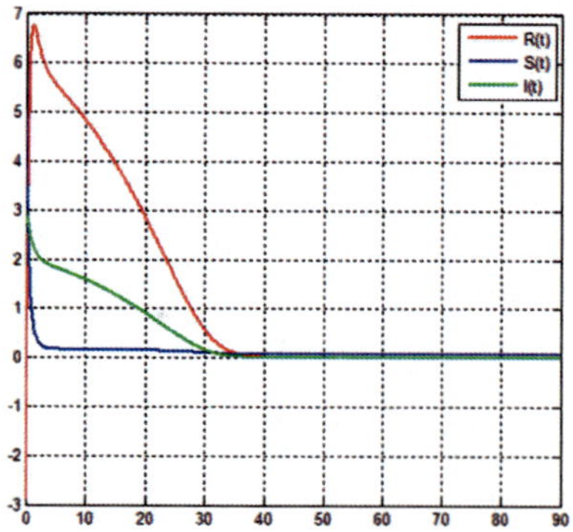

Fig. 1. The system (1) is stable with parameters $(r,d,\nu)=(3,-4.04,1)$, $(\alpha_1,\alpha_2,\alpha_3)=(9,8,7)$, and $(\Lambda,\beta)=(0.07,-6)$

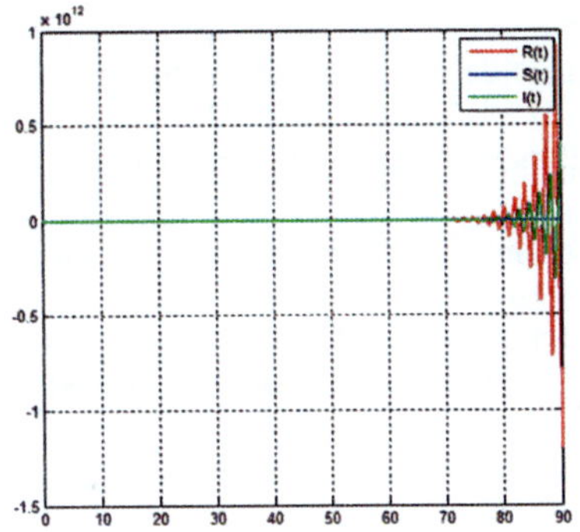

Fig. 2. The system (1) is unstable with parameters $(r, d, \nu) = (3, -2, 1)$, $(\alpha_1, \alpha_2, \alpha_3) = (5, 11, 9)$, and $(\Lambda, \beta) = (0.9, 8)$

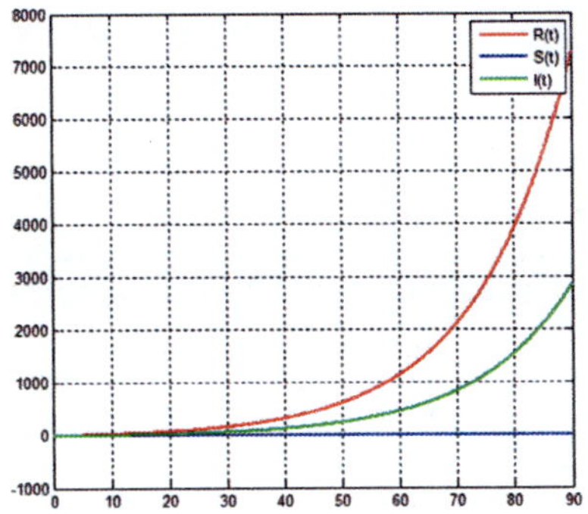

Fig. 3. The system (1) is unstable with parameters $(r, d, \nu) = (3, -4.18, 1.15)$, $(\alpha_1, \alpha_2, \alpha_3) = (5, 6, 7)$, and $(\Lambda, \beta) = (5, \frac{5}{2})$

6 Conclusion

In this work, we have studied the existence of solution to SIR epidemic model with nonlinear incidence rate involving Fractional Atangana derivative. Also we have prove the positivity and boundedness of the solutions. The stability is studied. Moreover, since the system (1) must be solved numerically to reconcile our results are given in Table 1, an appropriate numerical method should be selected.

References

1. Abel, N.H.: Solution de quelques problèmes á l'aide d'intégrales définies, Werke 1. Mag. Naturvidenkaberne 10–12 (1823)
2. Lin, W.: Global existence theory and chaos control of fractional differential equations. J. Math. Anal. Appl. **332**, 709–726 (2007)
3. Rossikhin, Y.A., Shitikova, M.V.: Application of fractional calculus to dynamic problems of linear and nonlinear hereditary mechanics of solids. Appl. Mech. Rev. **50**, 15–67 (1997)

4. Jia, G.L., Ming, Y.X.: Study on the viscoelasticity of cancellous bone based on higher-order fractional models. In: Proceeding of the 2nd International Conference on Bioinformatics and Biomedical Engineering (ICBBE 2008), pp. 1733–1736 (2006)
5. Magin, R.: Fractional calculus in bioengineering. Crit. Rev. Biomed. Eng. **32**(1), 13–77 (2004)
6. Scalas, E., Gorenflo, R., Mainardi, F.: Fractional calculus and continuous-time finance. Phys. A **284**(1–4), 376–384 (2000)
7. Capponetto, R., Dongola, G., Fortuna, L., Petras, I.: Fractional Order Systems: Modelling and Control Applications. World Scientific Series in Nonlinear Science, Series A, vol. 72 (2010)
8. Cole, K.S.: Electric conductance of biological systems. In: Cold Spring Harbor Symposia on Quantitative Biology, vol. 1, pp. 107–116 (1933)
9. Djordjevic, V.D., Jaric, J., Fabry, B., Fredberg, J.J., Stamenovic, D.: Fractional derivatives embody essential features of cell rheological behavior. Ann. Biomed. Eng. **31**, 692–699 (2003). https://doi.org/10.1114/1.1574026
10. Dos Santos, J.P.C., Monteiro, E., Vieira, G.B.: Global stability of fractional SIR epidemic model. Proc. Ser. Braz. Soc. Comput. Appl. Math. **5**(1), 1–7 (2017)
11. Okyere, E., Oduro, F.T., Amponsah, S.K., Dontwi, I.K., Frempong, N.K.: Fractional order SIR model with constant population. Br. J. Math. Comput. Sci. **14**(2), 1–12 (2016)
12. Guo, Y.: The stability of the positive solution for a fractional SIR model. Int. J. Biomath. **10**(1), 1–14 (2017)
13. Saeedian, M., Khalighi, M., Azimi-Tafreshi, N., Jafari, G.R., Ausloos, M.: Memory effects on epidemic evolution: the susceptible-infected-recovered epidemic model. Phys. Rev. E **95**, 022409 (2017)
14. Podlubny, I.: Fractional Differential Equations: An Introduction to Fractional Derivatives, Fractional Differential Equations, to Methods of Their Solution and Some of Their Applications, vol. 198. Academic Press, San Diego (1998)
15. Hattaf, K., Yousfi, N., Tridane, A.: Stability analysis of a virus dynamics model with general incidence rate and two delays. Appl. Math. Comput. **221**, 514–521 (2013)
16. Boukhouima, A., Hattaf, K., Yousfi, N.: Dynamics of a fractional order HIV infection model with specific functional response and cure rate. Int. J. Differ. Equ. **2017**, 8372140 (2017)
17. Petras, I.: Fractional-Order Nonlinear Systems: Modeling, Analysis and Simulation. Springer, Berlin (2011). https://doi.org/10.1007/978-3-642-18101-6

Fuzzy Local Ring

S. Achik(✉), M. Oukessou, and I. Bakhadach

LMACS, Laboratoire de Mathématiques Appliquées and Calcul Scientifique, Sultan Moulay Slimane University, PO Box 523, 23000 Beni Mellal, Morocco
sofyanebenabdou@gmail.com

Abstract. In this work we define a fuzzy divisible ring and fuzzy local ring using the notion of fuzzy point, Also we give a characterization of fuzzy local rings.

1 Introduction

In 1982 Liu developed the fuzzy abstract algebra, he discussed fuzzy subrings and fuzzy ideals [2], and that after the Zadeh's work [9] and Rosenfeld [6] and in 1984 Mukherjee, and Bhattacharya examined normal fuzzy subgroups and in 1997 Alkhamees and Mordeson characterize local rings in terms of certain fuzzy ideals [1]. They also characterize rings of fractions at a prime ideal in terms of fuzzy ideals, Then in 2009 Wang, Ruan and Kerre studied fuzzy subrings and fuzzy rings [8] as well as SwamyandSwamy defined and proved major theorems on fuzzy prime ideals of rings [7] and even Pu and Liu introduced the notion of fuzzy points [5], and in 2018 Melliani introduced the notion of a ring of fuzzy points *citeme* and based on this notion we introduce a fuzzy divisible ring and fuzzy local ring and give a characterization of this latter.

2 Preliminaries

2.1 FUZZY Sub-ring

Definition 1. [4] Let R be a ring, Then $\mu \subset R$ is a Fuzzy sub-ring if only if

i) $\mu(x-y) \geq \mu(x) \wedge \mu(y)$
ii) $\mu(xy) \geq \mu(x) \wedge \mu(y), \forall x, y \in R$
if moreover R is with identity we added the 3th condition
iii) $\mu(1) = 1$

Example 1. Let

$$\mu_B : \mathbb{Q} \to [0,1]$$

$$x \to \begin{cases} 1 & si\ x \in \mathbb{Z} \\ \frac{1}{2} & if\ not \end{cases}$$

S. Melliani and O. Castillo (Eds.): ICPAMS 2021, LNNS 476, pp. 37–40, 2023.
https://doi.org/10.1007/978-3-031-12416-7_4

μ is a Fuzzy sub-ring in $\mathbb{Q}$.

Proposition 1. *Let $\mu \subset R$ a fuzzy sub-ring of R. Then we have:*

i) $\mu(0) \geq \mu(x)$ *et* $\mu(x) = \mu(-x)$; $\quad \forall x \in R$
ii) Soit $x, y \in R$ *si* $\mu(x-y) = \mu(0)$, *alors* $\mu(x) = \mu(y)$.

Definition 2. [2] Let R be a ring, we say that $\mu \subset R$ is a fuzzy ideal if and only if:

i) $\mu(x-y) \geq \mu(x) \wedge \mu(y)$
ii) $\mu(xy) \geq \mu(x) \vee \mu(y)$,$\forall x, y \in R$.

Example: Let

$$\mu : \mathbb{Z}_4 \longrightarrow [0,1]$$
$$x \longmapsto \begin{cases} 1/2 & if \quad x = 0, 2 \\ \frac{1}{3} & if \quad x = 1, 3 \end{cases}$$

μ is a fuzzy ideal of $\mathbb{Z}_4$. With $\mathbb{Z}_4 = \{\overline{0}, \overline{1}, \overline{2}, \overline{3}\}$.

Theorem 1. *μ a fuzzy sub-ring of R if only if μ_t is a sub-ring of R $\forall t \in$* $[0, \mu(0)]$.

2.2 Set of Fuzzy Points

Definition 3. [5] Let A non-empty set. the Fuzzy sub-set A $x_\alpha : A \longrightarrow [0,1]$ with $x \in A$ and $\alpha \in (0,1]$ defined by:

$$x_\alpha(y) = \begin{cases} \alpha & si \quad x = y \\ 0 & si \quad x \neq y \end{cases}$$

is a fuzzy point.

Theorem 2. *[3] Let μ a fuzzy sub-ring of R, and x_t a fuzzy point of R. We have:*

i) $x_t \in \mu \Leftrightarrow \mu(x) \geq t$.
ii) $x_t + y_s = (x+y)_{t \wedge s}$
iii) $x_s . y_t = (x.y)_{t \wedge s}$.

3 Fuzzy Local Sub-rings

We consider μ a Fuzzy sub-ring in the ring with identity R.

Definition 4. A Fuzzy sub-ring μ is said to be divisible if every regular element (not a left or a right zero-divisor) is a unit (having a inverse).

Definition 5. A ring with identity is said to be a local ring if this ring has a unique maximal right ideal.

Theorem 3. *μ is a fuzzy sub-ring local of R* ***if only if*** *$M = \{x_t \in \mu;\ x_t$ is a unit$\}$ is an ideal of μ.*

Proof. $\Rightarrow$) Suppose that μ is a fuzzy local sub-ring of R.
and M_{max} the fuzzy maximal ideal of μ then show that M is included in a proper ideal.
It's clear that if $x_t \in M$ then $\forall y_s \in \mu;\ x_t.y_s \in M$.
Now, if $x_t, y_s \in M$ then $< x_t > \subset M_{max}$ and $< y_t > \subset M_{max}$.
Hence $x_t + y_s \in M_{max}$.
Thus $x_t + y_s \in M$. Finally M is a fuzzy proper ideal of μ.
Then we show the stability of the addition on M.
Therefor M is a fuzzy proper ideal of μ.
$\Leftarrow$) conversely suppose tha M is a fuzzy proper ideal of μ than it's easy to show that M is the unique fuzzy maximal ideal of μ.

Remark 1. μ is a fuzzy sub-ring local of R **if only if** the sum of any non-units is a non-unit.

Let $Z = \{x_t \in \mu / \exists y_s \in \mu$ *such that* $y_s \neq 0$ *and* $y_s.x_t = 0\}$ be a set of all fuzzy zero-divisors.

Theorem 4. *Let μ is a fuzzy divisible sub ring of R. Then μ is a fuzzy local sub-ring of R if and only if Z is proper ideal of μ.*

Proof. Let μ is a fuzzy divisible sub-ring of R.
$\Rightarrow$) Let's show that μ is local if Z is a proper ideal of μ.
Let ν a proper ideal so $\exists x_t \in \nu; x_t \notin Z$.
By the divisibility of μ; $\exists y_s \in \mu; y_s.x_t = 1_{s\wedge t}$.
Thus $1_{s\wedge t} \in \nu$ wich is impossible.
Therefore $\nu \subset \mu$.
By the divisibility of μ, we prove if ν is a fuzzy proper ideal of R then we have $\nu \subset \mu$.
$\Leftarrow$) Conversely, It's clear that Z is an ideal of μ and $1_\alpha \notin Z$ hence Z is a proper ideal of μ.

Theorem 5. *Let μ be a fuzzy sub-ring satisfying the following condition*

i) ideals containing J are principals.
ii) J is completely prime and is a principal ideal.

iii) $Z \subseteq J$.
iv) $J \neq 0$.

Then R is a local ring.

Proof. It suffices to prove that J is composed of all non-units. Let $J = Rx_t$ and $a_s \notin J$. Then by i) $J + a_sR = b_rR$ and $b_r \notin J$. Hence $x_t = by_{t_1}$. Since J is completely prime, $y_{t_1} \in J$. Therefore $y_{t_1} = c_{t_2}x_t$, i.e., $x_t = b_ry_{t_1} = b_rc_{t_2}x_t$. Thus $-b_rc_{t_2} \in Z$. This implies by iii) $1 - b_rc_{t_2} \in J$ and $b_rc_{t_2}$ is a unit and hence b_r is a unit. Consequently $J + a_sR = R$. Theref ore it follows that $1 = j_{t_3} + a_sc_{t_2}, j_{t_3} \in J$ and $1 - a_sc_{t_2} \in J$ and $a_sc_{t_2}$ is a unit. Thus we conclude that a_s is a unit.

Remark 2. = In the above theorem the condition that $J \neq 0$ is necessary as can be seen from the example of the ring of integers.

As an immediate consequence of 32 we have the following.

Corollary 1. *Let R be a principal ideal ring (every right and left ideal is principal) and* $Z \subseteq J$. *Then R is a local ring iff* $J \neq 0$ *and J is a completely primee ideal.*

Corollary 2. *Let R be a principal ideal ring which is not a domain. Then, if* $Z = J, R$ *is a local ring.*

References

1. Alkhamees, Y., Mordeson, J.N.: Fuzzy localized subrings. Inf. Sci. **99**, 183–193 (1997)
2. Liu, W.J.: Fuzzy invariant subgroups and fuzzy ideals. Fuzzy Sets Syst. **8**, 133–139 (1982)
3. Melliani, S., Bakhadach, I., Chadli, L.S.: Fuzzy rings and fuzzy polynomial rings. In: Badawi, A., Vedadi, M.R., Yassemi, S., Yousefian Darani, A. (eds.) SAA 2016. SPMS, vol. 228, pp. 89–98. Springer, Cham (2018). https://doi.org/10.1007/978-3-319-74195-6_8
4. Mordeson, J.N.: Fuzzy coefficient fields of fuzzy subrings. Fuzzy Sets Syst. **58**, 227–237 (1993)
5. Pu, P.M., Liu, Y.M.: Fuzzy topology. I. Neighborhood structure of a fuzzy point and Moore- Smith convergence. J. Math. Anal. Appl. **76**, 571–599 (1980)
6. Rosenfeld, A.: Fuzzy groups. J. Math. Anal. Appl. **35**, 512–517 (1971)
7. Swamy, U.M., Swamy, K.L.N.: Fuzzy prime ideals of rings. J. Math. Anal. Appl. **134**, 94–103 (1988)
8. Wang, X., Ruan, D., Kerre, E.E.: Mathematics of Fuzziness-Basic Issues. Springer, Heidelberg (2009). https://doi.org/10.1007/978-3-540-78311-4
9. Zadeh, L.A.: Fuzzy sets. Inf. Control **8**, 338–353 (1965)

Gevrey Class Regularity for the 2D Subcritical Dissipative Quasi-geostrophic Equation in Critical Fourier-Besov-Morrey Spaces

Achraf Azanzal(✉), Adil Abbassi, and Chakir Allalou

Laboratory LMACS, FST of Beni Mellal, Sultan Moulay Slimane University, Beni Mellal, Morocco
achraf0665@gmail.com

Abstract. The dissipative two dimensional Quasi-Geostrophic Equation (2D QGE) is studied. By using the Littlewood Paley theory, Fourier analysis and standard techniques we give the Gevrey class regularity results of this equation for small initial data v_0 belonging to the critical Fourier-Besov-Morrey spaces $\mathscr{F}\mathscr{N}^{1-2\alpha+\frac{2}{p'}+\frac{\lambda}{p}}_{p,\lambda,q}$.

1 Introduction

In this article, we consider the following Cauchy problem for the two-dimensional quasi-geostrophic equation with subcritical dissipation $\alpha > 1/2$.

$$\begin{cases} \partial_t v + k\Lambda^{2\alpha} v + K[v]\cdot\nabla v = 0, \quad x\in\mathbb{R}^2,\ t>0, \\ K[v] = (-\mathfrak{R}_2 v, \mathfrak{R}_1 v), \\ v(0,x) = v_0(x), \end{cases} \tag{1}$$

where $\mathfrak{R}_j = \partial_{x_j}(-\Delta)^{-1/2}$, $j = 1,2$, are the Riesz transforms, $\alpha > 1/2$ is a real number, $k > 0$ is a dissipative coefficient. Notice that (1) is called subcritical when $\alpha > 1/2$, critical when $\alpha = 1/2$ and supercritical when $\alpha < 1/2$. Λ is the operator defined by the fractional power of $-\Delta$:

$$\Lambda v = (-\Delta)^{1/2} v, \quad \mathscr{F}(\Lambda v) = \mathscr{F}((-\Delta)^{1/2} v) = |\xi|\mathscr{F}(v),$$

and more generally

$$\mathscr{F}(\Lambda^{2\alpha} v) = \mathscr{F}((-\Delta)^{\alpha} v) = |\xi|^{2\alpha}\mathscr{F}(v),$$

where $\mathscr{F}$ is the Fourier transform. The scalar function $v(x,t)$ represents the potential temperature, and $K[v]$ is the divergence free velocity which is determined by the Riesz transformation of v.

The 2D quasi-geostrophic equation is an important model in geophysical fluid dynamics, which represents the potential temperature dynamics of atmospheric and ocean flow. For further information on the physical background of this equation, see [13,20] and the references therein.

S. Melliani and O. Castillo (Eds.): ICPAMS 2021, LNNS 476, pp. 41–51, 2023.
https://doi.org/10.1007/978-3-031-12416-7_5

The mathematical analysis of the non-dissipative case ($k = 0$) has first been studied by Constantin, Majda and Tabak in [11] where it is shown to be an analogue to the 3D Euler equations. The subcritical dissipative case has then been studied by Constantin and Wu [12] where the authors obtained global existence of solutions in Sobolev spaces.

Now, we recall the scaling property of the equations:

if v solves (1) with initial data v_0, then v_γ with $v_\gamma(x,t) := \gamma^{2\alpha-1} v\left(\gamma x, \gamma^{2\alpha} t\right)$ is also a solution to (1) with the initial data

$$v_{0,\gamma}(x) := \gamma^{2\alpha-1} v_0(\gamma x). \tag{2}$$

Definition 1. A critical space for initial data of the Eq. (1) is any Banach space $E \subset \mathscr{S}'(\mathbb{R}^n)$ whose norm is invariant under the scaling (2) for all $\gamma > 0$, i.e.

$$\left\|v_{0,\gamma}(x)\right\|_E \approx \|v_0(x)\|_E.$$

In accordance with these scales, we can show that the space $\mathscr{F}\mathscr{N}_{p,\lambda,q}^{1-2\alpha+\frac{2}{p'}+\frac{\lambda}{p}}$ is critical for (1). The global well-posedness for small initial data in critical spaces was established by many researchers. The reader is referred to [2,3,6–9] and their references.

The analyticity of the solution is an important issue developed by numerous authors, notably in relation to the Navier-Stokes equations, see [1,23]. In this work, we will prove the spatial analyticity in the Fourier-Besov-Morrey space. Our main method is the Gevrey estimate, which was presented by Foias and Temam [17], since the Gevrey class approach has become an efficient method in the investigation of analytic solutions. Ferrari and Titi [15] obtained Gevrey regularity for several parabolic equations, see [21,22].

Throughout this paper, we use $\mathscr{F}\mathscr{N}_{p,\lambda,q}^{s}$ to denote the homogenous Fourier-Besov-Morrey spaces. Let A, B be Banach spaces, we denote $\|v\|_{A\cap B} := \|v\|_A + \|v\|_B$ and $\|(v,w)\|_A := \|v\|_A + \|w\|_A$, C will denote constants which can be different at different places, $U \lesssim V$ means that there exists a constant $C > 1$ such that $U \leq CV$, and p' is the conjugate of p satisfying $\frac{1}{p} + \frac{1}{p'} = 1$ for $1 \leq p \leq \infty$.

Since we are concerned with the dissipative case, we assume $k = 1$ for the sake of simplicity.

In order to obtain the Gevrey class regularity of the solution of the Eq. (1), we consider the following equivalent integral equation coming from Duhamel's principle

$$v(t) = \mathrm{S}_\alpha(t) v_0 + \mathscr{B}(v,v)(t), \tag{3}$$

where $\mathrm{S}_\alpha := e^{-t(-\Delta)^\alpha}$ denotes the fractional heat semigroup operator, which can be regarded as the convolution operator with the kernel $k_t(x) = \mathscr{F}^{-1}(e^{-t|\xi|^{2\alpha}})$, and

$$\mathscr{B}(v,\theta)(t) = -\int_0^t \mathrm{S}_\alpha(t-\tau)\left(K[v]\cdot\nabla\theta)\right)(\tau)d\tau. \tag{4}$$

Our results can be formulated as follows.

Theorem 1 *(Space Analyticity). Let* $1 \leq p < \infty$, $1 \leq q \leq \infty$, $0 \leq \lambda < 2$ *and* $\frac{1}{2} < \alpha < \min\{1, \alpha < 1 + \frac{\lambda}{2p} + \frac{1}{p'}\}$. *There exists a constant* $M > 0$ *such that for any* $v_0 \in \mathcal{F}\mathcal{N}_{p,\lambda,q}^{1-2\alpha+\frac{2}{p'}+\frac{\lambda}{p}}$ *satisfying* $\|v_0\|_{\mathcal{F}\mathcal{N}_{p,\lambda,q}^{1-2\alpha+\frac{2}{p'}+\frac{\lambda}{p}}} < M$. *The Eq. (1) has a unique analytic solution* v *in the sense that*

$$\|e^{\sqrt{t}|D|^{\alpha}} v\|_X \lesssim \|v_0\|_{\mathcal{F}\mathcal{N}_{p,\lambda,q}^{1-2\alpha+\frac{2}{p'}+\frac{\lambda}{p}}},$$

where $X := \mathcal{L}^{\infty}\left(\mathbb{R}^+, \mathcal{F}\mathcal{N}_{p,\lambda,q}^{1-2\alpha+\frac{2}{p'}+\frac{\lambda}{p}}\right) \cap \mathcal{L}^1\left(\mathbb{R}^+, \mathcal{F}\mathcal{N}_{p,\lambda,q}^{1+\frac{2}{p'}+\frac{\lambda}{p}}\right)$, *and* $e^{\sqrt{t}|D|^{\alpha}} v = \mathcal{F}^{-1}(e^{\sqrt{t}|\xi|^{\alpha}} \hat{v})$.

2 Preliminaries

In this section, we recall some basic properties of Fourier-Besov-Morrey spaces, the fractional heat semigroup, and other analysis tools that we will employ throughout this study.

The Fourier-Besov-Morrey spaces, presented in [16], are built by using a type of localization on Morrey spaces. The function spaces $\mathcal{M}_p^{\lambda}$ are defined as follows.

Definition 2 [18]. Let $1 \leq p \leq \infty$ and $0 \leq \lambda < n$. The homogeneous Morrey space $\mathcal{M}_p^{\lambda}$ is the set of all functions $f \in L^p(B(x_0, r))$ such that

$$\|f\|_{\mathcal{M}_p^{\lambda}} = \sup_{x_0 \in \mathbb{R}^n} \sup_{r>0} r^{-\frac{\lambda}{p}} \|f\|_{L^p(B(x_0,r))} < \infty, \tag{5}$$

where $B(x_0, r)$ is the open ball in $\mathbb{R}^n$ centered at x_0 and with radius $r > 0$.

The space $\mathcal{M}_p^{\lambda}$ endowed with the norm $\|f\|_{\mathcal{M}_p^{\lambda}}$ is a Banach space. When $\lambda = 0$, we have $\mathcal{M}_p^0 - L^p$.

The proofs of the results discussed in this work are based on a dyadic partition of unity in the Fourier variables, known as the homogeneous Littlewood-Paley decomposition. We present briefly this construction below. For more detail, we refer the reader to [5].

Let $f \in S'(\mathbb{R}^n)$. Define the Fourier transform as

$$\hat{f}(\xi) = \mathcal{F}f(\xi) = (2\pi)^{-\frac{n}{2}} \int_{\mathbb{R}^n} e^{-ix\cdot\xi} f(x)dx$$

and its inverse Fourier transform as

$$\check{f}(x) = \mathcal{F}^{-1}f(x) = (2\pi)^{-\frac{n}{2}} \int_{\mathbb{R}^n} e^{ix\cdot\xi} f(\xi)d\xi.$$

Let $\varphi \in S(\mathbb{R}^d)$ be such that $0 \leq \varphi \leq 1$ and $\text{supp}(\varphi) \subset \{\xi \in \mathbb{R}^d : \frac{3}{4} \leq |\xi| \leq \frac{8}{3}\}$ and

$$\sum_{j \in \mathbb{Z}} \varphi(2^{-j}\xi) = 1, \quad \text{for all } \xi \neq 0.$$

We denote

$$\varphi_j(\xi) = \varphi\left(2^{-j}\xi\right), \quad \delta_j(\xi) = \sum_{k\leq j-1} \varphi_k(\xi)$$

and

$$h(x) = \mathscr{F}^{-1}\varphi(x), \quad g(x) = \mathscr{F}^{-1}\delta(x).$$

We now present some frequency localization operators:

$$\dot{\Delta}_j f = \varphi_j(D)f = 2^{dj}\int_{\mathbb{R}^d} h\left(2^j y\right) f(x-y)dy$$

and

$$\dot{S}_j f = \sum_{k\leq j-1} \dot{\Delta}_k f = \delta_j(D)f = 2^{dj}\int_{\mathbb{R}^d} g\left(2^j y\right) f(x-y)dy.$$

From the definition, one easily derives that

$$\begin{aligned} \dot{\Delta}_j \dot{\Delta}_k f &= 0, \quad \text{if } |j-k| \geq 2 \\ \dot{\Delta}_j\left(\dot{S}_{k-1} f \dot{\Delta}_k f\right) &= 0, \quad \text{if } |j-k| \geq 5. \end{aligned}$$

The following Bony paraproduct decomposition will be applied throughout the paper.

$$uv = \dot{T}_u v + \dot{T}_v u + R(u,v)$$

where

$$\dot{T}_u v = \sum_{j\in\mathbb{Z}} \dot{S}_{j-1} u \dot{\Delta}_j v, \quad \dot{R}(u,v) = \sum_{j\in\mathbb{Z}} \dot{\Delta}_j u \tilde{\Delta}_j v, \quad \tilde{\Delta}_j v = \sum_{|j'-j|\leq 1} \dot{\Delta}_{j'} v.$$

Lemma 1 [16]. *Let* $1 \leq p_1, p_2, p_3 < \infty$ *and* $0 \leq \lambda_1,\ \lambda_2,\ \lambda_3 < n$.
(i) (Hölder's inequality) Let $\frac{1}{p_3} = \frac{1}{p_1} + \frac{1}{p_2}$ *and* $\frac{\lambda_3}{p_3} = \frac{\lambda_1}{p_1} + \frac{\lambda_2}{p_2}$, *then we have*

$$\|fg\|_{\mathscr{M}_{p_3}^{\lambda_3}} \leq \|f\|_{\mathscr{M}_{p_1}^{\lambda_1}} \|g\|_{\mathscr{M}_{p_2}^{\lambda_2}}. \tag{6}$$

(ii) (Young's inequality) If $\varphi \in L^1$ *and* $g \in \mathscr{M}_{p_1}^{\lambda_1}$, *then*

$$\|\varphi * g\|_{\mathscr{M}_{p_1}^{\lambda_1}} \leq \|\varphi\|_{L^1} \|g\|_{\mathscr{M}_{p_1}^{\lambda_1}}, \tag{7}$$

where $*$ *denotes the standard convolution operator.*

Now, we recall the Bernstein-type lemma in Fourier variables in Morrey spaces.

Lemma 2 [16]. *Let* $1 \leq q \leq p < \infty, 0 \leq \lambda_1, \lambda_2 < n$, $\frac{n-\lambda_1}{p} \leq \frac{n-\lambda_2}{q}$ *and let* γ *be a multi-index. If* $supp(\widehat{f}) \subset \{|\xi| \leq A2^j\}$, *then there is a constant* $C > 0$ *independent of* f *and* j *such that*

$$\|(i\xi)^\gamma \widehat{f}\|_{\mathscr{M}_q^{\lambda_2}} \leq C 2^{j|\gamma| + j(\frac{n-\lambda_2}{q} - \frac{n-\lambda_1}{p})} \|\widehat{f}\|_{\mathscr{M}_p^{\lambda_1}}. \tag{8}$$

Let us now recall the definition of Fourier-Besov-Morrey spaces $\mathscr{F}\mathscr{N}^{s}_{p,\lambda,q}(\mathbb{R}^n)$, see [16].

Definition 3 (Homogeneous Fourier-Besov-Morrey spaces). Let $1 \leq p,q \leq \infty$, $0 \leq \lambda < n$ and $s \in \mathbb{R}$. The homogeneous Fourier-Besov-Morrey space $\mathscr{F}\mathscr{N}^{s}_{p,\lambda,q}$ is defined as the set of all distributions $f \in \mathscr{S}'\backslash\mathscr{P}$, $\mathscr{P}$ is the set of all polynomials, such that the norm $\|f\|_{\mathscr{F}\mathscr{N}^{s}_{p,\lambda,q}}$ is finite, where

$$\|f\|_{\mathscr{F}\mathscr{N}^{s}_{p,\lambda,q}} \stackrel{\text{def}}{=} \begin{cases} \left(\sum_{j\in\mathbb{Z}} 2^{jsq} \|\varphi_j \hat{f}\|^q_{\mathscr{M}_p^{\lambda}}\right)^{\frac{1}{q}} & for \quad q<\infty \\ \sup_{j\in\mathbb{Z}} 2^{js} \|\varphi_j \hat{f}\|_{\mathscr{M}_p^{\lambda}} & for\ q=\infty. \end{cases} \tag{9}$$

Note that the space $\mathscr{F}\mathscr{N}^{s}_{p,\lambda,q}(\mathbb{R}^n)$ equipped with the norm (9) is a Banach space. Since $\mathscr{M}_p^0 = L^p$, we have $\mathscr{F}\mathscr{N}^{s}_{p,0,q} = \dot{FB}^{s}_{p,q}$, and $\mathscr{F}\mathscr{N}^{s}_{1,0,1} = \chi^s$ where χ^s is the Lei-Lin space [6].

The definition of mixed space-time spaces is given below.

Definition 4 [14]. Let $s \in \mathbb{R}, 1 \leq p < \infty, 1 \leq q,\rho \leq \infty, 0 \leq \lambda < n$, and $I = [0,T)$, $T \in (0,\infty]$. The space-time norm is defined on $u(t,x)$ by

$$\|u(t,x)\|_{\mathscr{L}^{\rho}(I,\mathscr{F}\mathscr{N}^{s}_{p,\lambda,q})} = \left\{ \sum_{j\in\mathbb{Z}} 2^{jqs} \|\widehat{\dot{\Delta}_j u}\|^q_{L^{\rho}(I,\mathscr{M}_p^{\lambda})} \right\}^{1/q},$$

and denote by $\mathscr{L}^{\rho}(I,\mathscr{F}]\mathscr{N}^{s}_{p,\lambda,q})$ the set of distributions in $\mathscr{S}'(\mathbb{R}\times\mathbb{R}^n)/\mathscr{P}$ with finite $\|.\|_{\mathscr{L}^{\rho}(I,\mathscr{F}\mathscr{N}^{s}_{p,\lambda,q})}$ norm.

According to Minkowski inequality, it is easy to verify that

$$L^{\rho}\left(I;\mathscr{F}\mathscr{N}^{s}_{p,\lambda,q}\right) \hookrightarrow \mathscr{L}^{\rho}\left(I,\mathscr{F}\mathscr{N}^{s}_{p,\lambda,q}\right), \quad \text{if } \rho \leq q, \tag{10}$$

$$\mathscr{L}^{\rho}\left(I,\mathscr{F}\mathscr{N}^{s}_{p,\lambda,q}\right) \hookrightarrow L^{\rho}\left(I;\mathscr{F}\mathscr{N}^{s}_{p,\lambda,q}\right), \quad \text{if } \rho \geq q, \tag{11}$$

where $\|u(t,x)\|_{L^{\rho}\left(I;\mathscr{F}\mathscr{N}^{s}_{p,\lambda,q}\right)} := \left(\int_I \|u(\tau,\cdot)\|^{\rho}_{\mathscr{F}\mathscr{N}^{s}_{p,\lambda,q}} d\tau\right)^{1/\rho}$.

At the end of this section, we will recall an existence and uniqueness result for an abstract operator equation in a Banach space that will be used to show Theorem 1 in the sequel. For the proof, we refer the reader to see [4, 19].

Lemma 3. *Let X be a Banach space with norm* $\|.\|_X$ *and* $B: X\times X \longmapsto X$ *be a bounded bilinear operator satisfying*

$$\|B(u,v)\|_X \leq C_0 \|u\|_X \|v\|_X$$

for all $u,v \in X$ *and a constant* $C_0 > 0$. *Then, if* $0 < \varepsilon < \frac{1}{4C_0}$ *and if* $y \in X$ *such that* $\|y\|_X \leq \varepsilon$, *the equation* $x := y + B(x,x)$ *has a solution* $\overline{x}$ *in X such that* $\|\overline{x}\|_X \leq 2\varepsilon$. *This solution is the only one in the ball* $\overline{B}(0,2\varepsilon)$. *Moreover, the solution depends continuously on y in the sense: if* $\|y'\|_X < \varepsilon$, $x' = y' + B(x',x')$, *and* $\|x'\|_X \leq 2\varepsilon$, *then*

$$\|\overline{x} - x'\|_X \leq \frac{1}{1-4\varepsilon C_0} \|y - y'\|_X.$$

3 Gevrey Class Regularity

In this section, we show the Gevrey class regularity for Eq. (1). For this aim, let $\bar{v} = e^{\sqrt{t}|D|^\alpha} v$ and using the integral Eq. (12), we obtain

$$v(t) = e^{\sqrt{t}|D|^\alpha} e^{-t(-\Delta)^\alpha} v_0 - \int_0^t e^{(\sqrt{t}-\sqrt{\tau})|D|^\alpha - \frac{1}{2}(t-\tau)(-\Delta)^\alpha} e^{-\frac{1}{2}(t-\tau)(-\Delta)^\alpha} e^{\sqrt{\tau}|D|^\alpha} (K[v]\cdot\nabla v))(\tau)d\tau$$
$$:= e^{\sqrt{t}|D|^\alpha} e^{-t(-\Delta)^\alpha} v_0 + \tilde{\mathscr{B}}(\bar{v},\bar{v}). \tag{12}$$

The following lemma is very useful in proving spatial analyticity (Gevrey regularity).

Lemma 4. *Let $0 < s \le t < \infty$ and $0 \le \alpha \le 1$. Then the following inequality holds*

$$t|\xi|^\alpha - \frac{1}{2}(t^2 - s^2)|\xi|^{2\alpha} - s|\xi - y|^\alpha - s|y|^\alpha \le \frac{1}{2} \tag{13}$$

for all ξ, $y \in \mathbb{R}^n$.

Now, we will establish some important linear estimates in Fourier-Besov-Morrey spaces.

Lemma 5. *Let $0 \le \lambda < 2$, $1 \le p < \infty$, $1 \le q \le \infty$ and $v_0 \in \mathscr{F}\mathscr{N}_{p,\lambda,q}^{1-2\alpha+\frac{2}{p'}+\frac{\lambda}{p}}(\mathbb{R}^2)$. Then there exists a constant $C > 0$ such that*

$$\|e^{\sqrt{t}|D|^\alpha} e^{-(-\triangle)^\alpha t} v_0\|_X \le C\|v_0\|_{\mathscr{F}\mathscr{N}_{p,\lambda,q}^{1-2\alpha+\frac{2}{p'}+\frac{\lambda}{p}}} \tag{14}$$

Proof. We have

$$e^{(\sqrt{t}|D|^\alpha - (-\Delta)^\alpha t)} v_0 = e^{-\frac{1}{2}(\sqrt{t}|D|^\alpha - 1)^2 + \frac{1}{2}} e^{-\frac{1}{2}(-\Delta)^\alpha t} v_0.$$

Using the Fourier transform, multiplying by φ_j and taking the $\mathscr{M}_p^\lambda$-norm we obtain

$$\|\varphi_j \widehat{e^{\sqrt{t}|D|^\alpha} e^{-(-\triangle)^\alpha t} v_0}\|_{\mathscr{M}_p^\lambda} \le C e^{-\frac{1}{2}t 2^{2j\alpha}(\frac{3}{4})^{2\alpha}} \|\varphi_j \widehat{v_0}\|_{\mathscr{M}_p^\lambda}.$$

Multiplying by $2^{(1-2\alpha+\frac{2}{p'}+\frac{\lambda}{p})j}$ and taking $l^q(\mathbb{Z})$-norm we get

$$\|e^{\sqrt{t}|D|^\alpha} e^{-(-\triangle)^\alpha t} v_0\|_{\mathscr{L}^\infty([0,\infty),\mathscr{F}\mathscr{N}_{p,\lambda,q}^{1-2\alpha+\frac{2}{p'}+\frac{\lambda}{p}})} \le C\|v_0\|_{\mathscr{F}\mathscr{N}_{p,\lambda,q}^{1-2\alpha+\frac{2}{p'}+\frac{\lambda}{p}}}.$$

Analoguesly,

$$\|e^{\sqrt{t}|D|^\alpha} e^{-(-\triangle)^\alpha t} v_0\|_{\mathscr{L}^1([0,\infty),\mathscr{F}\mathscr{N}_{p,\lambda,q}^{1+\frac{2}{p'}+\frac{\lambda}{p}})} \le C\|v_0\|_{\mathscr{F}\mathscr{N}_{p,\lambda,q}^{1-2\alpha+\frac{2}{p'}+\frac{\lambda}{p}}}.$$

Then

$$\|e^{\sqrt{t}|D|^\alpha} e^{-(-\triangle)^\alpha t} v_0\|_X \le C\|v_0\|_{\mathscr{F}\mathscr{N}_{p,\lambda,q}^{1-2\alpha+\frac{2}{p'}+\frac{\lambda}{p}}}.$$

Lemma 6 [10]. *Let $0 < T \leq \infty, s \in \mathbb{R}, 0 \leq \lambda < 3, 1 \leq p < \infty, 1 \leq q, \rho, r \leq \infty$ and $1 \leq r \leq \rho$. There exists a constant $C > 0$ such that*

$$\left\| \int_0^t e^{-(-\Delta)^\alpha (t-\tau)} h(\tau) d\tau \right\|_{\mathscr{L}^\rho([0,T), \mathscr{F}\mathscr{N}^s_{p,\lambda,q})} \leq C \|h\|_{\mathscr{L}^r([0,T), \mathscr{F}\mathscr{N}^{s-2\alpha-\frac{2\alpha}{\rho}+\frac{2\alpha}{r}}_{p,\lambda,q})},$$

for all $h \in \mathscr{L}^r([0,T), \mathscr{F}\mathscr{N}^s_{p,\lambda,q})$.

In the following proposition, we will establish the bilinear estimate which will be crucial in the proof of Theorem 1.

Proposition 1. *Under the hypothesis of Theorem 1, there exists a constant $C_0 > 0$ such that*

$$\|\tilde{\mathscr{B}}(\bar{v}, \bar{\theta})\|_X \leq C_0 \|\bar{v}\|_X \|\bar{\theta}\|_X,$$

for all $\bar{v}, \bar{\theta} \in X$.

Proof. First, using Lemma 6 and Lemma 4, we have

$$\begin{aligned} \|\mathscr{B}(\bar{v}, \bar{\theta})\|_X &\lesssim \|e^{\sqrt{\tau}|D|^\alpha} K[v] \cdot \nabla\theta\|_{\mathscr{L}^1\left([0,\infty[, \mathscr{F}\mathscr{N}^{1-2\alpha+\frac{2}{p'}+\frac{\lambda}{p}}_{p,\lambda,q}\right)} \\ &\lesssim \|e^{\sqrt{\tau}|D|^\alpha} \nabla\cdot(K[v]\theta)\|_{\mathscr{L}^1\left([0,\infty[, \mathscr{F}\mathscr{N}^{1-2\alpha+\frac{2}{p'}+\frac{\lambda}{p}}_{p,\lambda,q}\right)} \\ &\lesssim \|e^{\sqrt{\tau}|D|^\alpha} K[v]\theta\|_{\mathscr{L}^1\left([0,\infty[, \mathscr{F}\mathscr{N}^{2-2\alpha+\frac{2}{p'}+\frac{\lambda}{p}}_{p,\lambda,q}\right)} \end{aligned} \tag{15}$$

Then, the remainder of the proof is to show that

$$\|e^{\sqrt{\tau}|D|^\alpha} K[v]\theta\|_{\mathscr{L}^1\left([0,\infty[, \mathscr{F}\mathscr{N}^{2-2\alpha+\frac{2}{p'}+\frac{\lambda}{p}}_{p,\lambda,q}\right)} \lesssim \|\bar{v}\|_X \|\bar{\theta}\|_X. \tag{16}$$

Recalling that $K[v] = (-\mathfrak{R}_2 v, \mathfrak{R}_1 v)$. Since $\widehat{\mathfrak{R}_j v}(\xi) = -i\frac{\xi_j}{|\xi|}\hat{v}(\xi), j = 1,2$, then

$$\|\widehat{K[v]}\| \lesssim \|\hat{v}\|. \tag{17}$$

Applying Bony para-product decomposition and quasi-orthogonality property for Littlewood-Paley decomposition, for fixed j, we obtain

$$\begin{aligned} \dot{\Delta}_j e^{\sqrt{\tau}|D|^\alpha}(K[v]\theta) &= \sum_{|k-j|\leq 4} \dot{\Delta}_j e^{\sqrt{\tau}|D|^\alpha}(\dot{S}_{k-1}K[v]\dot{\Delta}_k\theta) + \sum_{|k-j|\leq 4} \dot{\Delta}_j e^{\sqrt{\tau}|D|^\alpha}(\dot{S}_{k-1}\theta\dot{\Delta}_k K[v]) \\ &\quad + \sum_{k\geq j-3} \dot{\Delta}_j e^{\sqrt{\tau}|D|^\alpha}(\dot{\Delta}_k K[v]\tilde{\dot{\Delta}}_k\theta) \\ &= R^1_j + R^2_j + R^3_j. \end{aligned}$$

Then, by the triangle inequalities in $\mathcal{M}_p^\lambda$ and in $l^q(\mathbb{Z})$, we have

$$
\begin{aligned}
\|e^{\sqrt{\tau}|D|^\alpha}K[v]\theta\|_{\mathscr{L}^1\left([0,\infty[,\mathscr{F}\mathscr{N}_{p,\lambda,q}^{2-2\alpha+\frac{2}{p'}+\frac{\lambda}{p}}\right)} &= \left\{\sum_{j\in\mathbb{Z}} 2^{j(2-2\alpha+\frac{2}{p'}+\frac{\lambda}{p})q}\|\dot{\Delta}_j(\widehat{e^{\sqrt{\tau}|D|^\alpha}K[v]\theta})\|^q_{L^1([0,\infty[,\mathscr{M}_p^\lambda)}\right\}^{\frac{1}{q}} \\
&\le \left\{\sum_{j\in\mathbb{Z}} 2^{j(2-2\alpha+\frac{2}{p'}+\frac{\lambda}{p})q}\|\widehat{R_j^1}\|^q_{L^1([0,\infty[,\mathscr{M}_p^\lambda)}\right\}^{\frac{1}{q}} \\
&+ \left\{\sum_{j\in\mathbb{Z}} 2^{j(2-2\alpha+\frac{2}{p'}+\frac{\lambda}{p})q}\|\widehat{R_j^2}\|^q_{L^1([0,\infty[,\mathscr{M}_p^\lambda)}\right\}^{\frac{1}{q}} \\
&+ \left\{\sum_{j\in\mathbb{Z}} 2^{j(2-2\alpha+\frac{2}{p'}+\frac{\lambda}{p})q}\|\widehat{R_j^3}\|^q_{L^1([0,\infty[,\mathscr{M}_p^\lambda)}\right\}^{\frac{1}{q}} \\
&:= \mathrm{I}_1+\mathrm{I}_2+\mathrm{I}_3.
\end{aligned}
$$

By Bernstein-type inequality (8) with $|\gamma|=0$, we have

$$
\left\|\varphi_m\widehat{K[v]}\right\|_{L^1} \le C2^{m(2-\frac{2-\lambda}{p})}\left\|\varphi_m\widehat{K[v]}\right\|_{\mathscr{M}_p^\lambda} \lesssim 2^{m(\frac{2}{p'}+\frac{\lambda}{p})}\|\varphi_m\hat{v}\|_{\mathscr{M}_p^\lambda}, \tag{18}
$$

where we have used (17).

Thus, using Young's inequality in Morrey spaces (7) and the estimate (18), we get

$$
\begin{aligned}
\|\widehat{R_j^1}\|_{L^1([0,\infty[,\mathscr{M}_p^\lambda)} &= \left\|\sum_{|k-j|\le 4}\varphi_j e^{\sqrt{\tau}|\xi|^\alpha}(\widehat{\dot{S}_{k-1}K[v]\dot{\Delta}_k\theta})\right\|_{L^1([0,\infty[,\mathscr{M}_p^\lambda)} \\
&= \left\|\sum_{|k-j|\le 4}\varphi_j e^{\sqrt{\tau}|\xi|^\alpha}\left[\left(\sum_{m\le k-2} e^{-\sqrt{\tau}|\xi|^\alpha}\varphi_m\widehat{K[\bar{v}]}\right)*e^{-\sqrt{\tau}|\xi|^\alpha}\varphi_k\widehat{\bar{\theta}}\right]\right\|_{L^1([0,\infty[,\mathscr{M}_p^\lambda)} \\
&= \left\|\sum_{|k-j|\le 4}\varphi_j\int_{\mathbb{R}^2} e^{\sqrt{\tau}(|\xi|^\alpha-|\xi-y|^\alpha-|y|^\alpha)}\left(\sum_{m\le k-2}\varphi_m\widehat{K[\bar{v}]}\right)(\xi-y)\varphi_k\widehat{\bar{\theta}}(y)dy\right\|_{L^1([0,\infty[,\mathscr{M}_p^\lambda)} \qquad (19) \\
&\lesssim \sum_{|k-j|\le 4}\|(\widehat{\dot{S}_{k-1}K[\bar{v}]\dot{\Delta}_k\bar{\theta}})\|_{L^1([0,\infty[,\mathscr{M}_p^\lambda)} \\
&\lesssim \sum_{|k-j|\le 4}\|\varphi_k\widehat{\bar{\theta}}\|_{L^1([0,\infty[,\mathscr{M}_p^\lambda)}\sum_{m\le k-2}\|\varphi_m\widehat{K[\bar{v}]}\|_{L^\infty([0,\infty[,L^1)} \\
&\lesssim \sum_{|k-j|\le 4}\|\varphi_k\widehat{\bar{\theta}}\|_{L^1([0,\infty[,\mathscr{M}_p^\lambda)}\sum_{m\le k-2}2^{(\frac{2}{p'}+\frac{\lambda}{p})m}\|\varphi_m\widehat{\bar{v}}\|_{L^\infty([0,\infty[,\mathscr{M}_p^\lambda)} \\
&\lesssim \sum_{|k-j|\le 4}\|\varphi_k\widehat{\bar{\theta}}\|_{L^1([0,\infty[,\mathscr{M}_p^\lambda)}\sum_{m\le k-2}2^{(1-2\alpha+\frac{2}{p'}+\frac{\lambda}{p})m}2^{(2\alpha-1)m}\|\varphi_m\widehat{\bar{v}}\|_{L^\infty([0,\infty[,\mathscr{M}_p^\lambda)} \\
&\lesssim \|\bar{v}\|_{\mathfrak{L}^\infty([0,\infty[,\mathscr{F}\mathscr{N}_{p,\lambda,q}^{1-2\alpha+\frac{2}{p'}+\frac{\lambda}{p}})}\sum_{|k-j|\le 4}\left(\sum_{m\le k-2}2^{m(2\alpha-1)q'}\right)^{\frac{1}{q'}}\|\varphi_k\widehat{\bar{\theta}}\|_{L^1([0,\infty[,\mathscr{M}_p^\lambda)} \\
&\lesssim \|\bar{v}\|_{\mathfrak{L}^\infty([0,\infty[,\mathscr{F}\mathscr{N}_{p,\lambda,q}^{1-2\alpha+\frac{2}{p'}+\frac{\lambda}{p}})}\sum_{|k-j|\le 4}2^{k(2\alpha-1)}\|\varphi_k\widehat{\bar{\theta}}\|_{L^1([0,\infty[,\mathscr{M}_p^\lambda)},
\end{aligned}
$$

where in (19) we have used Lemma 4.
Therefore, by using the Young inequality for series, one has

$$\begin{aligned}
\mathrm{I}_1 &\lesssim \|\bar{v}\|_{\mathfrak{L}^\infty([0,\infty[,\mathscr{F}\mathscr{N}_{p,\lambda,q}^{1-2\alpha+\frac{2}{p'}+\frac{\lambda}{p}})} \left\{\sum_{j\in\mathbb{Z}} 2^{j(2-2\alpha+\frac{2}{p'}+\frac{\lambda}{p})q}\Big(\sum_{|k-j|\le 4} 2^{k(2\alpha-1)}\|\varphi_k\widehat{\bar{\theta}}\|_{L^1([0,\infty[,\mathscr{M}_p^{\lambda})}\Big)^q\right\}^{\frac{1}{q}}\\
&\lesssim \|\bar{v}\|_{\mathfrak{L}^\infty([0,\infty[,\mathscr{F}\mathscr{N}_{p,\lambda,q}^{1-2\alpha+\frac{2}{p'}+\frac{\lambda}{p}})} \left\{\sum_{j\in\mathbb{Z}}\Big(\sum_{|k-j|\le 4} 2^{(j-k)(2-2\alpha+\frac{2}{p'}+\frac{\lambda}{p})q}2^{k(1+\frac{2}{p'}+\frac{\lambda}{p})}\|\varphi_k\widehat{\bar{\theta}}\|_{L^1([0,\infty[,\mathscr{M}_p^{\lambda})}\Big)^q\right\}^{\frac{1}{q}}\\
&\lesssim \|\bar{v}\|_{\mathfrak{L}^\infty([0,\infty[,\mathscr{F}\mathscr{N}_{p,\lambda,q}^{1-2\alpha+\frac{2}{p'}+\frac{\lambda}{p}})}\|\bar{\theta}\|_{\mathfrak{L}^1([0,\infty[,\mathscr{F}\mathscr{N}_{p,\lambda,q}^{1+\frac{2}{p'}+\frac{\lambda}{p}})}.
\end{aligned}$$

Similarly, we get

$$\mathrm{I}_2 \lesssim \|\bar{\theta}\|_{\mathfrak{L}^\infty([0,\infty[,\mathscr{F}\mathscr{N}_{p,\lambda,q}^{1-2\alpha+\frac{2}{p'}+\frac{\lambda}{p}})}\|\bar{v}\|_{\mathfrak{L}^1([0,\infty[,\mathscr{F}\mathscr{N}_{p,\lambda,q}^{1+\frac{2}{p'}+\frac{\lambda}{p}})}.$$

For I_3, first we use the Young inequality in Morrey spaces (7), the Bernstein-type inequality with $|\gamma|=0$ together with the Hölder inequality, to get

$$\begin{aligned}
\|\widehat{R_j^3}\|_{L^1([0,\infty[,\mathscr{M}_p^{\lambda})} &= \Big\|\sum_{k\ge j-3}\widehat{\dot{\Delta}_j e^{\sqrt{\tau}|D|^\alpha}(\dot{\Delta}_k K[v]\tilde{\dot{\Delta}}_k\theta)}\Big\|_{L^1([0,\infty[,\mathscr{M}_p^{\lambda})}\\
&= \Big\|\sum_{k\ge j-3}\varphi_j e^{\sqrt{\tau}|\xi|^\alpha}\widehat{\dot{\Delta}_k K[v]\tilde{\dot{\Delta}}_k\theta}\Big\|_{L^1([0,\infty[,\mathscr{M}_p^{\lambda})}\\
&= \Big\|\sum_{k\ge j-3}\varphi_j e^{\sqrt{\tau}|\xi|^\alpha}\Big[\Big(e^{-\sqrt{\tau}|\xi|^\alpha}\varphi_k\widehat{K[\bar{v}]}\Big)*\sum_{|m-k|\le 1}e^{-\sqrt{\tau}|\xi|^\alpha}\varphi_m\widehat{\bar{\theta}}\Big]\Big\|_{L^1([0,\infty[,\mathscr{M}_p^{\lambda})}\\
&= \Big\|\sum_{k\ge j-3}\varphi_j\int_{\mathbb{R}^2}e^{\sqrt{\tau}(|\xi|^\alpha-|\xi-y|^\alpha-|y|^\alpha)}\varphi_k\widehat{K[\bar{v}]}(\xi-y)\sum_{|m-k|\le 1}\varphi_m\widehat{\bar{\theta}}(y)dy\Big\|_{L^1([0,\infty[,\mathscr{M}_p^{\lambda})}\\
&\le \sum_{k\ge j-3}\|(\widehat{\dot{\Delta}_k K[\bar{v}]}*\widehat{\tilde{\dot{\Delta}}_k\bar{\theta}})\|_{L^1([0,\infty[,\mathscr{M}_p^{\lambda})}\\
&\le \sum_{k\ge j-3}\|\varphi_k\widehat{K[\bar{v}]}\|_{L^1([0,\infty[,\mathscr{M}_p^{\lambda})}\sum_{|m-k|\le 1}\|\varphi_m\widehat{\bar{\theta}}\|_{L^\infty([0,\infty[,L^1)}\\
&\lesssim \sum_{k\ge j-3}\|\varphi_k\widehat{\bar{v}}\|_{L^1([0,\infty[,\mathscr{M}_p^{\lambda})}\sum_{|m-k|\le 1}2^{(\frac{2}{p'}+\frac{\lambda}{p})m}\|\varphi_m\widehat{\bar{\theta}}\|_{L^\infty([0,\infty[,\mathscr{M}_p^{\lambda})}\\
&\lesssim \sum_{k\ge j-3}\|\varphi_k\widehat{\bar{v}}\|_{L^1([0,\infty[,\mathscr{M}_p^{\lambda})}\Big(\sum_{|m-k|\le 1}2^{m(2\alpha-1)q'}\Big)^{\frac{1}{q'}}\|\bar{\theta}\|_{\mathfrak{L}^\infty([0,\infty[,\mathscr{F}\mathscr{N}_{p,\lambda,q}^{1-2\alpha+\frac{2}{p'}+\frac{\lambda}{p}})}\\
&\lesssim \|\bar{\theta}\|_{\mathfrak{L}^\infty([0,\infty[,\mathscr{F}\mathscr{N}_{p,\lambda,q}^{1-2\alpha+\frac{2}{p'}+\frac{\lambda}{p}})}\sum_{k\ge j-3}2^{k(2\alpha-1)}\|\varphi_k\widehat{\bar{v}}\|_{L^1([0,\infty[,\mathscr{M}_p^{\lambda})}.
\end{aligned}$$

Then, applying the Young inequality for series, we obtain

$$\begin{aligned}
\mathrm{I}_3 &\lesssim \|\bar{\theta}\|_{\mathfrak{L}^\infty([0,\infty[,\mathscr{F}\mathscr{N}_{p,\lambda,q}^{1-2\alpha+\frac{2}{p'}+\frac{\lambda}{p}})}\left\{\sum_{j\in\mathbb{Z}}2^{j(2-2\alpha+\frac{2}{p'}+\frac{\lambda}{p})q}\Big(\sum_{k\ge j-3}2^{k(2\alpha-1)}\|\varphi_k\widehat{\bar{v}}\|_{L^1([0,\infty[,\mathscr{M}_p^{\lambda})}\Big)^q\right\}^{\frac{1}{q}}\\
&\lesssim \|\bar{\theta}\|_{\mathfrak{L}^\infty([0,\infty[,\mathscr{F}\mathscr{N}_{p,\lambda,q}^{1-2\alpha+\frac{2}{p'}+\frac{\lambda}{p}})}\left\{\sum_{j\in\mathbb{Z}}\Big(\sum_{k\ge j-3}2^{(j-k)(2-2\alpha+\frac{2}{p'}+\frac{\lambda}{p})}2^{k(1+\frac{2}{p'}+\frac{\lambda}{p})}\|\varphi_k\widehat{\bar{v}}\|_{L^1([0,\infty[,\mathscr{M}_p^{\lambda})}\Big)^q\right\}^{\frac{1}{q}}
\end{aligned}$$

$$\lesssim \|\bar{\theta}\|_{\mathfrak{L}^{\infty}([0,\infty[,\mathcal{FN}^{1-2\alpha+\frac{2}{p'}+\frac{\lambda}{p}}_{p,\lambda,q})} \|\bar{v}\|_{\mathfrak{L}^{1}([0,\infty[,\mathcal{FN}^{1+\frac{2}{p'}+\frac{\lambda}{p}}_{p,\lambda,q})} \sum_{i\leq 3} 2^{i(2-2\alpha+\frac{2}{p'}+\frac{\lambda}{p})}$$
$$\lesssim \|\bar{\theta}\|_{\mathfrak{L}^{\infty}([0,\infty[,\mathcal{FN}^{1-2\alpha+\frac{2}{p'}+\frac{\lambda}{p}}_{p,\lambda,q})} \|\bar{v}\|_{\mathfrak{L}^{1}([0,\infty[,\mathcal{FN}^{1+\frac{2}{p'}+\frac{\lambda}{p}}_{p,\lambda,q})}.$$

where the condition $\alpha < 1 + \frac{\lambda}{2p} + \frac{1}{p'}$ ensures that the series $\sum_{i\leq 3} 2^{i(2-2\alpha+\frac{2}{p'}+\frac{\lambda}{p})}$ converges. This finishes the proof of Proposition 1.

3.1 Proof of Theorem 1

From Proposition 1, we have

$$\|\mathcal{B}(\bar{v},\bar{\theta})\|_X \leq C_0 \|\bar{v}\|_X \|\bar{\theta}\|_X. \tag{20}$$

By Lemma 3, we know that if $\|e^{\sqrt{t}|D|^{\alpha}} e^{-(-\triangle)^{\alpha}t} v_0\|_X \leq \varepsilon$ with $\varepsilon = \frac{1}{4C_0}$, then the Eq. (12) has a unique solution in $\mathrm{B}(0,2\varepsilon) := \{y \in X : \|y\|_X \leq 2\varepsilon\}$.

Now, let us prove that $\|e^{\sqrt{t}|D|^{\alpha}} e^{-(-\triangle)^{\alpha}t} v_0\|_X \leq \varepsilon$. According to Lemma 5, we have

$$\|e^{\sqrt{t}|D|^{\alpha}} e^{-(-\triangle)^{\alpha}t} v_0\|_X \leq C\|v_0\|_{\mathcal{FN}^{1-2\alpha+\frac{2}{p'}+\frac{\lambda}{p}}_{p,\lambda,q}}. \tag{21}$$

If $\|v_0\|_{\mathcal{FN}^{1-2\alpha+\frac{2}{p'}+\frac{\lambda}{p}}_{p,\lambda,q}} \leq M$ with $M = \frac{1}{4CC_0}$, then (12) has a unique analytic solution.

This completes the proof of Theorem 1.

References

1. Abidin, M.Z., Chen, J.: Global well-posedness for fractional Navier-Stokes equations in variable exponent Fourier-Besov-Morrey spaces. Acta Math. Sci. **41**, 164–176 (2021). https://doi.org/10.1007/s10473-021-0109-1
2. Azanzal, A., Abbassi, A., Allalou, C.: On the Cauchy problem for the fractional drift-diffusion system in critical Fourier-Besov-Morrey spaces. Int. J. Optim. Appl. **1**(2), 28 (2021)
3. Bae, H.: Global well-posedness of dissipative quasi-geostrophic equations in critical spaces. Proc. Am. Math. Soc. **136**(1), 257–261 (2008)
4. Bahouri, H., Chemin, J.Y., Danchin, R.: Fourier Analysis and Nonlinear Partial Differential Equations, vol. 343. Springer, Cham (2011). https://doi.org/10.1007/978-3-642-16830-7
5. Bahouri, H.: The Littlewood-Paley theory: a common thread of many words in nonlinear analysis. Eur. Math. Soc. Newslett. (2019)
6. Benameur, J., Benhamed, M.: Global existence of the two-dimensional QGE with sub-critical dissipation. J. Math. Anal. Appl. **423**(2), 1330–1347 (2015)
7. Benameur, J., Blel, M.: Long-time decay to the global solution of the 2D dissipative quasi-geostrophic equation. In: Abstract and Applied Analysis, vol. 2012 (2012)
8. Benhamed, M., Abusalim, S.M.: Long time behavior of the solution of the two-dimensional dissipative QGE in Lei-Lin spaces. Int. J. Math. Math. Sci. **2020** (2020)
9. Chae, D., Lee, J.: Global well-posedness in the super-critical dissipative quasi-geostrophic equations. Commun. Math. Phys. **233**(2), 297–311 (2003). https://doi.org/10.1007/s00220-002-0750-z

10. Chen, X.: Well-posedness of the Keller-Segel system in Fourier-Besov-Morrey spaces. Zeitschrift für Analysis und ihre Anwendungen. **37**(4), 417–434 (2018)
11. Costantin, P., Majda, A., Tabak, E.: Formation of strong fronts in the 2D quasi-geostrophic thermal active scalar. Nonlinearity **7**, 1495–1533 (1994)
12. Constantin, P., Wu, J.: Behavior of solutions of 2D quasi-geostrophic equations. SIAM J. Math. Anal. **30**(5), 937–948 (1999)
13. Córdoba, A., Córdoba, D.: A maximum principle applied to quasi-geostrophic equations. Commun. Math. Phys. **249**(3), 511–528 (2004). https://doi.org/10.1007/s00220-004-1055-1
14. de Almeida, M.F., Ferreira, L.C., Lima, L.S.: Uniform global well-posedness of the Navier-Stokes-Coriolis system in a new critical space. Math. Z. **287**(3), 735–750 (2017). https://doi.org/10.1007/s00209-017-1843-x
15. Ferrari, A.B., Titi, E.S.: Gevrey regularity for nonlinear analytic parabolic equations. Commun. Partial Differ. Equ. **23**, 1–16 (1998)
16. Ferreira, L.C.F., Lidiane, S.M.L.: Self-similar solutions for active scalar equations in Fourier-Besov-Morrey spaces. Monatshefte für Mathematik **175**(4), 491–509 (2014)
17. Foias, C., Temam, R.: Gevrey class regularity for the solutions of the Navier-Stokes equations. J. Funct. Anal. **87**, 359–369 (1989)
18. Kato, T.: Strong solutions of the Navier-Stokes equation in Morrey spaces. Boletim da Sociedade Brasileira de Matemética-Bulletin/Braz. Math. Soc. **22**(2), 127–155 (1992). https://doi.org/10.1007/BF01232939
19. Lemariè-Rieusset, P.G.: Recent Developments in the Navier-Stokes Problem. CRC Press, Boca Raton (2002)
20. Pedlosky, J.: Geophysical Fluid Dynamics, vol. 710. Springer, New York (1987)
21. Toumlilin, M.: Global well-posedness and analyticity for generalized porous medium equation in critical Fourier-Besov-Morrey spaces (2019)
22. Wang, W.: Global well-posedness and analyticity for the 3D fractional magnetohydrodynamics equations in variable Fourier-Besov spaces. Z. Angew. Math. Phys. **70**, 1–16 (2019). https://doi.org/10.1007/s00033-019-1210-3
23. Zhou, X., Xiao, W.: Algebra properties in Fourier-Besov spaces and their applications. J. Funct. Spaces **2018** (2018)

Ambient Ozone Concentrations Modelling Using Feedforward Backpropagation Neural Networks: Spatial Modeling over the Agadir City (Morocco)

Anas Adnane[1](✉), Redouane Leghrib[1], Jamal Chaoufi[1], and Ahmed Chirmata[2]

[1] LETSMP, Department of Physics, Faculty of Sciences, University Ibn Zohr, Cité Dakhla BP8106, Agadir, Morocco
anas.adnane@edu.uiz.ac.ma, r.leghrib@uiz.ac.ma
[2] Department of Energy and Environment, Wilaya of Agadir, Boulevard Mohamed V, 80000 Agadir, Morocco

Abstract. The present study attempts to model ambient ozone (O_3) concentrations using the Feed-forward backpropagation neural network model as a function of meteorological conditions and to draw a map of ozone concentrations over the Agadir City and its neighboring regions. The data, used to train the network, was collected during a period of 30 days in Agadir City using a mobile monitoring station. The data was divided into two sets: 85% for the training and 15% for the testing. Different neural network architectures were tested during the training process. A network of two hidden layers with a [7-16-16-1] structure, Levenberg-Marquardt as a training algorithm and log-sigmoid as a transfer function was found to give the best predictions for both the training and testing data. Statistical agreement between predicted and observed values is evaluated by coefficient of correlation (CC), root mean square error (RMSE) and mean absolute error (MAE). Results show that artificial neural network modelling appears to be a promising method for modelling air-pollution. IDW Inverse distance method was used in order to interpolate the ozone concentrations over the city of Agadir and its neighboring areas.

Keywords: Air quality modelling · Air quality monitoring · Neural network model · Machine learning · Morocco · Pollution

1 Introduction

Air pollution is a major risk that affects the health of human beings in all over the world. The World Health Organization announced that almost 4.2 million people die prematurely every year as a result of exposure to ambient air pollution (World Health Organization 2016). Many studies have been carried out since the 1990s which highlights the strong dependence between urban pollution and short- and long-term health impacts

S. Melliani and O. Castillo (Eds.): ICPAMS 2021, LNNS 476, pp. 52–65, 2023.
https://doi.org/10.1007/978-3-031-12416-7_6

(ANSES 2012). Short-term effects are clinical, functional or biological manifestations that appear after exposure to air pollution in a short period of time (Kuenzli 2006). Currently, less research has been done on the long-term health effects of air pollution seeing that measuring the cumulative effects of lifelong exposure to air pollution is complicated and expensive (Tertre et al. 2002).

If we shed a light on Morocco, the gross domestic product GDP is remarkably affected by the environmental deterioration and especially the air quality degradation. Although it is not considered as an industrial country, Morocco has put serious interest in fighting air quality's deterioration. The Moroccan kingdom has also been protecting environment and health, considering them both as priorities. Thus, lots of measures and precautions were taken aiming to do so (Chirmata et al. 2017).

High ozone concentrations in the atmosphere near the ground are of concern because of potential effects on the environment and public health. Ozone is a secondary pollutant: it is not emitted directly into the air (Comrie 1997). It is the outcome of complex chemical reactions in the atmosphere. Ozone results when the primary pollutants nitrogen oxides (NOx) and non-methane hydrocarbons (NMHC) interact under the action of sunlight. The use of gasoline and other fuels is the main source of the ozone precursors. Furthermore, the generation of ozone is strongly connected to meteorological conditions (Abdul-Wahab and Al-Alawi 2002).

Due to the expensiveness of AQMN (Air Quality Monitoring Networks), observational data are not always available especially in non-developed countries. Hence, air quality modelling is an important tool for a better understanding of the situation. Nonetheless, air quality modelling remains a challenging task considering the following reasons:

- The lack of emission data in many countries worldwide.
- The large variety of species that influence pollutants such as ozone.
- The complexity of atmospheric processes (Abdallah et al. 2018).
- The complexity of the region's topography

Artificial Neural Networks (ANN) are a set of techniques in the field of machine learning, the principles of which are inspired by biological neural networks. The ANN has demonstrated a remarkable success in different engineering fields thanks to their ability to learn spontaneously from examples, over inexact and fuzzy data, and to provide adequate and rapid responses to new information not previously stored in memory (Elkamel et al. 2001). Due to the complex and non-linear ozone formation, we used the artificial neural networks (ANN) alongside with the Inverse Distance Weighted (IDW) method in this work in order to model and interpolate hourly ozone concentrations in the city of Agadir and its neighboring regions.

2 Methodology

2.1 Area Description

Our study area is the prefecture of Agadir-Ida-Outanane. It is the capital of the Souss Massa region. Bordered by the ocean to the west, the prefecture of Essaouira to the north, the prefecture of Inzegane Aït Melloul to the south and the province of Taroudant to the east. The Agadir-Ida-Outanane covers an area of 2,297 km^2. It is made up of 6 urban districts, 2 circles, an urban commune and 12 rural communes, with a population of 506517. Thanks to its climate, its diversity, its historical and cultural heritage and its extensive beaches, the city of Agadir is one of the major international tourist destinations. The region has significant economic potential based on its natural resources and its possibilities in the fields of agriculture, fishing and tourism. Fishing is a key sector in our study area and is an asset for the kingdom. It employs a large workforce and drains investment and foreign exchange. The city has a large offshore coastal fishing port, as well as a number of small fishing ports reserved for artisanal fishing boats.

2.2 Data Collection

In order to characterize the air quality in Agadir City, the Souss Massa Region has a regional mobile laboratory and one fixed ground station dedicated to air quality monitoring since 2010.

Hourly ozone concentrations and meteorological data (temperature, humidity, wind speed and wind direction) were collected in different sites over the city of Agadir (Table 1). During the data collection processes, Chirmata et al. took into consideration the representativeness of the station locations (Chirmata et al. 2017). That's why they chose three traffic sites (Al Massira Bus Station – Hassan II Hospital – International Camping of Agadir) and one industrial site (Anza).

Table 1. Monitoring sites in Agadir City and their locations (using the longitude and latitude coordinates)

Monitoring station	Location (Longitude; Latitude)
Anza	9,658485; 30,448091
International Camping of Agadir	9,608422; 30,423595
Al Massira Bus Station	9,565632; 30,415886
Hassan II Hospital	9,550661; 30,437560

Table 2 presents a statistical summary of the variables measured from 07/01/2016 to 07/30/2016. The values (maximum, minimum, average, variance, standard deviation and % missing values) are based on the hourly values in each of the four stations. Note that Anza represents the maximum of the O_3 concentrations recorded in all the other stations with a value of 183.97 $\mu g/m^3$. On the other hand, the average ozone concentrations in all stations do not exceed 54.71 $\mu g/m^3$.

Table 2. Characteristics of selected variables for the Anza, Al Massira Bus Station, Hassan II Hospital, International Camping of Agadir monitoring sites.

Station	Parameter	Unity	Range [Min; Max]	Average	% Missing values	Variance	Standrad deviation
Anza	Ozone	μg/m^3	[4,88; 183,97]	45,28	0	725,1	26,93
	Temperature	°C	[16,59; 27,2]	20,73	0	9,92	3,14
	Humidity	%	[0,27; 90,14]	84,85	0	283,12	16,83
	Wind direction	°	[29,23; 351,25]	211,99	0	4067,8	63,78
	Wind speed	m/s	[0,01; 0,52]	0,07	0	0,0049	0,07
International Camping of Agadir	Ozone	μg/m^3	[7,5; 132,58]	51,83	25,7	215,98	14,69
	Température	°C	[13,89; 27,59]	23,03	6,19	1,02	1,01
	Humidity	%	[0,02; 90,25]	80,75	6,19	432,47	20,79
	Wind direction	°	[4,48; 328,17]	165,09	6,73	5969,54	77,26
	Wind speed	m/s	[0,02; 4,15]	0,99	6,73	0,91	0,95
Al Massira Bus Station	Ozone	μg/m^3	[0,06; 84,52]	45,85	4,71	229,5	15,15
	Temperature	°C	[16,57; 30;33]	23,95	0	1,62	1,27
	Humidity	%	[0,22; 90,2]	78,01	0	653,5	25,56
	Wind direction	°	[4,27; 318,08]	165,09	0	4476,32	66,91
	Wind speed	m/s	[0,02; 3,7]	0,82	0	0,76	0,87
Hassan II Hospital	Ozone	μg/m^3	[0,98; 94,4]	54,71	19,24	420,2	20,23
	Temperature	°C	[16,43; 28,97]	23,27	19,24	2,86	1,69

(*continued*)

Table 2. (*continued*)

Station	Parameter	Unity	Range [Min; Max]	Average	% Missing values	Variance	Standrad deviation
	Humidity	%	[0,26; 90,2]	76,28	19,25	587,5	24,24
	Wind direction	°	[39,38; 334,35]	204,01	19,11	4308,11	65,64
	Wind speed	m/s	[0,02; 5,18]	0,67	19,24	0,69	0,83

In order to interpolate the O_3 concentrations with good resolution over the city of Agadir, we needed more observation points and therefore more data. As a result, we extracted weather data from the "Institut Agronomique et Vétérinaire Hassan II. Complexe Horticole Agadir" metrological database. Table 3 illustrates the different automatic measurement stations. These stations measure only meteorological parameters. In order to have also the ozone concentrations of these measurement points, we will use later in this work the trained neural networks from the previous stations to give the estimated concentrations.

Table 3. Sites near Agadir City and their locations (using the longitude and latitude coordinates).

Station	Coordinates (Longitude; Latitude)
Biougra Tin Hammou	9,3739; 30,1881
Ouled Teima - Lagfifat	9,1539; 30,1940
Temsia	9,2452; 30,2148
Ait Melloul	9,5040; 30,3379
Khmis Ait Amira – Tin Ali Mansour	9,5454; 30,2099
Sidi Bibi	9,5409; 30,2234

2.3 Data Treatment and Analysis

One of the main problems most often encountered is that of missing data. It should be noted that these are relatively linked to random or systematic errors. Therefore, it is important to take steps to ensure the recovery of missing data.

The lack of data in the series is a major problem. We can only go ahead and make analyzes on these series if the missing data reconstruction operation has already been completed. To estimate the missing values, the method of the mean of the neighboring points was used. This method is used when it is an isolated missing value. It consists of replacing a missing value with the average of days d 1 and d + 1.

We then calculated the values of the zonal wind and the meridional wind from the wind speed and the wind direction, using the following relationships:

$$Wind_x = w\cos\phi \tag{1}$$

$$Wind_y = -w\sin\phi \tag{2}$$

where $Wind_x$ represents the zonal wind, $Wind_y$ is the meridional wind, w denotes the wind speed and ϕ indicates the wind direction.

In order to compare the performance of the models, several numerical indices are used to explain the quality of the model. These indices will be useful to us in the selection of the best neural network for our case of study. The criteria used are as follows:

$$MAE = \frac{\sum_{i=1}^{n} |y_i - x_i|}{n} \quad \text{(Mean Absolute Error)} \tag{3}$$

$$RMSE = \sqrt{\frac{1}{n}\sum_{i=1}^{n} (y_i - x_i)^2} \quad \text{(Root - Mean - Square Deviation)} \tag{4}$$

$$CC = \frac{\sum_{i=1}^{n} (y_i - \overline{y}_i)(x_i - \overline{x}_i)}{\sqrt{\sum_{i=1}^{n} (y_i - \overline{y}_i)^2 \sum_{i=1}^{n} (x_i - \overline{x}_i)^2}} \quad \text{(Correlation coefficient)} \tag{5}$$

where y_i and x_i represent respectively the values of the observed and the predicted ozone concentrations, and $\overline{y}_i$ and $\overline{x}_i$ are respectively the mean values of the observed and the predicted ambient ozone concentrations.

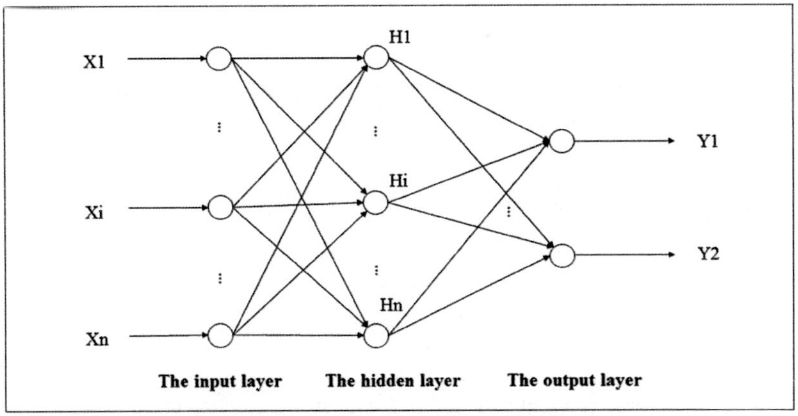

Fig. 1. The general structure of FFBP-NN.

3 Artificial Neural Networks

Like any other scientific field, the field of artificial neural networks has had a long history of development. This story begins in the early 1940s, but the real success only came about 30 years ago. Artificial neural networks are now considered among the best approaches for analyzing heterogeneous information. The use of neural networks has been developed in several fields and disciplines, namely meteorology, environment, medicine, etc. They are also applied to solve classification, forecasting or pattern recognition problems (Schmitt et al. 2001).

3.1 Feed-Forward Back Propagation Neural Network (FFBP-NN)

There exists a plenty of neural network architectures. The most frequently used are the ANNs based on a feed-forward configuration of the multilayer perceptron (MLP), also named FFBP-NN (Feed-forward Backpropagation neural networks). In the FFBP-NN architecture, the neurons are organized as layers and the information strictly flows forward: each neuron receives data, processes it, and provides an output, which becomes the input of subsequent neurons, to the output neurons of the model, and the errors of the network are propagated backwards. The MLP architecture consists of input layer, one or more hidden layers and output layer (Bai et al. 2016). Figure 1 shows a general structure of the three layers FFBP-NN.

The input layer receives the data to train the neural network, which is in different formats (images, videos, text, words, sounds or digital data). Its size is directly determined by the number of input variables. As for the hidden layers, they are not in direct contact with the outside. The activation functions of the hidden layers are nonlinear in general, but there is no general rule to follow. The choice of its size is not implicit and must be adjusted. The output layer presents the network results which are often obtained after compilation and training, especially to perform classifications or predictions, with the associated probabilities (Zhang and Ding 2017).

There is a plethora of activation functions. In this work, we used the log-sigmoid function in the hidden layers and a linear transfer function in the output layer. The mathematic expressions of the output of the hidden and output layers are expressed as follows (Trigo and Palutikof 1999).

$$\mathrm{x_j} = \mathbf{f}_{\text{hidden}}\left(\sum_{i=1}^{m} w_{ij}u_i\right), \text{ and } \mathrm{y_k} = \mathbf{f}_{\text{output}}\left(\sum_{i=1}^{m} w_{jk}\mathrm{x_j}\right), \tag{6}$$

where $u = (u_1, u_2, \ldots, u_i)$ $(i = 1, 2, \ldots, m)$ represents the inputs, $x = (x_1, x_2, \ldots, x_i)$ $(j = 1, 2, \ldots, n)$ represents the outputs of the hidden layer, $y = (y_1, y_2, \ldots, y_i)$ $(k = 1, 2, \ldots, p)$ represents the outputs of the network, w represents the weight matrix between two layers, and f_{hidden} and f_{output} are respectively the transfer functions of the hidden layer and the output layer.

In order to estimate the weights that can associate properly the predictors with the predictand, ANNs have to be trained, and there is a lot of algorithms to do so. In this work, we trained the ANNs using the Levenberg-Marquardt method, alongside with the Bayesian regularization and the scaled conjugate gradient back propagation.

3.2 Neural Network Development

Choosing the most suitable configuration of the model is not simple enough, given the multiplicity of combinations to be treated and the large number of parameters to be adjusted. In fact, there is no fixed rule for the choice of the number of neurons and layers to use, the activation function, or the learning algorithm.

Selecting model parameters is a crucial step in designing a learning model. Unfortunately, we do not have several meteorological and chemical parameters to test several combinations and choose the most optimal one. As a result, the parameters that we have chosen as input parameters of the neural network are: Hour of the day, longitude, latitude, temperature, humidity, zonal wind and meridional wind. Regarding the output layer, the concentration of ozone will be the only output parameter.

Like any other statistical model, ANNs should be trained and tested using two independent data sets. We divided our database into two groups: 85% of the data, or 612 cases, were used for training the model, the remaining 15%, or 107 cases, were used for validation and evaluation of the quality of the model. We trained, validated and evaluated the model using MATLAB Toolbox.

Table 4. FFBP-NNs used in this work and their learning algorithms, activation functions and number of hidden layers.

Structure	Algorithm	Activation function	Hidden layer
FFBP-NN1	Levenberg-Marquardt	Log-Sigmoid	1
FFBP-NN2	Bayesian Regularization	Log-Sigmoid	1
FFBP-NN3	Scaled Conjugate	Log-Sigmoid	1
FFBP-NN4	Levenberg-Marquardt	Log-Sigmoid	2

Different FFBP-NNs were trained on the data. We wanted to train different ANNs using different learning algorithms and also changing the number of hidden layers as shown in Table 4 and Table 5. The algorithms are written and ran under Matlab 2015 software.

Table 5. The best input-output structures of MLP neural networks and their CC, RMSE and MAE.

	Input-hidden-output structures	CC	RMSE ($\mu g/m^3$)	MAE ($\mu g/m^3$)
FFBP-NN1	7-10-1	0,62	23,85	18,54
FFBP-NN2	7-16-1	0,69	14,48	23,01
FFBP-NN3	7-12-1	0,41	18,51	19,13
FFBP-NN4	7-16-16-1	0,75	13,71	20,66

4 Results and Discussion

The objective of the calculations was to determine the effectiveness of the ANNs developed, in terms of its capacity to model ozone concentrations in the area of ongoing exploitation. The best network structures trained are shown in Table 5 alongside with their performance criteria. Three performance criteria were calculated in order to select the best FFBP-NN. From Table 5, we notice that CC, RMSE and MAE differ significantly from one structure to another.

By comparing between the structures with one hidden layer, we notice that FFBP-NN2 gives better results compared to FFBP-NN1 and FFBP-NN3 with the best CC with a value of 0,69 and the least RMSE with a value of 14,48 $\mu g/m^3$. In contrast, FFBP-NN2 represents the worst MAE worth 23,01 $\mu g/m^3$.

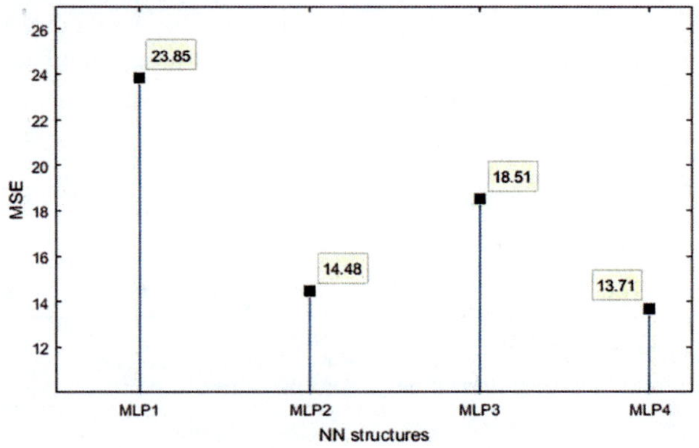

Fig. 2. Graph illustrating the evolution of MSE according to the different FFBP-NN structures.

Structures with two hidden layers have shown good performance compared to those with one single hidden layer. Figure 2 shows the evolution of RMSE according to the different FFBP-NN structures. Conforming to Table 5 and Fig. 2, FFBP-NN4 with a [7-16-16-1] structure is the best MLP trained, with a CC of 75%, RMSE of 13,71 $\mu g/m^3$ and MAE of 20,66 $\mu g/m^3$. The FFBP-NN4 with a [7-16-16-1] structure was selected for further analysis as the best of those presented in Table 5 and Fig. 2. These results are significant compared to other studies and models: a study was conducted in Agadir City to model ozone concentrations using CHIMERE transport model, the ozone concentrations obtained have a CC of 82%, a RMSE of 21,80 $\mu g/m^3$ and a MAE of 21,40 $\mu g/m^3$ (Ajdour et al. 2020).

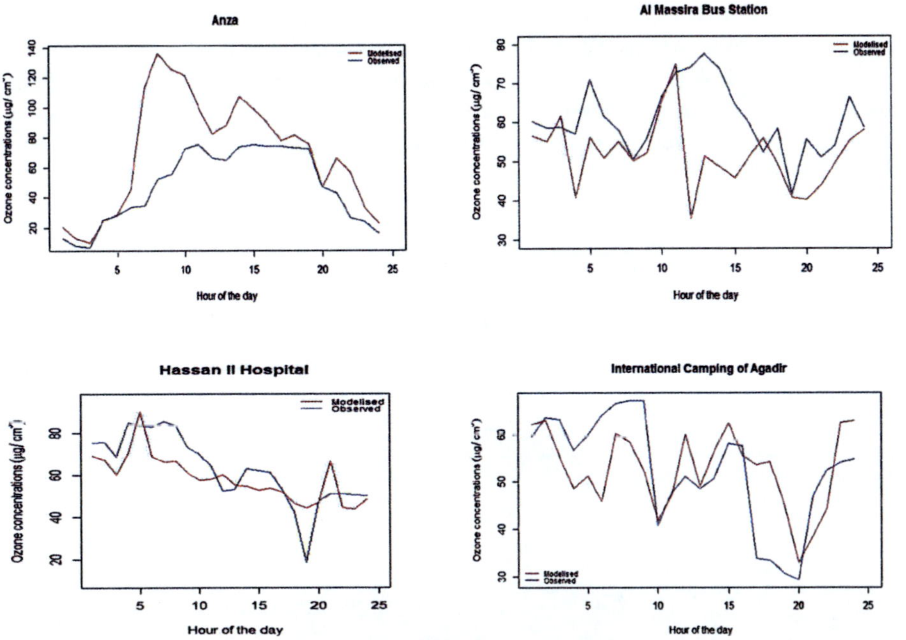

Fig. 3. Comparison of predicted and measured ozone concentrations for the validation data set in the four stations (Anza, International Camping of Agadir, Al Massira Bus Station and Hassan II Hospital).

In order to illustrate the performance of the FFBP-NN4 of [7-16-16-1] structure, a comparison of the changes between observed and predicted ozone concentrations in the four sites (Anza – Al Massira Bus Station – Hassan II Hospital – International Camping of Agadir) over a 24-h period is presented in Fig. 3. It can be noted that the predicted concentrations are sometimes overestimated and sometimes underestimated. In general, Artificial Neural Networks find it very challenging to estimate extreme values (Adnane et al. 2020; Bai et al. 2019). Despite their simple structures, the neural networks used in this paper allowed for a relatively accurate forecast of the ozone concentrations. However, the model requires calibration in order to further reduce the existing differences between the forecast and the observation.

In order to build a good resolution ozone map over the city of Agadir and its surrounding areas, we extracted weather data from the "Institut Agronomique et Vétérinaire Hassan II. Complexe Horticole Agadir" metrological database. The measurement stations, where the data were collected, do not have O_3 measurement sensors. It is for this reason that we injected the meteorological data of these stations into the neural network previously selected in this work, in order to produce the concentrations of O_3 in the six measurement stations indicated in Table 3. This allowed us to have a satisfactory number of points in order to interpolate the concentrations of ozone.

We used IDW method to interpolate the ozone concentrations over the city of Agadir and its neighboring regions. We chose IDW over the other interpolation methods thanks to the fact that IDW is easy to define and therefore easy to understand and interpret the results. Figure 4 shows a map that illustrates the interpolation of O_3 concentrations in the Agadir region using the IDW inverse distance method with the ARCGIS tool for the day of 10/18/2010 at 12 am.

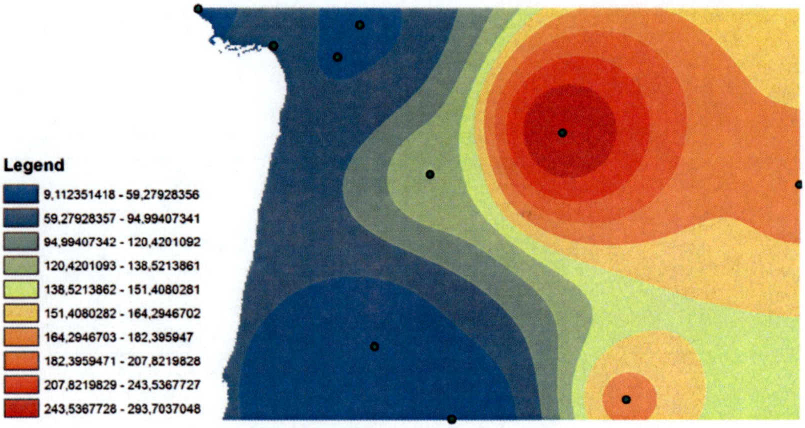

Fig. 4. Map of ozone concentrations interpolated over the Agadir region using the IDW method for the day of 01/07/2016 at 12 am.

5 Variable Importance Interpretation

A method for partitioning the neural network connection weights was proposed by (Garson 1991). This method aims to determine the relative importance the ANN's input variables. It should be noted that Garson's algorithm does not provide the direction of the relationship between the predictor and the predictand variables, because it adopts the absolute values of the connection weights when calculating variable contributions (Olden and Jackson 2002). The formula given by Garson to calculate the relative importance of the input parameters is as follows:

$$RI_i = \sum\nolimits_{i=1}^{n} \frac{\left|w_{ij}w_{jk}\right|}{\sum\nolimits_{j=1}^{m} \left|w_{ij}w_{jk}\right|} \tag{7}$$

where RI_i is the relative importance of neuron i, $\sum\nolimits_{j=1}^{m} w_{ij}w_{jk}$ is sum of product of final weights of the connections from input neurons to hidden neurons with the connection from hidden neurons to output neurons.

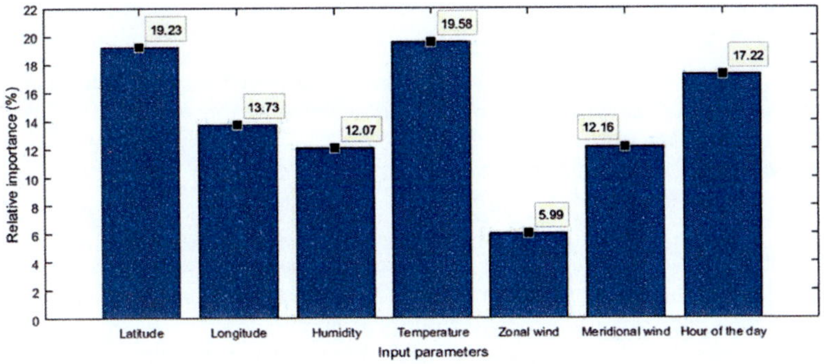

Fig. 5. Bar plots showing the relative importance of each input parameter for predicting ozone concentration based on Garson's algorithm (Garson et al. 1991).

Figure 5 illustrates the results featuring the relative importance of the ANN input parameters. The calculated contributions are ranged from 5,99 to 19,57%, with temperature and latitude expressing the strongest relationships with the predicted ozone concentrations, indicating that they should not be eliminated from the model. Although zonal and meridional wind are exhibiting the weakest relationship, they are important in our model because they give an indication of ozone dispersion (Ruiz-Suarez et al. 1995).

6 Conclusion and Perspectives

Our ANN models were developed for the Agadir City and its neighboring areas and should be considered specific to this selected site. The methodology is generalizable; however, it is probably not possible to extend the same exact model from the present study to other sites. It should be noted that works in this field are very rare in Morocco.

Despite the limited amount of data, the feed-forward backpropagation neural networks have been able to provide satisfactory results. The findings of this work show that the use of FFBP-NN in modelling air-pollution is promising. In general, the differences observed could be attributed to the absence of a sufficient number of data and input parameters to allow more advanced learning of neural networks, and to the scarcity of measurement stations to allow a broad comparison with the observation. The complex topography of the Agadir region is also cited as a reason for the deviation observed.

In order to improve the model, the next work will focus on the diversification of input data and input parameters of neural networks. More work is ongoing to model the other pollutants such as PM10 and NO2 over the city of Agadir.

Acknowledgements. We acknowledge the Wilaya and the Region of the Souss Massa for their collaboration and supporting the Research, under the leadership of the Wali and the president of the region of Souss Massa, to whom we present our great and special thanks.

A special acknowledge is dedicated also to AGROTECH association and the "Institut Agronomique et Vétérinaire Hassan II. Complexe Horticole Agadir".

References

Abdallah, C., Afif, C., El Masri, N., Öztürk, F., Keles, M., & Sartelet, K.: A first annual assessment of air quality modeling over Lebanon using WRF/Polyphemus. Atmos. Pollut. Res., March 2017 (2018). https://doi.org/10.1016/j.apr.2018.01.003

Abdul-Wahab, S.A., Al-Alawi, S.M.: Assessment and prediction of tropospheric ozone concentration levels using artificial neural networks. Environ. Model. Softw. **17**, 219–228 (2002)

Adnane, A., Leghrib, R., Chaoufi, J.: The use of a recurrent neural network for forecasting ozone concentrations in the city of Agadir (Morocco). J. Atomic Mol. Condens. Matter Nano Phys. **7**(3), 197–206 (2020). https://doi.org/10.26713/jamcnp.v7i3.1545

Ajdour, A., Leghrib, R., Chaoufi, J., Chirmata, A., Menut, L., Mailler, S.: Towards air quality modeling in Agadir City (Morocco). Mater. Today Proc. **24**, 17–23 (2020). https://doi.org/10.1016/j.matpr.2019.07.438

ANSES: Sélection des polluants à prendre en compte dans les évaluations des risques sanitaires réalisées dans le cadre des études d'impact des infrastructures routières (2012)

Bai, Y., Li, Y., Wang, X., Xie, J., Li, C.: Air pollutants concentrations forecasting using back propagation neural network based on wavelet decomposition with meteorological conditions. Atmos. Pollut. Res. **7**(3), 557–566 (2016). https://doi.org/10.1016/j.apr.2016.01.004

Bai, Y., Zeng, B., Li, C., Zhang, J.: Chemosphere an ensemble long short-term memory neural network for hourly PM 2. 5 concentration forecasting. Chemosphere **222**, 286–294 (2019). https://doi.org/10.1016/j.chemosphere.2019.01.121

Chirmata, A., Leghrib, R., Ichou, I.A.: Implementation of the air quality monitoring network at Agadir City in Morocco. J. Environ. Prot. **8**, 540–567 (2017). https://doi.org/10.4236/jep.2017.84037

Comrie, A.C.: Comparing neural networks and regression models for ozone forecasting. J. Air Waste Manag. Assoc. **47**(6), 653–663 (1997). https://doi.org/10.1080/10473289.1997.10463925

Elkamel, A., Abdul-Wahab, S., Bouhamra, W., Alper, E.: Measurement and prediction of ozone levels around a heavily industrialized area: a neural network approach. Adv. Environ. Res. **5**(1), 47–59 (2001). https://doi.org/10.1016/S1093-0191(00)00042-3

Garson, G.: Interpreting neural network connection weights. Artif. Intell. Expert **6**, 47–51 (1991)

Kuenzli, N.: Une longue route pour comprendre les effets à long terme de la pollution atmosphérique, pp. 5–8 (2006)

Schmitt, A., Le Blanc, B., Corsini, M., Lafond, C., Bruzek, J.: Artificial neural network : A relevant data processing tool for anthropology. Bulletins et Mémoires de La Société d'Anthropologie de Paris 13 (2001)

Tertre, A.L., et al.: Short-term effects of air pollution on mortality in nine French cities: a quantitative summary. Arch. Environ. Health **57**(4), 311–319 (2002). https://doi.org/10.1080/00039890209601414

Olden, J., Jackson, D.: Illuminating the 'black box': a randomization approach for understanding variable contribution in artifical neural networks. Ecol. Model. **154**, 135–150 (2002). https://doi.org/10.1016/S03033800(02)00064-9

Ruiz-Suarez, J., Mayora-Ibarra, O., Torres-Jimenez, J.: Short-term ozone forecasting by artificial neural networks. Adv. Eng. Softw. **23**, 143–149 (1995). https://doi.org/10.1016/0965-9978(95)00076-3

Trigo, R.M., Palutikof, J.P.: Simulation of daily temperatures for climate change scenarios over Portugal: a neural network model approach. Climate Res. **13**(1), 45–59 (1999). https://doi.org/10.3354/cr013045

World Health Organization, WHO (2016). http://www.who.int/airpollution

Zhang, J., Ding, W.: Prediction of air pollutants concentration based on an extreme learning machine: the case of Hong Kong. Int. J. Environ. Res. Public Health **14**(2), 1–19 (2017). https://doi.org/10.3390/ijerph14020114

Existence and Uniqueness of Solutions for Nonlinear Viscoelastic Plate Equation with $\overrightarrow{p}(x,t)-$ Laplacian Operator and Delay

Merah Ahlem and Mesloub Fatiha(✉)

Larbi Tebessi University, 12002 Tebessa, Algeria
ahlem.merah@univ-tebessa.dz, mesloubf@yahoo.fr

Abstract. In this work, we consider the Dirichlet–Neumann problem to the following initial nonlinear viscoelastic plate equation with a lower order perturbation of $\overrightarrow{p}(x,t)$-Laplacian operator and delay:

$$u_{tt} + \Delta^2 u\,(t) + \Delta_{\overrightarrow{p}(x,t)} u - \int_0^t g(t-s)\Delta^2 u\,(s)\,ds - \mu_1 \Delta u_t(t)$$
$$-\mu_2 \Delta u_t(t-\tau) + f(u) = 0.$$

Under suitable conditions on g, $f(u)$ and the variable exponent of $\overrightarrow{p}(x,t)$ – Laplacian operator, it is proved the local existence and the uniqueness of solution by the semi group method.

Keywords: Semi-group · Delay · Viscoelasticity plate equation · Nonstandard growth conditions · Anisotropy

1 Introduction

In this paper, we consider the Dirichlet–Neumann problem to the following initial nonlinear viscoelastic plate equation with a lower order perturbation of $\overrightarrow{p}(x,t)$-Laplacian operator and delay:

$$\begin{cases} u_{tt} + \Delta^2 u\,(t) - \Delta_{\overrightarrow{p}(x,t)} u - \int_0^t g(t-s)\Delta^2 u\,(s)\,ds - \mu_1 \Delta u_t(t) & \\ -\mu_2 \Delta u_t(t-\tau) + f(u) = 0, & x \in \Omega, t > 0, \\ u = \frac{\partial u}{\partial \nu} = 0, & x \in \partial\Omega, t > 0, \\ u\,(x,0) = u_0(x), u_t\,(x,0) = u_1(x), & x \in \Omega \\ u_t\,(x,t-\tau) = f_0\,(x,t-\tau)\,, & x \in \Omega, t \in (0,\tau) \end{cases} \tag{1}$$

where Ω is a bounded domain in $\mathbb{R}^N$, $n \geq 2$ with Lipschitz-continuous boundary $\Gamma = \partial\Omega$, and $q \geq 2$ is a positive constant

$$\Delta_{\overrightarrow{p}(x,t)} u = \sum_{i=1}^{n} \frac{\partial}{\partial x_i}\left(\left|\frac{\partial u}{\partial x_i}\right|^{p_i(x,t)-2} \frac{\partial u}{\partial x_i}\right), \overrightarrow{p} = (p_1, p_2, ..., p_n)$$

is the $\overrightarrow{p}(x,t)$-Laplacian operator. The constant μ_1 is positive and μ_2 is a real number, $\tau > 0$ represents the time delay, $g > 0$ is a memory kernel and f is

S. Melliani and O. Castillo (Eds.): ICPAMS 2021, LNNS 476, pp. 66–78, 2023.
https://doi.org/10.1007/978-3-031-12416-7_7

forcing term. In the absence of the viscoelastic term and delay term ($g = 0$ and $\mu_2 = 0$), and with the usual $p-$Laplacian operator $\Delta_p u = div\left(|\nabla u|^{p-2}\nabla u\right)$, ($p \geq 2$) the equation in (1.1) reduces to the fourth order wave equation

$$u_{tt} + \Delta^2 u(t) + div\left(|\nabla u|^{p-2}\nabla u\right) - \epsilon\Delta u_t = f(x,t,u,u_t), \tag{2}$$

which describes elastoplastic microstructure flows. The problem (1.2) has been extensively studied (see [3,4,19]) and results concerning existence, nonexistence and long-time behavior of solutions have been proved. The problem (1.2) without damping or forcing terms is related to the elastoplastic-microstructure models for longitudinal motion of an elastoplastic bar there arises the model equation

$$u_{tt} + u_{xxxx} = +a\left(u_x^2\right)_x + f(x),$$

where $a < 0$ is a constant (see [3]). I. Chueshov and I. Lasiecka in [6,7] discussed

$$u_{tt} + \Delta^2 u(t) + div\left(|\nabla u|^2 \nabla u\right) - ku_t = \sigma\Delta(u^2) + f(u),$$

and proved the existence of finite-dimensional global attractors

When $\epsilon = 0$ and in the presence of the viscoelastic term ($g \neq 0$) in (1.2), Jorge Silva and Ma [10], established exponential stability of solutions under the condition

$$g'(t) \leq -cg(t),\ \forall t \geq 0,\ c > 0.$$

Andrade and al. [1] proved exponential stability of solutions for the plate equation with finite memory and p-Laplacian. In the presence of the Kelvin–Voigt type dissipation ($\epsilon \neq 0$). In [4], Nakao obtained the existence of a global decaying solution for wave equation with Kelvin–Voigt dissipation and a derivative nonlinearity. Pukach et al. [17] established sufficient conditions of nonexistence of solution for a nonlinear hyperbolic equation with memory generalizing the Voigt–Kelvin model. Recently, Cavalcanti et al. [5] considered intrinsic decay rates for the energy of a nonlinear viscoelastic equation modeling the vibrations of thin rods with variable density.

In [2,8], the authors improved the results from [1] by establishing local and global existence, as well as the uniqueness of the weak solution $u(x,t)$ to problem (1.1). To be more important, the authors of [2] and [8] established the local and global existence, uniqueness of weak solutions and the asymptotic behaviour of solutions.

Time delays so often arise in many physical, chemical, biological, thermal, and economical phenomena because these phenomena depend not only on the present state but also on the past history of the system in a more complicated way. In recent years, there has been published much work concerning the wave equation with delay. Kafini and Messaoudi [11] considered the following nonlinear wave equation with delay

$$u_{tt}(t) - div(|\nabla u(t)|^{m-2}\nabla u(t)) + \mu_1 u_t(t) + \mu_2 u_t(t-\tau) = b|u(t)|^{p-2}u(t).$$

They proved the blow-up result of solutions with negative initial energy and $p > m$. LaterRecently, Kafini et al. [12] considered the blow up of solutions with negative initial energy for the second order abstract evolution system with delay. Motivated by previous works, we study the blow up of solutions. Recently, Shun-Tang WU [18] investigated the following nonlinear viscoelastic problem with delay

$$|u_t|^\rho u_{tt} - \Delta u - \Delta u_{tt} - \int_0^t g(t-s)\Delta u(s)\, ds + \mu_1 u_t(t) + \mu_2 u_t(t-\tau) = b|u|^{p-2}u.$$

He proved the blow-up result with nonpositive and positive initial energy by modifying the method in [11,12]. Motivated by previous works, in this paper, we investigate problem (1.1) and we prove a finite time blow-up result of solutions, we will see that the direct method introduced and developed by Georgiev and Todorova [9], in 1994 and Salim A. Messaoudi [6,7] is efficient in our case. Combining this method with some necessary modifications due the nature of the problem treated here. In this paper, we aim to prove that system (1) is well-posed The main features of this paper are summarized as follows:

- We set the functions space
- We define the anisotropic spaces of functions depending on x and t
- we adopt the semigroup method to obtain the well-posedness of system (1)

$$1 \leq p_i^- = const \leq p_i(x,t) = p(x) \leq p_i^+ = const < \infty, |p_{it}| \leq C_{p_i}, i = 1, ..., n \tag{3}$$

2 The Functions Space

Let $\Omega \subset \mathbb{R}^n, n \geq 2$ be a bounded domain with Lipschitz-continuous boundary $\Gamma = \partial\Omega, q \geq 2$ is a postive costant

We denote by $C_0^\infty(\Omega)$ the space of in nitely di erentiable functions with a compact support contained in Ω.The inner products and norms in $L^2(\Omega)$ and $H_0^1(\Omega)$ are represented by $(.,.)$, $\|.\|$ respectively and they are given by

$$(u,v)_\Omega = \int_\Omega u(x)v(x)dx \text{ and } \|u\|_{L^2(\Omega)}^2 = \|\nabla u\|_{2,\Omega}^2 = \int_\Omega u^2 dx,$$

$$\|u\|_{H_0^1(\Omega)}^2 = \|\nabla u\|_{2,\Omega}^2 = \int_\Omega |\nabla u|^2 dx$$

We recall some known facts from the theory of the Sobolev spaces with variable exponent (see [10,14]). Let $L^{p(\cdot)}(\Omega)$ be the set of measurable functions f on Ω such that

$$A_{p(.)}(f) = \int_\Omega |f(x)|^{p(x)} dx < \infty.$$

The set $L^{p(\cdot)}(\Omega)$ equipped with the Luxemburg norm

$$\|f\|_{p(\cdot),\Omega} = \|f\|_{p(\cdot)} = \inf\left\{\lambda > 0; A_{p(.)}\left(\frac{f}{\lambda}\right) \leq 1\right\}$$

is a Banach space.Let us list some basic properties of the spaces $L^{p(.)}(\Omega)$ used in the rest of this paper. It follows directly from the definition of the norm that

$$\min\left(\|f\|_{p(.)}^{p^-},\|f\|_{p(.)}^{p^+}\right)\le A_{p(.)}(f)\le \max\left(\|f\|_{p(.)}^{p^-},\|f\|_{p(.)}^{p^+}\right)$$

where

$$p^- = \inf_{\Omega} p(x), (p')^- = \inf_{\Omega} p'(x), p' = \frac{p(x)}{(p(x)-1)}.$$

We have following Holder-type inequality

$$\begin{aligned}\int_{\Omega}|fg|\,dx &\le \left(\frac{1}{p^-}+\frac{1}{(p')^-}\right)\|f\|_{p(.)}\|g\|_{p'(.)}\\ &\le 2\|f\|_{p(.)}\|g\|_{p'(.)},\end{aligned}$$

which holds for all $f\in L^{p(.)}(\Omega), g\in L^{p'(.)}(\Omega)$ with $p(x)\in(1,\infty)$. The Sobolev space $W_0^{1,p(.)}(\Omega)$ with $p(x)\in[p^-,p^+]\subset(1,\infty)$ is defined by

$$\begin{cases} W_0^{1,p(.)}(\Omega)=\left\{u\in L^{p(.)}(\Omega):|\nabla u|^{p(x)}\in L^1(\Omega), u=0 \text{ on } \partial\Omega\right\},\\ \|u\|_{W_0^{1,p(.)}(\Omega)}=\sum_i\|D_iu\|_{p(.),\Omega}+\|u\|_{p(.),\Omega}\end{cases} \tag{4}$$

Let $p(x)$ be log-continuous in $\Omega, \forall x,y\in\Omega$ such that $|x-y|<\frac{1}{2}$

$$|p(x)-p(y)|\le\omega(|x-y|) \text{ with } \overline{\lim}_{\tau\longrightarrow 0_+}\left(\omega(\tau)\ln\frac{1}{\tau}\right)=C<\infty \tag{5}$$

- Throughout the paper we use the following properties of the functions from the spaces $W_0^{1,p(.)}(\Omega)$:
- if condition (11) is ful lled, then $C_0^{\infty}(\Omega)$ is dense in $W_0^{1,p(.)}(\Omega)$ and the space $W_0^{1,p(.)}(\Omega)$ can be defined as the closure of $C_0^{\infty}(\Omega)$ with respect to the norm (10)-see [14, 30, 37, 38, 39];
- if $p(x)\in C^0(\overline{\Omega})$,the the space $W_0^{1,p(.)}(\Omega)$ is separable and reflexive;
- if $1<q(x)\le\sup_{\Omega}q(x)<\inf_{\Omega}p_+(x)$ with

$$p_+(x)=\begin{cases}\frac{p(x)}{n-p(x)} \text{ if } p(x)<n,\\ \infty \text{ if } p(x)<n,\end{cases}$$

then the embedding $W_0^{1,p(.)}(\Omega)\hookrightarrow L^{q(.)}(\Omega)$ is continuous and compact if $q<p_+(x)$.

3 Anisotropic Spaces of Functions Depending on x and t

Consider the cylinder

$$Q_T=\{z=(x,t):x\in\Omega, t\in[0,T]\}$$

of a nite height T. Wherever it doesn't cause a confusion we will use the notation $z=(x,t)$ for the points of the cylinder Q_T and drop the sub-index$T$. The lateral boundary of the cylinder Q is $\Gamma=\partial\Omega\times(0,T)$. If X is a Banach space, then we denote by $L^p(0,T,X)$, $1\le p\le\infty$ the Banach space of measurable vector valued functions $u:(0,T)\longrightarrow X$, such that

$$\|u(t)\|_{L^p(0,T,X)}=\left[\int_0^T\|u(t)\|_X^p\,dt\right]^{\frac{1}{p}},1\le p<\infty,$$

$$\|u(t)\|_{L^p(0,T,X)}=ess\sup_{0\le t\le T}\|u(t)\|_X,p=\infty.$$

We will use the following function spaces (see [6])

$$W=W(Q_T)=\left\{u:u\in L^2\left(0,T,H_0^2(\Omega)\right);u_t\in L^2\left(0,T,H_0^1(\Omega)\right)\right\},$$
$$W^\infty=W^\infty(Q_T)=\left\{u:u\in W(Q_T);u\in L^\infty\left(0,T,H_0^2(\Omega)\right);u_t\in L^\infty\left(0,T,L^2(\Omega)\right)\right\}$$

endowed with the norms

$$\|u\|_{W(Q)}=\|u\|_{W(Q_T)}+\|u\|_{L^\infty\left(0,T,H_0^2(\Omega)\right)}+\|u\|_{L^\infty(0,T,L^2(\Omega))}$$

Note that $\|u\|_{W(Q)}$ may be used in the equivalent form

$$\|u\|_{W(Q)}=\|u\|_{L^2(Q)}+\|\Delta u\|_{L^2(Q)}+\|\nabla u_t\|_{L^2(Q)}$$

Let $p(z)=\overrightarrow{p}=(p_1(z),...,p_n(z))$ be a vector-valued function defined on $Q=Q_T$. We assume that the components of $p(z)$ satisfy the conditions

$$\begin{cases} p_i(z) \text{ are measurable functions defined on } Q; \\ p_i(z):Q\longrightarrow(1,\infty), \\ \text{there exist constants } p_i^\pm,p^\pm \text{such that} \\ p_i(z)\in\left[p_i^-,p_i^+\right]\subseteq\left[p^-,p^+\right]\subset(1,\infty) \end{cases} \quad (6)$$

For every fixed $t\in(0,T)$; we introduce the anisotropic Banach space

$$V_t(\Omega)=\left\{u(x):u(x)\in L^2(\Omega)\cap W_0^{1,1}(\Omega),|D_iu(x)|^{p_i(x,t)}\in L^1(\Omega)\right\},$$

$$\|u\|_{V_t(\Omega)}=\|u\|_{2,\Omega}+\sum_{i=1}^n\|D_iu\|_{p_i(.,t),\Omega}+\|\Delta u\|_{2,\Omega}.$$

The elements of the space $V_t(\Omega)$ depend on $t\in(0,T)$as a parameter and the norms $\|u\|_{V_t(\Omega)}$ are functions of t By $V_t'(\Omega)$ we denote the dual space to $V_t(\Omega)$ with respect to the scalar product in $L^2(\Omega)$.For every $t\in(0,T)$ the inclusion.

$$V_t(\Omega)\subset X=W_0^{1,p^-}(\Omega)\cap L^2(\Omega)$$

holds. This is why $V_t(\Omega)$ is reflexive and separable as a closed subspace of X.By $W_{\overrightarrow{p}}(Q)$ we denote the Banach space

$$W_{\overrightarrow{p}}(Q)=\left\{u:(0,T)\longrightarrow V_t(\Omega)\left|u\in L^2(\Omega),|D_iu(x)|^{p_i(x,t)}\in L^1(\Omega),u=0\,\mathrm{on}\,\Gamma\right.\right\},$$

$$\|u\|_{W_{\overrightarrow{p}}(Q)}=\|u\|_{2,Q}+\sum_{i=1}^{n}\|D_iu\|_{p_i(.),Q}\,.$$

$\left(W_{\overrightarrow{p}}(Q)\right)'$ is the dual of $W_{\overrightarrow{p}}(Q)$ (the space of linear functionals over $W(Q)$). We have the following characterization

$$\omega\in\left(W_{\overrightarrow{p}}(Q)\right)'\Leftrightarrow\begin{cases}\exists\left(\omega_0,\omega_1,...,\omega_n\right),\omega_0\in L^2(\Omega),\omega_i\in L^{p_i'(.)}(Q),\\ \forall\phi\in W(Q)\quad\langle\langle\omega,\phi\rangle\rangle=\int_Q\left(\omega_0\phi+\sum_{i=1}^{n}\omega_iD_i\phi\right)dz.\end{cases}$$

The norm in $W'(Q)$ is defined by

$$\|u\|_{W'(Q)}=\sup\left\{\langle u,\phi\rangle\left|\phi\in W(Q)\ ,\|\phi\|_{W(Q)}\leq 1\right.\right\}.$$

Let $v=(v_1,...,v_n)$ be a vector-valued function de ned in Q. Assume that $p_i(z)$ satisfy conditions (12). Introduce the modular

$$A_{p(.)}(v)=\sum_{i=1}^{n}\int_Q|v_i|^{p_i(z)}\,dz.$$

For the elements of $W_{\overrightarrow{p}}(Q)$ the following inequality

$$\begin{aligned}&\min\left\{\sum_{i=1}^{n}\|D_iu\|_{p_i(.,.),Q}^{p^-},\sum_{i=1}^{n}\|D_iu\|_{p_i(.,.),Q}^{p^+}\right\}\\&\leq A_{p(.,.),Q}(\nabla u)\leq\max\left\{\sum_{i=1}^{n}\|D_iu\|_{p_i(.,.),Q}^{p^-},\sum_{i=1}^{n}\|D_iu\|_{p_i(.,.),Q}^{p^+}\right\}\end{aligned}\tag{7}$$

holds. We also use the space $W_{\overrightarrow{p}}^{\infty}(Q)$

$$W_{\overrightarrow{p}}^{\infty}(Q)=\left\{u:u\in W_{\overrightarrow{p}}(Q),|u_{x_i}|^{p_i(x,t)}\in L^{\infty}\left(0,T,L^1(\Omega)\right)\right\}.$$

Note that

$$W^{\infty}(Q)\subseteq W_{\overrightarrow{p}}^{\infty}(Q)\ \text{ if } p^+\leq\frac{2n}{n-2}.$$

We introduce also the functional space

$$U(Q)=W(Q)\cap W_{\overrightarrow{p}}(Q)$$

endowed with the norm

$$\|u\|_{U(Q)}=\|u\|_{W(Q)}+\|u\|_{W_{\overrightarrow{p}}(Q)}$$

and

$$U^{\infty}(Q) = W^{\infty}(Q) \cap W^{\infty}_{\overrightarrow{p}}(Q)$$

For the exponents $p_i(x,t)$ depending on $(x,t) \in Q$ we will use the notation $p_i \in C_{\log}(Q)$ if p_i satisfies condition (12) in the cylinder Q and:

$$C_{\log}(Q) := \left\{ p_i \in C^0\left(\overline{Q}\right) \middle| \begin{array}{c} \forall z = (x,t), \varsigma = (y,\tau) \in Q \\ \text{such that } |x-y| + |t-\tau| < \frac{1}{2}, \\ |p_i(z) - p_i(z)| \leq \omega(|z-\varsigma|) \end{array} \right\} \tag{8}$$

with a continuous function ω satisfying the condition

$$\overline{\lim}_{\tau \longrightarrow 0^+} \omega(\tau) \ln \frac{1}{\tau} = C < +\infty$$

4 Statement of the Problem

We consider a class of non linear viscoelastic plate equations with delay and with $\overrightarrow{p}(x,t)$-Laplacian type

$$\begin{cases} u_{tt} + \Delta^2 u(t) - \Delta_{\overrightarrow{p}(x,t)} u - \int_0^t g(t-s)\Delta^2 u(s)\, ds - \mu_1 \Delta u_t(t) \\ \qquad -\mu_2 \Delta z(1,t) + f(u) = 0, \ in\, \Omega \times \mathbb{R}^+, \end{cases} \tag{9}$$

$$\left\{ u = \tfrac{\partial u}{\partial \nu} = 0, \ in\, \partial\Omega \times \mathbb{R}^+ \right. \tag{10}$$

$$\begin{cases} u(x,0) = u_0(x), u_t(x,0) = u_1(x), \text{on}\, x \in \Omega \\ u_t(x, t-\tau) = f_0(x, t-\tau), \text{ on}\, x \in \Omega, t \in (0,\tau) \end{cases} \tag{11}$$

where Ω is a bounded domain on $\mathbb{R}^N$, $n \geq 2$ with Lipschitz-continuous boundary $\Gamma = \partial\Omega$. With μ_1 is a positive constant μ_2 is a real number,$\tau > 0$ represents the time delay, and $g > 0$ is a memory kernel f is a nonlinear function. We define the operator $\overrightarrow{p}(x,t)$-laplacian by the following formula

$$\Delta_{\overrightarrow{p}(x,t)} u = \sum_{i=1}^{n} \frac{\partial}{\partial x_i}\left(\left| \frac{\partial u}{\partial x_i} \right|^{p_i(x,t)-2} \frac{\partial u}{\partial x_i} \right), \overrightarrow{p} = (p_1, p_2, ..., p_n)$$

By using the direct calculations, we have

$$\begin{aligned} \int_0^t g(t-s)\left(\Delta u_t(t), \Delta u(s)\right) ds = & -\tfrac{1}{2}\tfrac{d}{dt}\left\{ (g \circ \Delta u)(t) - \left(\int_0^t g(s)ds\right) \|\Delta u\|^2 \right\} \\ & -\tfrac{1}{2} g(t) \|\Delta u\|^2 + \tfrac{1}{2}(g' \circ \Delta u)(t) \end{aligned} \tag{12}$$

where

$$(g \circ \Delta u)(t) = \int_0^t g(t-s) \|\Delta u(s) - \Delta u(t)\|^2 ds$$

With μ_1 and μ_2 are satisfied

$$|\mu_2| < \mu_1 \tag{13}$$

For the relaxation function g, we have the following assumptions

(A_1) $g : [0, +\infty) \longrightarrow [0, +\infty)$is a bounded differentiable function such that

$$g(0) > 0, g(s) \geq 0, 1 - \int_0^\infty g(s)ds := \gamma > 0$$

(A_2) There exists a nonincreasing differential function

$$\xi : [0, +\infty) \longrightarrow [0, +\infty) \text{ Such that}$$
$$g'(t) \leq -\xi g(t), \forall t \geq 0 \text{ and } \int_0^{+\infty} \xi(t)\, dt = +\infty$$

We refer the reader to the work of Nicaise and Pignotti[16] for the existence of solutions to nonlinear problems with delay. Let us introduce the function

$$z(x, \rho, t) = u_t(x, t - \tau\rho), x \in \Omega, \rho \in (0, 1), t > 0 \tag{14}$$

Then, we obtain

$$\tau z_t(x, \rho, t) + z_\rho(x, \rho, t) = 0, x \in \Omega, \rho \in (0, 1), t > 0 \tag{15}$$

The problem (9)–(11) is equivalent to

$$\begin{aligned} &\begin{matrix} u_{tt} + \Delta^2 u(t) - \Delta_{\overrightarrow{p}(x,t)} u - \int_0^t g(s)\Delta^2 u(s)\, ds - \mu_1 \Delta u_t(t) \\ -\mu_2 \Delta z(1, t) + f(u) = 0, \end{matrix} \quad on\ \Omega \times (0, 1) \times (0, \infty), \\ &\tau z_t(x, \rho, t) + z_\rho(x, \rho, t) = 0,\ on\ (0, 1) \times (0, \infty), \end{aligned} \tag{16}$$

$$\begin{aligned} z(0, t) &= u_t(t),\ on\ (0, \infty), \\ z(\rho, 0) &= f_0(x, -\tau\rho),\ on\ (0, 1), \end{aligned} \tag{17}$$

Let $\phi = u_t$ and denote by $U = (u, \phi, z)^T$, Therefore (16) can be rewritten as follows

$$\begin{cases} U' = AU \\ U(0) = (u_0, u_1, f_0(., -\tau))^T \end{cases} \tag{18}$$

where the linear operator A is defined by

$$A \begin{pmatrix} u \\ \phi \\ z \end{pmatrix} = \begin{pmatrix} \phi \\ -\Delta^2 u(t) + \Delta_{\overrightarrow{p}(x,t)} u + \int_0^t g(s)\Delta^2 u(s)\, ds + \mu_1 \Delta\phi(t) + \mu_2 \Delta z(1, t) - f(u) \\ -\frac{1}{\tau} z_\rho(x, \rho, t) \end{pmatrix} \tag{19}$$

With the domain

$$D(A) = \left\{(u, \phi, z)^T \in H, z(., 0) = \phi \text{ on } \Omega\right\}$$

Such that

$$H = \left\{u \in H^2(\Omega); \phi \in H_0^1(\Omega); z(., 1) \in L^2(\Omega); z, z_\rho \in L^2((0, 1) \times \Omega)\right\}$$

Now the energy space is defined by

$$K = H_0^1(\Omega) \times L^2(\Omega) \times L^2((0, 1) \times \Omega)$$

equipped with the inner product

$$\langle \varphi, \varkappa \rangle_{L^2((0,+\infty)\times H_0^1(\Omega))} = \int_\Omega \int_0^t g(s)\varphi_x(x,s)\,\varkappa_x(x,s)\,dsdx$$

Let

$$\left\{U = (u, \phi, z)^T \text{ and } \tilde{U} = (\tilde{u}, \tilde{\phi}, \tilde{z})^T\right.$$

Then, for a positive constant ζ satisfying

$$\tau\,|\mu_2| < \zeta < \tau\,(\mu_1 - |\mu_2|) \tag{20}$$

Now, we define the inner product in K as follows:

$$\begin{aligned}\left\langle U, \tilde{U}\right\rangle_K &= \int_\Omega \left\{\phi\tilde{\phi} - u_{xx}\tilde{u}_{xx}\right\} dx + \int_\Omega \sum_{i=1}^n |u_{x_i}|^{p_i}\,|\tilde{u}_{x_i}|^{p_i}\,dx \\ &+ \zeta \int_\Omega \int_0^1 z(x,\rho)\,\tilde{z}(x,\rho)\,dxd\rho + \int_\Omega \int_0^t g(s)\Delta u(s)\Delta\tilde{u}(s)dsdx\end{aligned}$$

The existence and uniqueness result is stated as follows;

Theorem 1. *For any $U_0 \in K$ there exists a unique solution $U \in C([0,+\infty[, K)$ of problem (18). Moreover, if $U_0 \in D(A)$ then $U \in C([0,+\infty[, D(A)) \cap C^1([0,+\infty[, K)$.*

Proof. In order to prove the result stated in Theorem 1, we use the semi-group theory, that is, we show that the operator A generates a C_0 semi-group in K. In this step, we concern ourselves to prove that the operator A is dissipative. Indeed, for$U = (u, \phi, z)^T \in D(A)$ and ζ is a positive constant, we have

$$\begin{aligned}\langle AU, U\rangle_K &= -\int_\Omega \Delta^2 u\phi(t)dx + \int_\Omega \Delta_{\overrightarrow{p}(x,t)} u\phi(t)dx + \int_\Omega\int_0^t g(s)\Delta^2 u(s)\,\phi(t)dsdx + \int_\Omega \mu_1\Delta\phi(t)\phi(t)dx \\ &+ \int_\Omega \mu_2\Delta z(x,1,t)\,\phi(t)\,dx - \int_\Omega f(u)\phi(t)\,dx - \frac{1}{\tau}\int_\Omega\int_0^1 z_\rho(x,\rho,t)\,\Delta z d\rho dx \\ &= 0\end{aligned} \tag{21}$$

By applying the integration by part for each term of (21) we find the results as following

$$-\int_\Omega \Delta^2 u\phi(t)dx = -\int_\Omega \Delta u(x,t)\Delta\phi(t)dx \tag{22}$$

$$\int_\Omega \Delta_{\overrightarrow{p}(x,t)} u\phi(t)dx = -\sum_{i=1}^n \int_\Omega |u_{x_i}|^{p_i-2} u_{x_i}\phi_{x_i}dx \tag{23}$$

$$\int_\Omega\int_0^t g(s)\Delta^2 u(s)\,\phi(t)dsdx = \int_\Omega\int_0^t g(s)\Delta u(s)\,\Delta\phi(t)dsdx \tag{24}$$

$$\mu_1\int_\Omega \Delta\phi(t)\phi(t)dx = -\mu_1\,\|\nabla\phi(t)\|_2^2 \tag{25}$$

$$\mu_2 \int_{\Omega} \Delta z(x,1,t)\,\phi(t)\,dx = -\mu_2 \int_{\Omega} \nabla z(x,1,t)\,\nabla\phi(t)\,dx \tag{26}$$

$$-\frac{1}{\tau}\int_{\Omega}\int_0^1 z_\rho(x,\rho,t)\,\Delta z d\rho dx = -\frac{1}{2\tau}\|\nabla z(x,1,t)\|_2^2 + \frac{1}{2\tau}\|\nabla\phi(t)\|_2^2 \tag{27}$$

By substituting (22)–(27) in (21) we find

$$\begin{aligned}\langle AU,U\rangle_K &= -\int_{\Omega}\Delta u(x,t)\Delta\phi(t)dx - \sum_{i=1}^{n}\int_{\Omega}|u_{x_i}|^{p_i-2}u_{x_i}\phi_{x_i}dx + \int_{\Omega}\int_0^t g(s)\Delta u(s)\,\Delta\phi(t)dsdx\\ &\quad -\mu_1\|\nabla\phi(t)\|_2^2 - \mu_2\int_{\Omega}\nabla z(x,1,t)\,\nabla\phi(t)\,dx\\ &\quad -\frac{1}{2\tau}\|\nabla z(x,1,t)\|_2^2 + \frac{1}{2\tau}\|\nabla\phi(t)\|_2^2 - \int_{\Omega} f(u)\phi(t)\,dx\\ &= 0\end{aligned} \tag{28}$$

Using Young's inequality, we obtain

$$-\mu_2\int_{\Omega}\nabla z(x,1,t)\,\nabla\phi(t)\,dx \le -\frac{|\mu_2|}{2}\|\nabla z(x,1,t)\|_2^2 - \frac{|\mu_2|}{2}\|\nabla\phi(t)\|_2^2 \tag{29}$$

Now by using (12) we get the following result

$$\begin{aligned}\int_{\Omega}\int_0^t g(s)\Delta u(s)\,\Delta\phi(t)dsdx &= \frac{1}{2}\left(g'\circ\Delta u\right)(t) - \frac{1}{2}\frac{d}{dt}\left\{(g\circ\Delta u)(t) - \left(\int_0^t g(s)ds\right)\|\Delta u\|_2^2\right\}\\ &\quad -\frac{1}{2}g(t)\|\Delta u\|_2^2\end{aligned} \tag{30}$$

If we replace the results obtained in (28) we find

$$\begin{aligned}\langle AU,U\rangle_K &\le -\int_{\Omega}\Delta u(x,t)\Delta\phi(t)dx - \sum_{i=1}^{n}\int_{\Omega}|u_{x_i}|^{p_i-2}u_{x_i}\phi_{x_i}dx - \int_{\Omega} f(u)\phi(t)\,dx\\ &\quad -\frac{1}{2}g(t)\|\Delta u\|_2^2 - \frac{1}{2}\frac{d}{dt}(g\circ\Delta u)(t) - \left(\frac{\zeta+1}{2\tau}\right)\|\nabla z(x,1,t)\|_2^2\\ &\quad +\frac{1}{2}(g'\circ\Delta u)(t) + \frac{1}{2}\frac{d}{dt}\left(\int_0^t g(s)ds\right)\|\Delta u\|_2^2\\ &\quad -\left(\mu_1 + \frac{\zeta-1}{2\tau}\right)\|\nabla\phi(t)\|_2^2\end{aligned}$$

keeping in mind the fact that

$$g'(t) \le -\xi g(t)$$

Then we have

$$\langle AU,U\rangle_K \le 0$$

Now to show that the operator A is maximal monotone, it is sufficient to show that the operator $\lambda I - A$ is surjective for a fixed $\lambda > 0$. Indeed, given $F = (f_1,f_2,f_3)^T \in H$, we seek $V = (u,\phi,z)^T \in D(A)$ solution of

$$\begin{cases} \lambda u - \phi = f_1 \\ \lambda\phi - (-\Delta^2 u\,(t) + \Delta_{\overrightarrow{p}(x,t)} u + \int_0^t g(s)\Delta^2 u\,(s)\,ds \\ +\mu_1 \Delta\phi(t) + \mu_2 \Delta z\,(1,t) - f(u)) = f_2 \\ \lambda z + \frac{1}{\tau} z_\rho\,(x,\rho,t) = f_3 \end{cases} \tag{31}$$

suppose we have find u with the appropriate regularity, then

$$\phi = \lambda u - f_1 \tag{32}$$

It is clear that $\phi \in H_0^1(\Omega)$. Furthermore, by (31), we can find z us$z\,(x,0) = \phi$, $x \in \Omega$. Using the approach as in Nicaise & Pignotti [16], we obtain,

$$z\,(x,1,t) = \phi(x)e^{-\lambda\tau} + \tau e^{-\lambda\tau} \int_0^1 e^{\lambda\tau\sigma} f_3\,(x,\sigma)\,d\sigma$$

from (32) we obtain

$$\begin{aligned} z\,(x,1,t) &= \lambda u(x)e^{-\lambda\tau} - f_1(x)e^{-\lambda\tau} + \tau e^{-\lambda\tau} \int_0^1 e^{\lambda\tau\sigma} f_3\,(x,\sigma)\,d\sigma \\ &= \lambda u(x)e^{-\lambda\tau} + z_0(x) \end{aligned} \tag{33}$$

With

$$z_0(x) = \tau e^{-\lambda\tau} \int_0^1 e^{\lambda\tau\sigma} f_3\,(x,\sigma)\,d\sigma - f_1(x)e^{-\lambda\tau} \tag{34}$$

So we find the second equation of (31) in the form

$$\begin{array}{c} \lambda u(x) - (-\Delta^2 u\,(x) + \Delta_{\overrightarrow{p}(x,t)} u + \int_0^t g(s)\Delta^2 u\,(s)\,ds \\ -\lambda f_1 + \mu_1 \Delta z\,(.,0) + \mu_2 \Delta z\,(1,t) - f(u)) = f_2 \end{array} \tag{35}$$

for $x \in \Omega$, we have

$$\begin{aligned} z\,(.,0) &= \lambda u(x) \\ z\,(1,t) &= \lambda u(x)e^{-\lambda\tau} + z_0(x) \end{aligned}$$

The system (31) can be reformulated as

$$\begin{array}{c} \lambda u(x) + \Delta^2 u\,(x) - \Delta_{\overrightarrow{p}(x,t)} u - \int_0^t g(s)\Delta^2 u\,(s)\,ds \\ -\lambda\mu_1 \Delta u\,(x) - \lambda e^{-\lambda\tau}\mu_2 \Delta u\,(x) + f(u) = f_2 + \lambda f_1 + \mu_2 \Delta z_0 \end{array} \tag{36}$$

We must now prove that (36) admits a solution $u \in H^2(\Omega) \cap H_0^1(\Omega)$ and replacing it in (32), (34) to get $V = (u,\phi,z)^T \in D(A)$. To solve the problem (31) we consider

$$\Phi\,(u,v) = \Psi\,(v)\,, \forall v \in H_0^1(\Omega) \tag{37}$$

where the bilinear form $\Phi : \left(H_0^1(\Omega)\right)^2 \longrightarrow \mathbb{R}$ and the linear form $\Psi : H_0^1(\Omega) \longrightarrow \mathbb{R}$ are defined by

$$\Phi\,(u,v) = \int_\Omega \left(\begin{array}{c} \lambda u(x) + \Delta^2 u\,(x) - \Delta_{\overrightarrow{p}(x,t)} u\,(x) - \int_0^t g(s)\Delta^2 u\,(s)\,ds \\ -\lambda\mu_1 \Delta u\,(x) - \lambda e^{-\lambda\tau}\mu_2 \Delta u\,(x) + f(u) \end{array} \right) v dx, \forall v \in H_0^1(\Omega)$$

And

$$\Psi(v) = \int_{\Omega} (f_2 + \mu_2 \Delta z_0 + \lambda f_1)\, v dx, \forall v \in H_0^1(\Omega)$$

it is clear that Φ is continuous and coercive, and Ψ is continuous. So applying the Lax-Milgram theorem, we deduce that for all $v \in H_0^1(\Omega)$ problem (37) admits a unique solution $u \in H_0^1(\Omega)$. It follows from (36) que $u \in H^2(\Omega) \cap H_0^1(\Omega)$. Therefore, the operator $\lambda I - A$is dissipative for any $\lambda > 0$ Then the result in Theorem 2.2.1 follows from the Hille-Yoshida theorem.

References

1. Andrade, D., Jorge Silva, M.A., Ma, T.F.: Exponential stability for a plate equation with p-Laplacian and memory terms. Math. Methods Appl. Sci. **35**, 417–426 (2012)
2. Antontsev, S., Ferreira, J.: On a Viscoleastic plate equation with strong damping and perturbation of $\overrightarrow{p}(x,t)$ -Laplacian: existence, and uniqueness. Differ. Integr. Equ. (2013, Submited)
3. An, L.J., Peirce, A.: The effect of microstructure on elastic-plastic models. SIAM J. Appl. Math. **54**, 708–730 (1994)
4. An, L.J., Peirce, A.: A weakly nonlinear analysis of elastoplastic-microstructure models. SIAM J. Appl. Math. **55**, 136–155 (1995)
5. Cavalcanti, M.M., Domingos Cavalcanti, V.N., Lasiecka, I., Webler, C.M.: Intrinsic decay rates for the energy of a nonlinear viscoelastic equation modeling the vibrations of thin rods with variable density. Adv. Nonlinear Anal. **6**(2), 121–145 (2017)
6. Chueshov, I., Lasiecka, I.: Existence, uniqueness of weak solutions and global attractors for a class of nonlinear 2D Kirchho-Boussinesq models. Discrete Contin. Dyn. Syst. **15**, 777–809 (2006)
7. Chueshov, I., Lasiecka, I.: Long-Time Behavior of Second Order Evolution Equations with Nonlinear Damping. Memories of American Mathematical Society, vol. 195. American Mathematical Society, Providence (2008)
8. Ferreira, J., Messaoudi, S.A.: On the general decay of a nonlinear viscoelastic plate equation with a strong damping and $\overrightarrow{p}(x,t)$-Laplacian. Nonlinear Anal. **104**, 40–49 (2014)
9. Georgiev, V., Todorova, G.: Existence of a solution of the wave equation with nonlinear damping and source terms. J. Differ. Equ. **109**, 295–308 (1994)
10. Jorge Silva, M.A., Ma, T.F.: On a viscoelastic plate equation with history setting and perturbation of p-Laplacian type. IMA J. Appl. Math. **78**, 1130–1146 (2013)
11. Kafini, M., Messaoudi, S.A.: A blow-up result in a nonlinear wave equation with delay. Mediterr. J. Math. **13**(1), 237–247 (2016)
12. Kafini, M., Messaoudi, S.A., Nicaise, S.: A blow-up result in a nonlinear abstract evolution system with delay. Nonlinear Differ. Equ. Appl. **23**(2), 1–14 (2016). https://doi.org/10.1007/s00030-016-0371-4
13. Liu, W.J., Yu, J.: On decay and blow-up of the solution for a viscoelastic wave equation with boundary damping and source terms. Nonlinear Anal. **74**(6), 2175–2190 (2011)
14. Messaoudi, S.A.: Blow-up of positive initial energy solutions of a nonlinear viscoelastic hyperbolic equation. J. Math. Anal. Appl. **320**(2), 902–915 (2006)

15. Nakao, M.: Global solutions to the initial-boundary value problem for the quasilinear viscoelastic equation with a derivative nonlinearity. Opusc. Math. **34**(3), 569–590 (2014)
16. Nicaise, S., Pignotti, C.: Stability and instability results of the wave equation with a delay term in the boundary or internal feedbacks. SIAM J. Control. Optim. **45**(5), 1561–1585 (2006)
17. Pukach, P., Il'kiv, V., Nytrebych, Z., Vovk, M.: On nonexistence of global in time solution for a mixed problem for a nonlinear evolution equation with memory generalizing the Voigt-Kelvin rheological model. Opusc. Math. **37**(5), 735–753 (2017)
18. Wu, S.-T.: Blow-up of solution for a viscoelastic wave equation with delay. Acta Math. Sci. Ser. B (Engl. Ed.) **39**(1), 329–338 (2019)
19. Yang, Z.: Longtime behavior for a nonlinear wave equation arising in elasto-plastic flow. Math. Methods Appl. Sci. **32**, 1082–1104 (2009)

Notes on the Existence Results for Nonlinear Fractional Differential Equations with Fuzzy Boundary Conditions

Ali El Mfadel, Said Melliani(✉), and Mhamed Elomari

LMACS, Sultan Moulay Slimane University, BP 523 - 23000 Beni Mellal, Morocco
{s.melliani,m.elomari}@usms.ma

Abstract. In this paper, we establish a new result concerning the existence and uniqueness of solutions for nonlinear fractional differential equations with fuzzy boundary conditions. As application, we give an illustrative example to show the effectiveness of the obtained result.

1 Introduction

Recently, fuzzy analysis and fuzzy differential equations were proposed to handle uncertainty due to incomplete information that appears in many mathematical or computer models of some deterministic real world phenomena.

In recent years, fractional differential equations have attracted a considerable interest both in mathematics and in applications as material theory, transport processes, fluid flow phenomena, earthquakes, solute transport, chemistry, wave Propagation, signal theory, biology, electromagnetic theory, thermodynamics, mechanics, geology, astrophysics, economics and control Theory (see [1,3]). The concept of fuzzy type Riemann-Liouville differentiability based on Hukuhara differentiability was initiated in [2] and using the Hausdorff measure of non compactness the authors established the existence to some fuzzy integral equations using appropriate compactness type conditions basic works related to the fuzzy fractional differential equations we refer the reader to [4,16,17].

Motivated by the above works, in this paper, we study the existence and uniqueness results of solutions for the following fuzzy fractional boundary value problem:

$$^cD^q u(t) = f(t, u(t)), \quad t \in [a,b], \tag{1}$$

$$u(a) = A \in E^1 \quad and \quad u(b) = B \in E^1. \tag{2}$$

where $^cD^q$ is the Caputo derivative of $u(t)$ at order $q \in [1,2]$ and E^1 is the collection of all fuzzy numbers.

The paper is organized as follows. In Sect. 2, we give some basic properties of fuzzy sets, operations of fuzzy numbers and some detailed definitions of fuzzy fractional integral and fuzzy fractional derivative which will be used in the rest of this paper. In Sect. 3, we introduce the existence and uniqueness results of solutions for fuzzy fractional boundary value problem. Illustrative example will be discussed in Sect. 4.

S. Melliani and O. Castillo (Eds.): ICPAMS 2021, LNNS 476, pp. 79–87, 2023.
https://doi.org/10.1007/978-3-031-12416-7_8

2 Preliminaries

Definition 1. [18] A fuzzy number is mapping $u : \mathbf{R}^n \longrightarrow [0,1]$ such that

1. u is upper semi-continuous,
2. u is normal, that is, there exist $x_0 \in \mathbf{R}^n$ such that $u(x_0) = 1$,
3. u is fuzzy convex, that is, $u(\lambda x + (1-\lambda)y) \geq \min\{u(x), u(y)\}$ for all $x, y \in \mathbf{R}^n$ and $\lambda \in [0,1]$,
4. $\overline{\{x \in \mathbf{R}^n,\ u(x) > 0\}}$ is compact.

The $\alpha - Cut$ of a fuzzy number u is defined as follows:

$$[u]^\alpha = \{x \in \mathbf{R}^n / u(x) \geq \alpha\}.$$

Moreover, we also can present the $\alpha - cut$ of fuzzy number u by $[u]^\alpha = [u_l(\alpha), u_r(\alpha)]$. We denote by E^n the collection of all fuzzy numbers.

Example 1. Let u be a fuzzy number defined by the following function:

$$\mu_u(x) = \begin{cases} x-1 & ; \quad x \in [1,2], \\ -x+3 & ; \quad x \in [2,3], \\ 0 & ; \quad elswhere. \end{cases}$$

Then we have $u^1 = \{2\}$.

Definition 2. Let $r \in [0,1]$.
A fuzzy number u in a parametric form is a pair of functions $(\underline{u}(r), \overline{u}(r))$, which satisfy the following requirements:

1. $\underline{u}(r)$ is a bounded non-increasing lef continuous function in $]0,1]$ and right continuous at 0,
2. $\overline{u}(r)$ is a bounded non-decreasing lef continuous function in $]0,1]$ and right continuous at 0,
3. $\underline{u}(r) \leq \overline{u}(r)\ \forall r \in [0,1]$.

Remark 1. Let $u \in E^1$.
We can present the $r - cut$ of u by

$$[u]^r = [\underline{u}(r), \overline{u}(r)].$$

Definition 3. [13] Let $r \in [0,1]$ and $u = (\underline{u}(r), \overline{u}(r)), v = (\underline{v}(r), \overline{v}(r)) \in E^1$, then the Hausdorf distance between u and v is given by

$$D(u,v) = \sup_{r \in [0,1]} d([u]^r, [v]^r) = \sup_{r \in [0,1]} \max\{|\underline{u}(r) - \underline{v}(r)|, |\overline{u}(r) - \overline{v}(r)|\}$$

Proposition 1. [10] *D is a metric on E^n and has the following properties:*

1. $(E^n; D)$ is a complete metric space.

2. $D(u+w, v+w) = D(u,v)$, $\forall u,v,w \in E^n$.
3. $D(ku, kv) = |k|D(u,v)$, $\forall u,v \in E^n$ *and* $k \in \mathbf{R}$.
4. $D(u+w, v+z) \leq D(u,v) + D(w,z)$, $\forall u,v,w,z \in E^n$.

We denote by $\mathscr{C}(J, E^n)$ space of all fuzzy-valued functions which are continuous on J, and $\mathscr{P}_c(\mathbf{R}^n)$ the collection of all the compact subset of $\mathbf{R}^n$.

Definition 4. [9] The generalized Hukuhara difference of two fuzzy numbers $u, v \in E^n$ is defined as follows:

$$u \ominus_{gH} v = w \Leftrightarrow \begin{cases} i) & u = v + w, \\ or & \\ ii) & v = u + (-1)w. \end{cases}$$

Proposition 2. [9] *If* $u \in E^1$ *and* $v \in E^1$, *then the following properties hold.*

1) $u \ominus_{gH}$ *exists and it is unique.*
2) $u \ominus_{gH} u = 0_{E^1}$.
3) $(u+v) \ominus_{gH} v = u$.
4) $u \ominus_{gH} v = 0_{E^1} \Leftrightarrow u = v$.

Definition 5. [18] According to the Zadeh's extension principle, the addition on E^1 is defined by:

$$(u \oplus v)(z) = \sup_{z=x+y} min\{u(x), v(y)\}.$$

And scalar multiplication of a fuzzy number is given by:

$$(k \odot u)(x) = \begin{cases} u(x/k) & ,k > 0, \\ \widetilde{0} & ,k = 0. \end{cases}$$

The following arithmetic operations on fuzzy numbers are well known and frequently used below(see [13]) If $u, v \in E^1$ and $\alpha \in [0,1]$, then we have:

$$[u+v]^\alpha = [u]^\alpha + [v]^\alpha,$$

$$[u-v]^\alpha = [u_1^\alpha - v_2^\alpha, u_2^\alpha - v_1^\alpha].$$

$$[ku]^\alpha = k[u]^\alpha = \begin{cases} [\lambda u_1^\alpha, \lambda u_2^\alpha] \; if & \lambda \geq 0, \\ [\lambda u_2^\alpha, \lambda u_1^\alpha] \; if & \lambda < 0. \end{cases}$$

$$[uv]^\alpha = [minu_1^\alpha v_1^\alpha, u_1^\alpha v_2^\alpha, u_2^\alpha v_1^\alpha, u_2^\alpha v_2^\alpha, maxu_1^\alpha v_1^\alpha, u_1^\alpha v_2^\alpha, u_2^\alpha v_1^\alpha, u_2^\alpha v_2^\alpha].$$

Definition 6. [7] Let $f : [a,b] \to E^n$ and $t_0 \in [a,b]$. We say that f is Hukuhara differentiable at t_0 if there exists $f'(t_0) \in E^n$ such that:

$$f'(t_0) = \lim_{h \to 0^+} \frac{f(t_0+h) \ominus_{gH} f(t_0)}{h} = \lim_{h \to 0^-} \frac{f(t_0) \ominus_{gH} f(t_0-h)}{h}.$$

Remark 2. Let $f : [a,b] \longrightarrow E^n$ be a fuzzy function such that $[f(x)]^\alpha = \left[\underline{f}(x;\alpha), \overline{f}(x;\alpha)\right]$ for each $\alpha \in [0,1]$ then

$$[f'(x)]^\alpha = \left[\underline{f}'(x;\alpha), \overline{f}'(x;\alpha)\right].$$

Definition 7. $F : [a,b] \longrightarrow E^n$ is strongly measurable if $\forall \alpha \in [0,1]$, the set-valued mapping $F_\alpha : [a,b] \longrightarrow \mathscr{P}_c(\mathbf{R}^n)$ defined by $F_\alpha(t) = [F(t)]^\alpha$ is Lebesgue measurable.
A function $F : [a,b] \longrightarrow E^n$ is called integrably bounded, if there exists an integrable function h such that, $|x| < h(t), \quad \forall x \in F_0(t)$.

Definition 8. Let $F : [a,b] \longrightarrow E^n$. The integral of F on $[a,b]$ denoted by $\int_I F(t)dt$, is given by

$$\left[\int_a^b F(t)dt\right]^\alpha = \int_J F_\alpha(t)dt = \left\{\int_a^b f(t)dt \mid f : [a,b] \longrightarrow \mathbf{R}^n \text{is a measurable selection for } F_\alpha\right\}.$$

for all $\alpha \in [0,1]$.

Proposition 3. *If $u \in E^1$ then the following properties hold*

1. $[u]^\beta \subset [u]^\alpha if 0 \leq \alpha \leq \beta$.
2. *If $\alpha_n \subset [0,1]$ is a nondecreasing sequence which converges to α, then*

$$[u]^\alpha = \bigcap_{n \geq 1} [u]^{\alpha_n}.$$

Conversely, if $A^\alpha = \{[u_1^\alpha, u_2^\alpha]; \alpha \in [0,1]\}$ is a family of closed real intervals verifying (1) and (2), then A^α defined a fuzzy number $u \in E^1$ such that $[u]^\alpha = A^\alpha$.

2.1 Fractional Integral and Fractional Derivative of Fuzzy Function

Let $q > 0$, the fractional integral of order q of a real function $g : [a,b] \rightarrow \mathbf{R}$ is given by

$$I^q g(t) = \frac{1}{\Gamma(q)} \int_a^t (t-s)^{q-1} g(s) ds.$$

Let $f(t) \in L\left([a,b], E^1\right)$ such that $f(t) = [f_1^\alpha(t), f_2^\alpha(t)]$. Suppose that $f_1^\alpha, f_2^\alpha \in L([a,b], \mathbf{R})$ for all $\alpha \in [0,1]$ and let

$$A^\alpha = \left[\frac{1}{\Gamma(q)} \int_a^t (t-s)^{q-1} f_1^\alpha(s) ds, \frac{1}{\Gamma(q)} \int_a^t (t-s)^{q-1} f_2^\alpha(s) ds\right]. \tag{3}$$

where $\Gamma(.)$ is the Euler gamma function.
We have the following lemma.

Lemma 1. [3] *The family $\{A^\alpha; \alpha \in [0,1]\}$ given by (3), defined a fuzzy number $u \in E^1$ such that $[u]^\alpha = A^\alpha$.*

Definition 9. [16] Let $f(t) \in L\left([a,b],E^1\right)$.
The fuzzy fractional integral of order $q \in [0,1]$ of f denoted by

$$I^q f(t) = \frac{1}{\Gamma(q)} \int_a^t (t-s)^{q-1} f(s) ds.$$

is defined by

$$[I^q f(t)]^\alpha = [I^q f_l(t;\alpha), I^q f_r(t;\alpha)].$$

Proposition 4. [16] *Let $f,g \in L([a,b],E^1)$ and $b \in E^1$, then we have:*

1. $I^q(bf)(t) = bI^q f(t)$.
2. $I^q(f+g)(t) = I^q f(t) + I^q g(t)$.
3. $I^{q_1} I^{q_2} f(t) = I^{q_1+q_2} f(t)$, *where* $(q_1,q_2) \in [0,1]^2$.

Example 2. Let $x : [a,b] \to E^1$ be a constant fuzzy function such that $x(t) = u \in E^1$. If $[u]^\alpha = [u_\alpha^1, u_\alpha^2]$, then

$$[I^q x(t)]^\alpha = \left[\frac{1}{\Gamma(q)} \int_a^t (t-s)^{q-1} u_\alpha^1(s) ds, \frac{1}{\Gamma(q)} \int_a^t (t-s)^{q-1} u_\alpha^2(s) ds\right].$$

$$[I^q x(t)]^\alpha = \frac{t^q}{\Gamma(\alpha+1)} [u_\alpha^1, u_\alpha^2].$$

$$[I^q x(t)]^\alpha = \frac{t^q}{\Gamma(q+1)} [u]^\alpha.$$

Definition 10. [16] Let $f \in C([a,b],E^1) \cap L([a,b],E^1)$.
The function f is called fuzzy Caputo fractional differentiable of order $0 < q < 1$ at t if there exists an element ${}^cD^q f(t) \in E^1$ such that

$${}^cD^q f(t) = \frac{1}{\Gamma(q)} \int_a^t (t-s)^{q-1} f'(s) ds.$$

Remark 3. [16] Since$[f(t)]^\alpha = [f_l(t;\alpha), f_r(t;\alpha)]$ for each $\alpha \in [0,1]$, then

$$[{}^cD^q f(t)]^\alpha = [{}^cD^q f_l(t;\alpha), {}^cD^q f_r(t;\alpha)].$$

Where

$${}^cD^q f_l(t;\alpha) = \frac{1}{\Gamma(q)} \int_a^t (t-s)^{q-1} f_l'(s,\alpha) ds.$$

$${}^cD^q f_r(t;\alpha) = \frac{1}{\Gamma(q)} \int_a^t (t-s)^{q-1} f_r'(s,\alpha) ds.$$

Example 3. Let $x : [a,b] \to E^1$ be a constant fuzzy function such that $x(t) = u \in E^1$. If $[u]^\alpha = [u_\alpha^1, u_\alpha^2]$, then

$$[{}^cD^q x(t)]^\alpha = \left[\frac{1}{\Gamma(q)} \int_a^t (t-s)^{q-1} (u_\alpha^1)' ds, \frac{1}{\Gamma(q)} \int_a^t (t-s)^{q-1} (u_\alpha^2)' ds\right].$$

$$[{}^cD^q x(t)]^\alpha = \{0\}.$$

$${}^cD^q x(t) = 0_{E^1}.$$

3 Main Results

Without loss of generality we solve the boundary value problem (1)–(2) for $A = 0_{E^1}$.

Definition 11. A function $u(t)$ is a solution of the system (1)–(2) if and only if it satisfes the differential Eq. 1 and the boundary conditions 2.

Lemma 2. *A fuzzy function $u(t)$ is a solution of the system (1)–(2) if and only if $u(t)$ satisfes the following requirements:*

1) $u(t)$ is continuous.

2) $u(t)$ satisfies the integral equation $u(t) = \frac{(t-a)^{q-1}}{(b-a)^{q-1}}B + \int_a^t G(t,s)f(t,u(s))ds.$

Where

$$G(t,s) = \frac{1}{\Gamma(q)}\begin{cases} \frac{(t-a)^{q-1}}{(b-a)^{q-1}}(b-s)^{q-1} - (t-s)^{q-1} & ,a \leq s \leq t \leq b \\ \frac{(t-a)^{q-1}}{(b-a)^{q-1}}(b-s)^{q-1} & ,a \leq t \leq s \leq b \end{cases}$$

Proof. By using the parametric form of fuzzy number we have $u(t) = (\underline{u}(r), \overline{u}(r))$, then the problem (1), (2) is equivalent to

$$\begin{cases} D^q\underline{x}(t;r) = \underline{f}(t,\underline{x}(t;r);r); & t \in [a,b] \\ \underline{x}(a;r) = \underline{0}(r), \quad \underline{x}(b;r) = \underline{B}(r). \end{cases} \tag{4}$$

And

$$\begin{cases} D^q\overline{u}(t;r) = \overline{f}(t,\overline{u}(t;r);r); & t \in [a,b] \\ \overline{u}(a;r) = \overline{0}(r), \quad \overline{u}(b;r) = \overline{B}(r). \end{cases} \tag{5}$$

It is well known that solving (3) is equivalent to solving the integral equation

$$\underline{u}(t;r) = c\frac{(t-a)^{q-1}}{\Gamma(q)} + d\frac{(t-a)^{q-2}}{\Gamma(q-2)} + \int_a^t G(t,s)\underline{f}(t,\underline{u}(t;r);r)ds.$$

where c and d are some real constants.
Now, $d = 0$ by the first boundary condition. On the other hand, $\underline{u}(b;r) = \underline{B}(r)$ wich implies that

$$\underline{B}(r) = c\frac{(b-a)^{q-1}}{\Gamma(q)} + \int_a^b (b-s)^{q-1}\underline{f}(s,\underline{u}(s;r);r)ds.$$

which implies that

$$c = \frac{\Gamma(q)}{(b-a)^{q-1}}\left(\underline{B}(r) - \int_a^b (b-s)^{q-1}\underline{f}(s,\underline{u}(s;r);r)ds\right)$$

$$\underline{u}(t;r) = \frac{\Gamma(q)}{(b-a)^{q-1}}\left(\underline{B}(r) - \int_a^b (b-s)^{q-1}\underline{f}(s,\underline{u}(s;r);r)ds\right)\frac{(t-a)^{q-1}}{\Gamma(q)} - \int_a^t (t-s)^{q-1}\underline{f}(t,\underline{u}(t;r);r)ds.$$

By the same way we can solve the problem (4) and this complete the proof.

Proposition 5. *Let G be the Green function given in Lemma* (2) *Then*

$$\int_a^b \mid G(t,s) \mid ds \leq \frac{(q-1)^{q-1}}{\Gamma(q)q^{q+1}}(b-a)^{q-1} \tag{6}$$

Proof. It is known [2], Lemma (2.2) that $G(t,s) \geq 0$ for all $t,s \in [a,b]$. Therefore

$$\begin{aligned}
\int_a^b \mid G(t,s) \mid ds = & \frac{1}{\Gamma(q)} \int_a^t \left(\frac{(t-a)^{q-1}}{(b-a)^{q-1}}(b-s)^{q-1} - (t-s)^{q-1} \right) ds, \\
& + \frac{1}{\Gamma(q)} \int_t^b \left(\frac{(t-a)^{q-1}}{(b-a)^{q-1}}(b-s)^{q-1} \right) ds, \\
& = \frac{1}{\Gamma(q)} \left(- \frac{(t-a)^{q-1}}{(b-a)^{q-1}} \frac{(b-t)^{q-1}}{q} + \frac{(t-a)^{q-1}}{(b-a)^{q}} \frac{(t-a)^{q-1}}{(b-a)^{q-1}} \frac{(b-a)^{q-1}}{q} \right), \\
& + \frac{1}{\Gamma(q)} \left(- \frac{(t-a)^{q-1}}{q} + \frac{(t-a)^{q-1}}{(b-a)^{q-1}} \frac{(b-t)^{q-1}}{q} \right), \\
& = \frac{1}{\Gamma(q)} \frac{(t-a)^{q-1}(b-t)}{q},
\end{aligned}$$

We define $g : [a,b] \rightarrow R$ by

$$g(t) = \frac{(t-a)^{q-1}(b-t)}{q}$$

Diferentiating the function g we immediately find that its maximum is achieved at the point

$$t^* = \frac{(q-1)b+a}{q},$$

Moreover

$$g(t^*) = \frac{(a-1)^{q-1}(b-a)^q}{q},$$

wich complete the proof.

Theorem 1. *Assume that* $f : [a;b] \times E^1 \longrightarrow R$ *is continuous and satisfes*

$$D(f(t,u(t)), f(t,v(t)) \leq KD(u,v); \quad such \quad that \quad K \in]0,1[.$$

If $\dfrac{K(q-1)^{q-1}}{\Gamma(q)q^{q+1}}(b-a)^{q-1} < 1$, *then the following fuzzy fractional boundary value problem (1)–(2) has a unique continuous solution.*

Proof. Let $C([a,b],E^1)$ be the complete metric space of all fuzzy continuous functions defined on $[a,b]$ with the distance D.

Let $u \in C([a,b],E^1)$ and by Lemma (2) u is a solution of (1)–(2) if and only if it is a solution of the integral equation

$$u(t) = \frac{(t-a)^{q-1}}{(b-a)^{q-1}}B + \int_a^b G(t,s)f(t,u(s))ds.$$

Let $T : C([a,b],E^1) \longrightarrow C([a,b],E^1)$ be an operator defined by

$$Tu(t) = \frac{(t-a)^{q-1}}{(b-a)^{q-1}}B + \int_a^b G(t,s)f(t,u(s))ds.$$

for $t \in [a,b]$. We will show that the operator T has a unique fixed point.
Let $u,v \in C([a,b],E^1)$, then

$$\begin{aligned} D(Tu(t),Tv(t)) &\leq \int_a^b |G(s,t)|D(f(u(s),s),f(v(s),s)))ds. \\ &\leq \int_a^b |G(s,t)|KD(u(s),v(s))ds. \\ &\leq \frac{K(q-1)^{q-1}}{\Gamma(q)q^{q+1}}(b-a)^{q-1}D(u,v). \end{aligned}$$

Thus we conclude that T is a contraction mapping on $C([a,b],E^1)$ and by using the Banach fixed point theorem we get the desired result.

3.1 Illustrative Example

Example 4. As an example we consider the fuzzy fractional boundary problem

$$D^{3/2}(\underline{u}(t;r),\overline{u}(t;r)) = (sin(\underline{u}(t;r)),sin(\overline{u}(t;r))) \quad t \in [0,1] \tag{7}$$

$$(\underline{u}(0;r),\overline{u}(0;r)) = (0,0) \quad (\underline{u}(1;r),\overline{u}(1;r)) = 0. \tag{8}$$

Here $f(t,u(t;r);r) = sin(\underline{u}(t;r))$.
And $|sin(u(t))| \leq 1 = K$.
Since $q = 3/2$, then we have

$$\frac{1(q-1)^{q-1}}{\Gamma(q)q^{q+1}}(1-0)^{q-1} = \frac{3}{4}\pi^{\frac{1}{3}}3^{2/3}.$$

Since the conditions of theorem 1 are satisfied, then the problem (7)–(8) has a unique solution.

4 Conclusion and Future Work

In this paper, we studied the existence results for a fuzzy fractional boundary problem by using the Banach fixed point theorem.

Our future works is to extend the results of this paper to generalized cases like the intuitionistic fuzzy fractional boundary value problem for parabolic and hyperbolic type equations.

Acknowledgement. The authors would like to express their sincere apreciation to the referees for their very helpful suggestions and many kind comments.

References

1. Agarwal, R.P., Belmekki, M., Benchohra, M.: A survey on semilinear differential equations and inclusions involving riemann-liouville fractional derivative. Adv. Differ. Equ. **2009**(1), 1–47 (2009). https://doi.org/10.1155/2009/981728
2. Agarwal, R.P., Lakshmikantham, V., Nieto, J.J.: On the concept of solution for fractional differential equations with uncertainty. Nonlinear Anal.: Theory Methods Appl. **72**, 2859–2862 (2010)
3. Agarwal, R.P., Zhou, Y., Wang, J., Luo, X.: Fractional functional differential equations with causal operators in banach spaces. Math. Comput. Model. **54**, 1440–1452 (2011)
4. Arshad, S., Lupulescu, V.: On the fractional differential equations with uncertainty. Nonlinear Anal. **74**, 3685–3693 (2011)
5. Diamond, P., Kloeden, P.E.: Metric spaces of fuzzy sets. Fuzzy Sets Syst. **35**, 241–250 (1990)
6. Dieudonn, J.: Acta. Sci. Math. Szeged Pars B (12), 38–40 (1950)
7. Dubois, D., Prade, H.: Fuzzy Sets and Systems: Theory and Applications. Academic Press, New York (1980)
8. Friedmana, M., Kandel, A.: Numerical solutions of fuzzy differential and integral equations. Fuzzy Sets Syst. **106**, 35–48 (1999)
9. Hoa, N.V., Lupulescu, V., O'Regan, D.: A note on initial value problems for fractional fuzzy differential equations. Fuzzy Sets Syst. **347**, 54–69 (2018)
10. Kaleva, O.: Fuzzy differential equations. Fuzzy Sets Syst. **24**, 301–317 (1987)
11. Kilbas, A., Srivastava, H.M., Trujillo, J.J.: Theory and Applications of Fractional Differential Equations. North-Holland Mathematical studies, vol. 204. Ed van Mill, Amsterdam (2006)
12. Lang, S.: Analysis, vol. 11. Addison-Wesley, Reading, MA (1969)
13. Nguyen, H.T.: A note on the extension principle for fuzzy sets. J. Math. Anal. Appl. **64**, 369–380 (1978)
14. Puri, M.L., Ralescu, D.A.: The concept of normality for fuzzy random variables. Ann. Probab. **13**, 1373–1379 (1985)
15. Puri, M.L., Ralescu, D.A.: Differential for fuzzy functions. J. Math. Anal. Appl. **91**, 552–558 (1983)
16. Salahshour, S., Allahviranloo, T., Abbasbandy, S., Baleanu, D.: Existence and uniqueness results for fractional differential equations with uncertainty. Adv. Diff. Equ. **112**, 1687–1847 (2012)
17. Salahshour, S., Allahviranloo, T., Abbasbandy, S.: Solving fuzzy fractional differential equations by fuzzy Laplace transforms. Commun. Nonlinear Sci. Numer. Simul. **17**, 1372–1381 (2012)
18. Zadeh, L.: Fuzzy sets. Inform. Control **3**, 338–356 (1965)

Numerical Method Combinations Assessment for Transport-Dominated Problems in the CHIMERE Model: A Case Study of Agadir (Morocco)

Amine Ajdour[1(✉)], Radouane Leghrib[1], Jamal Chaoufi[1], and Ahmed Chirmata[2]

[1] LETSMP, Department of Physics, Faculty of Sciences, University Ibn Zohr, Cité Dakhla, BP8106, Agadir, Morocco
amine.ajdour@edu.uiz.ac.ma, r.leghrib@uiz.ac.ma
[2] Department of Energy and Environment, Wilaya of Agadir, Boulevard Mohamed V, 80000 Agadir, Morocco

Abstract. The World Health Organization (WHO) estimates that outdoor and indoor air pollution effects cause approximately 7 million premature deaths each year. In Morocco's case, the economic cost of air quality degradation is about 1.62% of GDP. Several studies have been conducted in air quality modeling to address the transport equation, and various numerical methods have been proposed. As a result, the research aims to compare six numerical method combinations used in the CHIMERE model. Preliminary air quality modeling results have revealed a discrepancy between the model and ozone observations, which is more likely due to the accuracy of the input data, such as emissions, meteorology, or land use data. The fastest combination, adv00 (Upwind-Upwind), on the other hand, provides a significant improvement, especially when a high spatial resolution is used, as in this study. The results can be used to guide the selection of a perfect algebraic polynomial interpolation for transport-dominated problems (advection). Furthermore, it allows us to manipulate the CHIMERE model to make it as realistic as possible.

Keywords: Air quality modeling · Transport equation · Numerical method combinations · Transport-dominated problems · CHIMERE model

1 Introduction

For different reasons, including increasing urbanization, intensive industrial pollution, traffic emissions, agriculture, and energy consumption, increasing air pollution levels are a global issue (Avtar et al. 2019). There is a wealth of scientific information available today about air pollution and its effects on health (Rovira et al. 2020). Although there are still gaps and uncertainties in this database, it provides a solid foundation for the World Health Organization's (WHO) pollutant concentration guidelines (WHO 2020). The leading cause of fatal respiratory diseases was air pollution, both outside and

S. Melliani and O. Castillo (Eds.): ICPAMS 2021, LNNS 476, pp. 88–100, 2023.
https://doi.org/10.1007/978-3-031-12416-7_9

inside. The most serious environmental issue is pollution caused by high levels of ozone (Li et al. 2020). Photochemical reactions in the presence of light and precursor pollutants such as nitrogen oxides (NOx) and volatile organic compounds (VOC) produce ozone in the atmosphere. Significant health effects are expected if concentrations exceed 240 g/m3 for more than eight continuous hours. This conclusion is founded on the findings of numerous clinical inhalation and field studies. Healthy adults and asthmatics can expect significant reductions in pulmonary function and airway inflammation, resulting in significantly worsened health symptoms (Heberle et al. 2019).

Morocco has invested heavily in installing air quality control systems in several cities. In 2018, the air pollution's economic cost was estimated at 1.62% of the country's GDP, says Morocco's Secretary of State for Sustainable Development (Croitoru et al. 2017). This large number has prompted Morocco to make special efforts to combat air pollution. In this context, many new studies in Morocco, specifically over the city of Agadir, presented the evaluation of air quality. A survey was conducted in Agadir city to prioritize local air quality issues, highlighting the need to update the local Air Quality Management (LAQM) and the Air Quality Monitoring Network (AQMN) implementation regularly (Chirmata et al. 2017). We can point to studies that employ deterministic models, such as the CHIMERE model (Ajdour et al. 2019; Ajdour et al. 2020), as well as others that employ artificial intelligence to forecast air quality over Agadir (Adnane et al. 2021).

In the atmospheric science field, accurate numerical simulation of tropospheric air pollution phenomena has become a critical challenge (Todorov et al. 2020). Advection-dominated problems are a subset of partial differential equations (PDEs) in which other terms are relatively minor compared to advection (Appadu 2013). In the chemical species conservation equation, different numerical solvers are used for the advection equation, the properties of these solvers leading to a range of errors. It should be indicated that the advection term is to blame for all of the problems in computational processing. Mostly, the self-adjoint property in space is destroyed by this term (Cao et al. 2017). Typically, numerical methods for multidimensional problems are built by reducing them to a sequence of one-dimensional problems. Furthermore, these methods must preserve most aspects of the continuous model, such as the solution's positivity. Because most of the equations used do not have an analytical solution in closed form, it is critical to determine an accurate numerical approximation. Several studies on the CHIMERE model have been conducted, but only with the PPM numerical solver (Menut et al. 2013). Despite their rarity, it is also essential to mention some studies that have looked into this type of numerical comparison (Hutchison and Mitchell n.d.). The advection equation should be solved using an accurate numerical method to improve the model. Given the significance of the advection phenomenon, it is critical to identify the most appropriate numerical resolution method, particularly in the Agadir case.

The primary goal of this study is to improve the chemical transport module CHIMERE in the city of Agadir. As a result, we compared six numerical method combinations, three for horizontal and two for vertical transport. The results are evaluating using ozone observation data from the fixed station collected in July 2016. In addition, the stability and accuracy of each combination were investigated. This paper is organized as follows: in Sect. 2, we introduce the model theory, the transport equation, and the finite volume methods that have been evaluated. In Sect. 3, We will present the Chimere model configuration, and we will describe the results of this simulation. In Sect. 4, some conclusions are provided.

2 Methodology

2.1 CHIMERE Concept

The evolution equation of a chemical species' concentration fi can be decomposed as the number of individual factors, which can be determined directly or indirectly.

$$\frac{\partial f_i}{\partial t} = \left(\frac{\partial f_i}{\partial t}\right)_{Advection} + \left(\frac{\partial f_i}{\partial t}\right)_{Turbulence} + \left(\frac{\partial f_i}{\partial t}\right)_{Chemistry} + \left(\frac{\partial f_i}{\partial t}\right)_{Emissions} + \left(\frac{\partial f_i}{\partial t}\right)_{Depot} \tag{1}$$

With:

$$\left(\frac{\partial f_i}{\partial t}\right)_{Advection} = -\nabla(uf) \tag{2}$$

$$\left(\frac{\partial f_i}{\partial t}\right)_{Turbulence} = -\nabla(k\nabla f) \tag{3}$$

$$\left(\frac{\partial f_i}{\partial t}\right)_{Chemistry} = P_i - L_i.f_i \tag{4}$$

$$\left(\frac{\partial f_i}{\partial t}\right)_{Emissions} = E \tag{5}$$

$$\left(\frac{\partial f_i}{\partial t}\right)_{Depot} = v_d.\left(\frac{\partial f_i}{\partial z}\right)_{sol} \tag{6}$$

P and L represent output and loss terms due to chemical reactions, and νd is deposition speed. fi is a vector containing the concentrations of one chemical specie, u is the three dimensional wind vector, k is the tensor of eddy diffusivity.

There are two numerical methods: the first considers the problem as a whole and simultaneously integrates all of the trends; the second is the method of separation operators, which involves successively integrated the various trends in time. In the first equation, we are dealing with a stiff system of coupled differential equations. In the CHIMERE model, the first approach is used, and the time integration is based on the two-step algorithm. The calculation is performed on the application of a Gauss-Seidel iteration scheme (Blackledget 2006) to the two-step implicit backward differentiation (BDF2) formula (Emmrich 2009). We notice f, the vector containing the concentrations of all model species for every grid box.

$$f^{n+1} = \frac{4f^n}{3} - \frac{f^{n-1}}{3} + \frac{2}{3}\Delta t\left(P\left(f^{n+1}\right) - L\left(f^{n+1}\right)\cdot f^{n+1}\right) \tag{7}$$

It's worth noting that L is a diagonal matrix in this case. This equation reads as follows after rearranging it. A Gauss-Seidel approach can be used to solve the implicit nonlinear system obtained in this scheme (Verwer et al. 1996).

$$f^{n+1} = \left(I + \frac{2}{3}\Delta t L\left(f^{n+1}\right)\right)^{-1} \frac{4f^{n+1}}{3} - \frac{f^{n-1}}{3} + \frac{2}{3}\Delta t P\left(f^{n+1}\right) \tag{8}$$

In CHIMERE, the modified terms Pmod and Lmod replace the production and loss terms P, and L. Ph and Pv represent the temporal evolution of concentrations due to horizontal (only advection) and vertical (both advection and diffusion) inflow into a given grid box. L_h and L_v represent the temporal evolution due to the respective outflow divided by the concentration.

$$\begin{cases} P_{mod} = P + P_h + P_v \\ L_{mod} = L + L_h + L_v \end{cases} \tag{9}$$

2.2 The Transport Equation

For realistic description, we started with the three-dimensional scalar advection problem (Goyal and Kumar 2011). For each chemical species, the conservation equation is then numerically solved. (uf) indicating the mass flux corresponding to velocity u.

$$\frac{\partial f}{\partial t} = -\nabla(uf) \tag{10}$$

This equation can be discretized and solved separately for each of the three orthogonal directions: zonal, meridian, and vertical, using the operator splitting technique and the CHIMERE design, including parallelepiped structured grids. This technique is more computationally efficient and, in some cases, more stable and accurate than two- or three-dimensional approaches, notably when applied to high-order models. Generally, is widely used in meteorological and chemistry-transport modeling. We notice δα(f), the variation of (f) due to transport in the direction α and F is (uf). After time and space discretization, the discretized transport calculations are as follows:

$$\delta^{\alpha}(f) = \left(\frac{(F)^{\alpha}_{n+1/2} - (F)^{\alpha}_{n-1/2}}{\Delta x}\right)\Delta t \tag{11}$$

Since the inward and outward fluxes cancel out in each direction, this equation ensures mass conservation. From the initial concentration field, the concentration increments are determined sequentially for each direction. The calculation of fluxes at cell interfaces ($F^{\alpha}_{n+1/2}$) is a key problem in solving this equation. The characteristics of the transport scheme are determined by how these fluxes are numerically estimated. These numerical methods range from simple first order numerical to higher order methods.

Species concentrations and meteorological variables are defined on the same grid as in the CHIMERE model. The wind speeds at interfaces are interpolated linearly from wind speeds at the centers of the two grid cells separated by the interface. It is assumed that the grid cell length does not differ significantly from one grid cell to its neighbors in the horizontal directions. The thickness of the layers increases quasi-exponentially with altitude in the vertical direction to provide better vertical resolution in the lower model levels. At each grid cell, vertical mass fluxes are calculated to guarantee a zero flow divergence. The mass flux at the lower boundary of the lowest layer is zero, and at the top is determined in ascending order, from lowest to highest layer. After identifying the mass fluxes, the vertical transport method can be employed.

To simplify things, we'll continue with a one-dimensional scalar advection problem for one typical atmospheric pollutant. The following linear hyperbolic equation gives the transport equation, also known as the advection equation, to which we add an initial concentration f(x, 0) = f0, during the time interval [0, T]. The concentration of one atmospheric pollutant is denoted by f. As a result, the description is the general Cauchy problem.

$$\begin{cases} \frac{\partial f}{\partial t} = \frac{\partial (uf)}{\partial x} \\ f(x,0) = f_0 \end{cases} \quad \forall (x,t) \in \mathbb{R} \times [0,T] \tag{12}$$

2.3 Resolution Method

The equations governing the phenomenon of air pollution can only be resolved analytically for a few simple cases. However, to account for nonlinear and coupled differential equations that explain diffusion, advection, and chemical reactions, a numerical resolution that meets the boundary conditions is needed. The finite volume approach (Mazumder 2016) was used in the Chimere model. The need for stability and, as a result, a reduction in computation time drives the decision to use such a method. The conservative form of the transport equation in one dimension in space is solved using the finite volume method. First, we divide the spatial domain into cells called finite or controls volumes. This corresponds in one dimension to a partition of [0, L]. On the other hand, one built discrete equations from the integral form of the equation. A unidirectional grid is used to discretize the domain first. For time variable, let be $[t^n, t^{n+1}]$ a uniform division of [0, T], we designate time step as $\Delta t = t^{n+1} - t^n$. For space variable $[x_{j-1/2}, x_{j+1/2}]$ is a division of [0, L], we define space step as $\Delta x = x_{j+1/2} - x_{j-1/2}$, the control volume $\Omega_j = \left[x_{j-1/2}, x_{j+1/2}\right]$, and the center points $x_j = \frac{x_{j+1/2} - x_{j-1/2}}{2}$.

The integral form on each of the control volumes of the conservation law is given by

$$\int_{\Omega_j} \frac{\partial f}{\partial t} dx = Q(f(x_{j-1/2}, t)) - Q\left(f\left(x_{j+1/2}, t\right)\right) \tag{13}$$

We note $Q\left(f\left(x_{j-1/2}, t\right), Q\left(f\left(x_{j+1/2}, t\right)\right.\right.$,the fluxes inside the cell. By integrating in time:

$$\int_{\Omega_j} f(x, t+\Delta t) - \int_{\Omega_j} f(x,t) = \int_t^{t+\Delta t} Q\left(f\left(x_{j-1/2}, t\right)\right) dt - \int_t^{t+\Delta t} Q\left(f\left(x_{j+1/2}, t\right)\right) dt \tag{14}$$

$$\frac{1}{\Delta x} \int_{\Omega_j} f(x, t+\Delta t) = \frac{1}{\Delta x} \left(\int_{\Omega_j} f(x,t) + \int_t^{t+\Delta t} Q\left(f\left(x_{j-1/2}, t\right)\right) dt - \int_t^{t+\Delta t} Q\left(f\left(x_{j+1/2}, t\right)\right) dt \right) \tag{15}$$

This term can be expressed as:

$$\gamma_j^{n+1} = \gamma_j^n - \frac{1}{\Delta x}\left(Q_{j+1/2}^n - Q_{j-1/2}^n\right) \tag{16}$$

where, we define the exact flux Q and the average values of the exact solution γ at time t, on each cell, we note F, the approximation of the function by:

$$\begin{cases} Q^n_{j+1/2} = \frac{1}{\Delta x}\int_t^{t+\Delta t}(uf)(x_{j+1/2}, t)dt \\ \gamma_j^n = \frac{1}{\Delta x}\int_{x_{j-1/2}}^{x_{j+1/2}} F(x,t)dx \end{cases} \tag{17}$$

F is an interpolation function that is determined by the conservative algorithm. This paper employs three distinct F approximations, corresponding to constant, linear, and quadratic approximations of F over each of the central cells. By shifting the profile by u.Δt, the value of f in the (n + 1) step can be easily obtained. In the coming section, we'll go through the three transportation options available in the model.

$$f\left(x_{j+1}, t+\Delta t\right) = f\left(x_{j+1} - u\Delta t, t\right) = F_j^n\left(x_{j+1} - u\Delta t\right) \tag{18}$$

2.3.1 Upwind Method

The upwind method uses a condition on the velocity sign to determine the advection concentration into the neighboring cell via the considered surface (Falcone and Ferretti 2016). If u is positive, the moving wave solution of the equation above propagates to the right, with upwind refers to the left side, while downwind refers to the right side. If u is negative, the moving wave solution propagates to the left. The left hand is known as the downwind side, and the right side is known as the upwind side. The tracer concentration in each grid cell is assumed to be uniform in this scheme. Therefore, the interface mass flux is the product of the wind at the interface and the tracer concentration in the upwind cell. This method uses a constant function β, in a generic mesh element with Ωj as the boundaries. This constant is the function's value in the initial moment, when this value is known. For $\boldsymbol{x} \in \Omega_j = \left[\boldsymbol{x_{j+1/2} - u^n_{j+1/2}}\Delta t, \boldsymbol{x_{j+1/2}}\right]$:

$$\begin{cases} if\ u > 0\ F_j(x) = \beta_j \\ if\ u < 0\ F_j(x) = \beta_{j-1} \end{cases} \tag{19}$$

For u > 0, the flux value can be calculated as:

$$Q^n_{j+1/2} = \frac{1}{\Delta t}\int_{\Omega_j} F(x)dx = \frac{1}{\Delta t}\int_{\Omega_j} \beta_j dx \tag{20}$$

After calculation:

$$Q^n_{j+1/2} = u^n_{j+1/2}\beta_j \tag{21}$$

Following that, we can thoroughly describe the flux function Q for the Upwind method in terms of the average cell integral value γ as follows:

$$\begin{cases} if\ u(t)_{j+1/2} > 0\ Q_{j+1/2} = \gamma_j u_{j+1/2} \\ if\ u(t)_{j+1/2} < 0\ Q_{j+1/2} = \gamma_{j+1/2} u_{j+1/2} \end{cases} \tag{22}$$

For the CHIMERE model the function f is generally defined by f = ρ.c with c is the concentration, according to the sign of the wind speed at the cell interface, the fluxes at the cell interfaces are described by the following equations:

$$\begin{cases} if\ u(t)_{j+1/2} > 0\ F_{j\pm1/2} = \rho_j u_{j+1/2} c_j \\ if\ u(t)_{j+1/2} < 0\ Q_{j+1/2} = \rho_j u_{j+1/2} c_{j+1} \end{cases} \tag{23}$$

2.3.2 Van Leer Method

The Van Leer Method is a finite volume method for obtaining highly accurate numerical solutions for a given system, even when the solutions include shocks, discontinuities, or large gradients (Vanderheyden and Kashiwa 1998). The concept is to use reconstructed states derived from cell-averaged states from the previous time step to replace the constant approximation of the Upwind method. For x ∈ Ωj, and taking into account the velocity u < 0.

$$F_j(x) = \beta_j + \alpha_j \left(x - x_{j-1/2}\right) \tag{24}$$

From the initial condition of each cell, we can determine the coefficient βj:

$$\begin{cases} if\ u_{j+1/2} > 0\ F_j\left(x_{j-1/2}\right) = \beta_j = f_j \\ if\ u_{j+1/2} < 0\ F_j\left(x_{j-1/2}\right) = \beta_j = f_{j+1} \end{cases} \tag{25}$$

Based on monotonicity, to achieve a good relationship between the interpolation function and the original function. We have to specify a slope limiter to determine αj the slope term of the interpolation function. The main goal of flux limiter schemes is to keep spatial derivatives as practical as possible, which typically means physically achievable and relevant values. Limiter function Minmod provides a reasonable choice of the αj. We note $\left(f_j - f_{j-1}\right) = \delta f_j$

$$\begin{cases} if\ \left(\delta f_{j-1/2} - \delta f_{j-3/2}\right) > 0\ \ \alpha_j = f_j\ - f_{j-1} \\ if\ \left(\delta f_{j-1/2} - \delta f_{j-3/2}\right) < 0\ \ \alpha_j = f_{j-1}\ - f_{j-2} \end{cases} \tag{26}$$

The flux value can be calculated as:

$$Q_{j+1/2}^n = \frac{1}{\Delta t} \int_{\Omega_j} F(x)dx = \frac{1}{\Delta t} \int_{\Omega_j} \left(\beta_j + \alpha_j \left(x - x_{j-1/2}\right)\right)dx \tag{27}$$

In the monotonic case, the flux formulation can be calculated as follows, in terms of the average cell integral value γ, depending on the sign of the velocity. Generally, this formula is directly applied to determine the flux function.

$$\begin{cases} if\ u(t)_{j+1/2} > 0\ Q_{j+1/2} = \gamma_j u_{j+1/2} + \alpha_j \left(u_{j+1/2}\Delta x - u_j u_{j+1/2}\Delta t\right) \\ if\ u(t)_{j+1/2} < 0\ Q_{j+1/2} = \gamma_{j+1/2} u_{j+1/2} + \alpha_j \left(u_{j+1/2}\Delta x - u_j u_{j+1/2}\Delta t\right) \end{cases} \tag{28}$$

The concentration inside a grid cell is determined by a linear slope between the cell's two interfaces in the CHIMER model. This method is slightly more time-consuming than the first-order upwind method. Because of its higher numerical precision and lower scattering than the first-order upwind system. It is a good compromise solution for long-range transport in meteorology between numerical accuracy and computational efficiency.

$$c_j(x) = \beta_j + (x - x_j)\alpha_j \tag{29}$$

With $\beta_j = c_j$ and according to the following cases, the slope is calculated

$$if\ c_j \in [c_{j-1}, c_{j+1}]\ \beta_j = sign\,(c_{j+1} - c_{j-1}) \times min\left(\frac{c_{j+1} - c_{j-1}}{2\Delta x}, \frac{c_{j+1} - c_j}{\Delta x}, \frac{c_j - c_{j-1}}{\Delta x}\right) \tag{30}$$

2.3.3 Piecewise Parabolic Method (PPM)

Since PPM is a finite volume scheme, physical variables are represented as averages over a grid zone instead of single values at different points (Zhang et al. 2017). Then use the information from the average of the neighboring regions to fit a single monotonic parabola to the average area of each dependent variable. PPM is a computational technique developed for fluid flow modeling with heavy impacts and discontinuities. It can handle steep gradients in small meteorological flows. For x ∈ Ωj, and taking into account the velocity u < 0. We note $\overline{x_j} = (x - x_{j-1/2})/\Delta x_j$.

$$F_j(x) = \beta_j + \alpha_j x_j + \varphi_j x_j(1 - x_j) \tag{31}$$

The coefficients βj, αj and φj are obtained by the following formulas:

$$\beta_j = F_j(x_{j-1/2}) \tag{32}$$

$$\alpha_j = F_j(x_{j+1/2}) - F_j(x_{j-1/2}) \tag{33}$$

$$\varphi_j = 6\left(f_j - \frac{1}{2}(F_j(x_{j+1/2}) + F_j(x_{j-1/2}))\right) \tag{34}$$

We used the Colella and Woodward approach to measure the interface values. In general, slope modifications were added to the PPM method to address the spurious oscillations issue in numerical solutions caused by high-order interpolation. The average slope in cell Ω_j is noted by $\overline{\delta f}$, depending on the sign of $\gamma_{j+1} - \gamma_j$, and determined by $\overline{\delta f_j} = (\gamma_{j+1} - \gamma_j)/4$. The average slope and affect the dispersion errors of the numerical solutions are controlled by ε_1 and ε_2.

$$\begin{cases} if\,(\gamma_{j+1} - \gamma_j) > 0\ \overline{\delta f_j} = 2\min(\delta f, \varepsilon_1(\gamma_{j+1} - \gamma_j), \varepsilon_2(\gamma_j - \gamma_{j-1})) \\ if\,(\gamma_{j+1} - \gamma_j) < 0\ \overline{\delta f_j} = 2\max(\delta f, \varepsilon_1(\gamma_{j+1} - \gamma_j), \varepsilon_2(\gamma_j - \gamma_{j-1})) \end{cases} \tag{35}$$

The interface values are defined by

$$f_{j+1} = \frac{1}{2}(\gamma_{j+1} + \gamma_j) - \frac{1}{6}(\overline{\delta f_j} - \overline{\delta f_{j-1}}) \tag{36}$$

Generally, this formula is directly applied to determine the flux function.

$$\begin{cases} \textit{if } u(t)_{j+1/2} > 0 \; Q_{j+1/2} = u_{j+1/2}\left(F_j\left(x_{j+1/2}\right) - \frac{u_{j+1/2}\Delta t}{2\Delta x_j}\left(\Delta\beta_j - \left(1 - \frac{2u_{j+1}\Delta t}{3\Delta x_j}\right)\varphi_j\right)\right) \\ \textit{if } u(t)_{j+1/2} < 0 \; Q_{j+1/2} = u_{j+1/2}\left(F_{j+1}\left(x_{j+1/2}\right) - \frac{u_{j+1/2}\Delta t}{2\Delta x_j}\left(\Delta\beta_{j+1} - \left(1 - \frac{2u_{j+1}\Delta t}{3\Delta x_{j+1}}\right)\varphi_{j+1}\right)\right) \end{cases} \tag{37}$$

In the CHIMERE model, the piecewise parabolic method is used with $\varepsilon_1 = \varepsilon_2 = 0$. The errors caused by the neglect of the cross derivatives are approximately compensated due to the method's symmetry. As a result, because the PPM method implemented in CHIMERE treats transport in each horizontal dimension with 3rd order accuracy, it is much less diffusive than the simple 2d order Van Leer method.

3 Results and Discussion

Three models were used in the modeling process. The CHIMERE model, version 2017r3, calculate gas and aerosol concentrations based on WRF and Emis-surf data. Meteorological data is computed using the WRF (Weather Research Forecasting) model, version 4.0. Finally, the Emis-surf model, version 2016b, includes information on gas and aerosol emissions. CHIMERE is a multi-scale Eulerian chemistry transport model. It is designed to generate daily pollutant forecasts, reproduce long-term emission scenarios, and study typical cases. This model incorporates a wide range of data inputs, such as meteorological conditions, land use, and emissions. The chemical boundary conditions are from the three-dimensional global chemistry-climate model LMDz-INCA, while the aerosol boundary conditions are from the GOCART and LMDz-AERO global models (Folberth et al. 2006). Anthropogenic emissions are estimated using the 2010 EDGAR-HTAP v2 global emission inventory (Ferreyra et al. 2016). The horizontal resolution has been set at 0.02 degrees. The model is set up with 20 vertical levels in the troposphere ranging from 500 hPa to 200 hPa. The WRF and Chimere domains and the study domain are shown in Fig. 1. It should be noted that the position of the fixed station is at the center of the study domain.

Since 2010, the Souss Massa Region has had a regional mobile laboratory and a ground station for air quality monitoring to assess the air quality in Agadir. Both are outfitted with the Environment S.A. To detecting harmful pollutants, specific standardized analyzers and also meteorological variables are used. Data collected from the ground station are used as reference observational data in this study. The station is located in a residential area to assess urban background pollution. Souss Massa has a population of about 2.7 million people, with 33.7% living in the Agadir-Ida Outanane prefecture. The modeling ozone results are compared with monitoring data in this study. The performance of these models in Agadir is determined using four statistical indicators: The Mean Normalized Bias (MNE), the Correlation Coefficient (R), the Mean Bias (MB), and the Mean Squared Error (RMSE).

As illustrated in Table 1, six combinations were identified for this study based on the numerical method used during horizontal and vertical transport. Figure 2 shows the hourly ozone time series for each combination. A complete match can be seen in the night period, while differences in maximum values recorded in the middle of the day can be seen, which will affect the evening period. The adv11 combination makes a significant difference in the evening ozone values. Four statistical factors were calculated to accurately determine the discrepancies, as seen in Table 2. According to the MB significance values, CHIMERE overestimates the O3 concentration. This discrepancy is generally related to the low accuracy of the emission information at each time step derived from the annual totals of the HTAP inventory database, and also the model's horizontal resolution of 0.02°, while the stations record very local values of the measurements. The Mean Normalized Error shows a large percentage compared to the acceptable and recommended values by EPA. In addition to the low accuracy of the input leading to an increase in the accumulation of the errors. We can also see large proportions of typical ozone concentrations during the night, as shown in Fig. 2, which is not entirely consistent with the measurements. The correlation values indicate that the model maintains the measurement trend, which is a plus.

In general, the statistical results are close; adv11 has low statistical values, which corresponds to their position in Fig. 2. For adv00 and adv20, the best compromise between correlation and error is recorded. In terms of computation time, adv00 provides a significant improvement, particularly when a high spatial resolution is used, as in this study. In general, horizontal transport methods have an indirect effect on vertical transport and propagation, and the more prevalent horizontal methods tend to increase transport to the lower layers, which leads to an increase in surface concentrations, as seen when comparing the maximum values in the middle of the day, between adv00 and adv10 (Fig. 2).

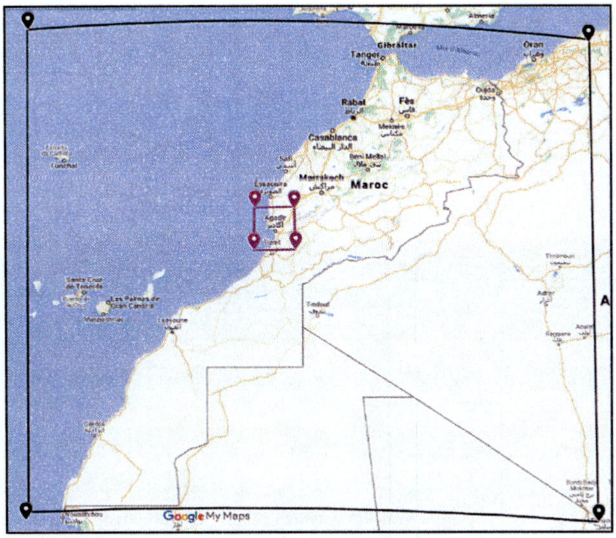

Fig. 1. WRF's domain (Black) and Chimere's domains (violate)

Table 1. The six combinations depending on the numerical method used during horizontal and vertical transport.

Horizontal	Vertical	
	Upwind	Van Leer
Upwind	adv00	adv01
PPM	adv10	adv11
Van Leer	adv20	adv21

Table 2. The Ozone evaluation of all combinations

	R	MB	MNE	RMSE
adv00	0.78	28.69	68%	28.40
adv01	0.77	29.10	69%	28.79
adv10	0.77	28.67	68%	28.37
adv11	0.76	29.12	69%	28.81
adv20	0.78	28.69	68%	28.39
adv21	0.78	29.12	69%	28.81

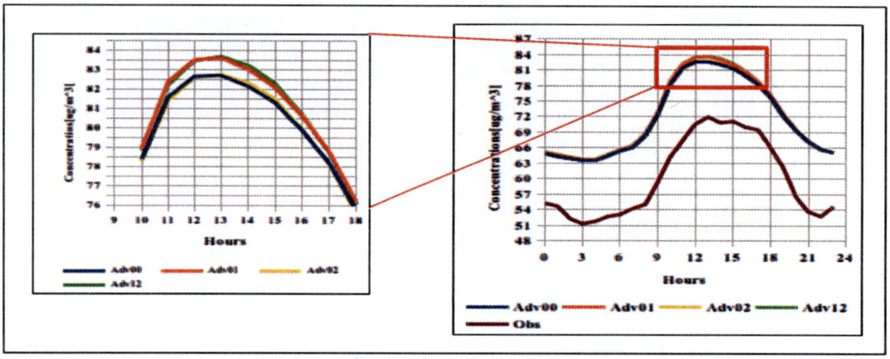

Fig. 2. Hourly ozone time series comparison between observations and four combinations

4 Conclusion

This study is based on the CHIMERE model and was conducted in Agadir in July of 2016. The goal is to improve the chemical transport module CHIMERE from the six combinations identified based on the numerical method used during horizontal and vertical transport. The results show that the ozone model-observation disagreement is more related to the accuracy of the input data, such as emissions, meteorology, or land use data. This finding is consistent with previous research on the subject (Derognat et al.

2003; Menut et al. 2013). If the results of the six combinations are similar, using the adv00 combination, which is the fastest, provides a significant improvement, especially, when a high spatial resolution is used, as in this study (Grylls et al. 2019; Tao et al. 2020). In conclusion, it appears that the type of algorithm used to simulate the transport phenomenon can be somewhat decisive, particularly in the case of long-time simulations with high spatial resolution (Gavete et al. 2012; Mazumder 2016). Even though only one month's data from the summer of 2016 was used in this study to assess air pollutants in Agadir, Morocco, this allows us to control in some way the CHIMERE model to align it as closely as possible with reality. Finally, it refers to a series of future studies.

Acknowledgements. We acknowledge the Wilaya and the Region of the Souss Massa for their collaboration in this research work, under the leadership of the Wali and the president of the region of Souss Massa, to whom we present our great and special thanks.

References

Adnane, A., Leghrib, R., Chaoufi, J.: The use of a recurrent neural network for forecasting ozone concentrations in the City of Agadir (Morocco). **7**(3), 197–206 (2021). https://doi.org/10.26713/jamcnp.v7i3.1545

Ajdour, A., Leghrib, R., Chaoufi, J., Chirmata, A., Menut, L., Mailler, S.: Towards air quality modeling in Agadir City (Morocco). Mater. Today: Proc. (2019). https://doi.org/10.1016/j.matpr.2019.07.438

Ajdour, A., Leghrib, R., Chaoufi, J., Fetmaoui, H.: Assessment of particulate matter (PM 10) using chemistry transport modeling in Agadir City, Morocco. **7**(3), 231–240 (2020). https://doi.org/10.26713/jamcnp.v7i3.1547

Appadu, A.R.: Numerical solution of the 1D advection-diffusion equation using standard and nonstandard finite difference schemes. J. Appl. Math. **2013**(3) (2013). https://doi.org/10.1155/2013/734374

Avtar, R., Tripathi, S., Aggarwal, A.K.: Population – urbanization – energy nexus: a review. 1–21 (2019). https://doi.org/10.3390/resources8030136

Blackledget, J.M.: Iterative methods of solution. Digit. Signal Process. i 237–254 (2006). https://doi.org/10.1533/9780857099457.2.237

Cao, Z.W., Liu, Z.F., Liu, Z.F., Wang, X.H.: A self-adaptive numerical method to solve convection-dominated diffusion problems. Math. Probl. Eng. **2017** (2017). https://doi.org/10.1155/2017/8379609

Chirmata, A., Leghrib, R., Ichou, I.A.: Implementation of the air quality monitoring network at Agadir City in Morocco. J. Environ. Prot. **08**(04), 540–567 (2017). https://doi.org/10.4236/jep.2017.84037

Croitoru, L., Sarraf, M.: Estimating the health cost of air pollution: the case of Morocco. 1087–1099 (2017). https://doi.org/10.4236/jep.2017.810069

Derognat, C., Beekmann, M., Baeumle, M., Martin, D., Schmidt, H.: Effect of biogenic volatile organic compound emissions on tropospheric chemistry during the Atmospheric Pollution Over the Paris Area (ESQUIF) campaign in the Ile-de-France region. J. Geophys. Res. D: Atmos. **108**(17) (2003). https://doi.org/10.1029/2001jd001421

Emmrich, E.: Convergence of the variable two-step BDF time discretisation of nonlinear evolution problems governed by a monotone potential operator. BIT Numer. Math. **49**(2), 297–323 (2009). https://doi.org/10.1007/s10543-009-0221-4

Falcone, M., Ferretti, R.: Numerical methods for Hamilton–Jacobi type equations. In: Handbook of Numerical Analysis, 1st edn., vol. 17. Elsevier B.V. (2016). https://doi.org/10.1016/bs.hna.2016.09.018

Ferreyra, M.F.G., Curci, G., Lanfri, M.: First implementation of the WRF-CHIMERE-EDGAR modeling system over Argentina. IEEE J. Sel. Top. Appl. Earth Obs. Remote Sens. **9**(12), 5304–5314 (2016). https://doi.org/10.1109/JSTARS.2016.2588502

Folberth, G.A., Hauglustaine, D.A., Lathièrel, J., Brocheton, F.: Interactive chemistry in the Laboratoire de Météorologie Dynamique general circulation model: model description and impact analysis of biogenic hydrocarbons on tropospheric chemistry. Atmos. Chem. Phys. **6**(8), 2273–2319 (2006). https://doi.org/10.5194/acp-6-2273-2006

Gavete, L., Vivanco, M.G., Molina, P., Gavete, M.L., Urena, F., Benito, J.J.: Implementation in CHIMERE of a conservative solver for the advection equation-cmmse10. J. Comput. Appl. Math. **236**(12), 3026–3033 (2012). https://doi.org/10.1016/j.cam.2011.04.003

Goyal, P., Kumar, A.: Mathematical modeling of air pollutants: an application to Indian Urban City. Air Qual.-Models Appl. (2011). https://doi.org/10.5772/16840

Grylls, T., Le Cornec, C.M.A., Salizzoni, P., Soulhac, L., Stettler, M.E.J., van Reeuwijk, M.: Evaluation of an operational air quality model using large-eddy simulation. Atmos. Environ.: X, **3**(November 2018), 100041 (2019). https://doi.org/10.1016/j.aeaoa.2019.100041

Heberle, S.M., da Costa, G.M., Barros, N., Rosa, M.S.G.: The effects of atmospheric pollution in respiratory health. In: Hussain, C.M. (ed.) Handbook of Environmental Materials Management, pp. 1271–1286. Springer, Cham (2019). https://doi.org/10.1007/978-3-319-73645-7_171

Hutchison, D., Mitchell, J.C.: (Lecture Notes in Computer Science 6785) Svajūnas Sajavičius (auth.), Beniamino Murgante, Osvaldo Gervasi, Andrés Iglesias, David Taniar, Bernady O. Apduhan (eds.) - Computational Science and I.pdf (n.d.)

Li, J., et al.: Ambient ozone pollution and years of life lost: association, effect modification, and additional life gain from a nationwide analysis in China. Environ. Int. **141**, 105771 (2020). https://doi.org/10.1016/j.envint.2020.105771

Mazumder, S.: Unstructured Finite Volume Method. In Numerical Methods for Partial Differential Equations, vol. M (2016). https://doi.org/10.1016/b978-0-12-849894-1.00007-x

Menut, L., et al.: CHIMERE 2013: a model for regional atmospheric composition modelling. Geosci. Model Dev. **6**(4), 981–1028 (2013). https://doi.org/10.5194/gmd-6-981-2013

Rovira, J., Domingo, J.L., Schuhmacher, M.: Air quality, health impacts and burden of disease due to air pollution (PM10, PM2.5, NO2 and O3): application of AirQ+ model to the Camp de Tarragona County (Catalonia, Spain). Sci. Total Environ. **703**, 135538 (2020). https://doi.org/10.1016/j.scitotenv.2019.135538

Tao, H., et al.: Impacts of improved modeling resolution on the simulation of meteorology, air quality, and human exposure to PM2.5, O3 in Beijing, China. J. Clean. Prod. **243**, 118574 (2020). https://doi.org/10.1016/j.jclepro.2019.118574

Todorov, V., Kandilarov, J., Dimov, I., Vulkov, L.: High-accuracy numerical methods for a parabolic system in air pollution modeling. Neural Comput. Appl. **32**(10), 6025–6040 (2020). https://doi.org/10.1007/s00521-019-04088-x

Vanderheyden, W.B., Kashiwa, B.A.: Compatible fluxes for van Leer advection. J. Comput. Phys. **146**(1), 1–28 (1998). https://doi.org/10.1006/jcph.1998.6070

Verwer, J.G., Blom, J.G., Van Loon, M., Spee, E.J.: A comparison of stiff ode solvers for atmospheric chemistry problems. Atmos. Environ. **30**(1), 49–58 (1996). https://doi.org/10.1016/1352-2310(95)00283-5

WHO. Air pollution (2020). https://www.who.int/health-topics/air-pollution#tab=tab_1

Zhang, W., Wang, T., Bai, J.S., Li, P., Wan, Z.H., Sun, D.J.: The piecewise parabolic method for Riemann problems in nonlinear elasticity. Sci. Rep. **7**(1), 1–17 (2017). https://doi.org/10.1038/s41598-017-13484-z

Generalized Solution of Transport Equation

Lalla Saadia Chadli, Abdelmjid Benmerrous, Abdelaziz Moujahid,
M'hamed Elomari, and Said Melliani(✉)

LMACS, Sultan Moulay Slimane University, BP 523 - 23000 Beni Mellal, Morocco
sa.chadli@yahoo.fr, m.elomari@usms.ma, s.melliani@usms.ma

Abstract. In this paper we proved some importance proprieties of Colombeau algebra, we proved the existence and uniqueness of solution of transport equation with variable speed and initial data in the Colombeau algebra $\mathcal{G}$. We proved the association of the generalized solution with the classical solution.

Keywords: Transport equation · Colombeau algebra · Generalized solution · Association

1 Introduction

The optimal solution for overcoming the problems that Schwartz theory of distributions is concerned with was offered by Colombeau (1984, 1985) [2] and [8]. He constructed an associative differential algebra of generalized functions $\mathcal{G}(\mathbb{R})$, which contains the space $\mathcal{D}'(\mathbb{R})$ of distributions as subspace and the algebra of $\mathcal{C}^\infty - functions$ as subalgebra. This theory of generalized functions of Colombeau actually generalizes the theory of Schwartz distributions: these new Colombeau generalized functions can be differentiated in the same way as distributions, but where multiplication and other nonlinear operations are concerned, it is significant that the result of these operations always exists in this algebra as Colombeau generalized function. These new generalized functions are very much related to the distributions, in the sense that their definition may be considered as a natural evolution of the Schwartz definition of distributions [2].

The notion of 'association' in $\mathcal{G}(\mathbb{R})$ is a faithful generalization of the equality of distributions, and again enables us to interpret results in terms of distributions.

Due to all these properties, Colombeau theory has found extensive application in different natural sciences and engineering, especially in fields where products of distributions with coinciding singularities are considered [8].

The transport equation describes how a scalar quantity is transported in a space. Usually, it is applied to the transport of a scalar field (e.g. chemical concentration, material properties or temperature) inside an incompressible flow. From the mathematical point of view, the transport equation is also called the convection-diffusion equation, which is a first-order PDE (partial differential equation). The convection-diffusion equation is the basis for the most common transportation models.

This paper solves a problem called non-regular transport problem on the domain $\Omega = \mathbb{R}^+ \times \mathbb{R}$.

$$\begin{cases} \partial_t u(t,x) + c(t,x)\partial_x u(t,x) = f(t,x)u(t,x) + a(t,x), & (t,x) \in \mathbb{R}^{*,+} \times \mathbb{R} \\ u(0,x) = u_0(x), & x \in \mathbb{R} \end{cases} \tag{1}$$

S. Melliani and O. Castillo (Eds.): ICPAMS 2021, LNNS 476, pp. 101–111, 2023.
https://doi.org/10.1007/978-3-031-12416-7_10

with c, f and a are discontinuous functions such that $c > 0$.

In Colombeau algebra of generalized functions $\mathcal{G}$ which allows multiplication of distributions and solution of nonlinear problems with singularities and proved the association of the solution.

The paper is organized as follows. After the introductory part, in the second section we give some basic preliminaries such as notations and definitions of the objects we shall work with. We also introduce different spaces of Colombeau algebra of generalized functions. In the third section we proved the existence and uniqueness of solution of transport equation with variable speed and initial data in the Colombeau algebra $\mathcal{G}$. Finally, in the fifth section we study the association.

2 Preliminaries

We use the following notations [2]:

$$\mathcal{A}_q = \left\{ \varphi \in \mathcal{D}(\mathbb{R}^n) / \int_{\mathbb{R}^n} \varphi(x)dx = 1, \int_{\mathbb{R}^n} x^\alpha \varphi(x)dx = 0 \text{ for } \quad 1 \le |\alpha| \le q \right\}$$

$q = 1, 2, \ldots$

$$\varphi_\varepsilon(x) = \frac{1}{\varepsilon^n} \varphi\left(\frac{x}{\varepsilon}\right) \quad \text{for} \quad \varphi \in \mathcal{D}(\mathbb{R}^n)$$

We denote by $\mathcal{E}(\mathbb{R}^n) = \{u : \mathcal{A}_1 \times \mathbb{R}^n \to \mathbb{C} /$ with $u(\varphi, x)$ is $\mathcal{C}^\infty$ to the second variable $x\}$

$$u(x, \varphi_\varepsilon) = u_\varepsilon(x) \quad \forall \varphi \in \mathcal{A}_1$$

$$\mathcal{E}_M(\mathbb{R}^n) = \{(u_\varepsilon)_{\varepsilon > 0} \subset \mathcal{E}(\mathbb{R}^n) / \forall K \subset\subset \mathbb{R}^n, \forall \alpha \in \mathbb{N}_0^n, \exists N \in \mathbb{N} \quad \text{such that} \quad \sup_{x \in K} |D^\alpha u_\varepsilon(x)| = \mathcal{O}\left(\varepsilon^{-N}\right) \text{ as } \varepsilon \to 0\}$$

$$\mathcal{N}(\mathbb{R}^n) = \{(u_\varepsilon)_{\varepsilon > 0} \subset \mathcal{E}(\mathbb{R}^n) / \forall K \subset\subset \mathbb{R}^n, \forall \alpha \in \mathbb{N}_0^n, \forall p \in \mathbb{N} \quad \text{such that} \quad \sup_{x \in K} |D^\alpha u_\varepsilon(x)| = \mathcal{O}(\varepsilon^p) \quad \text{as} \quad \varepsilon \to 0\}$$

The Colombeau algebra is defined as a factor set $\mathcal{G}(\mathbb{R}^n) = \mathcal{E}_M(\mathbb{R}^n) / \mathcal{N}(\mathbb{R}^n)$, where the elements of the set $\mathcal{E}_M(\mathbb{R}^n)$ are moderate while the elements of the set $\mathcal{N}(\mathbb{R}^n)$ are negligible [2].

Let $G_1, G_2 \in \mathcal{G}(\mathbb{R}^n)$ and $G_{1,\varepsilon}, G_{2,\varepsilon}$ their representatives respectively. We say that $G_1, G_2 \in \mathcal{G}(\mathbb{R}^n)$ are associated and we write $G_1 \approx G_2$, if for every $\varphi \in \mathcal{D}(\mathbb{R}^n)$

$$\lim_{\varepsilon \to 0} \int_{\mathbb{R}^n} (G_{1,\varepsilon} - G_{2,\varepsilon}) \varphi(x)dx = 0$$

Below is the statement of a problem called non-regular transport problem on the domain $\Omega = \mathbb{R}^+ \times \mathbb{R}$.

$$\begin{cases} \partial_t u(t,x) + c(t,x)\partial_x u(t,x) = f(t,x)u(t,x) + a(t,x), & (t,x) \in \mathbb{R}^{*,+} \times \mathbb{R} \\ u(0,x) = u_0(x), \quad x \in \mathbb{R} \end{cases} \tag{2}$$

with c, f and a are discontinuous functions such that $c > 0$.

3 Positive, Negative and Bounded Generalized Function

Definition 1 *[2,8].*
a- $U \in \mathcal{G}[\Omega]$ is said globally bound if it exists $c > 0$ and a representative $u \in \mathcal{E}_M[\Omega]$ of U and $N \in \mathbb{N}$ such as $\forall \phi \in \mathcal{A}_N$, we have

$$\sup_{y\in\Omega} |u(\phi_\varepsilon, y)| \leq c$$

when $\varepsilon \to 0$
b- $U \in \mathcal{G}[\Omega]$ is said to have local logarithmic growth if for any representative $u \in \mathcal{E}_M[\Omega]$ of U and for any compact K of Ω it exists $N \in \mathbb{N}$ such as $\forall \phi \in \mathcal{A}_N, \exists c > 0$, such as

$$\sup_{y\in\Omega} |u(\phi_\varepsilon, y)| \leq c \ln\left(\frac{1}{\varepsilon}\right)$$

when $\varepsilon \to 0$
c- $U \in \mathcal{G}[\Omega]$ is said to be strictly positive and we note $U > 0$ if for all compact K of Ω, there is a representative $u \in \mathcal{E}_M[\Omega]$ of $U, \exists N \in \mathbb{N}, c > 0, \forall \phi \in \mathcal{A}_N$, we have

$$u(\phi_\varepsilon, y) \geq c\varepsilon^N \quad \forall y \in K$$

when $\varepsilon \to 0$
d- $U \in \mathcal{G}[\Omega]$ is said to be strictly negative and we note $U < 0$ if for all compact K of Ω, there is a representative $u \in \mathcal{E}_M[\Omega]$ de $U, \exists N \in \mathbb{N}, c > 0, \forall \phi \in \mathcal{A}_N$, we have

$$u(\phi_\varepsilon, y) \leq -c\varepsilon^N \quad \forall y \in K$$

when $\varepsilon \to 0$

Proposition 1. *(c) and (d) of the previous definition do not depend on the chosen representative.*

Proof 1. *c- Let U an element of $\mathcal{G}[\Omega]$ strictly positive. Be a compact K of Ω. So there is a representative $u \in \mathcal{E}_M[\Omega]$ of $U, \exists N \in \mathbb{N}, \exists c > 0, \forall \phi \in \mathcal{A}_N$ such as*

$$u(\phi_\varepsilon, y) \geq c\varepsilon^N \quad y \in K$$

Let u_2 another representative of U. So

$$u - u_2 \in \mathcal{N}[\Omega]$$

i.e.

$$u(\phi_\varepsilon, y) - u_2(\phi_\varepsilon, y) < \varepsilon^q \quad \forall q$$

so

$$\begin{aligned} u_2(\phi_\varepsilon, y) &> -\varepsilon^q + u(\phi_\varepsilon, y) \\ &> -\varepsilon^q + c\varepsilon^N \\ &> c\varepsilon^N\left(1 - \frac{\varepsilon^{q-N}}{c}\right) \end{aligned}$$

By crossing the limit $q \longrightarrow +\infty$, so

$$u_2(\phi_\varepsilon, y) > c\varepsilon^N, \quad y \in K$$

d- Let U an element of $\mathcal{G}[\Omega]$, suppose that U is strictly negative. Be a compact K of Ω, So there is a representative $u \in \mathcal{E}_M[\Omega]$ of $U, \exists N \in \mathbb{N}, \exists c > 0, \forall \phi \in \mathcal{A}_N$ such that

$$U(\phi_\varepsilon, y) < -c\varepsilon^N \quad y \in K$$

Let u_2 another representative of U. So

$$u - u_2 \in \mathcal{N}[\Omega]$$

thus

$$u_2(\phi_\varepsilon, y) - u(\phi_\varepsilon, y) < \varepsilon^q$$

so

$$\begin{aligned} u_2(\phi_\varepsilon, y) &< \varepsilon^q + u(\phi_\varepsilon, y) \\ &< \varepsilon^q - c\varepsilon^N \\ &< -c\varepsilon^N\left(1 - \frac{\varepsilon^{q-N}}{c}\right) \end{aligned}$$

By crossing the limit $q \longrightarrow +\infty$, so

$$u_2(\phi_\varepsilon, y) < -c\varepsilon^N, \quad y \in K$$

Proposition 2. *(a) of the previous definition does not depend on the chosen representative.*

Proof 2. *Let u_2 another representative of U. So*

$$u - u_2 \in \mathcal{N}[\Omega]$$

then

$$\begin{aligned} |u_2(\phi_\varepsilon, y) - u(\phi_\varepsilon, y)| &< \varepsilon^q \quad \forall y \in \Omega \quad \forall q \\ |u_2(\phi_\varepsilon, y)| - |u(\phi_\varepsilon, y)| &< \varepsilon^q \\ |u_2(\phi_\varepsilon, y)| &< \varepsilon^q + |u(\phi_\varepsilon, y)| \end{aligned}$$

$|u_2(\phi_\varepsilon, y)| < \varepsilon^q + c$ *By crossing the limit $q \longrightarrow +\infty$, we find*

$$|u_2(\phi_\varepsilon, y)| < c \quad \forall y \in \Omega$$

4 Existence and Uniqueness of the Generalized Solution

Theorem 1. *We assume that c is globally bounded such that $c > 0, \partial_x c$ and f are with local logarithmic growth. So for an initial data $u_0 \in \mathcal{G}[\mathbb{R}]$ and a element of $\mathcal{G}[\Omega]$, the problem (1) admits a unique solution $u \in \mathcal{G}[\Omega]$.*

Proof 3. ***Existence:***
The transport problem with the representatives is as follows:

$$\begin{cases} \partial_t u_\varepsilon(t,x) + c_\varepsilon(t,x)\partial_x u_\varepsilon(t,x) = f_\varepsilon(t,x)u_\varepsilon(t,x) + a_\varepsilon(x,t), (t,x) \in \mathbb{R}^{+,*} \times \mathbb{R} \\ u_\varepsilon(0,x) = u_0(x), \quad x \in \mathbb{R} \end{cases} \tag{3}$$

This problem admits a unique class solution C^∞.
By hypothesis c is globally bounded, then $\exists M > 0$ such as:

$$|c_\varepsilon(x,t)| \leq M, \quad \forall (x,t) \in \Omega$$

so

$$\left|\frac{d\lambda_\varepsilon(x,t,s)}{ds}\right| \leq M, \quad \forall (x,t) \in \Omega$$

with $\lambda_\varepsilon(x,t,.)$ the characteristic curve corresponding to c_ε issue of the point (x,t), by drawing the lines passing through the point (x,t) and slope M and $-M$, we can determine a domain (compact of Ω) of determination of the solution that does not depend on ε.

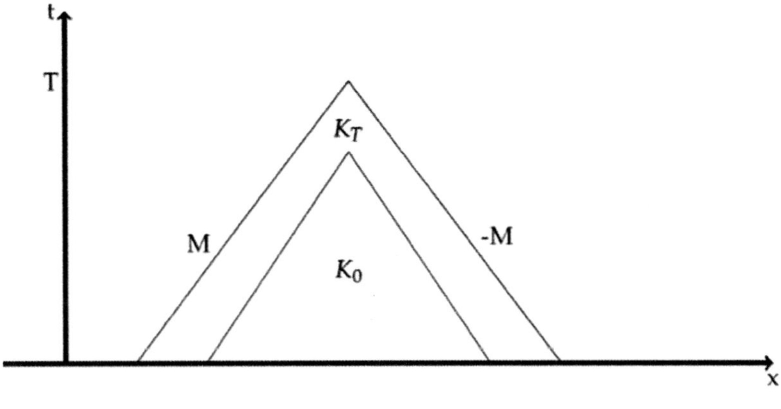

Fig. 1. .

In this case, for all $(x,t) \in K_T$, the characteristic curves resulting from this point remain in K_T, Furthermore:

$$u_{i,\varepsilon}(x,t) = u_{0,\varepsilon}(\lambda_\varepsilon(x,t,0)) + \int_0^t f_\varepsilon(\lambda_\varepsilon(x,t,s),s)\, u_\varepsilon(\lambda_\varepsilon(x,t,s),s)\, ds$$
$$+ \int_0^t a_{i,\varepsilon}(\lambda_\varepsilon(x,t,s),s)\, ds$$

$$|u_\varepsilon(x,t)| \leq \sup_{x \in K_0} |u_{0,\varepsilon}(x)| + \int_0^T \sup_{(t,x)\in K_T} |f_\varepsilon(t,x)| \sup_{K_s} |u_\varepsilon(t,x)|\, ds$$
$$+ \int_0^T \sup_{(t,x)\in K_T} |a_\varepsilon(t,x)|\, ds$$
$$\sup_{(t,x)\in K_T} |u_\varepsilon(t,x)| \leq \sup_{x\in K_0} |u_{0,\varepsilon}(x)| + T \sup_{(t,x)\in K_T} |a_\varepsilon(t,x)|$$
$$+ \int_0^t \sup_{(t,x)\in K_T} |f_\varepsilon(t,x)| \sup_{(t,x)\in K_s} |u_\varepsilon(t,x)|\, ds$$

Apply Gronwall's lemma to the function $s \to \sup_{(t,x)\in K_s} |u_\varepsilon(t,x)|$

$$\sup_{(t,x)\in K_T} |u_\varepsilon(t,x)| \leq \left[\sup_{x\in K_0} |u_{0,\varepsilon}(x)| + T \sup_{(t,x)\in K_T} |a_\varepsilon(t,x)|\right] \times \exp\left(T \sup_{(t,x)\in K_T} |f_\varepsilon(t,x)|\right)$$

As f has local logarithmic growth, then

$$\sup_{(t,x)\in K_T} |u_\varepsilon(t,x)| = \mathscr{O}\left(\varepsilon^{-N}\right), \quad N \in \mathbb{N}$$

Now let's apply the operator ∂_x on Eq. (2)

$$\begin{cases} \partial_t(\partial_x u_\varepsilon(t,x)) + c_\varepsilon(t,x)\partial_x\partial_x u_\varepsilon(t,x) = f_\varepsilon(t,x)\partial_x u_\varepsilon(t,x) - \partial_x c_\varepsilon(t,x)u_\varepsilon(t,x) \\ \qquad + \partial_x f_\varepsilon(t,x)u_\varepsilon(t,x) + \partial_x a_\varepsilon(x,t), (t,x) \in \mathbb{R}^{+,*} \times \mathbb{R} \\ \partial_x u_\varepsilon(0,x) = u_0'(x), x \in \mathbb{R} \end{cases} \tag{4}$$

then

$$\begin{aligned} \partial_x u_\varepsilon(t,x) = & u_{0,\varepsilon}'\left(\lambda_\varepsilon(x,t,0)\right) \\ & + \int_0^t \partial_x f_\varepsilon\left(\lambda_\varepsilon(x,t,s),s\right) u_\varepsilon\left(\lambda_\varepsilon(x,t,s),s\right) ds \\ & + \int_0^t f_\varepsilon\left(\lambda_\varepsilon(x,t,s),s\right) \partial_x u_\varepsilon\left(\lambda_\varepsilon(x,t,s),s\right) ds \\ & - \int_0^t \partial_x c_\varepsilon\left(\lambda_\varepsilon(x,t,s),s\right) u_\varepsilon\left(\lambda_\varepsilon(x,t,s),s\right) ds \\ & + \int_0^t \partial_x a_{i,\varepsilon}\left(\lambda_\varepsilon(x,t,s),s\right) ds \end{aligned}$$

$$|\partial_x u_\varepsilon(x,t)| \leq \sup_{x\in K_0} \left|u_{0,\varepsilon}'(x)\right| + \int_0^T \sup_{(t,x)\in K_T} |\partial_x f_\varepsilon(t,x)| \sup_{(t,x)\in K_T} |u_\varepsilon(t,x)| ds$$

$$\begin{aligned} & + \textstyle\int_0^T \sup_{(t,x)\in K_T} |f_\varepsilon(t,x)| \sup_{(t,x)\in K_s} |\partial_x u_\varepsilon(t,x)| ds \\ & + \textstyle\int_0^T \sup_{(t,x)\in K_T} |\partial_x c_\varepsilon(t,x)| \sup_{(t,x)\in K_T} |u_\varepsilon(t,x)| ds \\ & + \textstyle\int_0^T \sup_{(t,x)\in K_T} |\partial_x a_\varepsilon(t,x)| ds \end{aligned}$$

Apply Gronwall's lemma to the function $s \to \sup_{(t,x)\in K_s} |\partial_x u_\varepsilon(t,x)|$;

$$\begin{aligned} \sup_{K_T} |\partial_x u_\varepsilon(t,x)| \leq & \left[\sup_{x\in K_0} \left|u_{0,\varepsilon}'(x)\right| + T \sup_{(t,x)\in K_T} |\partial_x f_\varepsilon(t,x)| \sup_{(t,x)\in K_T} |u_\varepsilon(t,x)|\right. \\ & \left. + T \sup_{K_T} |\partial_x c_\varepsilon(t,x)| \sup_{(t,x)\in K_T} |u_\varepsilon(t,x)| + T \sup_{(t,x)\in K_T} |\partial_x a_\varepsilon(t,x)|\right] \\ & \times \exp\left(T \sup_{(t,x)\in K_T} |f_\varepsilon(t,x)|\right) \end{aligned}$$

As f has local logarithmic growth and (u_ε) is moderate, so

$$\sup_{(t,x)\in K_T} |\partial_x u_\varepsilon(t,x)| = \mathcal{O}\left(\varepsilon^{-N}\right), \quad N \in \mathbb{N}$$

By doing the same reasoning, we find that for everything $m \in \mathbb{N}$ and $m \geq 2$

$$\sup_{(t,x)\in K_T} |\partial_x^m u_\varepsilon(t,x)| = \mathcal{O}\left(\varepsilon^{-N}\right), \quad N \in \mathbb{N}$$

on the other hand, we have

$$\partial_t u_\varepsilon(t,x) = -c_\varepsilon(t,x)\partial_x u_\varepsilon(t,x) + f_\varepsilon(t,x)u_\varepsilon(t,x) + a_\varepsilon(x,t)$$

as f has local logarithmic growth and $(\partial_x u_\varepsilon)$ is moderate, then

$$\sup_{(t,x)\in K_T} |\partial_t u_\varepsilon(t,x)| = \mathcal{O}\left(\varepsilon^{-N}\right), \quad N \in \mathbb{N}$$

We also have

$$\begin{aligned}
\partial_t\partial_x u_\varepsilon(t,x) &= -\partial_x c_\varepsilon(t,x)\partial_x u_\varepsilon(t,x) - c_\varepsilon(t,x)\partial_x^2 u_\varepsilon(t,x)\\
&\quad + \partial_x f_\varepsilon(t,x)u_\varepsilon(t,x) + f_\varepsilon(t,x)\partial_x u_\varepsilon(t,x) + \partial_x a_\varepsilon(x,t)\\
\partial_t^2 u_\varepsilon(t,x) &= -\partial_t c_\varepsilon(t,x)\partial_x u_\varepsilon(t,x) - c_\varepsilon(t,x)\partial_t\partial_x u_\varepsilon(t,x)\\
&\quad + \partial_t f_\varepsilon(t,x)u_\varepsilon(t,x) + f_\varepsilon(t,x)\partial_t u_\varepsilon(t,x) + \partial_t a_\varepsilon(x,t)
\end{aligned}$$

etc. ....
Hence, for any derivative operator $\partial_t^n\partial_x^m, \exists N \in \mathbb{N}$

$$\sup_{(t,x)\in K_T} |\partial_t^n\partial_x^m u_\varepsilon(t,x)| = \mathcal{O}\left(\varepsilon^{-N}\right)$$

then

$$u \in \mathcal{G}(\Omega)$$

Uniqueness:
Suppose that problem (2) admits two solutions $u, v \in \mathcal{G}[\Omega]$ so they exist $d_{0,\varepsilon} \in \mathcal{N}[\mathbb{R}]$ such as:

$$\begin{cases}
\partial_t(u_\varepsilon(t,x) - v_\varepsilon(t,x)) + \lambda_{i\varepsilon}(x,t)\partial_x(u_\varepsilon(t,x) - v_\varepsilon(t,x)) = f_\varepsilon(x,t)(u_\varepsilon(x,t)\\
-v_\varepsilon(x,t))\\
u_\varepsilon(x,0) - v_\varepsilon(x,0) = d_{0,\varepsilon}(x)
\end{cases}$$

$$
\begin{aligned}
u_\varepsilon(t,x) - v_\varepsilon(t,x) =& d_{0,\varepsilon}\left(\lambda_\varepsilon(x,t,0)\right) \\
&+ \int_0^t f_\varepsilon\left(\lambda_\varepsilon(x,t,s),s\right)\left(u_\varepsilon\left(\lambda_\varepsilon(x,t,s),s\right) - v_\varepsilon\left(\lambda_\varepsilon(,,s),s\right)\right) ds \\
|u_\varepsilon(t,x) - v_\varepsilon(t,x)| \leq& \sup_{x\in K_0} |d_{0,\varepsilon}(x)| \\
&+ \int_0^T \sup_{(t,x)\in K_T} |f_\varepsilon(t,x)| \sup_{(t,x)\in K_s} |u_\varepsilon(t,x) - v_\varepsilon(t,x)| \, ds
\end{aligned}
$$

Apply Gronwall's lemma to the function $s \to \sup_{(t,x)\in K_s} |u_\varepsilon(t,x) - v_\varepsilon(t,x)|$

$$
\begin{aligned}
\sup\nolimits_{(t,x)\in K_T} |u_\varepsilon(t,x) - v_\varepsilon(t,x)| \leq& \left(\sup\nolimits_{x\in K_0} |d_{0,\varepsilon}(x)|\right) \\
&\times \exp\left(T \sup\nolimits_{(t,x)\in K_T} |f_\varepsilon(t,x)|\right)
\end{aligned}
$$

as f has local logarithmic growth, then

$$
\sup_{(t,x)\in K_T} |u_\varepsilon(t,x) - v_\varepsilon(t,x)| = \mathcal{O}(\varepsilon^q), \quad \forall q \in \mathbb{N}
$$

For the other derivatives, it is the same as the first part of the proof of the theorem. Consequently the problem (1) admits a unique solution $u \in \mathcal{G}[\Omega]$.

5 Application

We consider the following problem which presents the propagation of a wave in a discontinuous medium:

$$
\begin{cases} \partial_t u(t,x) + c(t,x)\partial_x u(t,x) = 0, & (t,x) \in \mathbb{R}_{+,*} \times \mathbb{R} \\ u(0,x) = u_0(x), & x \in \mathbb{R} \end{cases} \tag{5}
$$

with

$$
c(t,x) = \begin{cases} c_L, \ x \leq x_0 \\ c_R, \ x > x_0 \end{cases}
$$

and u_0 is a continuous function almost everywhere, and zero in the neighborhood of 0. If we set a condition of transition to x_0 (continuity of u on point x_0), then the solution of the problem (4) is given by:

$$
u(t,x) = u_0(\lambda(t,x,0))
$$

with λ the characteristic curve resulting from the point (t,x). is sloping c (Fig. 2)

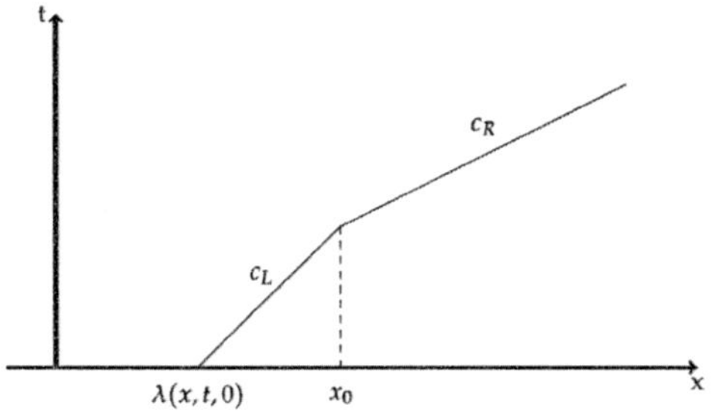

Fig. 2. .

$c \in L^\infty(\Omega)$, then there is $C \in \mathscr{G}[\Omega]$ such as $C \approx c$, c is globally bounded and $\partial_x C \in \mathscr{N}[\Omega]$ local logarithmic growth, and according to Theorem 1 problem 4 admits a unique solution $U = [(u_\varepsilon)] \in \mathscr{G}[\Omega]$, with:

$$\begin{cases} \partial_t u_\varepsilon(t,x) + c_\varepsilon(x)\partial_x u_\varepsilon(t,x) = 0, \ (t,x) \in \mathbb{R}^{*+} \times \mathbb{R} \\ u_\varepsilon(0,x) = u_{0,\varepsilon}(x), \qquad x \in \mathbb{R} \end{cases}$$

Take:

$$\lambda_\varepsilon = \lambda * \phi_{\eta_\varepsilon}$$

with $\phi \in \mathscr{D}(\mathbb{R}_+)$ such as:

$$\int_{\mathbb{R}_+} \phi(x)dx = 1 \quad \text{supp}\,(\phi_{\eta_\varepsilon}) \subset \ \ ]x_0 - \eta_\varepsilon, x_0 + \eta_\varepsilon[\quad \eta_\varepsilon = |\log\varepsilon|^{-1}$$

We pose:

$$u_\varepsilon(t,x) = u_{0,\varepsilon}(\lambda_\varepsilon(t,x,0))$$

To show that $U \approx u$, just prove that:

$$\lim_{\varepsilon\to 0} \int_\Omega (u_{0,\varepsilon}(\lambda_\varepsilon(t,x,0)) - u_0(\lambda(t,x,0)))\,\psi(t,x)dtdx = 0$$

for all $\psi \in \mathscr{D}(\Omega)$
We have

$$\begin{aligned} &\int_\Omega (u_{0,\varepsilon}(\lambda_\varepsilon(t,x,0)) - u_0(\lambda(t,x,0)))\,\psi(t,x)dtdx = \\ &\int_\Omega (u_{0,\varepsilon}(\lambda_\varepsilon(t,x,0)) - u_0(\lambda_\varepsilon(t,x,0)))\,\psi(t,x)dtdx \\ &+ \int_\Omega (u_0(\lambda_\varepsilon(t,x,0)) - u_0(\lambda(t,x,0)))\,\psi(t,x)dtdx \end{aligned}$$

But

$$\begin{aligned} &\int_\Omega (u_{0,\varepsilon}(\lambda_\varepsilon(t,x,0)) - u_0(\lambda_\varepsilon(t,x,0)))\,\psi(t,x)dtdx = \\ &\qquad \int_\Omega (u_{0,\varepsilon} - u_0)(\lambda_\varepsilon(t,x,0))\,\psi(t,x)dtdx \\ &\qquad \le \sup_{x\in\mathbb{R}} |u_0 * \phi_\varepsilon - u_0| \left| \int_{\text{supp}\,p(\phi)} \psi(t,x)dtdx \right| \end{aligned}$$

so

$$\lim_{\varepsilon\to 0} \int_\Omega (u_{0,\varepsilon}(\lambda_\varepsilon(t,x,0)) - u_0(\lambda(t,x,0)))\,\psi(t,x)dtdx = 0$$

To prove that

$$\lim_{\varepsilon\to 0}\int_{\Omega}(u_0(\lambda_\varepsilon(t,x,0))-u_0(\lambda(t,x,0)))\,\psi(t,x)dtdx=0$$

just show that

$$\lim_{\varepsilon\to 0}(\lambda_\varepsilon(t,x,0)-\lambda(t,x,0))=0$$

We know that c is globally bounded.
So there is $M>0$ such as

$$\sup_{(t,x)\in\Omega}|c_\varepsilon(t,x)|<M$$

So, we can frame the curve λ_ε between two broken curves (see Fig. 3), and we take the intersection of these two curves with the axis (ox), given by:

$$x_1=\tfrac{1}{c_L}[-2M\eta_\varepsilon+c_R(x-x_0-\eta_\varepsilon)]-\tfrac{t}{c_L}-\eta_\varepsilon+x_0$$
$$x_2=\tfrac{1}{c_L}[2M\eta_\varepsilon+c_R(x-x_0-\eta_\varepsilon)]-\tfrac{t}{c_L}-\eta_\varepsilon+x_0$$

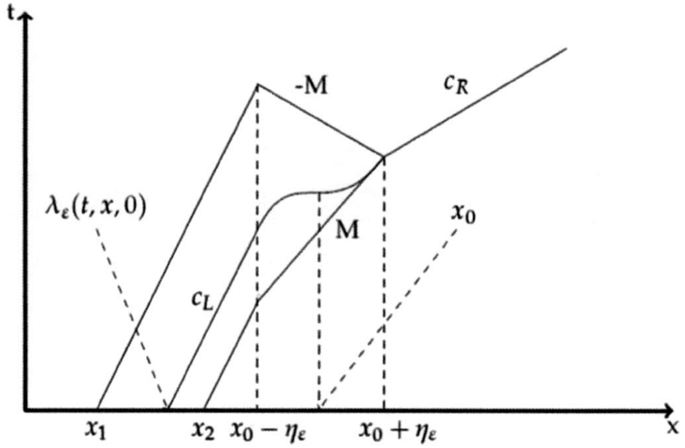

Fig. 3. .

such that

$$x_1\le\lambda_\varepsilon(t,x,0)\le x_2$$

Since

$$\lim_{\varepsilon\to 0}\lambda_\varepsilon(t,x,0)=\frac{c_R}{c_L}(x-x_0)-\frac{t}{c_L}+x_0$$
$$=\lambda(x,t,0)$$

so

$$U\approx u$$

References

1. Bourgain, J.: Global Solutions of Nonlinear Schrödinger Equations, vol. 46. Colloquium Publications, AMS (1999)
2. Colombeau, J.F.: Elementary Introduction in New Generalized Functions. North Holland, Amsterdam (1985)
3. Hermann, R., Oberguggenberger, M.: Ordinary differential equations and generalized functions. In: Grosser, M., Hörmann, G., Kunzinger, M., Oberguggenberger, M. (eds.) Proceedings of Workshop: Nonlinear Theory of Generalized Functions. Research Notes in Mathematical Series, October-December 1997. E. Schrödinger Inst, Chapman and Hall/CRC, Vienna (1999)
4. Nakamura, S.: Lectures on Schrödinger operators. Lectures given at the University of Tokyo, October 1992, February 1993
5. Oberguggenberger, M.: Generalized functions in nonlinear models a survey. Nonlinear Anal. **47**, 5029–5040 (2001)
6. Rajterc Ciric, D., Stojanovic, M.: Convolution-type derivatives and transforms of colombeau generalized stochastic processes. Integr. Transforms Spec. Funct. **22**(4–5), 319–326 (2011)
7. Reed, M., Simon, B.: Methods of Modern Mathematical Physics, II: Fourier Analysis, Self-adjointness. Academic Press, NewYork (1975)
8. Stojanovic, M.: Extension of Colombeau algebra to derivatives of arbitrary order D^{α}; $\alpha \in \mathbb{R}^{+} \cup \{0\}$: application to ODEs and PDEs with entire and fractional derivatives. Nonlinear Anal. **71**, 5458–5475 (2009)
9. Stojanovic, M.: Fondation of the fractional calculus in generalized function algebras. Anal. Appl. **10**(4), 439–467 (2012)
10. Stojanovic, M.: Nonlinear Schrödinger equation with singular potential and initial data. Nonlinear Anal. **64**, 1460–1474 (2006)

Modeling and Comparative Application of Fuzzy Logic and Artificial Neural Network in the Systemic Control of a Wind Turbine Using DFIG

Boaz Wadawa(✉), Youssef Errami, Abdellatif Obbadi, and Smail Sahnoun

Laboratory: Electronics, Instrumentation and Energy, Department of Physical, Faculty of Science, University Chouaib Doukkali, El-Jadida, Morocco
booz_wadawa@yahoo.fr, {errami.y,obbadi.a,sahnoun.s}@ucd.ac.ma

Abstract. This article discusses the applications of Artificial Neural Network (ANN) and Fuzzy Logic Control (FLC) strategies in a control context of a grid connected Wind Energy Conversion System (WECS), using a Doubly-Fed Induction Generator (DFIG). To rigorously explore the performance of the controllers, a systemic approach based on an ideal model and a nominal model of the wind turbine is used for the comparative analysis between PI, FLC and ANN. It turns out that the PI is less efficient than ANN and FLC. It turns out that the PI is less efficient than ANN and FLC. Whereas, for a high level of control, the FLC has better performance than the ANN. For a low level, the performance of the ANN is very slightly superior to the FLC. In addition, one notes a good rejection of disturbance and a good robustness with the FLC compared to the ANN. The latter has a Total Harmonic Distortion (THD) and cos $\phi \approx 1$ slightly better than the FLC. In short, the ANN and FLC present great advantages for the WECS and the results obtained are satisfactory. The platform used for modeling and simulation studies is MATLAB/Simulink.

Keywords: Wind Energy Conversion System · Fuzzy Logic Control (FLC) · Artificial Neural Network (ANN) · DFIG · Power grid

1 Introduction

Nowadays, in the renewable energy development sector, the wind power industry based on the use of DFIG is becoming more and more popular in the WECS [1, 2]. Compared to other types of generators, DFIG has the main advantages such as, multiple electrical configurations for a wide field of application, the production of specific or electrical power a little higher, and the reduction of the cost of conversion of investment engendered [1–4]. However, DFIG-based WECS are generally characterized by high non-linearity caused by disturbances from the randomness of the wind and the power grid; and/or interactions between different internal elements of the WECS [5]. This sensitivity to disturbances, when not properly controlled, can cause wind systems to shut down and/or

S. Melliani and O. Castillo (Eds.): ICPAMS 2021, LNNS 476, pp. 112–137, 2023.
https://doi.org/10.1007/978-3-031-12416-7_11

eventually malfunction. Therefore, it limits the use of control approaches based on linear models such as PI (Proportional Integral). As a result, several research works on wind turbines have been directed towards the synthesis of control family approaches that offer superior performance to conventional PI. Three main control families applied to wind turbines can be distinguished as follows:

- In the family of analog controls called optimal or advanced controls, we note: the sliding mode of the 1st or 2nd order, backstepping, H_2/H_∞ [6–8].
- Digital controls based on Artificial Intelligence (AI) controllers like, Particle Swarm Optimization (PSO), Artificial Neural Network (ANN), Fuzzy Logic (FLC) etc. [9–11].
- Mixed or adaptive controls. They are: Fuzzy-PI, PSO-GA and ANN-Fuzzy etc. [12–14].

Most of these controls show better performance than the PI in terms of robustness to parametric and model uncertainties, and good stability in the fast and accurate tracking of a given set point. However, these controls present problems such as the complexity of their mathematical models, the difficulty of theoretical and practical implementation, the additional cost generated and the computation time are relatively high. Furthermore, in the literature, the possible implementations of experimental or practical validation studies are very often omitted. In view of all these analyses of the control literature, controllers such as PI, FLC and ANN will receive our attention in this article, firstly because they are increasingly present in high-level research work, both in the laboratory and in the factory [13–16]. This can be justified, as the PI has advantages that allow it to dominate linear and non-linear servo systems, with relatively good performance and is simple to implement [17]. In addition, to validate the effectiveness of advanced models and/or controls, most of the work in the wind turbine sector uses the performance of the PI based system as a reference [6–16]. On the other hand, the strong non-linearity and the high level of complexity of the dynamic equation models of the systems on the one hand, and the difficulty to determine a precise criterion to be minimized on the other hand, have led several researchers to turn to the ANN and FLC strategies. The latter have the advantages of approximating any non-linear function without knowledge of the internal mathematical model of the system under study. Secondly, they adapt easily to changes in the system, and are less costly in terms of implementation [16, 18]. Moreover, behaviours related to the control of systems are less bad in case of a small amount of training data for the ANN, and similarly with a reduced number of inference rules for the FLC [10–18]. The ANN and FLC are also recognized as being robust to the PI [12, 14, 19, 20]. However, there is not yet a better methodology for the concrete design of FLC and ANN. While, in the FLC the difficulty to obtain high control accuracy or to eliminate the static error of the system; as well as the problem to predict the robustness and stability of the system to be controlled are more delicate [21–23]. In view of the minimal disadvantages for achieving increasingly better performance in the control of WECS with DFIG, the current trend is often to use ANN and FLC. In most cases, the implementation of controllers is based on the principle of vector control incorporating the power loops of the DFIG. Recently, some works prefer the joint use of ANN and FLC, or with other types of controllers. It can be shown that FL-PI and FL-ANN are superior to PI respectively [12, 14]. For these mixed controls, the difficulty increases when determining the optimal parameters or when setting or choosing the optimal ranges of variation of the parameters [16, 20, 22]. While in other works, systems with ANN or FL generally follow simplistic approaches and are not always studied especially the whole WECS. But we note nevertheless, the superiority of ANN over PI [24]. And a superiority of the FL over the PSO etc. [25]. Clearly, all these works can be criticized for insufficient exploitation of the great advantages offered by ANNs and LFs on the one hand. And on the other hand, the study and mastery of the dynamic behaviour of the wind system as a whole is still incomplete. However, the ANN and FL controllers have proved to be interesting and need to be sufficiently exploited in this paper, in order to make their applications more beneficial to the control of a wind turbine chain. To this end, the main objective of our work is to explore the relative effectiveness of ANN and FL, in order to obtain better performance in the systematic control of WECS. Specifically, this paper discusses in detail the application of ANN and FL strategies to a grid-connected wind power system of about 2 MW using a DFIG. All this is done in order to better evaluate and assess the dynamic and/or static performance of the system under study. In addition to ensuring that the system functions optimally, taking into account the uncertainties. Therefore, to obtain a quantity and quality of energy production to be injected into the electrical grid. As shown in the proposed wind power system in Fig. 1, the control strategy consists of ensuring:

- The maximum power extracted from the turbine by a Maximum Power Point Tracker (MPPT) control named Control (I). The latter uses the wind speed (V_{wind}) and the electromagnetic torque (W_{mec}) as input and output quantities respectively.
- Decoupling and control of active (P_s) and reactive (Q_s) power with two controllers (II) and (III) respectively. These controllers use P_s and Q_s as inputs and their respective references P_{sref} and Q_{sref}. Then, they deliver as output the quadratic (V'_{rq}) and direct (V'_{rd}) control voltages to be applied to Rotor Side Converter (RSC) to the DFIG.
- Control (IV), regulates the DC-BUS voltage V_{dc} around the capacitor C.
- The control of active (i_{fd}) and reactive (i_{fq}) filter currents via two controllers (V) and (VI) respectively. The direct control voltages V'_{fd} from (VI) and quadratic V'_{fq} from (V) are applied to the filter via the static converter on the grid side or GSC.

Indeed, taking into account the influences of disturbances due to the wind, the uncertainties of models and parameters on a proposed wind system, the performance evaluation criteria such as: rejection of disturbance and/or overrund, precision or static error, the speed, the amplitude of fluctuations or distortion, as well as the power factor will be considered in the observation of the quantities to be controlled. In addition, to better assess the efficiency of the controllers on the proposed wind power system, the performance of the FLC will be compared to that of the ANN and to conventional PI. MATLAB/Simulink Version 2013, is the platform used for the validation studies of the modeling and simulation results of the proposed wind power system. The document is structured as follows:

- The first part concerns the methodology: it details the mathematical simulation models of the different elements of the wind energy system such as the turbine, the DFIG, the Rotor Side Converter (RSC) and Grid Side Converter (GSC) Pulse-Width Modulation (PWM), the DC-BUS and the currents filter. In addition, the PI, FLC and ANN synthesis models are presented for each of the elements to be controlled in the system under consideration.

- The second part is the results and discussion phase of the simulation relating to the evaluation of the performance of the PI, FLC and ANN on the simplified DFIG model on the one hand and on the whole wind turbine chain with the real DFIG model on the other hand;
- The third part is devoted to conclusion.

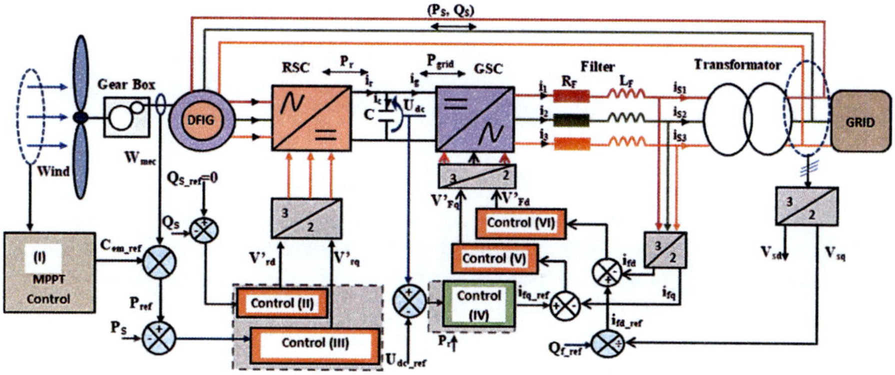

Fig. 1. Structure for controlling a wind chain on a DFIG-based electrical.

2 Mathematical Modeling of the Wind System

2.1 Wind Turbine Model

In general, a wind turbine is characterized by its mechanical efficiency, also known as the power coefficient C_p (λ, β). The latter is a function of the number of rotor blades, their geometric or aerodynamic shape, with an orientation angle or pitch angle (β); it also depends on the wind conditions called relative wind speed or specific speed (λ). Its expression is:

$$\lambda = \frac{\Omega_{\mathrm{turbine}} R}{V} \quad (1)$$

R: Radius of the rotor blades;
$\Omega_{\mathrm{turbine}}$: Turbine rotation speed;
V: Wind speed.

There are several mathematical approximation models of C_p (λ, β) [26]. The appropriate one for our study is:

$$C_p(\lambda, \beta) = C_1 \cdot \left(\frac{C_2}{\lambda_i} - C_3 \cdot \beta - C_4\right) \cdot \exp\left(-\frac{C_5}{\lambda_i}\right) + C_6 \cdot \lambda \quad (2)$$

With, $C_1 = 0.73$, $C_2 = 151$, $C_3 = 0.002$, $C_4 = 13.2$, $C_5 = 18.4$, $C_6 = 0$.
And,

$$\frac{1}{\lambda_i} = \frac{1}{\lambda + 0.08 \cdot \beta} \cdot \frac{0.035}{\beta^3 + 1} \tag{3}$$

The mechanical or aerodynamic power (Paero) extracted from the wind and transmitted to the turbine shaft is given as follows:

$$P_{aero} = \frac{1}{2} \cdot C_p(\lambda, \beta) \cdot \rho \cdot S \cdot V^3 \tag{4}$$

where, ρ: The density of the air ($\rho = 1.225\ kg/m^3$ at atmospheric pressure);
S: A surface swept by the propeller.
The model of the mechanical transmission can be summarized as follows:

$$J\frac{d\Omega_{mec}}{dt} = C_g - C_{em} - C_{vis} \tag{5}$$

With, C_{vis}: the viscous torque being proportional to the speed;
C_{em}: the electromechanical torque;
C_g: the total mechanical torque applied to the machine rotor;
Ω_{mec}: mechanical or angular speed of the rotor;
J: is the total inertia of the turbine [4].
In the case of the wind power system of Fig. 1, Eqs. 1–5 are used as in literature to model the turbine and establish its control model.

2.2 Generalized and Simplified Model of DFIG

The equivalent general dynamic model of DFIG in the system of axes (X, Y) that is linked to the rotating field is achieved by:

$$\begin{cases} U_{sx} = -r_s i_{sx} - L_s\frac{di_{sx}}{dt} + M\frac{di_{rx}}{dt} + \omega_{os} L_s i_{sy} - \omega_{os} M i_{ry} \\ U_{sy} = -r_s i_{sy} - L_s\frac{di_{sy}}{dt} + M\frac{di_{ry}}{dt} - \omega_{os} L_s i_{sx} - \omega_{os} M i_{rx} \\ U_{rx} = r_r i_{rx} + L_r\frac{di_{rx}}{dt} - M\frac{di_{sx}}{dt} - (\omega_{os} - \omega_r) L_r i_{ry} + (\omega_{os} - \omega_r) M i_{sy} \\ U_{ry} = r_r i_{ry} + L_r\frac{di_{ry}}{dt} - M\frac{di_{sy}}{dt} + (\omega_{os} - \omega_r) L_r i_{rx} - (\omega_{os} - \omega_r) M i_{sx} \end{cases} \tag{6}$$

where, U_{sx}, U_{sy} and U_{rx}, U_{ry} are respectively the expressions of the stator and rotor voltages of the reference frame (X, Y);
M_{sr} and M_{rs}: are the mutual inductances between the stator and rotor windings. The following relationship is generally considered: $M_{sr} = M_{rs} = M$;
r_r, r_s: The resistances of the rotor and stator windings;
L_r, L_s: The rotor and stator inductances;
ω_{0s} and ω_r: The angular rotational speeds of the field at the stator and rotor respectively;
i_{sx}, i_{sy} and i_{rx}, i_{ry}: The currents flowing in the stator and rotor respectively on the X and Y axes.

The electromagnetic torque of the machine is written [4]:

$$C_{em} = \frac{3}{2}p(\psi_{sx}i_{sy} - \psi_{sy}i_{sx}) \tag{7}$$

where, P is the number of pole pairs of the DFIG;

Ψ_{sx}, Ψ_{sy} are the expressions of the stator fluxes in the reference frame (X, Y).

The active and reactive stator and rotor powers are written as follows:

$$\begin{cases} P_s = U_{sx} \cdot i_{sx} + U_{sy} \cdot i_{sy} \\ Q_s = U_{sy} \cdot i_{sx} - U_{sx} \cdot i_{sy} \\ P_r = U_{rx} \cdot i_{rx} + U_{ry} \cdot i_{ry} \\ Q_r = U_{ry} \cdot i_{rx} - U_{rx} \cdot i_{ry} \end{cases} \tag{8}$$

The simplified DFIG model used in this document is based on the principle of vector control:

- For convenience, let us identify the notation (X, Y) with that of (d, q). Based on the orientation of the flow along the q-axis, we can write:

$$\begin{cases} \psi_{qs} = \psi_s = V_s\omega_s = cst \\ V_{ds} = 0 \end{cases} \tag{9}$$

- Assuming that the network is stable and of high power, we can write the simple voltage $V_s = cte$ and the frequency $f_s = cte$.

According to these assumptions, we obtain the so-called simplified mathematical model of the DFIG as follows:

$$\begin{cases} V_{rd} = \left[R_r + \left(L_r - \frac{M^2}{L_s}\right) \cdot s\right]i_{rd} - \omega_m\left(L_r - \frac{M^2}{L_s}\right)i_{rq} \\ V_{rq} = \left[R_r + \left(L_r - \frac{M^2}{L_s}\right) \cdot s\right]i_{rq} + \omega_m\left(L_r - \frac{M^2}{L_s}\right)i_{rd} + g\frac{MV_s}{L_s} \end{cases} \tag{10}$$

The objective being, among other things, to evaluate the dynamic performance of DFIG control in the context of the studied WECS, the generalized model of the DFIG of Eq. (8) is used in the form of a state matrix. Whereas, Eq. (10) of the simplified model of DFIG is generally associated with Eq. (8) for the study of the power control of an ideal wind system. The two case studies based on the generalized and simplified DFIG model are used in this work.

2.3 Inverter Modeling

The conversion of the energy from the alternating source to a continuous source and vice versa, on the rotor (RSC) and grid (GSC) side of the DFIG is provided here by an adjacently controlled inverter of the Pulse Width Modulation (PWM) type. The mathematical expression of the electrical model of the inverter is as follows [4]:

$$\begin{cases} V_{ab} = V_{an1} - V_{bn1} = E(f_1 - f_2) \\ V_{bc} = V_{bn1} - V_{cn1} = E(f_2 - f_3) \\ V_{ca} = V_{cn1} - V_{an1} = E(f_3 - f_1) \end{cases} \tag{11}$$

where, V_{ab}, V_{bc}, V_{ac}: the expressions of the voltages composed between the phases of the inverter;

Van, V_{bn}, V_{cn}: the simple voltages balanced between phase-to-neutral of the inverter;

f_1, f_2, f_3: the logic functions defining the switching states of the inverter;

E: the DC reference voltage at the input of the PWM converter.

2.4 DC Bus and Filter Models

Checking the DC bus voltage around the capacitance (C) in the Fig. 1 is based on the expression of the following power balance [4]:

$$P_R = P_c + P_g \tag{12}$$

where, P_c: the energy stored by the capacitor C;

P_g: the power transmitted or received to the grid by the inverter;

P_R: the active power of the bus on the rotor side of the DFIG.

Using the mesh law, the matrix of the voltages across the filter of R_F resistors and L_F inductors in Fig. 1 can be written as:

$$\begin{pmatrix} V_{F1} \\ V_{F2} \\ V_{F3} \end{pmatrix} = R_F \begin{pmatrix} i_1 \\ i_2 \\ i_3 \end{pmatrix} + L_F \frac{d}{dt} \begin{pmatrix} i_1 \\ i_2 \\ i_3 \end{pmatrix} + \begin{pmatrix} V_{ps1} \\ V_{ps2} \\ V_{ps3} \end{pmatrix} \tag{13}$$

V_{F123} and V_{ps123}: the three-phase input and output voltages at the filter terminals;

i_{123}: three-phase currents through the filter.

Applying Park's transform to Eq. (13), and then rearranging it, gives the equation for the control law of the active and reactive currents ifq and ifd of the filter.

$$\begin{cases} i_{fd} = \frac{1}{(R_F + L_F \cdot s)} \cdot \left(V_{Fd} + L_F \omega_s \cdot i_{fq} - V_{sd}\right) \\ i_{fq} = \frac{1}{(R_F + L_F \cdot s)} \cdot \left(V_{Fq} - L_F \cdot \omega_s \cdot i_{fd} - V_{sq}\right) \end{cases} \tag{14}$$

where, V_{fdq} and V_{sdq}, voltages at the filter terminals in reference frame (d, q).

The active and reactive powers are written as follows:

$$\begin{cases} P_f = V_s \cdot i_{fq} \\ Q_f = V_s \cdot i_{fd} \end{cases} \tag{15}$$

The power balance Eq. (12) helps us to establish around the capacitor (C) the coupling relationship between the machine and the electrical grid, as in Fig. 1. Among other things, the powers transmitted through the filter are defined by Eq. (15).

3 Controller Synthesis

In principle, the block diagram of a regulation law illustrated in Fig. 2, allows us to establish control relations between the transfer function of the PI corrector and the models of Eqs. (5), (10), (12) and (14) respectively. Therefore, the calculated values of the parameters of the proportional gain (K_P), the integral gain (K_I) and the response time (τ_r) of the PI are presented in the Table 3 in the appendix. However, for reasons of size of the article, these control relations relating to the various conventional PI will not be exposed.

3.1 Synthesis of the Fuzzy Regulator

Text The fuzzy logic, introduced by Lotfi Zadeh in 1965, will rapidly develop both in the modeling and in the control of complex systems [27]. This approach is mainly based on two concepts: the decomposition of a universe of discourse into one or more variables measured in the form of linguistic symbols, and rules derived from the expertise of the human operator. As early as 1974, E.H Mandani designed a fuzzy controller applied to a steam engine. To date, his model is the most widespread and will also be the subject of our choice in this article [28]. A fuzzy controller is usually represented by the structure of the Fig. 3.

The general principles of a fuzzy controller are defined as follows:

Fuzzification: this is the interface that allows you to convert physical input quantities into so-called fuzzy linguistic variables. The operation consists first of all in adapting the input ranges to the chosen universe of discourse. Then, in order to assign levels or degrees of linguistic variables to the input quantities, a choice are made among triangular, trapezoidal, Gaussian, etc. membership functions [29, 30];
Inference: this is the region where the decision-making rules are drawn up for each chosen membership function. A distinction is made between the methods of Mandani, Larsen and Sugeno etc. All of them are based on the use of programming operators such as AND, OR, IF, THEN etc. [29, 30];
Knowledge base: it is a database of fuzzy linguistic rules and data based on expert knowledge in the field of application and predictable results of the command [29, 30];
Defuzzification: it is the set of inverse operations of the fuzzification.

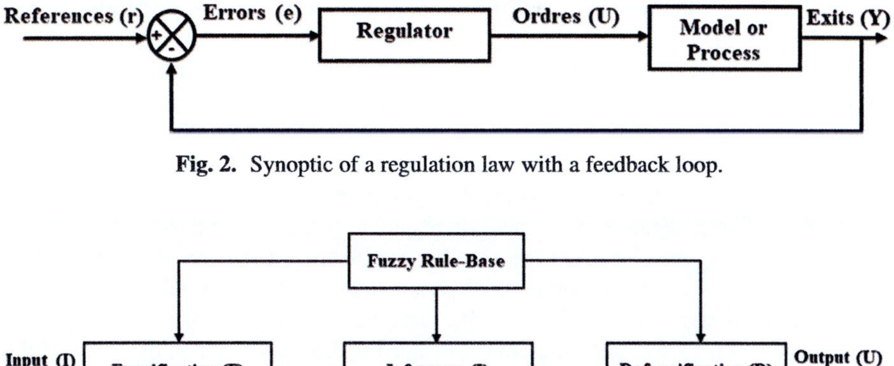

Fig. 2. Synoptic of a regulation law with a feedback loop.

Fig. 3. Configuration of general principles of fuzzy logic.

- ***Classic Fuzzy Controller Design Type1 (FLC)***

The Fig. 4, shows the main structure of the fuzzy controller type1, which is implemented for the control of the DFIG alone and for the control of the whole 2 MW wind generation system. As shown in Fig. 4, the Fuzzy Logic controller includes two normalized inputs, the error en (k) and its variation Δe_n (k), and a normalized variation output of the command U_n (k). The latter makes it possible to adjust at each instant (k) the command U (k) to be applied to the process or controlled system, in order to minimize the error e (k) between the set point Y * (k) and the response Y (k) of the system. Moreover, G_e and $G_{\Delta e}$ are the scaling or normalization factors of the error e (k) and its variation Δe (k) respectively. Whereas, G_n is also called the normalization gain of the control U (k) of the system. We define the error e (k) and its variation Δe (k) by the following relation [31, 32]:

$$\begin{cases} e(k) = Y^*(k) - Y(k) \\ \Delta e(k) = e(k) - e(k-1) \end{cases} \tag{16}$$

The three quantities e (k), Δe (k) and the command U (k) are standardized as follows:

$$\begin{cases} e_n(k) = G_e \cdot e(k) \\ \Delta e_n(k) = G_{\Delta e} \cdot \Delta e(k) \\ U_n(k) = U(k) \cdot G_u^{-1} \end{cases} \tag{17}$$

- ***Application of the FLC to DFIG control***

We consider the simplified DFIG model based on Eq. (10) to evaluate the efficiency of FLC, ANN and PI on the direct control of the active (P_s) and reactive (Q_s) power of the machine. The favorable study is that of an ideal wind power system based on the DFIG direct vector control structure shown in Fig. 5. The objective of the FLC control is to bring the input error (e) and the variation of the error (Δe) to an almost zero value; the degrees of membership likely to be occupied by intersections between the seven (07) rule levels of (e) and (Δe), give 49 combinations for 49 possible states. Depending on whether, for each of the possible membership states, there will be an automatic adjustment of the response (u) at the output of the fuzzy command. As shown in Fig. 6, the membership functions of (e) and (Δe) as input, are chosen identically and in a triangular shape in the interval [−1 1], with the seven (07) levels of rules or of degree belonging designated as follows: NG: big negative, NM: medium negative, NP: small negative, ZE: zero; PP: small positive, PM: medium positive and PG: big positive. Likewise, the variations of the controls (U) for the active and reactive powers are chosen in a triangular shape as in Fig. 6. Except that, the output of the fuzzy control (U_n (k)) will change according to the states of the inputs (e and Δe). In practice, to size the FLC of P_s and Q_s defined in Table 3 in the appendix, we based ourselves on the observations of the curves at the inputs (e) and (Δe), then of the response at the output of the regulation by PI on the quantities P_s and Q_s.

Therefore, we can choose a maximum value observed on the output response of the PI to assign it to U_n (k). The latter is then replaced in Eq. (17) to deduce the gain G_u, knowing that U (k) is set to 1 as the upper bound in Fig. 6. The same scenario will be repeated on the observations at the inputs (e) and (Δe) of the PI to determine G_e and $G_{\Delta e}$ respectively. Table 3 in the appendix gives a summary of the values of G_e, $G_{\Delta e}$, G_u, as well as those of the selected control intervals, $\Delta u \in [-U_{nmax}(k) + U_{nmax}(k)]$. The rule

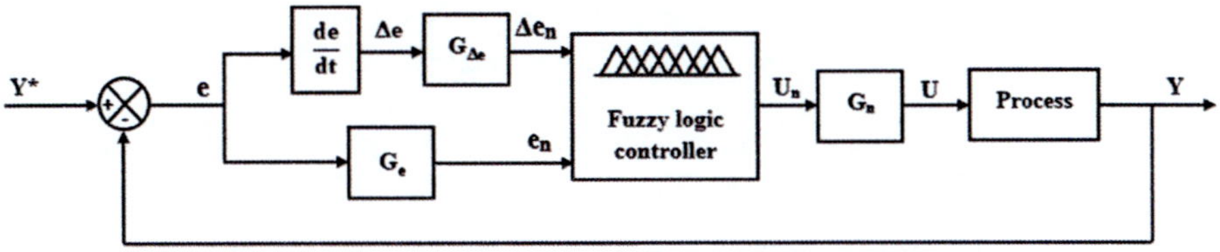

Fig. 4. Conventional fuzzy controller design structure.

table used for fuzzy interference is given below by Table 1. The filling of the latter is as follows, we align the seven (07) rule levels of (e) and (Δe) respectively, vertically at the extreme left and horizontally at the extreme top of Table 1. Then, the entire diagonal of Table 1 is filled with the zero level (ZE). Finally, going from left to right, we replace the boxes above (ZE) starting with NP towards NG. Then, below ZE, we start with PP towards PG.

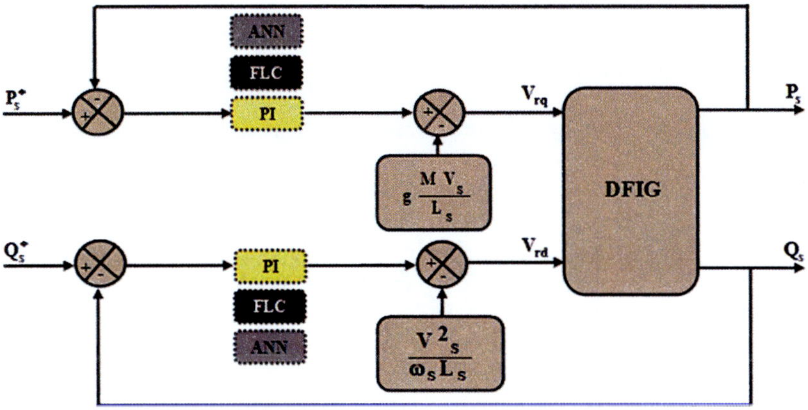

Fig. 5. DFIG direct vector power control model.

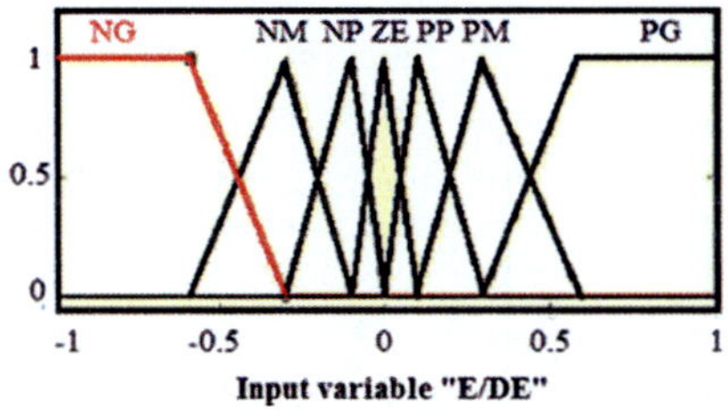

Fig. 6. The membership functions of the input variables e and Δe.

Table 1. Fuzzy inference rules of P_s and Q_s

P_s/Q_s		e(k)						
		NG	NM	NP	ZE	PP	PM	PG
Δe(k)	NG	NG	NG	NG	NG	NM	NP	ZE
	NM	NG	NG	NG	NM	NP	ZE	PP
	NP	NG	NG	NM	NP	ZE	PP	PM
	ZE	NG	NM	NP	ZE	PP	PM	PG
	PP	NM	NP	ZE	PP	PM	PG	PG
	PM	NP	ZE	PP	PM	PG	PG	PG
	PG	ZE	PP	PM	PG	PG	PG	PG

- ***Application of FLC to the 2 MW Wind Turbine Chain***

The entire wind energy system, comprising a total of six fuzzy controllers, leads us to search for a compromise between accuracy and the time needed to calculate or simulate the system. Consequently, the dimensioning of the FLC to control the quantities Ω_{mec}, P_s, Q_s, U_{dc}, i_{fd} and i_{fq} obeys the previous approach presented above in the section application of the FLC to the control of the DFIG. Except that here, we have limited the number of degrees of membership to three (03) noted, large negative (NG), zero (ZE) and large positive (PG). As a result, the membership functions (e) and (Δe) for these FLC are chosen identical as shown in Fig. 7. Then, the order variations (Δu) and the various normalization gains G_e, $G_{\Delta e}$ and G_u are determined for the six (06) FLC presented in Table 3 in the appendix. The rules used for the fuzzy inference of the six (06) FLC are given by the Table 2.

3.2 ANN Controller Synthesis

The concept of neural models dates back to the 1940s with the work of McCulloch and Pitts [16]. The latter defined the formal or artificial neuron model via simplified mathematical modeling, inspired by the functioning of the human brain [24]. In principle, as illustrated by Fig. 8, the mathematical representation of the biological neuron is assimilated to a threshold summing device with several inputs and a single output. This neural model proposed by Mc Culloch and Pitts is characterized by the following two main relationships:

$$Z_j(x) = -\theta_j + \sum\nolimits_{i=1}^{n} W_{ij} S_i(x) \tag{18}$$

$$S_j(x) = f\big(Z_j(x)\big) \tag{19}$$

where, S_i (x), S_j (x) and W_{ij} (x) are respectively, the ith input, the output of the neuron and the weight of the connection between neuron i and neuron j. Z_j (x) is the input function and θ_j is the bias or threshold of the neuron activation. f (.) is the activation function of the neuron, and can be step, sign and non-linear. Currently, ANN are among the major tools that have revolutionized AI. In various fields such as physics, chemistry, biology, finance etc. ANN have been successfully applied to model, predict, diagnose and for process control law synthesis [16].

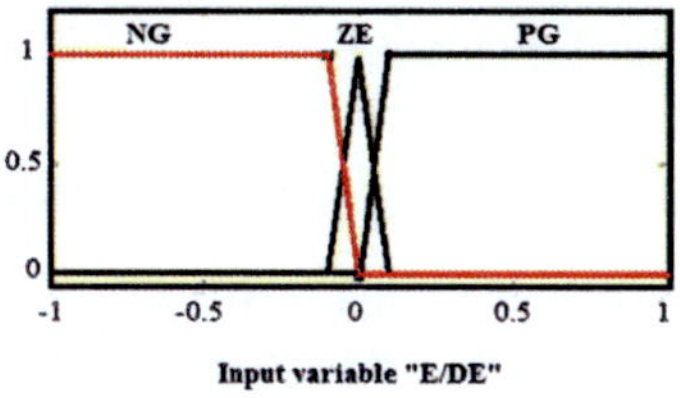

Fig. 7. The membership functions of the input variables e and Δe.

Table 2. Fuzzy inference rules of Ω_{mec}, P_s, Q_s, U_{dc} and i_{fdq}

Ω_{mec} /P_s/Q_s/U_{dc}/i_{fdq}		e(k)		
		NG	ZE	PG
Δe(k)	NG	NG	NG	ZE
	ZE	NG	ZE	PG
	PG	ZE	PG	PG

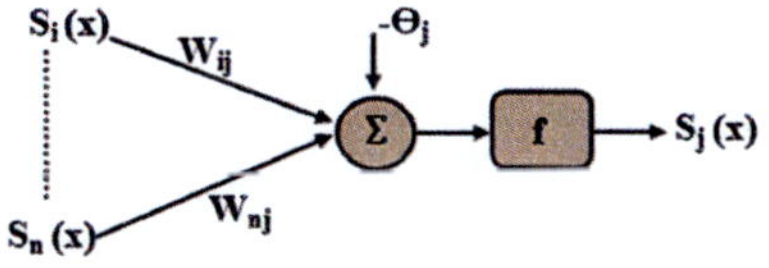

Fig. 8. Neural model proposed by Mc Culloch and Pitts.

- ***Application of ANN to DFIG and the wind system***

The ANN are applied to the active and reactive power controls of the DFIG of the ideal structure in Fig. 5, and on the whole wind power system in Fig. 1. In this work, the ANN implementations use the Leveuberg-Marquardt (LM) back propagation algorithm. The convergence of the ANNs has been obtained using the parameter values grouped in the attached Table 3. And, the ANNs are defined by on the internal logical structure of Fig. 9. The Number of neurons of the input and output layer are respectively 10 and 1.

4 Results and Discussions

In order to validate our study, we rely on the evaluation of the efficiency of the ANN controller by comparing its performance to that of the PI and the FLC. As a result, the structure of the simplified DFIG model and the modeling of the whole 2 MW wind turbine chain are used. The simulation parameters of all the elements of the wind system, including its controllers, are presented in Table 3 and 4 in the appendix. All these studies are carried out using the Matlab/Simulink software.

4.1 Case 1: Simulation of DFIG Power Control by PI, FLC and ANN

In order to highlight the different operating regimes that the DFIG can cope with, the reference active and reactive power levels are considered in the Figs. 10 and 11 as follows:

The time intervals (0 s to 0.15 s) and (0.3 s to 0.4 s) for $P_{sref} = 2$ MW and $Q_{sref} = 0$ MVAR corresponding to the normal operating state of the system. Whereas, in the interval (0.15 s to 0.3 s) we define respectively a sharp decrease in P_{sref} from 2 MW to 1.5 MW and a sharp increase in P_{sref} from 1.5 MW to 2 MW, at $Q_{sref} = 0$ MVAR. This reflects two large distinct disturbances, in order to also highlight the influence of model uncertainties or the coupling effect between P_s and Q_s in the DFIG. In general, you can clearly see in Figs. 10 and 11 that the curves of active power P_s and reactive power Q_s of the three controllers follow their reference value well with very insignificant overshoots, in particular for ANN and FLC. However, compared to PI, FLC and ANN allow controlled powers to reach their set points as quickly as possible. In addition, the FLC is very slightly precise and fast than the ANN. The static errors (SE) observed in the Figs. 10 and 11 are around the following values: 0.03% with the FLC, 0.048% with the ANN and 0% with the PI. For the robustness test presented in Figs. 12 and 13, we chose a 20% increase in the nominal value of the mutual inductance (M). In terms of robustness, Figs. 12 and 13 show that the curves of P_s and Q_s of the classical PI do not present interesting results. Because the PI curves are strongly more influenced than those of the FLC and ANN. While, ANN is less robust than the FLC, one observes on the ANN curves, a large static error and sensitivity to disturbances. Overall, although the FLC has a visible static error, its superiority to ANN and PI is remarkable. Moreover, compared to ANN and PI, the FLC offers good disturbance rejection, guaranteed decoupling of P_s and Q_s, good follow-up of its reference and good stability with respect to uncertainties. The ANN meanwhile, offers better performance than the classic PI.

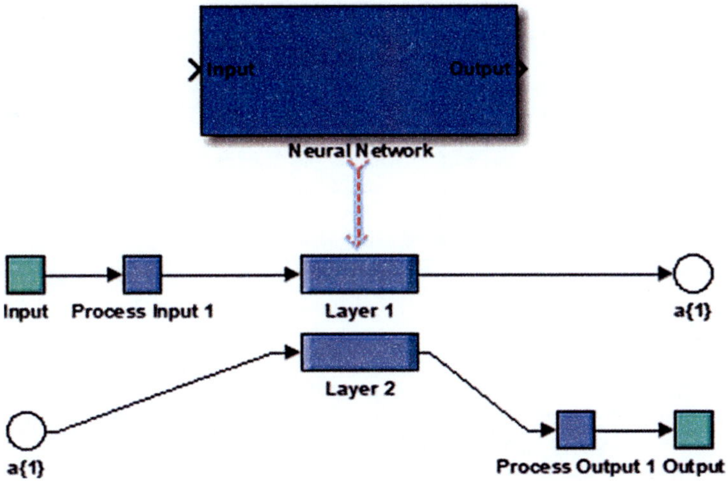

Fig. 9. Design structure of the ANN controller.

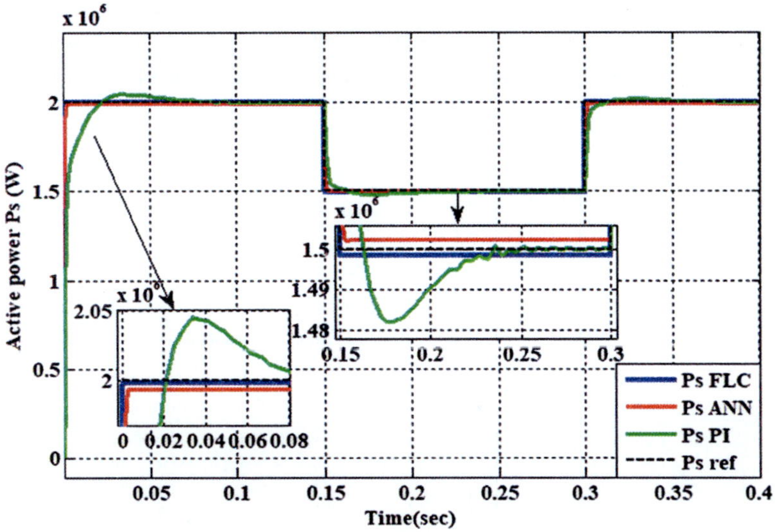

Fig. 10. Design structure of the ANN controller.

4.2 Case 2: Simulation of the Wind System Control with PI, FLC and FLSA-PI

The wind turbine which is studied in this work, can operate under optimal conditions, with a maximum coefficient of performance of $C_{pmax} = 0.44$, an optimal specific speed of $\lambda_{opt} = 5.841$ and a constant wind speed $V = 11.2$ m/s at $\beta = 0°$. In order to give an image of what could be observed or predicted experimentally or on site, the PI, ANN and FLC controllers are applied to a 2 MW wind system whose model is based on the

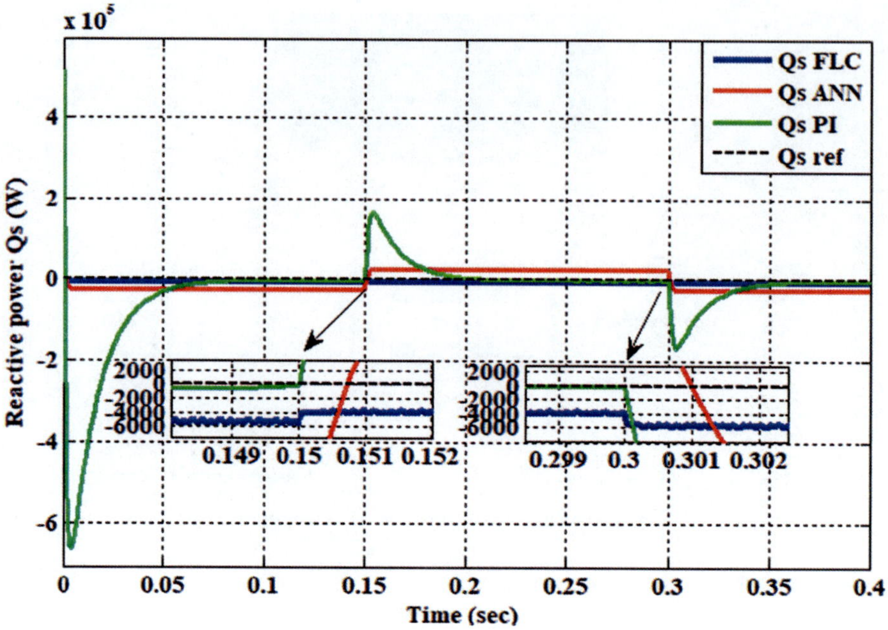

Fig. 11. Reactive power control by FLC, ANN and PI.

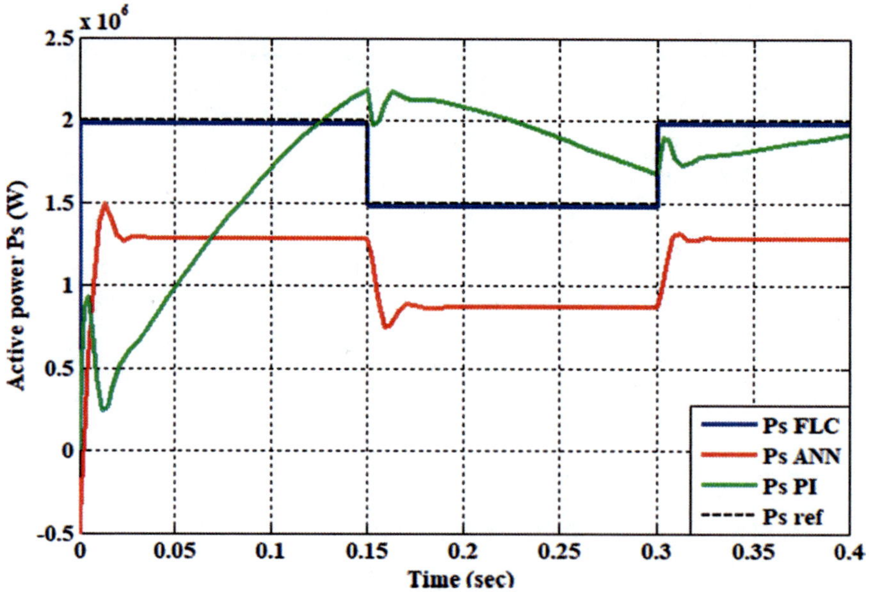

Fig. 12. Design structure of the ANN controller.

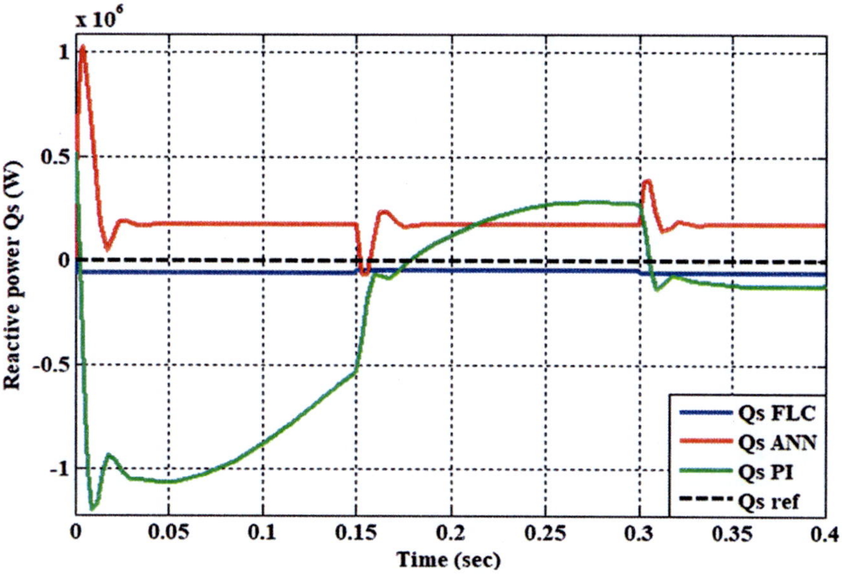

Fig. 13. Robustness reactive powers by FLC, ANN and PI.

equations: (5), (6), (11), (12), (13) with a random wind model illustrated in Fig. 14, based on the following equation [4]:

$$V = 11.4 + (0.2\sin(0.1047 \cdot t) + 0.2\sin(0.2665 \cdot t) + 0.2\sin(3.6645 \cdot t)) \quad (20)$$

The Fig. 15 shows us that the mechanical speed control is quickly well ensured by the ANN, followed by the FLC and finally the PI. The static error is almost 0% for the three controllers, while the overruns are 11% for the PI, 0% with the ANN and the FLC. From the active and reactive power curves in the Figs. 16, 17, 18, 19, 20, 21 it can also be seen that, in terms of sensitivity to overrunning, and rapid tracking of their set points, the ANN is higher than the FLC which is higher than the PI. Unlike the other two controllers, the FLC has the highest static error of around 4% in Fig. 18. The regulation of the DC bus voltage shown in Fig. 22 shows that the ANN is most reassuring in the rapid and precise pursuit of its reference value, and will be followed by the FLC in relation to the PI. The Figs. 23, 25 and 27 respectively show the quality of the currents to be fed into the grid by the wind system using the PI, ANN and FLC controls. Overall, the system is balanced for each control and the regime stabilizes around 0.5 s. This regime is established more quickly with the ANN, then with the PI and finally with the FLC. Indeed, it can be seen that the FLC has greater amplitudes and fluctuations at start-up than the other two controllers. Furthermore, with the same mains currents under the time interval conditions [0.5 s to 3 s], fundamental frequency (50 Hz), maximum frequency (1000 Hz) and number of cycles (125), it can be seen that the highest Total Harmonic Distortion (THD) is that of the FLC in the Fig. 28, followed by the PI in the Fig. 24, and finally the best ANN shown in the Fig. 26. On the other hand, cos$\phi \approx 1$ is the best with the ANN in Fig. 30, followed relatively by the FLC and the PI in Figs. 31

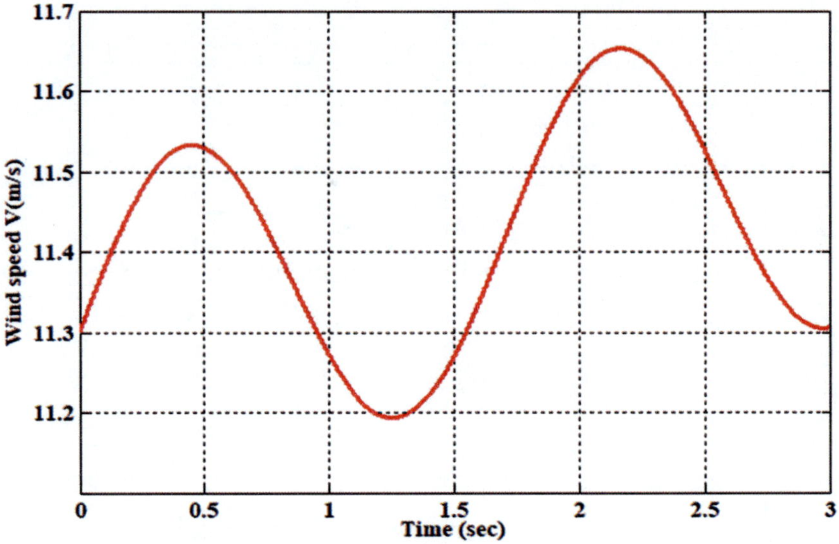

Fig. 14. Illustration of the random wind pattern.

and 29 respectively. In the case of controlling the whole wind system, we can say that decreasing the level of inference rules of the FLC to three, to lead to a slight decrease in the performance of the FLC compared to the ANN.

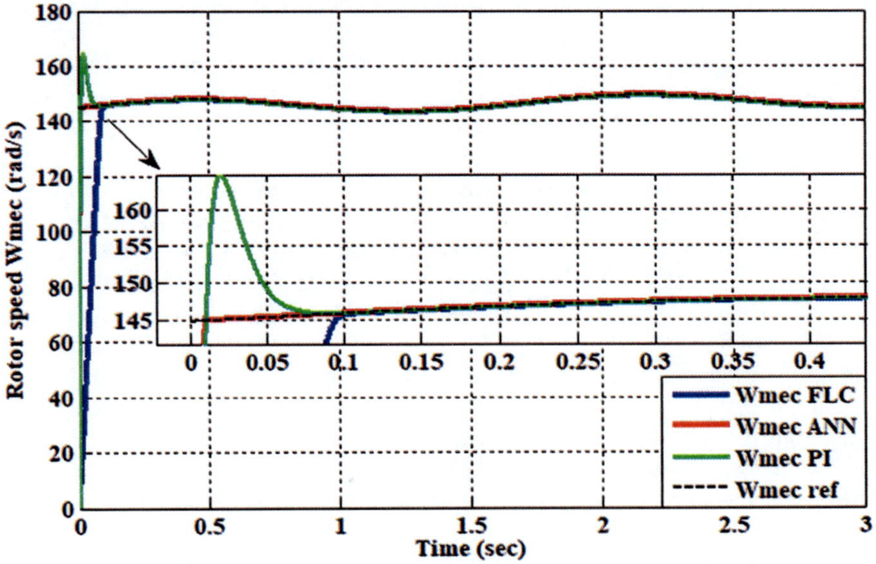

Fig. 15. Mechanical speed control by FLC, ANN and PI.

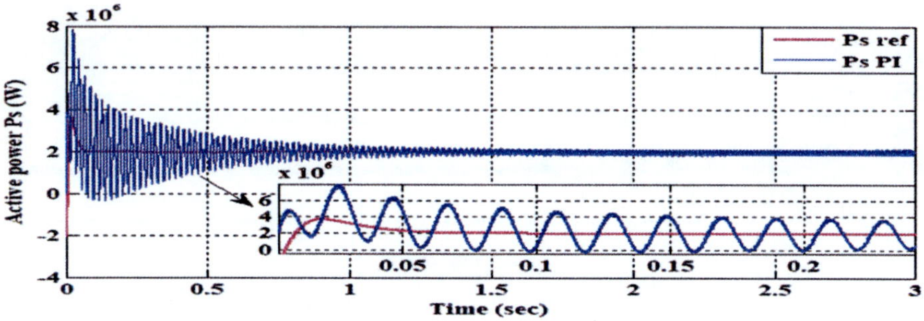

Fig. 16. Control of the active power with the PI action.

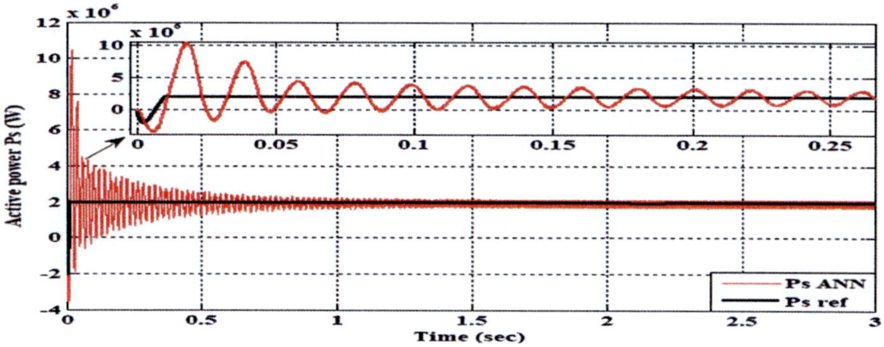

Fig. 17. Control of the active power with the ANN.

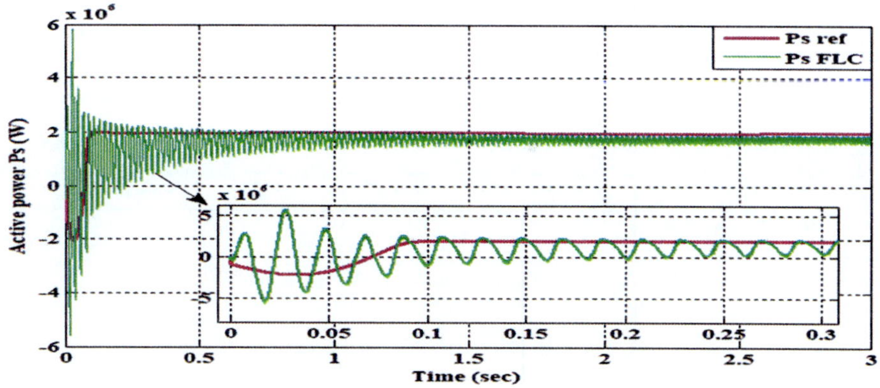

Fig. 18. Control of the active power with the FLC.

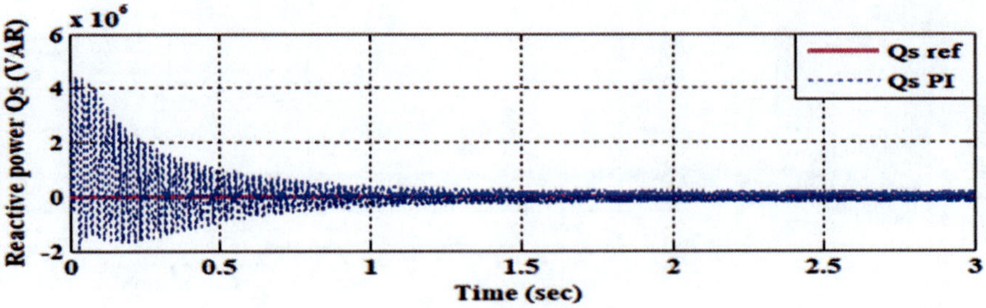

Fig. 19. Control of the reactive power with the PI.

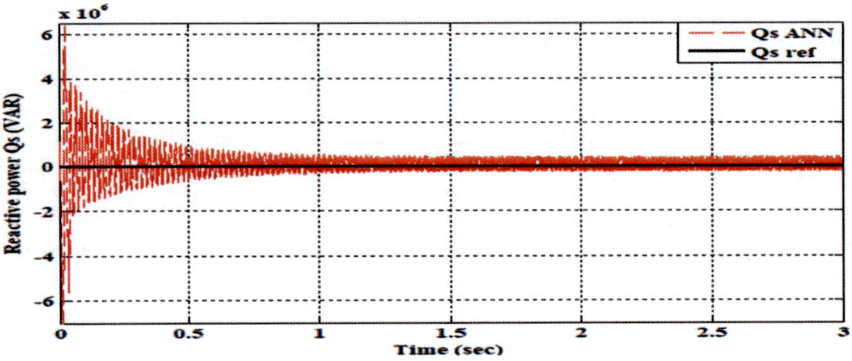

Fig. 20. Control of the reactive power with the ANN.

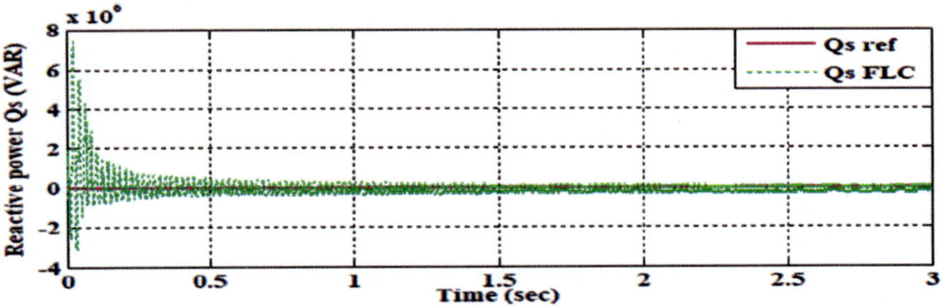

Fig. 21. Control of the reactive power with the FLC.

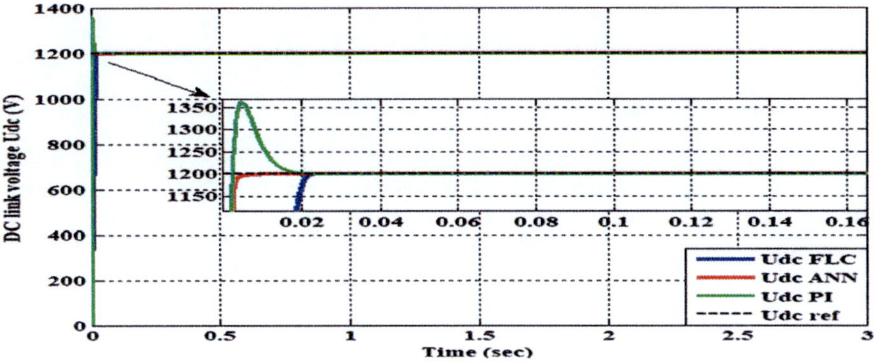

Fig. 22. Regulation of the DC bus voltage by FLC, ANN and PI.

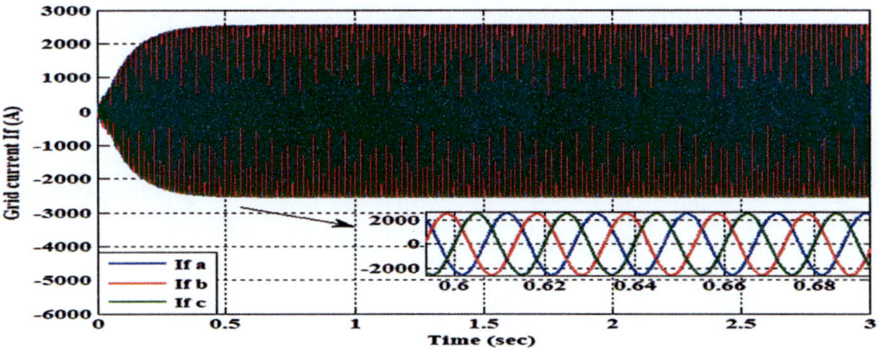

Fig. 23. Currents injected into the electrical grid with the PI.

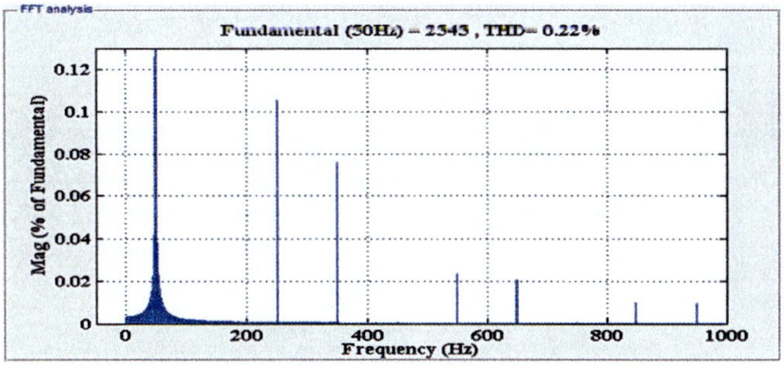

Fig. 24. THD of currents injected into the power grid with the PI.

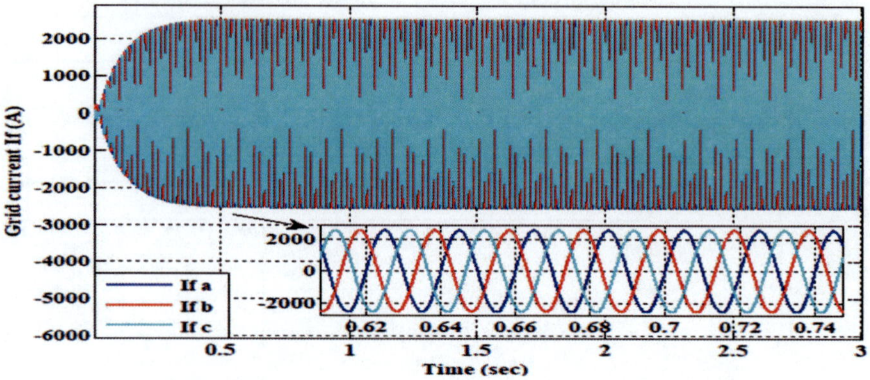

Fig. 25. Currents injected into the electrical grid with the ANN.

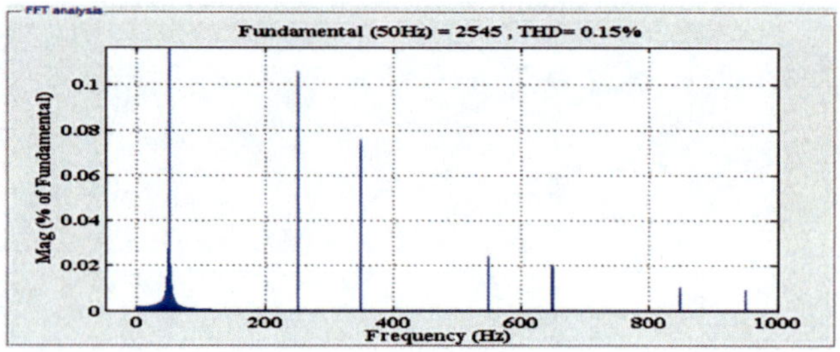

Fig. 26. THD of currents injected into the power grid with the ANN.

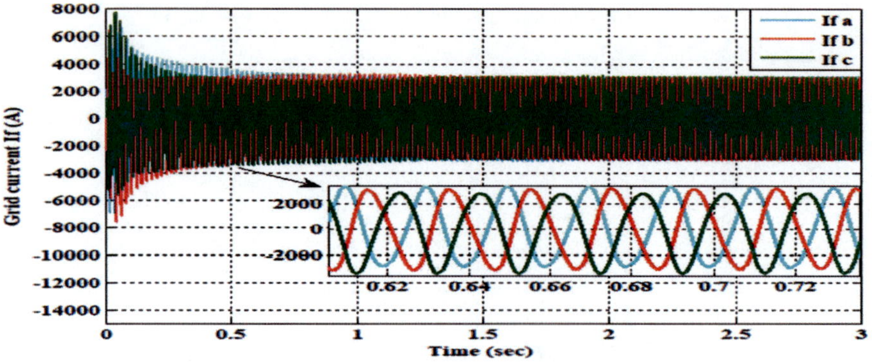

Fig. 27. Currents injected into the electrical grid with the FLC.

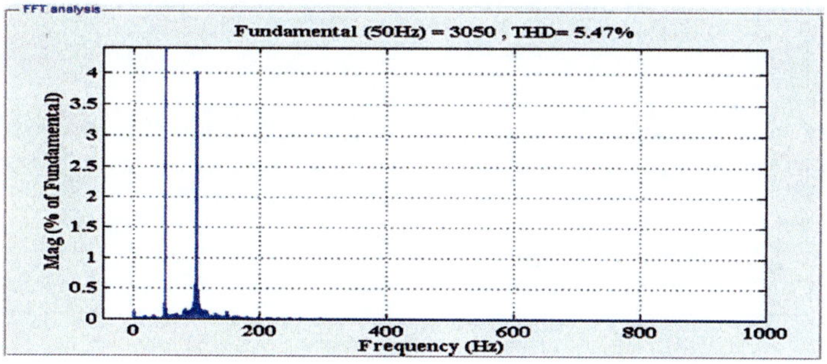

Fig. 28. THD of currents injected into the power grid with FLC.

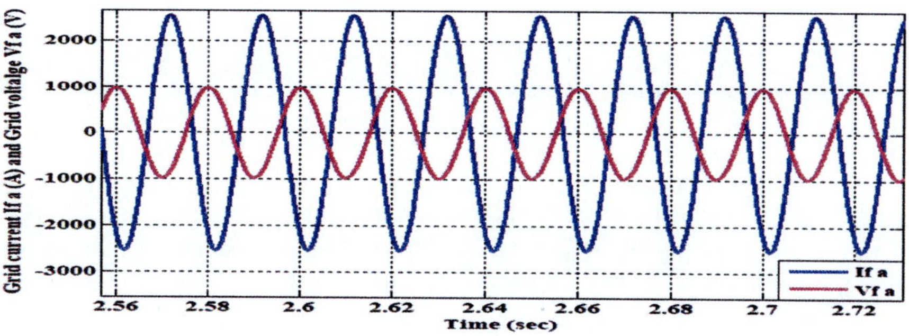

Fig. 29. Voltage and current phase shift PI to the power grid.

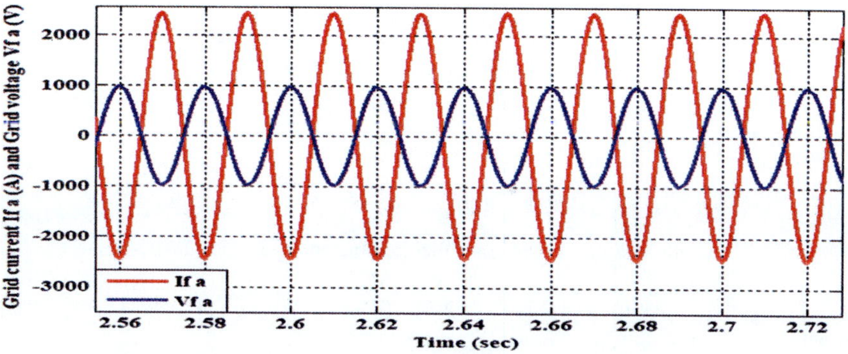

Fig. 30. Voltage and current phase shift ANN to the power grid.

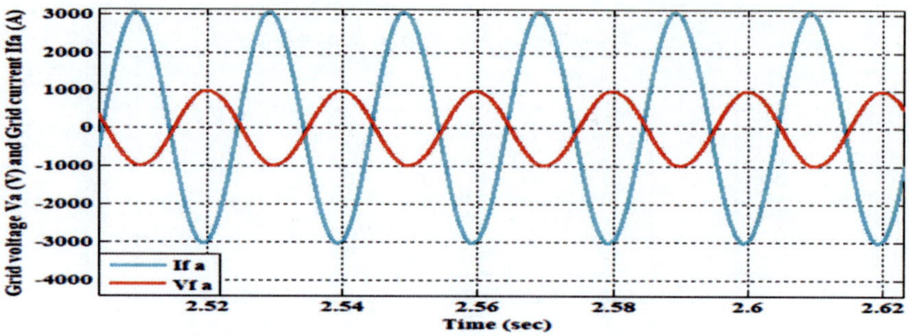

Fig. 31. Voltage and current phase shift FLC to the power grid.

5 Conclusion

This study proposes the use of Neural Network (ANN) and Fuzzy Logic Control (FLC) controllers, for the systemic control of a wind power system connected to the grid and using a Doubly-Fed Induction Generator (DFIG). To do this, comparative studies between Proportional Integral (PI), FLC and ANN were carried out in the case of an ideal system based on the simplified model of DFIG on the one hand; and in the case of a real wind system on the other hand. The results of the study show that, the PI performs less well than the ANN and FLC. Whereas, for a high level of fuzzy inference rules, the FLC performs better than the ANN. For low level, the performance of the ANN is very slightly superior to the FLC. In addition, one mainly notices a good disturbance rejection and a good robustness of the FLC compared to the ANN. The latter offers a static error, a standard deviation and a $\cos\phi \approx 1$, slightly better than the FLC. To this end, we can conclude that the ANN and FLC strategies are very promising for the control of wind systems.

Appendix

Table 3. Summary of controller parameters PI, FLC and ANN

Controlled Quantities of models		P_s/Q_s (DFIG simplified)	Ωmec (Wind turbine)	P_s/Q_s (DFIG widespread)	V_{dc} (DC-BUS)	I_{fd}/I_{fq} (Filter)
PI	K_p	4.871e−4	−136460	4.871e−4	30.4	24
	K_I	7.4439e−3	−6823e3	7.4439e−3	6080	240
	τ_r (ms)	1.2/0.6	40/39.4	1.2/0.6	10/9.4	2.5e−5/1.9e−5

(*continued*)

Table 3. (*continued*)

Controlled Quantities of models		P_s/Q_s (DFIG simplified)	Ωmec (Wind turbine)	P_s/Q_s (DFIG widespread)	V_{dc} (DC-BUS)	I_{fd}/I_{fq} (Filter)
FLC	G_e	5e−7	6.25e−3	1.6666e−7/1.43e−7	8.3333e−4	1.125e−4
	$G_{\Delta e}$	5.64e−10	2.9e−5	2e−18/2e−10	1e−6	1.125e−7
	$G_{\Delta U}$	1000	0.1	1	1	1
	ΔU	[−80 80]	[−2e7 2e7]	[−200 0]/[−350 0]	[−4000 4000]	[−1000 0]
ANN	Iterations	13/15	1000	500/789	1000	842/800
	Regression	0.9738/0.70	0.90	0.9738/0.70	0.9941	0.70/0.74

Table 4. Simulation parameters for DFIG, DC-Bus, Filter, Turbine [33]

Models	DFIG	DC-Link and Filter	Turbine
Parameters	$U_s = 690$ V/F = 50 Hz	C = 38000 μF	P = 2
	$R_s = 1.69$ mΩ	$U_{dc} = 1200$ V	G = 90
	$L_s = 2.95$ mH	$R_F = 6$ mΩ	$J_{turbine} = 50.105$ kg·m^2
	$R_r = 1.52$ mΩ	$L_F = 0.6$ mH	D = 82 m
	$L_r = 2.97$ mH/L_m(M) = 2.91 mH		f = 0.03/J_{rotor} = 682.3 kg·m^2

References

1. Moghadam, H.M., Gheisarnejad, M., Esfahani, Z., Khooban, M.-H.: A novel supervised control strategy for interconnected DFIG-based wind turbine systems: MiL validations. IEEE Trans. Emerg. Top. Comput. Intell. 1–10 (2020). https://doi.org/10.1109/tetci.2020.3010060
2. Hab, R.: Energies renouvelables en afrique: Enjeux, défis et opportunités. Renewable Energies in Africa: Issues, challenges and opportunities. Policy Paper (2019). https://media.africaportal.org/documents/energies_renouvelables_en_afrique
3. Shukla, R.D., Tripathi, R.K., Thakur, P.: K and DC grid/bus tied DFIG based wind energy system. Renew. Energy **108**, 179–193 (2017). https://doi.org/10.1016/j.renene.2017.02.064
4. Wadawa, B., Errami, Y., Obbadi, A., Sahnoun, S., Aoutoul, M.: Control of wind energy conversion system connected to the power grid and based on double fed induction generator. In: Proceedings of the 4th International Conference on Big Data and Internet of Things, pp. 1–8 (2019). https://doi.org/10.1145/3372938.3373004
5. Li, S., Wang, H., Tian, Y., et al.: Direct power control of DFIG wind turbine systems based on an intelligent proportional-integral sliding mode control. ISA Trans. **64**, 431–439 (2016). https://doi.org/10.1016/j.isatra.2016.06.003
6. Yang, B., Yu, T., Shu, H., Dong, J., Jiang, L.: Robust sliding-mode control of wind energy conversion systems for optimal power extraction via nonlinear perturbation observers. Appl. Energy **210**, 711–723 (2018). https://doi.org/10.1016/j.apenergy.2017.08.027

7. Bossoufi, B., Karim, M., Lagrioui, A., Taoussi, M., Derouich, A.: Observer backstepping control of DFIG-Generators for wind turbines variable-speed: FPGA-based implementation. Renew. Energy **81**, 903–917 (2015). https://doi.org/10.1016/j.renene.2015.04.013
8. Islam, K.S., Seyedmahmoudian, M., Stojcevski, A.: Decentralized robust mixed H_2/H_∞ reactive power control of DFIG cluster using SMES. Int. J. Electr. Power Energy Syst. **113**, 176–187 (2019). https://doi.org/10.1016/j.ijepes.2019.05.017
9. Tang, Y., Ju, P., He, H., Qin, C., Wu, F.: Optimized control of DFIG-based wind generation using sensitivity analysis and particle swam optimization. IEEE Trans. Smart Grid **4**(1), 509–520 (2013). https://doi.org/10.1109/tsg.2013.2237795
10. Ardjoun, S.A.E.M., Denai, M., Abid, M.: Robustification du contrôle des éoliennes pour une meilleure intégration dans un réseau déséquilibré. In: 2019 Algerian Large Electrical Network Conference, pp. 1–6. IEEE (2019). https://doi.org/10.1109/cagre.2019.8713286
11. Alnufaie, L.: DFIG control: a fuzzy approach. Int. J. Adv. Appl. Sci. **6**, 107–116 (2019). https://doi.org/10.21833/ijaas.2019.09.015
12. Chen, X., Wang, L., Sun, H., Chen, Y.: Fuzzy logic based adaptive droop control in multi-terminal HVDC for wind power integration. IEEE Trans. Energy Convers. **32**, 1200–1208 (2017). https://doi.org/10.1109/tec.2017
13. Bounar, N., Labdai, S., Boulkroune, A.: PSO–GSA based fuzzy sliding mode controller for DFIG-based wind turbine. ISA Trans. **85**, 177–188 (2019). https://doi.org/10.1016/j.isatra.2018.10.020
14. Shihabudheen, K.V., Pillai, G.N., Krishnama Raju, S.: Neuro-fuzzy control of DFIG wind energy system with distribution network. Electr. Power Compon. Syst. 1–16 (2019). https://doi.org/10.1080/15325008.2018.1499154
15. Debouza, M., Al-Durra, A., Errouissi, R., Muyeen, S.M.: Direct power control for grid-connected doubly fed induction generator using disturbance observer based control. Renew. Energy **125**, 365–372 (2018). https://doi.org/10.1016/j.renene.2018.02.121
16. Ata, R.: RETRACTED: artificial neural networks applications in wind energy systems: a review. Renew. Sustain. Energy Rev. **49**, 534–562 (2015). https://doi.org/10.1016/j.rser.2015.04.166
17. Raju, S.K., Pillai, G.N.: Design and real time implementation of type-2 fuzzy vector control for DFIG based wind generators. Renew. Energy **88**, 40–50 (2016). https://doi.org/10.1016/j.renene.2015.11.006
18. Matía, F., Marichal, G.N., Jiménez, E. (eds.): Fuzzy Modeling and Control: Theory and Applications. ACIS, vol. 9. Atlantis Press, Paris (2014). https://doi.org/10.2991/978-94-6239-082-9
19. Benbouhenni, H., Boudjema, Z., Belaidi, A.: Power ripple reduction of DPC DFIG drive using ANN controller. Acta Electrotechnica Informatica **20**, 15–22 (2020). https://doi.org/10.18280/mmep.070208
20. Azeem, B., et al.: Robust neural network scheme for generator side converter of doubly fed induction generator. In: International Symposium on Recent Advances in Electrical Engineering (RAEE) (2017). https://doi.org/10.1109/raee.2017.8246149
21. Rocha-Osorio, C.M., Solís-Chaves, J.S., Rodrigues, L.L., Puma, J.A., Sguarezi Filho, A.J.: Deadbeat–fuzzy controller for the power control of a doubly fed induction generator based wind power system. ISA Trans. **88**, 258–267 (2019). https://doi.org/10.1016/j.isatra.2018.11.038
22. Åström, K.J., Hägglund, T.: PID Controllers: Theory, Design, and Tuning, vol. 2. Instrument Society of America, Research Triangle Park (1995)
23. Xu, Q., Kan, J., Chen, S., Yan, S.: Fuzzy PID based trajectory tracking control of mobile robot and its simulation in Simulink. Int. J. Control Autom. **7**(8), 233–244 (2014). https://doi.org/10.14257/ijca.2014.7.8.20

24. Boudjellal, B., Benslimane, T.: Active and reactive powers control of DFIG based WECS using PI controller and artificial neural network based controller. Model. Measur. Control A **93**(1–4), 31–38 (2020). https://doi.org/10.18280/mmc_a.931-405
25. Mokhtari, M., Khazaei, J., Nazarpour, D.: Sub-synchronous resonance damping via doubly fed induction generator. Int. J. Electr. Power Energy Syst. **53**, 876–883 (2013). https://doi.org/10.1016/j.ijepes.2013.05.054
26. Rechka, S., Roy, G., Dennetiere, S., et al.: Modeling of multi-mass electromechanical systems based on asynchronous machines, using MATLAB and EMTP tools with application to wind turbines. Polypublie polytechnique montréal (2004). https://publications.polymtl.ca/2616/
27. Sánchez-Solano, S., Brox, M.: Hardware implementation of embedded fuzzy controllers on FPGAs and ASICs. In: Matía, F., Marichal, G.N., Jiménez, E. (eds.) Fuzzy Modeling and Control: Theory and Applications. ACIS, vol. 9, pp. 235–253. Atlantis Press, Paris (2014). https://doi.org/10.2991/978-94-6239-082-9_13
28. Haber, R., Toro, R.M.D., Godoy, J., Gajate, A.: Intelligent tuning of fuzzy controllers by learning and optimization. In: Matía, F., Marichal, G.N., Jiménez, E. (eds.) Fuzzy Modeling and Control: Theory and Applications. ACIS, vol. 9, pp. 135–158. Atlantis Press, Paris (2014). https://doi.org/10.2991/978-94-6239-082-9_7
29. Robyns, B., Francois, B., Degobert, P., Hautier, J.: Vector control of induction machines. In: Robyns, B., Francois, B., Degobert, P., Hautier, J. (eds.) Vector Control of Induction Machines, pp. 75–121. Springer, London (2012). https://doi.org/10.1007/978-0-85729-901-7_3
30. Sharma, J., Gupta, M.N., Nain, M.P.: Speed control of separately excited DC motor using fuzzy logic controller. Int. J. Tech. Res. (IJTR) **4**(2) (2015). https://doi.org/10.1109/iecon.1998.724140
31. Ta-Cao, M., Le-Huy, H.: Rotor resistance estimation using fuzzy logic for high performance induction motor drives. In: Proceedings of the 24th Annual Conference of the IEEE Industrial Electronics Society, IECON 1998 (Cat. No. 98CH36200), vol. 1, pp. 303–308 (1998). https://doi.org/10.1109/iecon.1998.724140
32. Mugisha, J.C., Munyazikwiye, B., Karimi, H.R.: Design of temperature control system using conventional PID and intelligent fuzzy logic controller. In: 2015 International Conference on Fuzzy Theory and its Applications (iFUZZY), pp. 50–55. IEEE (2015). https://doi.org/10.1109/iFUZZY.2015.7391893
33. Zhou, D., Blaabjerg, F.: Bandwidth oriented proportional-integral controller design for back-to-back power converters in DFIG wind turbine system. IET Renew. Power Gener. **11**(7), 941–951 (2017)

Bayesian Inference of a Discrete Fractional SEIRD Model

Mokhtar Zbair(✉), Abdelaziz Qaffou(✉), Fatima Cherkaoui(✉), and Khalid Hilal(✉)

Applied Mathematics and Scientific Computing Laboratory, Department of Applied Mathematics, Faculty of Sciences and Techniques, Sultan Moulay Slimane University-Beni-Mellal, C.P:523 Beni-Mellal, Morocco
{statzbair,hilalkhalid2005}@yahoo.fr, {aziz.qaffou,cherkaoui2310}@gmail.com

Abstract. This paper deals with the Bayesian estimation of the parameters of a discrete fractional epidemic model SEIRD as an extension of the classical SEIR model, describing the dynamics of disease propagation in a population. Equilibrium points are computed and the existence stability nature at these points are discussed. The basic reproduction number R_0 is calculated using next generation matrix method. The estimation of the parameters is based on Bayesian inference. The numerical simulations were used to illustrate the stability of the discrete fractional order SEIRD epidemic model and to evaluate the performance of the estimation method. The model introduced is applied to real data concerning pandemic COVID-19 in Morocco.

Keywords: Bayesian estimation · SEIRD epidemic model · Fractional order · Caputo · Stability · COVID 19 pandemic

AMS Subject Classification: 62F15 · 60E05 · 92D30

1 Introduction

At the begining of December 2019, a new viral disease was discovered in Wuhan, China. The scientists found that the cause of this infectious disease is the novel beta-coronavirus, which leads to the severe acute respiratory syndrome. This virus, which is now known as 2019-nCoV, SARS-CoV-2, and COVID-19, affects the lungs and has symptoms such as cough, fever, tiredness, and difficult breathing. Unfortunately, the spread of the 2019-nCoV was too rapid in Hubei Province and became an epidemic at the end of January 2020. Consequently, the Chinese government imposed the quarantine restrictions to prevent the outbreak. International travels were also declared. However, it was not successful, and the disease was spread in the whole globe. At the time, a large number of countries such as the USA, Italy, Spain, and Germany are affected by this disease and the governments try to defeat coronavirus by enforcing social distancing.

The infectious pandemics have substantial effects on the health and also on finance. Therefore, the study of the dynamics of transmission of the disease is of great importance. By the help of mathematical tools, it is a possibility to predict many real-world

S. Melliani and O. Castillo (Eds.): ICPAMS 2021, LNNS 476, pp. 138–146, 2023.
https://doi.org/10.1007/978-3-031-12416-7_12

time series in different fields such as economics, finance, and climate [1–5]. Mathematical models are an effective tool for understanding the dynamics of the outbreaks. These models are also useful in forecasting the spread of the disease and thus help the governments to be prepared and make necessary decisions [6]. The well-known and most used mathematical models for the spread of infections are the classical ordinary differential equations, such as SI, SIS, SIR, SEIR, SIRD, and SEIRD models.

In these models, each variable represents the number of individuals in different groups. From the discovery of the 2019-nCoV, several models have been proposed to study its dynamics [7–12]. Zhong et al. [13] proposed a simple SIR model for predicting novel coronavirus, according to China's first reported data. Yang and Wang [14] presented an extended SEIR model for COVID-19 with time-varying transmission rates by considering the environmental effects. Liang [15] described the growth propagation of three pandemic diseases, COVID-19, SARS, and MERS, by mathematical models and found that the growth rate of COVID-19 is much greater than SARS and MERS.

The fractional-order differential equations have been recently used for describing the behavior of the epidemics [16–18], [19–22]. The fractional derivatives are dependent on the historical states, in addition to the current state, and thus have memory properties [17, 18]. Therefore, they are a better choice for the epidemic's modeling. Furthermore, in the fractional model, the derivative order provides a degree of freedom in fitting data [18]. Due to these properties, the fractional differential equations have been used for various applications in different fields [23–29]. González-Parra et al. [18] presented a fractional-order SEIR model for explaining the outbreak of influenza A(H1N1). They showed that the fractional model agrees with the real data better than the classical model. Demirci et al. [16] proposed a fractional-order SEIR epidemic model with vertical transmission with considering that the death rate is dependent on the number of the total population. Area et al. [17] analyzed the data of the Ebola outbreak by both integer-order and fractional-order SEIR models. However, they reported that the classical model had better fitting results than the fractional one.

All of the studies in modeling the spread of COVID 19 have considered ordinary differential equations, while there are some claims that the fractional-order models have a better fitting to the real data. In [30], the authors have presented a SEIRD model for analyzing and predicting the COVID-19.

In this paper, we analyze this model with fractional-order derivatives. We compute the error of fitting the model to the real data, which refers to Italy from February 24 to April 7. Therefore, the optimum parameters are found for different derivative orders. It is observed that the model with fractional order has less fitting error than the integer model. Then, the model is tested by using the data of April 8 to May 16 (unseen by the model during parameter estimation). The results show that the fractional model provides a better prediction than the integer model.

In this paper, we propose a discrete fractional SEIRD model, where the parameters are estimated by Bayesian inference. Equilibrium points are computed and the existence stability nature at these points are discussed. The basic reproduction number R_0 is calculated using next generation matrix method. The numerical simulations were used to illustrate the stability of the discrete fractional order SEIRD epidemic model and to evaluate the performance of the estimation method. There are several definitions for fractional derivatives; in this paper we choose to work with the Caputo fractional derivative. One of the advantages of such derivative is allowing us to consider classical initial conditions to be included in the formulation of the problem. Also, the Caputo fractional derivative of a constant is zero, which is not true for other fractional derivatives. Also, since the fractional order can be any positive real α, we can choose the one that better fits the data. Therefore, we can adjust the model to real data and, thus, better predict the evolution of the disease.

The paper is outlined as follows. In Sect. 2, we first recall some definitions and results about fractional calculus and also present a new result that will be needed in what follows. In Sect. 3, we present the fractional discrete SEIRD epidemic model, and then, in Sect. 4, we give the dynamic properties of fractional calculus. Namely, we prove that the problem is well-posed and we evaluate the equilibrium points and the basic reproduction number. In Sect. 5, we estimate the parameters using Bayesian inference. Numerical simulations using R are given in Sect. 6. Finally, we conclude the paper.

2 Fractional Difference Operator

Fractional calculus is an extension of the ordinary calculus, by considering integrals and derivatives of arbitrary real or complex order [33, 59]. This subject is as old as calculus itself, and in the past decades it has proved to be applicable to real world phenomena. In some cases, considering the dynamic being modeled by a fractional derivative/integral, we obtain a more realistic model.

Let $\alpha > 0$ be a real number and $x : [a,b] \longrightarrow \mathbb{R}$ an integrable function. The Riemann-Liouville fractional integral of x of order α is given by the expression

$$I^{\alpha}_{a+}x(t) := \frac{1}{\Gamma(\alpha)} \int_a^t (t-\tau)^{\alpha-1} x(\tau) d\tau$$

where Γ denotes the Gamma function:

$$\Gamma(t) := \int_0^{+\infty} \tau^{t-1} \exp(-\tau) d\tau, \text{ for all } t > 0$$

We remark that $\Gamma(t+1) = t\Gamma(t), \forall t > 0$, and for positive integers n, we have $\Gamma(n+1) = n!$

Fractional calculus has a long history and there seems to be new and recent interest in the study of fractional calculus and fractional differential equations [2,3].

Let:

$$t^{(\alpha)} = \frac{\Gamma(t+1)}{\Gamma(t-\alpha+1)} \text{ and } \mathbb{N}_{\nu} = \nu + \mathbb{N}$$

From [1], the fractional sum is defined as:

$$\Delta^{-\nu}x(n) = \frac{1}{\Gamma(\nu)} \Sigma_{k=0}^{n-\nu} (n-k-1)^{(\nu-1)} x_k \text{ where } x : \mathbb{N} \to \mathbb{R}^p \text{ and } n \in \mathbb{N}_{\nu} \,.$$

For $m-1 < \nu < m$, the νth Caputo fractional difference is defined as:

$$\Delta^{\nu}_{*} = \Delta^{-(m-\nu)} \left(\Delta^m x\right) \text{ where } \Delta x_n = x_{n+1} - x_n$$

For $\nu \in]0,1]$:

$$\Delta_*^{\nu} x_n = \Delta^{-(1-\nu)} \Delta x_n = \frac{1}{\Gamma(1-\nu)} \Sigma_{k=0}^{n-(1-\nu)} (n-k-1)^{(-\nu)} \Delta x_k \text{ for } n \in \mathbb{N}_{1-\nu}.$$

Some properties:

$$\forall x : \mathbb{N} \to \mathbb{R}^p, \Delta^{-\nu_1} \Delta^{-\nu_2} x = \Delta^{-\nu_2} \Delta^{-\nu_1} x = \Delta^{-(\nu_1+\nu_2)} x$$

3 A Fractional Discrete SEIRD Epidemic Model

3.1 SEIRD Model

Following the classic SEIR model, we consider a compartment model with 6 compartments: Susceptible, Exposed (not yet infectious), Infected with mild symptoms, Infected with severe symptoms, Recovered and Death.

In Fig. 1, we present the compartments considered and the possible transitions between the compartments.

The Table 1 summarizes the compartments introduced.

Table 1. The compartments and their notation

Compartment	Notation
Susceptible	S_t
Exposed	E_t
Infected	I_t
Infected with mild symptoms	I_t^m
Infected with severe symptoms	I_t^s
Recovered	R_t
Death	D_t

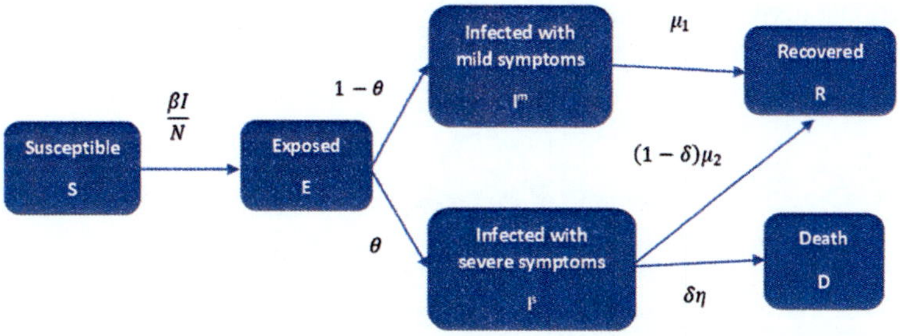

Fig. 1. Compartments and possible transitions between compartments

The parameters characterizing the interaction between the different compartments introduced are presented in the Table 2

Table 2. The parameters and their notation

Parameter	Meaning
β	Transmission rate
θ	Part of exposed to the virus with severe symptoms
$\frac{1}{\gamma}$	Average incubation duration
δ	Death rate due to epidemic with severe symptoms
$\frac{1}{\mu_1}$	Average duration of infection with mild symptoms
$\frac{1}{\mu_2}$	Average duration of infection with severe symptoms
$\frac{1}{\eta}$	Average time from onset of symptoms to death

The discrete SEIRD model is expressed by the following equations:

$$\begin{cases} \Delta S_t = -\frac{\beta}{N} S_t I_t \\ \Delta E_t = \frac{\beta}{N} S_t I_t - \gamma E_t \\ \Delta I_t^s = \theta \gamma E_t - ((1-\delta)\mu_2 + \delta\eta) I_t^s \\ \Delta I_t^m = (1-\theta)\gamma E_t - \mu_1 I_t^m \\ \Delta R_t = \mu_1 I_t^m + (1-\delta)\mu_2 I_t^s \\ \Delta D_t = \delta\eta I_t^s \end{cases} \tag{1}$$

where $\beta, \mu, \eta > 0$, $0 < \delta, \theta < 1$ and $N = S_t + E_t + I_t + R_t + D_t$ the total population.

3.2 Fractional SEIRD Model

There are several definitions for the fractional derivatives. Among them, the Caputo-type fractional derivative is more popular and used for real applications.

Now, we write the SEIRD model (Eq. 1) with fractional derivatives as:

$$\begin{cases} \Delta_*^\alpha S_t = -\frac{\beta}{N} S_t I_t \\ \Delta_*^\alpha E_t = \frac{\beta}{N} S_t I_t - \gamma E_t \\ \Delta_*^\alpha I_t^s = \theta \gamma E_t - ((1-\delta)\mu_2 + \delta\eta) I_t^s \\ \Delta_*^\alpha I_t^m = (1-\theta)\gamma E_t - \mu_1 I_t^m \\ \Delta_*^\alpha R_t = \mu_1 I_t^m + (1-\delta)\mu_2 I_t^s \\ \Delta_*^\alpha D_t = \delta\eta I_t^s \end{cases} \tag{2}$$

where $\beta, \mu, \eta > 0$, $0 < \delta, \theta < 1$ and $0 < \alpha < 1$.

4 Dynamic Properties

4.1 Disease Free Equilibrium (DFE) and the Basic Reproduction Number R_0

To evaluate the equilibria of the proposed model (2), we need to solve the following system:

$$\Delta_*^\alpha S_t = \Delta_*^\alpha E_t = \Delta_*^\alpha I_t^s = \Delta_*^\alpha I_t^m = \Delta_*^\alpha R_t = \Delta_*^\alpha D_t = 0$$

with the initials values: $E_t^* = I_t^{s*} = I_t^{m*} = 0$ and $S+E+I^s+I^m+R+D=N$. Then $F_0 = (N,0,0,0,0,0)$ is DFE.

The basic reproduction number noted R_0 represents the average number of secondary infections produced by an infected individual in a susceptible population.

We determine R_0 by using the next generation matrix approach. Then, R_0 is given by the formula

$$R_0 = \frac{(\theta\mu_1 + (1-\theta)\tilde{\mu}_2)\beta}{\mu_1\tilde{\mu}_2}$$

where $\tilde{\mu}_2 = \delta\eta + (1-\delta)\mu_2$.

4.2 Stability of DFE

As the dynamic of S, E, I^s and I^m does not depend on R and D, we consider only the first four equations. Then, the jacobian matrix J_{F_0} of the fractional model, evaluated around the DFE F_0 is

$$J_{F_0} = \begin{pmatrix} 0 & 0 & -\beta & -\beta \\ 0 & -\gamma & \beta & \beta \\ 0 & \theta\gamma & -\tilde{\mu}_2 & 0 \\ 0 & (1-\theta)\gamma & 0 & -\mu_1 \end{pmatrix}$$

The associated characteristic polynomial is

$$P(X) = X^4 + AX^3 + BX^2 + CX$$

with $A = \mu_1 + \tilde{\mu}_2 + \gamma$, $B = \mu_1\tilde{\mu}_2 + \gamma(\mu_1 + \tilde{\mu}_2 - \beta)$, $C = \gamma(\mu_1\tilde{\mu}_2 - (\theta\mu_1 + (1-\theta)\tilde{\mu}_2)\beta)$

Theorem 1. *if $C > 0$ and $AB > C$ then the DFE F_0 is stable.*

5 Numerical Simulations

We simulate the model for values $N = 1000000$, $\theta = 0.1$, $\delta = 0.3$, $\gamma = \frac{1}{5}\text{days}^{-1}$, $\mu_1 = \frac{1}{7}\text{days}^{-1}$, $\mu_2 = \frac{1}{14}$, $\eta = \frac{1}{18}\text{days}^{-1}$.

The Table 3 shows the values of R_0 as a function of β.

Table 3. The values of R_0 as a function of β

β	0.1	0.4	0.7	0.9
R_0	0.728	2.912	5.096	6.552

The figures below 2, 3, 4 and 5 show the evolution of the percentage of severely infected in the total population for $\beta = 0.1$, $\beta = 0.4$, $\beta = 0.7$ and $\beta = 0.9$ according to α . They show that the time needed in days to reach the peak of the number of severely infected individuals decreases as a function of fractional number α.

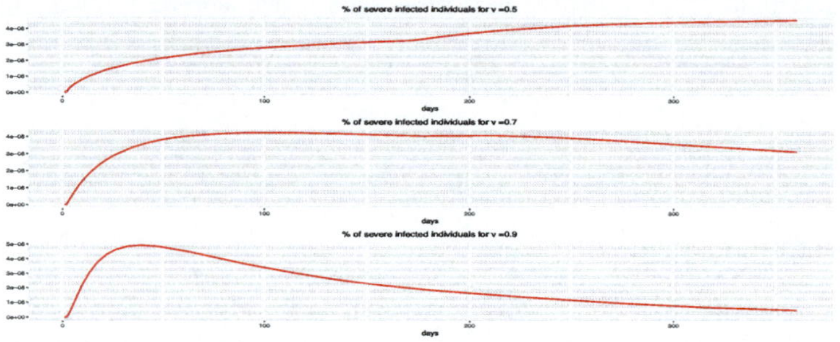

Fig. 2. Percentage of severely infected in the total population for $\beta = 0.1$, according to α

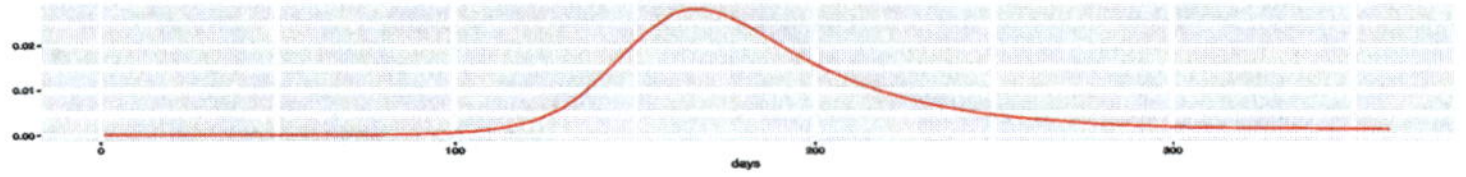

Fig. 3. Percentage of severely infected in the total population for $\beta = 0.4$, according to α

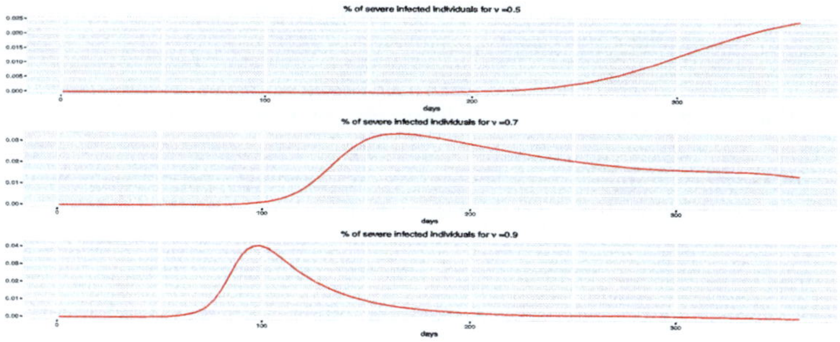

Fig. 4. Percentage of severely infected in the total population for $\beta = 0.7$, according to α

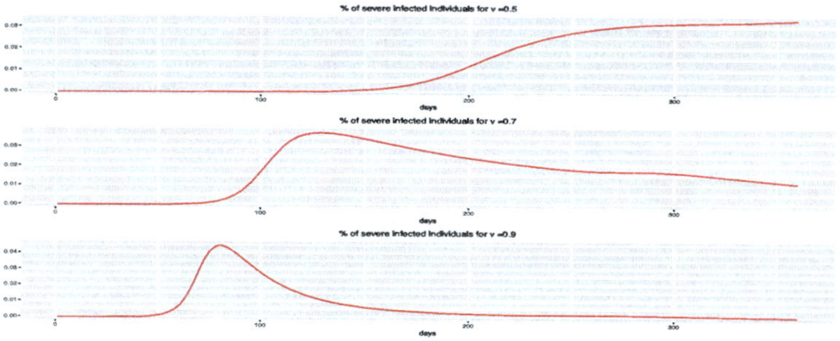

Fig. 5. Percentage of severely infected in the total population for $\beta = 0.9$, according to α

6 Conclusion

In this paper, we have proposed a discrete fractional SEIRD model that can be applied to any infectious disease. We proved that our fractional model has an equilibrium points. In addition, a sufficient condition for the stability of DFE was presented. The basic reproduction number was computed by next generation matrix method. Numerical simulations using R are given.

Finally, it is important to mention that mathematical models based on fractional derivatives are, in general, a more powerful approach to epidemiological models, not only because we can choose the order a of the fractional derivative, but also because of the memory properties of the fractional derivatives.

References

1. Barabási, A.-L., Albert, R.: Emergence of scaling in random networks. Science **286**(5439), 509–512 (1999)
2. Beaumont, M.A.: Approximate bayesian computation. Ann. Rev. Stat. Appl. **6**(2019), 2.1-2.25 (2018)
3. Bertozzi, A.L., Franco, E., et al.: Mathematical modeling of the spread of the coronavirus disease 2019 (COVID-19) taking into account the undetected infections. The case of China. Commun. Nonlinear Sci. Numer. Simul. **88**, 105303 (2020)

4. Bulut, C., Kato, Y.: Epidemiology of COVID-19. Turk. J. Med. Sci. **50**, 563–570 (2020)
5. Covidmaroc. http://www.covidmaroc.ma/
6. Fang, F., Zhao, D., Chen, Y., et al.: Recommendations for the diagnosis, prevention and control of the 2019 novel coronavirus infection in children. Chin. J. Pediatr. **58**(3), 169–174 (2020)
7. Giordano, G., Blanchini, F., Bruno, R., et al.: Modelling the COVID-19 epidemic and implementation of population-wide interventions in Italy. Nat Med. **26**, 855–860 (2020)
8. He, S., Peng, Y., Sun, K.: SEIR modeling of the COVID-19 and its dynamics. Nonlinear Dyn. **101**(3), 1667–1680 (2020). https://doi.org/10.1007/s11071-020-05743-y
9. Hethcote, H.W., Tudor, D.W.: Integral equation models for endemic infectious diseases. J. Math. Biology. **9**, 37–47 (1980). https://doi.org/10.1007/BF00276034
10. Kermack, W.O., McKendrick, A.G.: A contribution to the mathematical theory of epidemics. Proc. Roy. Soc. A. **115**(772), 700–721 (1927)
11. Li, J., et al.: Epidemiology of COVID-19: a systematic review and meta-analysis of clinical characteristics, risk factors, and outcomes. Med. Virol. J. **93**(3), 1449–1458 (2021)
12. Sisson, S.A., Fan, Y., Beaumont, M.: Handbook of Approximate Bayesian Computation. Chapman and Hall/CRC, 1st (ed.) (2018)
13. Toni, T., Welch, D., Strelkowa, N., Ipsen, A., Stumpf, M.P.: Approximate Bayesian computation scheme for parameter inference and model selection in dynamical systems. J. R. Soc. Interface **6**(31), 187–202 (2009)
14. Velavan, T.P., Meyer, C.G.: The COVID-19 epidemic; trop. Med. Int. Health **25**, 278–280 (2020)
15. Verity, R., Okell, L.C., et al.: Estimates of the severity of coronavirus disease 2019: a model-based analysis. Lancet. Infect. Dis. **20**(6), 669–677 (2020)
16. Wang, Y., Wang, Y., Chen, Y., Quin, Q.: Unique epidemiological and clinical features of the emerging 2019 novel coronavirus pneumonia (COVID-19) implicate special control measures. J. Med. Virol. **92**, 568–576 (2020)
17. Worldometers. https://www.worldometers.info/coronavirus/
18. Wu, Z., McGoogan, J.M.: Characteristics of and important lessons from the coronavirus disease 2019 (COVID-19) outbreak in China: summary of a report of 72 314 cases from the Chinese center for disease control and prevention. JAMA **323**(13), 1239–1242 (2020)

Existence and Uniqueness Results of Fuzzy Fractional Stochastic Differential Equations with Impulsive

E. Arhrrabi, M. Elomari, S. Melliani(✉), and L. S. Chadli

LMACS, Sultan Moulay Slimane University, BP 523 - 23000 Beni Mellal, Morocco
s.melliani@usms.ma

Abstract. In this paper, we study the existence and uniqueness of solution for a new kind of equations, namely fuzzy fractional stochastic differential equations (FFSDEs) under generalized Hukuhara differentiability using the principale of contraction mappings. An example is provided to illustrate the results.

1 Introduction

The theory of impulsive differential equations is emerging as a crucial area of investigation since such equations appear to represent a natural frame work for mathematical modeling of the many real processes and phenomena studied in electronics, optimal control and economics. On the other hand, the fuzzy stochastic differential equations could even be applicable within the investigation of various economic and engineering problems where the development are subjected to both fuzziness and randomness. Fuzziness dealt to unsharp boundaries of the parameters of the model and the randomness dealt to the stochastic variability of al feasible outcomes of a situation.

Recently, fractional differential equations are attracting more attentions of several researchers because they are more effective in modelling of many real-world problems.

The first paper on fractional differential equations (FDEs) with uncertainties have been given by Agarwal in [1]. For fuzzy fractional concepts, recently the authors [14–18] established, existence and uniqueness result of fuzzy fractional differential equation with nonlocal conditions, solving the fuzzy fractional differential wave equation by mean fuzzy Fourier Transform and proved existence, uniqueness and approximate solutions of fuzzy fractional differential equations. Abbasbandy et al. [2] studied a fuzzy FDEs with uncertainty under Hukuhara differentiability. In [3], M. Benchohra, et al. studied a Fuzzy solutions for impulsive differential equations. Combining the theories of impulsive differential equations Lakshmikantham et al. [4] initiated the study of fuzzy impulsive differential equations. Van and Vu Hoa [5] studied the existence and uniqueness of impulsive fuzzy functional differential equations.

S. Melliani and O. Castillo (Eds.): ICPAMS 2021, LNNS 476, pp. 147–163, 2023.
https://doi.org/10.1007/978-3-031-12416-7_13

Motivated by authors in paper, [4,6–13], we study the existence and uniqueness of solution of fuzzy impulsive fractional stochastic differential equations. The rest of this paper is organized as follows: In Sect. 2, we introduce some concepts, notations and basic results about fuzzy set, generalized Hukuhara differentiability and fuzzy fractional differential equations. The existence and uniqueness of solutions are discussed by using Banach fixed point Theorem in Sect. 3. Finally, in Sect. 4, we investigate an example to illustrate the results.

2 Preliminaries

Let $\mathbf{E}^n$ denote the set of fuzzy subsets of the real axis, if $\omega : \mathbb{R}^n \longrightarrow [0,1]$, satisfying the following properties:

(i) ω is normal, that is, there exists $z_0 \in \mathbb{R}^n$ such that $\omega(z_0) = 1$,
(ii) ω is fuzzy convex, that is, for $0 \leq \lambda \leq 1$

$$\omega(\lambda z_1 + (1-\lambda)z_2) \geq min\{\omega(z_1), \omega(z_2)\}, \ for\ any\ z_1, z_2 \in \mathbb{R}^n,$$

(iii) ω is upper semicontinous on $\mathbb{R}^n$,
(iv) $[\omega]^0 = cl\{z \in \mathbb{R}^n : \omega(z) > 0\}$ is compact, where cl denotes the closure in $(\mathbb{R}^n, |\,.\,|)$.
Then $\mathbf{E}^n$ is called the space of fuzzy number. For $r \in (0,1]$, we denote $[\omega]^r = \{z \in \mathbb{R}^n \mid \omega(z) \geq r\}$ and $[\omega]^0 = \{z \in \mathbb{R}^n \mid \omega(z) > 0\}$. From the conditions (i) to (iv), it follows that the $r-level$ set of ω, $[\omega]^r$, is a nonempty compact interval, for all $r \in [0,1]$ and any $\omega \in \mathbf{E}^n$.

The notation $[\omega]^r = [\underline{\omega}(r), \overline{\omega}(r)]$, denotes explicitly the $r-level$ set of ω, for $r \in [0,1]$. We refer to $\underline{\omega}$ and $\overline{\omega}$ as the lower and upper branches of ω, respectively. For $\omega \in \mathbf{E}^n$, we define the lengh of the $r-level$ set of ω as $len([\omega]^r) = \overline{\omega}(r) - \underline{\omega}(r)$. For addition and scalar multiplication in fuzzy set space $\mathbf{E}^n$, we have $[\omega_1 + \omega_2]^r = [\omega_1]^r + [\omega_2]^r$, $[\lambda\omega]^r = \lambda[\omega]^r$.

The Hausdorff distance between fuzzy numbers is given by

$$D[\omega_1, \omega_2] = \sup_{0 \leq r \leq 1} \{| \underline{\omega}_1(r) - \underline{\omega}_2(r) |, | \overline{\omega}_1(r) - \overline{\omega}_2(r) |\}.$$

The metric space $(\mathbf{E}^n, D)$ is complet metric space and the following properties of the metric D are valid.

$$D[\omega_1 + \omega_3, \omega_2 + \omega_3] = D[\omega_1, \omega_2],$$

$$D[\lambda\omega_1, \lambda\omega_2] = | \lambda | \, D[\omega_1, \omega_2],$$

$$D[\omega_1, \omega_2] \leq D[\omega_1, \omega_3] + D[\omega_3, \omega_2],$$

for all $\omega_1, \omega_2, \omega_3 \in \mathbf{E}^n$ and $\lambda \in \mathbb{R}^n$. Let $\omega_1, \omega_2 \in \mathbf{E}^n$, if there exists $\omega_3 \in \mathbf{E}^n$ such that $\omega_1 = \omega_2 + \omega_3$ then ω_3 is called the H-difference of ω_1, ω_2. We denote the ω_3 by $\omega_1 \ominus \omega_2$. Let us remark that $\omega_1 \ominus \omega_2 \neq \omega_1 + (-1)\omega_2$.

Definition 1. The generalized Hukuhara difference of two fuzzy numbers $\omega_1, \omega_2 \in \mathbf{E}^n$ (gH-difference for short) is defined as follows:

$$\omega_1 \ominus_{gH} \omega_2 = \omega_3 \Leftrightarrow \begin{cases} & (i)\ \ \omega_1 = \omega_2 + \omega_3, \\ or & (ii)\ \ \omega_2 = \omega_1 + (-1)\omega_3. \end{cases}$$

The generalized Hukuhara differentiability was introduced in [7].

Definition 2. Let $t \in (a, b)$ and h such that $t + h \in (a, b)$, then the generalized Hukuhara derivative of fuzzy-valued function $x : (a, b) \longrightarrow \mathbf{E}^n$ at t is defined as

$$D_{gH}x(t) = \lim_{h \longrightarrow 0} \frac{x(t + h) \ominus_{gH} x(t)}{h}. \tag{1}$$

If $D_{gH}x(t) \in \mathbf{E}^n$ satisfying (1) exists, we say that x is generalized Hukuhara differentiable (gH-differentiable for short) at t. Also, we say that x is $[(i) - gH]$-differentiable at t if (i) $[D_{gH}x(t)]^r = [\underline{x}'(t, r), \overline{x}'(t, r)]$, and that x is $[(ii) - gH]$-differentiable at t if (ii) $[D_{gH}x(t)]^r = [\overline{x}'(t, r), \underline{x}'(t, r)]$, $r \in [0, 1]$.

Theorem 1. *Let $x : [a, b] \longrightarrow \mathbf{E}^n$ be such that $[x(t)]^r = [\underline{x}'(t, r), \overline{x}'(t, r)]$ for $t \in [a, b]$, $r \in [0, 1]$. If the real-valued function $\underline{x}(t, r)$ and $\overline{x}(t, r)$ are differentiable at $t \in [a, b]$, then the function x is gH-differentiable at $t \in [a, b]$ and*

$$[D_{gH}x(t)]^r = [min\{\underline{x}'(t, r), \overline{x}'(t, r)\}, max\{\underline{x}'(t, r), \overline{x}'(t, r)\}]\,. \tag{2}$$

Let $I = [a, b] \subset \mathbb{R}^n$ be a compact interval we shall use the notation $C([a, b], \mathbf{E}^n) = \{x : I \longrightarrow \mathbf{E}^n \mid x \textit{ is continuous}\}$, where the continuous is one-sided at endpoints a, b. In the space $C([a, b], \mathbf{E}^n)$, we consider the following metric:

$$D_0^*[x, z] = \sup_{t \in [a,b]} D_0[x(t), z(t)].$$

It is known that $(C([a, b], \mathbf{E}^n), D_0^*)$ is a complete metric space. Also, we denote the space of all Lebesque integrable fuzzy-valued functions on $[a, b]$ by $L([a, b], \mathbf{E}^n)$. Let $x \in C([a, b], \mathbf{E}^n)$, we say that $x \in L(C([a, b], \mathbf{E}^n), D_0^*)$ if and only if $D_0[\int_a^b x(t)dt, \widehat{0}] < \infty$.

Definition 3. The Riemann-Liouville fractional integral operator of order $\alpha > 0$, of a real-valued function $\varphi \in L^1[a, b]$, is defined as

$$I_{a^+}^{\alpha}\varphi(t) = \frac{1}{\Gamma(\alpha)} \int_a^t (t - s)^{\alpha - 1}\varphi(s)ds,$$

where $\Gamma(.)$ is the Euler gamma function.

Definition 4. Let $\varphi : [a, b] \longrightarrow \mathbb{R}^n$, the Caputo fractional derivative of order $\alpha > 0$, $m - 1 < \alpha < m$, $m \in \mathbb{N}$, is defined as

$${}^{C}D_{a^+}^{\alpha}\varphi(t) = \frac{1}{\Gamma(m - \alpha)} \int_a^t (t - s)^{m - \alpha - 1}\varphi^{(m)}(s)ds,$$

where the function $\varphi(t)$ has absolutely continuous derivatives up to order $(m-1)$. If $\alpha \in (0,1)$, then

$$ {}^{C}D^{\alpha}_{a^+}\varphi(t) = \frac{1}{\Gamma(1-\alpha)}\int_a^t (t-s)^{-\alpha}\varphi'(s)ds. $$

Definition 5. Let $x : [a,b] \longrightarrow \mathbf{E}^n$, the fuzzy Rieman-Liouville integral of fuzzy-valued function x is defined as follows:

$$ (\mathcal{J}^{\alpha}_{a^+}x)(t) = \frac{1}{\Gamma(\alpha)}\int_a^t (t-s)^{\alpha-1}x(s)ds. \tag{3} $$

For $a \leq t$, and $0 < \alpha \leq 1$. For $= \alpha = 1$, we set $\mathcal{J}^1_a = I$, the identity operator.

Definition 6. Let $D_{gH} \in C([a,b],\mathbf{E}^n) \cap L([a,b],\mathbf{E}^n)$. The fuzzy gH-fractional Caputo diffentiability of fuzzy-valued function x ($[gH]^C_a - differentiable\ for\ short$) is defined as following:

$$ {}^{C}_{gH}\mathcal{D}^{\alpha}_{a^+}x(t) = \mathcal{J}^{1-\alpha}_{a^+}(D_{gH}x)(t) = \frac{1}{\Gamma(1-\alpha)}\int_a^t (t-s)^{-\alpha}(D_{gH}x)(s)ds, $$

where $0 < \alpha \leq 1$, $t > a$.

Lemma 1. *Suppose that $x : [a,b] \longrightarrow \mathbf{E}^n$ be a fuzzy function and $D_{gH}x(t) \in C([a,b],\mathbf{E}^n) \cap L([a,b],\mathbf{E}^n)$. Then*

$$ \mathcal{J}^{\alpha}_{a^+}({}^{C}_{gH}\mathcal{D}^{\alpha}_{a^+}x)(t) = x(t) \ominus_{gH} x(a). $$

Theorem 2. [4] *Let $D_{gH}x(t) \in C([a,b],\mathbf{E}^n) \cap L([a,b],\mathbf{E}^n)$ be such that $[x(t)]^r = [\underline{x}(t,r),\overline{x}(t,r)]$ for $0 \leq r \leq 1$, $t \in [a,b]$, then the function x is $[gH]^C_\alpha$-differentiable at $t \in [a,b]$ and*

$$ [{}^{C}_{gH}\mathcal{D}^{\alpha}_{a^+}x(t)]^r = \left[min\{{}^{C}\mathcal{D}^{\alpha}_{a^+}\underline{x}(t,r), {}^{C}\mathcal{D}^{\alpha}_{a^+}\overline{x}(t,r)\}, max\{{}^{C}\mathcal{D}^{\alpha}_{a^+}\underline{x}(t,r), {}^{C}\mathcal{D}^{\alpha}_{a^+}\overline{x}(t,r)\}\right]. \tag{4} $$

where ${}^{C}\mathcal{D}^{\alpha}_{a^+}\underline{x}(t,r)$ and ${}^{C}\mathcal{D}^{\alpha}_{a^+}\overline{x}(t,r)$ defined in Definition 4.

Lemma 2. *If $x(t) = (z_1(t), z_2(t), z_3(t))$ is a triangular fuzzy number valued function, then:*

(i) *If x is $[(i) - gH]$-differentiable at $t \in [a,b]$ then*

$$ ({}^{C}_{gH}\mathcal{D}^{\alpha}_{a^+}x)(t) = ({}^{C}\mathcal{D}^{\alpha}_{a^+}z_1(t), {}^{C}\mathcal{D}^{\alpha}_{a^+}z_2(t), {}^{C}\mathcal{D}^{\alpha}_{a^+}z_3(t)). $$

(ii) *If x is $[(ii) - gH]$-differentiable at $t \in [a,b]$ then*

$$ ({}^{C}_{gH}\mathcal{D}^{\alpha}_{a^+}x)(t) = ({}^{C}\mathcal{D}^{\alpha}_{a^+}z_3(t), {}^{C}\mathcal{D}^{\alpha}_{a^+}z_2(t), {}^{C}\mathcal{D}^{\alpha}_{a^+}z_1(t)). $$

Definition 7. Let $x : [a,b] \longrightarrow \mathbf{E}^n$ be $[gH]^C_\alpha$-differentiable at $t \in (a,b)$. We say x is $[(i) - gH]^C_\alpha$-differentiable at $t \in [a,b]$ if

$$ (i) \quad [({}^{C}_{gH}\mathcal{D}^{\alpha}_{a^+}x)(t)]^r = [{}^{C}\mathcal{D}^{\alpha}_{a^+}\underline{x}(t,r), {}^{C}\mathcal{D}^{\alpha}_{a^+}\overline{x}(t,r)], \; 0 \leq r \leq 1 \tag{5} $$

and that x is $[(ii) - gH]^C_\alpha$-differentiable at t if

$$(ii) \quad [({}^C_{gH}\mathcal{D}^\alpha_{a+}x)(t)]^r = [{}^C\mathcal{D}^\alpha_{a+}\overline{x}(t,r), {}^C\mathcal{D}^\alpha_{a+}\underline{x}(t,r)],\ 0 \leq r \leq 1 \tag{6}$$

where

$${}^C\mathcal{D}^\alpha_{a+}\underline{x}(t,r) = \frac{1}{\Gamma(1-\alpha)}\int_a^t (t-s)^{-\alpha}\frac{d}{ds}\underline{x}(s,r)ds,$$

$${}^C\mathcal{D}^\alpha_{a+}\overline{x}(t,r) = \frac{1}{\Gamma(1-\alpha)}\int_a^t (t-s)^{-\alpha}\frac{d}{ds}\overline{x}(s,r)ds.$$

Definition 8. Let $x : [a,b] \longrightarrow \mathbf{E}^n$ be a fuzzy function. A point $t \in (a,b)$ is said to be a switching point for the $[gH]^C_\alpha$-differentiable of x, if in any neighborhood V of t there exist points $t_1 < t < t_2$ such that:

type I at t_1 (5) holds while (6) does not hold and at t_2 (6) holds and (5) does not holds, or

type II at t_1 (6) holds while (5) does not hold and at t_2 (5) holds and (6) does not holds.

Let $\langle . \rangle : \mathbb{R}^n \longrightarrow \mathbf{E}^n$ denote the embedding of $\mathbb{R}^n$ into $\mathbf{E}^n$, i.e. for $r \in \mathbb{R}^n$ we have

$$\langle r\rangle(a) = \begin{cases} 1, & if \quad a = r, \\ 0, & if \quad a \neq r. \end{cases}$$

Notations: Let $(\Omega, \mathcal{A}, \mathbb{P})$ be a complete probability space with filtration $\{\mathcal{A}_t : t \in I := [0,T]\}$ satisfies the usual conditions. Let $B(t)$ be a one-dimentional Brownian motion defined on the probability space $(\Omega, \mathcal{A}, \mathbb{P})$ and $PC_0 := C([-\sigma, 0], \mathbf{E}^n)$ denote the space of continuous mapping $\varphi : [-\sigma, 0] \longrightarrow \mathbf{E}^n$ equipped with the metric

$$\sup_{t\geq 0} D[\varphi(t), \widehat{0}]e^{-a(s-t)} < \infty.$$

Denote by $L^2(\Omega, \mathcal{A}_t, \mathbf{E}^n)$ the Hilbert space of all square integrable random variable with values in $\mathbf{E}^n$ and $PC_T = PC([-\sigma, T], \mathbf{E}^n)$ the space of all function $x : [-\sigma, T] \longrightarrow \mathbf{E}^n$ which are continuous at $t \neq t_i$, $i = 1, .., n$ at which $x(t_i^+)$ and $x(t_i^-) = x(t_i)$ exist, where $x(t_i^+)$ and $x(t_i^-)$ stand for right and left limits of x at t_i respectively.

Let $\mathcal{B}_a := (I, L^2(\Omega, PC_T))$ denote the family of all bounded $\mathcal{A}_t$-measurable square integrable processes equipped with the following metric

$$D_a^2[x,y] = \sup_{t\in[-\sigma,T]} D^2[x(t), y(t)]e^{-a(s-t)}.$$

Note that $(\mathcal{B}_a, D_a)$ is a complete metric space.

3 Existance and Uniqueness of a Solution

In this section, we consider the following fuzzy impulsive fractional stochastic differential equations:

$$\begin{cases} {}_{gH}^{C}\mathcal{D}^{\alpha}x(t) = f(t,x_t) + \langle \int_0^t g(s,x_s)dB(s)\rangle, & t \in I, t \neq t_i, \\ x(t_k^+) = x(t_k) + I_i(t_i, x(t_i)), & i = 1,..,n, \\ x(t) = \varphi(t), & t \in [-\sigma, 0], \end{cases} \tag{7}$$

where $0 < \alpha < 1$ and ${}_{gH}^{C}\mathcal{D}^{\alpha}$ is the Caputo's generalized Hukuhara derivative, $0 = t_0 < t_1 < t_2...t_k < t_n < t_{n+1} = T$ and $f : I \times PC_0 \longrightarrow \mathbf{E}^n$, $g : I \times PC_0 \longrightarrow \mathbb{R}^n \times \mathbb{R}^n$ are continuous on T and for each $x_t \in PC_0$ we have $x_t(s) = x(t+s)$ for $s \in [-\sigma, 0]$, $I_i : I \times E \longrightarrow \mathbf{E}^n$ are continuous for $i = 1,..,n$. In this paper, we consider only (i)-differentiable type and (ii)-differentiable type solutions.

Lemma 3. *A fuzzy continuous mappiing $x : [-\sigma, T] \longrightarrow \mathbf{E}^n$ is a solution to the system (7) on $[-\sigma, T]$ if and only if x is a continuous fuzzy mapping and it satisfies to one of the following fuzzy integral equations:*

(I1) if x is (i)-differentiable, then

$$x(t) = \begin{cases} \varphi(t), & t \in [-\sigma, 0]; \\ \varphi(0) + \frac{1}{\Gamma(\alpha)}\int_0^t \frac{f(s,x_s)}{(t-s)^{1-\alpha}}ds + \frac{1}{\Gamma(\alpha)}\int_0^t \frac{\langle \int_0^s g(\theta,x_\theta)dB(\theta)\rangle}{(t-s)^{1-\alpha}}ds, & t \in [0, t_1]; \\ \cdot & \\ \cdot & \\ \cdot & \\ \varphi(0) + \sum_{i=1}^k \frac{1}{\Gamma(\alpha)}\int_{t_{i-1}}^{t_i} \frac{f(s,x_s)}{(t_i-s)^{1-\alpha}}ds + \frac{1}{\Gamma(\alpha)}\int_{t_k}^t \frac{f(s,x_s)}{(t-s)^{1-\alpha}}ds & \\ +\sum_{i=1}^k \frac{1}{\Gamma(\alpha)}\int_{t_{i-1}}^{t_i} \frac{\langle \int_0^s g(\theta,x_\theta)dB(\theta)\rangle}{(t_i-s)^{1-\alpha}}ds + \frac{1}{\Gamma(\alpha)}\int_{t_k}^t \frac{\langle \int_0^s g(\theta,x_\theta)dB(\theta)\rangle}{(t-s)^{1-\alpha}}ds & \\ +\sum_{i=1}^k I_i(t_i, x(t_i)), & t \in [t_k, t_{k+1}]. \end{cases} \tag{8}$$

(I2) if x is (ii)-differentiable, then

$$x(t) = \begin{cases} \varphi(t), & t \in [-\sigma, 0]; \\ \varphi(0) \ominus (-1)\left[\frac{1}{\Gamma(\alpha)}\int_0^t \frac{f(s,x_s)}{(t-s)^{1-\alpha}}ds + \frac{1}{\Gamma(\alpha)}\int_0^t \frac{\langle \int_0^s g(\theta,x_\theta)dB(\theta)\rangle}{(t-s)^{1-\alpha}}ds\right], & t \in [0, t_1]; \\ \cdot & \\ \cdot & \\ \cdot & \\ \varphi(0) \ominus (-1)[\sum_{i=1}^k \frac{1}{\Gamma(\alpha)}\int_{t_{i-1}}^{t_i} \frac{f(s,x_s)}{(t_i-s)^{1-\alpha}}ds + \frac{1}{\Gamma(\alpha)}\int_{t_k}^t \frac{f(s,x_s)}{(t-s)^{1-\alpha}}ds & \\ +\sum_{i=1}^k \frac{1}{\Gamma(\alpha)}\int_{t_{i-1}}^{t_i} \frac{\langle \int_0^s g(\theta,x_\theta)dB(\theta)\rangle}{(t_i-s)^{1-\alpha}}ds + \frac{1}{\Gamma(\alpha)}\int_{t_k}^t \frac{\langle \int_0^s g(\theta,x_\theta)dB(\theta)\rangle}{(t-s)^{1-\alpha}}ds & \\ +\sum_{i=1}^k I_i(t_i, x(t_i))], & t \in [t_k, t_{k+1}]. \end{cases} \tag{9}$$

Proof: We will prove only the case (i)-differentiable of solution because the other case of solution can be proved in similar way. If $x : [-\sigma, T] \longrightarrow \mathbf{E}^n$ satisfies (7), then it will be expressed as (8). Indeed, if $t \in [0, t_1]$, so we have:

$${}_{gH}^{C}\mathcal{D}^{\alpha}x(t) = f(t,x_t) + \langle \int_0^t g(s,x_s)dB(s)\rangle.$$

By Lemma 3.4 in [18], we get

$$x(t) = \varphi(0) + \frac{1}{\Gamma(\alpha)}\int_0^t \frac{f(s,x_s)}{(t-s)^{1-\alpha}}ds + \frac{1}{\Gamma(\alpha)}\int_0^t \frac{\langle \int_0^s g(\theta,x_\theta)dB(\theta)\rangle}{(t-s)^{1-\alpha}}ds,$$

so

$$x(t_1) = \varphi(0) + \frac{1}{\Gamma(\alpha)} \int_0^{t_1} \frac{f(s, x_s)}{(t_1 - s)^{1-\alpha}} ds + \frac{1}{\Gamma(\alpha)} \int_0^{t_1} \frac{\langle \int_0^s g(\theta, x_\theta) dB(\theta) \rangle}{(t_i - s)^{1-\alpha}} ds,$$

for $t \in [t_1, t_2]$, then Lemma 3.4 in [18] implies

$$\begin{aligned}
x(t) &= x(t_1^+) + \frac{1}{\Gamma(\alpha)} \int_{t_1}^{t} \frac{f(s, x_s)}{(t - s)^{1-\alpha}} ds + \frac{1}{\Gamma(\alpha)} \int_{t_1}^{t} \frac{\langle \int_0^s g(\theta, x_\theta) dB(\theta) \rangle}{(t - s)^{1-\alpha}} ds, \\
&= x(t_1) + I_1(t_1, x(t_1)) + \frac{1}{\Gamma(\alpha)} \int_{t_1}^{t} \frac{f(s, x_s)}{(t - s)^{1-\alpha}} ds + \frac{1}{\Gamma(\alpha)} \int_{t_1}^{t} \frac{\langle \int_0^s g(\theta, x_\theta) dB(\theta) \rangle}{(t - s)^{1-\alpha}} ds, \\
&= I_1(t_1, x(t_1)) + \varphi(0) + \frac{1}{\Gamma(\alpha)} \int_0^{t_1} \frac{f(s, x_s)}{(t_1 - s)^{1-\alpha}} ds + \frac{1}{\Gamma(\alpha)} \int_0^{t_1} \frac{\langle \int_0^s g(\theta, x_\theta) dB(\theta) \rangle}{(t_1 - s)^{1-\alpha}} ds \\
&+ \frac{1}{\Gamma(\alpha)} \int_{t_1}^{t} \frac{f(s, x_s)}{(t - s)^{1-\alpha}} ds + \frac{1}{\Gamma(\alpha)} \int_{t_1}^{t} \frac{\langle \int_0^s g(\theta, x_\theta) dB(\theta) \rangle}{(t - s)^{1-\alpha}} ds,
\end{aligned}$$

then, by mathematical induction, if $t \in (t_k, t_{k+1}]$, $k = 1, .., n$ and using again Lemma 3.4 in [18], we get

$$\begin{aligned}
x(t) &= x(t_k^+) + \frac{1}{\Gamma(\alpha)} \int_{t_k}^{t} \frac{f(s, x_s)}{(t_k - s)^{1-\alpha}} ds + \frac{1}{\Gamma(\alpha)} \int_{t_k}^{t} \frac{\langle \int_0^s g(\theta, x_\theta) dB(\theta) \rangle}{(t_k - s)^{1-\alpha}} ds, \\
&= x(t_k) + I_k(t_k, x(t_k)) + \frac{1}{\Gamma(\alpha)} \int_{t_k}^{t} \frac{f(s, x_s)}{(t_k - s)^{1-\alpha}} ds + \frac{1}{\Gamma(\alpha)} \int_{t_k}^{t} \frac{\langle \int_0^s g(\theta, x_\theta) dB(\theta) \rangle}{(t_k - s)^{1-\alpha}} ds, \\
&= I_k(t_k, x(t_k)) + \varphi(0) + \frac{1}{\Gamma(\alpha)} \int_0^{t_1} \frac{f(s, x_s)}{(t_1 - s)^{1-\alpha}} ds + \frac{1}{\Gamma(\alpha)} \int_0^{t_1} \frac{\langle \int_0^s g(\theta, x_\theta) dB(\theta) \rangle}{(t_1 - s)^{1-\alpha}} ds \\
&+ \frac{1}{\Gamma(\alpha)} \int_{t_1}^{t} \frac{f(s, x_s)}{(t - s)^{1-\alpha}} ds + \frac{1}{\Gamma(\alpha)} \int_{t_1}^{t} \frac{\langle \int_0^s g(\theta, x_\theta) dB(\theta) \rangle}{(t - s)^{1-\alpha}} ds + ... \\
&+ \frac{1}{\Gamma(\alpha)} \int_{t_{k-1}}^{t_k} \frac{f(s, x_s)}{(t_k - s)^{1-\alpha}} ds + \frac{1}{\Gamma(\alpha)} \int_{t_{k-1}}^{t_k} \frac{\langle \int_0^s g(\theta, x_\theta) dB(\theta) \rangle}{(t_k - s)^{1-\alpha}} ds \\
&+ \frac{1}{\Gamma(\alpha)} \int_{t_k}^{t} \frac{f(s, x_s)}{(t - s)^{1-\alpha}} ds + \frac{1}{\Gamma(\alpha)} \int_{t_k}^{t} \frac{\langle \int_0^s g(\theta, x_\theta) dB(\theta) \rangle}{(t - s)^{1-\alpha}} ds, \\
&= \sum_{i=1}^{k} I_i(t_i, x(t_i)) + \varphi(0) + \frac{1}{\Gamma(\alpha)} \sum_{i=1}^{k} \int_{t_{i-1}}^{t_i} \frac{f(s, x_s)}{(t_i - s)^{1-\alpha}} ds \\
&+ \frac{1}{\Gamma(\alpha)} \sum_{i=1}^{k} \int_{t_{i-1}}^{t_i} \frac{\langle \int_0^s g(\theta, x_\theta) dB(\theta) \rangle}{(t_i - s)^{1-\alpha}} ds + \frac{1}{\Gamma(\alpha)} \int_{t_k}^{t} \frac{f(s, x_s)}{(t - s)^{1-\alpha}} ds \\
&+ \frac{1}{\Gamma(\alpha)} \int_{t_k}^{t} \frac{\langle \int_0^s g(\theta, x_\theta) dB(\theta) \rangle}{(t - s)^{1-\alpha}} ds.
\end{aligned}$$

Conversely, assume that x satisfies the impulsive fractional integral equation (8). If $t \in [0, t_1]$, then $x(0) = \varphi(0)$ and using the fact ${}_{gH}^{C}\mathcal{D}^\alpha$ is the left inverse of I^α we get

$${}_{gH}^{C}\mathcal{D}^\alpha x(t) = f(t, x_t) + \langle \int_0^t g(s, x_s) dB(s) \rangle. \quad \text{for each} \quad t \in [0, t_1].$$

If $t \in (t_k, t_{k+1}]$, $k = 1,..,n$ and using the fact that ${}_{gH}^{C}\mathcal{D}^{\alpha}C = 0$, where C is a constant, we get

$$ {}_{gH}^{C}\mathcal{D}^{\alpha}x(t) = f(t, x_t) + \langle \int_0^t g(s, x_s)dB(s)\rangle. \quad \text{for each} \quad t \in [t_k, t_{k+1}]. $$

Also, we can show that

$$ x(t_k^+) = x(t_k) + I_k(t_k, x(t_k)), \quad k = 1, .., n. $$

□

Now, we will prove our result via principle of contraction mappings. Again, since the way of the proof is similar for two cases, we will consider only case (i)-differentiable of solution on $[0, T]$.

Theorem 3. *Assume that* $f : I \times PC_0 \longrightarrow \mathbf{E}^n$, $g : I \times PC_0 \longrightarrow \mathbb{R}^{n+1}$ *and* $I_i : I \times E \longrightarrow \mathbf{E}^n$ *are continuous and satisfy the following assumptions:*

($\mathcal{H}1$) *For all* φ, $\psi \in PC_0$, $\forall t \in I$, *there exist positive constant* p, q *such that*

$$ \mathbb{E}D^2\left[f(t,\varphi), f(t,\psi)\right] \le p\mathbb{E}D_{\sigma}^2[\varphi,\psi], $$

$$ \mathbb{E}\|g(t,\varphi) - g(t,\psi)\|^2 \le q\mathbb{E}D_{\sigma}^2[\varphi,\psi]. $$

($\mathcal{H}2$) *For all* $t \ge 0$, $\exists r, c > 0$ *and* $a > b \ge 0$ *such that*

$$ \mathbb{E}D^2[f(t,\widehat{0}),\widehat{0}] \le re^{b(s-t)}, $$

$$ \mathbb{E}D^2[g(t,\widehat{0}),\widehat{0}] \le ce^{b(s-t)}. $$

($\mathcal{H}3$) *For all* $\varphi, \psi \in PC_0$, $\forall t_i \in I$, $\exists l_i > 0$, $i = 1, .., n$ *such that*

$$ \mathbb{E}D^2[I_i(t_i,\varphi), I_i(t_i,\psi)] \le l_i D^2[\varphi,\psi]. $$

Then, the problem (7) *has a unique solution on* I, *provided that*

$$ H := \frac{5p}{a^{\alpha}} + \frac{5q}{a^{\alpha+1}} - \frac{5qT^{\alpha}e^{-as}}{\Gamma(\alpha+1)} + \frac{1}{2}k(k+1)\widehat{l} < 1, \tag{10} $$

where $\widehat{l} = \max\{l_1, ..., l_k\}$.

Proof: We transform the problem (7) into a fixed point problem. Consider the operator $T : \mathcal{B}_a \longrightarrow \mathcal{B}_a$, given by

$$ (Tx)(t) = \begin{cases} \varphi(t), & t \in [-\sigma, 0]; \\ \varphi(0) + \frac{1}{\Gamma(\alpha)}\int_0^t \frac{f(s,x_s)}{(t-s)^{1-\alpha}}ds + \frac{1}{\Gamma(\alpha)}\int_0^t \frac{\langle \int_0^s g(\theta,x_\theta)dB(\theta)\rangle}{(t-s)^{1-\alpha}}ds, & t \in [0, t_1]; \\ \cdot & \\ \cdot & \\ \cdot & \\ \varphi(0) + \sum_{i=1}^k \frac{1}{\Gamma(\alpha)}\int_{t_{i-1}}^{t_i} \frac{f(s,x_s)}{(t_i-s)^{1-\alpha}}ds + \frac{1}{\Gamma(\alpha)}\int_{t_k}^t \frac{f(s,x_s)}{(t-s)^{1-\alpha}}ds & \\ + \sum_{i=1}^k \frac{1}{\Gamma(\alpha)}\int_{t_{i-1}}^{t_i} \frac{\langle \int_0^s g(\theta,x_\theta)dB(\theta)\rangle}{(t_i-s)^{1-\alpha}}ds + \frac{1}{\Gamma(\alpha)}\int_{t_k}^t \frac{\langle \int_0^s g(\theta,x_\theta)dB(\theta)\rangle}{(t-s)^{1-\alpha}}ds & \\ + \sum_{i=1}^k I_i(t_i, x(t_i)), & t \in [t_k, t_{k+1}]. \end{cases} $$

Let $x \in \mathcal{B}_a$, i.e., there exists $\rho > 0$ such that $\mathbb{E}D^2[x(t), \widehat{0}] \le \rho e^{a(s-t)}$ for any $t \in [-\sigma, T]$. So, we have

$$\sup_{\theta\in[-\sigma,0]} \mathbb{E}D^2[x(t+\theta), \widehat{0}] \le \rho e^{a(s-t)}, \qquad \forall t \in I.$$

The proof will be given into two steps.

Step 1: The operator T is will defined. Indeed, for $t \in [0, t_1]$, we have:

$$\begin{aligned}
\mathbb{E}D^2[(Tx)(t), \widehat{0}] &= \mathbb{E}D^2[\varphi(0) + \frac{1}{\Gamma(\alpha)}\int_0^t \frac{f(s,x_s)}{(t-s)^{1-\alpha}}ds + \frac{1}{\Gamma(\alpha)}\int_0^t \frac{\langle \int_0^s g(\theta,x_\theta)dB(\theta)\rangle}{(t-s)^{1-\alpha}}ds, \widehat{0}],\\
&\le 3\mathbb{E}D^2[\varphi(0), \widehat{0}] + 3\mathbb{E}D^2[\frac{1}{\Gamma(\alpha)}\int_0^t \frac{f(s,x_s)}{(t-s)^{1-\alpha}}ds, \widehat{0}]\\
&+ 3\mathbb{E}D^2[\frac{1}{\Gamma(\alpha)}\int_0^t \frac{\langle \int_0^s g(\theta,x_\theta)dB(\theta)\rangle}{(t-s)^{1-\alpha}}ds, \widehat{0}],\\
&\le 3\mathbb{E}D^2[\varphi(0), \widehat{0}] + \frac{6}{\Gamma(\alpha)}\int_0^t \frac{\mathbb{E}D^2[f(s,x_s), f(s,\widehat{0})]}{(t-s)^{1-\alpha}}ds\\
&+ \frac{6}{\Gamma(\alpha)}\int_0^t \frac{\mathbb{E}D^2[f(s,\widehat{0}), \widehat{0}]}{(t-s)^{1-\alpha}}ds + \frac{6}{\Gamma(\alpha)}\int_0^t \frac{\langle \int_0^s \mathbb{E}D^2[g(\theta,x_\theta), g(\theta,\widehat{0})]dB(\theta)\rangle}{(t-s)^{1-\alpha}}ds\\
&+ \frac{6}{\Gamma(\alpha)}\int_0^t \frac{\langle \int_0^s \mathbb{E}D^2[g(\theta,\widehat{0}), \widehat{0}]dB(\theta)\rangle}{(t-s)^{1-\alpha}}ds,
\end{aligned}$$

using It$\widehat{o}$ isometric property, we get

$$\begin{aligned}
\mathbb{E}D^2[(Tx)(t), \widehat{0}] &\le 3\mathbb{E}D^2[\varphi(0), \widehat{0}] + \frac{6p}{\Gamma(\alpha)}\int_0^t \frac{\mathbb{E}D^2[x_s, \widehat{0}]}{(t-s)^{1-\alpha}}ds + \frac{6r}{\Gamma(\alpha)}\int_0^t \frac{e^{b(s-t)}}{(t-s)^{1-\alpha}}ds\\
&+ \frac{6q}{\Gamma(\alpha)}\int_0^t \frac{\int_0^s \mathbb{E}D^2[x_\theta, \widehat{0}]d\theta}{(t-s)^{1-\alpha}}ds + \frac{6c}{\Gamma(\alpha)}\int_0^t \frac{\int_0^s e^{b(\theta-s)}d\theta}{(t-s)^{1-\alpha}}ds,\\
&\le 3\mathbb{E}D^2[\varphi(0), \widehat{0}] + \frac{6p}{\Gamma(\alpha)}\int_0^t \frac{\sup_{\tau\in[-\sigma,0]}\mathbb{E}D^2[x(s+\tau), \widehat{0}]}{(t-s)^{1-\alpha}}ds + \frac{6r}{b^\alpha}\\
&+ \frac{6q}{\Gamma(\alpha)}\int_0^t \frac{\int_0^s \sup_{\tau\in[-\sigma,0]}\mathbb{E}D^2[x(\theta+\tau), \widehat{0}]d\theta}{(t-s)^{1-\alpha}}ds\\
&+ \frac{6c}{b\Gamma(\alpha)}\int_0^t \frac{(e^{b(s-t)} - e^{-bt})}{(t-s)^{1-\alpha}}ds,\\
&\le 3\mathbb{E}D^2[\varphi(0), \widehat{0}] + \frac{6p\rho}{\Gamma(\alpha)}\int_0^t \frac{e^{a(s-t)}}{(t-s)^{1-\alpha}}ds + \frac{6r}{b^\alpha} + \frac{6q\rho}{\Gamma(\alpha)}\int_0^t \frac{\int_0^s (e^{a(\theta-t)}d\theta)}{(t-s)^{1-\alpha}}ds\\
&+ \frac{6c}{b\Gamma(\alpha)}\int_0^t \frac{e^{b(s-t)}}{(t-s)^{1-\alpha}}ds - \frac{6ce^{-bt}}{b\Gamma(\alpha)}\int_0^t (t-s)^{\alpha-1}ds,\\
&\le 3\mathbb{E}D^2[\varphi(0), \widehat{0}] + \frac{6p\rho}{a^\alpha} + \frac{6r}{b^\alpha} + \frac{6q\rho}{a^{\alpha+1}} - \frac{6q\rho e^{-at}}{a\Gamma(\alpha+1)} + \frac{6c}{b^{\alpha+1}} - \frac{6ct^\alpha e^{-bt}}{b\Gamma(\alpha+1)}.
\end{aligned}$$

Thus

$$\begin{aligned}\sup_{t\in[0,t_1]} \mathbb{E}D^2[(Tx)(t),\widehat{0}]e^{-a(s-t)} &\le 3\mathbb{E}D^2[\varphi(0),\widehat{0}] + \frac{6p\rho}{a^\alpha} + \frac{6r}{b^\alpha} + \frac{6q\rho}{a^{\alpha+1}} - \frac{6q\rho e^{-as}}{a\Gamma(\alpha+1)}\\ &+ \frac{6c}{b^{\alpha+1}} - \frac{6ct^\alpha e^{-bs}}{b\Gamma(\alpha+1)},\\ &< \infty.\end{aligned}$$

Likewise, for $t \in (t_i, t_{i+1}]$, $i = 1, .., n$, we have

$$\begin{aligned}\mathbb{E}D^2[(Tx)(t),\widehat{0}] &= \mathbb{E}D^2[\varphi(0) + \sum_{i=1}^{k} \frac{1}{\Gamma(\alpha)} \int_{t_{i-1}}^{t_i} \frac{f(s,x_s)}{(t_i-s)^{1-\alpha}}ds + \frac{1}{\Gamma(\alpha)} \int_{t_k}^{t} \frac{f(s,x_s)}{(t-s)^{1-\alpha}}ds\\ &+ \sum_{i=1}^{k} \frac{1}{\Gamma(\alpha)} \int_{t_{i-1}}^{t_i} \frac{\langle \int_0^s g(\theta,x_\theta)dB(\theta)\rangle}{(t_i-s)^{1-\alpha}}ds + \frac{1}{\Gamma(\alpha)} \int_{t_k}^{t} \frac{\langle \int_0^s g(\theta,x_\theta)dB(\theta)\rangle}{(t-s)^{1-\alpha}}ds\\ &+ \sum_{i=1}^{k} I_i(t_i, x(t_i)),\widehat{0}],\\ &\le 6\mathbb{E}D^2[\varphi(0),\widehat{0}] + \sum_{i=1}^{k} \frac{12}{\Gamma(\alpha)} \int_{t_{i-1}}^{t_i} \frac{\mathbb{E}D^2[f(s,x_s),f(s,\widehat{0})]}{(t_i-s)^{1-\alpha}}ds\\ &+ \sum_{i=1}^{k} \frac{12}{\Gamma(\alpha)} \int_{t_{i-1}}^{t_i} \frac{\mathbb{E}D^2[f(s,\widehat{0}),\widehat{0}]}{(t_i-s)^{1-\alpha}}ds + \frac{12}{\Gamma(\alpha)} \int_{t_k}^{t} \frac{\mathbb{E}D^2[f(s,\widehat{0}),\widehat{0}]}{(t-s)^{1-\alpha}}ds\\ &+ \frac{12}{\Gamma(\alpha)} \int_{t_k}^{t} \frac{\mathbb{E}D^2[f(s,x_s),f(s,\widehat{0})]}{(t-s)^{1-\alpha}}ds\\ &+ \sum_{i=1}^{k} \frac{12}{\Gamma(\alpha)} \int_{t_{i-1}}^{t_i} \frac{\langle \int_0^s \mathbb{E}D^2[g(\theta,x_\theta),g(\theta,\widehat{0})]dB(\theta)\rangle}{(t_i-s)^{1-\alpha}}ds\\ &+ \sum_{i=1}^{k} \frac{12}{\Gamma(\alpha)} \int_{t_{i-1}}^{t_i} \frac{\langle \int_0^s \mathbb{E}D^2[g(\theta,\widehat{0}),\widehat{0}]dB(\theta)\rangle}{(t_i-s)^{1-\alpha}}ds\\ &+ \frac{12}{\Gamma(\alpha)} \int_{t_k}^{t} \frac{\langle \int_0^s \mathbb{E}D^2[g(\theta,x_\theta),g(\theta,\widehat{0})]dB(\theta)\rangle}{(t-s)^{1-\alpha}}ds\\ &+ \frac{12}{\Gamma(\alpha)} \int_{t_k}^{t} \frac{\langle \int_0^s \mathbb{E}D^2[g(\theta,\widehat{0}),\widehat{0}]dB(\theta)\rangle}{(t-s)^{1-\alpha}}ds + 6\sum_{i=1}^{k} l_i \mathbb{E}D^2[x,\widehat{0}],\end{aligned}$$

then

$$\mathbb{E}D^2[(Tx)(t),\widehat{0}] \le 6\mathbb{E}D^2[\varphi(0),\widehat{0}] + \sum_{i=1}^{k}\frac{12p}{\Gamma(\alpha)}\int_{t_{i-1}}^{t_i}\frac{\mathbb{E}D_\sigma^2[x_s,\widehat{0}]}{(t_i-s)^{1-\alpha}}ds + \sum_{i=1}^{k}\frac{12r}{\Gamma(\alpha)}\int_{t_{i-1}}^{t_i}\frac{e^{b(s-t)}}{(t_i-s)^{1-\alpha}}ds$$
$$+\frac{12p}{\Gamma(\alpha)}\int_{t_k}^{t}\frac{\mathbb{E}D_\sigma^2[x_s,\widehat{0}]}{(t-s)^{1-\alpha}}ds + \frac{12r}{\Gamma(\alpha)}\int_{t_k}^{t}\frac{e^{b(s-t)}}{(t-s)^{1-\alpha}}ds$$
$$+\sum_{i=1}^{k}\frac{12q}{\Gamma(\alpha)}\int_{t_{i-1}}^{t_i}\frac{\langle\int_0^s \mathbb{E}D_\sigma^2[x_\theta,\widehat{0}]dB(\theta)\rangle}{(t_i-s)^{1-\alpha}}ds + \sum_{i=1}^{k}\frac{12c}{\Gamma(\alpha)}\int_{t_{i-1}}^{t_i}\frac{\int_0^s e^{b(\theta-t)}d\theta}{(t_i-s)^{1-\alpha}}ds$$
$$+\frac{12q}{\Gamma(\alpha)}\int_{t_k}^{t}\frac{\langle\int_0^s \mathbb{E}D_\sigma^2[x_\theta,\widehat{0}]dB(\theta)\rangle}{(t-s)^{1-\alpha}}ds + \frac{12c}{\Gamma(\alpha)}\int_{t_k}^{t}\frac{\int_0^s e^{b(\theta-t)}d\theta}{(t-s)^{1-\alpha}}ds + 6\sum_{i=1}^{k} l_i\mathbb{E}D^2[x,\widehat{0}],$$
$$\le 6\mathbb{E}D^2[\varphi(0),\widehat{0}] + \sum_{i=1}^{k}\frac{12p}{\Gamma(\alpha)}\int_{t_{i-1}}^{t_i}\frac{\sup_{\tau\in[-\sigma,0]}\mathbb{E}D_\sigma^2[x(s+\tau),\widehat{0}]}{(t_i-s)^{1-\alpha}}ds$$
$$+\sum_{i=1}^{k}\frac{12r}{\Gamma(\alpha)}\int_{t_{i-1}}^{t_i}\frac{e^{b(s-t)}}{(t_i-s)^{1-\alpha}}ds + \frac{12p}{\Gamma(\alpha)}\int_{t_k}^{t}\frac{\sup_{\tau\in[-\sigma,0]}\mathbb{E}D_\sigma^2[x(s+\tau),\widehat{0}]}{(t-s)^{1-\alpha}}ds$$
$$+\frac{12r}{\Gamma(\alpha)}\int_{t_k}^{t}\frac{e^{b(s-t)}}{(t-s)^{1-\alpha}}ds + \sum_{i=1}^{k}\frac{12q}{\Gamma(\alpha)}\int_{t_{i-1}}^{t_i}\frac{\langle\int_0^s \sup_{\tau\in[-\sigma,0]}\mathbb{E}D_\sigma^2[x(\theta+\tau),\widehat{0}]dB(\theta)\rangle}{(t_i-s)^{1-\alpha}}ds$$
$$+\sum_{i=1}^{k}\frac{12c}{\Gamma(\alpha)}\int_{t_{i-1}}^{t_i}\frac{\int_0^s e^{b(\theta-t)}d\theta}{(t_i-s)^{1-\alpha}}ds + \sum_{i=1}^{k}\frac{12c}{\Gamma(\alpha)}\int_{t_{i-1}}^{t_i}\frac{\int_0^s e^{b(\theta-t)}d\theta}{(t_i-s)^{1-\alpha}}ds$$
$$+\frac{12q}{\Gamma(\alpha)}\int_{t_k}^{t}\frac{\langle\int_0^s \sup_{\tau\in[-\sigma,0]}\mathbb{E}D_\sigma^2[x(\theta+\tau),\widehat{0}]dB(\theta)\rangle}{(t-s)^{1-\alpha}}ds$$
$$+\frac{12c}{\Gamma(\alpha)}\int_{t_k}^{t}\frac{\int_0^s e^{b(\theta-t)}d\theta}{(t-s)^{1-\alpha}}ds + 6\sum_{i=1}^{k} l_i\mathbb{E}D^2[x,\widehat{0}],$$
$$\le 6\mathbb{E}D^2[\varphi(0),\widehat{0}] + \frac{12p\rho}{\Gamma(\alpha)}\int_0^t\frac{e^{a(s-t)}}{(t-s)^{1-\alpha}}ds + \frac{12r}{\Gamma(\alpha)}\int_0^t\frac{e^{b(s-t)}}{(t-s)^{1-\alpha}}ds$$
$$+\frac{12p\rho}{\Gamma(\alpha)}\int_0^t\frac{\int_0^s e^{a(\theta-t)}d\theta}{(t-s)^{1-\alpha}}ds + \frac{12c}{\Gamma(\alpha)}\int_0^t\frac{\int_0^s e^{b(\theta-t)}d\theta}{(t-s)^{1-\alpha}}ds + 6k\rho\widehat{l},$$

hence

$$\sup_{t\in[t_i,t_{i+1}]}\mathbb{E}D^2[(Tx)(t),\widehat{0}]e^{-a(s-t)} \le 6\mathbb{E}D^2[\varphi(0),\widehat{0}] + \frac{12p\rho}{a^\alpha} + \frac{12r}{b^\alpha} + \frac{12q\rho}{a^{\alpha+1}} - \frac{12q\rho e^{-as}}{a\Gamma(\alpha+1)}$$
$$+\frac{12c}{b^{\alpha+1}} - \frac{12st^\alpha e^{-bs}}{b\Gamma(\alpha+1)} + 6k\rho\widehat{l}e^{-a(s-T)}.$$
$$<\infty.$$

Thus, if we put $W = \sup_{\tau\in[-\sigma,0]}\mathbb{E}D^2[\varphi(\tau),\widehat{0}]$, we get

$$\sup_{t\in[-\sigma,t_1]}\mathbb{E}D^2[(Tx)(t),\widehat{0}]e^{-a(s-t)} \le 3W + \frac{6p\rho}{a^\alpha} + \frac{6r}{b^\alpha} + \frac{6q\rho}{a^{\alpha+1}} - \frac{6q\rho e^{-as}}{a\Gamma(\alpha+1)}$$
$$+\frac{6c}{b^{\alpha+1}} - \frac{6cT^\alpha e^{-bT}}{b\Gamma(\alpha+1)} < \infty,$$

and

$$\sup_{t\in[t_i,t_{i+1}]}\mathbb{E}D^2[(Tx)(t),\widehat{0}]e^{-a(s-t)} \le 6W + \frac{12p\rho}{a^\alpha} + \frac{12r}{b^\alpha} + \frac{12q\rho}{a^{\alpha+1}} - \frac{12q\rho e^{-as}}{a\Gamma(\alpha+1)}$$
$$+\frac{12c}{b^{\alpha+1}} - \frac{12cT^\alpha e^{-bs}}{b\Gamma(\alpha+1)} + 6k\rho\widehat{l}e^{-a(s-T)} < \infty.$$

On the other hand, for $t \in [-\sigma, 0]$, we have

$$\begin{aligned}\mathbb{E}D^2[(Tx)(t), \widehat{0}]e^{-a(s-t)} &= \mathbb{E}D^2[\varphi(0), \widehat{0}]e^{-a(s-t)},\\ &\leq We^{-a(s-t)} \leq We^{-as}.\end{aligned}$$

Then

$$\begin{aligned}\sup_{t\in[-\sigma,t_1]} \mathbb{E}D^2[(Tx)(t), \widehat{0}]e^{-a(s-t)} &\leq 3We^{-as} + \frac{6p\rho}{a^\alpha} + \frac{6r}{b^\alpha} + \frac{6q\rho}{a^{\alpha+1}} - \frac{6q\rho e^{-as}}{a\Gamma(\alpha+1)}\\ &+ \frac{6c}{b^{\alpha+1}} - \frac{6cT^\alpha e^{-bT}}{b\Gamma(\alpha+1)} < \infty,\end{aligned}$$

and

$$\begin{aligned}\sup_{t\in[t_i,t_{i+1}]} \mathbb{E}D^2[(Tx)(t), \widehat{0}]e^{-a(s-t)} &\leq 6We^{-as} + \frac{12p\rho}{a^\alpha} + \frac{12r}{b^\alpha} + \frac{12q\rho}{a^{\alpha+1}} - \frac{12q\rho e^{-as}}{a\Gamma(\alpha+1)}\\ &+ \frac{12c}{b^{\alpha+1}} - \frac{12cT^\alpha e^{-bs}}{b\Gamma(\alpha+1)} + 6k\rho\widehat{l}e^{-a(s-T)} < \infty.\end{aligned}$$

Hence, $Tx \in \mathcal{B}_a$.

Step 2: T is a contraction on $\mathcal{B}_a$. Indeed, let $x, y \in \mathcal{B}_a$. For $t \in [0, t_1]$,

$$\begin{aligned}\mathbb{E}D^2[(Tx)(t), (Ty)(t)] &\leq \frac{2p}{\Gamma(\alpha)}\int_0^t \frac{\mathbb{E}D^2[x(s), y(s)]}{(t-s)^{1-\alpha}}ds + \frac{2q}{\Gamma(\alpha)}\int_0^t \frac{\int_0^s \mathbb{E}D^2[x(\theta), y(\theta)]dB(\theta)}{(t-s)^{1-\alpha}}ds,\\ &\leq \frac{2p}{\Gamma(\alpha)}\int_0^t \frac{\sup_{r\in[s-\sigma,s]} \mathbb{E}D^2[x(r), y(r)]}{(t-s)^{1-\alpha}}ds\\ &+ \frac{2q}{\Gamma(\alpha)}\int_0^t \frac{\int_0^s \sup_{r\in[s-\theta,\theta]} \mathbb{E}D^2[x(r), y(r)]dB(\theta)}{(t-s)^{1-\alpha}}ds.\end{aligned}$$

In addition, we have

$$\mathbb{E}D^2[x(t), y(t)] \leq \mathbb{E}D_a^2[x, y]e^{a(s-t)}, \qquad \forall t \in [-\sigma, T].$$

So, we have

$$\begin{aligned}\mathbb{E}D^2[(Tx)(t), (Ty)(t)] &\leq \frac{2p}{\Gamma(\alpha)}\int_0^t \frac{\mathbb{E}D_a^2[x, y]e^{a(s-t)}}{(t-s)^{1-\alpha}}ds + \frac{2q}{\Gamma(\alpha)}\int_0^t \frac{\int_0^s \mathbb{E}D_a^2[x, y]e^{a(\theta-t)}d\theta}{(t-s)^{1-\alpha}}ds,\\ &\leq \frac{2p}{a^\alpha}\mathbb{E}D_a^2[x, y] + 2q\left(\frac{1}{a^{\alpha+1}} - \frac{t^\alpha e^{-at}}{\Gamma(\alpha+1)}\right)\mathbb{E}D_a^2[x, y],\\ &\leq \left(\frac{2p}{a^\alpha} + \frac{2q}{a^{\alpha+1}} - \frac{2qt^\alpha e^{-at}}{\Gamma(\alpha+1)}\right)\mathbb{E}D_a^2[x, y].\end{aligned}$$

Thus

$$\begin{aligned}\mathbb{E}D_a^2[Tx, Ty] &= \sup_{t\in[-\sigma,T]} \mathbb{E}D^2[(Tx)(t), (Ty)(t)]e^{-a(s-t)},\\ &\leq \left(\frac{2p}{a^\alpha} + \frac{2q}{a^{\alpha+1}} - \frac{2qT^\alpha e^{-as}}{\Gamma(\alpha+1)}\right)\mathbb{E}D_a^2[x, y]. \qquad (11)\end{aligned}$$

Similarly, for $t \in (t_i, t_{i+1}]$, $i = 1, .., n$, we have

$$\mathbb{E}D^2[(Tx)(t), (Ty)(t)] \leq \frac{5p}{\Gamma(\alpha)} \sum_{i=1}^{k} \int_{t_{i-1}}^{t_i} \frac{\mathbb{E}D^2[x(s), y(s)]}{(t_i - s)^{1-\alpha}} ds + \frac{5p}{\Gamma(\alpha)} \int_{t_k}^{t} \frac{\mathbb{E}D^2[x(s), y(s)]}{(t - s)^{1-\alpha}} ds$$
$$+ \frac{5q}{\Gamma(\alpha)} \sum_{i=1}^{k} \int_{t_{i-1}}^{t_i} \frac{\langle \int_0^s \mathbb{E}D^2[x(\theta), y(\theta)] dB(\theta) \rangle}{(t_i - s)^{1-\alpha}} ds$$
$$+ \frac{5q}{\Gamma(\alpha)} \int_{t_k}^{t} \frac{\langle \int_0^s \mathbb{E}D^2[x(\theta), y(\theta)] dB(\theta) \rangle}{(t - s)^{1-\alpha}} ds + 5 \sum_{i=1}^{k} l_i D^2[x(t_i), y(t_i)],$$
$$\leq \frac{5p}{\Gamma(\alpha)} \sum_{i=1}^{k} \int_{t_{i-1}}^{t_i} \frac{\sup_{\tau \in [-\sigma, 0]} \mathbb{E}D^2[x(s+\tau), y(s+\tau)]}{(t_i - s)^{1-\alpha}} ds$$
$$+ \frac{5p}{\Gamma(\alpha)} \int_{t_k}^{t} \frac{\sup_{\tau \in [-\sigma, 0]} \mathbb{E}D^2[x(s+\tau), y(s+\tau)]}{(t - s)^{1-\alpha}} ds$$
$$+ \frac{5q}{\Gamma(\alpha)} \sum_{i=1}^{k} \int_{t_{i-1}}^{t_i} \frac{\langle \int_0^s \sup_{\tau \in [-\sigma, 0]} \mathbb{E}D^2[x(\theta+\tau), y(\theta+\tau)] dB(\theta) \rangle}{(t_i - s)^{1-\alpha}} ds$$
$$+ \frac{5q}{\Gamma(\alpha)} \int_{t_k}^{t} \frac{\langle \int_0^s \sup_{\tau \in [-\sigma, 0]} \mathbb{E}D^2[x(\theta+\tau), y(\theta+\tau)] dB(\theta) \rangle}{(t - s)^{1-\alpha}} ds$$
$$+ 5 \sum_{i=1}^{k} l_i D^2[x(t_i), y(t_i)],$$

then

$$\mathbb{E}D^2[(Tx)(t), (Ty)(t)] \leq \frac{5p}{\Gamma(\alpha)} \sum_{i=1}^{k} \int_{t_{i-1}}^{t_i} \frac{\mathbb{E}D_a^2[x, y] e^{a(s-t)}}{(t_i - s)^{1-\alpha}} ds + \frac{5p}{\Gamma(\alpha)} \int_{t_k}^{t} \frac{\mathbb{E}D_a^2[x, y] e^{a(s-t)}}{(t - s)^{1-\alpha}} ds$$
$$+ \frac{5q}{\Gamma(\alpha)} \sum_{i=1}^{k} \int_{t_{i-1}}^{t_i} \frac{\langle \int_0^s \mathbb{E}D_a^2[x, y] e^{a(s-t)} dB(\theta) \rangle}{(t_i - s)^{1-\alpha}} ds$$
$$+ \frac{5q}{\Gamma(\alpha)} \int_{t_k}^{t} \frac{\langle \int_0^s \mathbb{E}D_a^2[x, y] e^{a(s-t)} dB(\theta) \rangle}{(t - s)^{1-\alpha}} ds + 5 \sum_{i=1}^{k} l_i e^{a(s-t_i)},$$
$$\leq \frac{5p}{a^\alpha} D_a^2[x, y] + \left(\frac{5q}{a^{\alpha+1}} - \frac{5qt^\alpha e^{-as}}{\Gamma(\alpha+1)} \right) D_a^2[x, y] + 5 \sum_{i=1}^{k} l_i e^{a(s-t)},$$

thus

$$\mathbb{E}D_a^2[Tx, Ty] = \sup_{t \in [-\sigma, T]} \mathbb{E}D^2[(Tx)(t), (Ty)(t)] e^{-a(s-t)},$$
$$\leq \left(\frac{5p}{a^\alpha} + \frac{5q}{a^{\alpha+1}} - \frac{5qT^\alpha e^{-as}}{\Gamma(\alpha+1)} + \frac{k}{2}(k+1)\widehat{l} \right) \mathbb{E}D_a^2[x, y]. \tag{12}$$

Then, from (11) and (12), we have

$$\mathbb{E}D_a^2[Tx, Ty] \leq \left(\frac{2p}{a^\alpha} + \frac{2q}{a^{\alpha+1}} - \frac{2qT^\alpha e^{-as}}{\Gamma(\alpha+1)} \right) \mathbb{E}D_a^2[x, y], \quad \text{for } t \in [0, t_1].$$

And

$$\mathbb{E}D_a^2[Tx, Ty] \leq \left(\frac{5p}{a^\alpha} + \frac{5q}{a^{\alpha+1}} - \frac{5qT^\alpha e^{-as}}{\Gamma(\alpha+1)} + \frac{k}{2}(k+1)\widehat{l} \right) \mathbb{E}D_a^2[x, y], \quad \text{for } t \in (t_i, t_{i+1}].$$

Consequently, by (10), T is a contraction. As a consequence of Banach fixed point Theorem, we deduce that T has a fixed point which is a solution of (7). □

4 Example

In this section, we will give an example to illustrat the usefulness of our main result. Let us consider the following impulsive stochastic differential equation with distributed delay. For $k \in \mathbb{N}$ and $0 < \sigma_1 < \sigma_2 < ... < \sigma_k < \sigma$. we have

$$\begin{cases} {}_{gH}^{C}\mathcal{D}^{\alpha}u(t) = \int_{-\sigma}^{0} f_0(s, u(t+s))ds + \sum_{i=1}^{k} f_i(s, u(t-\sigma)) \\ \quad + \int_{-\sigma}^{0} \langle \int_{0}^{s} g_0(\tau, u(\tau+s))dB(\tau)\rangle ds + \sum_{i=1}^{k} \langle \int_{0}^{s} g_i(\tau, u(\tau+s))dB(\tau)\rangle, \\ u(t_i^+) = \dfrac{u(t_i)}{2+u(t_i)}, \quad t = t_i. \\ u(t) = \varphi(t), \quad t \in [-\sigma, 0]. \end{cases} \tag{13}$$

Let $PC_\sigma := C([-\sigma, 0], \mathbf{E}^1)$ denote the space of fuzzy functions, $u : [-\sigma, T] \longrightarrow \mathbf{E}^1$, $f_i : I \times PC_\sigma \longrightarrow \mathbf{E}^1$ and $g_i : I \times PC_\sigma \longrightarrow \mathbb{R}^n$, $i = 1, .., k$ are fuzzy valued mappings which satisfies the following assumptions:
$(\mathcal{A}1)$ There exists $p_i, q_i > 0$ such that, $\forall u, v \in \mathbf{E}^1$ and $t \in I$:

$$\mathbb{E}D^2[f_i(t, u), f_i(t, v)] \leq p_i \mathbb{E}D_\sigma^2[u, v].$$

$$\mathbb{E}\|g_i(t, u) - g_i(t, v)\|^2 \leq q_i \mathbb{E}D_\sigma^2[u, v].$$

$(\mathcal{A}2)$ There exists $r_i, c_i > 0$ such that $\forall t \in I$:

$$\mathbb{E}D^2[f_i(t, \widehat{0}), \widehat{0}] \leq r_i e^{b(s-t)}.$$

$$\mathbb{E}D^2[g_i(t, \widehat{0}), \widehat{0}] \leq c_i e^{b(s-t)}.$$

$(\mathcal{A}3)$ There exist $l_i > 0$ such that $\forall u, v \in \mathbf{E}^1$ and $t \in I$:

$$\mathbb{E}D^2[I_i(t_i, u(t_i)), I_i(t_i, v(t_i))] \leq l_i \mathbb{E}D_\sigma^2[u, v].$$

Then, we define the fuzzy functions $f : I \times PC_0 \longrightarrow \mathbf{E}^1$, $g : I \times PC_0 \longrightarrow \mathbb{R}^n$ as follows:

$$f(t, u) = \int_{-\sigma}^{0} f_0(s, u(s))ds + \sum_{i=1}^{k} f_i(s, u(-\sigma)), \quad t \in I, \quad t \neq t_i,$$

$$g(t, u) = \int_{-\sigma}^{0} \langle \int_{0}^{s} g_0(\tau, u(\tau))dB(\tau)\rangle ds + \sum_{i=1}^{k} \langle \int_{0}^{s} g_i(\tau, u(\tau))dB(\tau)\rangle, \quad t \in I, \quad t \neq t_i.$$

Assume that the functions f_i and g_i $i = 1, .., k$ satisfy the assumptions $(\mathcal{A}1)$ – $(\mathcal{A}3)$. Then the system (13) has a unique solution on I.

Indeed, for $u, v \in \mathbf{E}^1$, $t \in I$, $t \neq t_i$, we have

$$\begin{aligned}
\mathbb{E}D^2[f(t,u), f(t,v)] &\leq 2\int_{-\sigma}^{0} \mathbb{E}D^2[f_0(s,u(s)), f_0(s,v(s))]ds \\
&\quad + 2\sum_{i=1}^{k} \mathbb{E}D^2[f_i(s,u(-\sigma_i)), f_i(s,v(-\sigma_i))], \\
&\leq 2p_0\sigma \mathbb{E}D_\sigma^2[u,v] + 2\sum_{i=1}^{k} p_i \mathbb{E}D_\sigma^2[u,v], \\
&\leq 2\left(p_0\sigma + \sum_{i=1}^{k} p_i\right)\mathbb{E}D_\sigma^2[u,v] := C_1\mathbb{E}D_\sigma^2[u,v].
\end{aligned}$$

And

$$\begin{aligned}
\mathbb{E}\|g(t,u) - g(t,v)\|^2 &\leq \int_{-\sigma}^{0} \langle \int_0^s \|g_0(\theta,u(\theta)), g_0(\theta,v(\theta))\|^2 dB(\theta)\rangle ds \\
&\quad + \sum_{i=0}^{k} \langle \int_0^s \|g_i(\theta,u(\theta)) - g_i(\theta,v(\theta))\|^2 dB(\theta)\rangle, \\
&\leq q_0\frac{\sigma^2}{2}\mathbb{E}D_\sigma^2[u,v] + \sum_{i=1}^{k} q_i T\mathbb{E}D_\sigma^2[u,v], \\
&\leq \left(q_0\frac{\sigma^2}{2} + T\sum_{i=1}^{k} q_i\right)\mathbb{E}D_\sigma^2[u,v] := C_2\mathbb{E}D_\sigma^2[u,v].
\end{aligned}$$

So, assumption $(\mathcal{H}1)$ is satisfied. On the other hand, we have

$$\begin{aligned}
\mathbb{E}D^2[f(t,\widehat{0}),\widehat{0}] &\leq 2\int_{-\sigma}^{0} \mathbb{E}D_\sigma^2[f_0(s,\widehat{0}),\widehat{0}]ds + 2\sum_{i=1}^{k} \mathbb{E}D_\sigma^2[f_i(s,\widehat{0}),\widehat{0}], \\
&\leq \left[\frac{2r_0}{b}\left(e^{-2bs-bt} - e^{-b\sigma}\right) + \sum_{i=1}^{k} r_i\right] e^{b(s-t)} := C_3 e^{b(s-t)}.
\end{aligned}$$

And

$$\begin{aligned}
\mathbb{E}D^2[g(t,\widehat{0}),\widehat{0}] &\leq 2\int_{-\sigma}^{0} \langle \int_0^s \mathbb{E}D^2[g_0(\theta,\widehat{0}),\widehat{0}]dB(\theta)\rangle ds + 2\sum_{i=1}^{k} \langle \int_0^s \mathbb{E}D^2[g_i(\theta,\widehat{0}),\widehat{0}]dB(\theta)\rangle, \\
&\leq \left[\frac{2s_0}{b^2}(e^{-b(s-t)} - e^{-b\sigma}) + 2e^{bT}\sum_{i=1}^{k} c_i\right] e^{b(s-t)} := C_4 e^{b(s-t)}.
\end{aligned}$$

Then, assumption $(\mathcal{H}2)$ is satisfied. Also, we have

$$\begin{aligned}\mathbb{E}D^2[I_i(t_i,u(t_i)),I_i(t_i,v(t_i))] &\leq \mathbb{E}D^2\left[\frac{u(t_i)}{2+u(t_i)},\frac{u(t_i)}{2+v(t_i)}\right],\\ &\leq \frac{1}{2}\sup_{\alpha\in[0,1]}\max\{|\underline{u}_\alpha(t_i)-\underline{v}_\alpha(t_i)|,|\overline{u}_\alpha(t_i)-\overline{v}_\alpha(t_i)|\},\\ &=\frac{1}{2}D_\sigma^2[u(t_i),v(t_i)].\end{aligned}$$

Thus, the assumption $(\mathcal{H}3)$ is satisfied with $l_i=\frac{1}{2}$. Hence, the problem (13) has a unique solution on $[0,T]$.

5 Conclusion

In this work, a class of fuzzy fractional stochastic differential equations with impulsive has been studied. By using the fixed point strategy, the existence and uniqueness of solutions have been proven. Eventually, an example is given to illustrat the effectiveness of result.

References

1. Agarwal, R.P., Lakshmikantham, V., Nieto, J.J.: On the concept of solution for fractional differential equations with uncertainty. Nonlinear Anal. Theor. Methods Appl. **72**, 2859–2862 (2010)
2. Abbasbandy, S., Allah, V.T., Balooch, S.M.R., Salahshour, S.: Fuzzy local fractional differential equations. Int. J. Ind. Math. **4**, 231–246 (2012)
3. Benchohra, M., Nieto, J.J., Ouahab, A.: Fuzzy solutions for impulsive differential equations. Commun. Appl. Anal. **11**, 379–394 (2007)
4. Lakshmikantham, V., McRae, F.A.: Basic results for fuzzy impulsive differential equations. Math. Inequalities Appl. **4**, 239–246 (2001)
5. Vu, H., Van Hao, N.: On impulsive fuzzy functional differential equations. Iran J. Fuzzy Syst. **13**(4), 79–94 (2016)
6. Allahviranloo, T., Abbasbandy, S., Salahshour, S., Hakimzadeh, A.: A new method for solving fuzzy linear differential equations. Computing **92**, 181–197 (2010). https://doi.org/10.1007/s00607-010-0136-6
7. Allahviranloo, T., Salahshour, S., Abbasbandy, S.: Explicit solutions of fractional differential equations with uncertainty. Soft. Comput. **16**, 297–302 (2011). https://doi.org/10.1007/s00500-011-0743-y
8. Allahviranloo, T., Abbasbandy, S., Sedaghgatfar, O., Darabi, P.: A new method for solving fuzzy integro-differential equation under generalized differentiability. Neural Comput. Appl. **21**, 191–196 (2011). https://doi.org/10.1007/s00521-011-0759-3
9. Guo, M., Xue, X., Li, R.: Impulsive functional differential inclusions and fuzzy population models. Fuzzy Sets Syst. **138**, 601–615 (2003)
10. Lupulescu, V.: On a class of fuzzy functional differential equations. Fuzzy Sets Syst. **160**, 1547–1562 (2009)

11. Nieto, J., Rodriguez-Lopez, R.: Periodic boundary value problem for non-Lipschitzian impulsive functional differential equations. J. Math. Anal. Appl. **318**, 593–610 (2006)
12. Vatsala, A.S.: Impulsive hybrid fuzzy differential equations. Facta. Univ. Ser Mech. Autom. Control Rob. **3**, 851–859 (2003)
13. Priyadharsini, J., Balasubramaniam, P.: Existence of fuzzy fractional stochastic differential system with impulses. Comput. Appl. Math. **39**(3), 1–21 (2020). https://doi.org/10.1007/s40314-020-01229-0
14. Harir, A., Melliani, S., Chadli, L.S.: Fuzzy conformable fractional differential equations. Int. J. Differ. Eqn. (2021)
15. Harir, A., Melliani, S., Chadli, L.S.: Existence, Uniqueness and Approximate Solutions of Fuzzy Fractional Differential Equations. Fuzzy Syst. (2020)
16. Melliani, S., Elomari, M., Chadli, L.S.: Solving the fuzzy fractional differential wave equation by mean fuzzy Fourier transform. In: Dutta, H., Hammouch, Z., Bulut, H., Baskonus, H.M. (eds.) CMES 2019. AISC, vol. 1111, pp. 133–147. Springer, Cham (2020). https://doi.org/10.1007/978-3-030-39112-6_9
17. Chadli, L.S., Melliani, S., Hilal, K., Kajouni, A.: Existence and uniqueness results of fuzzy fractional differential equation with nonlocal conditions. J. Univ. Math. **1**, 9–14 (2018)
18. Hoa, N.V., Tri, P.V., Dao, T.T., Zelinka, I.: Some global existence results and stability theorem for fuzzy functional differential equations. J. Intell. Fuzzy Syst. **28**, 393–409 (2015)

A New Smoothing Approach for Piecewise Smooth Functions: Application to Some Fundamental Functions

Elmehdi Amhraoui(✉) and Tawfik Masrour

Laboratory of Mathematical Modeling, Simulation and Smart Systems (L2M3S), ENSAM-Meknes, Moulay ISMAIL University, Meknes, Morocco
e.amhraoui@edu.umi.ac.ma, t.masrour@ensam.umi.ac.ma

Abstract. This article is about a new smoothing method for piecewise smooth functions. This smoothing method is based on formulating any piecewise smooth function as the expectation of a discrete random variable. By adopting this formulation, we show that smoothing apiecewise smooth function is equivalent to smooth a probability distribution. In addition, we propose to use the Boltzmann distribution as a smoothing approximation for this probability distribution. We apply our smoothing approach to smooth some fundamental functions. The theoretical and numerical results show the efficiency of our method.

1 Introduction

Piece-wise smooth functions (P-S) are a fundamental subclass of non-smooth functions. They have received significant attention for a long time as they appear in many areas including optimization, data modeling, and various equations [7,13,21–23]. The main problem with these functions is that they are not smooth over all the domain. As a result, their presence in a problem has many drawbacks. For example, in optimization, if the cost function is not at least continuously differentiable, gradient-based methods are not applicable. An important approach to solving this problem is to construct a smoothing approximation for the P-S function before the optimization procedure. The term smoothing approximation refers to a smooth function that approximates a nonsmooth function.

The first studies on smoothing approximations are proposed to solve the optimization problems where the objective function contains max or min operator in its formulation [6,28]. In [27], the authors proposed the hyperbolic smoothing method to solve the min-sum-min problem. The hyperbolic smoothing technique is also used in [3] to solve the essential problem known as the minimax problem. Smoothing approximations are also used in the context of penalty methods. In fact, most of the exact penalty functions are not smooth. Therefore, many papers studied the smoothing approximations for these exact penalty functions [11,14,17,18]. As a result, smoothing techniques of P-S functions play a crucial role in optimization and penalty methods.

Besides optimization and penalty methods, smoothing approximations are used in solving various equations [10,12,19,24], data modeling and compression [1,2,8,15,16, 20], blending surface problem [4,5,9], and regularization of neural networks [26].

S. Melliani and O. Castillo (Eds.): ICPAMS 2021, LNNS 476, pp. 164–178, 2023.
https://doi.org/10.1007/978-3-031-12416-7_14

It is clear that smoothing approximations of P-S functions have applications in a wide range of fields. This paper proposes a new approach of smoothing P-S functions. An extended version of our approach is presented in [29]. In [29], we defined a vector space of piecewise smooth functions and constructed an operator that can associate a smoothing function for each element of this space. Additionally, some complex partitions of the domain have been addressed. In Sect. 2, we reformulate any P-S function as the expectation of a random variable and show how this new formulation reduces the problem of smoothing a P-S function to the problem of smoothing a probability distribution. In Sect. 3, we propose to use the Boltzmann distribution to construct the smoothing approximations for P-S functions in the one-dimensional case. In Sect. 4, we extend our approach to the n-dimensional case. In Sect. 5, we apply our smoothing method to some fundamental functions that appear in nonsmooth optimization and penalty methods. In Sect. 6, we conclude this paper and discuss some of our future works.

2 A New Formulation for Piecewise Smooth Functions

Throughout this paper, $\Omega \neq \emptyset$ is an open subset of $\mathbb{R}^n$, $C^k(\Omega)$ $(k \geq 1)$ denotes the set of real-valued functions with continuous $k-th$ order partial derivatives on Ω. A real-valued function $f : \Omega \longrightarrow \mathbb{R}$ is considered to be smooth if it is at least of class $C^1(\Omega)$.

Now, we give the definition of a piecewise smooth function

Definition 1. A real-valued function $f : \Omega \longrightarrow \mathbb{R}$ is called piecewise smooth (P-S) function on Ω if it is continuous on Ω, there exists a finite collection of smooth functions $f_i : \Omega \longrightarrow \mathbb{R}$, $i = 1,2,\cdots,m$, and there exists $\{A_i\}_{i=1,\cdots,m}$ a partition of Ω, such that

$$f(x) = f_i(x) \ \forall x \in A_i, i = 1,\cdots,m.$$

Let $\tau = \{A_1,\cdots,A_m\}$ be a partition of Ω. Throughout this paper, we consider the following piccewisc smooth function

$$f(x) = \begin{cases} f_1(x), & x \in A_1, \\ f_2(x), & x \in A_2, \\ \vdots & \\ f_m(x), & x \in A_m. \end{cases} \tag{1}$$

Although the functions $f_i : \Omega \longrightarrow \mathbb{R}$, $i = 1,2,\cdots,r$ are smooth over Ω, the function f may not be differentiable over Ω; see Example 1. In this paper, we are interested in constructing a smooth function that approximates the P-S function f.

Example 1. Let $\tau_0 = \{(-\infty,0],(0,+\infty)\}$ and let $g(x) = \max(0,x) = \begin{cases} 0, & x \in (-\infty,0], \\ x, & x \in (0,+\infty). \end{cases}$

It is clear that $g_1(x) = 0$ and $g_2(x) = x$ are of class $C^\infty(\mathbb{R})$, but $g(x) = \max(0,x)$ is not differentiable at 0.

In the following, we will formulate any piecewise smooth function of the form 1 as the expectation of a discrete random variable. Let $(\mathscr{E}, \mathscr{F}, \mathbb{P})$ be a space of probability and let $X : \mathscr{E} \longrightarrow \Omega$ be a random variable with values on Ω. Let $E = \{1, \cdots, m\}$ and let $\pi(E)$ be the set of all possible probability distributions over E, that is

$$\pi(E) = \left\{ (p_1, \cdots, p_m) \in [0,1]^m \text{ such that } \sum_{i=0}^{m} p_i = 1 \right\} \tag{2}$$

Let $p : \Omega \longrightarrow \pi(E)$ be the function that associate to each $x \in \Omega$ a probability distribution $p(x) = (p_1(x), \cdots, p_m(x))$ over the set E.

For each possible outcome $x \in \Omega$ of the random variable X, we consider the conditional random variable $F(x) := F_{|X=x}$ which associates a piece $f_i(x)$ to x with probability $p_i(x)$, that is

$$\mathbb{P}(F = f_i(x) | X = x) = p_i(x), \forall i \in E.$$

We define $\hat{f}_p(x)$ as the expectation of the random variable $F(x)$ with respect to the probability $p(x)$, that is

$$\hat{f}_p(x) = \mathbb{E}_p[F(x)] = \sum_{i=1}^{m} p_i(x) f_i(x) \tag{3}$$

We define q_i as the indicator function of the set A_i, that is

$$q_i(x) = \begin{cases} 1, & x \in A_i, \\ 0, & x \notin A_i. \end{cases} \tag{4}$$

Since $\{A_1, \cdots, A_m\}$ is a partition of Ω, then we have $\sum_{i=1}^{m} q_i(x) = 1$ for each $x \in \Omega$. Thus, $q(x) = \big(q_1(x), \cdots, q_m(x)\big)$ can be seen as a probability distribution over the set E. We call q the **characteristic probability** of the partition τ. In addition, we have

$$\begin{aligned} \hat{f}_q(x) &= \mathbb{E}_q[F(x)] \\ &= \sum_{i=1}^{m} q_i(x) f_i(x) \\ &= \begin{cases} f_1(x), & x \in A_1, \\ f_2(x), & x \in A_2, \\ \vdots \\ f_m(x), & x \in A_m. \end{cases} \\ &= f(x). \end{aligned}$$

Therefore, we have $f(x) = \mathbb{E}_q[F(x)], \forall x \in \Omega$. Thus, we obtain a new formulation for all the piecewise smooth functions of the form 1. However, $q = \big(q_1, \cdots, q_m\big)$ is a nonsmooth function over Ω, which, firstly, explains the non-smoothness of f in most cases and, on the other hand, does not guarantee the smoothness of $\hat{f}_q$ despite the fact that $f_i, i = 1, \cdots, m$ are all smooth. As a result, if we construct a smooth probability distribution p that approximates the characteristic probability q, we will obtain a smoothing approximation $\hat{f}_p$ for each f of the form 1.

Theorem 1. *Take f and $\hat{f}_p$ as in (1) and (3), respectively. Assuming that $f_i, i = 1, \cdots, m$ are of class $C^k(\Omega)$ but f is only of class $C^r(\Omega)$, where $r < k$. If the function $p : \Omega \longrightarrow \pi(E)$ is of class $C^k(\Omega)$, then we have:*

(a) $\hat{f}_p$ is of class $C^k(\Omega)$,
(b) $\hat{f}_p \longrightarrow f$ pointwise as $p \longrightarrow q$ pointwise.

Proof. (a) Since $p_i, i = 1, \cdots, m$ and $f_i, i = 1, \cdots, m$ are of class $C^k(\Omega)$, then $\hat{f}_p(x) = \sum_{i=1}^m p_i(x) \times f_i(x)$ is of class $C^k(\Omega)$.
(b) Let $x \in \Omega$, we have

$$\begin{aligned}
\left|\hat{f}_p(x) - f(x)\right| &= \left|\hat{f}_p(x) - \hat{f}_q(x)\right| \\
&= \left|\sum_{i=1}^m p_i(x) f_i(x) - \sum_{i=1}^m q_i(x) f_i(x)\right| \\
&= \left|\sum_{i=1}^m \left(p_i(x) - q_i(x)\right) f_i(x)\right| \\
&\le \sum_{i=1}^m |p_i(x) - q_i(x)| \times |f_i(x)| \\
&\le \max_{i=1}^m |f_i(x)| \times \sum_{i=1}^m |p_i(x) - q_i(x)| \\
&= \max_{i=1}^m |f_i(x)| \times \|p(x) - q(x)\|_1
\end{aligned}$$

thus we obtain:

$$\left|\hat{f}_p(x) - f(x)\right| \le c(x) \times \|p(x) - q(x)\|_1, \forall x \in \Omega. \tag{5}$$

where $c(x) = \max_{i=1}^m |f_i(x)|$. As a result, $\hat{f}_p \longrightarrow f$ pointwise as $p \longrightarrow q$ pointwise.

Considering the formulation above, in what follows, we will try to construct a smooth function $p : \Omega \longrightarrow \pi(E)$ that approximates the characteristic probability of the partition τ. Once the smooth function p is available, the function $\hat{f}_p$ defined in (3) is a smoothing approximation for any piecewise smooth function of the form (1).

3 Smoothing Approximations for One-Dimensional Case

Let us consider the partition $\tau = \{A_0, A_1, \cdots A_m\}$ of $\mathbb{R}$, where

$$A_i = \begin{cases} (-\infty, a_1], & i = 0, \\ (a_i, a_{i+1}], & i = 1, \cdots, m-1, \\ (a_m, +\infty), & i = m. \end{cases} \tag{6}$$

and $a_1 < a_2 < \cdots < a_m$. In this case, the piecewise smooth function will have the following form

$$f(x) = \begin{cases} f_0(x), & x \in (-\infty, a_1], \\ f_2(x), & x \in (a_1, a_2], \\ \vdots & \\ f_m(x), & x \in (a_m, +\infty). \end{cases} \tag{7}$$

where $f_i, i = 0, \cdots, m$ are all smooth over $\mathbb{R}$.
We define the functions $h_i, i = 0, \cdots, m$ as follows

$$h_i(x) = \begin{cases} -(x-a_1), & i = 0, \\ -(x-a_i)(x-a_{i+1}), & i = 1, \cdots, m-1, \\ x - a_m, & i = m. \end{cases} \tag{8}$$

Proposition 1. *Take A_i and h_i as in (6) and (8), respectively.*
For $i = 0, \cdots, m$, we have:

$$(x \in \mathring{A}_i) \Longleftrightarrow \Big(h_i(x) > 0 \text{ and } h_k(x) < 0, \forall k \neq i\Big),$$

where $\mathring{A}_i$ is the interior of A_i.
Furthermore, for $i = 1, \cdots, m$, we have:

$$x = a_i \Leftrightarrow \Big(h_i(x) = h_{i-1}(x) = 0 \text{ and } h_k(x) < 0, \forall k \notin \{i, i-1\}\Big).$$

To transform the functions $h_i, i = 0, \cdots, m$ into a probability distribution for the condition random variable $F_{X=x}$, we propose the use of the Boltzmann distribution which is expressed in (9).

$$p_i^\alpha(x) = \frac{e^{\alpha h_i(x)}}{\sum_{k=0}^m e^{\alpha h_k(x)}}, \; i = 0, \cdots, m. \tag{9}$$

where $\alpha > 0$.
It is clear that for every $x \in \mathbb{R}$, we have $p_i(x) \in [0,1], i = 1, \cdots, m$ and $\sum_{i=1}^{m} p_i^\alpha(x) = 1$.
Thus, $p^\alpha(x) = \big(p_1^\alpha(x), \cdots, p_m^\alpha(x)\big)$ is a probability distribution.

To simplify, the smoothing approximation of the piecewise smooth function f is denoted $\hat{f}_\alpha$ instead of $\hat{f}_{p^\alpha}$.

Theorem 2. *Take f as in (7) such that $f_i, i = 1, \cdots, m$ are of class $C^k(\Omega)$ but f is only of class $C^r(\Omega)$, where $r < k$. If $p_i^\alpha, i = 0, \cdots, m$ are defined as in (9), then we have*

(a) $p_i^\alpha : \mathbb{R} \longrightarrow [0,1]$ is of class $C^\infty(\mathbb{R})$, $i = 0, \cdots, m$,
(b) $\hat{f}_\alpha \in C^k(\mathbb{R})$ $\forall \alpha \in \mathbb{R}$,
(c) $p^\alpha \longrightarrow q$ pointwise almost everywhere as $\alpha \longrightarrow +\infty$, where $p^\alpha = \big(p_0^\alpha, \cdots, p_m^\alpha\big)$ and $q = (q_0, \cdots, q_m)$ is the characteristic probability of the partition τ,
(d) $\hat{f}_\alpha \longrightarrow f$ pointwise as $\alpha \longrightarrow +\infty$.

Proof. (a) Easy to prove.
(b) Obtained immediately by applying the part (a) of Theorem 1.
(c)

$$\begin{aligned}
x \in \mathbb{R}\setminus\{a_1,a_2,\cdots,a_m\} &\Longrightarrow \exists!\ i \in \{0,1,\cdots,m\} \text{ such that } x \in \mathring{A}_i \\
&\Longrightarrow h_i(x) > 0 \text{ and } h_k(x) < 0, \forall k \neq i \ \big(\text{Proposition1}\big) \\
&\Longrightarrow h_k(x) - h_i(x) < 0, \forall k \neq i \\
&\Longrightarrow e^{\alpha\left(h_k(x)-h_i(x)\right)} \longrightarrow 0, \text{ as } \alpha \longrightarrow +\infty \\
&\Longrightarrow \frac{1}{1+\sum_{k\neq i} e^{\alpha\left(h_k(x)-h_i(x)\right)}} \longrightarrow 1, \text{ as } \alpha \longrightarrow +\infty \\
&\Longrightarrow p_i^\alpha(x) \longrightarrow 1, \text{ as } \alpha \longrightarrow +\infty \\
&\Longrightarrow p_i^\alpha(x) \longrightarrow 1 \text{ and } p_k^\alpha(x) \longrightarrow 0\ \forall k \neq i, \text{ as } \alpha \longrightarrow +\infty. \\
&\qquad \Big(\text{ because } \sum_{k=0}^{m} p_k^\alpha(x) = 1\Big)
\end{aligned}$$

Therefore, $\forall x \in \mathbb{R}\setminus\{a_1,a_2,\cdots,a_m\}$ we have, $p^\alpha \longrightarrow q$ as $\alpha \longrightarrow +\infty$.
(d) If $x \in \mathbb{R}\setminus\{a_1,a_2,\cdots,a_m\}$, we have $p^\alpha(x) \longrightarrow q(x)$ as $\alpha \longrightarrow +\infty$. By applying Theorem 1, we obtain:

$$\forall x \in \mathbb{R}\setminus\{a_1,a_2,\cdots,a_m\},\ \hat{f}_\alpha(x) \longrightarrow f(x) \text{ as } \alpha \longrightarrow +\infty. \tag{10}$$

If $x = a_i$, then we have:

$$\begin{aligned}
\Big|\hat{f}_\alpha(a_i) - f(a_i)\Big| &= \Big|\sum_{k=0}^{m} p_k^\alpha(a_i) f_k(a_i) - f(a_i)\Big| \\
&= \Big|\sum_{k=0}^{m} p_k^\alpha(a_i) f_k(a_i) - f_{i-1}(a_i)\Big| \ \Big(\text{because } f(a_i) = f_{i-1}(a_i)\Big) \\
&= \Big|\sum_{k=0}^{m} p_k^\alpha(a_i)\big(f_k(a_i) - f_{i-1}(a_i)\big)\Big| \\
&\leq \sum_{k=0}^{m} p_k^\alpha(a_i)\big|f_k(a_i) - f_{i-1}(a_i)\big| \\
&= p_{i-1}^\alpha(a_i)\big|f_{i-1}(a_i) - f_{i-1}(a_i)\big| + p_i^\alpha(a_i)\big|f_i(a_i) - f_{i-1}(a_i)\big| \\
&\quad + \sum_{\substack{k=0 \\ k\neq i \\ k\neq i-1}}^{m} p_k^\alpha(a_i)\big|f_k(a_i) - f_{i-1}(a_i)\big| \\
&= p_i^\alpha(a_i)\big|f_i(a_i) - f_{i-1}(a_i)\big| + \sum_{\substack{k=0 \\ k\neq i \\ k\neq i-1}}^{m} p_k^\alpha(a_i)\big|f_k(a_i) - f_{i-1}(a_i)\big|
\end{aligned}$$

Since f is assumed to be continuous, then we have $\left|f_i(a_i) - f_{i-1}(a_i)\right| = 0$. Therefore, we obtain:

$$\forall i \in \{0,1,\cdots,m\}, \left|\hat{f}_\alpha(a_i) - f(a_i)\right| \leq \sum_{\substack{k=0 \\ k\neq i \\ k\neq i-1}}^{m} p_k^\alpha(a_i)\left|f_k(a_i) - f_{i-1}(a_i)\right|. \tag{11}$$

Furthermore, we have

$$\begin{aligned}
p_k^\alpha(a_i) &= \frac{e^{\alpha h_k(a_i)}}{\sum_{j=0}^{m} e^{\alpha h_j(a_i)}} \\
&= \frac{1}{\sum_{j=0}^{m} e^{\alpha\left(h_j(a_i) - h_k(a_i)\right)}} \\
&= \frac{1}{1 + e^{\alpha\left(h_i(x) - h_k(x)\right)} + \sum_{\substack{j=0 \\ j\neq k \\ j\neq i}}^{m} e^{\alpha\left(h_j(x) - h_k(x)\right)}} \\
&\leq \frac{1}{1 + e^{\alpha\left(h_i(a_i) - h_k(a_i)\right)}}, \text{ (because } \sum_{\substack{j=0 \\ j\neq k \\ j\neq i}}^{m} e^{\alpha\left(h_j(x) - h_k(x)\right)} \geq 0) \\
&= \frac{1}{1 + e^{-\alpha h_k(a_i)}}, \text{ (because } h_i(a_i) = 0)
\end{aligned}$$

Since $h_k(a_i) < 0 \ \forall k \notin \{i, i-1\}$, then we have $\frac{1}{1+e^{-\alpha h_k(a_i)}} \longrightarrow 0$ as $\alpha \longrightarrow +\infty$ and $k \notin \{i, i-1\}$. Therefore, we obtain:

$$\forall i \in \{0,\cdots,m\}, \forall k \notin \{i, i-1\},\ p_k^\alpha(a_i) \longrightarrow 0 \text{ as } \alpha \longrightarrow +\infty. \tag{12}$$

From (11) and (12), we obtain

$$\forall i \in \{0,\cdots,m\},\ \left|\hat{f}_\alpha(a_i) - f(a_i)\right| \longrightarrow 0 \text{ as } \alpha \longrightarrow +\infty. \tag{13}$$

From (10) and (13), we obtain

$$\forall x \in \mathbb{R},\ \hat{f}_\alpha(x) \longrightarrow f(x) \text{ as } \alpha \longrightarrow +\infty.$$

To simplify and illustrate our approach, we consider the following partition of $\mathbb{R}$

$$\tau_a = \{(-\infty, a], (a, +\infty)\}. \tag{14}$$

The partition τ_a is a special case of (6), and it is studied extensively; see for instance [25].

For the partition τ_a, the P-S function f, the probability p^{α} and the smoothing approximation $\hat{f}_{\alpha}$ will take the form (15), (16) and (17), respectively.

$$f(x) = \begin{cases} f_0(x), & x \leq a, \\ f_1(x), & x > a. \end{cases} \tag{15}$$

$$p^{\alpha}(x) = \Big(\frac{1}{1+e^{\alpha(x-a)}}, \frac{1}{1+e^{-\alpha(x-a)}}\Big). \tag{16}$$

$$\hat{f}_{\alpha}(x) = \frac{1}{1+e^{\alpha(x-a)}} f_0(x) + \frac{1}{1+e^{-\alpha(x-a)}} f_1(x). \tag{17}$$

According to Theorem 2, $\hat{f}_{\alpha}$ is of class $C^k(\mathbb{R})$ and converges to f as α goes to infinity. Now, let us analyze the error estimates.

Theorem 3. *Take p^{α}, f, and $\hat{f}_{\alpha}$ as in (16), (15), and (17), respectively. We have*

$$\forall x \in \mathbb{R},\ \forall \alpha > 0,\ |\hat{f}_{\alpha}(x) - f(x)| \leq \frac{1}{2}|f_1(x) - f_0(x)|. \tag{18}$$

and for every compact C of $\mathbb{R}$, for every $\varepsilon > 0$, there exists $\alpha > 0$ such that

$$x \in C \Rightarrow |f_{\alpha}(x) - f(x)| \leq \varepsilon. \tag{19}$$

Proof. Let $x \in \mathbb{R}$, we have:

$$\begin{aligned}
|\hat{f}_{\alpha}(x) - f(x)| &= |p_0^{\alpha}(x) f_0(x) + p_1^{\alpha}(x) f_1(x) - f(x)| \\
&= |p_0^{\alpha}(x) f_0(x) + (1 - p_0^{\alpha}(x)) f_1(x) - f(x)| \\
&= |p_0^{\alpha}(x)\big(f_0(x) - f_1(x)\big) + f_1(x) - f(x)| \\
&= \begin{cases} \big(1 - p_0^{\alpha}(x)\big)|f_1(x) - f_0(x)|, & x \leq a, \\ p_0^{\alpha}(x)|f_1(x) - f_0(x)|, & x > a. \end{cases} \\
&= \begin{cases} \frac{1}{1+e^{-\alpha(x-a)}}|f_1(x) - f_0(x)|, & x \leq a, \\ \frac{1}{1+e^{\alpha(x-a)}}|f_1(x) - f_0(x)|, & x > a. \end{cases} \\
&= \frac{1}{1+e^{\alpha|x-a|}}|f_1(x) - f_0(x)|
\end{aligned}$$

Furthermore, we have:

$$\forall x \in \mathbb{R},\ \forall \alpha > 0, \frac{1}{1+e^{\alpha|x-a|}} \leq \frac{1}{2}.$$

Therefore, we obtain (18).

Let C be a compact of $\mathbb{R}$. Since $g(x) := f_1(x) - f_0(x)$ is continuous in a and $g(a) = 0$, then for every $\varepsilon > 0$, there exists $\delta > 0$ such that

$$\begin{aligned}
|x-a| \le \delta &\Rightarrow |g(x)-g(a)| \le \varepsilon \\
&\Rightarrow |f_1(x)-f_0(x)| \le \varepsilon \\
&\Rightarrow \frac{1}{1+e^{\alpha|x-a|}}|f_1(x)-f_0(x)| \le \frac{1}{1+e^{\alpha|x-a|}}\varepsilon, \forall \alpha > 0 \\
&\Rightarrow \frac{1}{1+e^{\alpha|x-a|}}|f_1(x)-f_0(x)| \le \varepsilon, \forall \alpha > 0 \\
&\Rightarrow |\hat{f}_\alpha(x)-f(x)| \le \varepsilon, \forall \alpha > 0.
\end{aligned}$$

Therefore, we have:

$$\forall \varepsilon > 0, \exists \delta > 0 \text{ such that } \big(|x-a| \le \delta \Rightarrow |\hat{f}_\alpha(x)-f(x)| \le \varepsilon, \forall \alpha > 0.\big) \tag{20}$$

Furthermore, for every $\beta > 0$ we have:

$$\begin{aligned}
|x-a| > \beta &\Rightarrow |\hat{f}_\alpha(x)-f(x)| < \frac{1}{1+e^{\alpha\beta}}|f_1(x)-f_0(x)|, \forall \alpha > 0 \\
&\Rightarrow |\hat{f}_\alpha(x)-f(x)| < \frac{M_C}{1+e^{\alpha\beta}}, \forall x \in C, \forall \alpha > 0 \\
&\quad \big(\text{ where } M_C = \max_{x\in C}|f_1(x)-f_0(x)|\big) \\
&\Rightarrow \forall \varepsilon > 0, \exists \hat{\alpha} > 0 \text{ such that } \big(|\hat{f}_{\hat{\alpha}}(x)-f(x)| < \varepsilon, \forall x \in C.\big) \\
&\quad \big(\text{ because } \lim_{\alpha\longrightarrow\infty} \frac{M_C}{1+e^{\alpha\delta}} = 0\big)
\end{aligned}$$

Therefore, we have:

$$\forall \beta > 0, \forall \varepsilon > 0, \exists \hat{\alpha} > 0 \text{ such that } \big(|x-a| > \beta \Rightarrow |\hat{f}_{\hat{\alpha}}(x)-f(x)| \le \varepsilon, \forall x \in C.\big) \tag{21}$$

For $\beta = \delta$ we have:

$$\forall \varepsilon > 0, \exists \hat{\alpha} > 0 \text{ such that } \big(|x-a| > \delta \Rightarrow |\hat{f}_{\hat{\alpha}}(x)-f(x)| \le \varepsilon, \forall x \in C.\big) \tag{22}$$

Since (20) is true for every $\alpha > 0$ and every $x \in \mathbb{R}$. Therefore, it is true for $\hat{\alpha}$ and for every $x \in C$, that is

$$\forall \varepsilon > 0, \exists \delta > 0 \text{ such that } \big(|x-a| \le \delta \Rightarrow |\hat{f}_{\hat{\alpha}}(x)-f(x)| \le \varepsilon, \forall x \in C.\big) \tag{23}$$

From (22) and (23), we obtain

$$\forall \varepsilon > 0, \exists \hat{\alpha} > 0 \text{ such that } \big(|\hat{f}_{\hat{\alpha}}(x)-f(x)| \le \varepsilon, \forall x \in C.\big)$$

This completes the proof of (19).

4 Smoothing Approximations for Multiple-dimensional Case

In this section, we aim to extend the above techniques to n-dimensional case. First, we consider the partition $\tau_G = \{A, A^c\}$ of $\mathbb{R}^n$, where

$$A = \{x \in \mathbb{R}^n |\ G(x) \le 0\},$$

and A^c is the complement of A. In this case, the P-S function f, the probability p^α and the smoothing approximation $\hat{f}_\alpha$ will take the form (24), (25) and (26), respectively.

$$f(x) = \begin{cases} f_0(x), & G(x) \leq 0, \\ f_1(x), & G(x) > 0. \end{cases} \tag{24}$$

$$p^\alpha(x) = \big(\frac{1}{1+e^{\alpha G(x)}}, \frac{1}{1+e^{-\alpha G(x)}}\big). \tag{25}$$

$$\hat{f}_\alpha(x) = \frac{1}{1+e^{\alpha G(x)}} f_0(x) + \frac{1}{1+e^{-\alpha G(x)}} f_1(x). \tag{26}$$

Theorem 4. *Take f and $\hat{f}_\alpha$ as in (24) and (26), respectively. If $G : \mathbb{R}^n \longrightarrow \mathbb{R}$ is of class $C^k(\mathbb{R}^n)$, then we have:*

(a) $\hat{f}_\alpha$ is of class $C^k(\mathbb{R}^n)$ $\forall \alpha$,
(b) $\hat{f}_\alpha \longrightarrow f$ as $\alpha \longrightarrow +\infty$.

Proof. (a) Obtained immediately.
(b) For each $x \in \mathbb{R}^n$, we have

$$\begin{aligned} \big|\hat{f}_\alpha(x) - f(x)\big| &= \big|p_0^\alpha(x) f_0(x) + p_1^\alpha(x) f_1(x) - f(x)\big| \\ &= \big|p_0^\alpha(x) f_0(x) + (1 - p_0^\alpha(x)) f_1(x) - f(x)\big| \\ &= \big|p_0^\alpha(x)\big(f_0(x) - f_1(x)\big) + f_1(x) - f(x)\big| \\ &= \begin{cases} \big(1 - p_0^\alpha(x)\big)|f_1(x) - f_0(x)|, & G(x) \leq 0 \\ p_0^\alpha(x)|f_1(x) - f_0(x)|, & G(x) > 0 \end{cases} \\ &= \begin{cases} \frac{1}{1+e^{-\alpha G(x)}}|f_1(x) - f_0(x)|, & G(x) \leq 0 \\ \frac{1}{1+e^{\alpha G(x)}}|f_1(x) - f_0(x)|, & G(x) > 0 \end{cases} \\ &= \frac{1}{1+e^{\alpha |G(x)|}}|f_1(x) - f_0(x)| \end{aligned}$$

If $G(x) \neq 0$, we have $\frac{1}{1+e^{\alpha|G(x)|}} \longrightarrow 0$ as $\alpha \longrightarrow +\infty$. Therefore, we have :

$$\forall x \in \mathbb{R}^n \setminus \{x \in \mathbb{R}^n |\ G(x) = 0\}, \hat{f}_\alpha(x) \longrightarrow f(x) \text{ as } \alpha \longrightarrow +\infty. \tag{27}$$

If $G(x) = 0$, then $|f_1(x) - f_0(x)| = 0$ (because f is assumed to be continuous). Therefore, we have:

$$\forall x \in \{x \in \mathbb{R}^n |\ G(x) = 0\},\ \hat{f}_\alpha(x) = f(x). \tag{28}$$

From (27) and (28), we obtain $\hat{f}_\alpha \longrightarrow f$ pointwise, as $\alpha \longrightarrow +\infty$.

Now, let us analyze the error estimates.

Theorem 5. *Take p^α, f , and $\hat{f}_\alpha$, as in (25), (24), and (26), respectively. We have*

$$\forall x \in \mathbb{R}^n,\ \forall \alpha > 0,\ |\hat{f}_\alpha(x) - f(x)| \leq \frac{1}{2}|f_1(x) - f_0(x)|, \tag{29}$$

and for every compact C of $\mathbb{R}^n$, for every $\varepsilon > 0$, there exists $\alpha > 0$ such that

$$x \in C \Rightarrow |\hat{f}_\alpha(x) - f(x)| \leq \varepsilon. \tag{30}$$

Proof. The proof is similar to the proof of Theorem 3.

5 Application to Some Fundamental Functions

In this section, our smoothing technique is applied to some fundamental functions that appear in the context of non-smooth optimization and penalty methods. In non-smooth optimization, the non-smoothness mostly stems from the presence of min, max, and $|.|$ in the formulation of the cost function. Therefore, to obtain a smooth version of the cost function, it suffices to smooth min, max and $|.|$.

Example 2. Consider the two functions $f(x) = |x|$ and $g(x) = \max(0,x)$. These functions are P-S function. However, they are not differentiable at 0 and their presence in any problem leads to limits in analysis. A smooth version of f and g, denoted $\hat{f}_\alpha$ and $\hat{g}_\alpha$, respectively, are obtained by applying our smoothing technique described above, that is

$$\hat{f}_\alpha(x) = \frac{1}{1+e^{\alpha x}} \times (-x) + \frac{1}{1+e^{-\alpha x}} \times x,$$

and

$$\hat{g}_\alpha(x) = \frac{1}{1+e^{-\alpha x}} \times x.$$

According to Theorem 2, $\hat{f}_\alpha$ and $\hat{g}_\alpha$ are of class $C^\infty(\mathbb{R})$ and converge to f and g, respectively, as α goes to infinity. The graphs of the function f, the smoothing approximation $\hat{f}_\alpha$, and the error of approximation are shown in Fig. 1. In Fig. 2, we show the graphs of the function g, the smoothing approximation $\hat{g}_\alpha$, and the error of the smoothing approximation.

Example 3. Let us consider the L_1-norm $\|(x,y)\|_1 = |x| + |y|$ which is not differentiable if $x = 0$ or $y = 0$. Therefore, the smoothing approximation of $\|(x,y)\|_1$, denoted $\|(x,y)\|_{\alpha,1}$ is obtained by applying our smoothing technique to the components $|x|$ and $|y|$. The graphs of the function $\|(x,y)\|_1$, the smoothing approximation $\|(x,y)\|_{\alpha,1}$, and the error of the smoothing approximation are shown in Fig. 3.

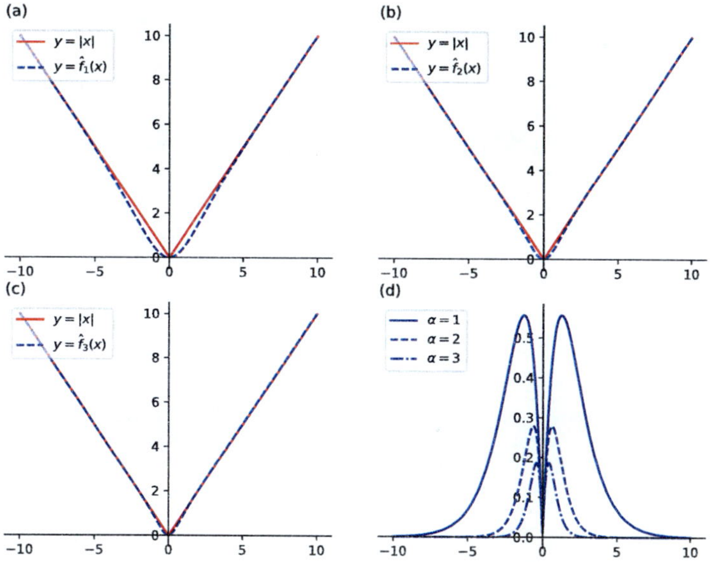

Fig. 1. (a), (b) and (c) contain the graphs of the function $f(x) = |x|$ and its smoothing approximation $\hat{f}_\alpha(x)$ for $\alpha = 1$, $\alpha = 2$, and $\alpha = 3$, respectively. (d) The error of the smoothing approximation for different values of α

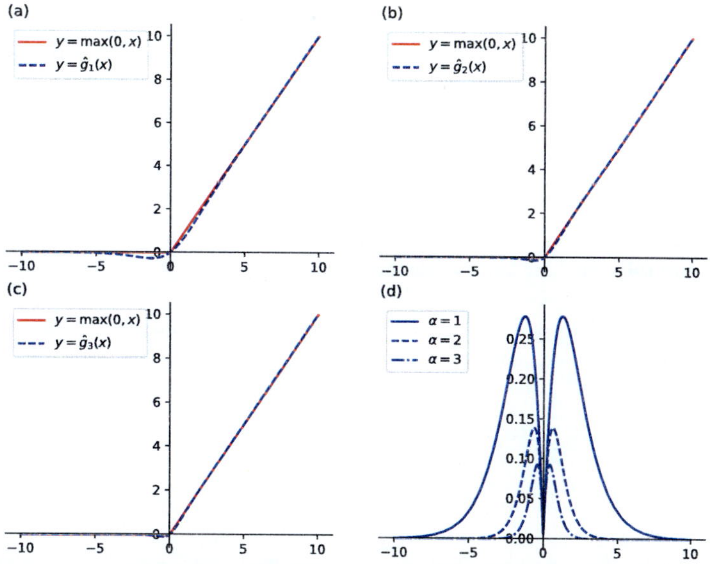

Fig. 2. (a), (b) and (c) contain the graphs of the function $g(x) = \max(0,x)$ and its smoothing approximation $\hat{g}_\alpha(x)$ for $\alpha = 1, \alpha = 2$ and $\alpha = 3$, respectively. (d) The error of the smoothing approximation for different values of α

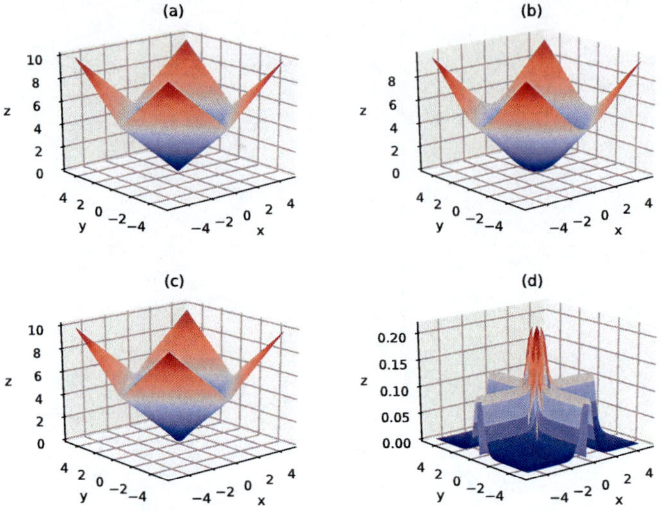

Fig. 3. (a) The graph of the function $\|(x,y)\|_1$. (b) The graph of the smoothing approximation $\|(x,y)\|_{\alpha,1}$ for $\alpha = 1$. (c) The graph of the smoothing approximation $\|(x,y)\|_{\alpha,1}$ for $\alpha = 5$. (d) The error of the smoothing approximation for $\alpha = 5$

6 Conclusion and Perspectives

Throughout this article, we have studied a new method of smoothing piecewise smooth functions. Therefore, we first propose a new formulation for piecewise smooth functions. Based on this formulation, we showed that smoothing a piecewise smooth function is equivalent to smooth a probability distribution. We have studied the properties of our smoothing method and applied it to important non-smooth functions that appear in many fields. This study shows that our approach has important properties and is easy to apply.

For future works, we intend to use our smoothing approach to solve non-smooth optimization problems and penalty methods.

Acknowledgements. The first author's work is supported by the national center for scientific and technical research, Morocco.

References

1. Amat, S., Busquier, S., Escudero, A., Trillo, J.C.: Lagrange interpolation for continuous piecewise smooth functions. J. Comput. Appl. Math. **221**, 47–51 (2008)
2. Aslam, M., Riemenschneider, S., Shen, L.: Smoothing transforms for wavelet approximation of piecewise smooth functions. IET Image Proc. **2**, 239–248 (2008)
3. Bagirov, A.M., Al Nuaimat, A., Sultanova, N.: Hyperbolic smoothing function method for minimax problems. Optimization **62**, 759–782 (2013)

4. Bai, G.Z., Liu, S.Y., Wang, H., Zhi, W., Yin, Z.J.: A novel method for smooth blending cylindrical surfaces whose axes are non-coplanar with intersecting line of two tapered surfaces. Appl. Mech. Mater. **687**, 1470–1473 (2014). Trans Tech Publ
5. Belkhatir, B., Zidna, A.: Construction of flexible blending parametric surfaces via curves. Math. Comput. Simul. **79**, 3599–3608 (2009)
6. Bertsekas, D.P.: Nondifferentiable optimization via approximation. In: Balinski, M.L., Wolfe, P. (eds.) Nondifferentiable Optimization. Mathematical Programming Studies, vol 3. Springer, Berlin, Heidelberg (1975). https://doi.org/10.1007/BFb0120696
7. Fukushima, M., Qi, L.: Reformulation: Nonsmooth, Piecewise Smooth, Semismooth and Smoothing Methods, vol. 22. Springer, Cham (2013). https://doi.org/10.1007/978-1-4757-6388-1
8. Han, X.: A degree by degree recursive construction of Hermite spline interpolants. J. Comput. Appl. Math. **225**, 113–123 (2009)
9. Hartmann, E.: Blending an implicit with a parametric surface. Comput. Aided Geom. Des. **12**, 825–835 (1995)
10. Hashemi, F., Ketabchi, S.: Numerical comparisons of smoothing functions for optimal correction of an infeasible system of absolute value equations. Numer. Algebra, Control Optim. **10**, 13 (2020)
11. Jiang, M., Shen, R., Xu, X., Meng, Z.: Second-order smoothing objective penalty function for constrained optimization problems. Numer. Funct. Anal. Optim. **35**, 294–309 (2014)
12. Jiang, X., Zhang, Y.: A smoothing-type algorithm for absolute value equations. J. Ind. Manag. Optim. **9**, 789 (2013)
13. Kuntz, L., Scholtes, S.: Qualitative aspects of the local approximation of a piecewise differentiable function. Nonlinear Anal. Theory Methods Appl. **25**, 197–215 (1995)
14. Lian, S., Duan, Y.: Smoothing of the lower-order exact penalty function for inequality constrained optimization. J. Inequalities Appl. **2016**, 1–12 (2016)
15. Lipman, Y., Levin, D.: Approximating piecewise-smooth functions. IMA J. Numer. Anal. **30**, 1159–1183 (2010)
16. Mazroui, A., Sbibih, D., Tijini, A.: A simple method for smoothing functions and compressing Hermite data. Adv. Comput. Math. **23**, 279–297 (2005)
17. Meng, Z., Dang, C., Yang, X.: On the smoothing of the square-root exact penalty function for inequality constrained optimization. Comput. Optim. Appl. **35**, 375–398 (2006)
18. Min, J., Meng, Z., Zhou, G., Shen, R.: On the smoothing of the norm objective penalty function for two-cardinality sparse constrained optimization problems. Neurocomputing **458**, 559–565 (2021)
19. Nguyen, C.T., Saheya, B., Chang, Y.-L., Chen, J.-S.: Unified smoothing functions for absolute value equation associated with second-order cone. Appl. Numer. Math. **135**, 206–227 (2019)
20. Prandoni, P., Vetterli, M.: Approximation and compression of piecewise smooth functions, Philosophical Transactions of the Royal Society of London. Ser. A: Math. Phys. Eng. Sci. **357**, 2573–2591 (1999)
21. Qi, L., Tseng, P.: On almost smooth functions and piecewise smooth functions. Nonlinear Anal. Theory Methods Appl. **67**, 773–794 (2007)
22. Ralph, D., Scholtes, S.: Sensitivity analysis of composite piecewise smooth equations. Math. Program. **76**, 593–612 (1997)
23. Rockafellar, R.T.: A property of piecewise smooth functions. Comput. Optim. Appl. **25**, 247–250 (2003)
24. Saheya, B., Yu, C.-H., Chen, J.-S.: Numerical comparisons based on four smoothing functions for absolute value equation. J. Appl. Math. Comput. **56**, 131–149 (2018)
25. Wu, H., Zhang, P., Lin, G.-H.: Smoothing approximations for some piecewise smooth functions. J. Oper. Res. Soc. China **3**, 317–329 (2015)

26. Wu, W., Fan, Q., Zurada, J.M., Wang, J., Yang, D., Liu, Y.: Batch gradient method with smoothing l1/2 regularization for training of feedforward neural networks. Neural Netw. **50**, 72–78 (2014)
27. Xavier, A.E.: The hyperbolic smoothing clustering method. Pattern Recogn. **43**, 731–737 (2010)
28. Zang, I.: A smoothing-out technique for min-max optimization. Math. Program. **19**, 61–77 (1980)
29. Amhraoui, E., Masrour, T.: Smoothing approximations for piecewise smooth functions: a probabilistic approach. Numer. Algebra Control Optim. https://doi.org/10.3934/naco.2021033

Modeling and Vector Control of a Variable Speed Power Plant System Based on 5 MW DFIG

Elmostafa Chetouani[1(✉)], Youssef Errami[2], Abdellatif Obbadi[2], and Smail Sahnoun[2]

[1] Electronics, Instrumentation and Energy, Faculty of Sciences, University of Chouaib Doukkali, El Jadida, Morocco
mschetouani@gmail.com

[2] Electronics, Instrumentation and Energy, El Jadida, Morocco
{errami.y,obbadi.a}@ucd.ac.ma

Abstract. This paper introduces a modeling and vector control study of a variable speed power plant system based on the Doubly Fed Induction Generator (DFIG). The generator is connected directly to the grid by its stator and fed by bidirectional converters at its wound rotor. The Direct and Indirect Field Oriented Control (D/IFOC) are designed and compared. The stator powers are regulated by using Proportional Integral (PI) controller based on the pole compensation technique. The generator stator magnetic field is oriented along the d-axis. The Maximum Power Point Tracking (MPPT) algorithm is applied to the variable-speed wind turbine. The performance of the whole system is verified by utilizing the Matlab/Simulink software. From the simulation results, we can observe that the indirect vector control gives better performance than the direct vector in terms of tracking reference with an improved error static and reduced overshoot.

1 Introduction

In recent decades, technological progress makes wind energy one of the most promising renewable energy sources that encourage governments to invest in wind energy and install more wind turbines. Wind energy is playing a significant role in satisfying the global energy demand and decreasing the environmental pollution caused by electricity production using fossil fuel energy sources in the world energy.

The high-power wind turbine, order of megawatts, can operate with different gen-erator technologies like Permanent Magnets Synchronous Generator (PMSG), Squir-rel Cage Induction Generator (SCIG), and Doubly-Fed Induction Generator (DFIG). Several advantages make the DFIG the widely utilized generator in variable speed wind energy systems over any other configuration [1]. Examples of such advantages are the ability to utilize a partial-sized converter in the rotor to control the power, reducing power losses and cost, reducing efforts on

S. Melliani and O. Castillo (Eds.): ICPAMS 2021, LNNS 476, pp. 179–194, 2023.
https://doi.org/10.1007/978-3-031-12416-7_15

me-chanical parts, noise reduction, power control, and controllable power factor [2,3]. The DFIG stator is directly connected to the grid, and two bidirectional converters connect the rotor to the electrical network via a DC-link voltage, as shown in Fig. 1. The Rotor Side Converter (RSC) is employed to control the powers exchanged with the grid and ensure the regulation of the Unit Power Factor (UPF). Yet, the Grid Side Converter (GSC) is utilized to maintain the DC-link voltage constant [4]. The DFIG is characterized by a multivariable non-linear mathematical model. The magnetic field and the electromagnetic torque are strongly coupled. So, the DFIG control is more complex and complicated compared to the DC machine [5]. There are several techniques to control the DFIG that have been proposed. The Field-Oriented Control (FOC) is commonly used in the power plant system based on the doubly fed-inductor generator and has shown satisfactory performance. In addition, this strategy permits to control the DFIG as a separately excited DC machine which makes it very popular in the industry. The Vector Control has two strategies: the Direct (DFOC) and Indirect Field-Oriented Control (IFOC). The latter has two types: the IFOC with open-loop and the IFOC with closed loop. Lamnadi et al. 2016 [2] have proposed a direct vector control for regulating the active and reactive power. Likewise, Bouderbala et al. 2018 [6] have suggested a comparative study between Direct and Indirect Field-Oriented Control (FOC) with a closed-loop for wind energy systems based on the doubly-fed induction machine. Similarly, Becheri et al. 2018 [7] have compared the direct and indirect vector control with a closed-loop for controlling the RSC converter. Direct vector control is simple and works only with two controllers regulating the powers by directly controlling the rotor voltages. However, the Indirect Vector Control with closed-loop is complex and needs four controllers to pilot the stator powers and the rotor currents. But, The IFOC offers satisfactory performance in terms of efficiency and robustness [5]. Most paper researches propose a comparison study between the IFOC with closed-loop and the DFOC. In this paper, we suggest a comparative analysis between the IFOC with open-loop and the direct vector control.

The extractable power from wind energy depends on the characteristics of each turbine and the wind variable speed. Tracking the maximum power generated is required when the wind turbine operates in region II [4]. This strategy is known as Maximum Power-Point Tracking (MPPT). To perform the MPPT many control schemes are developed and can be classified into two categories. The first one has required knowledge of the characteristic aerodynamic curve of wind turbine speed. The second one has not necessitated any information about wind speed to generate the optimal speed reference. Examples of these schemes are the MPPT with Optimal Tip Speed Ratio (OPTSR), the MPPT with Optimal Torque Control (OTC),the MPPT with Perturbation and Observation (P and O), and the MPPT with Power Signal Feedback (PSF) [8–10]. In this paper, a variable speed wind turbine based on a 5 MW doubly-fed induction generator is modeled and simulated by the Matlab/Simulink environment. The Rotor Side Converter (RSC) and Grid Side Converter (GSC) are controlled by a Pulse Width Modulation (PWM). The direct and indirect field-oriented controls (D/I-

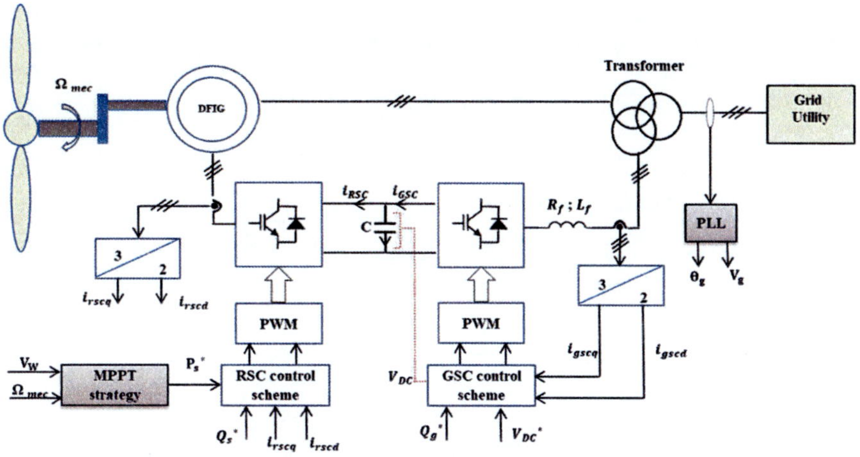

Fig. 1. Whole chain of power plant system

FOC) are proposed to control the injected powers (active and reactive) into the electrical network, consequently regulating the power factor to one.

This paper is structured as follows: Sect. 2 presents the modeling of wind energy conversion systems. Section 3 discusses the MPPT strategy established by the classical PI. Section 4 proposes the control of the active and reactive powers by applying the vector control technique. Section 5 discusses and interprets the simulation results, and the conclusion is presented in Sect. 6.

2 Power Plant System Components Modeling

2.1 Model of the Wind Turbine

The turbine mechanical power is expressed as [6,11]:

$$P_{Tu} = \frac{1}{2}.C_p(\alpha,\beta).\rho.\pi.R^2.V^3 \tag{1}$$

The mechanical torque TTu is written as below:

$$T_{Tu} = \frac{P_{Tu}}{\Omega_{Tu}} \tag{2}$$

where, ρ is the air density (kg/m3), R is the blade ray (m), Ω_{Tu} is the angular velocity of the turbine, and Cp represents the performance factor of the turbine. Cp can be formulated as [6]:

$$C_p(\alpha,\beta) = [0.5-0.0167.(\beta-2)].\sin(\frac{\pi(\lambda+1)}{(18.5-0.3(\beta-2)})-0.00184.(\lambda-3).(\beta-2) \tag{3}$$

This coefficient is calculated by the tip speed λ and the angle β of the blade pitch. The latter is fixed to $\beta = 2$ for having Cpmax. Equation (4) defines the expression of λ [11]:

$$\lambda = \frac{R.\Omega_{Tu}}{V} \tag{4}$$

2.2 Model of the Gearbox

Equation (5) presents the mechanical equation of the system, taking into account that the overall mechanical dynamics are brought back to the turbine shaft [12]:

$$J_{tot}.\frac{d\Omega_{mec}}{dt} + f.\Omega_{mec} = T_g - T_{Tem} \tag{5}$$

where

$$T_g = \frac{T_{Tu}}{G_B} \quad and \ G_B = \frac{\Omega_{mec}}{\Omega\Omega_{Tu}} \tag{6}$$

where J_{tot} is the overall inertia of wind energy system, T_{Tu} is the turbine torque, T_{Tem} is the electromagnetic torque of the generator, f is the overall viscous coefficient of friction, Ω_{mec} is the rotational speed at the rotor shaft of the gearbox (rad/s) and G_B is the gearbox multiplier.

2.3 DFIG Modeling

The Park transformation allows simplifying the general electrical model, which is determined by the following equations. Stator and rotor voltages equations are [2,6]:

$$\begin{aligned}
V_{sd} &= R_s.i_{sd} + \frac{d\varphi_{sd}}{dt} - \omega_s.\varphi_{sq} \\
V_{sq} &= R_s.i_{sq} + \frac{d\varphi_{sq}}{dt} + \omega_s.\varphi_{sd} \\
V_{rd} &= R_r.i_{rd} + \frac{d\varphi_{rd}}{dt} - \omega_r.\varphi_{rq} \\
V_{rq} &= R_r.i_{rq} + \frac{d\varphi_{rq}}{dt} - \omega_r.\varphi_{rd}
\end{aligned} \tag{7}$$

The stator and rotor field magnetic flux equations can be written as follows [13]:

$$\begin{aligned}
\varphi_{sd} &= L_s.i_{sd} + M.i_{rd} \\
\varphi_{sq} &= L_s.i_{sq} + M.i_{rq} \\
\varphi_{rd} &= L_r.i_{rd} + M.i_{sd} \\
\varphi_{rq} &= L_r.i_{rq} + M.i_{sq}
\end{aligned} \tag{8}$$

where, Vr and Vs are the voltages; is and ir are the currents; φ_s and φ_r are the flux; Rs and Rr are the resistances; ω_s and ω_r are the angular frequencies; Ls and Lr are the inductances; M is the mutual inductance. The r and s denote the rotor and stator, respectively. The electromagnetic torque can be expressed as follows [6]:

$$T_{em} = -p.\frac{M}{L_s}.(i_{rq} - i_{rd}.\varphi_{sq}) \tag{9}$$

where p is the number of generator pole pairs. The stator powers are expressed as follows [2,6]:

$$P_s = V_{sd}.i_{sd} + V_{sq}.i_{sq} \tag{10}$$

$$Q_s = V_{sd}.i_{sd} - V_{sd}.i_{sq} \tag{11}$$

3 MPPT Strategy

During the normal functioning of the wind turbine, the maximum power control method is developed to exploit the energy available in the wind as much as possible. The MPT method with mechanical speed control is established. This technique consists of maintaining the generator speed at its reference, which is maximized when the Cp is optimal. The electromagnetic torque developed by DFIG is equal to its reference value imposed by the control defined as [14]:

$$T_{em} = T_{em-opt} \tag{12}$$

The optimal electromagnetic torque T_{em-opt} for obtaining a rotation speed equal to the optimal speed is given as follows:

$$T_{em-opt} = [K_{pmppt} + K_{imppt}.\frac{1}{s}].[\Omega_{mec-opt} - \Omega_{mec}] \tag{13}$$

where, K_{pmppi} and K_{imppt} are the PI controller gains. The optimal speed ($\Omega_{mec-opt}$) is :

$$\Omega_{mec-opt} = G_B.\Omega_{Tu} \;\; ;with \;\; \Omega_{Tu} = \frac{V.\lambda_{opt}}{R} \tag{14}$$

Determination of the PI gains for MPPT

The PI controller parameters are determined by the pole compensation method. The time constant of the system (T_{sys}) is:

$$T_{sys} = \frac{J_{tot}}{f} \tag{15}$$

The gains of the controller are expressed as:

$$K_{imppt} = \frac{1}{\tau.f} \;\; and \;\; K_{pmppt} = \frac{-K_{imppt}.J_{tot}}{f} \;\; ; \;\; with \;\; \tau = \frac{T_{sys}}{1000} \tag{16}$$

4 Vector Control Strategy Applied to DFIG

4.1 Vector Control of Rotor Side Converter

Field Oriented Control (FOC) technique consists of aligning the stator flux along the d-axis of the rotating reference frame (Park Referee), as shown in Fig. 2. Vector con-trol has two types: Direct Field-Oriented Control (DFOC) and Indirect Field-Oriented Control (IFOC).

The stator flux is considered constant and is oriented according to d-axis. The stator winding resistance is neglected and the stator voltage equation can be simplified as follows [4–6]:

$$\varphi_{sd} = \varphi_s \ , \ \varphi_{sd} = 0 \ ; \ V_{sd} = 0 \ , \ V_{sq} = V_s = \omega_s.\varphi_s \tag{17}$$

The rotor voltages can be expressed as follows [4–6]:

$$V_{rd} = [R_r + (L_r - \frac{M^2}{L_s}).s].i_{rd} - g.\omega_s.(L_r - \frac{M^2}{L_s}).i_{rq} \tag{18}$$

$$V_{rq} = [R_r + (L_r - \frac{M^2}{L_s}).s].i_{rq} + g.\omega_s.(L_r - \frac{M^2}{L_s}).i_{rd} + g.\frac{V_s.M}{L_s} \tag{19}$$

From Eqs. (18,19), the rotor currents expressions are deduced as follows:

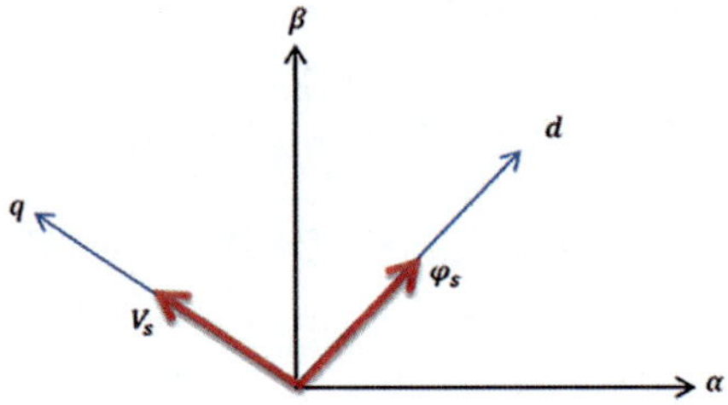

Fig. 2. Stator flux orientation along the d-axis

$$i_{rd} = [V_{rd} + g.\omega_s.(L_r - \frac{M^2}{L_s}).i_{rd}]/[R_r + (L_r - \frac{M^2}{L_s}).s] \tag{20}$$

$$i_{rq} = [V_{rq} - g.\omega_s.(L_r - \frac{M^2}{L_s}).i_{rd} - g.\frac{V_s.M}{L_s}]/[R_r + (L_r - \frac{M^2}{L_s}).s] \tag{21}$$

By using the vector control simplifications cited in Eq. (17) and from Eq. (8), the stator currents expressions are deduced as follows:

$$i_{sd} = -\frac{M}{L_s}.i_{rd} + \frac{\varphi_s}{L_s} \tag{22}$$

$$i_{sq} = -\frac{M}{L_s}.i_{rq} \tag{23}$$

The stator active and reactive powers can be expressed as follows [2]:

$$P_s = -V_s.\frac{M}{L_s}.i_{rq} \tag{24}$$

$$Q_s = .\frac{V_s^2}{\omega_s.L_s} - V_s.\frac{M}{L_s}.i_{rd} \tag{25}$$

The Eqs. (24,25) show that the powers are independently controlled by applying vector control. The active power is piloted by acting on the rotor q-component current. However, the reactive power is regulated by rotor d-component current.

4.1.1 Direct Vector Control

DFOC method consists of neglecting the coupling terms and uses the PI controller to regulate the active and reactive power in closed-loop, as shown in Fig. 3. The power regulators directly pilot the rotor voltages of the generator. The voltage references are generated by the following equations:

$$V_{rq}^{*} = [P_s^{*} - P_s].[K_{prsc1} + K_{irsc1}.\frac{1}{s}] + g.\frac{V_s.M}{L_s} \tag{26}$$

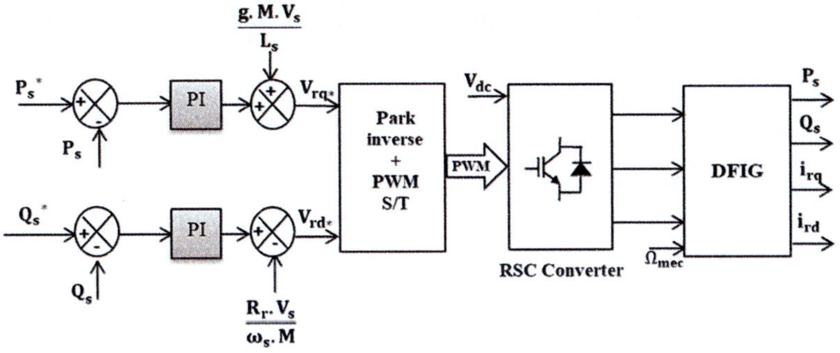

Fig. 3. Direct field oriented control of powers scheme

$$V_{rd}^{*} = [Q_s^{*} - Q_s].[K_{prsc1} + K_{irsc1}.\frac{1}{s}] + \frac{V_s.R_r}{\omega_s.M} \tag{27}$$

Determination of the gains of PI Controller

To set the PI parameters (K_{prsc1} , K_{irsc1}), the pole compensation method is utilized. The time constant of the system is:

$$T_{sys} = (L_r - \frac{M^2}{L_s})/R_r \tag{28}$$

The equations of PI parameters (K_{prsc1} , K_{irsc1}) are given as follows:

$$K_{prsc1} = \frac{1}{T_{rsc}}.\frac{L_s.(L_r - \frac{M^2}{L_s})}{M.V_s} \quad and \quad K_{irsc1} = \frac{1}{T_{rsc}}.\frac{R_r.L_s}{M.V_s} \; ; with \quad T_{rsc} = \frac{1}{100} \tag{29}$$

4.1.2 Indirect Vector Control

IFOC technique considers the coupling terms and compensates them by performing a system comprising two loops controlling the powers and the rotor currents. This strategy has two variants. The first one consists of regulating the powers and the rotor currents in the closed-loop. However, the second one governs the powers in an open-loop, and rotor currents are controlled in a closed-loop, as depicted in Fig. 4. In this paper, the second method is utilized to control the rotor side converter (RSC).

Considering the q-axis of the rotating reference frame is aligned to the stator voltage, as presented in Fig. 2. The current references can be deducted from Eqs. (24) and (25) as follows [2,6]:

$$i_{rq}^* = -\frac{L_s}{V_s.M}.P_s^* \tag{30}$$

$$i_{rd}^* = -\frac{L_s}{V_s.M}.(Q_s^* - \frac{V_s^2}{\omega_s.L_s}) \tag{31}$$

The voltage references are expressed easily from Eq. (18) and Eq. (19) as follows:

$$V_{rq}^* = [i_{rq}^* - i_{rq}].[K_{prsc2} + K_{irsc2}.\frac{1}{s}] + e_{rd} + g.\frac{V_s.M}{L_s} \tag{32}$$

$$V_{rd}^* = [i_{rd}^* - i_{rd}].[K_{prsc2} + K_{irsc2}.\frac{1}{s}] + e_{rq} \tag{33}$$

where

$$e_{rd} = g.\omega_s.(L_r - \frac{M^2}{L_s}].i_{rd} \quad and \quad e_{rd} = g.\omega_s.(L_r - \frac{M^2}{L_s}].i_{rq} \tag{34}$$

Determination of the gains of PI Controller

To set the PI parameters (K_{prsc}, K_{irsc}), the pole compensation method is utilized. The time constant of the system is:

$$T_s = (L_r - \frac{M^2}{L_s})/R_r \tag{35}$$

The equations of PI parameters (K_{prsc}, K_{irsc})are given as follows:

$$K_{prsc2} = \frac{1}{T_{rsc}}.(L_r - \frac{M^2}{L_s}) \; and \; K_{prsc2} = \frac{K_{prsc2}.R_r}{(L_r - \frac{M^2}{L_s})} \; ; \; with \; T_{rsc} = \frac{T_s}{1000} \tag{36}$$

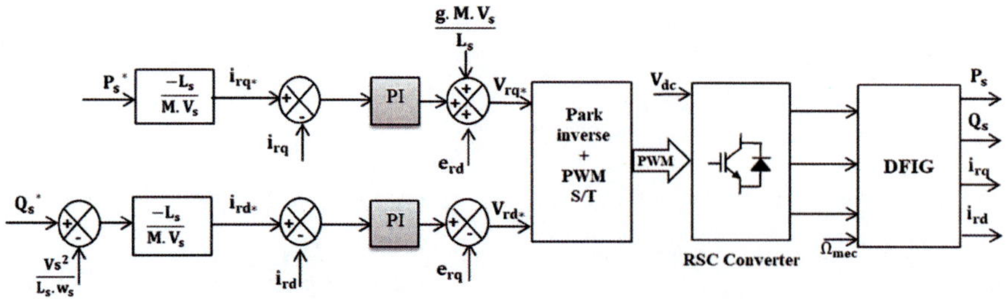

Fig. 4. Indirect field oriented control of powers scheme

4.2 Grid Side Converter (GSC) Control

The objective of the GSC is to regulate the DC-link voltage and ensure that the power factor is equal to one. The GSC control scheme is given in Fig. 5. The powers exchanged equations between the grid, and the converter can be expressed by applying Park Transformation and considering a receptor convention of the converters, as shown in Fig. 1, as follows [15]:

$$P_g = V_{gd}.i_{fd} + V_{gq}.i_{fq} \tag{37}$$

$$Q_g = V_{gq}.i_{fd} - V_{gd}.i_{fd} \tag{38}$$

The equations used to model the bus continuous are given below:

$$i_c = i_{GSC} - i_{RSC} \tag{39}$$

$$V_{DC} = \frac{1}{c.s}.i_c \tag{40}$$

Considering the grid voltage is oriented along the q-axis, as shown in Fig. 2. So:

$$V_{gd} = 0 \ ; \ V_{gd} = V_g \tag{41}$$

The output voltage of the GSC can be written as [15,16]:

$$V_{gd-gsc} = -[R_f + L_f.s].i_{fd} + \omega_g.L_f.i_{fq} \tag{42}$$

$$V_{gq-gsc} = -[R_f + L_f.s].i_{fq} - \omega_g.L_f.i_{fd} + V_g \tag{43}$$

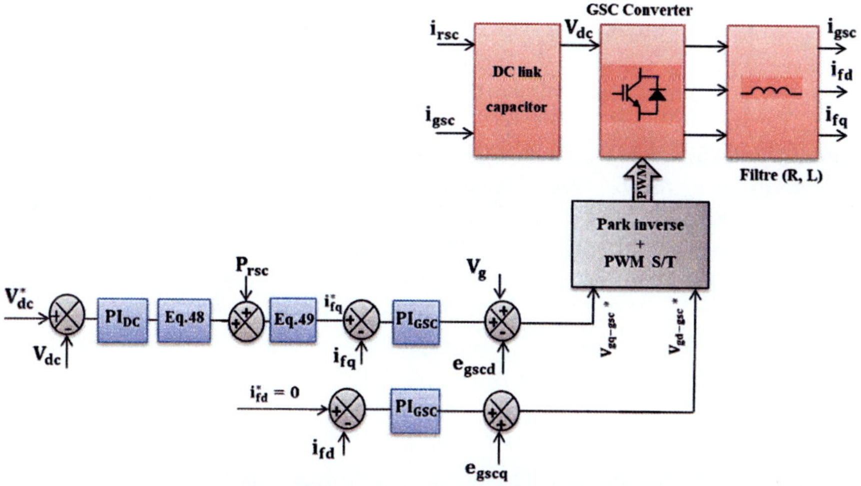

Fig. 5. Grid side converter (GSC) control scheme

Therefore, the Eqs. (37,38) can be simplified as follows [14]:

$$P_g = V_{gq}.i_{fq} \ and \ Q_g = V_{gq}.i_{fd} \tag{44}$$

Assuming the Back-to-Back converter is lossless and neglecting the losses in the inductor resistor, the relation between powers can be expressed as follows:

$$V_{DC}.i_c = P_{GSC} - P_{RSC} \tag{45}$$

$$P_g = P_{GSC} = V_{DC}.i_c + P_{RSC} \tag{46}$$

$$P_{RSC} = V_{DC}.i_{RSC} \tag{47}$$

$$P_c^* = V_{DC}.i_c^* \tag{48}$$

From the above equations, the references of i_{fq}^* can be derived as follows:

$$i_{fq}^* = \frac{1}{V_g}.(V_{DC}^*.i_c^* + P_{RSC}) \tag{49}$$

The voltage references are expressed as follows:

$$V_{gd-gsc}^* = [i_{fd}^* - i_{fd}].[K_{pgsc+K_{igsc}}.s] + e_{gscq} \tag{50}$$

$$V_{gq-gsc}^* = [i_{fq}^* - i_{fq}].[K_{pgsc+K_{igsc}}.s] - e_{gscd} + V_g \tag{51}$$

where

$$e_{gscq} = \omega_g.L_f.i_{fq} \ and \ e_{gscd} = \omega_g.L_f.i_{fd} \tag{52}$$

From Eq. (19) and Eq. (20) the expressions of the grid currents can be deducted as follows:

$$i_{fq} = -\frac{1}{[R_f + L_f.s]}.(V_{gq-gsc}^* - \omega_g.L_f.i_{fd} - V_{DC}.S_q) \tag{53}$$

$$i_{fd} = -\frac{1}{[R_f + L_f.s]}.(V_{gd-gsc}^* + \omega_g.L_f.i_{fq} - V_{DC}.S_d) \tag{54}$$

where Sd and Sq are the switching states computed by Park transformation.
Determination of the DC-link controller parameters

The PI controller is used to maintain the DC-link voltage constant, as shown in Fig. 6. From Eq. (40) the current flowing through the DC bus is deduced as below:

$$i_c = C.s.V_{DC} \tag{55}$$

From the block diagram shown in Fig. 6:

$$C.s.V_{DC} = \frac{K_{pDC}.\,s + K_{iDC}}{s}.(V_{DC}^* - V_{DC}) \tag{56}$$

The transfer function in closed loop can be expressed as follows:

$$\frac{V_{DC}}{V_{DC}^*} = \frac{\frac{1}{C}(K_{pDC}\,.s + K_{iDC})}{s^2 + s.\frac{K_{pDC}}{C} + \frac{K_{iDC}}{C}} \tag{57}$$

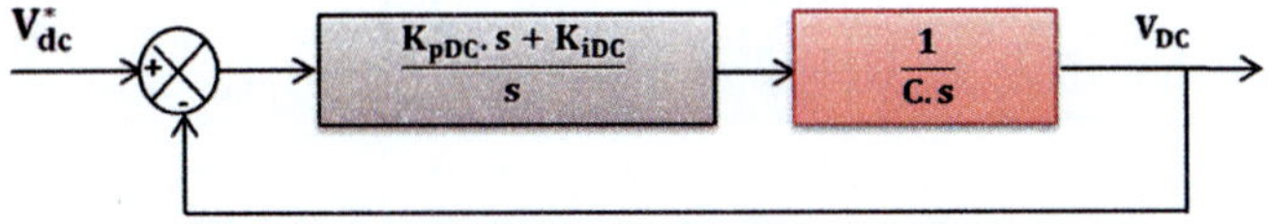

Fig. 6. Regulation loop of DC- link voltage

Comparing denominator of Eq. (57) with the canonical second order polynomial given below:

$$F(s) = \frac{K.\omega_0}{s^2 + 2.\xi.\omega_0 + \omega_0^2} \tag{58}$$

PI controller gains are:

$$K_{pDC} = 2.\xi.\omega.C \text{ ; } With\ \xi\ is\ the\ damping\ coefficient \tag{59}$$

$$K_{iDC} = \omega^2.C \tag{60}$$

Determination of the GSC controller gains

The currents ifq and ifd are controlled by the same regulation loop, as shown in Fig. 7. The GSC is considered as a unit gain (G_{GSC}=1). The PI controllers compensate the coupling terms (e_{gscq} and e_{gscd}).

The time constant of the controlled system is given in Eq. (61), which is divided by 10 to have a satisfactory system dynamic.

$$T_s = \frac{L_f}{R_f} \tag{61}$$

The transfer function in open loop (TFO) is expressed as follows:

$$TFO = K_{igsc}.\frac{\frac{K_{pgsc}}{K_{igsc}}.s+1}{s}.\frac{\frac{1}{R_f}}{\frac{L_f}{R_f}.s+1} = \frac{K_{igsc}}{R_f.s} \text{ ; } With\ \frac{K_{pgsc}}{K_{igsc}} = \frac{L_f}{R_f} \tag{62}$$

The transfer function in closed-loop (TFC) can be written as follows:

$$TFC = \frac{TFO}{TFO+1} = \frac{1}{\frac{R_f}{K_{igsc}}.s+1} \tag{63}$$

By applying the pole compensator method, the PI controller gains (K_{pgsc}, K_{igsc}) are given as follows:

$$K_{pgsc} = \frac{L_f}{R_f}\ and\ K_{igsc} = \frac{R_f}{T} \text{ ; } With\ T = \frac{T_s}{10} \tag{64}$$

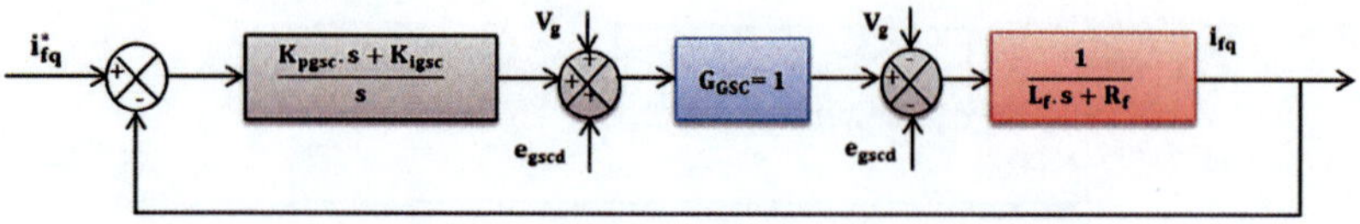

Fig. 7. Control loop of currents flowing through the filter RL

5 Simulation Results and Discussion

With help of the before-mentioned mathematical models, the wind chain is modeled and simulated employing the parameters presented in Tables (1,2) under Matlab/Simulink software. The wind profile used in this work is presented in Fig. 8, which is considered a variable. However, Fig. 9 illustrates the produced mechanical power, which is pursuing the evolution of the wind speed. From 0.2 (s) to 0.7 (s), the mechanical power is equal to 5 MW, which corresponds to the rated value of the wind velocity that is 12.5 (m/s). This power will be utilized as a reference for driving the system. Figure 10 shows the mechanical speed computed by the MPPT strategy based on the PI controller. Figures 11 and 12 show the efficiency of the MPPT technique for making constant the tip speed ratio at its best value, which is $\lambda_{opt} = 9.19$, and the coefficient of power is preserved at its maximum (Cpmax= 0.5) regardless of the wind speed change. So, maximal power is achieved. The rotor side converter is primarily controlled by the direct field-aligned control, and then it is piloted by the indirect vector control. These variants of the vector control use two PI controllers. The first one controls the stator active power, and the second one governs the reactive power. The comparison analysis is established by considering the same condition. The response time of the system is divided by 103 for having a rapid dynamic system. As you can see in Fig. 13, the stator active power is computed by two techniques: the direct and indirect field-aligned control based on the two PI controllers, where their gains are adjusted by using the pole compensation method. Notably, these two methods offer approximately a similar time response, which is 2.1 (ms). Besides, the static error is improved by the IFOC from 0.84 (percent) to 0.16 (percent)of the rated value. Besides, the overshot is also improved. The desired reactive power is zero ($Q_s^* = 0$) to ensure a power factor equal to 1, as shown in Fig. 14. It is observa-ble that the static error is also improved, and the stability is guaranteed when especially wind changes. Figure 15 shows the DC-link voltage, which is regulated by a PI controller at its reference value ($V_{DC}^* = 1200\,\text{V}$). This Figure shows the influence of the evolution of the wind speed on the stability of the voltage. An overshot is observed when the wind change. So, the indirect field-oriented control with open-loop control gives better performances than the direct vector control in terms of the static error amelioration and the overshot reduction.

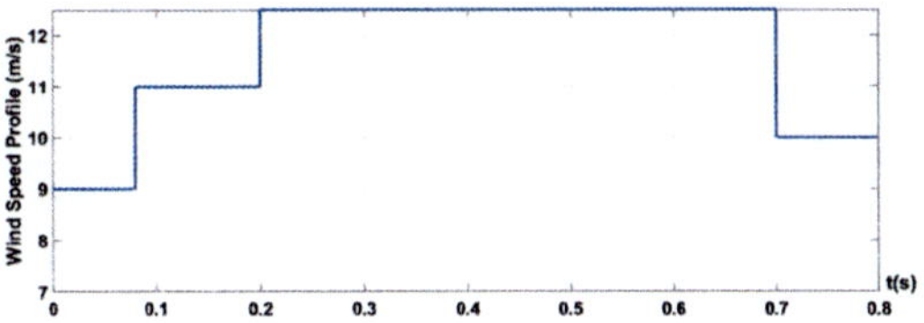

Fig. 8. Wind speed profile (m/s)

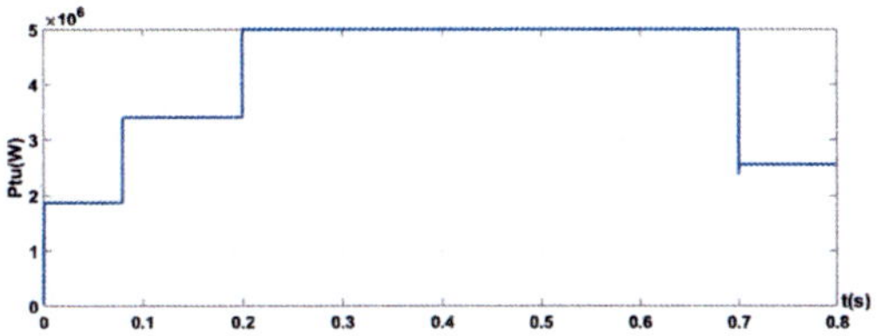

Fig. 9. Extracted mechanical power by MPPT strategy

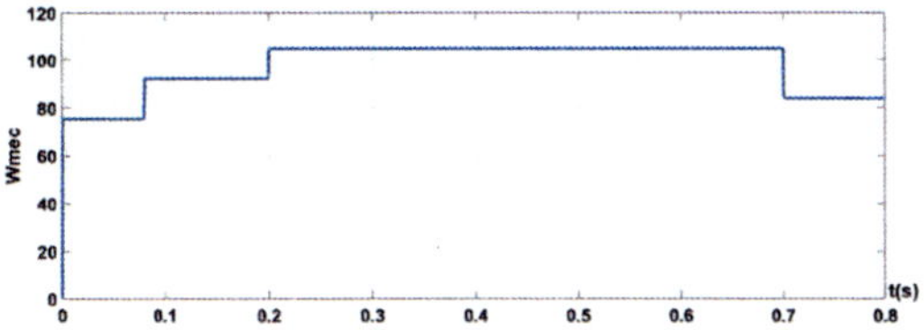

Fig. 10. Mechanical rotation speed of the DFIG (rad/s)

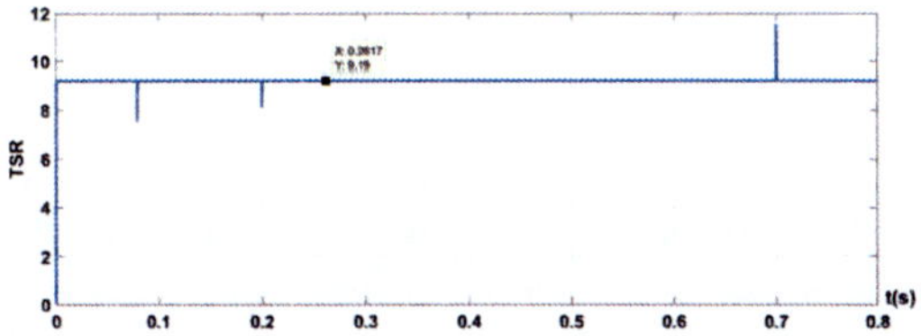

Fig. 11. Tip speed ratio (TSR)

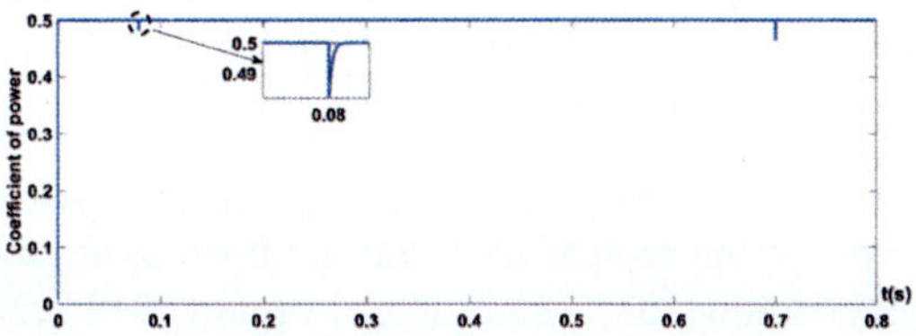

Fig. 12. Coefficient of power (Cp)

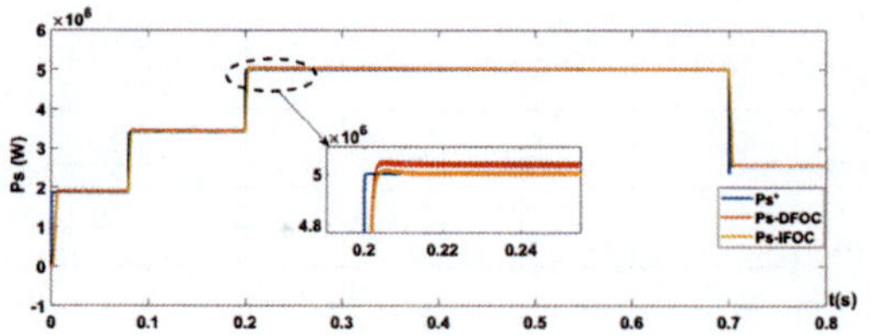

Fig. 13. Stator active power (W)

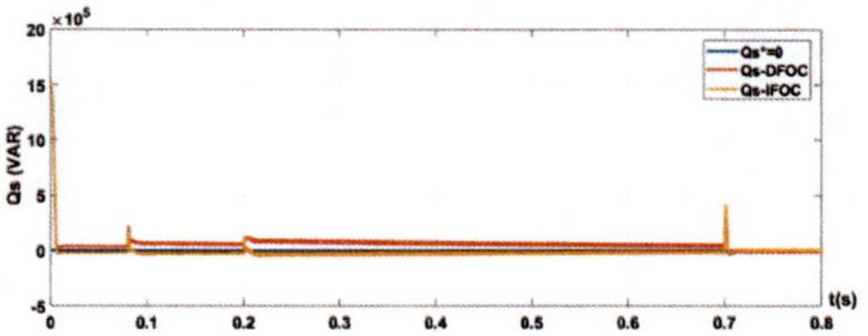

Fig. 14. Stator reactive power (VAR)

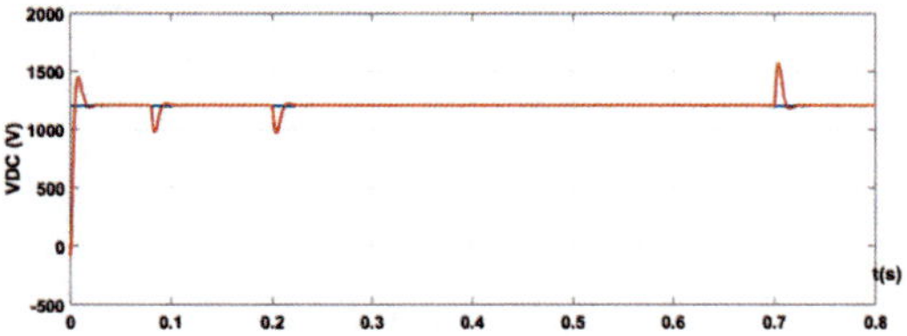

Fig. 15. DC link voltage (V)

6 Conclusion

In this paper, the wind energy conversion system is modeled and simulated based on variable wind speed. The DFIG stator is directly linked to the electrical network, and the rotor is coupled via two bidirectional converters. The Maximum Power Tracking with speed regulation is established. Then, the control of the rotor side converter is designed by employing the IFOC with open loop and DFOC based on the classical PI controller, in which the gains are calculated by the pole compensation method. The results of the two strategies are compared and analyzed. It is observed that the IFOC with open loop brings satisfactory results than the direct vector control (DFOC) under wind speed variation.

The future works will focus on:

- Proposing and comparing another type of algorithms with the indirect vector control like Backstepping control, Sliding mode control, and Neural network Controller for controlling the WECS based on the DFIG.
- Testing the robustness of the proposed control under the grid faults.

Appendix

Table 1. Proportional and Integral (PI) gains for controllers

MPPT controller	RSC controller		DC link controller	GSC controller
	DFOC strategy	IFOC strategy		
Kp = −7.2e+6 Ki = 51.87	Kprsc1 = 4.3*e* − 3 Kirsc1 = 7*e* − 3	Kprsc2 = 1.446 Kirsc2 = 2.376	Kpdc = 1.848 Kidc = 396	Kpgsc = 200 Kigsc = 5e+4

Table 2. Set of parameters used in the simulation

Component	Parameter	Symbol	Value
Turbine	Ray of blade	**R**	51,583 m
	Coefficient of multiplier	$\mathbf{G}_B$	47,23
	Total moment of inertia	$\mathbf{J}_{tot}$	1000 kg.m^2
DFIG	DFIG rated power	**Ps**	5 MW
	Stator inductance	**Ls**	1,2721 mH
	Rotor resistance	**Rr**	1,446 mΩ
	Rotor inductance	**Lr**	1,1194 mH
	Mutual inductance	**M**	0,55182 mH

References

1. Li, S., Haskew, T.: Energy capture, conversion, and control study of DFIG wind turbine under weibull wind distribution. In: 2009 IEEE Power & Energy Society General Meeting. Calgary, AB. IEEE, vol. 2009, pp. 1–9 (2009)
2. Lamnadi, M., Trihi, M., Bossoufi, B., Boulezhar, A.: Modeling and control of a doubly-fed induction generator for wind turbine-generator systems. Int. J. Power Electron. Drive Syst. (IJPEDS) **7**(3), 973–985 (2016). https://doi.org/10.11591/ijpeds.v7.i3.pp982-994
3. Ouassaid, M., Elyaalaoui, K., Cherkaoui, M.: Sliding mode control of induction generator wind turbine connected to the grid. In: Vaidyanathan, S., Volos, C. (eds.) Advances and Applications in Nonlinear Control Systems. SCI, vol. 635, pp. 531–553. Springer, Cham (2016). https://doi.org/10.1007/978-3-319-30169-3_23
4. Rouabhi, R., Djerioui, A.: Control of the power generated by variable speed wind turbine driving a doubly fed induction generator. J. Electr. Eng. **14**(3), 1–7 (2014)
5. Ihedrane, Y., Bekkali, C.E., Bossoufi, B., Bouderbala, M.: Control of power of a DFIG generator with MPPT technique for wind turbines variable speed. In: Derbel, N., Zhu, Q. (eds.) Modeling, Identification and Control Methods in Renewable Energy Systems. GET, pp. 105–129. Springer, Singapore (2019). https://doi.org/10.1007/978-981-13-1945-7_5

6. Bouderbala, M., Bossoufi, B., Lagrioui, A., Taoussi, M., Alami, H., Ihedrane, Y.: Direct and indirect vector control of a doubly fed induction generator based in a wind energy conversion system. Int. J. Electr. Comput. Eng. (IJECE) **9**(3), 1531–1540 (2018)
7. Becheri, M., Bousserhane, I.K., Bouchiba, B., Harrouz, A., Belbekri, T.: Vector control of wind turbine conversion chain variable speed based on DFIG using MPPT strategy. Int. J. Appl. Eng. Res. **13**(7), 5404–5410 (2018)
8. Ananth, D., Kumar, N.: Tip speed ratio based MPPT algorithm and improved field oriented control for extracting optimal real power and independent reactive power control for grid connected doubly fed induction generator. Int. J. Electr. Comput. Eng. (IJECE) **6**(3), 1319–1331 (2016). https://doi.org/10.11591/ijece.v6i3.9306
9. Chavero-Navarrete, E., Trejo-Perea, M., Jáuregui-Correa, J.C., Carrillo-Serrano, R.V., Ríos-Moreno, J.G.: Expert control systems for maximum power point tracking in a wind turbine with PMSG: state of the art. Appl. Sci. **9**(12), 1–24 (2019). https://doi.org/10.3390/app9122469
10. Cortes-Vega, D., Ornelas-Tellez, F., Anzurez-Marin, J.: Comparative analysis of MPPT techniques and optimal control for a PMSG-based WECS. In: 2019 IEEE 4th Colombian Conference on Automatic Control (CCAC), Medellín, Colombia, pp. 1–6 (2019). https://doi.org/10.1109/CCAC.2019.8921183
11. Errami, Y., Obbadi, A., Sahnoun, S.: Control of PMSG wind electrical system in network context and during the MPP tracking process. Int. J. Syst. Control Commun. **11**(2), 200–225 (2020). https://doi.org/10.1504/IJSCC.2020.106578
12. Errami, Y., Obbadi, A., Sahnoun, S., Benhmida, M., Ouassaid, M., Maaroufi, M.: Design of a nonlinear backstepping control strategy of grid interconnected wind power system based PMSG. In: AIP Conference Proceeding, vol. 1758, p. 030053 (2016). https://doi.org/10.1063/1.4959449
13. Mensou, S., Essadki, A., Minka, I., Nasser, T., Bououlid Idrissi, B.: Backstepping Controller for a Variable Wind Speed Energy Conversion System Based on a DFIG. In: International Journal of Electrical and Computer Engineering, vol. 12, no. 9 (2018)
14. Baran, J., Jaderko, A.: An MPPT control of a PMSG-based WECS with disturbance compensation and wind speed estimation. Energies **13**(23), 6344 (2020)
15. Elazzaoui, M.: Modeling and Control of a Wind System Based Doubly Fed Induction Generator: Optimization of the Power Produced. J. Electr. Electron. Syst. **4**(1), 1–8 (2015). https://doi.org/10.4172/2332-0796.1000141
16. Qian, K., Gao, G., Sheng, Z.: A maximum current control method for three-phase PWM rectifier for the ITER in-vessel vertical stability coil power supply. IEEE Trans. Plasma Sci. **46**(5), 1689–1693 (2018)

Stability for Delay SEIR Epidemic Models with Saturated Incidence Rates and Diffusion

C. Allalou(✉), A. Abbassi, and E. Anaama

Laboratory LMACS, FST of Beni-Mellal, Sultan Moulay Slimane University, Beni Mellal, Morocco
chakir.allalou@yahoo.fr

Abstract. In this paper, we investigate the effect of spatial diffusion and delay on the dynamical behavior of the SEIR epidemic model. The introduction of the delay in this model makes it more realistic and modelizes the latency period. In addition, the consideration of an SEIR model with diffusion aims to better understand the impact of the spatial heterogeneity of the environment and the movement of individuals on the persistence and extinction of disease. First, we determined a threshold value $\mathscr{R}_0$ of the delayed SEIR model with diffusion. Next, By using the theory of partial functional differential equations, we have shown that the unique disease-free equilibrium and the endemic equilibrium are asymptotically stable. Moreover,we determine, using Lyapunov functionals, conditions by which the disease-free equilibrium and the endemic equilibrium are globally asymptotically stable. Also some numerical simulations are given to illustrate the theoretical results.

1 Introduction

Epidemiological models with latent or incubation period have been studied by many authors, because many diseases have a latent or incubation period, during which the individual is said to be infected but not infectious. This period can be modeled by incorporating it as a delay effect [2], or by introducing an exposed class [7]. Therefore, it is an important subject to compare this two types of modeling incubation period.

The models mentioned above have concentrated only on the temporal dimension with out diffusion. But, in many cases the spatial variation of population plays an important role in the disease spreading model and the time variation governs the dynamical behavior of the disease spreading, see [11]. However, the spatial effects cannot be neglected in studying the spread of epidemics; because due to the large mobility of people within a country or even worldwide, spatially uniform models are not sufficient to give a realistic picture of disease diffusion. Focusing on the influence of space on the qualitative behavior of the SIR epidemic model, several improvements are made (see, e.g., [15,16] and references cited therein).

In this paper, we generalize all the DDE and DDEs models PDE presented in [1,18] by proposing the following delayed SEIR epidemic model with spatial diffusion and saturated incidence function:

S. Melliani and O. Castillo (Eds.): ICPAMS 2021, LNNS 476, pp. 195–209, 2023.
https://doi.org/10.1007/978-3-031-12416-7_16

$$\begin{cases} \frac{\partial S(x,t)}{\partial t} = d\Delta S(x,t) + A - \mu S(x,t) - \frac{\beta S(x,t)E(x,t)}{1+\alpha_1 S(x,t)} \\ \frac{\partial E(x,t)}{\partial t} = d\Delta I(x,t) + \frac{\beta e^{-\mu\tau} S(x,t-\tau)E(x,t-\tau)}{1+\alpha_1 S(x,t-\tau} - (\mu+\gamma)I(x,t) \\ \frac{\partial I(x,t)}{\partial t} = d\Delta I(x,t) + \gamma E(x,t) - (\mu+\alpha+\delta)I(x,t) \\ \frac{\partial R(x,t)}{\partial t} = d\Delta R(x,t) + \delta I(x,t) - \mu R(x,t) \end{cases} \tag{1}$$

where Δ denotes the Laplacian operator, $S(x,t), E(x,t), I(x,t), R(x,t)$ are the numbers of susceptible, infectious but not yet symptomatic, infected and recovered individuals at location x and time t, respectively.

The recruitment rate of new individuals into the susceptible class is A, μ is positive constants representing the natural mortality rate of the population. α is a positive constants representing the death rate due to disease.The positive constant d indicates the diffusion rate, the transmission rate is β, and α_1 is the parameters that measure the inhibitory effect. The exposed individuals develop symptoms at a rate γ, so $1/\gamma$ is the latent period, as many γE exposed will be infected. The number of exposed increases as many $\beta \frac{S(x,t)E(x,t)}{1+\alpha_1 S(x,t)}$ individuals after direct contact between susceptible and exposed.

Likewise the infected symptomatically can be recover at rates δ.
Throughout this paper, we consider the system 1 with initial conditions

$$\begin{cases} S(x,t) = \psi_1(x,t) \geq 0, \\ E(x,t) = \psi_2(x,t) \geq 0, \\ I(x,t) = \psi_3(x,t) \geq 0, \\ R(x,t) = \psi_5(x,t) \geq 0, \end{cases} for\ (x,t) \in \bar{\Omega} \times [-\tau, 0], \tag{2}$$

and zero-flux boundary conditions

$$\frac{\partial S}{\partial \nu} = \frac{\partial E}{\partial \nu} = \frac{\partial I}{\partial \nu} = \frac{\partial R}{\partial \nu} = 0, t \geq 0, x \in \partial\Omega, \tag{3}$$

where Ω is a bounded domain in $\mathbf{R}^n$ with a smooth boundary $\partial\Omega$ and $\frac{\partial}{\partial \nu}$ represents the outside normal derivative on $\partial\Omega$. The boundary condition in 3 implies that susceptible,exposed, infectious, quarantined and recovered individuals do not across the boundary $\partial\Omega$.

The paper is organized as follows. In next section, we study the well-posedness for model 1. Section 3 is devoted to investigate to the local stability of the disease-free equilibrium and the endemic through the study of associated characteristic equations. equilibrium. Next, in Sect. 4, we prove the global asymptotical stability of the disease-free equilibrium. In Sect. 5, to support our theoretical predictions, some numerical simulations are given.

2 The Well-Posedness

In this section, we focus on the well-posedness of solutions for 1 by establishing the global existence, uniqueness, nonnegativity and boundedness of solutions. In the following, we need some notations. Let $\mathbf{X} = \mathscr{C}\left(\overline{\Omega}, \mathbf{R}^4\right)$ be the Banach space of continuous functions from $\overline{\Omega}$ into $\mathbf{R}^4$, and $\mathscr{C}_{\mathbf{X}} = \mathscr{C}([-\tau, 0], \mathbf{X})$ denotes the Banach space of continuous $\mathbf{X}$ -valued functions on $[-\tau,0]$ equipped with the supremum norm. For any real numbers $a \leq b, t \in [a,b]$ and any continuous function $u : [a-\tau,b] \to \mathbf{X}$, u_t is the element of $\mathscr{C}_{\mathbf{X}}$ given by $u_t(\theta) = u(t+\theta)$ for $\theta \in [-\tau,0]$.

Moreover, we identify any element $\psi \in \mathscr{C}_{\mathbf{X}}$ as a function from $\overline{\Omega} \times [-\tau,0]$ in $\mathbf{R}^4$ defined by $\psi(x,t) = \psi(t)(x)$ The next theorem gives us the existence and uniqueness of the global positive solution.

Theorem 1. *For any given initial condition $\psi \in \mathscr{C}_{\mathbf{X}}$ satisfying (2) and (3), the system (1) admits a unique nonnegative solution. Moreover, this solution is global and remains nonnegative.*

Proof. Let $\psi = (\psi_1, \psi_2, \psi_3, \psi_4) \in \mathscr{C}_{\mathbf{X}}$ and $x \in \overline{\Omega}$. We define $f = (f_1, f_2, f_3, f_4) : \mathscr{C}_{\mathbf{X}} \to \mathbf{X}$ by

$$\begin{aligned}
f_1(\psi)(x) &= A - \mu\psi_1(x,0) - \tfrac{\beta\psi_1(x,0)\psi_2(x,0)}{1+\alpha_1\psi_1(x,0)} \\
f_2(\psi)(x) &= \tfrac{\beta e^{-\mu\tau}\psi_1(x,-\tau)\psi_2(x,-\tau)}{1+\alpha_1\psi_1(x,-\tau)} - (\mu+\gamma)\psi_2(x,0) \\
f_3(\psi)(x) &= \gamma\psi_2(x,0) - (\mu+\alpha+\delta)\psi_3(x,0) \\
f_4(\psi)(x) &= \delta\psi_3(x,0) - \mu\psi_4(x,0)
\end{aligned}$$

Then the system 1–3 can be rewritten as an abstract differential equation in the phase space $\mathscr{C}_{\mathbf{X}}$ in the form

$$\begin{cases} \dot{u} = Bu + f(u_t), & t \geq 0 \\ u(0) = \psi \in \mathscr{C}_{\mathbf{X}}, \end{cases} \tag{4}$$

where $u(t) = (S(.,t), E(.,t), I(.,t), R(.,t))^{\top}$, $\psi = (\psi_1, \psi_2, \psi_3, \psi_4)$ and $Bu = (d\Delta S, d\Delta E, d\Delta I, d\Delta R)$. We can easily show that f is locally Lipschitz in $\mathscr{C}_{\mathbf{X}}$. According to [4,9,10, 13,17], we deduce that the system 4 admits a unique local solution on its maximal interval of existence $[0, t_{\max})$.

Since $0 = (0,0,0,0)$ is a lower-solution of the problem 1–3, we have $S(x,t) \geq 0$, $E(x,t) \geq 0$, $I(x,t) \geq 0$, and $R(x,t) \geq 0$.

In the following, our goal is to show that the maximum solution of the problem 1–3, is global. Let's first consider the first equation of the system 1, then we have

$$\begin{cases} \frac{\partial S(x,t)}{\partial t} - d\Delta S(x,t) \leq A - \mu S(x,t), \\ \frac{\partial S}{\partial v} = 0, \\ S(x,0) = \psi_1(x,0) \geq 0. \end{cases} \tag{5}$$

By the comparison principle [12], we have $S(x,t) \leq \tilde{S}(t)$, where $\tilde{S}(t) = \tilde{S}(0)e^{-\mu t} + \frac{A}{\mu}(1 - e^{-\mu t})$ is the solution of the following ordinary equation:

$$\begin{cases} \frac{d\tilde{S}}{dt} = A - \mu\tilde{S}, \\ \tilde{S}(0) = \max_{x\in\overline{\Omega}}(\psi_1(x,0)). \end{cases} \tag{6}$$

Hence,

$$S(x,t) \leq \max\left\{\frac{A}{\mu}, \max_{x\in\overline{\Omega}}(\psi_1(x,0))\right\}, \forall (x,t) \in \overline{\Omega} \times [0.t_{\max}),$$

this implies that S is bounded.
Let $\quad L(x,t)=e^{-\mu\tau}S(x,t-\tau)+E(x,t)+I(x,t)+R(x,t),$
thus,

$$\frac{\partial L(x,t)}{\partial t}=e^{-\mu\tau}d\Delta S(x,t-\tau)+d\Delta E(x,t)+d\Delta I(x,t)+d\Delta R(x,t)+e^{-\mu\tau}A-\mu L(x,t)-\alpha I(x,t).$$

Then, we have

$$\begin{cases}\frac{\partial L(x,t)}{\partial t}-d\Delta L(x,t)\leq e^{-\mu\tau}A-\mu L(x,t),\\ \frac{\partial L}{\partial v}=0,\\ L(x,0)=e^{-\mu\tau}\psi_1(x,-\tau)+\psi_2(x,0)+\psi_3(x,0)+\psi_4(x,0).\end{cases}\tag{7}$$

Applying the comparison principle to the system 7 we obtain

$$L(x,t)\leq\max\left\{\frac{e^{-\mu\tau}A}{\mu},\max_{x\in\overline{\Omega}}L(x,0)\right\},\forall(x,t)\in\overline{\Omega}\times[0.t_{\max}).$$

Therefore, E,I and R are bounded. So, we proved that S,E,I and R are bounded on $\overline{\Omega}\times[0,t_{\max})$. By the standard theory for semilinear parabolic systems [6], we deduce that $t_{\max}=+\infty$. This completes the proof.

3 Basic Reproduction Number and Existence of Equilibirum

In this section we determine the equilibrium of the SEIR models, for that we solve the following system

$$\begin{cases}A-\mu S-\frac{\beta SE}{1+\alpha_1 S}=0\\ \frac{\beta e^{-\mu\tau}SE}{1+\alpha_1 S}-(\mu+\gamma)I=0\\ \gamma E-(\mu+\alpha+\delta)I=0\\ \delta I-\mu R=0\end{cases}\tag{8}$$

Then the disease-free equilibrium is given by

$$\mathscr{P}_0=\left(S^0,E^0,I^0,R^0\right)$$

where $E^0=I^0=R^0=0$ and $S^0=\frac{A}{\mu}$.

Furthermore, the system 1 has a unique endemic equilibirum

$$\mathscr{P}^*=\left(S^*,E^*,I^*,R^*\right),$$

where

$$\begin{cases}S^*=\frac{\mu+\gamma}{\beta e^{-\mu\tau}-\alpha_1(\mu+\gamma)},\\ E^*=\frac{1}{\beta}\left(1+\alpha_1 S^*\right)\left(\frac{A}{S^*}-\mu\right),\\ I^*=\frac{\gamma}{\mu+\delta+\alpha}E^*,\\ R^*=\frac{\delta\gamma}{(\mu+\delta+\alpha)\mu}E^*.\end{cases}\tag{9}$$

Now let's determine the expression of basic reproduction number denoted by $\mathscr{R}_0$, by using the method presented in [14], using the same notations, the matrix F and V are given by

$$F = \begin{pmatrix} \frac{\beta e^{-\mu\tau} S^0}{1+\alpha_1 S^0} & 0 \\ 0 & 0 \end{pmatrix}, \quad and \quad V = \begin{pmatrix} \mu+\gamma & 0 \\ -\gamma & \mu+\alpha+\delta \end{pmatrix}.$$

Thus

$$FV^{-1} = \begin{pmatrix} A & 0 \\ 0 & 0 \end{pmatrix}$$

where

$$A = \frac{\beta A e^{-\mu\tau}}{(\mu+\alpha_1 A)(\mu+\gamma)}.$$

If ρ is the spectral radius of FV^{-1}, then the expression of the basic reproduction number is as follows

$$\mathscr{R}_0 = \rho\left(FV^{-1}\right) = \frac{\beta A e^{-\mu\tau}}{(\mu+\alpha_1 A)(\mu+\gamma)}.$$

So, we can rewrite 9 as follows

$$\begin{cases} S^* = \frac{A}{\mu\mathscr{R}_0+\alpha_1 A(\mathscr{R}_0-1)}, \\ E^* = \frac{\mathscr{R}_0(\mathscr{R}_0-1)(\mu+\alpha_1 A)^2}{(\mu\mathscr{R}_0+\alpha_1 A(\mathscr{R}_0-1))}, \\ I^* = \frac{\gamma\mathscr{R}_0(\mathscr{R}_0-1)(\mu+\alpha_1 A)^2}{(\mu+\delta+\alpha)(\mu\mathscr{R}_0+\alpha_1 A(\mathscr{R}_0-1))}, \\ R^* = \frac{\gamma\delta\mathscr{R}_0(\mathscr{R}_0-1)(\mu+\alpha_1 A)^2}{\mu(\mu+\delta+\alpha)(\mu\mathscr{R}_0+\alpha_1 A(\mathscr{R}_0-1))}. \end{cases} \tag{10}$$

Then, $\mathscr{P}^* = (S^*, E^*, I^*, R^*)$ exist if $\mathscr{R}_0 > 1$.

4 Local Stability of the Equilibria for the SEIR Models

In this section, by analyzing the corresponding characteristic equations of equilibria $\mathscr{P}$ and $\mathscr{P}^*$ of 1, we study the local stability of them, respectively.

Let $\tilde{S} = S - S^*, \tilde{E} = E - E^*, \tilde{I} = I - I^*$ and $\tilde{R} = R - R^*$, where $(S^*, E^*, I^*, R^*)^\top$ is an arbitrary equilibrium point, and drop bars for simplicity. Then the system 1 can be transformed into the following form

$$\begin{cases} \frac{\partial S(x,t)}{\partial t} = d\Delta S(x,t) + A - \mu\left(S(x,t)+S^*\right) - \frac{\beta(S(x,t)+S^*)(E(x,t)+E^*)}{1+\alpha_1(S(x,t)+S^*)} \\ \frac{\partial E(x,t)}{\partial t} = d\Delta E(x,t) + \frac{\beta e^{-\mu t}(S(x,t-\tau)+S^*)(E(x,t)+E^*)}{1+\alpha_1(S(x,t-\tau)+S^*)} - (\mu+\gamma)\left(E(x,t)+E^*\right), \\ \frac{\partial I(x,t)}{\partial t} = d\Delta I(x,t) + \gamma(E(x,t)+E^*) - (\mu+\alpha+\delta)\left(I(x,t)+I^*\right). \\ \frac{\partial R(x,t)}{\partial t} = d\Delta R(x,t) + \delta\left(I(x,t)+I^*\right) - \mu\left(R(x,t)+R^*\right). \end{cases} \tag{11}$$

Thus, the arbitrary equilibrium point $\mathscr{P}^* = (S^*, E^*, I^*, R^*)^\top$ of the system 1 is transformed into the zero equilibrium point $(0,0,0,0)^\top$ of the system 11. In the following, we will analyze stability of the zero equilibrium point of the system 11. Denote $u(t) = (S(.,t), E(.,t), I(.,t), R(.,t))^\top$ and $\psi = (\psi_1, \psi_2, \psi_3, \psi_4) \in \mathscr{C}_X$, then the system

11 can be rewritten as an abstract differential equation in the phase space $\mathscr{C}_{\mathbf{X}}$ of the form

$$\dot{u}(t) = D\Delta u(t) + L(u_t) + g(u_t), \tag{12}$$

where $D = diag(d,d,d,d)$, $L : \mathscr{C}_{\mathbf{X}} \to \mathbf{X}$ and $g : \mathscr{C}_{\mathbf{X}} \to \mathbf{X}$ are given, respectively, by

$$L(\psi)(x) = \begin{pmatrix} -\left(\mu + \frac{\beta e^*}{(1+\alpha_1 S^*)}\right)\psi_1(x,0) - \frac{\beta S^*}{(1+\alpha_1 S^*)}\psi_2(x,0) \\ \frac{\beta E^* e^{-\mu\tau}}{(1+\alpha_1 S^*)}\psi_1(x,-\tau) + \frac{\beta S^* e^{-\mu\tau}}{(1+\alpha_1 S^*)}\psi_2(x,-\tau) - (\mu+\gamma)\psi_2(x,0) \\ \gamma\psi_2(x,0) - (\mu+\alpha+\delta)\psi_3(x,0) \\ \delta\psi_3(x,0) - \mu\psi_4(x,0) \end{pmatrix} \tag{13}$$

and

$$g(\psi)(x) = \begin{pmatrix} g_1(\psi)(x) \\ g_2(\psi)(x) \\ g_3(\psi)(x) \\ g_4(\psi)(x) \end{pmatrix},$$

where

$$\begin{cases} g_1(\psi)(x) = \frac{\beta E^*}{(1+\alpha_1 S^*)}\psi_1(x,0) + \frac{\beta S^*}{(1+\alpha_1 S^*)}\psi_2(x,0) + A - \frac{\beta(\psi_1(x,0)+S^*)(\psi_2(x,0)+E^*)}{1+\alpha_1(\psi_1(x,0)+S^*)} - \mu S^* \\ g_2(\psi)(x) = -\frac{\beta E^* e^{-\mu\tau}}{(1+\alpha_1 S^*)}\psi_1(x,-\tau) - \frac{\beta S^* e^{-\mu\tau}}{(1+\alpha_1 S^*)}\psi_2(x,-\tau) + \frac{\beta e^{-\mu\tau}(\psi_1(x,-\tau)+S^*)(\psi_2(x,-\tau)+E^*)}{1+\alpha_1(\psi_1(x,-\tau)+S^*)} \\ \quad -(\mu+\gamma)E^* \\ g_3(\psi)(x) = \gamma E^* - (\mu+\alpha+\delta)I^* \\ g_4(\psi)(x) = \delta I^* - \mu R^* \end{cases} \tag{14}$$

For $\psi = u_t, \psi = (\psi_1, \psi_2, \psi_3, \psi_4)^\top \in \mathscr{C}_{\mathbf{X}}$, the linearized system of 12 at the zero equilibrium point is

$$\dot{u} = D\Delta u(t) + L(u_t),$$

and its characteristic equation is

$$\lambda\omega - D\Delta\omega - L\left(e^{\lambda \cdot}\omega\right) = 0, \tag{15}$$

where $\omega \in dom(\Delta)$, and $\omega \neq 0, dom(\Delta) \subset \mathbf{X}$.

Let $0 = \eta_0 < \eta_1 < \cdots$ be the sequence of eigenvalues for the elliptic operator $-\Delta$ subject to the Neumann boundary condition on Ω, and $E(\eta_i)$ be the eigenspace corresponding to η_i in $L^2(\Omega)$.

Let $0 = \eta_0 < \eta_1 < \cdots$ be the sequence of eigenvalues for the elliptic operator $-\Delta$ subject to the Neumann boundary condition on Ω, and $E(\eta_i)$ be the eigenspace corresponding to η_i in $L^2(\Omega)$.

Let $\{\phi_{ij}, j = 1, \ldots, dimE(\eta_i)\}$ be an orthonormal basis of $E(\eta_i)$, and $\mathbf{Y}_{ij} = \{a\phi_{ij}, a \in \mathbf{R}\}$.
Then

$$L^2(\Omega) = \bigoplus_{i=0}^{+\infty} \mathbf{Y}_i \text{ and } \mathbf{V}_i = \bigoplus_{j=1}^{dimE(\eta_i)} \mathbf{Y}_{ij}.$$

Moreover, we put for $i=0,1,2,\ldots, j=1,2,\ldots, dimE(\eta_i)$

$$\beta_{ij}^1=\begin{pmatrix}\phi_{ij}\\0\\0\\0\end{pmatrix},\quad \beta_{ij}^2=\begin{pmatrix}0\\\phi_{ij}\\0\\0\end{pmatrix},\quad \beta_{ij}^3=\begin{pmatrix}0\\0\\\phi_{ij}\\0\end{pmatrix},\quad \beta_{ij}^4=\begin{pmatrix}0\\0\\0\\\phi_{ij}\end{pmatrix}. \tag{16}$$

Clearly, the family $\left(\beta_{ij}^1,\beta_{ij}^2,\beta_{ij}^3,\beta_{ij}^4\right)$ is a basis of $\left(L^2(\Omega)\right)^4$. Therefore, any element ω of $\mathbf{X}$ can be written in the in the following form

$$\begin{aligned}\omega &= (\omega_1,\omega_2,\omega_3,\omega_4)\\ &= \sum_{i=0}^{+\infty}\sum_{j=1}^{dimE(\eta_i)}\langle\omega_1,\phi_{ij}\rangle\beta_{ij}^1+\langle\omega_2,\phi_{ij}\rangle\beta_{ij}^2+\langle\omega_3,\phi_{ij}\rangle\beta_{ij}^3+\langle\omega_4,\phi_{ij}\rangle\beta_{ij}^4.\end{aligned} \tag{17}$$

Next, from a straightforward analysis and using 16 and 17 we show that 15 is equivalent to

$$(\lambda I_5+\eta_i D-M)\begin{pmatrix}\langle\omega_1,\phi_{ij}\rangle\\\langle\omega_2,\phi_{ij}\rangle\\\langle\omega_3,\phi_{ij}\rangle\\\langle\omega_4,\phi_{ij}\rangle\end{pmatrix}=\begin{pmatrix}0\\0\\0\\0\end{pmatrix}, i=0,1,2,\ldots,\quad j=1,2,\ldots,dimE(\eta_i), \tag{18}$$

where M is given by

$$M=\begin{pmatrix}-\mu-\frac{\beta E^*}{(1+\alpha_1 S^*)} & -\frac{\beta S^*}{(1+\alpha_1 S^*)} & 0 & 0\\ \frac{\beta E^* e^{-\mu\tau}}{(1+\alpha_1 S^*)}e^{-\lambda\tau} & -(\mu+\gamma)+\frac{\beta S^* e^{-\mu\tau}}{(1+\alpha_1 S^*)}e^{-\lambda\tau} & 0 & 0\\ 0 & \gamma & -(\mu+\alpha+\delta) & 0\\ 0 & 0 & \delta & -\mu\end{pmatrix}.$$

Thus the characteristic equation is

$$(\lambda+d\eta_i+\mu+\alpha+\delta)(\lambda+d\eta_i+\mu)\left(\lambda^2+a\lambda+b+(c\lambda+d)e^{-\lambda\tau}\right)=0,\quad i=0,1,\ldots, \tag{19}$$

where

$$\begin{aligned}a&=2(\eta_i d+\mu)+\gamma+\tfrac{\beta E^*}{1+\alpha_1 S^*},\\ b&=\left(\eta_i d+\mu+\tfrac{\beta E^*}{1+\alpha_1 S^*}\right)(\eta_i d+\mu+\gamma),\\ c&=-\tfrac{\beta S^* e^{-\mu\tau}}{1+\alpha_1 S^*},\\ d&=-\tfrac{\beta S^* e^{-\mu\tau}}{1+\alpha_1 S^*}(\eta_i d+\mu).\end{aligned}$$

4.1 Stability of Disease-Free Equilibirum $\mathscr{P}$

Using the above analysis, in this subsection, we take $(S^*,E^*,I^*,R^*)=\mathscr{P}=\left(\frac{A}{\mu},0,0,0\right)$. Thus, the characteristic Eq. 19 becomes for $i=0,1,\ldots$

$$(\lambda+d\eta_i+\mu+\alpha+\delta)(\lambda+d\eta_i+\mu)^2\left(\lambda+\eta_i d+\mu+\gamma-\frac{\beta A e^{-\mu\tau}}{\mu+\alpha_1 A}e^{-\lambda\tau}\right)=0, \tag{20}$$

Theorem 2. *If $\mathscr{R}_0 \leq 1$, then the disease-free equilibrium $\mathscr{P}$ is locally asymptotically stable for all $\tau \geq 0$.*

Proof. For $\tau = 0$, the Eq. 20 is equivalent to the following cubic equation

$$(\lambda + d\eta_i + \mu + \alpha + \delta)(\lambda + d\eta_i + \mu)^2 [\lambda + \eta_i d - (\mu + \gamma)(\mathscr{R}_0 - 1)]. \tag{21}$$

Clearly, 21 has three roots $\lambda_1 = -d\eta_i - \mu - \delta - \alpha < 0$, $\lambda_1 = -d\eta_i - \mu < 0$ and $\lambda_2 = -d\eta_i + (\mu + \gamma)(\mathscr{R}_0 - 1)$. According to the Routh-Hurwitz criteria, if $\mathscr{R}_0 \leq 1$, all the roots of Eq. 21 have negative real parts. Therefore, when $\tau = 0$, the disease-free equilibrium point $\mathscr{P}$ is locally asymptotically stable.

Next we discuss the effect of the delay τ on the stability of disease-free equilibrium $\mathscr{P}$. Assume that 20 has a purely imaginary root $i\omega$, with $\omega > 0$. Then ω should satisfy the following equation for η_i.

$$\begin{cases} \mu + \gamma + d\eta_i = \frac{\beta A e^{-\mu\tau}}{\mu + \alpha_1 A} \cos(\omega\tau) \\ \omega = -\frac{\beta A e^{-\mu\tau}}{\mu + \alpha_1 A} \sin(\omega\tau) \end{cases} \tag{22}$$

Taking square on both sides of the Eq. of 22 and summing them up, we obtain

$$\omega^2 = \left[(\mu + \gamma + d\eta_i) + \frac{\beta A e^{-\mu\tau}}{\mu + \alpha_1 A}\right] [(\mu + \gamma)(\mathscr{R}_0 - 1) - d\eta_i]. \tag{23}$$

Therefore, For $\mathscr{R}_0 \leq 1$ Eq. 23 has no positive roots and characteristic Eq. 20 does not admit any purely imaginary root for all τ. Since $\mathscr{P}$ is asymptotically stable for $\tau = 0$, it remains asymptotically stable for all $\tau \geq 0$.

4.2 Stability of Endemic Equilibirum $\mathscr{P}^*$

In this subsection, we will discuss the local stability of the endemic equilibrium $\mathscr{P}^*$. First, we take $(S^*, E^*, I^*, R^*) = \mathscr{P}^*$. Thus, the characteristic Eq. 19 becomes for $i = 0, 1, \ldots$

$$(\lambda + d\eta_i + \mu + \alpha + \delta_q)(\lambda + d\eta_i + \mu)\left(\lambda^2 + a\lambda + b + (\lambda c + d)e^{-\lambda\tau}\right) = 0, \tag{24}$$

where

$$\begin{aligned} a &= 2(\eta_i d + \mu) + \gamma + \tfrac{\beta E^*}{1 + \alpha_1 S^*}, \\ b &= \left(\eta_i d + \mu + \tfrac{\beta E^*}{1 + \alpha_1 S^*}\right)(\eta_i d + \mu + \gamma), \\ c &= -\tfrac{\beta S^* e^{-\mu\tau}}{1 + \alpha_1 S^*}, \\ d &= -\tfrac{\beta S^* e^{-\mu\tau}}{1 + \alpha_1 S^*}(\eta_i d + \mu). \end{aligned}$$

Theorem 3. *If $\mathscr{R}_0 > 1$ then the endemic equilibrium $\mathscr{P}^*$ is locally asymptotically stable for all $\tau \geq 0$.*

Proof. For $\tau = 0$, the characteristic Eq. 24 is transformed into the following form

$$(\lambda + d\eta_i + \mu + \alpha + \delta_q)(\lambda + d\eta_i + \mu)\left(\lambda^2 + (a+c)\lambda + b + d\right), \tag{25}$$

where

$$\begin{aligned} a &= 2(\eta_i d + \mu) + \gamma + \tfrac{\beta E^*}{1+\alpha_1 S^*}, \\ b &= \left(\eta_i d + \mu + \tfrac{\beta E^*}{1+\alpha_1 S^*}\right)(\eta_i d + \mu + \gamma), \\ c &= -\tfrac{\beta S^* e^{-\mu\tau}}{1+\alpha_1 S^*}, \\ d &= -\tfrac{\beta S^* e^{-\mu\tau}}{1+\alpha_1 S^*}(\eta_i d + \mu). \end{aligned}$$

As $\mathscr{R}_0 > 1$, and$-\frac{\beta S^* e^{-\mu\tau}}{1+\alpha_1 S^*} = -(\mu+\gamma)$ we deduce that

$$\begin{aligned} a + c &= 2(\eta_i d + \mu) + \gamma + \tfrac{\beta E^*}{1+\alpha_1 S^*} - \tfrac{\beta S^* e^{-\mu\tau}}{1+\alpha_1 S^*} \\ &= 2\eta_i d + \mu + \tfrac{\beta E^*}{1+\alpha_1 S^*} > 0, \end{aligned}$$

and

$$\begin{aligned} b + d &= \left(\eta_i d + \mu + \tfrac{\beta E^*}{1+\alpha_1 S^*}\right)(\eta_i d + \mu + \gamma) - \tfrac{\beta S^* e^{-\mu\tau}}{1+\alpha_1 S^*}(\eta_i d + \mu) \\ &= \eta_i d(\eta_i d + \mu) + (\eta_i d + \mu + \gamma)\tfrac{\beta E^*}{1+\alpha_1 S^*} > 0, \end{aligned}$$

According to the Routh-Hurwitz criteria, all the roots of Eq. 25 have negative real parts. Therefore, when $\tau = 0$, the endemic equilibrium point $\mathscr{P}^*$ is locally asymptotically stable.

Next, Since all the roots of Eq. 25 have negative real parts for $\tau = 0$. it follows that if instability occurs for a particular value of the delay τ, a characteristic root of 24 must intersect the imaginary axis. If 24 has a purely imaginary root $i\omega$, with $\omega > 0$, then, by separating real and imaginary parts in 24, we have

$$\begin{cases} c\omega\sin(\omega\tau) + d\cos(\omega\tau) = \omega^2 - b, \\ c\omega\cos(\omega\tau) - d\sin(\omega\tau) = \ \ -a\omega \end{cases} \tag{26}$$

Taking square on both sides of the Eq. of 26 and summing them up, we obtain

$$\omega^4 + \left(a^2 - c^2 - 2b\right)\omega^2 + b^2 - d^2 = 0. \tag{27}$$

As $\mathscr{R}_0 > 1$, , we obtain

$$b - d = \left(\eta_i d + \mu + \tfrac{\beta E^*}{1+\alpha_1 S^*}\right)(\eta_i d + \mu + \gamma) + \tfrac{\beta S^* e^{-\mu\tau}}{1+\alpha_1 S^*}(\eta_i d + \mu) > 0,$$

we deduce that $\qquad b^2 - d^2 = (b+d)(b-d) > 0.$

Moreover, as $-\frac{\beta S^* e^{-\mu\tau}}{1+\alpha_1 S^*} = -(\mu+\gamma)$, we have

$$\begin{aligned} a^2 - c^2 + 2b &= \left(2(\eta_i d + \mu) + \gamma + \tfrac{\beta E^*}{1+\alpha_1 S^*}\right)^2 - 2\left(\eta_i d + \mu + \tfrac{\beta E^*}{1+\alpha_1 S^*}\right)(\eta_i d + \mu + \gamma) \\ &\quad - \left(\tfrac{\beta S^* e^{-\mu\tau}}{1+\alpha_1 S^*}\right)^2 \\ &= (\eta_i d)^2 + 2\eta_i d(\mu + \gamma) + \left(\eta_i d + \mu + \tfrac{\beta E^*}{1+\alpha_1 S^*}\right)^2 > 0. \end{aligned}$$

Therefore, Eq. 27 has no positive roots and characteristic Eq. 24 does not admit any purely imaginary root for all η_i. Since $\mathscr{P}^*$ is asymptotically stable for $\tau = 0$, it remains asymptotically stable for all $\tau \geq 0$.

5 Global Stability of Disease-Free Equilibirum $\mathscr{P}$

In this section, by constructing an appropriate Lyapunov function, we investigate the global stability of the disease-free equilibrium $\mathscr{P}$ of 1 when $\mathscr{R}_0 \leq 1$.

Theorem 4. *If $\mathscr{R}_0 \leq 1$, then the disease-free equilibrium $\mathscr{P}$ of system 1–3 is globally asymptotically stable for all $\tau \geq 0$.*

Proof. We consider the following Lyapunov functional

$$V = \int_\Omega \left[e^{-\mu\tau} \int_{\frac{A}{\mu}}^{S(x,t-\tau)} \left(1 - \frac{A(1+\alpha_1 u)}{(\mu+\alpha_1 A)u}\right) du + E(x,t) + (\mu+\gamma) \int_0^\tau E(x,t-u)du \right] dx.$$

Calculating the time derivative of V along solution of system 1–3,we get

$$\begin{aligned}\frac{dV(t)}{dt} &= \int_\Omega \Big\{ e^{-\mu\tau} \left(1 - \frac{A(1+\alpha_1 S(x,t-\tau)}{(\mu+\alpha_1 A)S(x,t-\tau)}\right) (d\Delta S(x,t-\tau) + A - \mu S(x,t-\tau) \\ &\quad - \frac{\beta S(x,t-\tau)E(x,t-\tau)}{1+\alpha_1 S(x,t-\tau)}\Big) + d\Delta E(x,t) + \frac{e^{-\mu\tau}\beta S(x,t-\tau)E(t-\tau)}{1+\alpha_1 S(x,t-\tau)} \\ &\quad - (\mu+\gamma)E(x,t) \quad + (\mu+\gamma)\left[E(x,t) - E(x,t-\tau)\right]\Big\} dx.\end{aligned}$$

Then

$$\begin{aligned}\frac{dV(t)}{dt} &= \int_\Omega \Big\{ e^{-\mu\tau} \left(1 - \frac{A(1+\alpha_1 S(x,t-\tau))}{(\mu+\alpha_1 A)S(x,t-\tau)}\right) (d\Delta S(x,t-\tau) + A - \mu S(x,t-\tau)) + d\Delta E(x,t) \\ &\quad + (\mu+\gamma)\left(\frac{A\beta e^{-\mu\tau}}{(\mu+\gamma)(\mu+\alpha_1 A)} - 1\right) E(x,t-\tau)\Big\} dx \\ &= \int_\Omega \Big\{ e^{-\mu\tau} \left(1 - \frac{A(1+\alpha_1 S(x,t-\tau))}{(\mu+\alpha_1 A)S(x,t-\tau)}\right) (d\Delta S(x,t-\tau)) - \frac{e^{-\mu\tau}(A-\mu S(x,t-\tau))^2}{(\mu+\alpha_1 A)S(x,t-\tau)} \\ &\quad + d\Delta E(x,t) + (\mu+\gamma)(\mathscr{R}_0 - 1)E(x,t-\tau)\Big\} dx\end{aligned}$$

Recall that $\int_\Omega \Delta S(x,t-\tau)dx = 0$, $\int_\Omega \Delta E(x,t)dx = 0$ and using Green's formula, we have

$$\begin{aligned}\frac{dV(t)}{dt} &= \int_\Omega \Big\{ -e^{-\mu\tau} \frac{dA}{(\mu+\alpha_1 A)} \frac{\|\nabla S(x,t-\tau)\|^2}{S^2(x,t-\tau)} - \frac{e^{-\mu\tau}(A-\mu S(x,t-\tau))^2}{(\mu+\alpha_1 A)S(x,t-\tau)} \\ &\quad + (\mu+\gamma)(\mathscr{R}_0 - 1)E(x,t-\tau)\Big\} dx\end{aligned}$$

Then $\mathscr{R}_0 \leq 1$ ensures $\frac{dV}{dt} \leq 0$ for all $t \geq 0$. Moreover, it can be shown that the largest compact invariant set in $\left\{(S,E,I,R) \mid \frac{dV}{dt} = 0\right\}$ is the singleton $\{\mathscr{P}\}$. Therefore, it follows from LaSalle's invariant principle [5] that $\mathscr{P}$ is globally asymptotically stable when $\mathscr{R}_0 \leq 1$. This completes the proof.

6 Numerical

In this section, we perform some numerical simulations to illustrate the theoretical results. For the sake of simplicity, we consider a one-dimensional bounded spatial domain $\Omega = [0,1]$. Thus, we propose system 1 with Neumann boundary conditions

$$\frac{\partial S}{\partial \nu} = \frac{\partial E}{\partial \nu} = \frac{\partial I}{\partial \nu} = \frac{\partial R}{\partial \nu}, \quad t \geq 0, x = 0,1$$

and initial conditions

$$S(x,t) = |\sin(2\pi x)| \geq 0,\ E(x,t) = |\sin(3\pi x)| \geq 0,\ I(x,t) = |\sin(3\pi x)| \geq 0,$$
$$R(x,t) = |\sin(2\pi x)| \geq 0, \quad (x,t) \in [0,1] \times [-\tau, 0].$$

In addition, to solve system 1 using a numerical algorithm, we must discretize each equation of system 1 as a finite difference equation. The Crank-Nicolson method [3] is a finite difference method used for numerically solving a partial differential equation. It is a second-order method in time and space, and is numerically stable. Thereafter, a brief description of the Crank-Nicolson method applied to our problem will be provided below. We first start by partitioning the spatial interval $[0, 1]$ and temporal interval $[0, t_f]$ into respective finite grids as follows.

$$\begin{cases} x_i = (i-1)\Delta x,\ i = 1, 2, \ldots, N_x + 1\, where\ \Delta x := \frac{1}{N_x}, \\ t_j = (j-1)\Delta t,\ j = 1, 2, \ldots, N_t + 1\ where\ \Delta t := \frac{t_f}{N_t}. \end{cases}$$

Therefore, using discretization, we can describe $S(x,t)$ as $S_{i,j}\,(i = 1, \ldots, N_x + 1, j = 1, \ldots, N_t + 1)$, $E(x,t)$ as $E_{i,j}\,(i = 1, \ldots, N_x + 1, j = 1, \ldots, N_t + 1)$, $I(x,t)$ as $I_{i,j}\,(i = 1, \ldots, N_x + 1, j = 1, \ldots, N_t + 1)$ and $R(x,t)$ as $R_{i,j}\,(i = 1, \ldots, N_x + 1, j = 1, \ldots, N_t + 1)$.

Moreover, we can discretize the system 1 as follows:

$$\begin{cases} \frac{S_{i,j+1} - S_{i,j}}{\Delta t} = \frac{d}{2}\left\{\frac{S_{i+1,j+1} - 2S_{i,j+1} + S_{i-1,j+1}}{\Delta x^2} + \frac{S_{i+1,j} - 2S_{i,j} + S_{i-1,j}}{\Delta x^2}\right\} \\ \qquad + A - \mu S_{i,j} - \frac{\beta S_{i,j} I_{i,j}}{1 + \alpha_1 S_{i,j}} \\ \frac{E_{i,j+1} - E_{i,j}}{\Delta t} = \frac{d}{2}\left\{\frac{E_{i+1,j+1} - 2E_{i,j+1} + E_{i-1,j+1}}{\Delta x^2} + \frac{E_{i+1,j} - 2E_{i,j} + E_{i-1,j}}{\Delta x^2}\right\} \\ \qquad + \frac{e^{-\mu\tau}\beta S_{i,j-\tau/\Delta t} E_{i,j-\tau/\Delta t}}{1 + \alpha_1 S_{i,j-\tau/\Delta t}} - (\mu + \gamma) E_{i,j}, \\ \frac{I_{i,j+1} - I_{i,j}}{\Delta t} = \frac{d}{2}\left\{\frac{I_{i+1,j+1} - 2I_{i,j+1} + I_{i-1,j+1}}{\Delta x^2} + \frac{I_{i+1,j} - 2I_{i,j} + I_{i-1,j}}{\Delta x^2}\right\} \\ \qquad + \gamma E_{i,j} - (\mu + \alpha + \delta) I_{i,j} \\ \frac{R_{i,j+1} - R_{i,j}}{\Delta t} = \frac{d}{2}\left\{\frac{R_{i+1,j+1} - 2R_{i,j+1} + R_{i-1,j+1}}{\Delta x^2} + \frac{R_{i+1,j} - 2R_{i,j} + R_{i-1,j}}{\Delta x^2}\right\} \\ \qquad + \delta I_{i,j} - \mu R_{i,j}. \end{cases} \tag{28}$$

Applying the central difference formula to approximate the Neumann boundary condition 3, we see that 28 yields the following system:

$$\begin{cases} MS_{j+1} = NS_j + U_j, \\ ME_{j+1} = NE_j + V_j, \\ MI_{j+1} = NI_j + W_j, \\ MR_{j+1} = NR_j + Y_j, \end{cases} \tag{29}$$

where

$$S_j = \begin{bmatrix} S_{1,j} \\ S_{2,j} \\ \vdots \\ S_{N_x,j} \\ S_{N_x+1,j} \end{bmatrix}, \quad E_j = \begin{bmatrix} E_{1,j} \\ E_{2,j} \\ \vdots \\ E_{N_x,j} \\ E_{N_x+1,j} \end{bmatrix}, \quad I_j = \begin{bmatrix} I_{1,j} \\ I_{2,j} \\ \vdots \\ I_{N_x,j} \\ I_{N_x+1,j} \end{bmatrix}, \quad R_j = \begin{bmatrix} R_{1,j} \\ R_{2,j} \\ \vdots \\ R_{N_x,j} \\ R_{N_x+1,j} \end{bmatrix},$$

$$U_j = 2\Delta t \cdot \begin{bmatrix} A-\mu S_{1,j} - \frac{\beta S_{1,j} I_{1,j}}{1+\alpha_1 S_{1,j}} \\ A-\mu S_{2,j} - \frac{\beta S_{2,j} I_{2,j}}{1+\alpha_1 S_{2,j}} \\ \vdots \\ A-\mu S_{N_x,j} - \frac{\beta S_{N_x,j} I_{N_x,j}}{1+\alpha_1 S_{N_x,j}} \\ A-\mu S_{N_x+1,j} - \frac{\beta S_{N_x+1,j} I_{N_x+1,j}}{1+\alpha_1 S_{N_x+1,j}} \end{bmatrix},$$

$$V_j = 2\Delta t \cdot \begin{bmatrix} \frac{e^{-\mu\tau}\beta S_{1,j-\tau/\Delta t} E_{1,j-\tau/\Delta t}}{1+\alpha_1 S_{1,j-\tau/\Delta t}} - (\mu+\gamma) E_{1,j} \\ \frac{e^{-\mu\tau}\beta S_{2,j-\tau/\Delta t} E_{2,j-\tau/\Delta t}}{1+\alpha_1 S_{2,j-\tau/\Delta t}} - (\mu+\gamma) E_{2,j} \\ \vdots \\ \frac{e^{-\mu\tau}\beta S_{N_x,j-\tau/\Delta t} E_{N_x,j-\tau/\Delta t}}{1+\alpha_1 S_{N_x,j-\tau/\Delta t}} - (\mu+\gamma) E_{N_x,j} \\ \frac{e^{-\mu\tau}\beta S_{N_x+1,j-\tau/\Delta t} E_{N_x+1,j-\tau/\Delta t}}{1+\alpha_1 S_{N_x+1,j-\tau/\Delta t}} - (\mu+\gamma) E_{N_x+1,j} \end{bmatrix}$$

$$W_j = 2\Delta t \cdot \begin{bmatrix} \gamma E_{1,j} - (\mu+\alpha+\delta) I_{1,j} \\ \gamma E_{2,j} - (\mu+\alpha+\delta) I_{2,j} \\ \vdots \\ \gamma E_{N_x,j} - (\mu+\alpha+\delta) I_{N_x,j} \\ \gamma E_{N_x+1,j} - (\mu+\alpha+\delta) I_{N_x+1,j} \end{bmatrix} \quad \textit{and} \quad Y_j = 2\Delta t \cdot \begin{bmatrix} \delta I_{1,j} - \mu R_{1,j} \\ \delta I_{2,j} - \mu R_{2,j} \\ \vdots \\ \delta I_{N_x,j} - \mu R_{N_x,j} \\ \delta I_{N_x+1,j} - \mu R_{N_x+1,j} \end{bmatrix},$$

and we take $r := d\frac{\Delta t}{\Delta x^2}$, then the tridiagonal matrices M and N are given by:

$$M = \begin{bmatrix} 2+2r & -2r & 0 & 0 & \cdots & 0 \\ -r & 2+2r & -r & 0 & \ddots & \vdots \\ 0 & -r & \ddots & \ddots & \ddots & 0 \\ 0 & \ddots & \ddots & \ddots & -r & 0 \\ \vdots & \ddots & 0 & -r & 2+2r & -r \\ 0 & \cdots & 0 & 0 & -2r & 2+2r \end{bmatrix}, \quad N = \begin{bmatrix} 2-2r & 2r & 0 & 0 & \cdots & 0 \\ r & 2-2r & r & 0 & \ddots & \vdots \\ 0 & r & \ddots & \ddots & \ddots & 0 \\ 0 & \ddots & \ddots & \ddots & r & 0 \\ \vdots & \ddots & 0 & r & 2-2r & r \\ 0 & \cdots & 0 & 0 & 2r & 2-2r \end{bmatrix}.$$

Consequently, it follows from 29 that

$$\begin{cases} S_{j+1} = M^{-1}\left\{NS_j + U_j\right\}, \\ E_{j+1} = M^{-1}\left\{NE_j + V_j\right\}, \\ I_{j+1} = M^{-1}\left\{NI_j + W_j\right\}, \\ R_{j+1} = M^{-1}\left\{NR_j + Z_j\right\}. \end{cases}$$

Therefore, we get a recursive schema, with is numerically stable. The parameters employed in the numerical simulations are summarized in Table 1.

Table 1. List of parameters and their values used in numerical simulations

Parameter	Description	Value
A	Recruitment rate of the population	*varied*
μ	Natural death of the population	0.0402
γ	Rate of exposed individuals to the infected	0.04
β	Transmission rate	*varied*
δ	Recovery rate	*varied*
α_1	The parameters that measure the inhibitory effect	0.01
d	Rate of diffusion	0.00008
τ	Time incubation	8

Now, if we choose the values from Table 1 , with $A = 0.08, \delta = 0.05$ and $\beta = 0.04$, then we get $\mathscr{R}_0 = 0.4746$. By Theorem 2, the disease-free equilibrium $\mathscr{P}(1.005,0,0,0,0)$ is locally asymptotically stable. This means that the disease dies out (see Fig. 1).

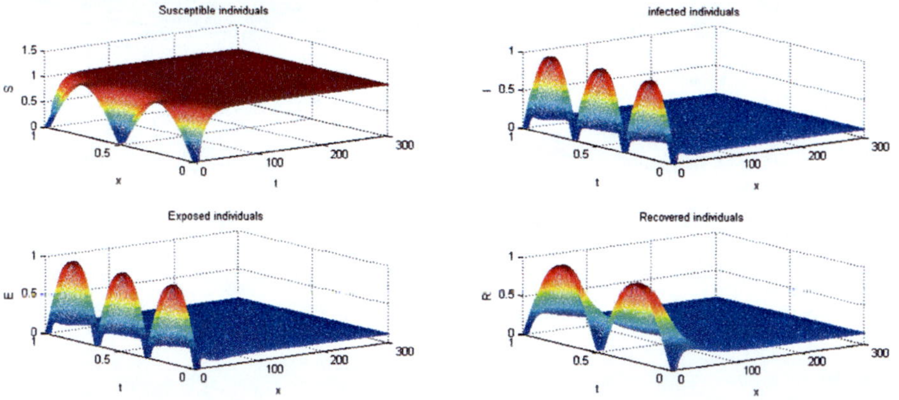

Fig. 1. Spatiotemporal solution found by numerical integration of system 1 under conditions 2 and 3 when $\mathscr{R}_0 = 0.4746$

to better understand Fig. 1 we propose Fig. 2 where we present the curve of S in the case of $x = \frac{1}{4}$ and $x = \frac{1}{2}$.

Next, if we choose $A = 0.2$, $\delta = 0.4$ and $\beta = 0.07$, then we get $\mathscr{R}_0 = 3.1985$. It follows from Theorem 3 that the endemic equilibrium $P^*(1.5113, 1.126667, 0.0874, 1.1728)$ is locally asymptotically stable (see Fig. 3).

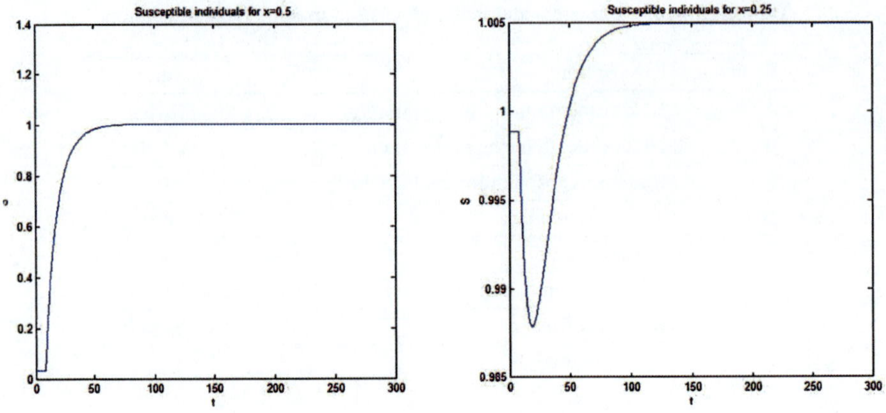

Fig. 2. The curve of S in the case of $x = \frac{1}{2}$ and $x = \frac{1}{4}$.

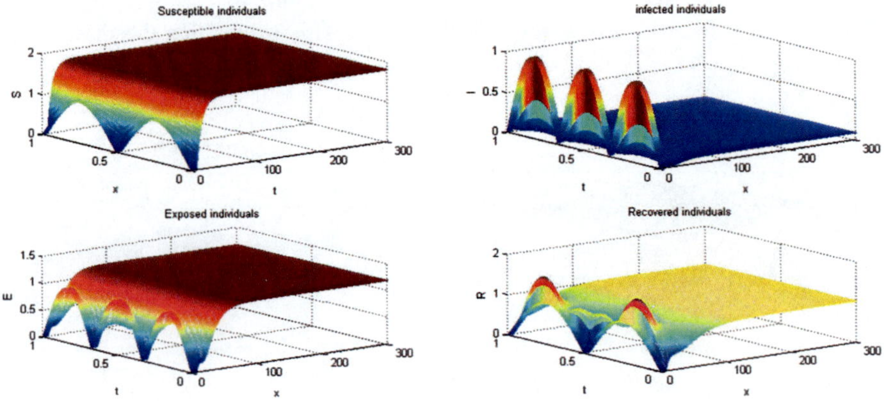

Fig. 3. Spatiotemporal solution found by numerical integration of system 1 under conditions 2 and 3 when $\mathscr{R}_0 = 3.1985$

7 Conclusion

By comparing the results in Theorems 2 and 3 with the propositions 1, 2 of [8] and the proposition 2 of [1], we affirm that we have obtained the same results, but for a more general class of population models. In reality, we have extended these results to contain our model of reaction-diffusion epidemic. Firstly, by analyzing the corresponding characteristic equations, we discussed the local stability of the disease-free equilibrium $\mathscr{P}$ and the endemic equilibrium $\mathscr{P}^*$ of system 1 under homogeneous Neumann boundary conditions. Since $\mathscr{R}_0$ has no relation to the diffusion coefficient d, we have shown in Theorem 2 and Theorem 3 that spatial diffusion has no effect on the local stability of the steady states of our SEIR model. Which indicates that, whatever the choice of the diffusion coefficient d, the stability of the equilibrium points remains invariant when the system passes from the dynamics governed by the ordinary differential equations ODE [1] to that governed by the partial differential equations PDE.

References

1. Abta, A., Kaddar, A., Alaoui, H.T.: Global stability for delay SIR and SEIR epidemic models with saturated incidence rates. Electron. J. Differ. Equat. **2012**(23), 1–13 (2012)
2. Cooke, K.L.: Stability analysis for a vector disease model. Rocky Mt. J. Math. **9**(1), 31–42 (1979)
3. Crank, J., Nicolson, P.: A practical method for numerical evaluation of solutions of partial differential equations of the heat-conduction type. In: Mathematical Proceedings of the Cambridge Philosophical Society, vol. 43, no. 1, pp. 50–67. Cambridge University Press (1947)
4. Fitzgibbon, W.E.: Semilinear functional differential equations in Banach space. J. Differ. Equat. **29**(1), 1–14 (1978)
5. Hale, J.K., Lunel, S.M.V.: Introduction to Functional Differential Equations. Springer-Verlag, New York (1993). https://doi.org/10.1007/978-1-4612-4342-7
6. Henry, D.: Neighborhood of a periodic solution. In: Geometric Theory of Semilinear Parabolic Equations. LNM, vol. 840, pp. 247–274. Springer, Heidelberg (1981). https://doi.org/10.1007/BFb0089656
7. Hethcote, H.W., Stech, H.W., van den Driessche, P., Busenberg, S., Cooke, K.L.: Differential Equations and Applications in Ecology, Epidemics and Population Problems (1981)
8. Kaddar, A., Abta, A., Alaoui, H.T.: A comparison of delayed SIR and SEIR epidemic models. Nonlinear Anal. Model. Control **16**(2), 181–190 (2011)
9. Martin, R.H., Smith, H.L.: Abstract functional-differential equations and reaction-diffusion systems. Trans. Am. Math. Soc. **321**(1), 1–44 (1990)
10. Smith, H.L., Martin, R.J.: Reaction-diffusion systems with time delays: monotonicity, invariance, comparison and convergence (1991)
11. Murray, J.D. (ed.): Mathematical Biology. IAM, vol. 18. Springer, New York (2003). https://doi.org/10.1007/b98869
12. Weinberger, H.F.: Maximum Principles in Differential Equations (1967)
13. Travis, C.C., Webb, G.: Existence and stability for partial functional differential equations. Trans. Am. Math. Soc. **200**, 395–418 (1974)
14. Van den Driessche, P., Watmough, J.: Reproduction numbers and sub-threshold endemic equilibria for compartmental models of disease transmission. Math. Biosci. **180**(1–2), 29–48 (2002)
15. Wang, W., Cai, Y., Wu, M., Wang, K., Li, Z.: Complex dynamics of a reaction-diffusion epidemic model. Nonlinear Anal. Real World Appl. **13**(5), 2240–2258 (2012)
16. Wang, X.S., Wang, H., Wu, J.: Traveling waves of diffusive predator-prey systems: disease outbreak propagation. Discrete Continuous Dyn. Syst.-A **32**(9), 3303 (2012)
17. Wu, J.: Theory and Applications of Partial Functional Differential Equations, vol. 119. Springer Science Business Media, New York (2012). https://doi.org/10.1007/978-1-4612-4050-1
18. Yang, J., Liang, S., Zhang, Y.: Travelling waves of a delayed SIR epidemic model with nonlinear incidence rate and spatial diffusion. PLoS One **6**(6), e21128 (2011)

Generalized Solution of Non-homogeneous Wave Equation

S. Melliani(✉), A. Moujahid , F. Gueradi, and M. Elomari

LMACS, Sultan Moulay Slimane University, BP 523 - 23000 Beni Mellal, Morocco
s.melliani@usms.ma

Abstract. In this paper, we are interested to study the non-homogeneous wave equation in generalized function algebra, we give a result of existence and uniqueness of generalized solution with initial data are distributions (elements of the Colombeau algebra). Then we study the association concept with the classical solution.

Keywords: Colombeau algebra, non-homogeneous wave equation, Generalized solution, association.

1 Introduction

The algebras of Colombeau are constructed by J.F. Colombeau [1] and [2], as factor algebras of infinite powers of the space C^∞ modulo a particular class of ideals. Enjoying a list of optimal properties, These algebras contain the space of distributions D' as a subspace with an embedding realized through convolution with a suitable mollifier. Elements of these algebras are classes of nets of smooth functions. This theory was been used for solving the linear and nonlinear partial differential equations with singularities, and in the last few years it was developed and also applied in different domains.

On the other hand these problems have been studied by some authors [6–8] in different cases, for example M. Oberguggenberger and Y.G. Wang, studied the Delta-waves for semi linear hyperbolic Cauchy problems [4], also Nonlinear stochastic wave equations have been studied by M. Oberguggenberger and F. Russo [5]. This paper in first part we introduce the Colombeau algebras and we give some properties and tools, after that in the second part we study the existence and uniqueness of generalized solution of non homogeneous wave equation with the initial data are singular, finally we proved the association of generalized solution with classical solution.

2 Preliminaries

In this section we will introduce basic notations and definitions from Colombeau theory.

S. Melliani and O. Castillo (Eds.): ICPAMS 2021, LNNS 476, pp. 210–218, 2023.
https://doi.org/10.1007/978-3-031-12416-7_17

For $q \in \mathbb{N}_0$ with $\mathbb{N}_0 = \mathbb{N} \cup 0$

$$\mathcal{A}_q = \{\varphi \in \mathcal{D}(\mathbb{R}^n) / \int_{\mathbb{R}^n} \varphi(x)dx = 1, \quad \int_{\mathbb{R}^n} x^\alpha \varphi(x)dx = 0 \quad \text{for} \quad 1 \leq |\alpha| \leq q\}$$

$q = 1, 2, ...$
where $D(\mathbb{R}^n)$ is the space of all C^∞ functions $\varphi : \mathbb{R}^n \to \mathbb{C}$ with compact support. The elements of the set $\mathcal{A}_q$ are called test functions.
It is simple too see that $A1 \supset A2 \supset A3....$ Also, $A_i \neq \emptyset$; for all $i \in \mathbb{N}$

$$\varphi_\varepsilon(x) = \frac{1}{\varepsilon^n}\varphi(\frac{x}{\varepsilon}) \quad \text{for} \quad \varphi \in \mathcal{D}(\mathbb{R}^n)$$

We denote by

$$\mathcal{E}(\mathbb{R}^n) = \{u : \mathcal{A}_1 \times \mathbb{R}^n \to \mathbb{C} / \quad \text{with} \quad u(\varphi, x) \quad \text{is} \quad \mathcal{C}^\infty \quad \text{to the second variable} \quad x\}$$

$$u(x, \varphi_\varepsilon) = u_\varepsilon(x) \quad \forall \varphi \in \mathcal{A}_1$$

$$\mathcal{E}_M(\mathbb{R}^n) = \{(u_\varepsilon)_{\varepsilon>0} \subset \mathcal{E}(\mathbb{R}^n) / \forall K \subset\subset \mathbb{R}^n, \forall \alpha \in \mathbb{N}_0^n, \exists N \in \mathbb{N} \quad \text{such that} \quad \sup_{x\in K} |D^\alpha u_\varepsilon(x)| = \mathcal{O}(\varepsilon^{-N}) \quad \text{as} \quad \varepsilon \to 0\}$$

$$\mathcal{N}(\mathbb{R}^n) = \{(u_\varepsilon)_{\varepsilon>0} \subset \mathcal{E}(\mathbb{R}^n) / \forall K \subset\subset \mathbb{R}^n, \forall \alpha \in \mathbb{N}_0^n, \forall p \in \mathbb{N} \quad \text{such that} \quad \sup_{x\in K} |D^\alpha u_\varepsilon(x)| = \mathcal{O}(\varepsilon^{p}) \quad \text{as} \quad \varepsilon \to 0\}$$

The Colombeau algebra is defined as a factor set $\mathcal{G}(\mathbb{R}^n) = \mathcal{E}_M(\mathbb{R}^n)/\mathcal{N}(\mathbb{R}^n)$, where the elements of the set $\mathcal{E}_M(\mathbb{R}^n)$ are moderate while the elements of the set $\mathcal{N}(\mathbb{R}^n)$ are negligible.
The Colombeau algebra $\mathcal{G}(\mathbb{R}^n)$ contains the distributions space as subspace by the map :

$$i : D'(\mathbb{R}^n) \to \mathcal{G}(\mathbb{R}^n)$$
$$u \to [u] = u * \varphi_\varepsilon + \mathcal{N}(\mathbb{R}^n)$$

where $*$ denotes the convolution product of two distributions and is given by:

$$u * \varphi = < u(y), \varphi(y - x) >$$

Definition 1. Let $G_1, G_2 \in \mathcal{G}(\mathbb{R}^n)$ and $G_{1,\varepsilon}, G_{2,\varepsilon}$ their representatives respectively.

We say that $G_1, G_2 \in \mathcal{G}(\mathbb{R}^n)$ are associated and we write $G_1 \approx G_2$, if for every $\varphi \in \mathcal{D}(\mathbb{R}^n)$

$$\lim_{\varepsilon\to 0} \int_{\mathbb{R}^n} (G_{1,\varepsilon} - G_{2,\varepsilon})\varphi(x)dx = 0$$

3 Application

This section is devoted to solving the non-homogeneous wave equation in Colombeau algebra
$\mathcal{G}(\mathbb{R}^+ \times \mathbb{R})$. Recall first that if u is a classical solution of the following problem we consider the following problem:

$$\begin{cases} \frac{d^2}{dt^2}u(t,x) - c^2\frac{d^2}{dx^2}u(t,x) = F(t,u(t,x)) \quad x \in \mathbb{R}, \quad t \geqslant 0 \\ u(0,x) = a(x) \\ \partial_t u(0,x) = b(x) \end{cases} \tag{1}$$

With $a, b \in D'(\mathbb{R})$.
Then it solves the integral equation

$$u(x,\ t) = \frac{1}{2}(a(x-ct) + a(x+ct)) + \frac{1}{2c}\int_{x-ct}^{x+ct} b(y)dy + \frac{1}{2c}\int_0^t \int_{x-c(t+s)}^{x+c(t-s)} F\big(s, u(s,y)\big)dy\ ds. \tag{2}$$

We define the domains:

$$\Delta_s = \{(s,y) \in \mathbb{R} \times [0,\ \infty)\ /\ 0 \leq s \leq t,\ y \in I_s\}.$$

$$I_s = \{z \in \mathbb{R}/\ \ x - cs \leq z \leq x + cs\}$$

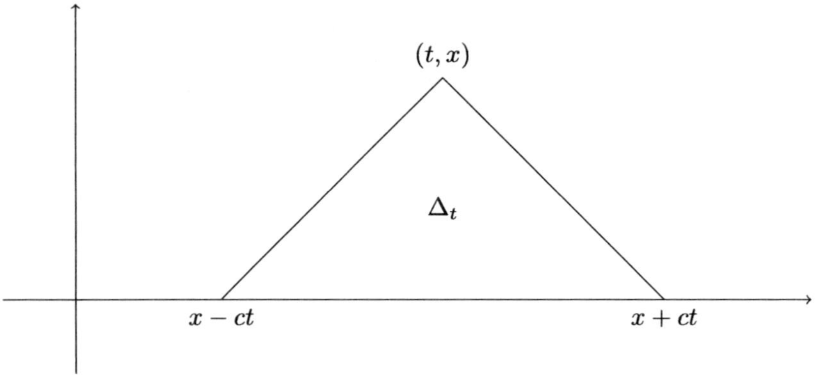

Using (3.2), the following estimates are easily deduced $(0 \leq t \leq T)$

$$\|u\|_{L^\infty(\Delta_T)} \leq \|a\|_{L^\infty(I_0)} + T\|b\|_{L^\infty(I_0)} + T\int_0^T \|F(s,u(s,.))\|_{L^\infty(\Delta_s)}ds, \tag{3}$$

$$\|u(t,.)\|_{L^\infty(I_t)} \leq \|a\|_{L^\infty(I_0)} + T\|b\|_{L^\infty(I_0)} + T\int_0^T \|F(s,u(s,.))\|_{L^\infty(I_s)}ds. \tag{4}$$

Definition 2. An element $F \in \mathcal{G}[R^n]$ is of L^∞ logarithmic type if it has a representative $(F_\varepsilon)_\varepsilon \in \mathcal{E}_M[\mathbb{R}^+ \times \mathbb{R}]$ such that

$$\|F_\varepsilon\|_{L^\infty(\mathbb{R}^+\times\mathbb{R})} = \mathcal{O}(\log(\frac{1}{\varepsilon})) \qquad \text{as} \qquad \varepsilon \to 0$$

Proposition 1. *If $F \in \mathcal{G}[R^n]$ is L^∞ log-type then for any representative is satisfied $(F_\varepsilon)_\varepsilon \in \mathcal{E}_M[\mathbb{R}^+ \times \mathbb{R}]$ we have*

$$\|F_\varepsilon\|_{L^\infty(\mathbb{R}^+\times\mathbb{R})} = \mathcal{O}(\log(\frac{1}{\varepsilon})) \qquad \textit{as} \qquad \varepsilon \to 0$$

Proof. $F \in \mathcal{G}[R^n]$ is L^∞ log-type then there is a representative $(F_{1,\varepsilon})_\varepsilon \in \mathcal{E}_M[\mathbb{R}^+ \times \mathbb{R}]$ such that

$$\|F_{1,\varepsilon}\|_{L^\infty(\mathbb{R}^+\times\mathbb{R})} = \mathcal{O}(\log(\frac{1}{\varepsilon})) \qquad \text{as} \qquad \varepsilon \to 0$$

let $(F_{2,\varepsilon})_\varepsilon$ be another representative of F , then $(F_{2,\varepsilon} - F_{1,\varepsilon})_\varepsilon \in \mathcal{N}(\mathbb{R}^n)$

$$\Rightarrow \qquad \|F_{2,\varepsilon} - F_{1,\varepsilon}\|_{L^\infty(\mathbb{R}^+\times\mathbb{R})} = \mathcal{O}(\varepsilon^q) \quad \forall q \in \mathbb{N}$$

$$\Rightarrow \exists c_1 \geqslant 0 \qquad \|F_{2,\varepsilon} - F_{1,\varepsilon}\|_{L^\infty(\mathbb{R}^+\times\mathbb{R})} \le c_1\varepsilon^q \quad \forall q \in \mathbb{N}$$

$$\Rightarrow \exists c_1, c_2 \geqslant 0 \quad \|F_{2,\varepsilon}\|_{L^\infty(\mathbb{R}^+\times\mathbb{R})} \le c_1\varepsilon^q + c_2\log(\frac{1}{\varepsilon}) \quad \forall q \in \mathbb{N}$$

$$\Rightarrow \exists c_1, c_2 \geqslant 0 \quad \|F_{2,\varepsilon}\|_{L^\infty(\mathbb{R}^+\times\mathbb{R})} \le c_1\log(\frac{1}{\varepsilon}) + c_2\log(\frac{1}{\varepsilon}) \quad \forall q \in \mathbb{N}$$

$$\Rightarrow \quad \exists c' \geqslant 0 \qquad \|F_{2,\varepsilon}\|_{L^\infty(\mathbb{R}^+\times\mathbb{R})} \le c'\log(\frac{1}{\varepsilon}) \quad \forall q \in \mathbb{N}$$

$$\Rightarrow \qquad \|F_{2,\varepsilon}\|_{L^\infty(\mathbb{R}^+\times\mathbb{R})} = \mathcal{O}(\log(\frac{1}{\varepsilon}))$$

this completes the proof of the proposition.

Theorem 1. *Let $a, b \in \mathcal{G}(\mathbb{R})$, ∇F is L^∞ log-type, then the problem (3.2) has a unique solution $u \in \mathcal{G}(\mathbb{R} \times [0,\infty))$.*

Proof.
existence :
To prove the existence of a solution, we transform the problem in the Colombeau algebra, then we obtain:

$$\begin{cases} \frac{d^2}{dt^2}u_\varepsilon(t,x) - c^2\frac{d^2}{dx^2}u_\varepsilon(t,x) = F_\varepsilon(t, u_\varepsilon(t,x)) \quad x \in \mathbb{R}, \quad t \geqslant 0 \\ u_\varepsilon(0,x) = a_\varepsilon(x) \\ \partial_t u_\varepsilon(0,x) = b_\varepsilon(x) \end{cases} \tag{5}$$

With $a_\varepsilon, b_\varepsilon$ and F_ε are the Representative of a, b, F respectively.
The integral solution is :

$$u_\varepsilon(t,x) = \frac{1}{2}(a_\varepsilon(x-ct) + a_\varepsilon(x+ct)) + \frac{1}{2c}\int_{x-ct}^{x+ct} b_\varepsilon(y)dy + \frac{1}{2c}\int_0^t \int_{x-c(t+s)}^{x+c(t-s)} F_\varepsilon(s, u_\varepsilon(s,y))dy\, ds.$$

We apply the estimate (3.3) successively to all the derivatives,

$$\|u\|_{L^\infty(\Delta_T)} \leq \|a\|_{L^\infty(I_0)} + T\|b\|_{L^\infty(I_0)} + T\int_0^T \|F(s, u(s,.))\|_{L^\infty(\Delta_s)}ds,$$

The first approximation of F_ε:

$$F_\varepsilon(s, u_\varepsilon(s,.)) = F_\varepsilon(s,0) + |\nabla F_\varepsilon| u_\varepsilon(s,.) + N_\varepsilon(s,.)$$

with $N_\varepsilon \in \mathcal{N}(\mathbb{R}^+ \times \mathbb{R})$.
Then,

$$\begin{aligned}\|u_\varepsilon\|_{L^\infty(\Delta_T)} &\leq \|a_\varepsilon\|_{L^\infty(I_0)} + T\|b_\varepsilon\|_{L^\infty(I_0)} \\ &+ T\int_0^T \Big[||F_\varepsilon(.,0)||_{L^\infty([0,T])} + |\nabla F_\varepsilon|\; ||u_\varepsilon(s,.)||_{L^\infty(\Delta_s)} + ||N_\varepsilon||_{L^\infty(\Delta_T)}\Big] ds.\end{aligned}$$

Therefore,

$$\begin{aligned}\|u_\varepsilon\|_{L^\infty(\Delta_T)} &\leq \|a_\varepsilon\|_{L^\infty(I_0)} + T\|b_\varepsilon\|_{L^\infty(I_0)} + T^2||F_\varepsilon(.,0)||_{L^\infty([0,T])} + T^2||N_\varepsilon||_{L^\infty(\Delta_T)} \\ &+ T\int_0^T |\nabla F_\varepsilon|\; ||u_\varepsilon||_{L^\infty(\Delta_s)} ds.\end{aligned}$$

By the Granwall's inequality, we have

$$\begin{aligned}\|u_\varepsilon\|_{L^\infty(\Delta_T)} \leq \Big(&\|a_\varepsilon\|_{L^\infty(I_0)} + T\|b_\varepsilon\|_{L^\infty(I_0)} + T^2||F_\varepsilon(.,0)||_{L^\infty([0,T])} \\ &+ T^2||N_\varepsilon||_{L^\infty(\Delta_T)}\Big) \times \exp\Big(T^2|\nabla F_\varepsilon|\Big).\end{aligned}$$

As $a \in \mathcal{G}(\mathbb{R})$, $b \in \mathcal{G}(\mathbb{R})$ and ∇F is $L^\infty - logtype$ there exist $M \in \mathbb{N}$ such that

$$\|u_\varepsilon\|_{L^\infty(\Delta_T)} = \mathcal{O}(\varepsilon^{-M}) \qquad as \qquad \varepsilon \to 0$$

Uniqueness:
To prove uniqueness, we consider representatives $u_\varepsilon, v_\varepsilon \in \mathcal{E}(\mathbb{R}^+ \times \mathbb{R})$ of two solutions u and v. Their difference satisfies

$$\begin{cases} \frac{d^2}{dt^2}\Big(u_\varepsilon(t,x) - v_\varepsilon(t,x)\Big) - c^2\frac{d^2}{dt^2}\Big((u_\varepsilon(t,x) - v_\varepsilon(t,x)\Big) = F_\varepsilon\Big(t, u_\varepsilon(t,x)\Big) - F_\varepsilon\Big(t, v_\varepsilon(t,x)\Big) + n_\varepsilon(t,x) \quad x \in \mathbb{R}, \quad t \geqslant 0 \\ u_\varepsilon(0,x) - v_\varepsilon(0,x) = n_{1,\varepsilon}(x) \\ \partial_t u_\varepsilon(0,x) - \partial_t v_\varepsilon(0,x) = n_{2,\varepsilon}(x) \end{cases} \tag{6}$$

$$\begin{aligned}u_\varepsilon(x,t) - v_\varepsilon(t,x) =& \frac{1}{2}(n_{1,\varepsilon}(x-ct) + n_{1,\varepsilon}(x+ct)) + \frac{1}{2c}\int_{x-ct}^{x+ct} n_{2,\varepsilon}(y)dy \\ &+ \frac{1}{2c}\int_0^t \int_{x-c(t+s)}^{x+c(t-s)} F_\varepsilon\Big(t, u_\varepsilon(t,x)\Big) - F_\varepsilon\Big(t, v_\varepsilon(t,x)\Big) + n_\epsilon(s,x)\, dyds.\end{aligned}$$

$$\|u_\varepsilon - v_\varepsilon\|_{L^\infty(\Delta_T)} \leq \|n_{1,\varepsilon}\|_{L^\infty(I_0)} + T\|n_{2,\varepsilon}\|_{L^\infty(I_0)} + T\int_0^T \|F_\varepsilon(s, u_\varepsilon(s,.)) - F_\varepsilon(s, v_\varepsilon(s,.)) - n_\varepsilon(s,.)\|_{L^\infty(\Delta_s)} ds,$$

The first approximation of F_ε:

$$F_\varepsilon(s, u_\varepsilon(s,.)) = F_\varepsilon(s,0) + |\nabla F_\varepsilon| u_\varepsilon(s,.) + N_{1,\varepsilon}(s,.)$$

$$F_\varepsilon(s, v_\varepsilon(s,.)) = F_\varepsilon(s,0) + |\nabla F_\varepsilon| v_\varepsilon(s,.) + N_{2,\varepsilon}(s,.)$$

with $N_1, N_2 \in \mathcal{N}(\mathbb{R} \times \mathbb{R}^+)$.

Then there exist $N_3 \in \mathcal{N}(\mathbb{R} \times \mathbb{R}^+)$ such that,

$$\|u_\varepsilon - v_\varepsilon\|_{L^\infty(\Delta_T)} \leq \|n_{1,\varepsilon}\|_{L^\infty(I_0)} + T\|n_{2,\varepsilon}\|_{L^\infty(I_0)} + T\int_0^T \| \, |\nabla F_\varepsilon|\big(u_\varepsilon(s,.) - v_\varepsilon(s,.)\big) + N_{3,\varepsilon}(s,.)\|_{L^\infty(\Delta_s)} ds,$$

So,

$$\begin{aligned}\|u_\varepsilon - v_\varepsilon\|_{L^\infty(\Delta_T)} \leq& \|n_{1,\varepsilon}\|_{L^\infty(I_0)} \\ &+ T\|n_{2,\varepsilon}\|_{L^\infty(I_0)} + T\int_0^T \| \, |\nabla F_\varepsilon|\big(u_\varepsilon(s,.) - v_\varepsilon(s,.)\big)\|_{L^\infty(\Delta_s)} \\ &+ T\int_0^T \|N_{3,\varepsilon}(s,.)\|_{L^\infty(\Delta_s)} ds,\end{aligned}$$

Then,

$$\|u_\varepsilon - v_\varepsilon\|_{L^\infty(\Delta_T)} \leq \|n_{1,\varepsilon}\|_{L^\infty(I_0)} + T\|n_{2,\varepsilon}\|_{L^\infty(I_0)} + T^2\|N_{3,\varepsilon}\|_{L^\infty(\Delta_T)} + T\int_0^T \| \, |\nabla F_\varepsilon|\big(u_\varepsilon(s,.) - v_\varepsilon(s,.)\big)\|_{L^\infty(\Delta_s)},$$

we apply Granwall's inequality on the function $s \mapsto \|u_\varepsilon - v_\varepsilon\|_{L^\infty(\Delta_s)}$ we obtain:

$$\|u_\varepsilon - v_\varepsilon\|_{L^\infty(\Delta_T)} \leq \Big(\|n_{1,\varepsilon}\|_{L^\infty(I_0)} + T\|n_{2,\varepsilon}\|_{L^\infty(I_0)} + T^2\|N_{3,\varepsilon}\|_{L^\infty(\Delta_T)}\Big) \times \exp\Big(T\int_0^T |\nabla F_\varepsilon| ds\Big),$$

As $n_1, n_2 \in \mathcal{N}(\mathbb{R})$, $N_3 \in \mathcal{N}(\mathbb{R}^+ \times \mathbb{R})$ and ∇F is $L^\infty - logtype$

$$\|u_\varepsilon - v_\varepsilon\|_{L^\infty(\Delta_T)} = \mathcal{O}(\varepsilon^q) \quad as \quad \varepsilon \to 0 \quad \forall q$$

Then the problem have a unique solution in $\mathcal{G}(\mathbb{R}^+ \times \mathbb{R})$.

4 Association with classical solution

Let v the classical solution to

$$\begin{cases} \frac{d^2}{dt^2}v(t,x) - c^2\frac{d^2}{dx^2}v(t,x) = 0 \quad x \in \mathbb{R}, \quad t \geqslant 0 \\ v(0,x) = a(x) \\ \partial_t v(0,x) = b(x) \end{cases} \tag{7}$$

And w the classical solution to

$$\begin{cases} \frac{d^2}{dt^2}w(t,x) - c^2\frac{d^2}{dx^2}w(t,x) = F(w+v) \quad x \in \mathbb{R}, \quad t \geqslant 0 \\ w(0,x) = 0 \\ \partial_t w(0,x) = 0 \end{cases} \tag{8}$$

Proposition 2. *The generalized solution of* (3.5) *is associated with* $v+w$.

Proof. Let v_ϵ the solution to

$$\begin{cases} \frac{d^2}{dt^2}v_\varepsilon(t,x) - c^2\frac{d^2}{dx^2}v_\varepsilon(t,x) = 0 \quad x \in \mathbb{R}, \quad t \geqslant 0 \\ v_\varepsilon(0,x) = a_\varepsilon(x) \\ \partial_t v_\varepsilon(0,x) = b_\varepsilon(x) \end{cases} \tag{9}$$

w_ϵ the solution to

$$\begin{cases} \frac{d^2}{dt^2}w_\epsilon(t,x) - c^2\frac{d^2}{dx^2}w_\epsilon(t,x) = F_\epsilon(w_\epsilon + v_\epsilon) \quad x \in \mathbb{R}, \quad t \geqslant 0 \\ w_\epsilon(0,x) = 0 \\ \partial_t w_\epsilon(0,x) = 0 \end{cases} \tag{10}$$

And u_ϵ the solution to

$$\begin{cases} \frac{d^2}{dt^2}u_\varepsilon(t,x) - c^2\frac{d^2}{dx^2}u_\varepsilon(t,x) = F_\varepsilon(u_\varepsilon(t,x)) \quad x \in \mathbb{R}, \quad t \geqslant 0 \\ u_\varepsilon(0,x) = a_\varepsilon(x) \\ \partial_t u_\varepsilon(0,x) = b_\varepsilon(x) \end{cases} \tag{11}$$

Then, we have

$$\begin{cases} (\frac{d^2}{dt^2} - c^2\frac{d^2}{dx^2})(u_\epsilon(t,x) - v_\epsilon(t,x) - w_\varepsilon(t,x)) = F_\epsilon(t,u_\epsilon(t,x)) - F_\epsilon(t,w_\epsilon(t,x) + v_\epsilon(t,x)) + n_\epsilon(t,x) \quad x \in \mathbb{R}, \quad t \geqslant 0 \\ (u_\epsilon - v_\epsilon - w_\epsilon)(0,x) = n_{1,\epsilon}(x) \\ \partial_t(u_\epsilon - v_\epsilon - w_\epsilon)(0,x) = n_{2,\epsilon}(x) \end{cases} \tag{12}$$

With $n_1, n_2 \in \mathcal{N}(\mathbb{R})$.
The integral solution is:

$$\begin{aligned} u_\varepsilon(x,t) - v_\varepsilon(t,x)w_\varepsilon(t,x) =& \frac{1}{2}(n_{1,\varepsilon}(x-ct) + n_{1,\varepsilon}(x+ct)) \\ &+ \frac{1}{2c}\int_{x-ct}^{x+ct} n_{2,\varepsilon}(y)dy + \frac{1}{2c}\int_0^t \int_{x-c(t+s)}^{x+c(t-s)} F_\varepsilon\Big(t, u_\varepsilon(t,x)\Big) \\ &- F_\varepsilon\Big(t, w_\varepsilon(t,x) + v_\varepsilon(t,x)\Big) + n_\epsilon(t,x)\ dyds. \end{aligned}$$

Using (3.3) we have,

$$\begin{aligned}\|u_\varepsilon - v_\varepsilon - w_\varepsilon\|_{L^\infty(\Delta_T)} \leq &\|n_{1,\varepsilon}\|_{L^\infty(I_0)} \\ &+ T\|n_{2,\varepsilon}\|_{L^\infty(I_0)} + T\int_0^T \|F_\varepsilon(s, u_\varepsilon(s,.)) \\ &- F_\varepsilon(s, w_\varepsilon(s,.) + v_\varepsilon(s,.)) - n_\varepsilon(s,.)\|_{L^\infty(\Delta_s)} ds,\end{aligned}$$

The first approximation of F_ε:

$$F_\varepsilon(s, u_\varepsilon(s,.)) = F_\varepsilon(s,0) + |\nabla F_\varepsilon| u_\varepsilon(s,.) + N_{1,\varepsilon}(s,.)$$

$$F_\varepsilon(s, v_\varepsilon(s,.) + w_\varepsilon(s,.)) = F_\varepsilon(s,0) + |\nabla F_\varepsilon|\big(v_\varepsilon(s,.) + w_\varepsilon(s,.)\big) + N_{2,\varepsilon}(s,.)$$

with $N_1, N_2 \in \mathcal{N}(\mathbb{R}\times\mathbb{R}^+)$.

Then there exist $N_3 \in \mathcal{N}(\mathbb{R}\times\mathbb{R}^+)$ such that,

$$\begin{aligned}\|u_\varepsilon - v_\varepsilon - w_\varepsilon\|_{L^\infty(\Delta_T)} \leq &\|n_{1,\varepsilon}\|_{L^\infty(I_0)} \\ &+ T\|n_{2,\varepsilon}\|_{L^\infty(I_0)} + T\int_0^T \|\,|\nabla F_\varepsilon|\big(u_\varepsilon(s,.) - v_\varepsilon(s,.) \\ &- w_\varepsilon(s,.)\big) + N_{3,\varepsilon}(s,.)\|_{L^\infty(\Delta_s)} ds,\end{aligned}$$

So,

$$\begin{aligned}\|u_\varepsilon - v_\varepsilon - w_\varepsilon\|_{L^\infty(\Delta_T)} \leq &\|n_{1,\varepsilon}\|_{L^\infty(I_0)} \\ &+ T\|n_{2,\varepsilon}\|_{L^\infty(I_0)} + T\int_0^T \|\,|\nabla F_\varepsilon|\big(u_\varepsilon(s,.) - v_\varepsilon(s,.) \\ &- w_\varepsilon(s,.)\big)\|_{L^\infty(\Delta_s)} + T\int_0^T \|N_{3,\varepsilon}(s,.)\|_{L^\infty(\Delta_s)} ds,\end{aligned}$$

Then,

$$\begin{aligned}\|u_\varepsilon - v_\varepsilon - w_\varepsilon\|_{L^\infty(\Delta_T)} \leq &\|n_{1,\varepsilon}\|_{L^\infty(I_0)} + T\|n_{2,\varepsilon}\|_{L^\infty(I_0)} + T^2\|N_{3,\varepsilon}\|_{L^\infty(\Delta_T)} \\ &+ T\int_0^T \|\,|\nabla F_\varepsilon|\big(u_\varepsilon(s,.) - v_\varepsilon(s,.) - w_\varepsilon(s,.)\big)\|_{L^\infty(\Delta_s)},\end{aligned}$$

We apply Granwall's inequality

$$\begin{aligned}\|u_\varepsilon - v_\varepsilon - w_\varepsilon\|_{L^\infty(\Delta_T)} \leq &\Big(\|n_{1,\varepsilon}\|_{L^\infty(I_0)} + T\|n_{2,\varepsilon}\|_{L^\infty(I_0)} + T^2\|N_{3,\varepsilon}\|_{L^\infty(\Delta_T)}\Big) \\ &\times \exp\big(T\int_0^T |\nabla F_\varepsilon| ds\big),\end{aligned}$$

As $n_1, n_2 \in \mathcal{N}(\mathbb{R})$, $N_3 \in \mathcal{N}(\mathbb{R}^+\times\mathbb{R})$ and ∇F is $L^\infty - logtype$

$$\|u_\varepsilon - v_\varepsilon - w_\varepsilon\|_{L^\infty(\Delta_T)} = \mathcal{O}(\varepsilon^q) \quad as \quad \varepsilon \to 0 \quad \forall q$$

Then, u is associated to $w + v$

References

1. Colombeau, J.F.: Elementary introduction to new generalized functions. North-Holland
2. Colombeau, J.F.: Multiplication of Distributions. LNM, vol. 1532. Springer, Heidelberg (1992). https://doi.org/10.1007/BFb0088952
3. Colombeau, J.F.: New generalized functions and multiplication of distributions. Notas de matematica 90 North-Holland mathematics studies 84. North-Holland (1984)
4. Oberguggenberger, M., Wang, Y.G.: Delta-waves for semi linear hyperbolic Cauchy problems. Math. Nachr. **166**, 317–327 (1994)
5. Oberguggenberger, M., Russo, F.: Nonlinear stochastic wave equations. Integr. Transform Spec. Functions **6**, 71–83 (1998)
6. Hurd, A.E., Sattinger, D.H.: Questions of existence and uniqueness for hyperbolic equations with discontinuous coefficients. Trans. Amer. Math. Soc. **132**, 159–174 (1968)
7. Gramchev, T.: Semilinear hyperbolic systems with singyular initial data. Monarsh. Math. **112**, 99–113 (1991). https://doi.org/10.1007/BF01525797
8. Grosser, M., Kunzinger, M., Oberguggenberger, M., Steinbauer, R.: Geometric Theory of Generalized Functions with Applications to General Relativity, Mathematics and its Applications. Kluwer Acad. Publ, Dordrecht (2001)
9. Nedeljkov, M., Oberguggenberger, M., Pilipovic, S.: Generalized solutions to a semilinear wave equation, non-linear. Analysis **61**, 461–475 (2005)
10. Oberguggenberger, M.: Generalized solutions to semilinear hyperbolic systems. Monatshefte Math. **103**, 133–144 (1987). https://doi.org/10.1007/BF01630683

Fuzzy Fractional Equation with Derivative of Atangana-Baleanu and Fuzzy Semigroup

Fouziya Zamtain, Said Melliani(✉), Mohamed Elomari, and Lalla Saadia Chadli

LMACS, Sultan Moulay Slimane University, BP 523 - 23000 Beni Mellal, Morocco
s.melliani@usms.ma

Abstract. In this work, we are going to solve the initial fuzzy fractional problem under the fuzzy derivative of Atangana-Baleanu in sense of Caputo and via the generalized Hukuhara difference. More precisely, we will discuss the existence of a mild solution for the fuzzy fractional equation of Atangana-Baleanu by using the notion of fuzzy semi-group, and the fixed point theorem.

1 Introduction

The fuzzy set theory was created by L. Zadeh to describe the uncertainties that exist in many real phenomena in 1965 (see [10]). It is also an extension of classical theory, in which if we notice $\mathbb{E}$ the space of all fuzzy number then, we have $\mathbb{R} \subset \mathbb{E}$. In particular, the notions of continuous, convergence, differentiability, and integrability of fuzzy function are studied by many researchers [4,12–14]. Moreover, the concept of fuzzy metric space was studied by P. Diamond and P. E. Kloeden [11]. On the other hand in classical theory, the Mankowski difference doesn't satisfy $A - A = 0$, for A be a subset of $\mathbb{R}$ unless A is equal a simple element of A. But in fuzzy set theory and thanks to Hukuhara difference we have $A \ominus_H A = 0$. Unfortunately, this last difference has again a problems for that Stefanini generalize the Hukuhara difference to solve the problem facing searchers using Hukuhara difference (for more details see [8,15]).

Fuzzy fractional calculus is a generalization of differential ordinary calculus, it is used in many areas to be specified and defined in FFDEs as well. In books [7,16, 17] we find the works that studied accurately the subject of fractional calculus. The fractional derivative of Atangana-Baleanu was created by Abdon Atangana and Dermitu Baleanu in 2016 to develop the theory of fractional calculus [3]. Then D. Baleanu, A. Fernandez studied some proprieties of this derivative. Also, there are many problems and many equations solved by using this derivative [1]. The concept of fuzzy semi-group generated by linear operators of a fuzzy-valued function was introduced by Gal and Gal [6]. M. Elomari et all studied the existence and uniqueness of the mild solution to the problem with strong semi-group and using the conformable derivative [5]. The notion of semi-group was

S. Melliani and O. Castillo (Eds.): ICPAMS 2021, LNNS 476, pp. 219–227, 2023.
https://doi.org/10.1007/978-3-031-12416-7_18

not limited to fuzzy set but S. Melliani et all proved the existence and uniqueness of mild solution with strong semi-group and using the fuzzy derivative of Caputo in a fuzzy intuitionistic set.

The rest of this paper is organized as follows. The next section bears some notions, definitions of fuzzy set concepts and the fractional theory. The definition of fuzzy semigroup and its proprieties is presented in Sect. 3. The existence and the uniqueness of a mild solution are proved in Sect. 4.

2 Preliminaries

In this section, we present some necessary concepts about fuzzy set theory, then we introduce some definitions from fractional calculus theory, which are used all around this article.

Definition 1. [1] Let consider $\mathbb{E} = \{u : \mathbb{R} \longrightarrow [0,1]\}$ the space of all fuzzy subsets on $\mathbb{R}$ satisfying the following expressions:

1. u is normal, i.e., there exists an $x_0 \in \mathbb{R}$, such that $u(x_0) = 1$.

2. u is fuzzy convex, i.e., for $x, y \in \mathbb{R}$ and $0 \leq \lambda \leq 1$,

$$u(\lambda x + (1-\lambda)y) \geq \min[u(x), u(y)].$$

3. u is upper semi-continuous.

4. $[u]^0 = \mathrm{cl}\{x \in \mathbb{R} \mid u(x) > 0\}$ is compact.

For $0 < \alpha \leq 1$, denote $[u]^\alpha = \{x \in \mathbb{R}^n / u(x) \geq \alpha\}$.

Definition 2. [1] The parametric interval form of a fuzzy number u is shown as,

$$u[r] = [\underline{u}(r), \overline{u}(r)], \quad 0 \leq r \leq 1.$$

Where,

- $\underline{u}(r)$ is a left continuous and non-decreasing function with respect to r.
- $\overline{u}(r)$ is a left continuous and non-increasing function with respect to r.
- For each $r \in [0,1]$, we have $\underline{u}(r) \leq \overline{u}(r)$.

Definition 3. [2] Let $u, v : X \longrightarrow [0,1]$ be the fuzzy sets. Then, $u = v$ if and only if $[u]^\alpha = [v]^a$ for all $\alpha \in [0,1]$. The following arithmetic operations on fuzzy numbers are well known and frequently used below. If $u, v \in \mathbb{E}$, then

$$\begin{aligned}[u+v]^\alpha &= [\underline{u}^\alpha + \underline{v}^a, \overline{u}^\alpha + \overline{v}^a] \\ [u-v]^\alpha &= [\underline{u}^\alpha - \overline{v}^a, \overline{u}^\alpha - \underline{v}^a]\end{aligned}$$

$$[\lambda u]^\alpha = \lambda [u]^a = \begin{cases} [\lambda \underline{u}^\alpha, \lambda \overline{u}^\alpha], & \text{if } \lambda \geq 0 \\ [\lambda \overline{u}^\alpha, \lambda \underline{u}^\alpha], & \text{if } \lambda < 0. \end{cases}$$

Definition 4. [1] The Hausdorff distance is defined as the following,

$$d : \mathbb{E} \times \mathbb{E} \rightarrow \mathbb{R}_{+} \cup \{0\}$$
$$d(u,v) = \sup_{r\in[0,1]} \max\{|u_l(r) - v_l(r)|, |u_u(r) - v_u(r)|\}$$

Where $\mathbb{E}$ is the set of all fuzzy numbers on real numbers and $(\mathbb{E}, d)$ is a complete metric space and the following properties

1. $d(u \oplus w, v \oplus w) = d(u,v), \quad \forall u,v,w \in \mathbb{E}.$
2. $d(u \oplus w, 0) = d(u,0) + d(v,0), \quad \forall u,v,w \in \mathbb{E}.$
3. $d(u \oplus v, u \oplus w) = d(v,w), \quad \forall u,v,w \in \mathbb{E}.$
4. $d(u \oplus v, w \oplus z) \leq d(u,w) + d(v,z), \quad \forall u,v,w \in \mathbb{E}.$
5. $d(u \ominus v, w \ominus z) \leq d(u,w) + d(v,z), \quad \forall u,v,w \in \mathbb{E},\ u \ominus v, w \ominus z$ exist.
6. $d(\lambda \odot u, \lambda \odot v) = |\lambda| d(u,v), \quad \forall u,v \in \mathbb{E}, \lambda \in \mathbb{E}.$

Definition 5. [9] For $u, v \in \mathbb{E}$, the gH difference of u and v, denoted by $u \ominus_{gH} v$, is defined as the element $w \in \mathbb{E}$ such that

$$u \ominus_{gH} v = w \Leftrightarrow \begin{cases} i) & u = v \oplus w, \\ ii) & v = u \oplus (-1)w. \end{cases}$$

Definition 6. [1] (Generalized derivative) The gH-derivative of function $y(\tau)$ can be defined in the following form,

$$y'_{gH}(\tau) = \lim_{h\rightarrow 0+} \frac{y(\tau+h) \ominus_{gH} y(\tau)}{h} = \lim_{h\rightarrow 0+} \frac{y(\tau) \ominus_{gH} y(\tau+h)}{h}.$$

Remark 1. By consideration of gH difference and by using the generalized derivative we say

case 1: y is i-differentiable if the gH difference exists in the generalized derivative definition coincide with the case (i) in definition 5.
case 2: y is ii-differentiable if the gH difference exists in the generalized derivative definition coincide with the case (ii) in definition 5.

Definition 7. [4] Suppose that $f : [a,b] \rightarrow X \subset \mathbb{E}$ is a fuzzy-valued function with parametric form $f(t) = \left(\underline{f}(t), \overline{f}(t)\right), t \in [a,b]$, and $\underline{f}, \overline{f}$ are measurable and Lebesgue integrable on $[a,b]$. Then we define $\int f(t)dt$ by the parametric form

$$\int_a^b f(t)dt = \left(\int_a^b \underline{f}(t)dt, \int_a^b \overline{f}(t)dt\right).$$

This means

$$\left[\int_a^b f(t)dt\right]^{\alpha} = \left[\int_a^b \underline{f}_{\alpha}(t)dt, \int_a^b \overline{f}_{\alpha}(t)dt\right] \quad \text{for all} \quad \alpha \in [0,1].$$

Then we say that f is Lebesgue integrable on $[a,b]$.

Definition 8. [1] The AB fractional derivative in the sense of Caputo of the function $f(t)$ is defined as follow,

$$^{ABC}D^{\alpha}f(t)=\frac{B(\alpha)}{1-\alpha}\int_0^t f'(\tau)E_{\alpha}\left(-\frac{\alpha}{1-\alpha}(t-\tau)^{\alpha}\right)d\tau.$$

With $B(\alpha)=1-\alpha+\frac{\alpha}{\Gamma(\alpha)}$.

Theorem 1. *[1] The AB fractional derivative in the sense of Caputo is defined in two cases as follow,*

$$\left[^{ABC}D^{\alpha}y(t)\right]^r=\left[^{ABC}D^{\alpha}\underline{y}(t,r),^{ABC}D^{\alpha}\overline{y}(t,r)\right],\qquad case(1).$$

$$\left[^{ABC}D^{\alpha}y(t)\right]^r=\left[^{ABC}D^{\alpha}\overline{y}(t,r),^{ABC}D^{\alpha}\underline{y}(t,r)\right],\qquad case(2).$$

where

$$^{ABC}D^{\alpha}\underline{y}(t,r)=\frac{B(\alpha)}{1-\alpha}\int_0^t (_{i-gH})\underline{y}'(s)E_{\alpha}\big(-\frac{\alpha}{1-\alpha}(t-s)^{\alpha}\big)ds,$$

$$^{ABC}D^{\alpha}\overline{y}(t,r)=\frac{B(\alpha)}{1-\alpha}\int_0^t (_{ii-gH})\overline{y}'(s)E_{\alpha}\big(-\frac{\alpha}{1-\alpha}(t-s)^{\alpha}\big)ds.$$

Definition 9. [1] The AB fractional integral of the function $f(t)$ is given by

$$^{AB}I^{\alpha}f(t)=\frac{1-\alpha}{B(\alpha)}f(t)+\frac{\alpha}{B(\alpha)\Gamma(\alpha)}\int_0^t (t-\tau)^{\alpha-1}f(\tau)d\tau.$$

Definition 10. [1] The fractional integral operator associated to the ABC fractional derivative operator on fuzzy valued functions in the interval parametric form is denoted as,

$$[^{AB}I^{\alpha}(f(t))]^r=\left[^{AB}I^{\alpha}\left(\underline{f}(t,r)\right),^{AB}I^{\alpha}\left(\overline{f}(t,r)\right)\right].$$

Where

$$^{AB}I^{\alpha}\underline{f}(t)=\frac{1-\alpha}{B(\alpha)}\underline{f}(t,r)+\frac{\alpha}{B(\alpha)\Gamma(\alpha)}\int_0^t (t-\tau)^{\alpha-1}\underline{f}(\tau,r)d\tau.$$

$$^{AB}I^{\alpha}\overline{f}(t)=\frac{1-\alpha}{B(\alpha)}\overline{f}(t,r)+\frac{\alpha}{B(\alpha)\Gamma(\alpha)}\int_0^t (t-\tau)^{\alpha-1}\overline{f}(\tau,r)d\tau.$$

Lemma 1. *[1] The composition of fractional derivative and fractional integral of Atangana-Baleanu in Caputo sense of a fuzzy function is given as*

$$^{AB}I^{\alpha}\left(^{AB}D^{\alpha}f(t)\right)=f(t)\ominus_{gH}f(0).$$

Definition 11. Let $f : [0,\infty) \to X \subset \mathbb{E}$ be a continuous function such that $e^{-st} \odot f(t)$ is integrable. Then the fuzzy Laplace transform of f, denoted by $\mathbf{L}[f(t)]$, is

$$\mathbf{L}[f(t)] := F(s) = \int_0^{\infty} e^{-st} \odot f(t)dt, \quad s > 0.$$

Theorem 2. *[3] The fuzzy Laplace transform of the Atangana-Baleanu derivative in Caputo sense is given as*

$$\mathbf{L}\left(D^{ABC} f(t)\right)(s) = \frac{B(\alpha)}{1-\alpha} \frac{s^{\alpha} F(s) - s^{\alpha-1} f(0)}{s^{\alpha} + \frac{\alpha}{1-\alpha}}.$$

3 Fuzzy Strongly Continuous Semigroup

Definition 12. By a fuzzy semigroup on E we mean a family $\{T(t), t \geq 0\}$ of operators from E into itself satisfying the following conditions:

1. $T(0) = I$, the identity mapping on E;

2. $T(t+s) = T(t)T(s)$ for all $t, s \geq 0$;

3. the function $g : [0,\infty[\to E$ defined by $g(t) = T(t)(x)$ is continuous at $t = 0$ for all $x \in E$, that is, $\lim_{t \to 0^+} T(t)(x) = x$

$\{T(t), t \geq 0\}$ is also called a fuzzy $\mathbb{C}_0$ -semigroup.

Definition 13. Let $\{T(t)\}_{t \geq 0}$ be a strongly continuous semigroup on $\mathbb{E}$. The infinitesimal generator $A : D(A) \subset \mathbb{E} \to \mathbb{E}$ is defined by

$$Ax = \lim_{h \to 0^+} \frac{1}{h} \odot (T(h)x \ominus_{gH} x).$$

On the domain

$$D(A) := \left\{ x \in \mathbb{E} / \lim_{h \to 0^+} \frac{1}{h} \odot (T(h)x \ominus_{gH} x) \text{ exists in } \mathbb{E} \right\}.$$

Theorem 3. *If $(T(t))_t \geq 0$ and $(S(t))_t \geq 0$ are two fuzzy C_0-semigroup having the same infinitesimal generator A. Then*

$$T(t) = S(t), \qquad \forall t \geq 0.$$

4 Main Results

In this part, we show the existence of the mild solution to the following problem in Banach space.

$$\begin{cases} {}^{ABC}D^{\alpha}y(t) = Ay(t) + f(t, y(t)), \quad t \in [0, T] \\ y(0) = y_0 \in \mathbb{E}. \end{cases} \tag{1}$$

With A generate a C_0 semigroup $T(t)$.

Theorem 4. *Let $y \in C([0,T], \mathbb{E})$ be a mild solution of the problem (1) if y satisfies the following equation*
Case 1

$$y(t) = S_{\alpha}(t)y_0 \oplus \int_0^t T_{\alpha}(t-s)f(s, y(s))ds. \tag{2}$$

Case 2

$$y(t) = S_{\alpha}(t)y_0 \ominus (-1) \int_0^t T_{\alpha}(t-s)f(s, y(s))ds. \tag{3}$$

Where,

$$S_{\alpha}(t) = \mathbf{L}^{-1}\left[\left(\frac{B(\alpha)t^{\alpha}}{(1-\alpha)t^{\alpha}+\alpha}Id - A\right)^{-1}\frac{B(\alpha)t^{\alpha-1}}{(1-\alpha)t^{\alpha}+\alpha}\right].$$

$$T_{\alpha}(t) = \mathbf{L}^{-1}\left[\left(\frac{B(\alpha)t^{\alpha}}{(1-\alpha)t^{\alpha}+\alpha}Id - A\right)^{-1}\right].$$

Proof. We consider the equation of problem (1)

$${}^{ABC}D^{\alpha}y(t) = Ay(t) + f(t, y(t)) \tag{4}$$

Then we apply the fuzzy fractional integral of Atangana-Baleanu in sense of Caputo on both sides of the equality (4), we obtain

$$y(t) \ominus_{gH} y(0) =^{AB} I\Big(Ay(t) + f(t, y(t))\Big)$$

From the linearity of ${}^{AB}I^{\alpha}$, we get

$$y(t) \ominus_{gH} y(0) = A^{AB}Iy(t) +^{AB} I^{\alpha}f(t, y(t))$$

Then, by using the Remark 1 and by applying the fuzzy Laplace transform, we get
For **case 1**

$$\mathbf{L}(y(t))(s) = \mathbf{L}(y(0))(S) + \mathbf{L}(A^{AB}Iy(t))(s) + \mathbf{L}({}^{AB}I^{\alpha}f(t, y(t)))(s)$$

Then, we obtain

$$Y(s) = \frac{1}{s}y_0 + A\left(\frac{(1-\alpha)s^{\alpha}+\alpha}{s^{\alpha}B(\alpha)}\right)Y(s) + \left(\frac{(1-\alpha)s^{\alpha}+\alpha}{s^{\alpha}B(\alpha)}\right)F(s).$$

So,

$$Y(s) = \left(\frac{s^{\alpha}B(\alpha)}{(1-\alpha)s^{\alpha}+\alpha}I_d - A\right)^{-1}\left[\frac{s^{\alpha-1}B(\alpha)}{(1-\alpha)s^{\alpha}+\alpha}y_0 + F(s)\right]$$

By applying the fuzzy inverse of Laplace transform we get

$$y(t) = S_{\alpha}(t)y_0 \oplus \int_0^t T_{\alpha}(t-s)f(s, y(s))ds.$$

The same for **case 2**.
This completes the proof.

Theorem 5. *Let the following assumption hold*

1. (H_1) $f : [0, T] \times \mathbb{E} \to \mathbb{E}$ *is continuous and Lipschitz with respect to the second argument, that is, there exists a constant* $L > 0$ *such that*

$$d(f(t, x(t)), f(t, y(t))) \leq Ld(x(t), y(t)) \qquad \text{for all } t \in [0, T], x, y \in \mathbb{E}$$

2. (H_2) *There exists* $M \geq 1$ *and* $\omega > 0$ *such that*

$$d\left(T_{\alpha}u(t), T_{\alpha}v(t)\right) \leq Me^{\omega t}d(u(t), v(t)), \quad t \in [0, T].$$

3. (H_2) $\quad Me^{\omega T} < (LT)^{-1}$.

Then the problem (1) has a unique mild solution.

Proof. We first define the following mapping $G : C([0, T], \mathbb{E}) \longrightarrow C([0, T], \mathbb{E})$ as

$$Gy(t) = S_{\alpha}(t)y_0 \oplus \int_0^t T_{\alpha}(t-s)f(s, y(s))ds.$$

- The mapping G is well define. In fact

  Let $x \in C([0, T], \mathbb{E})$ and $t \in [0, T]$ and $\epsilon > 0$ sufficiently small,

$$d_{\infty}(Gy(t+\epsilon), Gy(t)) = d_{\infty}\bigg(S_{\alpha}(t+\epsilon)y_0 \oplus \int_0^{t+\epsilon} T_{\alpha}(t-s+\epsilon)f(s, y(s))ds, S_{\alpha}(t)y_0 \oplus \int_0^t T_{\alpha}(t-s)f(s, y(s))ds\bigg)$$

  Then

$$\begin{aligned} d_{\infty}(Gy(t+\epsilon), Gy(t)) &\leq d_{\infty}\Big(S_{\alpha}(t+\epsilon)y_0, S_{\alpha}(t)y_0\Big) \\ &\quad \oplus d_{\infty}\Big(\int_0^{t+\epsilon} T_{\alpha}(t-s+\epsilon)f(s, y(s))ds, \int_0^t T_{\alpha}(t-s)f(s, y(s))ds\Big) \\ &\leq d_{\infty}\Big(S_{\alpha}(t+\epsilon)y_0, S_{\alpha}(t)y_0\Big) \\ &\quad \oplus d_{\infty}\Big(\int_0^t T_{\alpha}(t-s)f(s-\epsilon, y(s-\epsilon))ds, \int_0^t T_{\alpha}(t-s)f(s, y(s))ds\Big) \\ &\leq d_{\infty}\Big(S_{\alpha}(t+\epsilon)y_0, S_{\alpha}(t)y_0\Big) \oplus M\int_0^t d_{\infty}\Big(f(s-\epsilon, y(s-\epsilon)), \int_0^t f(s, y(s))\Big)ds \end{aligned}$$

By using the limit and the dominated convergence theorem, we obtain

$$d_\infty(Gy(t-\epsilon), Gy(t)) \to 0 \quad \text{as} \quad \epsilon \to 0^+.$$

The same we show that $d_\infty(Gy(t-\epsilon), Gy(t)) \to 0$ as $\epsilon \to 0^+$.
Consequently, Gy is continuous at each $t \in [0, T]$.
Hence $Gy \in \mathbb{C}([0, T], \mathbb{E})$.

- The mapping G has a unique fixed point. In fact :

Let $x(t)$ and $y(t)$ are in $C([0, T], \mathbb{E})$, then we have

$$\begin{aligned} d(Gx(t), Gy(t)) &= d\Big(S_\alpha(t)y_0 \oplus \int_0^t T_\alpha(t-s)f(s, x(s))ds, S_\alpha(t)y_0 \oplus \int_0^t T_\alpha(t-s)f(s, y(s))ds\Big) \\ &\leq \int_0^t d\Big(T_\alpha(t-s)f(s, x(s))ds, T_\alpha(t-s)f(s, y(s))ds\Big) \\ &\leq M \int_0^t e^{\omega(t-s)} d\Big(f(s, x(s)), f(s, y(s))\Big) ds. \end{aligned}$$

Then, we can deduce that

$$d_\infty(Gx, Gy) \leq MLTe^{\omega T} d_\infty(x, y).$$

Which implies that G is a contraction mapping.
By the fixed point theorem the problem (1) has a unique fixed point which is the mild solution of (1).

The same for **case 2**.

References

1. Allahviranloo, T., Ghanbari, B.: On the fuzzy fractional differential equation with interval 85 Atangana-Baleanu fractional derivative approach. Chaos Solit. Fractals **130**, 109397 (2020)
2. Allaoui, A., Melliani, S., Chadli, L.: Fuzzy dynamical systems and invariant attractor sets for fuzzy strongly continuous semigroups. J. Fuzzy Set Valued Anal. **2**, 148–155 (2016)
3. Atangana, A., Baleanu, D.: arXiv preprint arXiv:1602.03408. arxiv.org (2016)
4. Bede, B., Gal, S.G.: Generalization of the differentiability of fuzzy-number-valued functions with applications to fuzzy differential equations. Fuzzy Sets Syst. **151**, 581–599 (2005)
5. Elomari, M., Melliani, S., Chadli, L.S.: Conformable fractional cosine families of operators. J. Math. Sci. Model. **2**(2), 112–116 (2019)
6. Gal, C.G., Gal, S.G.: Semigroups of Operators on Spaces of Fuzzy-Number-Valued Functions with Applications to Fuzzy Differential Equations, arXiv:1306.3928v1, 17 June 2013
7. Harir, A., Melliani, S., Chadli, L.S.: Fuzzy fractional evolution equations and fuzzy solution operators. In: Advances in Fuzzy Systems, vol. 2019, Article ID 5734190, 10 p. Hindawi (2019)

8. Son, N.T.K.: A foundation on semigroups of operators defined on the set of triangular fuzzy numbers and its application to fuzzy fractional evolution equations. Fuzzy Sets Syst. **347**, 1–28 (2018)
9. Stefanini, L., Bede, B.: Generalized hukuhara diferentiability of interval-valued functions and interval diferential equations. Nonlinear Anal. **71**(34), 1311–1328 (2009)
10. Zadeh, L.A.: Fuzzy sets. Inf. Control **8**, 338–353 (1965)
11. Diamond, P., Kloeden, P.E.: Metric Spaces of Fuzzy Sets: Theory and Applications. World Scientific, Singapore (1994)
12. Goo, H.Y., Park, J.S.: On the continuity of the Zadeh extensions. Chungcheong Math. Soc. **20**(4), 525–533 (2007)
13. Puri, M.L., Ralescu, D.A.: Differentials of fuzzy functions. J. Math. Anal. Appl. **91**, 552–558 (1983)
14. Zhong, Q.: On fuzzy measure and fuzzy integral on fuzzy set. Fuzzy Sets Syst. **37**, 77–92 (1990)
15. Hukuhara, M.: Integration des Applicaitons Mesurables dont la Valeur est un Compact Convexe. Funkcialaj Ekvacioj **10**, 205–223 (1967)
16. Sabatier, J., Agrawal, O.P., Tenreiro Machado, J.A.: Advances in Fractional Calculus. Springer, Dordrecht (2007). https://doi.org/10.1007/978-1-4020-6042-7
17. Kochubei, A., Luchko, Y.: Handbook of Fractional Calculus with Applications, Volume 1: Basic Theory (2019)

Pricing and Hedging of Swaptions: Setting up a Pricer of Interest Rate Swaptions

Ghizlane Kouaiba[1(✉)], Ali Malloum Goni[2], Ahmed Doghmi[2], and Driss Mentagui[1]

[1] Faculty of Sciences, Ibn Tofail University, Kenitra, Morocco
kouaiba.ghizlane@gmail.com
[2] National Institute of Statistics and Applied Economics (INSEA), Rabat, Morocco
a.doghmi@insea.ac.ma

Abstract. The objective of the project, carried out within the middle office of the Attijariwafa Bank trading room, is to develop a swaption pricer under vba excel. The establishment of this pricer within the trading room does not come at random. Indeed, the implementation of such a pricer meets the needs of the bank. For this purpose, the bank should have an internal pricer for each product processed by its commercial catalog, which means that the bank should be independent regarding the external pricer. For the valuation of swaptions, two models are used, namely: the Black model founded in 1976, which aims to value options on rates, this model remains the most used in the financial market and the Hull model -White which consists first of valuing short rates, and then to evaluate the options on rates, taking into account the evolution of volatility over time, this does not appear in Black's model. Our project is not just about setting up this pricer. Nonetheless, our work sheds light on the different hedging strategies that an investor must adopt to deal with several market risks by using swaptions.

1 Introduction

In the derivatives markets, the investor do not buy a product directly, but there are possibilities of forward buying or selling with predefined conditions. Therefore, they allow you to take large buying or selling positions with a limited down payment. The interest rate derivatives market is one of the largest and liquid derivative markets. Given that in the financial market, the interest rate is a risk whether for a loan or a borrowing, hence the birth of derivatives on interest rates, such as CAP, FLOR, COLLAR, SWAP, SWAPTION.... The swap is defined as a derivative product that allows the exchange of future cash flows between two parties. Generally, it is the exchange of a variable rate against a fixed rate. The latter is considered as a firm-hedging instrument that generates a risk in case of bad anticipation, hence the use of the swaption. Thus, a swaption is an option on an interest rate swap, i.e. the right to set up the interest rate swap whose characteristics are

S. Melliani and O. Castillo (Eds.): ICPAMS 2021, LNNS 476, pp. 228–239, 2023.
https://doi.org/10.1007/978-3-031-12416-7_19

fixed in advance by paying a premium if it is a buyer. From this point of view, our end of studies project will consist in setting up a tool necessary to determine and verify the premium of the swaption buyer for the account of the Middle Office entity of the trading room of Attijariwafa Bank, in order to verify the positions taken by the Front Office Traders. We will first define swaps in this report, and then we will discuss the different types of swaps. Second, we will present options in a general way. Then we will discuss swap options, also known as swaptions. What is a swaption? What are the different types of swaption valuations used in this report? and we'll see at the end, when should we use swaption to hedge against the interest rate risk?

2 Swaps

The swap [1,2] is a derivative that allows future cash flows to be exchanged between two parties according to an amount called the notional amount and a fixed schedule. Generally, the exchange is between a fixed rate and a variable rate, but sometimes any other type of exchange can be considered for example the variable rate against the variable.

The most widely used type of swap is the standard rate swap, also known as the "Vanilla" swap rate. In this case, a company undertakes to pay cash flows at fixed interest rates on a given principal for a certain number of years. And in return, it receives the product of variable interest rate (Libor, Euribor ...) on the same principal with the same duration. We talk about [5]:

- payer swap: We pay the fixed rate and receive the variable rate.
- receiver swap: The operation is carried out in the opposite direction, in this case we pay the variable rate and we receive the fixed rate.

3 Interest Rate Swaps Valuation

3.1 The Swap as a Bond Portfolio

This method consists in valuing the swap in term of bond prices, that is to say, considering the swap as a structured product composed of two bonds, the first with a fixed interest rate and the other with a variable interest rate. In this case, the value of the swap is given as follows [8]:

- In the case of a fixed payer swap:

$$V = B_{var} - B_{fix} \tag{1}$$

- In the case of a fixed receiver swap:

$$V = B_{fix} - B_{var} \tag{2}$$

where

- B_{fix}: value of the fixed rate bond associated to the swap.
- B_{var}: The value of the floating rate bond associated to the swap.
- B_{fix} is given by:

$$\sum_{i=1}^{n} Ke^{-r_i t_i} + Le^{-r_n t_n} \tag{3}$$

With:
- t_i : the duration until the i-th payment.
- L: The notional principal.
- r_i: The zero LIBOR/Swap rate for each maturity t_i.
- K: The payment of the fixed leg at each date t_i.

In return, the floating rate bond is considered a bond with a shorter maturity. Just before the payment, the price of this bond is thereafter worth $L + K^*$, with K^*being the payment made on the variable leg of the swap. If we consider t_1 the delay until the next payment we will have:

$$B_{var} = (L + K^*)e^{-r_1 t_1} \tag{4}$$

3.2 Valuation of Swaps in Terms of FRA (Forward Rate Agreement)

The FRA [8] can be estimated assuming that forward rates [5] will actually be future spot rates. We then proceed as follows:

- For each LIBOR rate determining the swap chashflows, the corresponding fordward rate is calculated by the following relation $R_F = (R_2T_2 - R_1T_1)/(T_2 - T_1)$ knowing that R_2 and R_1 are the zero LIBOR/Swap rates respectively observed in T_2 and T_1.
- Each cash flow is estimated assuming that the future LIBOR rate will be the forward rate observed today.
- The value of the swap is the sum of these discounted cash flows with the LIBOR/swap curve.

3.3 Applications

By using"the swap as a bond portfolio" method:

			term	zero libor rate
fixed rate	8%		3	10%
Frenquency	2		9	10,50%
notional principal	100 000 000		15	11%
LIBOR rate	10,20%			
Frenquency	2			
	fixed payer swap	4 267 175,85		
	fixed receiver swap	- 4 267 175,85		

Fig. 1. The swap as a bond portfolio

By using the "FRA "method:

			term	zero libor rate
fixed rate	8%		3	10%
Frenquency	2		9	10,50%
notional principal	100 000 000		15	11%
LIBOR rate	10,20%			
Frenquency	2			
	Payeur de fix	4 267 175,85		
	Receveur de fix	- 4 267 175,85		

Fig. 2. The"FRA "method

The amount of the debt is 4 267 175.85 Euro for the fixed-payer swap and -4 267 175.85 Euro for the opposite case.

P.S

In the case of a speculative swap, the operator is exposed to unlimited risk in the event of an unfavorable movement of rates. An initially expensive debt will remain so even once the swap is concluded, because it cannot act on the past. Hence the resort to swaption.

4 Swaptions

4.1 Concept and Operations

The interest rate swap option, or swaption [1], is a contract between the seller and the buyer that gives the buyer the right but not the obligation to enter a swap on a particular date. The interest rate swap characteristics are set in advance. In return, the buyer pays a premium to the seller [3].

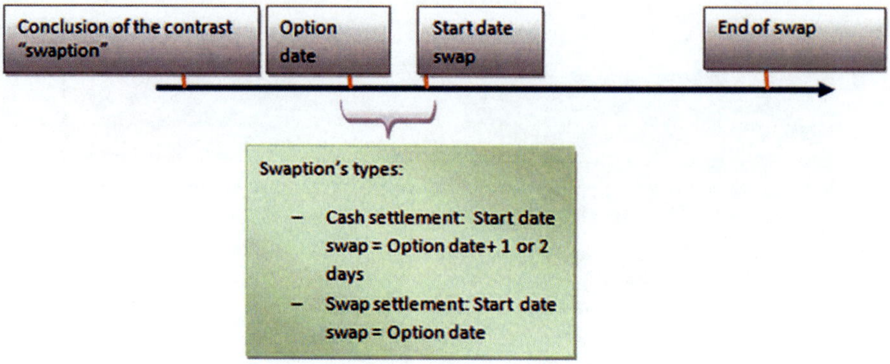

Fig. 3. The swaption process

5 Valuation of Interest Rate Swaption

5.1 Black 76 Model

Initially, the formula of Black and Scholes (1973) [4,12] was used to value the options on financial index, shares ... Thus, in 1976 Fischer Black [11] extended this model to value the options of the interest rate and which will be used mainly to value the cap, floor, swaption... The diffusion equation of this model is given by $dr_f = r_f \sigma dw$ knowing that w represents the Brownian motion. Remember that a payer swaption is a contrast where the holder pays the fixed leg if the option is exercised, i.e. the forward swap rate is greater than r_k, the strike price or Strike. And we note his payoff as follows [7]:

$$\frac{L}{m} max(r_f(t) - r_k, 0) \tag{5}$$

With

- L: the notional principal amount
- m: frequency of payments per year
- r_f: The forward swap rate
- r_k: Strike.

The log-normality assumption swap rate using the Black model gives us the cash flow value received in time T_i is given by:

$$\frac{L}{m} P(0, T_i)[r_f^0 N(d_1) - r_k N(d_2)] \tag{6}$$

Knowing that:

- r_f^0: The forward swap rate is expressed as follows $r_f^0 * \sum_{i=1}^{m*n} a_i * P(0, T_i) = P(0, T_0) - P(0, T_{m*n})$ and a_i is the accrual fraction corresponding to the time period between T_i and T_{i+1}.

- N(d): The distribution function of the normal reduced center law given by:

$$\int_{-\infty}^{d} \frac{1}{2\pi} e^{-x^2/2} dx. \tag{7}$$

- $d_1 = \frac{ln(\frac{r_f^0}{r_k})+\frac{\sigma^2 T}{2}}{\sigma\sqrt{T}}$ and $d_2 = d_1 - \sigma\sqrt{T}$.
- $P(0, T_i)$: the price of a zero-coupon bond covering the period $[0\,, T_i]$ So the total value of the payout swaption after n years and m frequency of payments is:

$$\sum_{i=1}^{m*n} \frac{L}{m} P(0, T_i)[r_f^0 N(d_1) - r_k N(d_2)] \tag{8}$$

5.2 Hull-White (One-Factor) Model

The previous model, namely the Black model used for swaption valuation, is based on the assumption of log-normality of the underlying which is our swap rate. However, their interest is limited by the fact that it provides no description of the evolution of rates over time because it considers the volatility is constant. Hence the interest of stochastic models those are part of this evolution. The rate process in this model is written as follows [8]:

$$dr = a(\theta(t)/a - r)dt + \sigma dw \tag{9}$$

With a and σ are constant.

The bond prices at time t by using the Hull-White model are given by:

$$P(t, T) = A(t, T)e^{-B(t,T)r(t)} \tag{10}$$

where $B(t, T) = \frac{1-e^{a(T-t)}}{a}$

and $ln(A(t, T)) = ln(\frac{P(0,T)}{P(0,t)}) + B(t, T)F(0, T) - \frac{1}{4a^3}\sigma^2(e^{-aT} - e^{-at})^2(e^{-2at} - 1)$

With: F (0, t) is the instantaneous forward rate for horizon t, seen on date 0.

The price at time zero of a call option that mature at time T on a zero-coupon bond which maturity at time s is given by

$$LP(0, s)N(h) - KP(0, T)N(h - \sigma_p) \tag{11}$$

where $h = \frac{1}{\sigma_p} ln(\frac{LP(0,s)}{KP(0,T)}) + \frac{\sigma_p}{2}$ And $\sigma_p = \frac{\sigma}{a}[1 - e^{-a(s-T)}]\sqrt{\frac{1-e^{-2aT}}{2a}}$

To obtain a and σ, we need a calibration, this consists in a first step to look for parameters that minimize:

$$\sum_{i=1}^{n} = (U_i - V_i)^2 \tag{12}$$

where U_i is the market price of instrument i and V_i the price calculated by the model. It is assumed here that n liquid assets are used for calibration.

Assets used for calibration should be chosen as close as possible to the assets that will be valued by the model.

Application. Take an example of a swaption carried out in Attijariwafa bank's trading room:

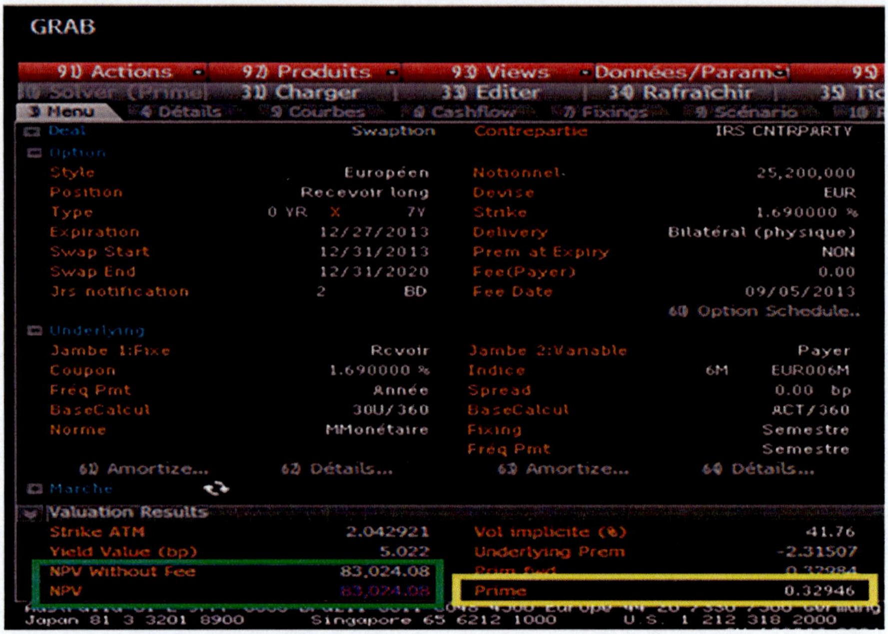

Fig. 4. Bloomberg's pricer (black 76 model)

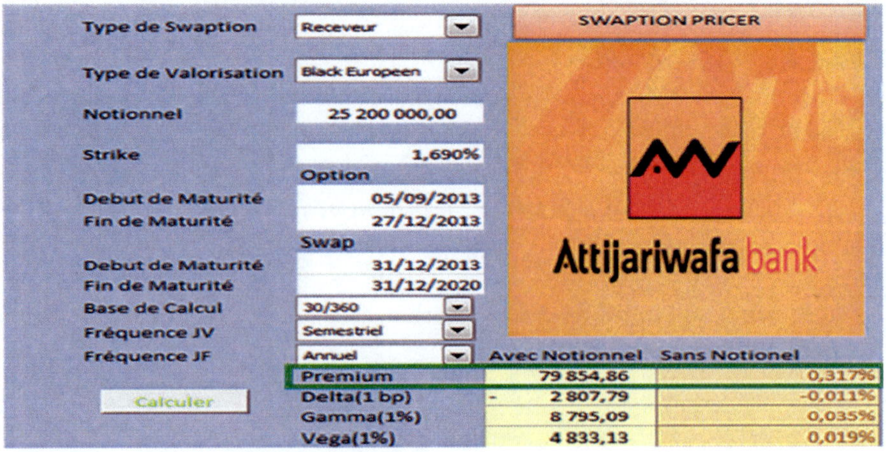

Fig. 5. VBA's pricer (black 76 model)

This pricer gives a premium of 0.317% or 79 854.86 EUR. This difference in value with that of Bloomberg is often explained by the interpolation method used for the yield curve.

Fig. 6. Bloomberg's pricer (hull-white (one factor) model)

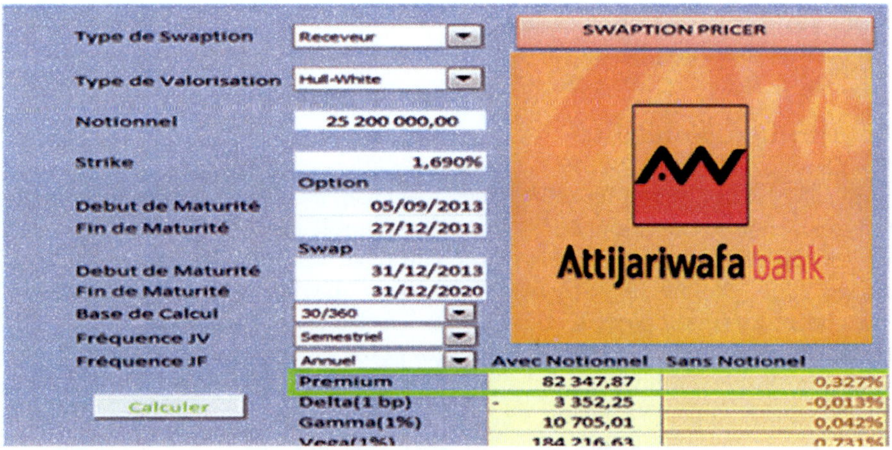

Fig. 7. VBA's pricer (hull-white (one factor) model)

It is clear that the value obtained by the Hull-White model of our pricer (0.327%) is very close to that obtained by Bloomberg's pricer (0.32850%).

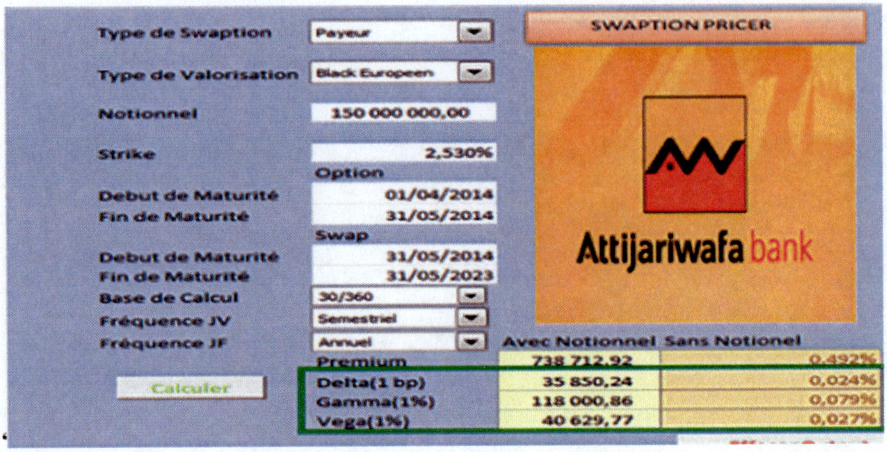

Fig. 8. Greek sensitivity parameters

Note also that for the Greek sensitivity [10] parameters we get the following values:

Delta (1 bps) = 0.024%, i.e. if the underlying varies by one basis point (0.01%) then this implies a variation of 0.024% of the price of the swaption, i.e. an increase of 35 850. 24 EUR whose principal amount is equal to 150 000 000 EUR.

Gamma (1%) = 0.079%, which means that when the Delta varies by 1%then the price varies by 0.079% or 118 000.86 EUR.

Vega (1%) = 0.027%, that means when the volatility of the swaption varies by 1% implies that the price of the swaption varies 0.027% i.e. 40 629.77 EUR.

6 Implied Volatility

Implied volatility [9] can be defined as the value of σ, such that if we replace it in the expression of the call deduced from the example of the Black Scholes model, we will find the numerical value of the call or of the put given by the market. However, it remains very interesting to know, having the price, what would have been the volatility that would have been necessary to enter in a theoretical model to obtain the price as it is on the market. Thus, we can calculate the volatility. Since the inverse of the expression of a European option defined by Black Scholes appears difficult, we refer to an iterative search method such as the Newton-Raphson algorithm. This algorithm allows to have the numerical value of the implied volatility as follow:

$$\sigma_{n+1} = \sigma_n + \frac{P^{market} - P^{Black}}{\frac{\partial P^{Black}}{\partial \sigma_n}} \tag{13}$$

The algorithm stops at iteration i once $|\sigma_{i+1} - \sigma_i|$ is less than a certain precision that it is fixed in advance and finally retains the value σ_{i+1}. Similarly and at the level of our swaption price, the option to calculate the implied volatility (Fig. 9) by knowing the price of the swaption is availabale, for example if the price is 750 000 euros with the precise characteristics then the corresponding volatility is 38,85%.

We will establish a volatility matrix (Fig. 10) according to the maturity of the option (1MO, 3MO, ...) and that of the swap "Tenor" (1Yr, 2Yr....), this matrix will allow us to plot the surface of the volatility according to the maturity of the option and that of the swap. It should also be noted that for almost all strikes, implied volatility decreases with the strike, which is called the skew phenomenon. However, for very large strikes, we sometimes observe a slight rise in implied volatility, which is called the smile phenomenon.

7 Cover Strategies with Swaption

Generally, swaption is used when holding bonds [3] or eurobonds (bonds denominated in a currency other than that of the issuer's country). And these bonds are valued by a zero coupon curve. The latter will depend on the currency with which the bonds are issued, which can be for example a ZC Euribor curve if it is the Euro and ZC Libor is the dollar. When constraining rate risks for these bonds, we can set up a swaption for protection. Thus, the interest rate swaption contract has a double advantage, which allows the buyer of the option to hedge against an unfavorable market trend while benefiting from a possible favorable evolution. So we can summarize the interest of the use of swaption for the case of a lending and borrowing and with its strategy adopted in the following table:

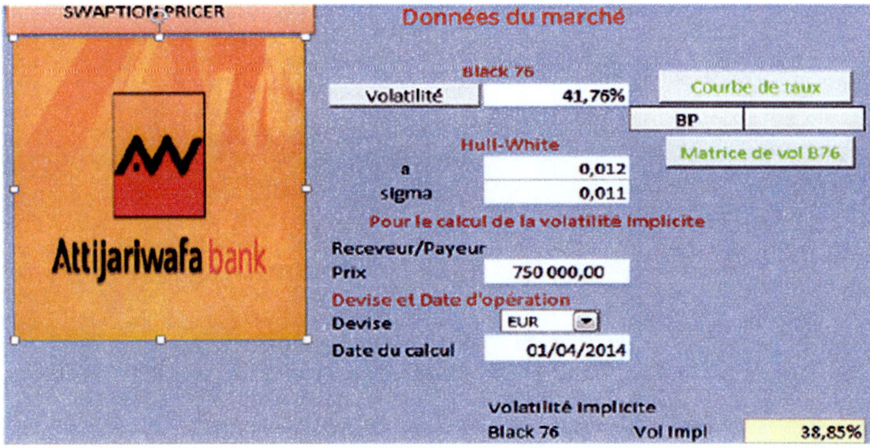

Fig. 9. The implied volatility calculation

Table 1. Cover strategies with swaption

Swaption operation	Protection sought	Strategy of the swaption
Floating Rate borrowing or Fixed rate lending	Rising interest rates	Purchase of a right to pay a fixed rate
Floating Rate lending or Fixed rate borrowing	Falling interest rates	Purchase of a right to receive a fixed rate

8 Conclusion

In light of the above, we can say that swaptions are necessary tools to hedge interest rate risk better than swaps, as these could generate counterparty risk in the event of poor anticipation.

However, two models were used to value the swap options: the black 76 model and the Hull-White model. Our two models used converge almost to the same results as those of Bloomberg. The Hull-White model gives us the best result compared to that of the Black, even if this Hull-White model is not a reference pattern of the financial market. So the VBA's pricer is a tool that could serve for the Middle Office entity to undoubtedly control the premium given by front office traders. And also a major utility in the context of mark to market, to evaluate the swaption during its period of life to see the opportune and risky moment of swaption.

The Hull-White model, despite being theoretically sound, uses the market data offered by the black model, as it is the most used model on the market.

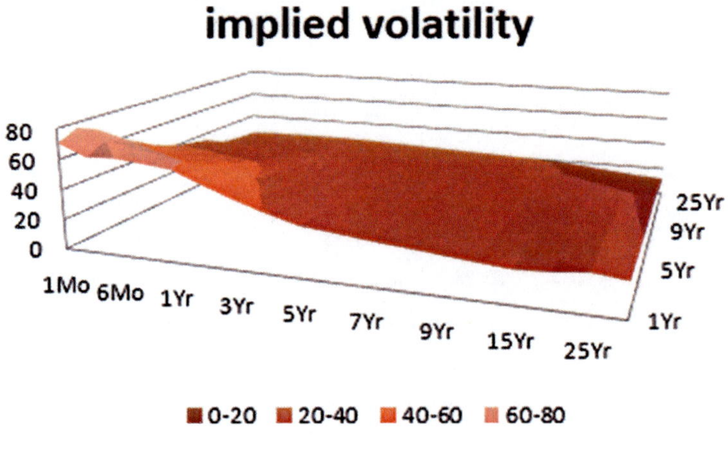

Fig. 10. Area of implied volatility as a function of the tenor and maturity of the option

References

1. Fauth, A.: Modèle de taux, surface de volatilité et introduction au risque de crédit, *Université Lille I*, working paper, pp. 10–19 (2014)
2. Bouarafa, A.: Etude et mise en place d'un outil d'aide à la décision pour l'évaluation et le pricing des produits de couverture du risque et taux d'intérêt: Swap de taux (2009)
3. Lyonnais, C.: Les produits dérivés de Taux, working paper (1995)
4. Akume, D., Luderer, B., Weber, G.-W.: Pricing and hedging of swaptions, *TOM 8*, N 4 (2003)
5. Wilhemly, F.: Analyse des modèles de taux d'intérêts pour la gestion actif-passif. *Claude Bernard University - Lyon 1*, France (2010)
6. Jauvion, G.: Adaption de la structure par des Taux aux conditions de marché Actuelles. Mémoire d'Actuariat, ISUP - Ecole Centrale Paris, France (2009)
7. Jean-Babtiste, D.: Valorisation des options, Université d'Atrois,working paper, pp. 4–6 (2019)
8. Hull, J.: Options futures et autres actifs dérivés,*Pearson Education*, 6ème édition (2007)
9. Saimi, N.: Estimation de la volatilité et filtrations non linéaire. Québec à Trois-Rivières University, Canada (2004)
10. Portait, R., de Marché, P.P.F.: Instrument de base, produits dérivés, portefeuille de risque,*Dalloz*, 3ème édition
11. Burgess, N.: Interest rate swaptions: a review and derivation of swaption pricing formulae. J. Econ. Financ. Anal. **2**(2), 87–103 (2008). https://ojs.tripaledu.com/index.php/jefa/article/view/38/44
12. Black, F., Scholes, M.: The pricing of options and corporate liabilities. J. Polit. Econ. **81**, 637–59 (1973)
13. Cox, J.C., Ingersoll, J.E., Jr., Ross, S.A.: An intertemporal general equilibrium model of asset prices. Econometrica **53**, 363–84 (1985)
14. Crépey, S.: Calibration of the local volatility in a generalized Black-Scholes model using Tikhonov regularization. SIAM J. Math. Anal. **34**(5), 1183–1206 (2003)
15. Bouchaud, J.-P., Sornette, D.: The Black-Scholes option pricing problem in mathematical finance: generalization and extensions for a large class of stochastic processes. J. de Phys. I, Fr. EDP Sci. **4**(6), 863–881 (1994)
16. Kouaiba, G., Saikak, M.-D.: The pricing of interest rate derivatives: caps/floors and the construction of the yield curve. In: Ben Ahmed, M., Boudhir, A.A. (eds.) SCAMS 2017. LNNS, vol. 37, pp. 997–1005. Springer, Cham (2018). https://doi.org/10.1007/978-3-319-74500-8_89
17. Rosenberg, J.V.: Implied volatility functions. J. Derivatives **7**(3), 51–64 (2010). https://doi.org/10.3905/jod.2000.319124

Non-linear Age-Dependent Population Dynamics with Spatial Diffusion

Khalid Hilal(✉), Ahmed Kajouni, and Hiba El Asraoui

Sultan Moulay Slimane University, BP 523, 23000 Beni Mellal, Morocco
hilalkhalid2005@yahoo.fr

Abstract. Several models in population dynamics are governed by reaction-diffusion equations or parabolic equations. In this work, we present a population model containing both age-structure and spatial diffusion.

Our model is:

$$\begin{cases} \dfrac{\partial u}{\partial t} + \dfrac{\partial u}{\partial a} = \Delta u - \mu_n(a)u(x,t,a) - \mu_e(P(x,t))u(x,t,a). \\ P(x,t) = \displaystyle\int_0^\infty u(x,t,a)da. \end{cases}$$

where $u(x,t,a)$ is a positive function which represents the density in both age (a) and space (x). $P(x,t)$ represents the total population at position x.

Existence and uniqueness results are obtained, and also the asymptotic behavior of the solution is studied.

1 Introduction

The simplest model of population dynamics is the Malthusian model [5],

$$\frac{dP}{dt} = rP.$$

where $P(t)$ is the total population at time t and r is the growth rate. This law cannot be applicable to situations in which population competes for resources like space and food.

To overecome this problem, Verhulst [6] proposed the following model:

$$\frac{dP}{dt} = r(1-\frac{P}{K})P.$$

where r is as before the growth rate and K is a carrying capacity of the environment related to resources. The disadventage of the Malthus and Verhulst models is that they are age independent.

Consider a population dispersing in a bounded domain of $\mathbb{R}^n$ with $n \leq 3$. Let $u(x,t,a)$ be the density of population having at time $t > 0$ the age a, and located at the geographic position $x \in \Omega$. We are going to study existence, uniqueness and the asymptotic behavior of the solution of this equation:

$$\frac{\partial u}{\partial t} + \frac{\partial u}{\partial a} = \Delta u - \mu(a)u(x,t,a) - \mu_e(P(x,t))u(x,t,a). \tag{1}$$

S. Melliani and O. Castillo (Eds.): ICPAMS 2021, LNNS 476, pp. 240–254, 2023.
https://doi.org/10.1007/978-3-031-12416-7_20

where,

$$P(x,t) = \int_0^{\infty} u(x,t,a)da. \tag{2}$$

- $P = P(x,t)$ is the total population at time t and at position x.
- $\mu_n(a)$ is the probability of dying of natural causes at age a. $\mu_n(a) \geq 0$.
- One defines the function π in the interval $[0,\infty[$, by

$$\pi(a) = \exp(-\int_0^{a} \mu(\sigma)d\sigma).$$

  $\pi(a)$ is the probability of living to age a.One remarks that: $\pi(0) = 1$, π is decreasing and $\lim_{a \longrightarrow +\infty} \pi(a) = 0$.
- $\mu_e(P)$ is the probability of death due to environmental factors. μ_e is a function of the total population, such that $0 \leq \mu_e(P) \leq \overline{\mu}$, where $\overline{\mu}$ is a minimum mortality rate.
- The birth low is given by

$$u(x,t,0) = \int_0^{\infty} \beta(a)u(x,t,a)da. \tag{3}$$

- β is called the birth-modulus which represents the fertility of the population. One assumes that β has a compact support in $[0,\infty[$, so that:

$$Supp\beta \subset [0,A].$$

  Where

$$A = \max\{a; \beta(a) > 0\} < \infty,$$

  is the minopause age.

We can consider several types of boundary value problems:

- Dirichlet boundary conditions:

$$u(x,t,a) = 0 \qquad \text{for } x \in \partial\omega.$$

  This type of boundary conditions corresponds to uninhabited spatial area.
- Neumann boundary condition:

$$\frac{\partial u}{\partial n} = 0 \qquad x \in \partial\omega, \tag{4}$$

  which corresponds to spatial area without extrenel exchange, around which the population can be dense but through which there is almost no population flow.

Here we analyze a model of the second type.

We give the initial condition:

$$u(x,0,a) = u_0(x,a). \tag{5}$$

$u_0(x,a)$ is the initial distribution of the population at position $x \in \Omega$ and age a. The purpose of this work is to study the asymptotic behavior of the solution of the parabolic problem 1, 2, 3, 4, 5.

The behavior will be expressed in term of the parameter λ_1 which represents the pure diffusion effect, it is the smallest eigenvalue of the Neumann problem for $-\Delta$ in Ω, $(\lambda_1 \geq 0)$. And the parameter $r^\star$ which represents the pure population growth effect, which is root of the characteristic equation:

$$\int_0^\infty \beta(a)\pi(a)\exp(-r^\star a)da = 1. \tag{6}$$

In the following, we will establish some conditions on μ_n and μ_e to have the existence of $r^\star$.

2 Assumptions and Notations

2.1 Assumptions

(H1) We assume that

$$0 \leq \beta(a) \leq \beta_1 < \infty \text{ on } [0,\infty[.$$

where β_1 is a constant real. And,

$$\int_0^\infty \beta^2(a)da < \infty. \tag{7}$$

(H2) We assume that,

$$0 \leq \mu_n(a) \leq \mu_1 < \infty \text{ on } [0,\infty[.$$

where μ_1 is a constant real.
(H3) μ_e is a positive, continuous and locally lipschitz function on $[0,\infty[$. We assume also that it is an increasing function on $[0,\infty[$.
(H4) We assume that,

$$0 \leq u_0(x,a) \leq \hat{u}_0(a) \text{ with } \hat{u}_0 \in L^2(0,\infty),$$

and $u_0 \in L^2(\Omega \times (0,\infty)) \cap L^1(\Omega \times (0,\infty))$.
(H5) $0 \leq P_0(x) \leq M < \infty$.

2.2 Notations

Let $X = L^2(\Omega \times \mathbb{R})$,and $A = -\Delta : D(A) \longrightarrow X$,
where

$$D(A) = \{\phi \in H^2(\Omega \times \mathbb{R}); \frac{\partial \phi}{\partial n} = 0 \text{ sur } \partial\Omega \times \mathbb{R}\}.$$

Let α be a real number such that $\frac{3}{4} < \alpha < 1$ and X^α the fractional space associated with A, defined by $X^\alpha = D(A^\alpha)$, and $D(X^\alpha) = D(A^\alpha)$.

3 Main Results

Proposition 1 [1]. *For $n \leq 3$ and $\frac{3}{4} < \alpha < 1$, we have,*

$$X^{\alpha} \subset H^{1}(\Omega) \cap L^{\infty}(\Omega).$$

With continuous injection.

Proof. According to Nirenberg-Gagliardo's inequality [3], for $u \in H^2(\Omega)$ and $\frac{1}{2} < \theta < 1$, we have

$$\|u\|_{H^1} \leq c\|u\|_{H^2}^{\theta}\|u\|_{L^2}^{1-\theta}. \tag{8}$$

And from ([1] p. 177), for $v \in D(A)$, we have

$$\|v\|_{H^1} \leq c_1\|Av\|_{L^2}^{\theta}\|v\|_{L^2}^{1-\theta}. \tag{9}$$

where c and c_1 are positive constants.

Let $v \in X^{\alpha}$, then there exists $w \in L^2(\Omega)$ such that $v = A^{-\alpha}w$.
From the definition of $A^{-\alpha}$, we can write:

$$v = \frac{1}{\Gamma(\alpha)}\int_0^{\infty} t^{\alpha-1}T(t)w dt.$$

By replacing w with $A^{\alpha}v$, we obtain:

$$A^{-1}v = \frac{1}{\Gamma(\alpha)}\int_0^{\infty} t^{\alpha-1}T(t)A^{\alpha-1}v dt.$$

From the analiticity of the semigroup $(T(t))_{t\geq 0}$, we deduce that, $\exists K > 0$, such that

$$\|A^{-1}v\|_{H^1} \leq K\|A^{\alpha-1}v\|_{H^1}.$$

And according to 9, $\exists K' > 0$, such that

$$\|A^{-1}v\|_{H^1} \leq K'\|A^{\alpha}v\|_{L^2}^{\theta}\|A^{\alpha-1}v\|_{L^2}^{1-\theta}.$$

If $u \in X^{\alpha+1}$, then, $Au \in X^{\alpha}$, and as a result

$$\|u\|_{H^1} = \|A^{-1}Au\|_{H^1} \leq K'\|A^{\alpha}Au\|_{L^2}^{\theta}\|A^{\alpha-1}Au\|_{L^2}^{1-\theta}.$$

Which means,

$$\|u\|_{H^1} \leq K'\|A^{\alpha+1}u\|_{L^2}^{\theta}\|A^{\alpha}u\|_{L^2}^{1-\theta},$$

then,

$$\|u\|_{H^1} \leq K'\|u\|_{X^{\alpha+1}}^{\theta}\|u\|_{X^{\alpha}}^{1-\theta},$$

however, $X^{\alpha+1} \hookrightarrow X^{\alpha}$. Then, there exists a constant $k > 0$, such that

$$\|u\|_{X^{\alpha}} \leq k\|u\|_{X^{\alpha+1}}.$$

Finally, we have,

$$\|u\|_{H^1} \leq C\|u\|_{X^\alpha}.$$

Where C is a constant, as a result,

$$X^\alpha \hookrightarrow H^1(\Omega).$$

Let us now show that, $X^\alpha \hookrightarrow L^\infty$.

Let $u \in X^\alpha$, then, there exists $w \in L^2(\Omega)$ such that $u = A^{-\alpha}w$.

According to ([1], p. 100), we can write

$$u = \frac{\sin \pi\alpha}{\pi} \int_0^\infty t^{-\alpha}(tId + A)^{-1} w dt. \tag{10}$$

$-A$ is a dissipative operator in $L^2(\Omega)$, then for every $(tI+A)^{-1}w \in D(A)$, we have

$$\|A(tI+A)^{-1}w\| \leq \|w\|. \tag{11}$$

And $\forall t \geq 0$, $\exists M \geq 0$;

$$\|(tI+A)^{-1}w\| \leq (t+M)^{-1}\|w\|. \tag{12}$$

From Sobolev's interpolation inequalities in the spaces $L^p(\Omega)$, we can write for $\theta = \frac{1}{4}$:

$$\|(tI+A)^{-1}w\|_\infty \leq c\|A(tI+A)^{-1}w\|_{L^2}^{\frac{3}{4}} \times \|(tI+A)^{-1}x\|_{L^2}^{\frac{1}{4}}. \tag{13}$$

From 10 and 13, we deduce that

$$\|A^{-\alpha}w\|_\infty \leq c\left|\frac{\sin \pi\alpha}{\pi}\right| \int_0^\infty t^{-\alpha}\|w\|_{L^2}^{\frac{3}{4}} \times (t+M)^{-\frac{1}{4}}\|w\|_{L^2}^{\frac{1}{4}} dt,$$

then,

$$\|A^{-\alpha}w\|_\infty \leq k\int_0^\infty t^{-\alpha} \times (t+M)^{-\frac{1}{4}} dt \times \|w\|_{L^2},$$

where k is a positive constant, however,

$$\int_0^\infty t^{-\alpha} \times (t+M)^{-\frac{1}{4}} dt \leq \int_0^\infty t^{-(\alpha+\frac{1}{4})} dt = \left[\frac{t^{-(\alpha+\frac{1}{4})+1}}{-(\alpha+\frac{1}{4})+1}\right]_0^\infty.$$

As $\alpha > \frac{3}{4}$, then the integral $\int_0^\infty t^{(-\alpha+\frac{1}{4})dt}$ converges.

So there exists $K > 0$,

$$\|A^{-\alpha}w\|_\infty \leq K\|w\|_{L^2}.$$

i.e.

$$\|u\| \leq K\|A^\alpha u\|_{L^2}.$$

Finally,

$$X^\alpha \hookrightarrow L^\infty(\Omega).$$

3.1 Separable Solution

We first define the notion of separable solution of the problem 1, 2, 3, 4, 5, which is a solution that can be written as,

$$u(x,t,a) = \psi(a)P(x,t). \tag{14}$$

With the normalisation,

$$\int_0^\infty \phi(a)da = 1. \tag{15}$$

3.1.1 Necessary and Sufficient Condition for the Existence of Separable Solution

For the existence of $r^\star$, the root of the characteristic equation 6; we assume that the hypothesis **(H1)** and **(H2)** are satisfied. Consider the function

$$\phi(r) = \int_0^\infty \beta(a)\pi(a)e^{-ra}da,$$

this function is continuous, and satisfies,
$\lim_{r \longrightarrow -\infty} \phi(r) = +\infty$ and $\lim_{r \longrightarrow +\infty} \phi(r) = 0$.
And according to the intermediate value theorem, there exists a unique real number $r^\star$, such that $\phi(r^\star) = 1$.

Lemma 1. *Let $r^\star$ be the root of 6. There exists a nontrivial separable solution of the problem 1–5, if and only if*

$$\int_0^\infty \pi(a)e^{-r^\star a}da < +\infty. \tag{16}$$

Remark 1. $\pi(a) = \exp(-\int_0^a \mu_n(\sigma)d\sigma)$, thus,

$$\int_0^\infty \pi(a)\exp(-r^\star a)da = \int_0^\infty \exp\left[-(r^\star a + \int_0^a \mu_n(\sigma)d\sigma)\right] da.$$

If $r^\star \leq 0$, then the condition 16 is satisfied.

Remark 2. If for every $a > 0$, $0 < \overline{\mu} \leq \mu_n(a)$ ($\mu_n(a)$ can represent a minimum natural mortality rate). Then, if $r^\star > -\overline{\mu}$, the condition 16 is satisfied.
Indeed,

$$\int_0^\infty \pi(a)\exp(r^\star a)da \leq \int_0^\infty exp\left[-(r^\star + \overline{\mu})\right] da < +\infty.$$

Proof. Let $u(x,t,a) = \psi(a)P(x,t)$ be a separable solution, then $\int_0^\infty \phi(a)da = 1$.
However, $\psi(a) = \psi_0\pi(a)\exp(-r^\star a)$, thus,

$$\psi_0 \times \int_0^\infty \pi(a)\exp(-r^\star a)da = 1.$$

Terefore,

$$\psi_0^{-1} = \int_0^\infty \pi(a)\exp(-r^\star a)da < \infty$$

Inversely, if $\int_0^{\infty} \pi(a)\exp(-r^{\star}a)da < \infty$. Let

$$\psi_0 = (\int_0^{\infty} \pi(a)\exp(-r^{\star}a)da)^{-1},$$

and,

$$\psi(a) = \psi_0 \pi(a)\exp(-r^{\star}a).$$

It remains to show the existence of a function $P(x,t)$, such that the function $u(x,t,a) = \psi(a)P(x,t)$ will be a separable solution. We will define this function in the following paragraph.

3.1.2 Equations Equivalent to the Problem 1, 2, 3, 4, 5

Let u be a separable solution to the problem 1, 2, 3, 4, 5. By replaciny u with $\psi \times P$, in the Eq. 1, we obtain,

$$-\frac{1}{\psi}\frac{d\psi}{da} - \mu_n(a) = -\frac{1}{P}\Delta P + \frac{1}{P}\frac{\partial p}{\partial t} + \mu_e(P).$$

In this equality the left term depends only on a, while the right term depends only on (x,t). Then these two terms are constant and equal to a constant r. This leads to the following two equations.

$$-\frac{1}{\psi}\frac{d\psi}{da} - \mu_n(a) = r, \tag{17}$$

and,

$$-\frac{1}{P}\Delta P + \frac{1}{P}\frac{\partial p}{\partial t} + \mu_e(P) = r. \tag{18}$$

The Eq. 17 can be written as:

$$\frac{d\psi}{da} = -\psi(a)(\mu_n(a) + r), \text{ on } [0, \infty[.$$

This equation has the solution,

$$\psi(a) = \psi_0 \pi(a)\exp(-ra). \tag{19}$$

From 15, we deduce that

$$\psi_0 = (\int_0^{\infty} \pi(a)\exp(-ra)da)^{-1}. \tag{20}$$

The condition 3

$$u(x,t,0) = \int_0^{\infty} \beta(a)u(x,t,a)da,$$

can be written as

$$\begin{aligned}\psi(0)P(x,t) &= P(x,t)\int_0^{\infty} \beta(a)\psi(a)da \\ &= P(x,t)\int_0^{\infty} \beta(a)\psi(a)\pi(a)\exp(-ra)da\end{aligned}$$

Thus, $\int_0^{\infty} \beta(a)\pi(a)e^{-ra}da = 1$, and by uniqueness of the solution of the characteristic equation 6,

$$r = r^{\star}.$$

Hence, the solution of the Eq. 17 is:

$$\psi(a) = \psi_0 \pi(a) e^{-r^{\star}a},$$

where ψ_0 is given by 20.

3.1.3 The Study of the Equation 18

We take $r = r^{\star}$. The Eq. 18 can be written as

$$\begin{cases} \frac{\partial P}{\partial t} = \Delta P + P(r^{\star} - \mu_e(P)) \text{ in } \Omega \times]0,\infty[. \\ P(x,0) = P_0(x) \text{ in } \Omega \\ \frac{\partial P}{\partial \nu} = 0 \text{ for } (x,t) \in \partial\Omega \times]0,\infty[. \end{cases} \tag{21}$$

Theorem 1. *Under the assumptions made on $r^{\star}$ and μ_e, the parabolic problem 21 has a unique solution P.*

Proof. Let f be the function defined by

$$f(P) = P(r^{\star} - \mu_e(P)).$$

As the function μ_e is locally lipschitz in $[0,\infty[$, then f is also locally lipschitz. $A = -\Delta$ is sectoriel. In addition f is boundning (i.e. it transforms a bounded set to a bounded set). The conditions of the theorem (3.1.5) [4] are satisfied. Hence, the parabolic equation 21 has a unique solution.

In what follows, we will denote by $I = [0, \frac{3}{2}\mu_e^{-1}(r^{\star})]$ and by f the function

$$f(P) = P(r^{\star} - \mu_e(P)).$$

Theorem 2. *We assume that $P_0 \in X^{\alpha}$ $(\frac{3}{4} < \alpha < 1)$ and for every $x \in \Omega$, $P_0(x) \in I$. Then:*

There exist two constants $k > 0$ and $C > 0$; such that the solution $P(.,t,P_0)$ of the Eq. 21 and passing through P_0, satisfies the following property:

$$\|P(.,t,P_0) - \overline{P}(t)\|_{X^{\alpha}} \leq Ke^{-Ct}, \textit{for } t \geq 0. \tag{22}$$

Where

$$\overline{P}(t) = \frac{1}{|\Omega|}\int_{\Omega} P(x,t,P_0)dx. \tag{23}$$

The function $\overline{P}$ satisfies the equation,

$$\frac{d\overline{P}(t)}{dt} = f(\overline{P}(t)) + g(t). \tag{24}$$

With,

$$|g(t)| \leq Ke^{-Ct}, \textit{for } t \geq 0. \tag{25}$$

Remark 3. We have $X^\alpha \hookrightarrow H^1 \cap L^\infty$, as $\frac{3}{4} < \alpha < 1$. For this reason we assumed that $\forall x \in \Omega,\ P_0(x) \in I$.

Proof. We consider the following decomposition of the space X^α,

$$X^\alpha = \mathbf{R} \oplus W.$$

where R is the set of constant functions (in space), and W is the orthogonal complement of $\mathbf{R}$ in X^α, with respect to the scalar product induced by L^2.

If $P = v + w \in X^\alpha$, with $v \in \mathbf{R}$ and $w \in W$, then,

$$v = \frac{1}{|\Omega|} \int_\Omega P(x,t)dx \text{ and } \int_\Omega w(x)dx = 0.$$

Let $P(t,.)$ be the solution of the Eq. 21 passing from P_0. We have,

$$P(.,t) = v(t) + w(.,t).$$

where v and w verify the following system of equations,

$$\begin{cases} \frac{\partial v(t)}{\partial t} = R(v(t), w(.,t)) \text{ for } t \geq 0 \\ \frac{\partial w(t,x)}{\partial t} = \Delta w + S(v(t); w(x,t)), \text{ for } t \geq 0,\ x \in \Omega \end{cases} \tag{26}$$

where for $v \in \mathbf{R}$ and $w \in W$,

$$R(v,\phi) = \frac{1}{|\Omega|} \int_\Omega f(v + \phi(x))dx, \tag{27}$$

and,

$$S(v,\phi)(x) = f(v + \phi(x)) - R(v,\phi). \tag{28}$$

we will need the following lemma.

Lemma 2. *Under the assumptions of the Theorem 2, the function v defined by the system 26 satisfies*

$$v(t) \in I = [0, \frac{3b}{2}],\ \forall t \geq 0.$$

where $b = \mu_e^{-1}(r^\star)$.

Proof. For $\alpha > \frac{3}{4}$, $X^\alpha \hookrightarrow L^\infty$. Thus there exists $K > 0$, such that

$$\|\phi\|_{L^\infty} \leq K\|\phi\|_{X^\alpha} \text{ for } \phi \in X^\alpha. \tag{29}$$

We consider the operator A, defined as,

$$\begin{cases} Au = -\Delta u \text{ in } \Omega, \\ \frac{\partial u}{\partial v} = 0 \text{ in } \partial\Omega. \end{cases}$$

Let $(T(t))_{t\geq 0}$ be the analytic semigroup generated by A in X^α. From the analiticity of $(T(t))_{t\geq 0}$, there exists a constant $K_1 > 0$ such that,

$$\|T(t)Z\|_{X^\alpha} \leq K_1 e^{-\lambda_1 t}\|Z\|_{X^\alpha}. \tag{30}$$

We define the constants M and L by,

$$M = \sup_{v \in I} |f'(v)|. \tag{31}$$

and,

$$L = \int_0^{\infty} u^{-\alpha} e^{-(\lambda_1 - \delta)y} dy. \tag{32}$$

where $\delta < \lambda_1$ is chosen in such away that,

$$1 - 2KK_1ML|\Omega|^{\frac{1}{2}} > 0. \tag{33}$$

Let $b = \mu_e^{-1}(r^\star)$ and $K_{2b} = \frac{K_1}{1-2KK_1ML|\Omega|^{\frac{1}{2}}}$.

In the assumptions of the lemma, we supposed that $\forall x \in \Omega,\ P_0(x) \in I$.

Then, $v_0 < \frac{3b}{2}$ and $\|w\|_\infty < \frac{b}{2K_{2b}}$.

Let w be a solution of the equation,

$$\begin{cases} \frac{\partial w}{\partial t} = \Delta w + S(v,w) \text{ in } \Omega \times [0,\infty[. \\ \frac{\partial w}{\partial v} = 0 \text{ in } \partial\Omega \times [0,\infty[. \\ w(x,0) = w_0(x) \text{ in } \Omega. \end{cases}$$

If $\|w(.,t)\|_\infty \leq K_{2b}\|w_0\|_\infty$ for every $t \geq 0$ such that $\max_{0\leq s\leq t} v(s) < 2b$.

Then $\|w\|_\infty \leq \frac{b}{2}$ as long as $v(t) < 2b$.

Thus

$$v < 2b \Rightarrow \|w\|_\infty \leq \frac{b}{2}.$$

Assume that,

$$\frac{3b}{2} < \max_{0\leq s\leq t} v(s) < 2b,\ t \geq 0. \tag{34}$$

Then, there exists $t_0 \geq 0$, such that,

$$\frac{3b}{2} < v(t_0) = \max_{0\leq s\leq t} v(s) < 2b,\ t \geq 0.$$

On remarks that $t_0 > 0$ since $v(0) < \frac{3b}{2}$. Hence $v(t_0) + w(t_0) \geq b$, which implies that, $\mu_e(v(t_0) + w(t_0)) > r^\star$ since μ_e is increasing.

$\int_\Omega w = 0$, so $\int_\Omega \frac{\partial w}{\partial t} = 0$. $P = v + w$, and $\int_\Omega \Delta P = 0$. Thus

$$\int_\Omega \frac{\partial P}{\partial t} = \int_\Omega \frac{\partial v}{\partial t} = \int_\Omega P(r^\star - \mu_e(P))dx,$$

hence,

$$\frac{dv}{dt} = \frac{1}{|\Omega|}\int_\Omega P(r^\star - \mu_e(P))dx.$$

However, $\mu_e(v(t_0) + w(t_0)) > r^\star$. Then,

$$\frac{dv}{dt}(t_0) < 0.$$

To conclude, we need the following lemma,

Lemma 3. *(Zygmund's Lemma [2])*
If v is a continuous function in an interval $J=[a,b]$ and there exists $M\geq 0$ such that:

$$\lim_{h\longrightarrow 0}\sup\frac{v(t+h)-v(t)}{h}\leq 0,$$

in every point t, where $v(t)\geq M$. And $v(a)\leq M$. Then, we have, $v(t)\leq M$ for every $t\geq 0$.

By applying Zygmund's Lemma, we obtain $v(t)<\frac{3b}{2}$, for every $t\geq 0$, which contradicts the assumption 34. Finally $v(t)\in I$, $\forall t\geq 0$.

Let us go back to the proof of the theorem.

Let $\psi\in w$ and $v\in I$, then,

$$(\frac{\partial S}{\partial\phi}(v,\phi)\psi)(x)=f'(v+\phi(x))\psi(x)-\frac{1}{|\Omega|}\int_{\Omega}f'(v+\phi(y))\psi(y)dy.$$

Thus, for $\phi\in W$, such that,

$$v+\phi(x)\in I,\ \forall x\in\Omega.$$

From 31 and (29),

$$\Big\|\frac{\partial S}{\partial\phi}(v,\phi)\Big\|_{X^\alpha}\leq 2MK|\Omega|^{\frac{1}{2}}\|\psi\|_{X^\alpha}.$$

Hence,

$$\Big\|\frac{\partial S}{\partial\phi}(v,\phi)\psi\Big\|_{X^\alpha}\leq 2MK|\Omega|^{\frac{1}{2}}. \tag{35}$$

Using the formula of variation of contants, we can write,

$$w(t)=T(t)w_0+\int_0^t T(t-s)[S(v(s),w(s))]ds.$$

From 30 and 35, we obtain,

$$\|w(t)\|_{X^\alpha}\leq K_1e^{-\lambda_1 t}\|w_0\|_{X^\alpha}+2KK_1M|\Omega|^{\frac{1}{2}}\int_0^t e^{-\lambda_1(t-s)}(t-s)^{-\alpha}\|w(s)\|_{X^\alpha}ds.$$

We consider the function $z(t)$, defined as,

$$z(t)=\sup_{0\leq s\leq t}e^{\delta s}\|w(s)\|_{X^\alpha}.$$

Thus, from 32, we have

$$e^{\delta t}\|w(t)\|_{X^\alpha}\leq K_1e^{-(\lambda_1-\delta)t}\|w_0\|_{X^\alpha}+2KK_1LM|\Omega|^{\frac{1}{2}}\times z(t).\ \forall t\geq 0.$$

According to 33, we have

$$z(t)\leq\frac{K_1}{1-2KK_1ML|\Omega|^{\frac{1}{2}}}\|w_0\|_{X^\alpha}.$$

We deduce that,

$$\|w(t)\|_{X^\alpha} \leq \frac{K_1}{1-\chi} e^{-\delta t} \|w_0\|_{X^\alpha}.$$

where $\chi = 2KK_1ML|\Omega|^{\frac{1}{2}}$.

With the choice made on v_0, and w_0, we have $\|w(t)\|_\infty \leq K_{2b}\|w_0\|_{X^\alpha}$, and $\lim_{t \longrightarrow \infty} \|w(t)\|_{X^\alpha} = 0$, exponentially.

In conclusion,

$$\|P(.,t,P_0) - v(t)\|_{X^\alpha} = \|w(t)\|_{X^\alpha} \leq Ke^{-\delta t}, \ \forall t \geq 0.$$

with,

$$v(t) = \frac{1}{|\Omega|} \int_\Omega P(.,t,P_0)dx.$$

and

$$\frac{dv(t)}{dt} = f(v(t)) + [R(v(t),w(t)) - f(v(t))].$$

and:

$$\|R(v(t),w(t)) - f(v(t))\| = \|S(v(t),w(t))\| \leq 2KM|\Omega|^{\frac{1}{2}} \|w(t)\|_{X^\alpha}.$$

This expression tends to 0 when $t \longrightarrow \infty$. Which ends the proof.

Proposition 2. *Let $P(t,.) = v(t) + w(t)$ be a non null solution of the Eq. 20. Then, we have:*

$$\lim_{t \longrightarrow +\infty} v(t) = b = \mu_e^{-1}(r^\star).$$

Proof. According to the Theorem 2, the function w satisfies: $\lim_{t \longrightarrow +\infty} w(t,.) = 0$ exponentially.

The function v is bounded in $[0,+\infty[$.

Assume that:

$$\lim_{t \longrightarrow \infty} \sup v(t) > b. \tag{36}$$

Then, $\exists \varepsilon > 0$, $\exists t_n \longrightarrow \infty$, such that, $v(t_n) \geq b + \varepsilon$.

However, $\lim_{t \longrightarrow +\infty} w(t,.) = 0$, then,

$$\exists n_0 \in \mathbb{N}, \ \forall n \geq n_0, \ \|w(t_n,.)\| \leq \varepsilon.$$

Then, for every $x \in \Omega$,

$$v(t_n) + w(t_n,x) \geq b.$$

As μ_e is increasing, these for every $x \in \Omega$, $\mu_e(v(t_n) + w(t_n,x)) > r^\star$. However,

$$\frac{dv}{dt} = \frac{1}{|\Omega|} \int_\Omega (v + w(x,.))(r^\star - \mu_e(v+w))dx.$$

Thus,

$$\forall n \geq n_0, \ \frac{dv}{dt}(t_n) < 0.$$

According to Zygmund's Lemma,

$$\lim_{t\longrightarrow\infty} \sup v(t_n) \leq b.$$

Which contradicts the assumption 36. Finally,

$$\lim_{t\longrightarrow\infty} \sup v(t) \leq b.$$

Let us show that:

$$\lim_{t\longrightarrow\infty} \inf v(t) \geq b.$$

Assume that $\lim_{t\longrightarrow\infty} \inf v(t) < b$. Then,

$$\exists \varepsilon > 0,\ \exists t_n > 0,\ t_n \longrightarrow \infty,\ v(t_n) \leq b - \varepsilon.$$

Applying the Zygmund's Lemma leads to a contradiction. Then $\lim_{t\longrightarrow\infty} \inf v(t) \geq b$. Hence,

$$\lim v(t) = b = \mu_e^{-1}(r^\star).$$

Proposition 3. *If $r^\star < \lambda_1 + \mu_e(0)$, and if P is a solution of the parabolic problem 21. Then, P satisfies,*

$$\lim_{t\longrightarrow\infty} P(.,t,P_0) = 0,$$

uniformly in Ω.

Proof. Let $P(t,.,P_0)$ be a solution of the problem:

$$\begin{cases} \frac{\partial P}{\partial t} = \Delta P + P(r^\star - \mu_e(P)) \text{ in } \Omega\times]0,\infty[. \\ P(x,0) = P_0(x) \text{ in } \Omega \\ \frac{\partial P}{\partial \nu} = 0 \text{ for } (x,t) \in \partial\Omega\times]0,\infty[. \end{cases}$$

This equation can be written as:

$$\frac{\partial P}{\partial t} = \Delta P + P(r^\star - \mu_e(0)) - P(\mu_e(P) - \mu_e(0)) \text{ in } \Omega\times]0,\infty[.$$

As $\mu_e(P) - \mu_e(0) \geq 0$, then,

$$0 \leq P \leq \tilde{P},$$

where $\tilde{P}$ is the solution of the equation

$$\begin{cases} \frac{\partial P}{\partial t} = \Delta P + P(r^\star - \mu_e(0)), \\ \frac{\partial P}{\partial \nu} = 0 \text{ for } (x,t) \in \partial\Omega\times]0,\infty[. \end{cases} \tag{37}$$

Let ϕ be an eigenfunction associated to λ_1, with the Neumann's condition. We recall that $\lambda_1 > 0$, and that ϕ has a constant sign, we will chose $\phi > 0$.

We define the function $U(t)$ by,

$$U(t) = \int_\Omega \tilde{P}(t,x)\phi(x)dx.$$

Then, we have

$$\frac{dU}{dt}(t) = \int_{\Omega} \frac{\partial \tilde{P}}{\partial t}(x,t)\phi(x)dx.$$

As $\tilde{P}$ is the solution of the Eq. 37, we deduce that:

$$\frac{dU}{dt}(t) = \int_{\Omega} \Delta \tilde{P}(x,t)\phi(x)dx + \int_{\Omega} (r^{\star} - \mu_e(0))\tilde{P}\phi(x)dx.$$

Which means

$$\frac{dU}{dt}(t) = [\lambda_1 + r^{\star} - \mu_e(0)]U(t). \tag{38}$$

Hence, if $r^{\star} < \lambda_1 + \mu_e(0)$, then the solution of the Eq. 38 is the function:

$$U(t) = c\exp(-\lambda_1 + r^{\star} - \mu_e(0))t.$$

This function tends to 0 when $t \longrightarrow \infty$.

As $\phi > 0$, the solution $\tilde{P}$ is such that

$$\lim_{t \longrightarrow \infty} \tilde{P}(x,t) = 0.$$

Finally,

$$\lim_{t \longrightarrow \infty} P(x,t) = 0.$$

Proposition 4. *The previous proposition is still valid for* $r^{\star} = \lambda_1 + \mu_e(0)$.

Proof. The Eq. 37 becomes:

$$\frac{\partial P}{\partial t} = \Delta P + \lambda_1 P.$$

Its solution $\tilde{P}$ is then bounded. However, $0 \leq P \leq \tilde{P}$, thus P is bounded.

In addition, as λ_1 is the smallest eigenvalue of $-\Delta$, then, the other eigenvalues of $\Delta + \lambda_1 I$ are negative. And hence,

$$\lim_{t \longrightarrow \infty} \tilde{P}(x,t) = 0.$$

And P tends also to 0 when $t \longrightarrow \infty$.

Corollary 1. *For* $r^{\star} > \lambda_1 + \mu_e(0)$, *the solution of the Eq. 20 satisfies:*

$$\lim_{t \longrightarrow \infty} P(x,t) = b = \mu_e^{-1}(r^{\star}).$$

4 Conclusion

In this work, we proved the existence and uniqueness of the solution of the problem 1, 2, 3, 4, 5, and we studied the asymptotic behavior of the solution.

References

1. Friedman, A.: Partial Differential Equations. Holt, Rinehart and Winston, New York (1969)
2. Ladzs, G.E., Lakshmikantham, V.: Differential Equations in Abstract Spaces. Academic Press (1972)
3. Brezis, H.: Analyse fonctionnelle. Collect. Math. Appl. Masson (1983)
4. Hilal, K.: Contribution à l'étude des équations parabpliques à retard. Application à la dynamique de population (1994)
5. Malthus, T.R.: An Essay on the Principle of Population (1878)
6. Verhulst, P.-F.: Notice sur la loi que la population poursuit dans son accroissement. Corresp. Math. Phys. **10**, 113–126 (1838)

An Epidemiological Model "Covid 19 British Variant"

Khadija Channan, Khalid Hilal(✉), and Ahmed Kajouni

Laboratory of Applied Mathematics Scientific Calculus, Sultan Moulay Slimane University, BP 523, 23000 Beni Mellal, Morocco
hilalkhalid2005@yahoo.fr

Abstract. Everyone is talking about coronavirus from last of months due to its exponential spread throughout the globe. So, in this work we are interested in the study of an epidemiological model "covid 19 british variant". More precisely, we made the Numerical simulation of our model.

1 Introduction

The COVID-19 (coronavirus disease of 2019) pandemic represents a global public health emergency unprecedented in recent history. Since the release of the first World Health Organization report describing the COVID-19 outbreak centered in Wuhan, China, a few months ago, the number of confirmed cases has risen sharply in all regions of the world. More than that, The emergence of mutations in the disease is much more serious and so far there are two mutations, the Indian mutation and the English variant mutation. And in this article, we will study this latter.

A variant much more contagious than the classic covid 19, according to a new English study published Wednesday, March 10, the English variant, which we already knew to be more contagious, would also be 64% more deadly. For 1000 cases detected, the English vary causes 4.1 deaths, against 2.5 for the original virus, conclude the authors of this work published in the medical journal BMJ.

This work aims to present a model describing a transmissible and infectious disease in epidemiology called Covid-19 british variant and its transmission factors.

2 Diagram Transmission of Covid-19 British Variant Between Human

Currently our community is divided into two parts, a category that is already had covid-19, so it has more immunity than the other class of individuals who did not get it.

S. Melliani and O. Castillo (Eds.): ICPAMS 2021, LNNS 476, pp. 255–258, 2023.
https://doi.org/10.1007/978-3-031-12416-7_21

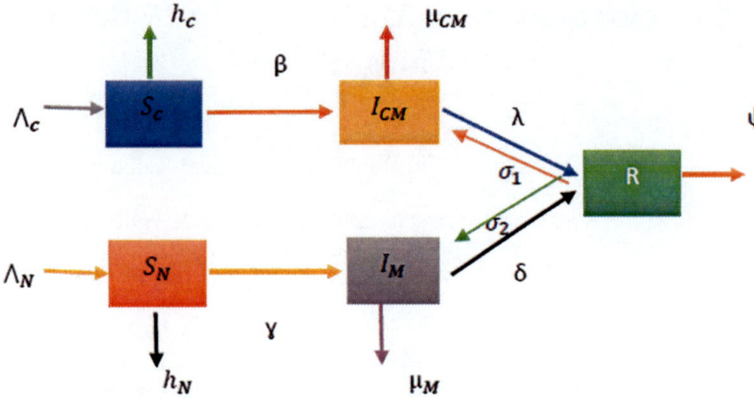

Description of Biological Parameters:

- S_C : The susceptible individuals who have already had covid-19
- S_N : The susceptible individuals who have not had the covid-19
- I_{CM} : The individuals who have previously infected by covid-19 and are now infected by the British variant of covid-19.
- I_M : The individuals infected by the covid-19 British variant and have not previously infected with covid-19.
- R : The individuals withdrawn (healed or dead).
- β : the rate of individuals who become infected by the covid-19 British variant and who had already contacted by the classical covid-19.
- γ: the rate of individuals who become infected by the covid-19 British variant.
- λ : recovery rate (for the people who are already immune by covid-19).
- δ : recovery rate (for the people who are not already infected by the classical covid-19).
- σ_1, σ_2 : the re-injury rate (for both compartments I_{CM} and I_M respectively).
- $h_C, h_N, \mu_{CM}, \mu_M, \psi$: Natural mortality rate (for the compartments S_C, S_N, I_{CM} , I_M *and* R respectively).
- Λ_N, Λ_C : Birth rate (for the compartments S_C, S_N respectively).

The dynamics of this model is given by the following system:

$$(1)\begin{cases} \dfrac{dS_C(t)}{dt} = \Lambda_C - \beta S_C(t)I_{CM}(t) - h_C S_C(t) \\ \dfrac{dS_N(t)}{dt} = \Lambda_N - \gamma S_N(t)I_M(t) - h_N S_N(t) \\ \dfrac{dI_{CM}(t)}{dt} = \beta S_C(t)I_{CM}(t) + \sigma_1 R(t) - \lambda I_{CM}(t) - \mu_{CM} I_{CM}(t) \\ \qquad - \psi I_{CM} \\ \dfrac{dI_M(t)}{dt} = \gamma S_N(t)I_M(t) + \delta R(t) - \sigma_2 I_M(t) - \mu_M I_M(t) \\ \qquad - \psi I_M(t) \\ \dfrac{dR(t)}{dt} = \lambda I_{CM}(t) + \delta I_M(t) - \sigma_1 I_{CM}(t) - \sigma_2 I_M(t) - \psi R(t) \end{cases}$$

The system (1) is provided with the initial conditions:

$$S_C(0) = S_{C0} \geq 0, \quad S_N(0) = S_{N0} \geq 0 \ \ , , I_{CM}(0) = I_{CM0} \geq 0,$$
$$I_M = I_{M0} \geq 0 \quad and \quad R(t) = R_0 \geq 0$$

3 Numerical Results and Discussion:

In this section, we present the numerical simulation of our model of the dynamics of "the British variant of covid-19" in a given population. In the following, we set a period of three months to study the evolution of the disease.

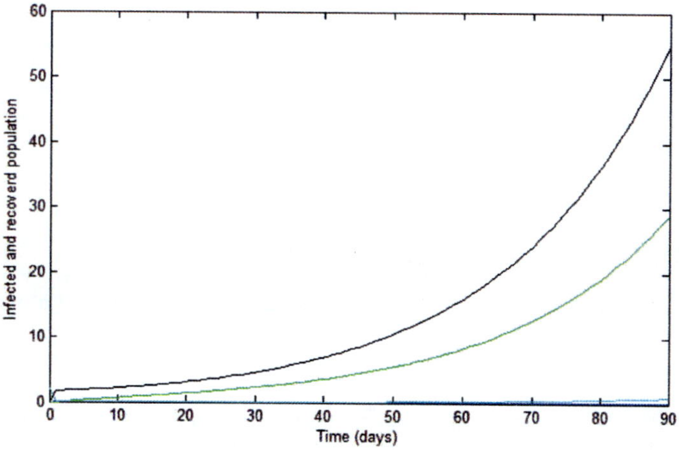

The plots of I_M, I_{CM} and R with respect to time t.

In this figure we see the evolution of infected cases of the covid-19 British variant. We notice that the curve which is in black (which represents I_M) is higher than the green curve (which represents I_{CM}) it comes down to acquired immunity following recovery from covid-19 disease. Moreover, we know that nowadays and thanks to the vaccination which plays a very important role in the reduction of the disease. On the other hand, the curve which is in blue represents the individuals which are retrofitted.

We studied the evolution of infected cases of the British variant of covid-19 using The parameter estimates are given in the following table:

!t

The settings	Estimate
β	0.004
Λ_C	0.01
Λ_N	0.01
γ	0.005
μ_{CM}	0.09
μ_M	0.05
σ_1	0.02
σ_2	0.003
h_C	0.3
h_N	0.2
ψ	0.0002
λ	1
δ	0.05

References

1. Bichara, D.: Study of epidemiological models: stability, observation and estimation of parameters. University of Lorraine (2013)
2. Hilal, K., Kajouni, A.: Boundary value problems for hybrid differential equations with fractional order. Adv. Differ. Eqns. **2015**, 1–19 (2015)
3. Hilal, K., Kajouni, A.: Boundary value problems for hybrid differential equations. Math. Theor. Model. (2015)
4. Hakiki Toubeish, K.: Simulation numérique par les ondelettes des modèles fractionnaires en épidémiologie (2018)

Existence and Uniqueness of Solutions for a Second-Order Iterative Fractional Conformable Boundary Value Problem

Khalid Hilal(✉), Manal Menchih, and Ahmed Kajouni

Sultan Moulay Slimane University, BP 523, 23000 Beni Mellal, Morocco
hilalkhalid2005@yahoo.fr

Abstract. Motivated by some papers treating the iterative differential equations, we disscuss the existence and uniqueness of solution of the sencond-order iterative fractional boundary value problem

$$\begin{cases} D^{\alpha}x(t) = f(t, x(t), x^{[2]}(t)), & t \in]0,1[, \\ x(0) = 0, \quad x(1) = 1. \end{cases}$$

where, $x^{[2]}(t) = x(x(t))$, $1 < \alpha < 2$ and D^{α} is the conformable derivative. The main tools employed to establish our results are the Schauder and Banach fixed point theorems. We give an example in order to illustrate this situation.

1 Introduction

During the past decades, fractional differential equations have attracted many authors [4–7, 11, 13]. The differential equations involving fractional derivatives in time, compared with those of integer order in time, are more realistic to describe many phenomena in nature (for instance, to describe the memory and hereditary properties of various materials and processes), the study of such equations has become an object of extensive study during recent years.

The notion of the conformable derivative was introduced by Khalil and Al [16]. This new fractional derivatives quickly becomes the subject of many contributions in sevral areas of science [5–7, 11, 13].

Iteration is a repeated action of the same operation. It exists extensively in mathematics, in science, in the nature and in the world.

The iterative differential equations are the equations involving derivative and iterate of unkown function. They can be used to describe infective disease processes and the motions of charged particles with retarded interactions [1], but since iterative differential equations are quite different from the usual differential equations, therefore the standard existence and uniqueness theorems cannot be applied.

The study of this equations can be traced back to papers by Petuhov [18] and Eder [1].

In 1965, Petuhov [18] considered the existence of solutions of the functional differential equation

$$x''(t) = \lambda x(x(t)),$$

S. Melliani and O. Castillo (Eds.): ICPAMS 2021, LNNS 476, pp. 259–266, 2023.
https://doi.org/10.1007/978-3-031-12416-7_22

under the condition that $x(t)$ maps the interval $[-T,T]$ into itself and that $x(0)=x(T)=\alpha$. He obtained conditions on λ and α for existence and uniqueness of solutions.

In 1984, Eder [1] considered the functional differential equation

$$x'(t) = x(x(t))$$

and proved that every solutions either vanishes identically or is strictly monotonic.

In 1990 and 1993, Fečkan and Wang [8,12] studied the equation

$$x'(t) = f(x(x(t))),$$

and they showed the existence of solutions by using the fixed-point theorems of Schauder and Banach.

In 2018, Kaufmann [2] established the existence and uniqueness of solutions of the functional differential equation

$$x''(t) = f(t,x(t),x(x(t))),$$

by using Shauder's fixed point.

Inspired by the above works and the fact that many dynamical systems are better characterized by the fractional derivatives, in this paper we consider the existence and uniqueness of solutions to the second-order iterative fractional boundary value problem

$$\begin{cases} D^{\alpha}x(t) = f(t,x(t),x^{[2]}(t)), & t \in]0,1[, \\ x(0)=0, \quad x(1)=1. \end{cases} \tag{1}$$

where, $x^{[2]}(t) = x(x(t))$, $f : [0,1]\times\mathbb{R}\times\mathbb{R}\to\mathbb{R}$ is a continuous map and the derivative is the conformable fractional derivative, with $1<\alpha<2$.

Due to the iterative term $x^{[2]}(t)$, in order for solution to be well-defined, we require that the image of x be in the interval $[0,1]$; that is, we need $0\le x(t)\le 1$, for all $t\in[0,1]$.

We arrange the rest of the paper as follows. In Sect. 2, we present some definitions of fractional calculus related to our work. Section 3 contains the main results, while example illustrating the main results are given in Sect. 4. Some interesting observations are presented in the concluding section.

2 Preliminary Results and Definitions

This section recalls some fundamental definitions of the conformable fractional derivatives and integrals and investigates their main characteristics as well.

Definition 1. [17] Let $\alpha\in(1,2]$, the conformable fractional derivative of a function $f : [0,+\infty)\to\mathbb{R}$, where $f'(t)$ exists, is defined by

$$f^{(\alpha)}(t) = \lim_{\varepsilon\to 0}\frac{f'(t+\varepsilon t^{2-\alpha}) - f'(t)}{\varepsilon}; \quad f^{(\alpha)}(0) = \lim_{t\to 0} f^{(\alpha)}(t). \tag{2}$$

Remark 1. [17]

1. *If the conformable fractional derivative of f of order α exists, then we say that f is α differentiable.*
2. *If f is 2-time differentiable, then*

$$f^{(\alpha)}(t) = t^{2-\alpha} f''(t). \tag{3}$$

In [11], we find the following properties

Proposition 1. [11] *Let f et g two α times differentiable function on $t > 0$, then we have*

1. $(af+bg)^{(\alpha)}(t) = af^{(\alpha)}(t) + bg^{(\alpha)}(t)$,
2. $(fg)^{(\alpha)}(t) = f^{(\alpha)}(t)g(t) + f(t)g^{(\alpha)}(t)$,
3. $(t^p)^{(\alpha)} = pt^{p-\alpha}$,
4. $\left(\frac{f(t)}{g(t)}\right)^{(\alpha)} = \frac{f^{(\alpha)}(t)g(t) - f(t)g^{(\alpha)}(t)}{g^2(t)}$,
5. *If $c \in \mathbb{R}$, then $c^{(\alpha)} = 0$.*

Next we give the definition of the fractional integral of any order $\alpha > 0$.

Definition 2. [17] Let $\alpha \in (1,2]$, then the fractional integral of a function $f : [0,+\infty) \to \mathbb{R}$ of order α is defined by

$$(I^\alpha f)(t) = \int_0^t (t-s)s^{\alpha-2} f(s)ds. \tag{4}$$

Lemma 1. [17] *Let $\alpha \in (1,2]$, and $f : (0,+\infty] \to \mathbb{R}$ be a continuous function. Then we have*

$$(I^\alpha f(t))^{(\alpha)} = f(t). \tag{5}$$

Lemma 2. [17] *Let $\alpha \in (1,2]$, and $f : (0,+\infty] \to \mathbb{R}$ be 2 times differentiable, Then we have*

$$I^\alpha \left(f^{(\alpha)}(t)\right) = f(t) - f(0) - f'(0)t. \tag{6}$$

3 Main Results

We first convert the boundary value problem (1) to a fixed point problem. To do, applying the conformable fractional integral to the both sides of the first equation of (1), we get

$$x(t) = x(0) + x'(0)t + \int_0^t (t-s)s^{\alpha-2} f(s,x(s),x^{[2]}(s))ds. \tag{7}$$

Using the value conditions, we find that

$$x'(0) = 1 - \int_0^1 (1-s)s^{\alpha-2} f(s,x(s),x^{[2]}(s))ds. \tag{8}$$

Intersecting (8) in (7), we find

$$x(t) = t + \int_0^t (t-1)s^{\alpha-2} f(s, x(s), x^{[2]}(s))ds$$
$$- \int_t^1 t(1-s)s^{\alpha-2} f(s, x(s), x^{[2]}(s))ds.$$

Then, x satisfies the integral solution.

$$x(t) = t + \int_0^1 G(t,s) f(s, x(s), x^{[2]}(s))ds, \tag{9}$$

where,

$$G(t,s) = \begin{cases} s^{\alpha-1}(t-1), & 0 \le s \le t \le 1, \\ -t(1-s)s^{\alpha-2}, & 0 \le t \le s \le 1. \end{cases} \tag{10}$$

Motivate by the above calculus, we can introduce the following lemma.

Lemma 3. *x is a solution of the problem* (1) *if only and if x is a solution of* (9).

Proof. Suppose that x is a solution of the integral solution (9). It's clear that $x(0) = 0$ and $x(1) = 1$. Since $f \in C([0,1] \times \mathbb{R} \times \mathbb{R}, \mathbb{R})$, then by application of D^{α}, we obtain

$$D^{\alpha}x(t) = D^{\alpha}\left(\int_0^t (t-1)s^{\alpha-1} f(s, x(s), x^{[2]}(s))ds\right) + D^{\alpha}\left(\int_t^1 t(s-1)s^{\alpha-2} f(s, x(s), x^{[2]}(s))ds\right)$$
$$= f(t, x(t), x^{[2]}(t)).$$

□

We donote $X = C([0,1], \mathbb{R})$ the Banach space of all continuous functions from $[0,1]$ into $\mathbb{R}$ with the norm

$$\|y\| = \sup\{|y(t)|; t \in [0,1]\}.$$

Define the operator $T : X \to X$ by

$$(Tx)(t) = t + \int_0^1 G(t,s) f(s, x(s), x^{[2]}(s))ds. \tag{11}$$

We have the following lemma.

Lemma 4. *The function x is a solution of the problem* (1) *if only and if $0 \le Tx(t) \le 1$ and x is a fixed point of T.*

Proof. Just use the Lemma 3 and the fact that the solution is well defined.

To establish our existence and uniqueness results, we will need the following results concerning the Green function.

Lemma 5. *The Green's function $G(t,s)$ given by* (10) *satisfies the following properties*

1. $G(t,s) \in C([0,1] \times [0,1])$,

2. $|G(t,s)| \leq |G(s,s)|$, *for all* $t,s \in [0,1]$,
3. $\int_0^1 |G(t,s)| < 1$.

Proof. 1. By expression of G, this function is belong to $C([0,1] \times [0,1])$.
2. • If $0 \leq s \leq t \leq 1$, we have $|G(t,s)| = s^{\alpha-1}(1-t)$. Then

$$\frac{\partial |G(t,s)|}{\partial t} = -s^{\alpha-1} \leq 0.$$

Which implies that

$$|G(t,s)| \leq |G(s,s)|.$$

• Now if $0 \leq t \leq s \leq 1$, we have $|G(t,s)| = t(1-s)s^{\alpha-2}$. Then

$$\frac{\partial |G(t,s)|}{\partial t} = (1-s)s^{\alpha-2} \geq 0.$$

Which implies that

$$|G(t,s)| \leq |G(s,s)|.$$

Which complete the proof.
3. We have

$$\begin{aligned}\int_0^1 G(s,s)ds &= \int_0^1 s^{\alpha-1}(s-1)ds \\ &= B(\alpha,2) < 1.\end{aligned}$$

□

We set the following assumptions
(H1) There exists $A > 0$ such that

$$|f(t,u,v)| < A,$$

for all $t \in [0,1], u, v \in \mathbb{R}$, and $1 - A\left(\frac{1}{\alpha} - \frac{1}{\alpha-1}\right) > 0$.
(H2) There exists $M, N > 0$ such that

$$|f(t,u_1,v_1) - f(t,u_2,v_2)| < |u_1 - u_2| + |v_1 - v_2|,$$

for all $t \in [0,1], u_1, u_2, v_1, v_2 \in \mathbb{R}$.
We are now ready to state our first result which will poved by Shauder fixed point.

Theorem 1. *Assume that hypothesis **(H1)** holds. Then there exist a solution of the fractional second-iterative boundary value problem* (1).

Proof. We define a subset S by

$$S = \{x \in X; \|x\| \leq N\}.$$

where $\dfrac{N}{A+1} \geq 1.$

Obviously, S is a closed convex subset.
We split the proof in two steep.
Step 1: Let $t \in [0,1]$. Since **(H1)** holds,

$$(Tx)'(t) = 1 + \int_0^t s^{\alpha-1} f(s,x(s),x^{[2]})ds - \int_0^t s^{\alpha-2} f(s,x(s),x^{[2]})ds.$$
$$\geq 1 - A[\tfrac{1}{\alpha} - \tfrac{1}{\alpha-1}] > 0.$$

Consequently Tx is increasing, and since $Tx(0) = 0$ and $Tx(1) = 1$, then

$$0 \leq Tx(t) \leq 1, \quad \text{for all} \quad t \in [0,1].$$

Step 2: We show that $T(S) \subset S$.
Let $t \in [0,1]$ and $x \in S$,

$$|Tx(t)| \leq \sup\{t + \int_0^1 G(t,s) f(s,x(s),x^{[2]})ds\}$$

$$\leq 1 + A \int_0^1 |G(s,s)| ds$$

$$\leq 1 + A \leq N,$$

which, on taking the taking the norm of $t \in [0,1]$, yields

$$\|Tx\| \leq N.$$

An application of Schauder's theorem yields a fixed point of the operator T. □
By lemma (4), the function x is the solution of (1).

Now, we consider uniqueness of solutions of problem (1). To this end, we need the Banach fixed point to show our second result.

Theorem 2. *Suppose that* **(H1)** *and* **(H2)** *hold. Assume that* $M + N < 1$. *Then there exists a unique solution of the problem* (1).

Proof. Since **(H1)** holds, we have $T(S) \subset S$.
Let consider $x, y \in X$. Then for all $t \in [0,1]$, we obtain

$$|Tx(t) - Ty(t)| \leq \int_0^1 |G(t,s)||f(s,x(s),x^{[2]}) - f(s,y(s),y^{[2]})| ds$$

$$\leq \int_0^1 |G(t,s)|[M\|x-y\| + N\|x-y\|]$$

$$\leq (M+N)\|x-y\|,$$

which shows that T is a contraction. Hence we deduce by Banach contraction mapping principle that the operator has a unique fixed point which correspond to a unique solution of problem (1) on $[0,1]$. this complete the proof. □

4 Illustrated Example

Consider the following fractional differential equation

$$D^{\frac{3}{2}}x(t) = f(t,x(t),x^{[2]}(t)), \quad t \in]0,1[, \tag{12}$$

associated to the boundary value

$$x(0) = 0, \quad x(1) = 1. \tag{13}$$

Where

$$f(t,u,v) = \frac{1}{\sqrt{t^2+144}}\left\{\sin(v) + e^{-t}u\right\}, \quad t \in [0,1]. \tag{14}$$

which obviously continuous on $[0,1]$. Hence there exist $A > 0$ such that

$$|f(t,u,v)| < A.$$

Thus f satisfies the hypothesis **(H1)**

By the mean value theorem, we have

$$|f(t,u_1,v_1)) - f(t,u_2,v_2)| \leq \frac{1}{12}\left[|v_1 - v_2| + |u_1 - u_2|\right].$$

Which implies that the hypothesis **(H2)** holds. Thus all the asumptions of theorem (1) and (2) are satisfied. Hence there exist a unique solution for the problem (12)–(13), with f is given by (14) on $[0,1]$.

5 Conclusion

We have discussed the solvability of a iterative fractional differential equation involving conformable fractional derivatives with boundary conditions $x(0) = 0$ and $x(1) = 1$. The results obtained in this paper are new and strengthen the literature on iterative fractional differential equations. Moreover, the results in this paper can be extended to boundary-value problems of the form

$$\begin{cases} D^{\alpha}x(t) = f(t,x(t),x^{[2]}(t),\ldots,x^{[n]}(t)), & t \in]0,1[, \\ x(0) = 0, \quad x(1) = 1. \end{cases}$$

References

1. Eder, E.: The functional-differential equation $x'(t) = x(x(t))$. J. Differ. Equ. **54**(3), 390–400 (1984)
2. Kaufmann, E.R.: Existence and uniqueness of solutions for a second-order iterative boundary-value problem. Electron. J. Differ. Equ. (2018)

3. Zhao, H.: Smooth solutions of a class of iterative functional differential equations. Abstr. Appl. Anal. (2012)
4. Guida, K., Hilal, K., Ibnelazyz, L.: Existence results for a class of coupled Hilfer fractional pantograph differential equations with nonlocal integral boundary value conditions. Adv. Math. Phys. (2020)
5. Hilal, K., Kajouni, A.: Boundary value problems for hybrid differential equations. Math. Theor. Model (2015)
6. Hilal, K., Kajouni, A.: Existence of the solution for system of coupled hybrid differential equations with fractional order and nonlocal conditions. Int. J. Differ. Equ. (2016)
7. Hilal, K., Guida, K., Ibnelazyz, L., Oukessou, M.: Existence results for an impulsive fractional integro-differential equations with a non-compact semigroup. J. Univ. Math. (2018)
8. Wang, K.: On the Equation $x'(t) = f(x(x(t)))$. Funkcial, Ekvac (1990)
9. Bouaouid, M., Hannabou, M. Hilal, K.: Nonlocal conformable-fractional differential equations with a measure of noncompactness in Banach spaces. J. Math. (2020)
10. Bouaouid, M., Hilal, K., Melliani, S.: Nonlocal conformable fractional Cauchy problem with sectorial operator. Indian J. Pure Appl. Math. **50**(4), 999–1010 (2019). https://doi.org/10.1007/s13226-019-0369-9
11. Elomari, M., Melliani, S., Chadli, L.S.: Conformable fractional cosine families of operators. J. Math. Sci. Model **2**(2), 112–116 (2019)
12. Fečkan, M.: On a certain type of functional-differential equations. Math. Slovaca **43**(1), 39–43 (1993)
13. Hannabou, M., Hilal, K., Kajouni, A.: Existence results of hybrid fractional sequential integro-differntial equations. Eng. Math. Lett. (2020)
14. Zhang, P.: Analytic solutions for iterative functional differential equations. Electron. J. Differ. Equ. **180**, 1–7 (2012)
15. Zhang, P., Gong, X.: Existence of solutions for iterative differential equations. Electron. J. Differ. Equ. (2014)
16. Khalil, R., Al Horani, M., Yousef, A.: Sababhehm: a new definition of fractional derivative. J. Comput. Appl. Math. **264**, 65–70 (2014)
17. Abdeljawad, T.: On a conformable calculus. J. Comput. Appl. Math. **279**, 57–66 (2015)
18. Petuhov, V.R.: On a boundary value problem, (Russian. English summary), Trudy Sem. Teor. Differencial. Uravneniis Otklon. Argumentom Univ. Druzby Narodov Patrisa Limumby (1965)

A System Coupled of Nonlinear Impulsive Differential Equations

M. El Hassnaoui, S. Melliani(✉), L. S. Chadli, and M. Oukessou

LMACS, Sultan Moulay Slimane University, BP 523 - 23000 Beni Mellal, Morocco
s.melliani@usms.ma

Abstract. In this paper, we discuss the existence and uniqueness of solutions for the new system of nonlinear mixed type impulsive fuzzy differential equation (1) in the metric space of normal fuzzy convex sets with distance given by maximum of the Hausdorff distance between level sets. Our analysis is fundamentally based on fixed point theory. We obtain some new existence and uniqueness theorems of solutions for this system of first-order impulsive fuzzy differential equations by using Banach fixed point theorem.

1 Introduction

There has been a lot of research into the equations of existence and uniqueness of solutions to differential equations boundary value problems in recent years [9]. Meanwhile, impulsive differential equations [2-6] have received much interest as a useful method for studying practical problems in biology, engineering, and physics. Many researchers have studied differential equations with applications in Banach spaces and fuzzy differential equations under various initial and boundary conditions. Chang and Zadeh [3] were the first to present the definition of fuzzy derivative in 1972. Following that, a framework for studying fuzzy differential equations was developed, as well as the basic properties of fuzzy differential equation solutions (see, for example, [1,2,4–8,11] and the references therein).

Recently, Nieto [10] proved a variant of the classical Peano existence theorem for initial value problems for a fuzzy differential equation in the metric space of normal fuzzy convex sets with the distance given by the maximum of the Hausdorff distance between level sets for a fuzzy differential equation in the metric space of normal fuzzy convex sets with the distance given by the maximum of the Hausdorff distance between level sets. The effects of Nieto [10] add to Kaleva's [7] existence and uniqueness result.

We are concerned with the existence and uniqueness of solution to the following nonlinear implicit impulsive fuzzy differential equations

$$\begin{cases} x'(t) = f(t, x(t), y'(t)), \ t \neq t_i \\ y'(t) = g(t, \lambda y(t), \gamma x'(t)), \ t \neq t_i \\ \Delta x \mid_{t=t_i} = I_i(x(t_i)), \ i = 1, 2, \cdots, m \\ \Delta y \mid_{t=t_i} = \widehat{I}_i(x(t_i)), \ i = 1, 2, \cdots, m \\ x(t_0) = x_0, \ \ y(t_0) = y_0, \end{cases} \tag{1}$$

S. Melliani and O. Castillo (Eds.): ICPAMS 2021, LNNS 476, pp. 267–274, 2023.
https://doi.org/10.1007/978-3-031-12416-7_23

where $f, g : J \times E^n \times E^n \longrightarrow E^n$ are continuous, $J = [t_0, t_0 + a] \subset \mathbb{R}$ and E^n is the family of all fuzzy sets $u : \mathbb{R}^n \longrightarrow [0,1]$, $x_0,\ y_0 \in E^n$, $I_i, \widehat{I_i} \in C(E^n, E^n)$ for $i = 1, 2, \cdots, m$, and $\lambda,\ \gamma \geq 0$.

The rest of this paper is organized as follows: In the next section we introduce the functional setting and gather some preliminary facts in connection with the problem. The existence and uniqueness of the problem (Theorems 1 and 2) by the Banach fixed point theorem are stated in Sect. 3.

2 Notations and Preliminaries

In this section we shall fix the notations and introduce the functional framework, which will be used through this paper. Let $P_k(\mathbb{R}^n)$ denote the family of non-empty compact, convex subsets of $\mathbb{R}^n$. If $\alpha, \beta \in \mathbb{R}$ and $A, B \in P_k(\mathbb{R}^n)$

$$\alpha(A + B) = \alpha A + \alpha B,$$

$$\alpha(\beta A) = (\alpha\beta)A, \quad 1.A = A.$$

For $A, B \in P_k(\mathbb{R}^n)$, the Hausdorff metric is defined as

$$d(A, B) = \max\{\sup_{x \in A} \inf_{y \in B} \|x - y\|,\ \sup_{y \in B} \inf_{x \in A} \|x - y\|\}.$$

A fuzzy set in $\mathbb{R}^n$ is a function with domain $\mathbb{R}^n$ and values in $[0,1]$, i.e., an element of $[0,1]^{\mathbb{R}^n}$.

Let $u, v \in [0,1]^{\mathbb{R}^n}$. Then we have

a) u is contained in v denoted by $u \leq v$ if and only if $u(x) \leq v(x)$ for all $x \in \mathbb{R}^n$;
b) $u \wedge v \in [0,1]^{\mathbb{R}^n}$ by $(u \wedge v)(x) = \min\{u(x), v(x)\}$ for all $x \in \mathbb{R}^n$ (intersection);
c) $u \vee v \in [0,1]^{\mathbb{R}^n}$ by $(u \vee v)(x) = \max\{u(x), v(x)\}$ for all $x \in \mathbb{R}^n$ (union);
d) $u^c \in [0,1]^{\mathbb{R}^n}$ by $u^c(x) = 1 - u(x)$ for all $x \in \mathbb{R}^n$.

Definition 1. Denote by $E^n = \{u : \mathbb{R}^n \longrightarrow [0,1]$ such that u satisfies (1) to (4) mentioned below $\}$:

1. u is normal, that is, there exists an $x_0 \in \mathbb{R}^n$ such that $u(x_0) = 1$;
2. u is fuzzy convex, that is, for $x, y \in \mathbb{R}^n$ and $0 \leq \nu \leq 1$, $u(\nu x + (1 - \nu)y) \geq \min\{u(x), u(y)\}$;
3. u is upper semi continuous;
4. $[u]^0 = \overline{\{x \in \mathbb{R}^n \ : \ u(x) > 0\}}$ is compact.

If $u \in E^n$, then it follows from (1)-(4) that, for each $\alpha \in (0,1]$, the α-level set

$$[u]^\alpha = \{x \in \mathbb{R}^n \ : \ u(x) \geq \alpha\}.$$

is a non empty compact convex subset of $\mathbb{R}^n$, that is, $[u]^\alpha \in P_k(\mathbb{R}^n)$ for all $0 \leq \alpha \leq 1$. Moreover, if $u, v, w \in E^n$ and $\alpha > 0$, then the addition and (positive) scalar multiplication in E^n are defined in terms of the λ-level sets by

$$[u + v]^\alpha = [u]^\alpha + [v]^\alpha, \ [\lambda.u]^\alpha = \lambda[u]^\alpha, \ \forall \alpha \in [0,1].$$

We define $D : E^n \times E^n \longrightarrow [0, +\infty)$ as

$$D(u,v) = \sup\{d([u]^\alpha, [v]^\alpha) \; : \; \alpha \in [0,1]\}.$$

D is a metric in E^n and (E^n, D) is a complete metric space. Moreover, D has a linear structure in the sense that

$$D(u+w, v+w) = D(u,v) \text{ and } D(\lambda u, \lambda v) = \lambda D(u,v).$$

Let $J = [t_0, t_0 + a]$ with $a > 0$ and $x, y \in E^n$. A mapping $F : J \longrightarrow E^n$ is differentiable at $t \in J$ if there exists a $F'(t) \in E^n$ such that the limits

$$\lim_{h \to 0^+} \frac{F(t+h) \ominus F(t)}{h}$$

and

$$\lim_{h \to 0^-} \frac{F(t) \ominus F(t-h)}{h}$$

exist and are equal to $F'(t)$

The integral of F over J denoted by $\int_J F(t)dt$, is defined levelwise by

$$\begin{aligned}\left[\int_J F(t)dt\right]^\alpha &= \int_J [F(t)]^\alpha dt = \int_J F_\alpha(t)dt \\ &= \left\{\int_J F(t)dt \mid F : J \longrightarrow \mathbb{R}^n \text{ is a measurable selection for } F_\alpha\right\}.\end{aligned}$$

Lemma 1. *If $F : J \longrightarrow E^n$ is continuous, then it is integrable and the function*

$$G(t) = \int_{t_0}^t F(s)ds, \; \forall t \in J$$

is differentiable and $G'(t) = F(t)$. Furthermore,

$$F(t) - F(t_0) = \int_{t_0}^t F'(s)ds.$$

Let $J = [t_0, t_0 + a]$ $(a > 0)$, $t_0 < t_1 < \cdots < t_m < t_0 + a < +\infty$, $J_0 = [t_0, t_1]$, $J_1 = (t_1, t_2]$, $\cdots, J_i = (t_i, t_{i+1}], \cdots, J_m = (t_m, t_0 + a]$. We define

$$\begin{aligned}PC^1(J, E^n) = \{x : \; & x \text{ is a map from J into } E^n \text{ such that is is continuously} \\ & \text{differential on } (t_i, t_{i+1}), \text{left continuous at } t_i, \\ & \text{and } x(t_i^+), x'(t_i^-), x'(t_i^+) \text{ exists }, i = 1, 2, \cdots, m\},\end{aligned}$$

where $x(t_i^+)$ represents the right limits of $x(t)$ at $t = t_i$, and $x'(t_i^-), x'(t_i^+)$ represent, respectively, the left and right derivatives of $x(t)$ at $t = t_i$.

We define $H(x,y)$ by

$$H(x,y) = \sup_{t \in J}\{D(x(t), y(t)) + D(x'(t), y'(t))\},$$

for all $x, y \in PC^1(J, E^n)$. Moreover $(PC^1(J, E^n), H)$ is a complete metric space.

The following lemma is important in our result.

Lemma 2. *For any $p \in PC^1(J;B)$, $u_0 \in B$ and $v_i \in B$, $i = 1,2,\cdots,n$, the following initial value problem*

$$\begin{cases} u'(t) + \alpha u(t) = f(t), \\ \Delta u\,|_{t=t_i} = v_i, \quad i = 1,2,\cdots,n \\ u(t_0) = u_0 \end{cases} \tag{2}$$

has a unique solution $u \in PC^1(J;B)$ given by

$$u(t) = u_0 e^{-\alpha(t-t_0)} + \int_{t_0}^{t} e^{-\alpha(t-s)} f(s) + \sum_{t_0<t_i<t} e^{-\alpha(t-t_i)} v_i,$$

where $\alpha \geq 0$ is a constant.

3 Main Results

Now, we are in a position to prove our main results concerning the solutions of the nonlinear mixed type impulsive differential equation system (1) in Banach spaces.

Lemma 3. *Assume that $f,g : J \times E^n \times E^n \longrightarrow E^n$ is continuous. Then a mapping (x,y) : $J \times J \longrightarrow E^n \times E^n$ is a solution of the problem (1) in $PC^1(J,E^n)$ if and only if x and y satisfies the following impulsive integral equation*

$$\begin{aligned} x(t) = x_0 e^{-M(t-t_0)} &+ \int_{t_0}^{t} e^{-M(t-s)} [f(s,x(s),y'(s)) + Mx(s)]ds \\ &+ \sum_{t_0<t_i<t} e^{-M(t-t_i)} I_i(x(t_i)), \end{aligned}$$

$$\begin{aligned} y(t) = y_0 e^{-M(t-t_0)} &+ \int_{t_0}^{t} e^{-M(t-s)} [g(s,\lambda y(s),\gamma x'(s)) + My(s)]ds \\ &+ \sum_{t_0<t_i<t} e^{-M(t-t_i)} \widehat{I_i}(x(t_i)), \end{aligned}$$

where $M > 0$ is a constant.

Now, let us first list the following assumptions:

$H1$: f and g are continuous, and there exist non negative constants ρ, μ, ϖ and $\underline{w}$ such that

$$\begin{cases} D(f(t,x_1,y_1), f(t,x_2,y_2)) \leq \rho D(x_1,x_2) + \mu D(y_1,y_2) \\ D(g(t,z_1,v_1), g(t,z_2,v_2)) \leq \varpi D(z_1,z_2) + \underline{w} D(v_1,v_2), \ \forall t \in J \end{cases}$$

$H2$: there exist non negative constants b_1 and b_2 such that

$$\begin{cases} D(I_i(x_1(t)), I_i(x_2(t)) \leq b_1 D(x_1(t), x_2(t)) \\ D(\widehat{I}_i(y_1(t)), \widehat{I}_i(y_2(t)) \leq b_2 D(y_1(t), y_2(t)), \ \forall i = 1, 2, \cdots, m \end{cases}$$

$H3$

$$\begin{aligned} \sigma = \max\{ & mb_1 + h(\rho + M) + M(1 - mb_1 - h(\rho + M); h(\lambda\varpi + M) \\ & + mb_2 + M(1 - h(\lambda\varpi + M) - mb_2) + \lambda\varpi); \\ & h\underline{w}\gamma + \gamma\underline{w}(1 - hM); h\eta + \eta(1 - hM)\} < 1, \end{aligned}$$

where

$$h = \frac{e^{\delta} - 1}{M}.$$

Theorem 1. *Suppose that the conditions* $(H1) - (H3)$ *hold. Then problem* (1) *has a unique solution.*

Proof. Let $(\tau_1, x(t), y(t)) \in J \times E^n \times E^n$ be arbitrary and $\delta > 0$ be a constant. We will first show that the initial value problem

$$\begin{cases} x'(t) = f(t, x(t), y'(t)), \ t \neq t_i \\ y'(t) = g(t, \lambda y(t), \gamma x'(t)), \ t \neq t_i \\ \Delta x \mid_{t=t_i} = I_i(x(t_i)), \ i = 1, 2, \cdots, m \\ \Delta y \mid_{t=t_i} = \widehat{I}_i(x(t_i)), \ i = 1, 2, \cdots, m \\ x(\tau_1) = \overline{x}, \ y(\tau_1) = \overline{y}, \end{cases} \tag{3}$$

has a unique solution on $J_1 = [\tau_1, \tau_1 + \delta]$.

For any $x, y \in PC^1(J, E^n)$, define $F(x, y) = (Ax, Gy)$ on $J_1 \times J_1$ by

$$\begin{aligned} Ax(t) = \overline{x}e^{-M(t-\tau_1)} + & \int_{\tau_1}^{t} e^{-M(t-s)}[f(s, x(s), y'(s)) + Mx(s)]ds \\ & + \sum_{\tau_1 < t_i < t} e^{-M(t-t_i)} I_i(x(t_i)), \end{aligned}$$

$$\begin{aligned} Gy(t) = \overline{y}e^{-M(t-\tau_1)} + & \int_{\tau_1}^{t} e^{-M(t-s)}[g(s, \lambda y(s), \gamma x'(s)) + My(s)]ds \\ & + \sum_{\tau_1 < t_i < t} e^{-M(t-t_i)} \widehat{I}_i(x(t_i)), \end{aligned}$$

and for any given$(x, y) \in PC^1(J, E^n) \times PC^1(J, E^n)$ and $t \neq t_i$, $i = 1, 2, \cdots, m$, we have

$$(Ax)'(t) = -MA(x(t)) + Mx(t) + f(t, x(t), y'(t)),$$

$$(Gy)'(t) = -MG(y(t)) + My(t) + g(t, \lambda y(t), \gamma x'(t)).$$

In the sequel, we prove that $F : PC^1(J,E^n) \times PC^1(J,E^n) \longrightarrow PC^1(J,E^n) \times PC^1(J,E^n)$ is a contraction mapping. for any $(x_1,y_1),(x_2,y_2) \in PC^1(J,E^n) \times PC^1(J,E^n)$, we have

$$D(A(x_1(t)),A(x_2(t))) \leq [mb_1 + h(\rho+M)]D(x_1(t),x_2(t)) + h\mu D(y_1'(t),y_2'(t)),$$

$$\begin{aligned} D(A(x_1(t)),A(x_2(t))) \leq& \int_{\tau_1}^{\tau_1+\delta} e^{-M(t-s)}\left[D\left(f(s,x_1(s),y_1'(s)),f(s,x_2(s),y_2'(s))\right) + MD(x_1(s),x_2(s)\right] ds \\ &+ \sum_{\tau_1<t_i<t} e^{-M(t-t_i)} D(I_i(x_1(t_i)),I_i(x_2(t_i))) \\ &\leq [M(1-mb_1-h(\rho+M))]D(x_1(t),x_2(t)) + \mu(1-Mh)D(y_1'(t),y_2'(t)),\end{aligned}$$

and we have

$$\begin{aligned} D((Ax_1)'(t)),(Ax_2)'(t))) \leq& [M(1-mb_1-h(\rho+M))]D(x_1(t),x_2(t)) \\ &+\mu(1-Mh)D(y_1'(t),y_2'(t)).\end{aligned}$$

Using de same calculus, we obtain

$$\begin{aligned} D(G(y_1(t)),G(y_2(t))) \leq& \int_{\tau_1}^{\tau_1+\delta} e^{-M(t-s)}\left[D\left(g(s,y_1(s),\alpha x_1'(s)),g(s,y_2(s),\alpha x_2'(s))\right) + MD(y_1(s),y_2(s)\right] ds \\ &+ \sum_{\tau_1<t_i<t} e^{-M(t-t_i)} D(\widehat{I}_i(y_1(t_i)),\widehat{I}_i(y_2(t_i))) \\ &\leq e^{-Mt}\int_{\tau_1}^{\tau_1+\delta} e^{Ms}\left[(\varpi\lambda+M)D(y_1(s),y_2(s)) + \underline{w}\alpha D(x_1'(s),x_2'(s)\right] ds \\ &+mb_2 D(y_1(t),y_2(t)) \\ &\leq h\underline{w}\gamma D(x_1'(t),x_2'(t)) + (h(\lambda\varpi+M)+mb_2)D(y_1(t),y_2(t))\end{aligned}$$

Using de same calculus, we obtain

$$\begin{aligned} D((Gx_1)'(t)),(Gx_2)'(t))) \leq& \gamma\underline{w}(1-Mh)D(x_1'(t),x_2'(t)) \\ &+[M(1-h(\lambda\varpi+M)-mb_2)+\lambda\varpi]D(y_1(t),y_2(t))\end{aligned}$$

Therefore, we have

$$\begin{aligned} H(F(x_1,y_1),F(x_2,y_2)) =& \sup_{t\in J_1}\left\{D(F(x_1(t),y_1(t)),F(x_2(t),y_2(t)) + D\left(F'(x_1(t),y_1(t)),F'(x_2(t),y_2(t))\right)\right\} \\ &\leq D((Ax_1)(t)),(Ax_2)(t))) + D((Gx_1)(t)),(Gx_2)(t))) \\ &+D((Ax_1)'(t)),(Ax_2)'(t))) + D((Gx_1)'(t)),(Gx_2)'(t))).\end{aligned}$$

Then

$$H(F(x_1,y_1),F(x_2,y_2)) \leq \sigma H((x_1,y_1),(x_2,y_2)),$$

where

$$\begin{aligned} \sigma = \max\{&mb_1 + h(\rho+M) + M(1-mb_1-h(\rho+M); h(\lambda\varpi+M) \\ &+ mb_2 + M(1-h(\lambda\varpi+M)-mb_2)+\lambda\varpi); \\ &h\underline{w}\gamma + \gamma\underline{w}(1-hM); h\eta+\eta(1-hM)\}.\end{aligned}$$

Therefore, by Banach fixed point theorem, F has a unique fixed point, which by Lemma 3 is the desired solution to the problem (1). Express J as a union of a finite family of intervals J_i with the length of each interval less than δ this completes the proof.

Theorem 2. *Let $f,g : J \times E^n \times E^n \longrightarrow E^n$ be continuous mappings. Assume that for all $x_k, y_k, z_k, v_k : J \longrightarrow E^n$ $(k = 1,2)$, there exist non negative constants ρ, ϖ, b_1 and b_2 such that, for all $t \in J$, $\alpha \in [0,1]$ and $i = 1,2,\cdots,m$,*

$$\begin{cases} d([f(t,x_1,y_1)]^\alpha, [f(t,x_2,y_2)]^\alpha) \leq \rho d([x_1]^\alpha, [x_2]^\alpha) + \rho d([y_1]^\alpha, [y_2]^\alpha) \\ d([g(t,z_1,v_1)]^\alpha, [g(t,z_2,v_2)]^\alpha) \leq \varpi d([z_1]^\alpha, [z_2]^\alpha) + \varpi d([v_1]^\alpha, [v_2]^\alpha), \end{cases}$$

$$\begin{cases} d([I_i(x_1(t))]^\alpha, [I_i(x_2(t)]^\alpha) \leq b_1 d([x_1(t)]^\alpha, [x_2(t)]^\alpha) \\ d([\widehat{I_i}(y_1(t))]^\alpha, [\widehat{I_i}(y_2(t)]^\alpha) \leq b_2 d([y_1(t)]^\alpha, [y_2(t)]^\alpha). \end{cases}$$

Then the problem ((1) has a unique solution.

Proof. For $\alpha \in [0,1]$, We have

$$\begin{aligned} D(f(t,x_1,y_1), f(t,x_2,y_2)) &= \sup\{d([f(t,x_1,y_1)]^\alpha, [f(t,x_2,y_2)]^\alpha)\} \\ &\leq \rho \sup\{d([x_1]^\alpha, [x_2]^\alpha) + \rho d([y_1]^\alpha, [y_2]^\alpha)\} \\ &= \rho D(x_1,x_2) + \rho D(y_1,y_2), \end{aligned}$$

$$\begin{aligned} D(g(t,z_1,v_1), g(t,z_2,v_2)) &= \sup\{d([g(t,z_1,v_1)]^\alpha, [g(t,z_2,v_2)]^\alpha)\} \\ &\leq \varpi \sup\{d([z_1]^\alpha, [z_2]^\alpha) + \rho d([v_1]^\alpha, [v_2]^\alpha)\} \\ &= \varpi D(z_1,z_2) + \varpi D(v_1,v_2), \end{aligned}$$

$$\begin{aligned} D(I_i(x_1(t)), I_i(x_2(t))) &= \sup\{d([I_i(x_1(t))]^\alpha, [I_i(x_2(t)]^\alpha)\} \\ &\leq b_1 \sup\{d([x_1(t)]^\alpha, [x_2(t)]^\alpha)\} \\ &= b_1 D(x_1(t), x_2(t)), \end{aligned}$$

and

$$\begin{aligned} D(\widehat{I_i}(y_1(t)), \widehat{I_i}(y_2(t))) &= \sup\{d([\widehat{I_i}(y_1(t))]^\alpha, [\widehat{I_i}(y_2(t)]^\alpha)\} \\ &\leq b_2 \sup\{d([y_1(t)]^\alpha, [y_2(t)]^\alpha)\} \\ &= b_2 D(y_1(t), y_2(t)). \end{aligned}$$

By Theorem 1, we know that the problem (1) has a unique solution.

4 Conclusion

In the present work, we studied a system of nonlinear mixed type implicit impulsive fuzzy differential equation, We obtained some new existence and uniqueness theorems of solutions for a system of nonlinear first-order impulsive fuzzy differential equations in Banach spaces under some weaker conditions.

References

1. Abbasbandy, S., Viranloo, T.A., López-Pouso, Ó., Nieto, J.J.: Numerical methods for fuzzy differential inclusions. Comput. Math. Appl. **48**, 1633–1641 (2004)
2. Chalco-Cano, Y., Roman-Flores, H.: On new solutions of fuzzy differential equations. Chaos, Solitons Fractals (2006). In Press, Corrected Proof
3. Chang, S.S.L., Zadeh, L.A.: On fuzzy mapping and control. IEEE Trans. Syst. Man Cybernet. **2**, 30–34 (1972)
4. Georgiou, D.N., Kougias, I.E.: On Cauchy problems for fuzzy differential equations. Int. J. Math. Math. Sci. **15**, 799–805 (2004)
5. Bhaskar, T.G., Lakshmikantham, V., Devi, V.: Revisiting fuzzy differential equations. Nonlinear Anal. **58**, 351–358 (2004)
6. Kaleva, O.: Fuzzy differential equations. Fuzzy Sets Syst. **24**, 301–317 (1987)
7. Kaleva, O.: The Cauchy problem for fuzzy differential equations. Fuzzy Sets Syst. **35**, 389–396 (1990)
8. Kloeden, P.E.: Remarks on Peano-like theorems for fuzzy differential equations. Fuzzy Sets Syst. **44**, 161–163 (1991)
9. Ladde, G.S., Lakshmikantham, V., Vatsala, A.S.: Monotone Iterative Techniques for Nonlinear Differential Equation. Pitman Advanced Publishing Program, Pitman, London (1985)
10. Nieto, J.J.: The Cauchy problem for continuous fuzzy differential equations. Fuzzy Sets Syst. **102**, 259–262 (1999)
11. Stefanini, L., Sorini, L., Guerra, M.L.: Parametric representation of fuzzy numbers and application to fuzzy calculus. Fuzzy Sets Syst. **157**, 2423–2455 (2006)

Existence and Uniqueness of Weak Solution for a Class of Nonlinear Degenerate Elliptic Problems in Weighted Sobolev Spaces

Mohamed El Ouaarabi, Ail Abbassi, and Chakir Allalou(✉)

LMACS, Laboratory of Applied Mathematics and Scientific Calculus, Sultan Moulay Sliman University, Beni Mellal, Morocco
chakir.allalou@yahoo.fr

Abstract. This work is devoted to study the existence and uniqueness of weak solution for a class of nonlinear degenerate elliptic problems of the following form

$$-\text{div}\Big[\omega_1\mathcal{A}(x,\nabla u)+\omega_2\mathcal{B}(x,u,\nabla u)\Big]+\omega_3\mathcal{H}(x,u)=\phi(x),$$

where ω_1, ω_2 and ω_3 are A_p-weight functions and the operators $\mathcal{A}$, $\mathcal{B}$ and $\mathcal{H}$ are Caratéodory functions that satisfy some conditions and the right-hand side term $\phi\in L^{p'}(\Omega,\omega_1^{1-p'})$. Our technical approach is based on the Browder-Minty Theorem and the weighted Sobolev spaces theory.

Keywords: Nonlinear degenerate elliptic problems · Dirichlet problem · Browder-Minty Theorem · Weighted Sobolev spaces · Weak solution

1 Introduction

Our aim in this work is to prove the existence and uniqueness of weak solution in the weighted Sobolev space $W_0^{1,p}(\Omega,\omega_1)$ (p is not necessarily equal 2) for the Dirichlet problem associated to the nonlinear degenerate elliptic equation of the form

$$\begin{cases} -\text{div}\Big[\omega_1 a(x,\nabla u)+\omega_2 b(x,u,\nabla u)\Big]+\omega_3 g(x,u)=\phi(x) & \text{in } \Omega,\\ u(x)=0 & \text{on } \partial\Omega, \end{cases} \tag{1}$$

where, Ω is a bounded open set in $\mathbb{R}^N$, ω_1, ω_2 and ω_3 are A_p-weight functions, and the functions $b:\Omega\times\mathbb{R}\times\mathbb{R}^N\longrightarrow\mathbb{R}^N$, $a:\Omega\times\mathbb{R}^N\longrightarrow\mathbb{R}^N$ and $g:\Omega\times\mathbb{R}\longrightarrow\mathbb{R}$ are Caratéodory functions that satisfy the assumptions of growth, ellipticity and monotonicity, with $\phi\in L^{p'}(\Omega,\omega_1^{1-p'})$.

Problems such as (1) have been studied by several authors in the case $\omega_1\equiv\omega_2\equiv\omega_3\equiv 1$. For $\omega_1\equiv\omega_2\equiv\omega_3\equiv 1$ (the non weighted case) and the term

S. Melliani and O. Castillo (Eds.): ICPAMS 2021, LNNS 476, pp. 275–290, 2023.
https://doi.org/10.1007/978-3-031-12416-7_24

$a(x, \nabla u)$ is equal to zero, existence results for problem (1) have been shown in [3] (see also [6]). The degenerate case with various conditions conditions have been studied by many authors, we mention some works in this direction [1,2,15–17].

In [5] the author proved the existence of solution, when the term $g(x, u)$ is equal to zero and in [4] when the term $a(x, \nabla u)$ is equal to zero.

The paper is divided into five sections. In Sect. 2 we present the definitions and basic results. In Sect. 3, we make precise all the assumptions on a, b, g and we introduce the notion of weak solution for the problem (1). Our main result and his proof are collected in Sect. 4. In Sect. 5, we give an example.

2 Mathematical Preliminaries

In what follows, we recall some definitions and basic properties of Lebesgue and Sobolev spaces with weight. Complete expositions can be found in the monographs by J. Garcia-Cuerva and J. L. Rubio de Prancia [9] and A. Torchinsky [18].

Let ω be a weight function in $\mathbb{R}^N$, that is ω measurable and strictly positive a.e. in $\mathbb{R}^N$. For $0 < p < \infty$, we denote by $L^p(\Omega, \omega)$ the space of measurable functions f on Ω such that

$$||f||_{L^p(\Omega,\omega)} = \left(\int_\Omega |f(x)|^p \omega(x) dx\right)^{\frac{1}{p}} < \infty,$$

where ω is a weight, and Ω be open in $\mathbb{R}^N$. It is a well-known fact that the space $L^p(\Omega, \omega)$, endowed with this norm is a Banach space. We also have that the dual space of $L^p(\Omega, \omega)$ is the space $L^{p'}(\Omega, \omega^{1-p'})$.

We now determine conditions on the weight ω that guarantee that functions in $L^p(\Omega, \omega)$ are locally integrable on Ω.

Proposition 1 [12,13]. *Let $1 \leq p < \infty$. If the weight ω is such that*

$$\omega^{\frac{-1}{p-1}} \in L^1_{loc}(\Omega) \qquad if \quad p > 1,$$

$$ess \sup_{x \in B} \frac{1}{\omega(x)} < +\infty \qquad if \quad p = 1,$$

for every ball $B \subset \Omega$. Then,

$$L^p(\Omega, \omega) \subset L^1_{loc}(\Omega).$$

A class of weights, which is particularly well understood, is the class of A_p-weight. These classes have found many useful applications in harmonic analysis [18].

Definition 1. Let $1 \leq p < \infty$. A weight ω is said to be an A_p-weight, or ω belongs to the Muckenhoupt class, if there exists a positive constant $\theta = \theta(p,\omega)$ such that, for every ball $B \subset \mathbb{R}^n$

$$\left(\frac{1}{|B|}\int_B \omega(x)dx\right)\left(\frac{1}{|B|}\int_B (\omega(x))^{\frac{-1}{p-1}}\,dx\right)^{p-1} \leqslant \theta \qquad \text{if } \ p>1,$$

$$\left(\frac{1}{|B|}\int_B \omega(x)dx\right) ess \sup_{x\in B}\frac{1}{\omega(x)} \leqslant \theta \qquad \text{if } \ p=1,$$

where $|.|$ denotes the n-dimensional Lebesgue measure in $\mathbb{R}^N$.

The infimum over all such constants θ is called the A_p constant of ω. We denote by A_p, $1 \leq p < \infty$, the set of all A_p-weights. We refer to [10,11,19] for more informations about A_p-weights.

Example 1. (Example of A_p-weights)

(i) If $C \leq \omega(y) \leq D$ for a.e. $y \in \mathbb{R}^n$, such that C and D two positive constants, then $\omega \in A_p$ for $1 < p < \infty$.
(ii) Suppose that $\omega(y) = |y|^\eta$, $y \in \mathbb{R}^N$. Then $\omega \in A_p$ iff $-n < \eta < N(p-1)$ for $1 < p < \infty$ (see Corollary 4.4 in [18]).
(iii) $\omega(x) = e^{\lambda v(y)} \in A_2$, with $v \in W^{1,N}(\Omega)$ and λ is sufficiently small (see Corollary 2.18 in [14]).

The weighted Sobolev space $W^{1,p}(\Omega,\omega)$ is defined as follows.

Definition 2. Let $\Omega \subset \mathbb{R}^n$ be open, and let ω be an A_p-weight, $1 < p < \infty$. We define the weighted Sobolev space $W^{1,p}(\Omega,\omega)$ as the set of functions $u \in L^p(\Omega,\omega)$ with $D_k u \in L^p(\Omega,\omega)$, for $k = 1, ..., n$. The norm of u in $W^{1,p}(\Omega,\omega)$ is

$$||u||_{W^{1,p}(\Omega,\omega)} = \left(\int_\Omega |u(x)|^p\omega(x)dx + \sum_{k=1}^{N}\int_\Omega |D_k u(x)|^p\omega(x)dx\right)^{\frac{1}{p}}. \tag{2}$$

We also define $W_0^{1,p}(\Omega,\omega)$ as the closure of $C_0^\infty(\Omega)$ in $W^{1,p}(\Omega,\omega)$ with respect to the norm $||.||_{W^{1,p}(\Omega,\omega)}$. Note that $C_0^\infty(\Omega)$ is dense in $W_0^{1,p}(\Omega,\omega)$.

The spaces $W^{1,p}(\Omega,\omega)$ and $W_0^{1,p}(\Omega,\omega)$ are separable and reflexive Banach spaces with respect to the norm (2) (see Proposition 2.1.2. in [12] and see [11,13] for more informations about the spaces $W^{1,p}(\Omega,\omega)$). The dual of space $W_0^{1,p}(\Omega,\omega)$ is the space $W_0^{-1,p'}(\Omega,\omega^{1-p'})$, where

$$W_0^{-1,p'}(\Omega,\omega^{1-p'}) = \left\{f_0 - \sum_{k=1}^{N} D_k f_k \, : \, \frac{f_k}{\omega} \in L^{p'}(\Omega,\omega),\ k = 0, ..., n\right\}.$$

In this paper we use the following results.

Theorem 1 [8]. *Let $\omega \in A_p$, $1 \leqslant p < \infty$, and let Ω be a bounded open set in $\mathbb{R}^N$. If $u_k \longrightarrow u$ in $L^p(\Omega,\omega)$, then there exist a subsequence (u_{k_i}) and $\psi \in L^p(\Omega,\omega)$ such that*

(i) $u_{k_i}(x) \longrightarrow u(x)$, $k_i \longrightarrow \infty$, *$\omega$-a.e. on Ω.*
(ii) $|u_{k_i}(x)| \leqslant \psi(x)$, *$\omega$-a.e. on Ω.*

Theorem 2 [7]. *Let $\omega \in A_p$, $1 \leqslant p < \infty$, and let Ω be a bounded open set in $\mathbb{R}^N$. There exist constants C_Ω and ε positive such that for all $v \in W_0^{1,p}(\Omega,\omega)$ and all κ satisfying $1 \leqslant \kappa \leqslant \frac{N}{N-1} + \varepsilon$,*

$$||v||_{L^{\kappa p}(\Omega,\omega)} \leqslant C_\Omega ||\nabla v||_{L^p(\Omega,\omega)},$$

where C_Ω depends only on n, p, the A_p constant of ω and the diameter of Ω.

The Browder-Minty Theorem is stated as follows

Theorem 3 [20]. *Let $A : W \longrightarrow W^*$ be a coercive, hemicontinuous and monotone operator on the real, reflexive and separable Banach space W. Then the following assertions hold:*

(a) For each $T \in W^*$, the equation $Av = T$ has a solution $v \in W$.
(b) If the operator A is strictly monotone, then equation $Av = T$ has a unique solution $v \in W$.

3 Basic Assumptions and Notion of Solutions

In the sequel, we make precise all the assumptions on a, b, g and we introduce the notion of weak solution for the problem (1).

3.1 Basic Assumptions

Let us now give the precise hypotheses on the problem (1), we assume that the following assumptions: Ω be a bounded open subset of $\mathbb{R}^N$ $(N \geq 2)$, $1 < q < p < \infty$, let ω_1, ω_2 and ω_3 are a weights functions, and let $a_j : \Omega \times \mathbb{R}^N \longrightarrow \mathbb{R}$, $b_j : \Omega\times\mathbb{R}\times\mathbb{R}^N \longrightarrow \mathbb{R}$ $(j = 1, ..., N)$, with $b(x,\eta,\xi) = \Big(b_1(x,\eta,\xi), ..., b_N(x,\eta,\xi)\Big)$ and $a(x,\xi) = \Big(a_1(x,\xi), ..., a_N(x,\xi)\Big)$ and $g : \Omega\times\mathbb{R} \longrightarrow \mathbb{R}$ satisfying the following assumptions:

(A1) For $j = 1, ..., N$, b_j, a_j and g are Caratéodory functions.
(A2) There are positive functions $h_1, h_2, h_3, h_4 \in L^\infty(\Omega)$ and $f_1 \in L^{p'}(\Omega,\omega_1)$ $\Big(\text{with}\frac{1}{p} + \frac{1}{p'} = 1\Big)$, $f_2 \in L^{q'}(\Omega,\omega_2)$ $\Big(\text{with}\frac{1}{q} + \frac{1}{q'} = 1\Big)$ and $f_3 \in L^{s'}(\Omega,\omega_3)$ $\Big(\text{with}\frac{1}{s} + \frac{1}{s'} = 1\Big)$ such that:

$$|a(x,\xi)| \leq f_1(x) + h_1(x)|\xi|^{p-1},$$

$$|b(x,\eta,\xi)| \leq f_2(x) + h_2(x)|\eta|^{q-1} + h_3(x)|\xi|^{q-1},$$

and

$$|g(x,\eta)| \leq f_3(x) + h_4(x)|\eta|^{s-1}.$$

(A3) There exists a constant $\alpha > 0$ such that:

$$\langle a(x,\xi) - a(x,\xi'),\xi - \xi'\rangle \geqslant \alpha|\xi - \xi'|^p,$$

$$\langle b(x,\eta,\xi) - b(x,\eta',\xi'),\xi - \xi'\rangle \geqslant 0,$$

and

$$\Big(g(x,\eta) - g(x,\eta')\Big)\Big(\eta - \eta'\Big) \geqslant 0,$$

whenever $(\eta,\xi), (\eta',\xi') \in \mathbb{R} \times \mathbb{R}^N$ with $\eta \neq \eta'$ and $\xi \neq \xi'$.

(A4) There are constants β_1, β_2, $\beta_3 > 0$ such that:

$$\langle a(x,\xi),\xi\rangle \geqslant \beta_1|\xi|^p,$$

$$\langle b(x,\eta,\xi),\xi\rangle \geqslant \beta_2|\xi|^q + \beta_3|\eta|^q,$$

and

$$g(x,\eta)\eta \geqslant 0.$$

3.2 Notions of Solutions

The definition of a weak solution for problem (1) can be stated as follows.

Definition 3. A function $u \in W_0^{1,p}(\Omega,\omega_1)$ is said to be a weak solution to (1) if

$$\int_\Omega \langle a(x,\nabla u),\nabla v\rangle \omega_1 dx + \int_\Omega \langle b(x,u,\nabla u),\nabla v\rangle \omega_2 dx + \int_\Omega g(x,u)v\omega_3 dx = \int_\Omega \phi v dx,$$

for all $v \in W_0^{1,p}(\Omega,\omega_1)$.

Remark 1. For all ω_1, ω_2, $\omega_3 \in A_p$ we have

(i) If $\frac{\omega_2}{\omega_1} \in L^{r_1}(\Omega,\omega_1)$ where $= \frac{p}{p-q}$ and $1 < q < p < \infty$, then, by Hölder inequality we obtain

$$||u||_{L^q(\Omega,\omega_2)} \leqslant C_{p,q}||u||_{L^p(\Omega,\omega_1)},$$

where $C_{p,q} = ||\frac{\omega_2}{\omega_1}||^{1/q}_{L^{r_1}(\Omega,\omega_1)}$.

(ii) Analogously, if $\frac{\omega_3}{\omega_1} \in L^{r_2}(\Omega,\omega_1)$ where $r_2 = \frac{p}{p-s}$ and $1 < s < p < \infty$, then

$$||u||_{L^s(\Omega,\omega_3)} \leqslant C_{p,s}||u||_{L^p(\Omega,\omega_1)},$$

where $C_{p,s} = ||\frac{\omega_3}{\omega_1}||^{1/s}_{L^{r_2}(\Omega,\omega_1)}$.

4 Main General Result

The main result of this article is given in the next theorem.

Theorem 4. *Let $\omega_i \in A_p$ for $i = 1,2,3$, $1 < q, s < p < \infty$ and assume that the conditions **(A1)–(A4)** holds. Then the problem (1) has exactly one solution $u \in W_0^{1,p}(\Omega,\omega_1)$.*

Proof. The essential concept of our proof is to abate the (1) to an operator problem $\mathbf{A}u = \mathbf{T}$, in order to apply the Theorem 3.

We define

$$\mathbf{O} : W_0^{1,p}(\Omega,\omega_1) \times W_0^{1,p}(\Omega,\omega_1) \longrightarrow \mathbb{R}$$

and

$$\mathbf{T} : W_0^{1,p}(\Omega,\omega_1) \longrightarrow \mathbb{R},$$

where $\mathbf{O}$ and $\mathbf{T}$ are defined below.

Then $u \in W_0^{1,p}(\Omega,\omega_1)$ is a weak solution of (1) if and only if

$$\mathbf{O}(u,v) = \mathbf{T}(v), \qquad \text{for all } v \in W_0^{1,p}(\Omega,\omega_1).$$

The proof of Theorem 4 is divided into several steps.

Step 1: Equivalent Operator Equation

In this step, we use the some tools and the condition **(A2)** to prove an existence the operator $\mathbf{A}$ such that the problem (1) is equivalent to the operator equation $\mathbf{A}u = \mathbf{T}$.

Using Hölder inequality and Theorem 2, we obtain

$$\begin{aligned} |\mathbf{T}(v)| &\leq \int_\Omega \frac{|\phi|}{\omega_1} |v|\omega_1 \, dx \\ &\leq ||\phi/\omega_1||_{L^{p'}(\Omega,\omega_1)} ||v||_{L^p(\Omega,\omega_1)} \\ &\leq C_\Omega ||\phi/\omega_1||_{L^{p'}(\Omega,\omega_1)} ||v||_{W_0^{1,p}(\Omega,\omega_1)}. \end{aligned}$$

Since $\phi \in L^{p'}(\Omega,\omega_1^{1-p'})$, then $\mathbf{T} \in W_0^{-1,p'}(\Omega,\omega_1^{1-p'})$.

The operator $\mathbf{O}$ is broken down into the from

$$\mathbf{O}(u,v) = \mathbf{O}_1(u,v) + \mathbf{O}_2(u,v) + \mathbf{O}_3(u,v),$$

where

$$\begin{aligned} &\mathbf{O}_1 : W_0^{1,p}(\Omega,\omega_1) \times W_0^{1,p}(\Omega,\omega_1) \longrightarrow \mathbb{R} \\ &\mathbf{O}_1(u,v) = \int_\Omega \langle a(x,\nabla u), \nabla v\rangle \omega_1 dx, \end{aligned}$$

$$\begin{aligned} &\mathbf{O}_2 : W_0^{1,p}(\Omega,\omega_1) \times W_0^{1,p}(\Omega,\omega_1) \longrightarrow \mathbb{R} \\ &\mathbf{O}_2(u,v) = \int_\Omega \langle b(x,u,\nabla u), \nabla v\rangle \omega_2 dx, \end{aligned}$$

$$\mathbf{O}_3 : W_0^{1,p}(\Omega,\omega_1) \times W_0^{1,p}(\Omega,\omega_1) \longrightarrow \mathbb{R}$$
$$\mathbf{O}_3(u,v) = \int_\Omega g(x,u)v\omega_3 dx.$$

Then, we have

$$|\mathbf{O}(u,v)| \leq |\mathbf{O}_1(u,v)| + |\mathbf{O}_2(u,v)| + |\mathbf{O}_3(u,v)|. \tag{3}$$

On the other hand, we get by using **(A2)**, Hölder inequality, Remark 1 (i) and Theorem 2,

$$\begin{aligned}
|\mathbf{O}_1(u,v)| &\leq \int_\Omega |a(x,\nabla u)||\nabla v|\omega_1 dx \\
&\leq \int_\Omega \left(f_1 + h_1|\nabla u|^{p-1}\right)|\nabla v|\omega_1 dx \\
&\leq ||f_1||_{L^{p'}(\Omega,\omega_1)}||\nabla v||_{L^p(\Omega,\omega_1)} + ||h_1||_{L^\infty(\Omega)}||\nabla u||^{p-1}_{L^p(\Omega,\omega_1)}||\nabla v||_{L^p(\Omega,\omega_1)} \\
&\leq \left(||f_1||_{L^{p'}(\Omega,\omega_1)} + ||h_1||_{L^\infty(\Omega)}||u||^{p-1}_{W_0^{1,p}(\Omega,\omega_1)}\right)||v||_{W_0^{1,p}(\Omega,\omega_1)},
\end{aligned}$$

and

$$\begin{aligned}
|\mathbf{O}_2(u,v)| &\leq \int_\Omega |b(x,u,\nabla u)||\nabla v|\omega_2 dx \\
&\leq \int_\Omega \left(f_2 + h_2|u|^{q-1} + h_3|\nabla u|^{q-1}\right)|\nabla v|\omega_2 dx \\
&\leq ||f_2||_{L^{q'}(\Omega,\omega_2)}||\nabla v||_{L^q(\Omega,\omega_2)} + ||h_2||_{L^\infty(\Omega)}||u||^{q-1}_{L^q(\Omega,\omega_2)}||\nabla v||_{L^q(\Omega,\omega_2)} \\
&\quad + ||h_3||_{L^\infty(\Omega)}||\nabla u||^{q-1}_{L^q(\Omega,\omega_2)}||\nabla v||_{L^q(\Omega,\omega_2)} \\
&\leq \Big[C_{p,q}||f_2||_{L^{q'}(\Omega,\omega_2)} + C^q_{p,q}\Big(C_\Omega^{q-1}||h_2||_{L^\infty(\Omega)} \\
&\quad + ||h_3||_{L^\infty(\Omega)}\Big)||u||^{q-1}_{W_0^{1,p}(\Omega,\omega_1)}\Big]||v||_{W_0^{1,p}(\Omega,\omega_1)}.
\end{aligned}$$

Analogously, using **(A2)**, Hölder inequality, Remark 1 (ii) and Theorem 2, we obtain

$$\begin{aligned}
|\mathbf{O}_3(u,v)| &\leq \int_\Omega |g(x,u)||v|\omega_3 dx \\
&\leq \left[C_\Omega C_{p,s}||f_3||_{L^{s'}(\Omega,\omega_3)} + C^s_{p,s}C^s_\Omega||h_4||_{L^\infty(\Omega)}||u||^{s-1}_{W_0^{1,p}(\Omega,\omega_1)}\right]||v||_{W_0^{1,p}(\Omega,\omega_1)}.
\end{aligned}$$

Hence, in (3) we obtain, for all $u,v \in W_0^{1,p}(\Omega,\omega_1)$

$$\begin{aligned}
|\mathbf{O}(u,v)| \leq \Big[&||f_1||_{L^{p'}(\Omega,\omega_1)} + ||h_1||_{L^\infty(\Omega)}||u||^{p-1}_{W_0^{1,p}(\Omega,\omega_1)} + C_\Omega C_{p,s}||f_3||_{L^{s'}(\Omega,\omega_3)} \\
+ \ &C_{p,q}||f_2||_{L^{q'}(\Omega,\omega_2)} + C^s_{p,s}C^s_\Omega||h_4||_{L^\infty(\Omega)}||u||^{s-1}_{W_0^{1,p}(\Omega,\omega_1)} \\
+ \ &C^q_{p,q}\left(C_\Omega^{q-1}||h_2||_{L^\infty(\Omega)} + ||h_3||_{L^\infty(\Omega)}\right)||u||^{q-1}_{W_0^{1,p}(\Omega,\omega_1)}\Big]||v||_{W_0^{1,p}(\Omega,\omega_1)}.
\end{aligned}$$

Then $\mathbf{O}(u, .)$ is linear and continuous, for each $u \in W_0^{1,p}(\Omega, \omega_1)$. Thus, there exists a linear and continuous operator on $W_0^{1,p}(\Omega, \omega_1)$ denoted by $\mathbf{A}$ such that

$$\langle \mathbf{A}u, v\rangle = \mathbf{O}(u, v), \quad \text{for all } u, v \in W_0^{1,p}(\Omega, \omega_1).$$

Moreover, we have

$$\begin{aligned}\|\mathbf{A}u\|_* \leq & \, ||f_1||_{L^{p'}(\Omega,\omega_1)} + ||h_1||_{L^\infty(\Omega)}||u||^{p-1}_{W_0^{1,p}(\Omega,\omega_1)} + C_\Omega C_{p,s}||f_3||_{L^{s'}(\Omega,\omega_3)} \\ & + \, C_{p,q}||f_2||_{L^{q'}(\Omega,\omega_2)} + C^s_{p,s} C^s_\Omega ||h_4||_{L^\infty(\Omega)}||u||^{s-1}_{W_0^{1,p}(\Omega,\omega_1)} \\ & + \, C^q_{p,q}\left(C^{q-1}_\Omega ||h_2||_{L^\infty(\Omega)} + ||h_3||_{L^\infty(\Omega)}\right)||u||^{q-1}_{W_0^{1,p}(\Omega,\omega_1)},\end{aligned}$$

where

$$\|\mathbf{A}u\|_* := \sup\Big\{|\langle \mathbf{A}u, v\rangle| = |\mathbf{O}(u, v)| : v \in W_0^{1,p}(\Omega, \omega_1), \|v\|_{W_0^{1,p}(\Omega,\omega_1)} = 1\Big\},$$

is the norm in $W_0^{-1,p'}(\Omega, \omega_1^{1-p'})$. Hence, we obtain the operator

$$\begin{aligned}\mathbf{A} : W_0^{1,p}(\Omega, \omega_1) &\longrightarrow W_0^{-1,p'}(\Omega, \omega_1^{1-p'}) \\ u &\longmapsto \mathbf{A}u.\end{aligned}$$

However, the problem (1) is equivalent to the equation of the operator

$$\mathbf{A}u = \mathbf{T}, \quad u \in W_0^{1,p}(\Omega, \omega_1).$$

Step 2: Monotonicity of the Operator A

We will we prove that the operator $\mathbf{A}$ is strictly monotone.
Let $u_1, u_2 \in W_0^{1,p}(\Omega, \omega_1)$ with $u_1 \neq u_2$. We have

$$\begin{aligned}&\langle \mathbf{A}u_1 - \mathbf{A}u_2, u_1 - u_2\rangle \\ &= \mathbf{O}(u_1, u_1 - u_2) - \mathbf{O}(u_2, u_1 - u_2) \\ &= \int_\Omega \langle a(x, \nabla u_1), \nabla(u_1 - u_2)\rangle \omega_1 dx - \int_\Omega \langle a(x, \nabla u_2), \nabla(u_1 - u_2)\rangle \omega_1 dx \\ &\quad + \int_\Omega \langle b(x, u_1, \nabla u_1), \nabla(u_1 - u_2)\rangle \omega_2 dx - \int_\Omega \langle b(x, u_2, \nabla u_2), \nabla(u_1 - u_2)\rangle \omega_2 dx \\ &\quad + \int_\Omega g(x, u_1)(u_1 - u_2)\omega_3 dx - \int_\Omega g(x, u_2)(u_1 - u_2)\omega_3 dx \\ &= \int_\Omega \langle a(x, \nabla u_1) - a(x, \nabla u_2), \nabla(u_1 - u_2)\rangle \omega_1 dx \\ &\quad + \int_\Omega \Big(g(x, u_1) - g(x, u_2)\Big)\Big(u_1 - u_2\Big)\omega_3 dx \\ &\quad + \int_\Omega \langle b(x, u_1, \nabla u_1) - b(x, u_2, \nabla u_2), \nabla(u_1 - u_2)\rangle \omega_2 dx.\end{aligned}$$

Thanks to **(A3)**, we obtain

$$\begin{aligned}\langle \mathbf{A}u_1 - \mathbf{A}u_2, u_1 - u_2\rangle &\geq \int_\Omega \alpha|\nabla(u_1 - u_2)|^p \omega_1 dx \\ &\geq \alpha\|\nabla(u_1 - u_2)\|^p_{L^p(\Omega,\omega_1)},\end{aligned}$$

and by Theorem 2(with $\kappa = 1$), we conclude that

$$\langle \mathbf{A}u_1 - \mathbf{A}u_2, u_1 - u_2\rangle \geq \frac{\alpha}{(C^p_\Omega + 1)}\|u_1 - u_2\|^p_{W^{1,p}_0(\Omega,\omega_1)}.$$

Therefore, the operator **A** is strictly monotone.

Step 3: Coercivity of the Operator A

In this step, we prove that the operator **A** is coercive. To this purpose let $u \in W^{1,p}_0(\Omega, \omega_1)$, we have

$$\begin{aligned}\langle \mathbf{A}u, u\rangle &= \mathbf{O}(u, u) \\ &= \int_\Omega \langle a(x, \nabla u), \nabla u\rangle \omega_1 dx + \int_\Omega \langle b(x, u, \nabla u), \nabla u\rangle \omega_2 dx + \int_\Omega g(x, u)u\, \omega_3 dx.\end{aligned}$$

Moreover, from **(A4)** and Theorem 2(with $\kappa = 1$), we obtain

$$\begin{aligned}\langle \mathbf{A}u, u\rangle &\geq \beta_1 \int_\Omega |\nabla u|^p \omega_1 dx + \beta_2 \int_\Omega |\nabla u|^q \omega_2 dx + \beta_3 \int_\Omega |u|^q \omega_2 dx \\ &\geq min(\beta_1, \lambda_4)\left[\int_\Omega |\nabla u|^p \omega_1 dx + \int_\Omega |u|^p \omega_1 dx\right] \\ &\quad + min(\beta_2, \beta_3)\left[\int_\Omega |\nabla u|^q \omega_2 dx + \int_\Omega |u|^q \omega_2 dx\right] \\ &= min(\beta_1, \lambda_4)\|u\|^p_{W^{1,p}_0(\Omega,\omega_1)} + min(\beta_2, \beta_3)\|u\|^q_{W^{1,q}_0(\Omega,\omega_2)} \\ &\geq min(\beta_1, \lambda_4)\|u\|^p_{W^{1,p}_0(\Omega,\omega_1)}.\end{aligned}$$

Hence, we obtain

$$\frac{\langle \mathbf{A}u, u\rangle}{\|u\|_{W^{1,p}_0(\Omega,\omega_1)}} \geq min(\beta_1, \lambda_4)\|u\|^{p-1}_{W^{1,p}_0(\Omega,\omega_1)}.$$

Therefore, since $p > 1$, we have

$$\frac{\langle \mathbf{A}u, u\rangle}{\|u\|_{W^{1,p}_0(\Omega,\omega_1)}} \longrightarrow +\infty \text{ as } \|u\|_{W^{1,p}_0(\Omega,\omega_1)} \longrightarrow +\infty,$$

that is, **A** is coercive.

Step 4: Continuity of the Operator A

We need to show that the operator **A** is continuous. To this purpose let $u_k \longrightarrow u$ in $W^{1,p}_0(\Omega, \omega_1)$ as $k \longrightarrow \infty$. Then $u_k \longrightarrow u$ in $L^p(\Omega, \omega_1)$ and $\nabla u_k \longrightarrow \nabla u$ in

$(L^p(\Omega,\omega_1))^n$. Hence, thanks to Theorem 1, there exist a subsequence (u_{k_i}), functions $\psi_1 \in L^p(\Omega,\omega_1)$ and $\psi_2 \in L^p(\Omega,\omega_1)$ such that

$$\begin{array}{ll} u_{k_i}(x) \longrightarrow u(x), & a.e. \text{ in } \Omega \\ |u_{k_i}(x)| \leq \psi_1(x), & a.e. \text{ in } \Omega \\ \nabla u_{k_i}(x) \longrightarrow \nabla u(x), & a.e. \text{ in } \Omega \\ |\nabla u_{k_i}(x)| \leq \psi_2(x), & a.e. \text{ in } \Omega. \end{array} \tag{4}$$

We will show that $\mathbf{A}u_k \longrightarrow \mathbf{A}u$ in $W_0^{-1,p'}(\Omega,\omega_1^{1-p'})$. In order to prove this convergence we proceed in several steps.

Step 1:

We define the operator

$$\begin{array}{l} B_j : W_0^{1,p}(\Omega,\omega_1) \longrightarrow L^{p'}(\Omega,\omega_1) \\ (B_j u)(x) = a_j(x,\nabla u(x)). \end{array}$$

We now show that

$$B_j u_k \longrightarrow B_j u \quad \text{in} \quad L^{p'}(\Omega,\omega_1).$$

(i) Let $u \in W_0^{1,p}(\Omega,\omega_1)$. Using **(A2)** and Theorem 2(with $\kappa = 1$), we obtain

$$\begin{aligned} \|B_j u\|^{p'}_{L^{p'}(\Omega,\omega_1)} &= \int_\Omega |B_j u(x)|^{p'} \omega_1 dx = \int_\Omega |a_j(x,\nabla u)|^{p'} \omega_1 dx \\ &\leq \int_\Omega \left(f_1 + h_1|\nabla u|^{p-1}\right)^{p'} \omega_1 dx \\ &\leq C_p \int_\Omega \left(f_1^{p'} + h_1^{p'}|\nabla u|^p\right) \omega_1 dx \\ &\leq C_p \left[\|f_1\|^{p'}_{L^{p'}(\Omega,\omega_1)} + \|h_1\|^{p'}_{L^\infty(\Omega)} \|\nabla u\|^p_{L^p(\Omega,\omega_1)}\right] \\ &\leq C_p \left[\|f_1\|^{p'}_{L^{p'}(\Omega,\omega_1)} + \|h_1\|^{p'}_{L^\infty(\Omega)} \|u\|^p_{W_0^{1,p}(\Omega,\omega_1)}\right], \end{aligned}$$

where the constant C_p depends only on p.

(ii) By **(A2)** and (4), we obtain

$$\begin{aligned} \|B_j u_{k_i} - B_j u\|^{p'}_{L^{p'}(\Omega,\omega_1)} &= \int_\Omega |B_j u_{k_i}(x) - B_j u(x)|^{p'} \omega_1 dx \\ &\leq \int_\Omega \left(|a_j(x,\nabla u_{k_i})| + |a_j(x,\nabla u)|\right)^{p'} \omega_1 dx \\ &\leq C_p \int_\Omega \left(|a_j(x,\nabla u_{k_i})|^{p'} + |a_j(x,\nabla u)|^{p'}\right) \omega_1 dx \\ &\leq C_p \int_\Omega \left[\left(f_1 + h_1|\nabla u_{k_i}|^{p-1}\right)^{p'} + \left(f_1 + h_1|\nabla u|^{p-1}\right)^{p'}\right] \omega_1 dx \\ &\leq C_p \int_\Omega \left[\left(f_1 + h_1\psi_2^{p-1}\right)^{p'} + \left(f_1 + h_1\psi_2^{p-1}\right)^{p'}\right] \omega_1 dx \\ &\leq 2C_p C_p' \int_\Omega \left(f_1^{p'} + h_1^{p'}\psi_2^p\right) \omega_1 dx \\ &\leq 2C_p C_p' \left[\|f_1\|^{p'}_{L^{p'}(\Omega,\omega_1)} + \|h_1\|^{p'}_{L^\infty(\Omega)} \|\psi_2\|^p_{L^p(\Omega,\omega_1)}\right]. \end{aligned}$$

Hence, thanks to **(A1)**, we get, as $k \longrightarrow \infty$

$$B_j u_{k_i}(x) = a_j(x, \nabla u_{k_i}(x)) \longrightarrow a_j(x, \nabla u(x)) = B_j u(x), \qquad \text{for almost all } x \in \Omega.$$

Therefore, by Lebesgue's theorem, we obtain

$$\|B_j u_{k_i} - B_j u\|_{L^{p'}(\Omega,\omega_1)} \longrightarrow 0,$$

that is,

$$B_j u_{k_i} \longrightarrow B_j u \quad \text{in} \quad L^{p'}(\Omega, \omega_1).$$

Finally, in view to convergence principle in Banach spaces, we have

$$B_j u_k \longrightarrow B_j u \quad \text{in} \quad L^{p'}(\Omega, \omega_1). \tag{5}$$

Step 2:

We define the operator

$$G_j : W_0^{1,p}(\Omega, \omega_1) \longrightarrow L^{q'}(\Omega, \omega_2)$$
$$(G_j u)(x) = b_j(x, u(x), \nabla u(x)).$$

We also have that

$$G_j u_k \longrightarrow G_j u \quad \text{in} \quad L^{q'}(\Omega, \omega_2). \tag{6}$$

In fact,

(i) Let $u \in W_0^{1,p}(\Omega, \omega_1)$. Using **(A2)**, Remark 1 **(i)** and Theorem 2 (with $\kappa = 1$), we obtain

$$\begin{aligned}
&\|G_j u\|^{q'}_{L^{q'}(\Omega,\omega_2)} \\
&= \int_\Omega |b_j(x, u, \nabla u)|^{q'} \omega_2 dx \\
&\leq \int_\Omega \left(f_2 + h_2|u|^{q-1} + h_3|\nabla u|^{q-1}\right)^{q'} \omega_2 dx \\
&\leq C_q \int_\Omega \left[f_2^{q'} + h_2^{q'}|u|^q + h_3^{q'}|\nabla u|^q\right] \omega_2 dx \\
&\leq C_q \left[\|f_2\|^{q'}_{L^{q'}(\Omega,\omega_2)} + \|h_2\|^{q'}_{L^\infty(\Omega)}\|u\|^q_{L^q(\Omega,\omega_2)} + \|h_3\|^{q'}_{L^\infty(\Omega)}\|\nabla u\|^q_{L^q(\Omega,\omega_2)}\right] \\
&\leq C_q \left[\|f_2\|^{q'}_{L^{q'}(\Omega,\omega_2)} + C^q_{p,q}\left(C^q_\Omega\|h_2\|^{q'}_{L^\infty(\Omega)} + \|h_3\|^{q'}_{L^\infty(\Omega)}\right)\|u\|^q_{W_0^{1,p}(\Omega,\omega_1)}\right],
\end{aligned}$$

where the constant C_q depends only on q.

(ii) According to **(A2)**, Remark 1 (i) and the same arguments used in Step 1 **(ii)**, we obtain analogously,

$$G_j u_k \longrightarrow G_j u \quad \text{in} \quad L^{q'}(\Omega, \omega_2). \tag{7}$$

Step 3:

We define the operator

$$\begin{aligned} &H : W_0^{1,p}(\Omega, \omega_1) \longrightarrow L^{s'}(\Omega, \omega_3) \\ &(Hu)(x) = g(x, u(x)). \end{aligned}$$

In this step, we will show that

$$Hu_k \longrightarrow Hu \quad \text{in} \quad L^{s'}(\Omega, \omega_3). \tag{8}$$

(i) Let $u \in W_0^{1,p}(\Omega, \omega_1)$. Using **(A2)** and Remark 1 (ii), we obtain

$$\begin{aligned} \|Hu\|^{s'}_{L^{s'}(\Omega,\omega_3)} &= \int_\Omega |g(x,u)|^{s'} \omega_3 dx \\ &\leq \int_\Omega \left(f_3 + h_4|u|^{s-1}\right)^{s'} \omega_3 dx \\ &\leq C_s \int_\Omega \left(f_3^{s'} + h_4^{s'}|u|^s\right) \omega_3 dx \\ &\leq C_s \left[\|f_3\|^{s'}_{L^{s'}(\Omega,\omega_3)} + \|h_4\|^{p'}_{L^\infty(\Omega)} \|u\|^s_{L^s(\Omega,\omega_3)}\right] \\ &\leq C_s \left[\|f_3\|^{s'}_{L^{s'}(\Omega,\omega_3)} + C^s_{p,s}\|h_4\|^{p'}_{L^\infty(\Omega)} \|u\|^s_{L^p(\Omega,\omega_1)}\right] \\ &\leq C_s \left[\|f_3\|_{L^{s'}(\Omega,\omega_1)} + C^s_{p,s}C^s_\Omega\|h_4\|^{s'}_{L^\infty(\Omega)} \|u\|^s_{W_0^{1,p}(\Omega,\omega_1)}\right], \end{aligned}$$

where the constant C_s depends only on s.

(ii) By using **(A2)** and Remark 1 (ii), we get

$$\begin{aligned} \|Hu_{k_i} - Hu\|^{s'}_{L^{s'}(\Omega,\omega_3)} &= \int_\Omega |Hu_{k_i}(x) - Hu(x)|^{p'} \omega_3 dx \\ &\leq \int_\Omega \left(|g(x,u_{k_i})| + |g(x,u)|\right)^{s'} \omega_3 dx \\ &\leq C_s \int_\Omega \left(|g(x,u_{k_i})|^{s'} + |g(x,u)|^{s'}\right) \omega_3 dx \\ &\leq C_s \int_\Omega \left[\left(f_3 + h_4|u_{k_i}|^{s-1}\right)^{s'} + \left(f_3 + h_4|u|^{s-1}\right)^{s'}\right] \omega_3 dx \\ &\leq C_s \int_\Omega \left[\left(f_3 + h_4|\psi_1|^{s-1}\right)^{s'} + \left(f_3 + h_4\psi_1^{s-1}\right)^{s'}\right] \omega_3 dx \\ &\leq 2C_sC'_s\left[\|f_3\|^{s'}_{L^{s'}(\Omega,\omega_3)} + \|h_4\|^{s'}_{L^\infty(\Omega)} \|\psi_1\|^s_{L^s(\Omega,\omega_3)}\right] \\ &\leq 2C_sC'_s\left[\|f_3\|^{s'}_{L^{s'}(\Omega,\omega_3)} + C^s_{p,s}\|h_4\|^{s'}_{L^\infty(\Omega)} \|\psi_1\|^s_{L^p(\Omega,\omega_1)}\right], \end{aligned}$$

then, using condition **(A1)**, we deduce, as $k \longrightarrow \infty$

$$Hu_{k_i}(x) = g(x, u_{k_i}(x)) \longrightarrow g(x, u(x)) = Hu(x), \qquad \text{a.e. } x \in \Omega.$$

Therefore, by the Lebesgue's theorem, we obtain

$$\|Hu_{k_i} - Hu\|_{L^{s'}(\Omega,\omega_3)} \longrightarrow 0,$$

that is,

$$Hu_{k_i} \longrightarrow Hu \quad \text{in} \quad L^{s'}(\Omega, \omega_3).$$

We conclude, from the convergence principle in Banach spaces, that

$$Hu_k \longrightarrow Hu \quad \text{in} \quad L^{s'}(\Omega, \omega_3). \tag{9}$$

Finally, let $v \in W_0^{1,p}(\Omega, \omega_1)$ and using Hölder inequality, Theorem 2(with $\kappa = 1$) and Remark 1, we obtain

$$\begin{aligned}
|\mathbf{O}_1(u_k, v) - \mathbf{O}_1(u, v)| &= |\int_\Omega \langle a(x, \nabla u_k) - a(x, \nabla u), \nabla v\rangle \omega_1 dx| \\
&\leq \sum_{j=1}^{n} \int_\Omega |a_j(x, \nabla u_k) - a_j(x, \nabla u)||D_j v|\omega_1 dx \\
&= \sum_{j=1}^{n} \int_\Omega |B_j u_k - B_j u||D_j v|\omega_1 dx \\
&\leq \sum_{j=1}^{n} \|B_j u_k - B_j u\|_{L^{p'}(\Omega,\omega_1)} \|D_j v\|_{L^p(\Omega,\omega_1)} \\
&\leq \left(\sum_{j=1}^{n} \|B_j u_k - B_j u\|_{L^{p'}(\Omega,\omega_1)}\right) \|v\|_{W_0^{1,p}(\Omega,\omega_1)},
\end{aligned}$$

$$\begin{aligned}
|\mathbf{O}_2(u_k, v) - \mathbf{O}_2(u, v)| &= |\int_\Omega \langle b(x, u_k, \nabla u_k) - b(x, u, \nabla u), \nabla v\rangle \omega_2 dx| \\
&\leq \sum_{j=1}^{n} \int_\Omega |b_j(x, u_k, \nabla u_k) - b_j(x, u, \nabla u)||D_j v|\omega_2 dx \\
&= \sum_{j=1}^{n} \int_\Omega |G_j u_k - G_j u||D_j v|\omega_2 dx \\
&\leq \left(\sum_{j=1}^{n} \|G_j u_k - G_j u\|_{L^{q'}(\Omega,\omega_2)}\right) \|\nabla v\|_{L^q(\Omega,\omega_2)} \\
&\leq C_{p,q} \left(\sum_{j=1}^{n} \|G_j u_k - G_j u\|_{L^{q'}(\Omega,\omega_2)}\right) \|\nabla v\|_{L^p(\Omega,\omega_1)} \\
&\leq C_{p,q} \left(\sum_{j=1}^{n} \|G_j u_k - G_j u\|_{L^{q'}(\Omega,\omega_2)}\right) \|v\|_{W_0^{1,p}(\Omega,\omega_1)},
\end{aligned}$$

$$\begin{aligned}
|\mathbf{O}_3(u_k, v) - \mathbf{O}_3(u, v)| &\leq \int_\Omega |g(x, u_k) - g(x, u)||v|\omega_3 dx \\
&= \int_\Omega |Hu_k - Hu||v|\omega_3 dx \\
&\leq \|Hu_k - Hu\|_{L^{s'}(\Omega,\omega_3)} \|v\|_{L^s(\Omega,\omega_3)} \\
&\leq C_{p,s} \|Hu_k - Hu\|_{L^{s'}(\Omega,\omega_3)} \|v\|_{L^p(\Omega,\omega_1)} \\
&\leq C_{p,s} C_\Omega \|Hu_k - Hu\|_{L^{s'}(\Omega,\omega_3)} \|v\|_{W_0^{1,p}(\Omega,\omega_1)}.
\end{aligned}$$

Hence, for all $v \in W_0^{1,p}(\Omega,\omega_1)$, we have

$$\begin{aligned}|\mathbf{O}(u_k,v)-\mathbf{O}(u,v)| &\le |\mathbf{O}_1(u_k,v)-\mathbf{O}_1(u,v)|+|\mathbf{O}_2(u_k,v)-\mathbf{O}_2(u,v)|\\ &\quad +|\mathbf{O}_3(u_k,v)-\mathbf{O}_3(u,v)|\\ &\le \Big[\sum_{j=1}^{n}\Big(\|B_ju_k-B_ju\|_{L^{p'}(\Omega,\omega_1)}+C_{p,q}\|G_ju_k-G_ju\|_{L^{q'}(\Omega,\omega_2)}\Big)\\ &\quad +C_{p,s}C_\Omega\|Hu_k-Hu\|_{L^{s'}(\Omega,\omega_3)}\Big]\|v\|_{W_0^{1,p}(\Omega,\omega_1)}.\end{aligned}$$

Then, we get

$$\begin{aligned}\|\mathbf{A}u_k-\mathbf{A}u\|_* &\le \sum_{j=1}^{n}\Big(\|B_ju_k-B_ju\|_{L^{p'}(\Omega,\omega_1)}+C_{p,q}\|G_ju_k-G_ju\|_{L^{q'}(\Omega,\omega_2)}\Big)\\ &\quad +C_{p,s}C_\Omega\|Hu_k-Hu\|_{L^{s'}(\Omega,\omega_3)}.\end{aligned}$$

Combining (5), (7) and (9), we deduce that

$$\|\mathbf{A}u_k-\mathbf{A}u\|_* \longrightarrow 0 \text{ as } m\longrightarrow\infty,$$

that is, $\mathbf{A}u_k \longrightarrow \mathbf{A}u$ in $W_0^{-1,p'}(\Omega,\omega_1^{1-p'})$. Which implies $\mathbf{A}$ is continuous.

Thus, the operator $\mathbf{A}$ is hemicontinuous, coercive and strictly monotone and $\mathbf{T}\in W_0^{-1,p'}(\Omega,\omega_1^{1-p'})$, so all the conditions of Theorem 3 are satisfied.

Consequently, by Theorem 3, the operator equation $\mathbf{A}u=\mathbf{T}$ has exactly one solution $u\in W_0^{1,p}(\Omega,\omega_1)$ and it follows that u is the unique weak solution of problem (1). Thus, the proof of Theorem 4 is finished.

5 Example

We get $\Omega=\{(x,y)\in\mathbb{R}^2 : x^2+y^2<1\}$, and consider the weight functions $\omega_1(x,y)=(x^2+y^2)^{-1/2}$, $\omega_2(x,y)=(x^2+y^2)^{-1/3}$ and $\omega_3(x,y)=(x^2+y^2)^{-1}$ (we have that $\omega_1,\omega_2,\omega_3\in A_4$, $p=4$, $q=3$ and $s=2$), and the functions $b:\Omega\times\mathbb{R}\times\mathbb{R}^2\longrightarrow\mathbb{R}^2$, $a:\Omega\times\mathbb{R}^2\longrightarrow\mathbb{R}^2$ and $g:\Omega\times\mathbb{R}\longrightarrow\mathbb{R}$ defined by

$$a\Big((x,y),\xi\Big)=h_1(x,y)|\xi|^3 sgn(\xi),$$

where $h_1(x,y)=2e^{(x^2+y^2)}$, and

$$b\Big((x,y),\eta,\xi\Big)=|\xi|^2 sgn(\xi),$$

and

$$g\Big((x,y),\eta\Big)=h_4(x,y)|\eta| sgn(\eta),$$

where $h_4(x,y)=2-cos^2(xy)$.

Let us consider the problem

$$\begin{cases} Lu(x,y) = cos(x+y) & \text{in } \Omega, \\ u(x,y) = 0 & \text{on } \partial\Omega, \end{cases} \tag{10}$$

where,

$$Lu(x,y) = -\operatorname{div}\Big[\omega_1 a\Big((x,y), \nabla u(x,y)\Big) + \omega_2 b\Big((x,y), u(x,y), \nabla u(x,y)\Big)\Big] + \omega_3 g\Big((x,y), u(x,y)\Big).$$

Hence, by Theorem 4, the problem (10) has exactly one solution $u \in W_0^{1,4}(\Omega, \omega_1)$.

6 Conclusion

In this work, we use the main theorem on monotone operators and the weighted Sobolev spaces theory to prove the existence and uniqueness of weak solution of weak solution in the space $W_0^{1,p}(\Omega, \omega_1)$. We hope in a future work to solve other similar problems in spaces with weight and variable exponents.

References

1. Abbassi, A., Allalou, C., Kassidi, A.: Topological degree methods for a Neumann problem governed by nonlinear elliptic equation. Moroccan J. Pure Appl. Anal. **6**(2), 231–242 (2020)
2. Abbassi, A., Allalou, C., Kassidi, A.: Existence results for some nonlinear elliptic equations via topological degree methods. J. Elliptic Parabol. Equ. **7**(1), 121–136 (2021). https://doi.org/10.1007/s41808-021-00098-w
3. Bensoussan, A., Boccardo, L., Murat, F.: On a non linear partial differential equation having natural growth terms and unbounded solution. Annales de l'Institut Henri Poincare (C) Non Linear Anal. **5**(4), 347–364 (1988)
4. Cavalheiro, A.C.: Existence results for degenerate quasilinear elliptic equations in weighted Sobolev spaces. Bull. Belgian Math. Soc. Simon Stevin **17**(1), 141–153 (2010)
5. Cavalheiro, A.C.: Existence results for a class of nonlinear degenerate elliptic equations. Moroccan J. Pure Appl. Anal. **5**(2), 164–178 (2019)
6. Chiarenza, F.: Regularity for solutions of quasilinear elliptic equations under minimal assumptions. Potential Theory Degener. Partial Diff. Oper. **4**, 325–334 (1995)
7. Drabek, P., Kufner, A., Mustonen, V.: Pseudo-monotonicity and degenerated or singular elliptic operators. Bull. Aust. Math. Soc. **58**(2), 213–221 (1998)
8. Fucik, S., John, O., Kufner, A.: Function Spaces. Noordhoff International Publishing, Leyden. Academia, Publishing House of the Czechoslovak Academy of Sciences, Prague (1977)
9. Garcia-Cuerva, J., Rubio de Francia, J.L.: Weighted Norm Inequalities and Related Topics. Elsevier, Amsterdam (2011)
10. Heinonen, J., Kilpelainen, T., Martio, O.: Nonlinear Potential Theory of Degenerate Elliptic Equations. Oxford Math. Monographs. Clarendon Press, Oxford (1993)

11. Kufner, A.: Weighted Sobolev Spaces, vol. 31. Wiley, New York (1985)
12. Kufner, A., Opic, B.: How to define reasonably weighted Sobolev spaces. Commentationes Mathematicae Universitatis Carolinae **25**(3), 537–554 (1984)
13. Muckenhoupt, B.: Weighted norm inequalities for the Hardy maximal function. Trans. Am. Math. Soc. **165**, 207–226 (1972)
14. Muckenhoupt, B.: The equivalence of two conditions for weight functions. Studia Math **49**(1), 101–106 (1974)
15. Ouaarabi, M.E., Abbassi, A., Allalou, C.: Existence result for a Dirichlet problem governed by nonlinear degenerate elliptic equation in weighted Sobolev spaces. J. Elliptic Parabol. Equ. **7**(1), 221–242 (2021). https://doi.org/10.1007/s41808-021-00102-3
16. Ouaarabi, M.E., Allalou, C., Abbassi, A.: On the Dirichlet problem for some nonlinear degenerated elliptic equations with weight. In: 2021 7th International Conference on Optimization and Applications (ICOA), pp. 1–6 (2021)
17. Ouaarabi, M.E., Abbassi, A., Allalou, C.: Existence result for a general nonlinear degenerate elliptic problems with measure datum in weighted Sobolev spaces. Int. J. Optim. Appl. **1**(2), 1–9 (2021)
18. Torchinsky, A.: Real-Variable Methods in Harmonic Analysis. Academic Press, Cambridge (1986)
19. Turesson, B.O.: Nonlinear Potential Theory and Weighted Sobolev Spaces, vol. 1736. Springer, Heidelberg (2000). https://doi.org/10.1007/BFb0103908
20. Zeidler, E.: Nonlinear Functional Analysis and its Applications, vol. II/B. Springer, New York (1990). https://doi.org/10.1007/978-1-4612-0985-0

Unbounded Optimal Control of a Class of Distributed Bilinear Systems

El Hassan Zerrik[1], Abderrahman Ait Aadi[2](✉), and Mohamed Ouhafsa[1]

[1] MACS Team, Department of Mathematics, Moulay Ismail University, Meknes, Morocco
zerrik3@yahoo.fr, mohamed.ouhafsa@edu.umi.ac.ma

[2] Department of Sciences, ENS, Moulay Ismail University, Meknes, Morocco
abderrahman.aitaadi@gmail.com

Abstract. In this paper, we consider optimal control of infinite dimensional bilinear systems evolving in a spatial domain $\Omega \subset \mathbb{R}^n, n \geq 1$ using unbounded control. Then, we minimize a functional cost constituted of the deviation between the desired state and the final one at time T, the effort term and the energy one. The purpose of this study is to prove that an optimal control exists, and characterized as a solution to an optimality system. Numerical algorithm for the computation of an optimal control is given and successfully illustrated through simulations for the heat equation.

1 Introduction

Bilinear systems constitute an important subclass of nonlinear systems. The nonlinearity in mathematical models of bilinear systems appears in the multiplication of state and control in the dynamical process. The major importance of bilinear systems indeed lies in their applications to real word systems, and represent many physical processes, for example: the basic law of mass action [13] and dynamics of heat exchanger with controlled flow [12].

Controllability of a class of infinite dimensional bilinear systems has been studied by many authors using different control techniques. In [3], the authors considered the approximate controllability of the bilinear beam equation in the mono-dimensional case. In [8], the author proved the multiplicative controllability of various parabolic and hyperbolic equations of semilinear type using asymptotic qualitative methods. He studied the controllability of bilinear parabolic equations with the reaction-diffusion term satisfying Newton law [7].

Optimal control of infinite dimensional bilinear systems has been developed in many works. In [9], the author proved the existence and gave characterization of an optimal control of a convective-diffusive fluid bilinear equation. In [6], the author discussed optimal control of a bilinear heat equation with distributed bounded control. In [1], the authors discussed unbounded optimal control for a class of bilinear systems. In [10], the authors studied optimal control problem of a wave equation excited by boundary controls. In [4], the authors considered optimal control of Kirchoff plate equation using distributed bounded controls. In [11], the author proved optimal control problem

S. Melliani and O. Castillo (Eds.): ICPAMS 2021, LNNS 476, pp. 291–301, 2023.
https://doi.org/10.1007/978-3-031-12416-7_25

of the bilinear wave equation using bounded controls, and in [5], the authors considered optimal control of the bilinear Kirchoff equation using bounded spatial controls. In [15], the authors discussed optimal control of a class of infinite dimensional bilinear systems with distributed bounded and unbounded controls and in [2], the authors studied constrained optimal control of the bilinear plate equation with distributed bounded controls and acting on the velocity term. The used approach is based essentially on the differentiability in the Gateaux sense.

The present work deals with optimal control of a large class of infinite dimensional bilinear systems where the functional cost is constituted of the deviation between the desired state and the final one at time T, the effort term and the energy one. Then, we prove the existence of a solution of such problem and we give characterization of an optimal control. Moreover, we develop a numerical approach that leads to an algorithm and simulations. The used approach is mainly based on the Frechet differentiability and the characterization of the normal vector to the set of admissible controls.

More precisely, let Ω be an open bounded domain of $\mathbb{R}^n (n \geq 1)$, and we consider the following bilinear system

$$\begin{cases} \dot{z}(t) = Az(t) + u(t)Bz(t) & 0 < t < T \\ z(0) = z_0 \in L^2(\Omega) \end{cases} \tag{1}$$

where $A : D(A) \subset L^2(\Omega) \longrightarrow L^2(\Omega)$ is the infinitesimal generator of strongly continuous semigroup $(S(t))_{t\geq 0}$ on state space $L^2(\Omega)$, whose norm and scalar product are denoted, respectively, by $\|.\|$ and $\langle .,. \rangle$, $u \in U_{ad}$ is the control function, where U_{ad} is the set of admissible controls, assumed to be a nonempty, closed and convex subset of $L^2(0,T)$, and $B : L^2(\Omega) \longrightarrow L^2(\Omega)$ is a linear bounded operator.

For all $u \in L^2(0,T)$ and $z_0 \in L^2(\Omega)$, the system (1) has a unique weak solution in the space $C(0,T;L^2(\Omega))$ (see [14]) and z is the solution of

$$z(t) = S(t)z_0 + \int_0^t S(t-s)u(s)Bz(s)ds. \tag{2}$$

Consider the following functional

$$\mathscr{J}(u) = \frac{\alpha}{2}\|z(T) - z_d\|^2 + \frac{\beta}{2}\int_0^T \|z(t) - z_d\|^2 dt + \frac{\varepsilon}{2}\|u\|^2_{L^2(0,T)} \tag{3}$$

where α, β, and ε are nonnegative constants, and $z_d \in L^2(\Omega)$ indicates the desired state.

Our problem consists in finding a control u that steers the state close to z_d, over the time interval $[0,T]$ with a reasonable amount of energy. This may be stated as the following minimization problem

$$\begin{cases} \min \mathscr{J}(u) \\ u \in U_{ad} \end{cases} \tag{4}$$

The paper is organized as follows: in Sect. 2, we prove the existence of an optimal control solution of problem (4). In Sect. 3, we establish a characterization of an optimal control solution of problem (4). In Sect. 4, we develop a numerical algorithm for the computation of such optimal control. The approach is successfully illustrated by simulations.

2 Existence of an Optimal Control

This section is devoted to the existence of an optimal control solution of problem (4).

Theorem 1. *There exists an optimal control $u \in U_{ad}$, solution of problem (4).*

Proof. The set $\{\mathcal{J}(u)\ ;\ u \in U_{ad}\}$ is non-empty and nonnegative, and thus it has a nonnegative infimum.

Let $(u_n)_{n\in\mathbb{N}}$ be a minimizing sequence in U_{ad}, as $\varepsilon > 0$ we have

$$\|u_n\|^2_{L^2(0,T)} \leq \frac{2}{\varepsilon}\mathcal{J}(u_n),\ \ \forall n \in \mathbb{N}$$

It follows that $(u_n)_{n\in\mathbb{N}}$ is bounded. Then, there exists a subsequence, still denoted by $(u_n)_{n\in\mathbb{N}}$ that converges weakly to a limit u^*. Since U_{ad} is closed and convex, it is closed for the weak topology, which implies that $u^* \in U_{ad}$.

Let z_n and z^* be the unique solutions of systems (1), associated to u_n and u^* respectively. From (2), we have

$$\begin{aligned} z_n(t) - z^*(t) &= \int_0^t S(t-s)[u_n(s)Bz_n(s) - u^*(s)Bz^*(s)]ds \\ &= \int_0^t S(t-s)[(u_n(s) - u^*(s))Bz^*(s) + u_n(s)(Bz_n(s) - Bz^*(s))]ds \end{aligned}$$

Then

$$\begin{aligned} \|z_n(t) - z^*(t)\| \leq& \left\| \int_0^t S(t-s)(u_n(s) - u^*(s))Bz^*(s)ds \right\| \\ &+ \int_0^t \|S(t-s)\| |u_n(s)| \|B\| \|z_n(s) - z^*(s)\| ds \end{aligned}$$

Using Gronwall Lemma, the above inequality becomes

$$\|z_n(t) - z^*(t)\| \leq \left\| \int_0^t S(t-s)(u_n(s) - u^*(s))Bz^*(s)ds \right\| e^{\|B\| \int_0^t \|S(t-s)\| |u_n(s)| ds}$$

There exist constant $M \geq 1$ and $\rho \in \mathbb{R}$ such that $\|S(t)\| \leq Me^{|\rho|t}$, then

$$\int_0^t \|S(t-s)\| |u(s)| ds \leq Me^{|\rho|T} T^{\frac{1}{2}} \left(\int_0^T |u(s)|^2 ds \right)^{\frac{1}{2}}$$

The above inequality yields

$$\|z_n(t) - z^*(t)\| \leq \|\Lambda_t(u_n - u^*)\| e^{Me^{|\rho|T}(\|B\| T^{\frac{1}{2}} \mu)},\ \ \forall n \in \mathbb{N} \tag{5}$$

where $\mu = \sup \|u_n\|_{L^2(0,T)}$, and the operator $\Lambda_t : L^2(0,T) \longrightarrow L^2(\Omega)$ is defined by

$$\Lambda_t u = \int_0^t S(t-s)u(s)Bz^*(s)ds,\ \ \forall u \in L^2(0,T)$$

Let us prove that for all $t \in [0,T]$, Λ_t is compact.

To this end, we prove that for any sequence $(\varphi_n)_n$ in $L^2(\Omega)$ that weakly converges to 0 in $L^2(\Omega)$, $(\Lambda_t^* \varphi_n)_n$ converge with norm to 0 in $L^2(\Omega)$.

Let us calculate the adjoint of the operator Λ_t.

Let $u \in L^2(0,T)$ and $y \in L^2(\Omega)$, we have

$$< \Lambda_t(u), y > = \int_0^t < S(t-s)u(s)Bz^*(s), y(s) > ds$$

$$= \int_\Omega < u(s), S^*(t-s)y(s)B^*z^*(s) >_{L^2(0,T)} dx$$

$$= < u(.), \int_\Omega S^*(t-s)y(x,.)B^*z^*(x,.)dx >_{L^2(0,T)}$$

Thus $\Lambda_t^* : L^2(\Omega) \to L^2(0,T)$ is given by

$$[\Lambda_t^* z](s) = \begin{cases} < S(t-s)Bz^*(.), z > & if \quad 0 \le s < t \\ 0 & if \quad t < s \le T \end{cases}$$

Let $(\phi_l)_{l \in \mathbb{N}}$ be a sequence in $L^2(\Omega)$ such that $\phi_l \underset{l\to\infty}{\rightharpoonup} 0$. Without loss of generality, we can assume that $\|\phi_l\|_{L^2(\Omega)} \le l, \ \forall l \in \mathbb{N}$, then

$$\phi_l \underset{l\to\infty}{\rightharpoonup} 0 \Longrightarrow \|[\Lambda_t^* \phi_l](s)\| \underset{l\to\infty}{\longrightarrow} 0 \ a.e \text{ on } [0,T]$$

Thus, for all $s \in [0,T]$, we have

$$|[\Lambda_t^* \phi_l](s)| \le \|S(t-s)\| \|B\| \|z^*(s)\| \|\phi_l(s)\| \le Me^{|\rho|T} \|B\| \|z^*(s)\|, \ \forall s \in [0,T]$$

Applying the dominated convergence theorem allows

$$\|\Lambda_t^* \phi_l(s)\| \underset{l\to\infty}{\longrightarrow} 0, \text{ in } L^2(0,T)$$

We conclude that Λ_t is compact.

It follows from the weak convergence $(u_n - u^*) \underset{n\to\infty}{\rightharpoonup} 0$ that

$$\lim_{n\to\infty} \|\Lambda_t(u_n - u^*)\|_{L^2(\Omega)} = 0.$$

Therefore, by the inequality (5), we obtain

$$\lim_{n\to\infty} \|z_n(t) - z^*(t)\|_{L^2(\Omega)} = 0, \ a.e \text{ on } t \in [0,T]$$

We have $\|z^*(T) - z_d\|^2_{L^2(\Omega)} = \liminf_{n\to\infty} \|z_n(T) - z_d\|^2_{L^2(\Omega)}$.

Applying Fatou Lemma, gives

$$\int_0^T \|z^*(T) - z_d\|^2 dt \le \liminf_{n\to\infty} \int_0^T \|z_n(T) - z_d\|^2 dt$$

Since the norms are lower semi-continuous for the weak topology, it follows that the weak convergence of $(u_n) \to u^*$ yields that $\|u^*\|_{L^2(0,T)} \le \liminf_{n\to\infty} \|u_n\|_{L^2(0,T)}$.

Then

$$\mathscr{J}(u^*) \le \liminf_{n\to\infty} \mathscr{J}(u_n) = \mathscr{J}^*.$$

We conclude that u^* is solution of problem (4).

3 Characterization of an Optimal Control

In this section, we give characterization of an optimal control solution of problem (4).

First, we characterize the differential of the functional cost (3).

Proposition 1. *The functional (3) is differentiable in the Frechet sense, and its differential for all $u \in L^2(0,T)$ is given by*

$$\begin{cases} d_u \mathscr{J} \cdot h = \langle \mathscr{J}'(u), h \rangle_{L^2(0,T)} \\ \mathscr{J}'(u)(t) = < \varphi(t), Bz(t) > + \varepsilon u(t) \end{cases} \tag{6}$$

where z is the solution of system (1), and φ is the solution of the following adjoint equation

$$\begin{cases} \dot{\varphi}(t) = -A^* \varphi(t) - u(t) B^* z(t) \varphi(t) - \beta(z(T) - z_d) \\ \varphi(T) = \alpha(z(T) - z_d) \end{cases} \tag{7}$$

Proof. The adjoint operator A^* generates a $\mathscr{C}_0$-semigroup $(S^*(t))_{t>0}$ on $L^2(\Omega)$.

For $u \in U_{ad}$, the Eq. (7) has a unique weak solution $\varphi \in \mathscr{C}([0,T], L^2(\Omega))$, φ is also the solution of the following equation

$$\varphi(t) = S^*(T-t)\alpha(z(T) - z_d) + \int_t^T S^*(s-t)[u(t)B^* z(t)(\varphi(t)) - \beta(z(T) - z_d)]ds \tag{8}$$

Let us show that $\mathscr{J}$ is Frechet differentiable.

The mapping $u \longmapsto z_u$ is Frechet differentiable in $L^2(0,T)$ and $d_u z(t)$ its differential at u. We have $z_h(t) = d_u z(t) h$ is the solution of the integral equation

$$z_h(t) = \int_0^t S(t-s)h(s)Bz_u(s) + u(s)Bz_u(s)z_h(s)ds \tag{9}$$

For all $u, u+h \in U_{ad}$, we have

$$\begin{aligned} \|z_{u+h}(T) - z_d\|^2_{L^2(\Omega)} &= < z_{u+h}(T) - z_d, z_{u+h}(T) - z_d >_{L^2(\Omega)} \\ &= < z_u(T) + z_h(T) - z_d, z_u(T) + z_h(T) - z_d >_{L^2(\Omega)} \\ &= < z_u(T) - z_d, z_u(T) + z_h(T) - z_d >_{L^2(\Omega)} + < z_h(T), z_u(T) > \\ &+ z_h(T) - z_d >_{L^2(\Omega)} \\ &= < z_u(T) - z_d, z_u(T) - z_d >_{L^2(\Omega)} + < z_h(T), z_u(T) - z_d >_{L^2(\Omega)} \\ &+ < z_h(T), z_u(T) - z_d >_{L^2(\Omega)} + < z_h(T), z_h(T) >_{L^2(\Omega)} \\ &= \|z_u(T) - z_d\|^2_{L^2(\Omega)} + 2 < z_h(T), z_u(T) - z_d >_{L^2(\Omega)} + o\|h\|_{L^2(0,T)} \end{aligned}$$

Then

$$\|z_{u+h}(T) - z_d\|^2_{L^2(\Omega)} - \|z_u(T) - z_d\|^2_{L^2(\Omega)} = 2 < z_h(T), z_u(T) - z_d >_{L^2(\Omega)} + o\|h\|_{L^2(0,T)}$$

Using similar calculations to those given above, we obtain

$$\int_0^T \|z_{u+h}(t) - z_d\|^2_{L^2(\Omega)} - \|z_u(t) - z_d\|^2_{L^2(\Omega)} dt = 2\int_0^T < z_h(t), z_u(t) - z_d >_{L^2(\Omega)} dt + o\|h\|$$

and it is easy to see that $\|u+h\|^2_{L^2(0,T)} - \|u\|^2_{L^2(0,T)} = 2 < u,h >_{L^2(0,T)} + o\|h\|_{L^2(0,T)}$. Then, $\mathscr{J}$ is Frechet differentiable over U_{ad}, and its derivative at u is given by

$$\mathscr{J}'(u)\cdot h = < z_h(T), \alpha z(T) - z_d >_{L^2(\Omega)} + \int_0^T < z_h(t), \beta(z(T) - z_d) >_{L^2(\Omega)} dt + \varepsilon \int_0^T u(t)h(t)dt$$

where z_h solution of (9) is the solution of the equation

$$\begin{cases} \dot{z}_h(t) = Az_h(t) + h(t)Bz(t) + u(t)Bz(t)z_h(t) \\ z_h(0) = 0 \end{cases} \tag{10}$$

For $n \in \rho(A)$ (the resolving set of A), let $A_n = nA(nI - A)^{-1}$, and $A_n^* = nA^*(nI - A^*)^{-1}$ be the Yosida's approximations of A and A^*.

Let φ_n and z_n the solutions of Eqs. (7) and (10), with A_n and A_n^* instead of A and A^*. Since A_n and A_n^* are bounded, we have $\dot{z}_n \in L^2(0,T;L^2(\Omega))$ and $\dot{\varphi}_n \in L^2(0,T;L^2(\Omega))$. Then

$$\begin{aligned} &\int_0^T < \beta(z(T) - z_d), z_n(t) > dt = -\int_0^T < \dot{\varphi}_n(t) + (A_n^* + u(t)[B(z(t))]^*)\varphi_n(t), z_n(t) > dt \\ &= -\int_0^T [< \dot{\varphi}_n(t), z_n(t) > + < \varphi_n(t), A_n z_n(t) u(t) B(z(t)) z_n(t) >] dt \\ &= -\int_0^T < \dot{\varphi}_n(t), z_n(t) > dt - \int_0^T < \varphi_n(t), \dot{z}_n(t) - h(t)Bz(t) > dt \end{aligned}$$

Integrating by part,as $\dot{z}_n \in L^2(0,T;L^2(\Omega)), \dot{\varphi} \in L^2(0,T;L^2(\Omega))$, gives

$$\int_0^T < \dot{\varphi}_n(t), z_n(t) > dt + \int_0^T < \varphi_n(t), \dot{z}_n(t) > dt = < \varphi_n(T), z_n(T) > - < \varphi(0), z_n(0) >$$

Since $z_n(0) = 0$ and $\varphi_n(T) = \alpha(z(T) - z_d)$, the above equality becomes

$$\int_0^T < \dot{\varphi}_n(t), z_n(t) > dt + \int_0^T < \varphi_n(t), \dot{z}_n(t) > dt = < \alpha(z(T) - z_d), z_n(T) >$$

Thus

$$\int_0^T < \beta(z(t) - z_d), z_n(t) > dt = - < \alpha(z(t) - z_d), z_n(T) > + \int_0^T < \varphi_n(t), h(t)Bz(t) > dt.$$

Or again

$$< \alpha(z(T) - z_d), z_n(T) > + \int_0^T < \beta(z(t) - z_d), z_n(t) > dt = \int_0^T < \varphi_n(t), h(t)Bz(t) >_{L^2(\Omega)} dt.$$

Let φ be the solution of the adjoint equation (7), we have $\lim_{n\to+\infty} \|z_n - z_h\|_{\mathscr{C}([0,T],L^2(\Omega))} = 0$ and $\lim_{n\to+\infty} \|\varphi_n - \varphi\|_{\mathscr{C}([0,T],L^2(\Omega))} = 0$.

The above calculation allows

$$< \alpha(z(T)-z_d), z_h(T) > + \int_0^T < \beta(z(t)-z_d), z_h(t) > dt = < B^*z(t)\varphi(t), h(t) >_{L^2(0,T)}$$

and

$$< \alpha(z(T)-z_d), z_h(T) > + \int_0^T < \beta(z(t)-z_d), z_h(t) > dt = \int_0^T < \varphi_n(t), Bz(t) >_{L^2(\Omega)} dt.$$

Therefore the derivative of $\mathscr{J}$ is given by

$$\mathscr{J}'(u)\cdot h = \int_0^T [\langle \varphi(t), Bz(t) \rangle_{L^2(\Omega)} + \varepsilon u(t)]h(t)dt.$$

Hence $d_u \mathscr{J} \cdot h = < \mathscr{J}'(u), h >_{L^2(0,T)}$ ·

Now, we give the necessary optimality conditions.

Proposition 2. *Let u^* be an optimal control solution of problem (4). Then u^* satisfies*

$$< \mathscr{J}'(u^*), v-u^* >_{L^2(0,T)} \geq 0, \ \forall\, v \in U_{ad} \tag{11}$$

where $\mathscr{J}'(u^*)$ is the Frechet derivative of $\mathscr{J}$ at u^*, given by (6).

Proof. Let u^* be an optimal control, and $v \in U_{ad}$, the convexity of U_{ad} implies that

$$u^* + \lambda(v-u^*) \in U_{ad}, \ \forall \lambda \in]0,1[.$$

Then

$$\mathscr{J}(u^*) \leq \mathscr{J}(u^* + \lambda(v-u^*)), \tag{12}$$

So

$$\mathscr{J}(u^*) \leq \mathscr{J}(u^*) + \lambda < \mathscr{J}'(u^*), v-u^* > + \|\lambda(v-u^*)\|\theta(\lambda(v-u^*))$$

where $\lim_{\|z\|\to 0} \theta(z) = 0$.

$< \mathscr{J}'(u^*), v-u^* > \geq -\|v-u^*\|\theta(\lambda(v-u^*)), \ \forall\, \lambda \in]0,1[.$

Thus

$$< \mathscr{J}'(u^*), v-u^* > \geq -\|v-u^*\| \lim_{\lambda\to 0} \theta(\lambda(v-u^*))$$

As $\lim_{\lambda\to 0} \theta(\lambda(v-u^*)) = 0$ we deduce that $< \mathscr{J}'(u^*), v-u^* > \geq 0$.

Proposition 3. *Let u^* be an optimal control, solution of problem (4), then*

$$\begin{cases} \mathscr{J}'(u^*) = 0 \\ or\ \ u^* \in \partial U_{ad}\ and\ -\dfrac{\mathscr{J}'(u^*)}{\|\mathscr{J}'(u^*)\|}\ is\ a\ normal\ vector\ to\ U_{ad}\ at\ u^* \end{cases} \tag{13}$$

Proof. Let u^* be an optimal control. We suppose that $\mathscr{J}'(u^*) \neq 0$, then $u^* \in \partial U_{ad}$. Indeed, if $u^* \in U_{ad}$, then there exists $r > 0$ such that $B(u^*, r) \subset U_{ad}$, where $B(u^*, r)$ the open ball with centre u^* and radius r.

Applying the inequality (11) to the elements of $B(u^*, r)$, we obtain $\mathscr{J}'(u^*) = 0$ which is absurd. Then $u^* \in \partial U_{ad}$.

Using (11), we deduce that $-\dfrac{\mathscr{J}'(u^*)}{\|\mathscr{J}'(u^*)\|}$ is the normal vector to U_{ad} at u^*.

We apply the optimality condition (13), to characterize the optimal control.

Theorem 2. *The optimal control solution of problem (4) is given by the formula*

$$u^*(t) = -\frac{1}{\varepsilon} < \varphi(t), Bz(t) > . \tag{14}$$

Proof. Let u^* be an optimal control, then $u^* \in Int(U_{ad}) = U_{ad}$, so, the condition (13) becomes $\mathscr{J}'(u^*) = 0$. We deduce that $u^*(t) = -\dfrac{1}{\varepsilon} < \varphi(t), Bz(t) >$.

Remark 1. For $\alpha = 0$, we retrieve the result established in [15] and in [2] for the plate equation.

4 Numerical Approach and Simulations

We have seen that an optimal control solution of problem (4) is solution of the Eq. (13) that gives by the formula (14). The computation of such a control may be achieved considering the following algorithm.

- **Step 1:** Initialization: y_0 initial state, $u_0 \in U_{ad}$, $\kappa > 0$ the precision.
- **Step 2:**
  Solving the system (1) gives z_n.
  Solving the adjoint system (7) gives φ_n.
- **Step 3:**
  The control $u_{n+1}(t)$ is computed by the formula (14).
- **Step 4:**
  If $\| u_{n+1} - u_n \| \leq \kappa$, stop.
  else n = n + 1, go to step 2.

Example: Heat equation

On $]0,1[$, we consider the following bilinear heat equation

$$\begin{cases} \dfrac{\partial}{\partial t} z(x,t) = c\dfrac{\partial^2}{\partial x^2} z(x,t) + u(t) z(x,t) & x \in]0,1[,\ t \in]0,T[, \\ z(0,t) = z(1,t) = 0 & t \in]0,T[\\ z(x,0) = z_0 & x \in]0,1[\end{cases} \tag{15}$$

where $A = c\dfrac{\partial^2}{\partial x^2}$, $c \in \mathbb{R}$ with domain $D(A) = H^2(0,1) \cap H_0^1(0,1)$ and control operator $B = I$. The operator A generates a C_0 semi-group $(S(t))_{t \geq 0}$ on $L^2(\Omega)$ and is given by

$$S(t)z = \sum_{n=0}^{\infty} e^{-(n\pi)^2 ct} \langle \phi_n, z \rangle \phi_n$$

where $\phi_0 = 1$, $\phi_n(x) = \sqrt{2}\sin(n\pi x)$, $\forall n \in \mathbb{N}^*$.

Consider the problem (4) with $U_{ad} = L^2(0,T)$ and

$$\mathscr{J}(u) = \frac{\alpha}{2}\|z(.,T) - z_d\|^2_{L^2(0,1)} + \frac{\beta}{2}\int_0^T \|z(.,t) - z_d\|^2_{L^2(0,1)}dt + \frac{\varepsilon}{2}\int_0^T |u(t)|^2 dt \quad (16)$$

The differential of the functional cost is given by

$$\mathscr{J}'(u)(t) = <\varphi(t), z(t)>_{L^2(0,1)} + \varepsilon u(t) \quad (17)$$

where φ is the solution of

$$\begin{cases} \dot{\varphi}(t) = [-c\frac{\partial^2}{\partial x^2} - u(t)z(t)]\varphi(t) - \beta(z(T) - z_d) \\ \varphi(T) = \alpha(z(T) - z_d) \end{cases} \quad (18)$$

We consider the following data $z_0(x) = x(1-x^3), z_d(x) = 0.2x(1-x)(x^2+x+1), c = 0.001$, $T = 1$, $\alpha = 10^3$, $\beta = 10^2$, $\varepsilon = 1$, $\kappa = 10^{-3}$.

The optimal control is given by

$$u^*(t) = -\frac{1}{\varepsilon} < \varphi(t), z(t) >_{L^2(0,1)}$$

and the simulations give the following figures:

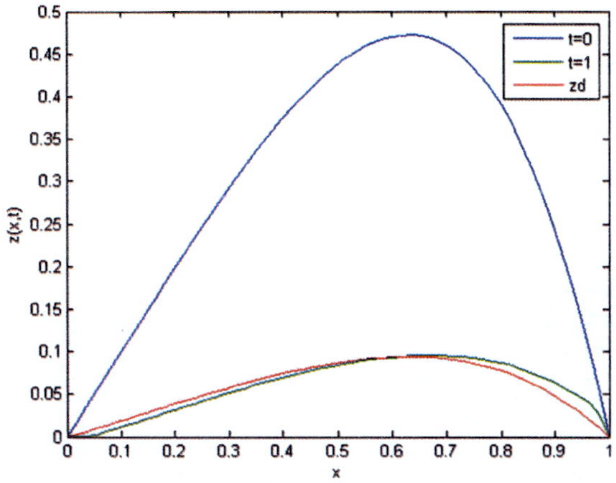

Fig. 1. Final state on $]0,1[$

Figure 1 shows that the final state is very close to the desired one with error $\|z(T) - z_d\|_{L^2(\Omega)} = 4.4133 \times 10^{-6}$ and the evolution of control is given by Fig. 2.

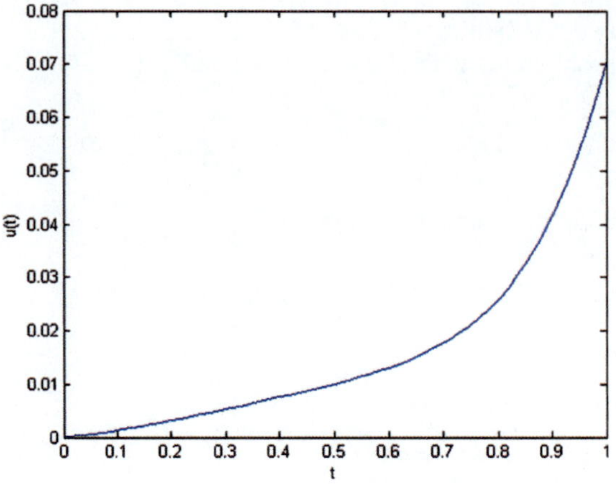

Fig. 2. Evolution of control function

5 Conclusion

Optimal control of a class of infinite dimensional bilinear systems is considered using unbounded control. The existence of an optimal control is proved and characterized by optimality conditions. The obtained results are illustrated by examples and successfully tested through simulations. Questions are still open, for instance the case of optimal control of linear systems acted by bilinear boundary controls.

Acknowledgments. This work was carried out with the help of the Academy Hassan II of Sciences and Technology.

References

1. Addou, A., Benbrik, A.: Existence and uniqueness of optimal control for a distributed parameter bilinear systems. J. Dyn. Control Syst. **8**, 141–152 (2002)
2. Aadi, A.A., Zerrik, E.: Constrained regional control problem of a bilinear plate equation. Int. J. Control (2020). https://doi.org/10.1080/00207179.2020.1834144
3. Ball, J.M., Marsden, J.E., Slemrod, M.: Controllability for distributed bilinear systems. SIAM J. Control Optim. **20**(4), 575–597 (1982)
4. Bradley, M.E., Lenhart, S.: Bilinear optimal control of a Kirchhoff plate. J. Syst. Control Lett. **22**(1), 27–38 (1994)
5. Bradley, M.E., Lenhart, S.: Bilinear spatial control of the velocity term in a Kirchhoff plate equation. Electron. J. Diff. Equ. **22**(1), 1–15 (2001)
6. Joshi, H.R.: Optimal control of the convective velocity coefficient in a parabolic problem. Nonlinear Anal. Theory Methods Appl. **63**, 1383–1390 (2005)
7. Khapalov, A.K.: On bilinear controllability of the parabolic equation with the reaction-diffusion term satisfying Newton's law. J. Comput. Appl. Math. **21**(1), 1–23 (2002)
8. Khapalov, A.K.: Controllability of Partial Differential Equations Governed by Multiplicative Controls. Springer, Heidelberg (2010). https://doi.org/10.1007/978-3-642-12413-6

9. Lenhart, S.: Optimal control of a convective-diffusive fluid problem. J. Math. Models Methods Appl. Sci. **5**(2), 225–237 (1995)
10. Lenhart, S., Protopopescu, V., Yong, J.: Optimal control of a reflection boundary coefficient in an acoustic wave equation. J. Appl. Anal. **68**, 179–194 (1998)
11. Liang, M.: Bilinear optimal control for a wave equation. Math. Models Methods Appl. Sci. **9**, 45–68 (1999)
12. Mohler, R.R.: Bilinear Control Processes: With Applications to Engineering, Ecology, and Medicine. Academic Press, New York (1973)
13. Mohler, R.R., Ruberti, A.: Theory and Applications of Variable Structure Systems. Academic Press, New York (1972)
14. Pazy, A.: Semi-groups of Linear Operators and Applications to Partial Differential Equations. Springer, New York (1983). https://doi.org/10.1007/978-1-4612-5561-1
15. Zerrik, E., El Kabouss, A.: Regional optimal control problem of a class of infinite dimensional bi-linear systems. Int. J. Control **90**(7), 1495–1504 (2017)

Approximate Bayesian Estimation of Parameters of an Agent-Based Model in Epidemiology

Mokhtar Zbair(✉), Abdelaziz Qaffou(✉), and Khalid Hilal(✉)

Applied Mathematics and Scientific Computing Laboratory, Department of Applied Mathematics, Faculty of Sciences and Techniques, C.P: 523, Sultan Moulay Slimane University -Beni -Mellal, Beni Mellal, Morocco
statzbair@yahoo.fr, aziz.qaffou@gmail.com, hilalkhalid2005@yahoo.fr

Abstract. This paper is concerned with the Bayesian estimation of the parameters of an epidemic agent-based model. The proposed model simulates the classic SEIR epidemiological model at the agent level. The estimation of parameters of the introduced model is based on Approximate Bayesian Computation (ABC). Simulated datasets are used to test the proposed methodology.

Keywords: Approximate Bayesian Computation · Agent-based models · Seir epidemic model

AMS Subject Classification: 62F15 · 60E05 · 60J10 · 60J60 · 92D30 · 65C05

1 Introduction

The modeling of complex systems generates models that are difficult to estimate by classical inference methods. The likelihood for this kind of models is often computationally intractable. Free likelihood methods offer a promising way to deal with this kind of problem. Approximate Bayesian Computation (see Sisson et al. [16]) is a family of free likelihood algorithms widely applied in disciplines that rely on genetics, ecology ...etc. This paper aims to propose an approximate Bayesian computation algorithm to estimate the parameters of an agent-based model in epidemiology. The paper is structured as follows. Section 2 formalises the agent-based model; Sect. 3 present an MC simulation study of the dynamics of the introduced model; Sect. 4 is devoted to the presentation of an approximate Bayesian estimation proposed to estimate the parameter of the agent based model introduced; the evaluation of the estimation algorithm is given in Sect. 5; finally, Sect. 6 conclude the paper.

S. Melliani and O. Castillo (Eds.): ICPAMS 2021, LNNS 476, pp. 302–314, 2023.
https://doi.org/10.1007/978-3-031-12416-7_26

2 Agent-Based Model Proposed

Agent-based models date back to the work of modeling complex systems in the years 1969–1985 (see, Axelrod and Hamilton [1], Langton [14] and Thomas C. Schelling [15] among others). The literature on the application of agent-based models in epidemiology shows two main classes of models (see, Hunter et al. [9]). The first class contains general models whose objective is to show how diseases spread according to the different parameters of the models. The second class includes models specific to given disease.

As noted by Hunter et al. in [9], four components are distinguished in agent-based models applied in epidemiology: the structure of agents, the environment of agents, the mobility of agents and the transmission dynamics of disease.

Hence, the specification of an agent-based model requires the precision of the following elements (see, Hunter et al. [9]): the structure of the agents, the space in which the agents evolve (discrete or continuous), time (discrete or continuous), the mobility of the agents, the environment and the dynamics of contagion.

The model proposed here is a general type model. The time is assumed to be discrete and divided into days.

The agents live in households of random size. We assume that the household size follows a Poisson probability distribution of parameter λ. The choice of this probability distribution is motivated by [11] and [12], more precisely, the household size n_h is generated as follows:

$$n_h = 1 + \epsilon_h \text{where} \epsilon_h \sim Poisson(\lambda).$$

We modeled the household size by a Poisson probability distribution shifted by 1 to ensure that each household contains at least one agent with probability equal to 1.

Here we use $X \sim P$ to indicate that X is a random variable following the probability distribution P.

From this specification, the number of agents is a random variable with mean $N(1+\lambda)$ and variance $N\lambda$ where N is the number of households.

Each agent is characterized by a state with 4 possible values: 0 (susceptible), 1 (exposed), 2 (infected) and 3 (recovered).

Every day, agents make contacts modeled by Erdős-Rényi random network (see [7] and [5]).

The dynamic of contagion is based on the classic SEIR model (see, [2] and [13]). SEIR model is a compartmental model used to model infectious diseases with a latent phase during which the individual is infected but not yet infectious. In this model, the population is affected to 4 compartments: Susceptible, Exposed, Infectious and Recovered (noted respectively: S, E, I and R). Each member of the population transits from susceptible to exposed to infectious and finally to recovered as shown in the Fig. 1 below:

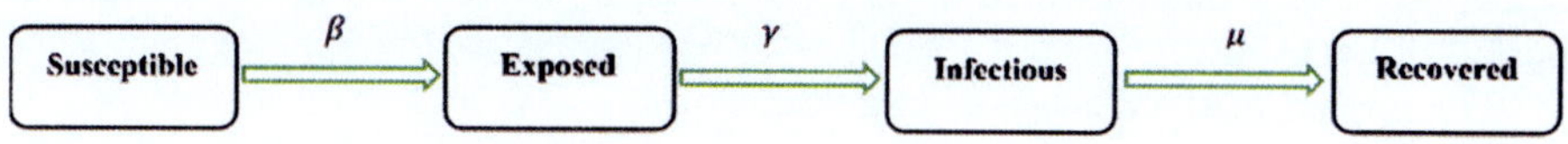

Fig. 1. SEIR model

The parameter β is the infectious rate, it controls the rate of spread of epidemic which is the probability of transmitting disease between a susceptible and an infectious individual. The rate γ is the incubation rate, it is the rate of latent individuals becoming infectious. The average duration of incubation is $\frac{1}{\gamma}$. The parameter μ is the recovery rate, this rate expresses the average duration of infection.

The model is expressed by the following set of ordinary differential equations:

$$\frac{dS_t}{dt} = -\frac{\beta S_t I_t}{M}$$
$$\frac{dE_t}{dt} = \frac{\beta S_t I_t}{M} - \gamma E_t$$
$$\frac{dI_t}{dt} = -\gamma E_t - \mu I_t$$
$$\frac{dR_t}{dt} = -\mu I_t$$

where $M = S_t + E_t + I_t + R_t$ the total population.

Inspired from this model, we assume that a susceptible agent in contact with an infected agent can be infected with a probability β. The incubation time is assumed to be random with a Poisson probability distribution of mean $\frac{1}{\gamma}$. An infected agent recovers after a random period with a Poisson probability distribution of mean $\frac{1}{\mu}$.

More formally, the model is expressed as follows:

Let A be a set of agents of size N. For $a \in A$ and $t \in \mathbb{N}$ (time is represented by natural numbers), let $x_a(t)$ the state of agent a on day t.

Let $\mathcal{R}_t = (A, E_t, p)$ the random network of contacts on day t. The vertices are the set of agents and the edges are specified by the adjacency matrix E_t defined as follow:

$$E_t(a, a') = \begin{cases} 1, & \text{if agents} a \text{ and } a' \text{ come from the same househould} \\ u \sim \text{Binomial}(1, p), & \text{otherwise.} \end{cases}$$

Here, p is the probability that a contact will be established between two agents who do not come from the same household. We assume that the probability p is independent of time.

2.1 Mechanism of Disease Spread

In this subsection we present how the epidemic state of an agent evolves over time.

Initially, only one agent is infected. For an agent $a \in A$, let $\eta_a = \inf\{t \in \mathbb{N}, x_a(t) = 1\}$ the day on which the agent a is contaminated. Let τ_a the incubation period of agent a.

The spread of the disease is as follows:

Let $a, a' \in A$

- if $x_a(t) = 0$, $x_{a'}(t) = 2$ and $E_t(a,a') = 1$, then a becomes infected with probability β.
- if $x_a(t) = 1$, let v be a realization of the Poisson distribution of parameter $\frac{1}{\gamma}$, if $t - \tau_a > v$ then $x_a(t+1) = 2$.
- if $x_a(t) = 2$, let v be a realization of the Poisson distribution $\frac{1}{\mu}$, if $t - \tau_a > v$ then $x_a(t+1) = 3$.

Proposition 2.1. *The probability that a susceptible agent on day t will become infected knowing his contacts and the status of his contacts and the epidemic states of his contacts, is given as follows:*

Let $I_{a,t} = \{x_a(t); (x_{a'}(t))_{a' \neq a}; (E_t(a,a'))_{a' \in \mathbb{N}}\}$

$$P[x_a(t+1) = 1 | I_{a,t}] = 1 - (1-\beta)^{l_a(t)}$$

where $l_a(t) = card\{a' \in \mathbb{N}, E_t(a,a') = 1 \text{ and } x_a(t) = 2\}$ *the number of infectious contacts of agent a at t.*

Proof. Let $a \in A$ susceptible at t, the agent remains susceptible at $t+1$ if it isn't infected by its infectious contacts.

As every $a' \in \{a' \in \mathbb{N}, E_t(a,a') = 1$ and $x_a(t) = 2\}$, can infect the agent with probability β. Then

$$P[x_a(t+1) = 0 \mid I_{a,t}] = (1-\beta)^{l_a(t)}$$

as the agent is in state 0 (susceptible), at $t+1$ it cannot transit to states 2 (infected) and 3 (recovered).
So,

$$P[x_a(t+1) = 2 \mid I_{a,t}] = P[x_a(t+1) = 3 \mid I_{a,t}] = 0$$

and

$$P[x_a(t+1) = 0 \mid I_{a,t}] + P[x_a(t+1) = 1 \mid I_{a,t}] + P[x_a(t+1) = 2 \mid I_{a,t}] + P[x_a(t+1) = 3 \mid I_{a,t}] = 1$$

then

$$P[x_a(t+1) = 1 \mid I_{a,t}] = 1 - (1-\beta)^{l_a(t)}$$

□

In the following subsection, we propose an algorithm to simulate the evolution of agents states.

2.2 Simulation Algorithm

To simulate the evolution of agents states over time we propose the following algorithm.

Algorithm 1. Simulation algorithm

```
Input:
N number of households
p, β, γ, μ and λ
T=Simulation duration
(τ_a)_a, (η_a)_a: vectors initialized to 0
A: agents set
Output:
(x_a(t))_{a,t}
Initialization:
at t = 0,
Generate the agents given the number of households and λ
Choose a_0 a random in A, put x_{a_0}(0) = 2 and η_{a_0} = 0
Disease spread:
for t = 0 : T − 1
 for a ∈ A
  case (x_a(t) = 0)
   compute l_a
   r = random
   if (r < 1 − (1 − β)^{l_a(t)})
    x_a(t + 1) = 1
    τ_a = t + 1
   end if
  case (x_a(t) = 1)
   draw v ~ Poisson(1/γ)
   if ((t + 1) − τ_a > v)
    x_a(t + 1) = 2
    η_a = t + 1
   end if
  case (x_a(t) = 2)
   draw v ~ Poisson(1/μ)
   if ((t + 1) − η_a > u), x_a(t + 1) = 3
 end for
end for
```

3 Dynamic of the Introduced Model: Simulation Study

We study the model introduced in the cases of the A/H1N1 influenza pandemic (see [3]) and Covid 19 (see [10]). The behavior of the system is simulated for several values of β.

The Table 1 shows the parameters γ and μ considered for Poisson distributions.

Table 1. The parameters γ and μ

	A/H1N1 influenza	Covid 19
γ	$1/2\text{days}^{-1}$ [3]	$1/6\text{days}^{-1}$ [10]
μ	$1/4\text{days}^{-1}$ [3]	$1/14\text{days}^{-1}$ [10]

The simulations are conducted for a population of 1000 households with $\lambda = 2$, which gives a household of average size equal to 3, and $p = \frac{log(N')}{N'}$ where N' is the number of agents.

In the Table 2, we summarize for some beta values, the estimate of the maximum number of infectious agents reached noted MI and the duration in days to reach the maximum of active cases (infected cases minus recovered cases) noted DMA. These results are obtained after 100 simulations.

Table 2. Simulations result

			MC mean	MC confidence interval
Covid 19	$\beta = 0.01$	MI(%)	0.79%	[0.02%, 5.13%]
		DMA (days)	44.31	[1,172.15]
	$\beta = 0.02$	MI(%)	90%	[89%, 92%]
		DMA (days)	94	[92,116]
	$\beta = 0.05$	MI(%)	99.67%	[99.51%, 99.80%]
		DMA (days)	65.87	[60,74]
	$\beta = 0.1$	MI(%)	99.99%	[99.98%, 100%]
		DMA (days)	48.5	[45.96,52.03]
	$\beta = 0.5$	MI(%)	100%	[100%, 100%]
		DMA (days)	29.8	[28.98,31]
A/H1N1 influenza	$\beta = 0.01$	MI(%)	0.02%	[0.02%, 1%]
		DMA (days)	1	[1,10]
	$\beta = 0.05$	MI(%)	45.15%	[0.02%, 80.71%]
		DMA (days)	36.57	[1.72]
	$\beta = 0.06$	MI(%)	91.21%	[90.7%, 92.76%]
		DMA (days)	42.77	[37.4,47.6]
	$\beta = 0.1$	MI(%)	95.35%	[94.97%, 98.07%]
		DMA (days)	33.10	[29.28,40]
	$\beta = 0.5$	MI(%)	100%	[100%, 100%]
		DMA (days)	17.27	[16.73,18]

These results suggest the existence of a critical value of beta, from which the infectious disease can spread in population. For Covid 19, the critical value is about 0.02. Concerning A/H1N1 influenza, it is about 0.05.

4 Estimation of Model Parameters

Usually, agents level data are not available, only aggregates data are available. Inference on parameters from theses aggregates is very complicated. It is almost impossible to compute the probability law governing the model. To overcome this problem, we propose the use of Approximate Bayesian Computational methods.

Approximate Bayesian Computation methods are a family of algorithms developed to perform Bayesian inference in the case of computationally intractable likelihood function. The steps of the basic ABC algorithm (see[17]) are as follows:

Algorithm 2. Basic ABC algorithm

Input: Observed data D_0, prior distribution π, threshold error e and discrepancy measure Δ

Output: a sample of size n from approximate posterior $P_a(\theta|D_0)$

1. Draw θ from the prior distribution $\pi(\theta)$.
2. Simulate D from the model $p(D|\theta)$, using parameters θ to get data D.
3. If $\Delta(D, D_0) < e$ accept θ otherwise reject.
4. Repeat 1.,2. and 3. until n values θ are accepted.

To improve the efficiency of the basic algorithm, several algorithms have been developed such as ABC-MCMC, ABC-SMC (see [3] and [16]).

The simulation of agent-based models is very expensive in terms of computation. This limits the number of simulations that can be generated. To remedy this problem, we propose an algorithm inspired by the ABC methods based on the modeling of the discrepancy measure (see, [8]).

Let $\Delta_\theta = \Delta(D, D_0)$ where D the simulate data under $p(D|\theta)$ and D_0 the observed data. We model $(\log(\Delta_\theta))_\theta$ as a Gaussian process with mean function $m(\theta)$ and covariance function $c(\theta, \theta')$. The standard choice of m and c are (see, [8]):

$$m(\theta) = \sum_{i=1}^{p} a_i \theta_i^2 + b_i \theta_i + c$$

and

$$c(\theta, \theta') = \sigma^2 \exp\left(-\sum_{i=1}^{p} \frac{(\theta_i - \theta_i^{'})^2}{l_i^2}\right) + \sigma_0^2 1_{\{\theta=\theta'\}} \text{ for } \theta = \{\theta_1, ..., \theta_p\}$$

The estimation of the parameters of m and c can be done by the maximum likelihood method. From a sample $(\theta_i, \Delta_{\theta_i})_{i=1,...,K}$, let $\hat{m}$ and $\hat{c}$ the estimates obtained.

For new value of θ, the prediction of Δ_θ is done as follow:

$$\hat{\Delta}_\theta = \hat{m}(\theta) + \hat{b}(\theta)'\hat{V}^{-1}\hat{e} \ (\star)$$

where $\hat{b}(\theta) = (c(\theta, \theta'), i = 1, ..., K)'$, $\hat{V} = (\hat{c}(\theta_i, \theta_j), i, j = 1, ..., K)$ and $\hat{e} = (\Delta_{\theta_i} - \hat{m}(\theta_i), i = 1, ..., K)'$.

The estimation algorithm consists of two steps. In the first step (the most expensive part in terms of computation) we apply the basic ABC algorithm without rejecting any generated value. We thus obtain a sample of size K of parameters θ and the associated discrepancies. From this sample we estimate a Gaussian process modeling the logarithm of the discrepancy depending on θ.

In the second step, we apply Basic algorithm without simulating the model. Discrepancies are predicted using the Gaussian process estimated in the first step.

So, the proposed algorithm is as follow:

Algorithm 3

Input: Observed data D_O, prior distribution π, threshold error e and discrepancy measure Δ, K, m

Output: a sample of size n from approximate posterior $P_a(\theta|D_O)$

1. Draw a sample $(\theta_i)_{i=1,..,K}$ from the prior distribution $\pi(\theta)$.
2. For each θ_i, simulate data from the model, using parameters θ_i to get data D_i and $\Delta_i = \Delta(D_i, D_O)$.
3. From the sample $(\theta_i, \Delta_i)_{i=1,..,K}$, estimate a Gaussian process to obtain $\hat{m}$ and $\hat{c}$.
4. Draw θ from the prior distribution $\pi(\theta)$.
5. Compute $\hat{\Delta}_\theta$ according to (*)
6. If $\hat{\Delta}_\theta < e$ accept θ otherwise reject.
7. Repeat 4,5,6 m times.

---

5 Evaluation of the Estimation Algorithm

We use a Gaussian process with polynomial mean function on θ, $m(\theta) = P(\theta)$ and covariance function $c(\theta, \theta') = \sigma^2 \exp\left(\frac{\|\theta_i - \theta_i'\|}{l}\right) + \sigma_0^2 1_{\{\theta=\theta'\}}$. The degree of the polynomial m is chosen so as to minimize the Bayesian information criterion of the model.

Two discrepancy measures were tested, namely:

$$\Delta_\infty = \max_t \left(| I_t - I_t^0 |, | R_t - R_t^0 |\right)$$

$$\Delta_1 = \sum_t | I_t - I_t^0 | + \sum_t | R_t - R_t^0 |$$

where I_t^0 and R_t^0 are respectively: observed infected cases, and observed recovered cases. We perform the approximate Bayesian computation with a noninformative prior on β: the uniform probability distribution on [0,1]. The error was set to accept a fraction α of the generated sample of θ. Several values of α have been tested. We perform the algorithm with $K = 100$ and $m = 10000$.

The estimates obtained are close to the true values. Indeed, the average absolute error compared to the true value of β varies between 1.3% (value reached for $\beta = 0.9$) and 22% (value reached for $\beta = 0.1$). By decreasing the alpha acceptance rate the performance of the estimation algorithm improves. Indeed, by going from $\alpha = 15\%$ to $\alpha = 5\%$, the gain in terms of average absolute error varies between 55% and 73%.

By comparing the two discrepancy measure adopted, it turns out that Δ_∞ perform better.

Tables 3 and 4 summarize the results obtained according to the various values of α and the discrepancy measures adopted.

Table 3. Results obtained with discrepancy Δ_∞

	True value of β	Estimation	Mean absolute error	Credible interval
Δ_∞ and $\alpha = 5\%$	$\beta = 0.05$	0.045	0.012	$[0.023, 0.068]$
	$\beta = 0.1$	0.108	0.014	$[0.085, 0.134]$
	$\beta = 0.3$	0.284	0.018	$[0.261, 0.308]$
	$\beta = 0.6$	0.602	0.013	$[0.579, 0.628]$
	$\beta = 0.9$	0.912	0.012	$[0.877, 0.924$
Δ_∞ and $\alpha = 10\%$	$\beta = 0.05$	0.072	0.030	$[0.024, 0.119]$
	$\beta = 0.1$	0.113	0.027	$[0.065, 0.159]$
	$\beta = 0.3$	0.275	0.032	$[0.228, 0.324]$
	$\beta = 0.6$	0.564	0.038	$[0.517, 0.610]$
	$\beta = 0.9$	0.911	0.026	$[0.864, 0.960]$
Δ_∞ and $\alpha = 15\%$	$\beta = 0.05$	0.081	0.044	$[0.009, 0.153$
	$\beta = 0.1$	0.122	0.040	$[0.050, 0.192]$
	$\beta = 0.3$	0.28	0.040	$[0.208, 0.349]$
	$\beta = 0.6$	0.564	0.046	$[0.492, 0.635]$
	$\beta = 0.9$	0.906	0.036	$[0.838, 0.974]$

Table 4. Results obtained with discrepancy Δ_1

	True value of β	Estimation	Mean absolute error	Credible interval
Δ_1 and $\alpha = 5\%$	$\beta = 0.05$	0.051	0.012	$[0.027, 0.074]$
	$\beta = 0.1$	0.129	0.029	$[0.105, 0.153]$
	$\beta = 0.3$	0.280	0.020	$[0.256, 0.305]$
	$\beta = 0.6$	0.717	0.117	$[0.692, 0.742]$
	$\beta = 0.9$	0.974	0.074	$[0.951, 0.998]$
Δ_1 and $\alpha = 10\%$	$\beta = 0.05$	0.055	0.024	$[0.010, 0.101]$
	$\beta = 0.1$	0.134	0.036	$[0.089, 0.179]$
	$\beta = 0.3$	0.282	0.028	$[0.234, 0.329]$
	$\beta = 0.6$	0.717	0.117	$[0.666, 0.766]$
	$\beta = 0.9$	0.951	0.051	$[0.905, 0.997]$
Δ_1 and $\alpha = 15\%$	$\beta = 0.05$	0.074	0.041	$[0.005, 0.143$
	$\beta = 0.1$	0.146	0.052	$[0.072, 0.216]$
	$\beta = 0.3$	0.287	0.039	$[0.214, 0.357]$
	$\beta = 0.6$	0.715	0.115	$[0.643, 0.788]$
	$\beta = 0.9$	0.921	0.041	$[0.846, 0.996]$

Figs. 2, 3, 4, 5 and 6 represent the approximate posterior distributions according to the different values of beta. The dashed line refers to the approximate posterior mean. We note that these probability distributions are concentrated around the true values of the parameters beta.

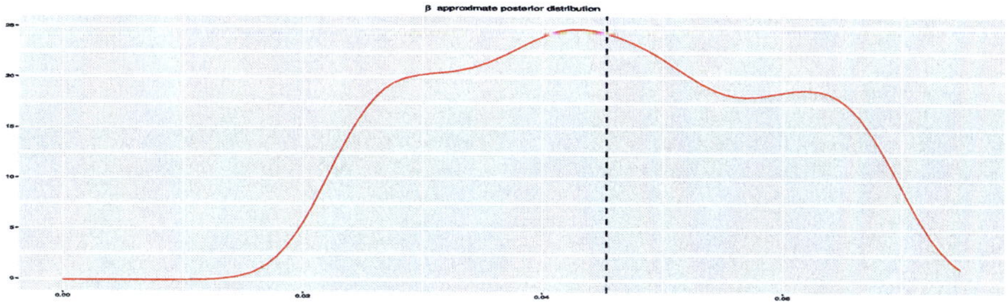

Fig. 2. Case $\beta = 0.05$ (acceptancy rate $= 5\%$) and dicrepancy mesure Δ_∞

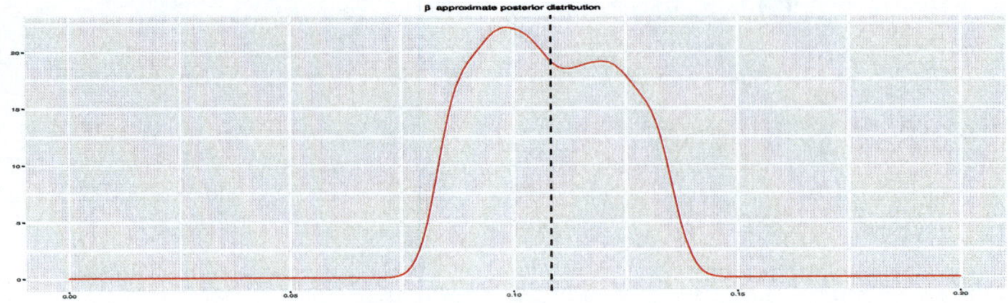

Fig. 3. Case $\beta = 0.1$ (acceptancy rate = 5%) and dicrepancy measure Δ_∞

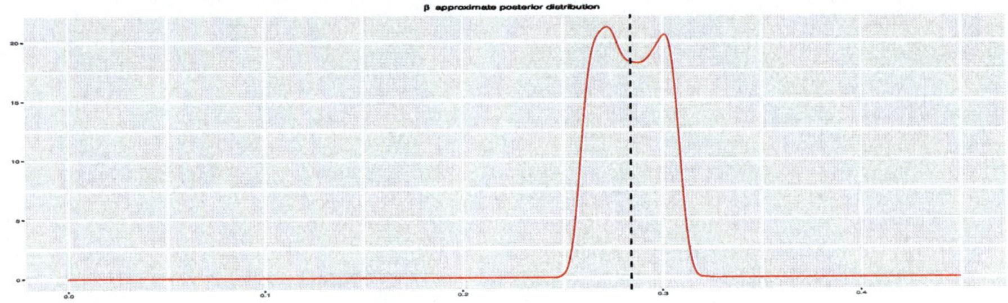

Fig. 4. Case $\beta = 0.3$ (acceptancy rate = 5%) and dicrepancy measure Δ_∞

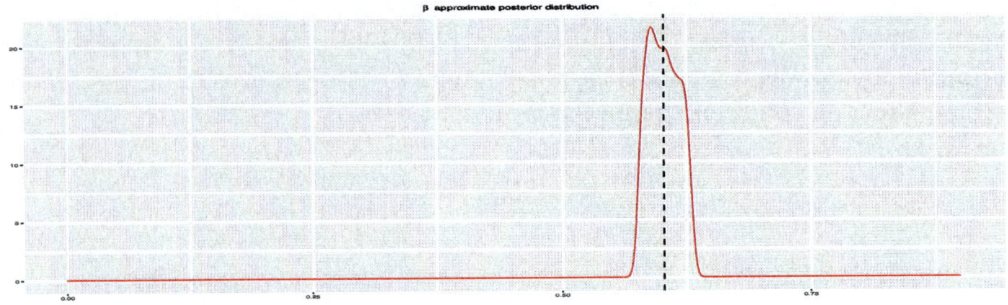

Fig. 5. Case $\beta = 0.6$ (acceptancy rate = 5%) and dicrepancy measure Δ_∞

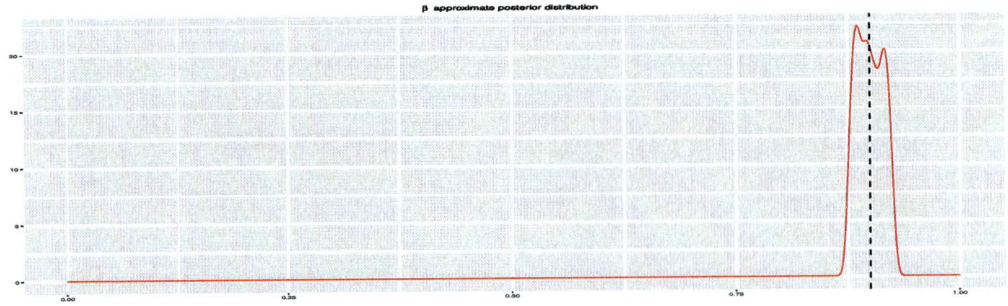

Fig. 6. Case $\beta = 0.9$ (acceptancy rate = 5%) and dicrepancy measure Δ_∞

6 Conclusion

This paper aimed to apply Bayesian approximation methods to estimate the parameters of an agent-based model in epidemiology. First, an agent-based model was formalized, and its dynamic properties were studied by simulation. Then, we proposed an estimation algorithm based on Bayesian approximation methods to infer parameters of the introduced model. The algorithm has been tested on simulated data and has given very satisfactory results.

For the future studies, a possible extension of the introduced agent based model is to take into account the heterogeneities of agents, for example: age, morbidity, etc. These characteristics will influence the contact network and the parameters of the epidemic model: β, μ, γ.

References

1. Axelrod, R., Hamilton, W.D.: The evolution of cooperation. Science **211**(4489), 1390–1396 (1981)
2. Brauer, F., Castillo-Chavez, C., Feng, Z.: Mathematical Models in Epidemiology. TAM, vol. 69. Springer, New York (2019). https://doi.org/10.1007/978-1-4939-9828-9
3. Beaumont, M.A.: Approximate bayesian computation. Ann. Rev. Stat. Appl. **6**, 2.1–2.25 (2018)
4. Boëlle, P.-Y., Ansart, S., Cori, A., Valleron, A.-J.: Transmission parameters of the A/H1N1 (2009) influenza virus pandemic: a review. Influenza Other Respir. Viruses **5**(5), 306–316 (2011)
5. Bollobás, B.: Random Graphs, 2nd edn. Cambridge University Press, Cambridge (2001)
6. Buzbas, E.O., Rosenberg, N.A.: AABC: approximate approximate Bayesian computation for inference in population-genetic models. Theor. Popul. Biol. **99**, 31–42 (2015)
7. Erdős, P., Rényi, A.: On random graphs. I. Publicationes Mathematicae (1959)
8. Gutmann, M.U., Corander, J.: Bayesian optimization for likelihood-free inference of simulator-based statistical models. J. Mach. Learn. Res. **17**(125), 1–47 (2016)
9. Hunter, E., Mac Namee, B., Kelleher, J.D.: A Taxonomy for Agent-Based Models in Human Infectious Disease Epidemiology. J. Artif. Soc. Soc. Simul. **20**(3) (2017)
10. Izadi, N., Taherpour, N., Mokhayeri, Y., Ghorbani, S.S., Rahmani, K., Nazari, S.S.H.: The epidemiologic parameters for COVID-19: a Systematic review and meta-analysis. Epidemiology and Infection **148** (2020)
11. Jarosz, B.: Poisson distribution: a model for estimating households by household size. Popul. Res. Policy Rev. **40**(2), 149–162 (2020). https://doi.org/10.1007/s11113-020-09575-x
12. Jennings, V., Lloyd-Smith, B., Ironmonger, D.: Household size and the poisson distribution. J. Aust. Popul. Assoc. **16**(1), 65–84 (1999). https://doi.org/10.1007/BF03029455
13. Kermack, W.O., McKendrick, A.G.: A Contribution to the Mathematical Theory of Epidemics. Proc. R. Soc. A. **115**(772), 700–721 (1927)
14. Langton, C.G.: Self reproduction in cellular automata. Phys. D **10**, 135–144 (1984)
15. Schelling, T.C.: Models of segregation. Am. Econ. Rev. **59**(2), 488–493 (1969)

16. Sisson, S.A., Fan, Y., Beaumont, M.: Handbook of Approximate Bayesian Computation, 1st edn. Chapman and Hall/CRC, Boca Raton (2018)
17. Toni, T., Welch, D., Strelkowa, N., Ipsen, A., Stumpf, M.P.: Approximate Bayesian computation scheme for parameter inference and model selection in dynamical systems. J. R. Soc. Interface **6**(31), 187–202 (2009)

Numerical Simulation of CdS/GaSe Solar Cell Using SCAPs Simulation Software

Mohamed Al-Hattab[1](✉), L'houcine Moudou[1], Lhouceine Moulaoui[1], Younes Chrafih[2], Abdelhafid Najim[6], Mohammed Khenfouch[3,4,5], Omar Bajjou[6], and Khalid Rahmani[1]

[1] ERPTM, Polydisciplinary Faculty, Sultan Moulay Slimane University, B.P 592, 23000 Beni Mellal, Morocco
med.alhattab@gmail.com
[2] LPHE-Modeling and Simulation Faculty of Science, Mohammed V University, Rabat, Morocco
[3] Faculty of Applied Sciences, Ibn Zohr University, Ait-Melloul, Morocco
[4] EPTHE, Department of Physics, Faculty of Sciences, Ibn Zohr University, Agadir, Morocco
[5] UNESCO UNISA Africa Chair in Nanosciences and Nanotechnology (U2ACN2), College of Graduate Studies, University of South Africa (UNISA), Pretoria, South Africa
[6] Material Physics Laboratory, Faculty of Sciences and Technics, Sultan Moulay Slimane University, BP 523, 23000 Beni Mellal, Morocco

Abstract. In this work a GaSe/CdS thin layers structure of a solar cell was digitally processed by the SCAPS-1D simulation program where GaSe is the absorbent material and CdS is the buffer layer; the study of this cell mainly consists in analyzing the effect of the different parameters of the two GaSe and CdS layers such as thickness, concentration career, temperature and R_s, R_{sh} resistances. The P-N junction is covered by an ITO layer acting as an OTC; the simulation results led to good values photovoltaic parameters: more than 1 V for the open circuit voltage, more than 12 mA/cm^2 for J_{sc}, an excellent fill factor which exceeds 90%, with an efficiency around 16%.

1 Introduction

The production of energy is a challenge of great importance in the future. Today, a large part of the world's energy production comes from fossil sources. The sun is a green, clean, renewable and sustainable source of energy, it has a very high potential to meet the world's electricity demand, and the sunlight can be directly converted into electricity through the solar cell using the principle known as the photovoltaic effect. Researchers are working on many different solar materials such as: Si, CdTe, CIGS, CGS, CZTS and organic resources [1–10], with the goal of obtaining low cost and high efficiency solar cells.

The GaSe compound (III-VI), is a more promising semiconductor because of its optical and electrical properties, it has a lot of capacities in the field of optoelectronic devices. Studies concerning technological applications of crystals have shown that GaSe has potentials for the use in optoelectronics [11, 12], non-linear optics [13], portable devices,

S. Melliani and O. Castillo (Eds.): ICPAMS 2021, LNNS 476, pp. 315–325, 2023.
https://doi.org/10.1007/978-3-031-12416-7_27

energy storage [14, 15], photovoltaics [16, 17] (Has a high absorption coefficient in the visible range [18]), the objective of this work is the simulation of a GaSe-based solar cell with the CdS emitter by the SCAPS simulator, and analysis their photovoltaic parameters as a function of variation of the parameters, thickness, concentration, temperature and resistance of the GaSe and CdS layers.

2 Methodology

In this work the results were obtained by SCAPS-1D simulator, it is a one dimensional solar cell simulation software developed at the Department of Electronics and Information Systems (EIS), University of Ghent, Belgium, their purpose is the digital analysis of solar cells [19]. To achieve the desired structure, several physical properties of the different layers composing the cell can be changed such as band gap, electron affinity, dielectric permittivity, etc.

The necessary specification of the working point can be indicated in the action panel. The software supports the classification of all physical parameters and also the specification of the properties of the front and rear contact. A large number of AC and DC electrical measurements, including short-circuit current density J_{sc}, fill factor FF, open-circuit voltage V_{oc}, conversion efficiency η, quantum efficiency QE, response spectral, generation and recombination profile [20–22]. Figure 1 shows the structure of the solar cell used in this study.

The illumination of the light was through the side of the ITO layer with a light output of $1000\,W/m^2$ at room temperature. The values of the physical parameters used in this study are illustrated in Table 1 [17, 23–28].

The observation of the effect of thickness and concentration of GaSe and CdS layers on the performance of GaSe-based solar cell was carried out by the following method: the thickness (concentration) of the buffer layer was modified from 0.05 μm to 0.15 μm ($10^{17}\,cm^{-3}$ to $10^{21}\,cm^{-3}$) while keeping the absorbent layer set at 2 μm ($10^{17}\,cm^{-3}$). In the following, the buffer layer was set at a thickness (concentration) in which PCE is maximum and the GaSe absorbent layer thickness was changed in the range of 0.2 μm to 2.0 μm ($10^{13}\,cm^{-3}$ to $10^{17}\,cm^{-3}$). To observe the temperature effect of this structure, the simulation was done in a range of temperature going from 280 to 400 K with a step of 20 K. the effect of resistors R_s and R_{sh} also was simulated.

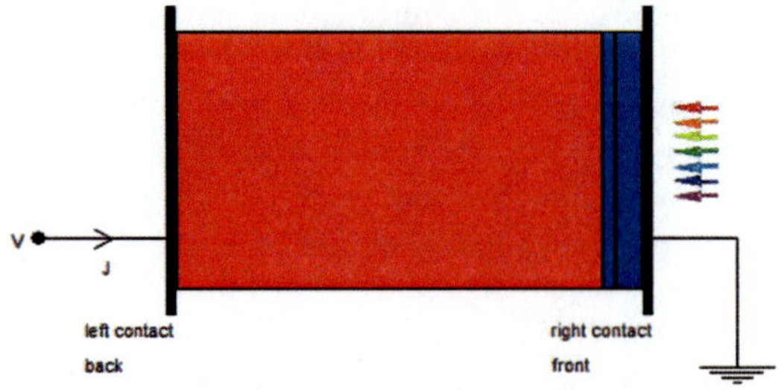

Fig. 1. Solar cell structure of ITO/CdS/GaSe/Mo layer

Table 1. Physical parameters used in the simulation

Parameters	p-GaSe	n-CdS	n-ITO
Thickness(μm)	0.2–2	0.05–0.15	0.1
E_g(eV)	2.1	2.4	3.6
χ(eV)	4.6	4.5	4.1
ε_r	8	10	10
$N_C(cm^{-3})$	2.2×10^{18}	2.2×10^{18}	2.2×10^{18}
$N_V(cm^{-3})$	1.8×10^{19}	1.8×10^{19}	1.8×10^{19}
$\mu_n(cm^2/Vs)$	250	100	50
$\mu_p(cm^2/Vs)$	25	25	75
$N_D(cm^{-3})$	–	$10^{17} - 10^{21}$	10^{18}
$N_A(cm^{-3})$	$10^{13} - 10^{17}$	-	-
V_{eth}(cm/s)	10^7	10^7	10^7
V_{pth}(cm/s)	10^7	10^7	10^7
$N_t(cm^{-3})$	10^{14}	10^{14}	10^{14}

3 Results and Discussions

3.1 Effect of Variation of Buffer Layer Thickness and Carrier Concentration

An N-type buffer layer was used between the absorbent layer and the optical window layer so that the generated charge carriers would be directed to the electrode before recombination occurred. Figure 2 shows the influence of the thickness and the concentration of the CdS layer on the photovoltaic characteristics of the simulated solar cell. The concentration and thickness of the CdS were varied from 10^{17} cm^{-3} to 10^{21} cm^{-3} and from 0.05 to 0.15 μm, respectively. The increase in the number of donors results in a slight decrease in V_{oc}, and a slight increase in J_{sc}. Figure 2 also shows that the photovoltaic parameters of the simulated structure is independent of the thickness of CdS.

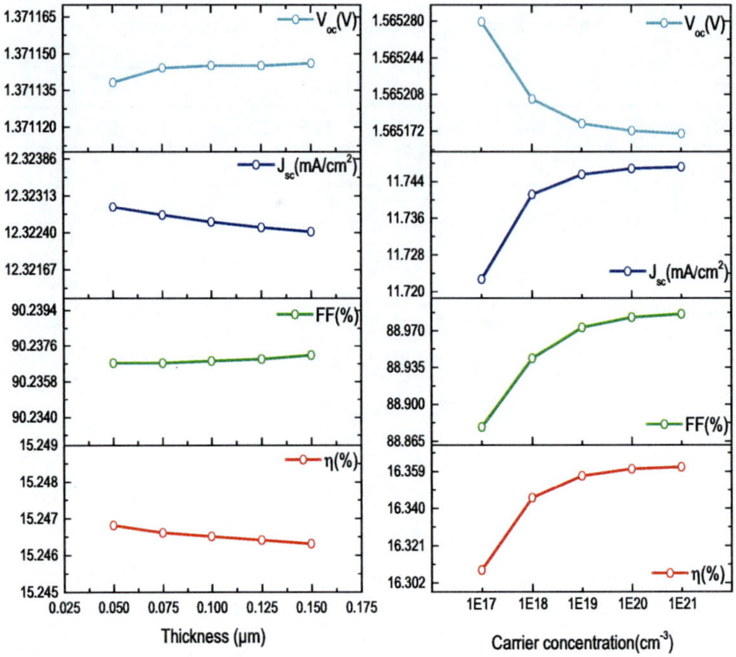

Fig. 2. Thickness and carrier concentration variation curve for CdS buffer layer

3.2 Effect of Variation of Absorption Layer Thickness and Carrier Concentration

In this work Fig. 3 shows that the increase in the thickness of the GaSe absorber, increased the parameters PCE, J_{sc} and V_{oc} and reach up to 15.24%, 12.32 mA/cm^2 and 1.37 V respectively at 2 μm. The FF decreases due to the increase in series resistance with an increase in the thickness of the absorber [26]. We varied the acceptor concentration value between 10^{13} cm^{-3} and 10^{17} cm^{-3} in the p-GaSe absorbent layer, here PCE increased from 15.04% to 16.46%, which reveals that proper doping of the absorbent layer would produce greater efficiency. Likewise, V_{oc} also increase with increasing values of N_A.

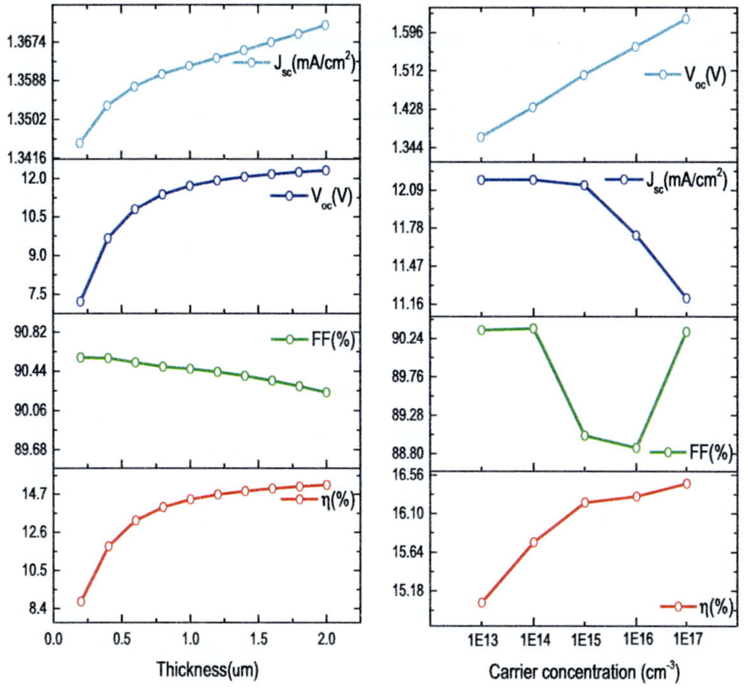

Fig. 3. Thickness and carrier concentration variation curve for GaSe absorption layer

Figure 4 shows the illuminated current density-voltage and quantum efficiency (EQ) for the simulated solar cell. Figure 4-a presents the current-voltage density curves for different thicknesses of absorbent layer varying from 0.2 μm to 2 μm. As a result of the increase in the thickness of the absorbent layer, photons with longer wavelengths were also absorbed. This effect is associated with improving the collection of photo-generated carriers. Figure 4-b shows the current-voltage density curves with a variable GaSe doping concentration varying from 10^{13} cm^{-3} and 10^{17} cm^{-3}. According to this figure, the higher concentrations produce the best charge transport.

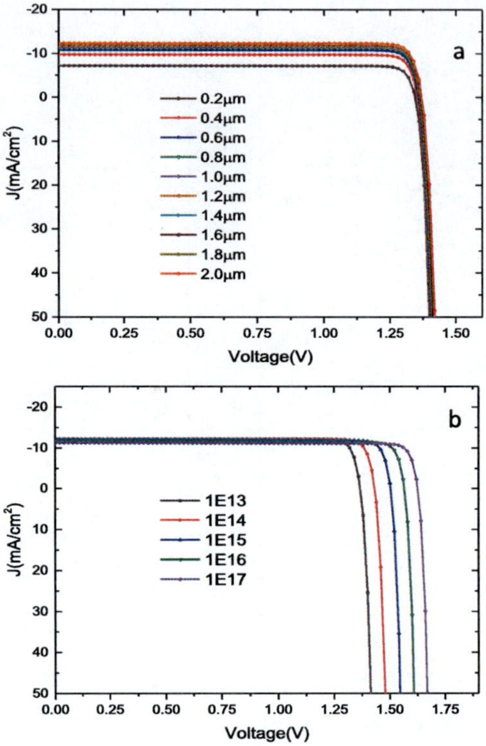

Fig. 4. J-V characteristics curve of a) Thickness b) carrier concentration variation of GaSe absorption layer

Quantum efficiency (Fig. 5) is the ratio of the number of charge carriers passing through the external circuit to the number of incident carriers. For a given wavelength, the external quantum efficiency is equal to 1 if each photon generates a pair of hole electron. The effect of the thicknesses of the GaSe layer on the quantum efficiency was also analyzed. Figure 5-a displays the quantum efficiency (QE) of the structure of the solar cell (ITO/CdS/GaSe/Mo) with different thicknesses of GaSe layer. Note that the QE of this structure increases with increasing thickness of the absorbent layer. In fact, more photons are absorbed when the thickness of the absorbent layer is increased. The quantum efficiency is also observed to be reduced for higher acceptor densities due to enhanced recombination process (Fig. 5-b).

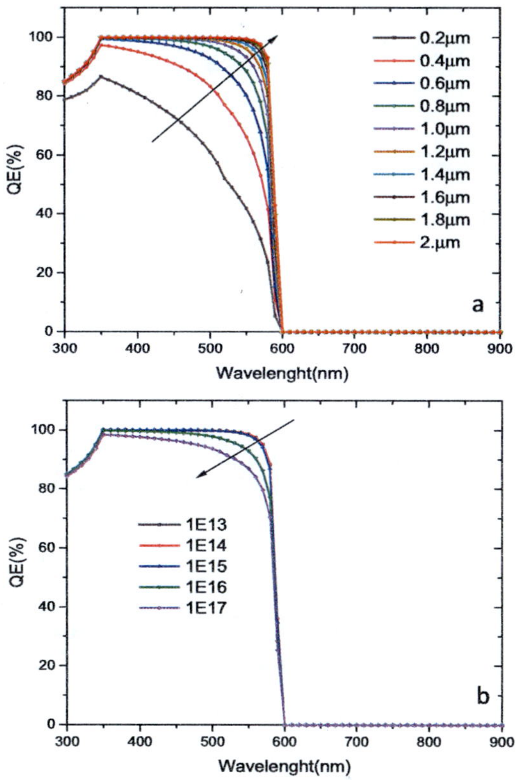

Fig. 5. QE characteristics curve of a) Thickness b) carrier concentration variation of GaSe absorption layer

3.3 Effect of Variation of Temperature

We analyzed the device for various working temperatures from 280 K to 400 K, for a better understanding of their behavior in the environment. Figure 6 shows that with increasing operating temperature in the range from 280 K to 400 K, deterioration occurs for V_{oc} from 1.60 V to 1.40 V. This is caused by increasing saturation current and the rate of recombination with increasing temperature [29, 30]. On the other hand, J_{sc} increases from 11.64 mA/cm^2 to 12.10 mA/cm^2 due to thermal generation of carriers. the FF goes from 88.93% to 88.19% and therefore PCE decreases from 16.60% to 14.97%.

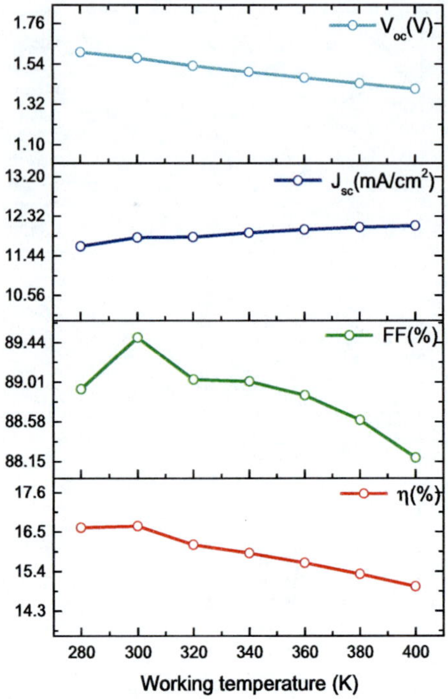

Fig. 6. Cell performance of ITO/CdS/GaSe/Mo QE as a function ofworking temperature.

3.4 Effect of Variation of Resistance R_s and R_{sh}

The performance of the cell was observed by varying the R_s from 2 to 20 $\Omega \cdot \text{cm}^2$ as shown in Fig. 7-a indicates that the V_{oc} and the J_{sc} are independent of R_s. It can also be seen from the figure that the FF and the PCE decrease with the increase in R_s due to the increase in power loss [26]. For the R_{sh} it was varied from 100 to 1000 $\Omega \cdot \text{cm}^2$ to explore its effect on the performance of the cell as shown in Fig. 7-b. We observe from the figure that the V_{oc} decreases with the decrease of the R_{sh}, due to the loss of current towards R_{sh}. Figure 7-b shows that low R_{sh} has a massive impact on cell performance. We can conclude that good performance of solar cells is obtained for low values of R_s and high values of R_{sh}.

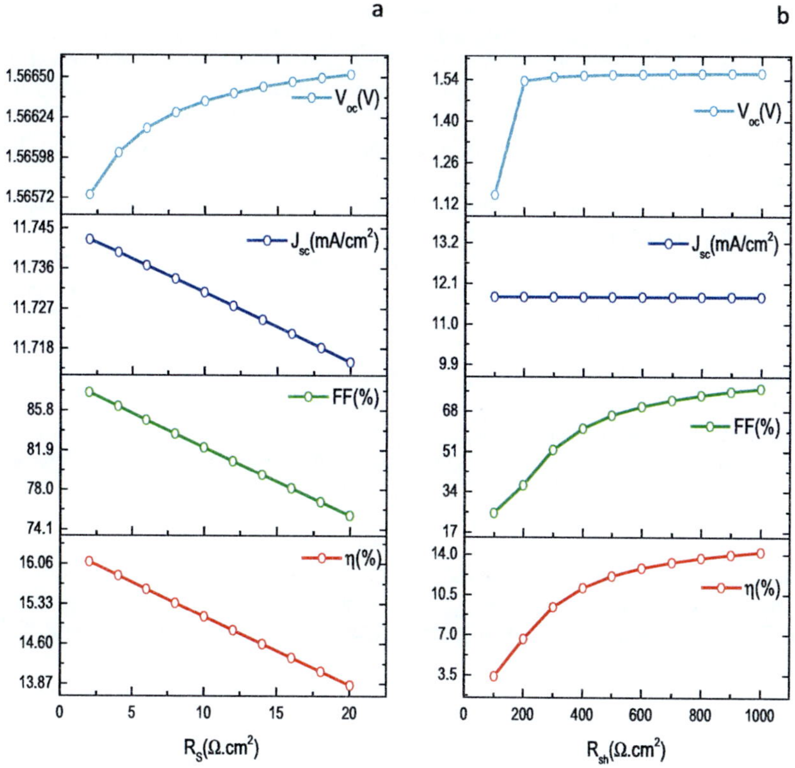

Fig. 7. Cell performance of ITO/CdS/GaSe/Mo as a function of a) series resistance b) shunt resistance.

4 Conclusion

In the present study, thin films of GaSe, CdS, ITO, were used to simulate the solar cell. Efficiencies in the order of 15% and more, and fill factors exceeding 90% were recorded, the study began with the analysis of the effect of thickness and concentration of the dopants of the first two layers on cell performance which showed that increasing career concentration positively affects cell function for both layers, however the variation in thickness did not cause a big change in the case of CdS; then the change in temperature and resistors R_s and R_{sh} was also treated.

Acknowledgements. The authors gratefully acknowledge Prof. Marc Burgelman, University of Gent, Belgium, for providing SCAPS 1-D simulation software.

References

1. Ahmmed, S., Aktar, A., Hossain, J., Ismail, A.B.M.: Enhancing the open circuit voltage of the SnS based heterojunction solar cell using NiO HTL. Sol. Energy **207**, 693–702 (2020). https://doi.org/10.1016/j.solener.2020.07.003

2. Al-Hattab, M., Moudou, L., Chrafih, Y., Khenfouch, M., Bajjou, O., Rahmani, K.: The anisotropic optical properties of different polytypes (ε, β, δ, γ) of GaSe lamellar materials. Eur. Phys. J. Appl. Phys. **91**, 30102 (2020). https://doi.org/10.1051/epjap/2020200136
3. Bacewicz, R.: Polycrystalline thin film solar cells. Opto-electronics Rev. **1997**, 167–173 (1997)
4. Balkanski, M., Julien, C., Emery, J.Y.: Integrable lithium solid-state microbatteries. J. Power Sour. **26**, 615–622 (1989). https://doi.org/10.1016/0378-7753(89)80189-2
5. Basol, B.M.: Low cost techniques for the preparation of Cu (In, Ga) (Se, S)2 absorber layers. Thin Solid Films **361–362**, 514–519 (2000)
6. Burgelman, M., Nollet, P., Degrave, S.: Modelling polycrystalline semiconductor solar cells. Thin Solid Films **361**, 527–532 (2000). https://doi.org/10.1016/S0040-6090(99)00825-1
7. Burgelman, M., Verschraegen, J., Degrave, S., Nollet, P.: Modeling thin-film PV devices. Prog. Photovoltaics Res. Appl. **12**, 143–153 (2004). https://doi.org/10.1002/pip.524
8. Chen, H., et al.: Electrodeposited CZTS solar cells from Reline electrolyte. Green Chem. **16**, 3841–3845 (2014). https://doi.org/10.1039/c4gc00142g
9. Cuculescu, E., Evtodiev, I., Caraman, M.: Non-equilibrium charge carriers generation - recombination mechanisms at the interface of the SnO2/GaSe heterojunction. Thin Solid Films **517**, 2515–2518 (2009). https://doi.org/10.1016/j.tsf.2008.11.066
10. Drapak, S.I., Katerynchyk, V.M., Kovalyuk, Z.D.: Electrical, luminescent and photovoltaic properties of the indium tin oxide-GaSe heterojunctions with a thin layer of gallium oxide. Thin Solid Films **515**, 6356–6359 (2007). https://doi.org/10.1016/j.tsf.2006.11.066
11. Elbar, M., Tobbeche, S., Merazga, A.: Effect of top-cell CGS thickness on the performance of CGS/CIGS tandem solar cell. Sol Energy **122**, 104–112 (2015). https://doi.org/10.1016/j.solener.2015.08.029
12. Ertap, H.: Nonlinear absorption, SHG behavior and carrier dynamics of Nd and Pr doped GaSe single crystals. Opt. Mater. (Amst) **83**, 99–103 (2018). https://doi.org/10.1016/j.optmat.2018.05.073
13. Ferrer-Roca, C., Bouvier, J., Segura, A., Andrés, M.V., Muñoz, V.: Light-induced transmission nonlinearities in gallium selenide. J. Appl. Phys. **85**, 3780–3785 (1999). https://doi.org/10.1063/1.369748
14. Huo, C., Yan, Z., Song, X., Zeng, H.: 2D materials via liquid exfoliation: a review on fabrication and applications. Sci. Bull. **60**(23), 1994–2008 (2015). https://doi.org/10.1007/s11434-015-0936-3
15. Jeong, J.H., Jung, D.W., Oh, E.S.: Lithium storage characteristics of a new promising gallium selenide anodic material. J. Alloys Compd. **613**, 42–45 (2014). https://doi.org/10.1016/j.jallcom.2014.06.017
16. Jhuma, F.A., Shaily, M.Z., Rashid, M.J.: Towards high-efficiency CZTS solar cell through buffer layer optimization. Mater. Renew. Sustain. Energy **8**(1), 1–7 (2019). https://doi.org/10.1007/s40243-019-0144-1
17. Kaelin, M., Rudmann, D., Tiwari, A.N.: Low cost processing of CIGS thin film solar cells. Sol. Energy **77**, 749–756 (2004). https://doi.org/10.1016/j.solener.2004.08.015
18. Lang, O., Klein, A., Pettenkofer, C., Jaegermann, W., Chevy, A.: Band lineup of lattice mismatched InSe/GaSe quantum well structures prepared by van der Waals epitaxy: absence of interfacial dipoles. J. Appl. Phys. **80**, 3817–3821 (1996). https://doi.org/10.1063/1.363335
19. Li, Y., Pullerits, T., Zhao, M., Sun, M.: Theoretical characterization of the PC60BM:PDDTT model for an organic solar cell. J. Phys. Chem. C **115**, 21865–21873 (2011). https://doi.org/10.1021/jp2040696
20. Lin, P., Lin, L., Yu, J., Cheng, S., Lu, P., Zheng, Q.: Numerical simulation of Cu2ZnSnS4 based solar cells with In2S3 buffer layers by SCAPS-1D. J. Appl. Sci. Eng. **17**, 383–390 (2014). https://doi.org/10.6180/jase.2014.17.4.05

21. Mazhari, B.: An improved solar cell circuit model for organic solar cells. Sol. Energy Mater. Sol. Cells **90**, 1021–1033 (2006). https://doi.org/10.1016/j.solmat.2005.05.017
22. Mebarkia, C., Dib, D., Zerfaoui, H., Belghit, R.: Energy efficiency of a photovoltaic cell based thin films CZTS by SCAPS. J. Fundam. Appl. Sci. **8**, 363 (2016). https://doi.org/10.4314/jfas.v8i2.13
23. Nykyruy, L.I., Yavorskyi, R.S., Zapukhlyak, Z.R., Wisz, G., Potera, P.: Evaluation of CdS/CdTe thin film solar cells: SCAPS thickness simulation and analysis of optical properties. Opt. Mater. (Amst) **92**, 319–329 (2019). https://doi.org/10.1016/j.optmat.2019.04.029
24. Sites, J., Pan, J.: Strategies to increase CdTe solar-cell voltage. Thin Solid Films **515**, 6099–6102 (2007). https://doi.org/10.1016/j.tsf.2006.12.147
25. Sobayel, K., et al.: Numerical modeling on prospective buffer layers for tungsten di-sulfide (WS2) solar cells by scaps-1D. Chalcogenide Lett. **15**, 307–315 (2018)
26. Sun, S., Fan, Z., Wang, Y., Haliburton, J.: Organic solar cell optimizations. J. Mater. Sci. **40**, 1429–1443 (2005). https://doi.org/10.1007/s10853-005-0579-x
27. Tang, L., Zhao, Z., Yuan, S., Yang, T., Zhou, B., Zhou, H.: Self-catalytic VLS growth one dimensional layered GaSe nanobelts for high performance photodetectors. J. Phys. Chem. Solids **118**, 186–191 (2018). https://doi.org/10.1016/j.jpcs.2018.03.015
28. Verschraegen, J., Burgelman, M.: Numerical modeling of intra-band tunneling for heterojunction solar cells in scaps. Thin Solid Films **515**, 6276–6279 (2007). https://doi.org/10.1016/j.tsf.2006.12.049
29. Wada, T., Kohara, N., Nishiwaki, S., Negami, T.: Characterization of the Cu(In, Ga)Se2/Mo interface in CIGS solar cells. Thin Solid Films **387**, 118–122 (2001). https://doi.org/10.1016/S0040-6090(00)01846-0
30. Zhao, W., Zhou, W., Miao, X.: Numerical simulation of CZTS thin film solar cell. In: 2012 7th IEEE International Conference on Nano/Micro Engineered and Molecular Systems, NEMS 2012, pp. 502–505 (2012). https://doi.org/10.1109/NEMS.2012.6196826

Amalgamation over a Graded Ring

Moussa Thiaw(✉)

African Institute for Mathematical Sciences (AIMS-SENEGAL) and Gaston Berger University, Km2 route de Joal centre IRD, Saint-Louis, Senegal
moussa.thiaw@aims-senegal.org, thiawskr@gmail.com

Abstract. Let A and B be commutative rings, let J be an ideal of B, let $f : A \longrightarrow B$ be a ring homomorphism and we note by $J(B)$ the Jacobson radical of B. The purpose of this paper is to study the graduation of the amalgamation $A \bowtie^f J$ and $A \bowtie I$ introduced by D'Anna and Fontana in 2007. We investigate also some homological properties of the amalgamation over a graded ring and we show that if (A, m) is a Noetherian local ring with dimension d, $f : A \longrightarrow B$ a ring homomorphism and $0 \neq J \subseteq J(B)$ an ideal such that J is a finitely generated A-module, then $H^d_{m\bowtie^f J}(A \bowtie^f J)$ is $FP - gr$-injective if and only if $H^d_m(A)$ and $H^d_m(J)$ are $FP - gr$-injective.

Keywords: Amalgamation · Primary ideal · Graded ring · Graded module · $FP - gr$-injective

1 Introduction

The notion of amalgamation of A with B along J with respect to f was introduced by D'Anna et al. see [12]. Let $f : A \longrightarrow B$ be a ring homomorphism. The subring of $A \times B$: $A \bowtie^f J = \{(a, f(a) + j) \mid a \in A, j \in J\}$ is called the amalgamation of A with B along J with respect to f. Certain properties of the rings (resp. ideals) have been transferred to the amalgamation $A \bowtie^f J$ (resp.$I \bowtie^f J$) see [1,12]. This notion of amalgamation of A with B along J with respect to f is a generalization of the notion of amalgamated duplication defined by D'Anna and Fontana as follows $A \bowtie I = \{(a, a+i) \mid a \in A, i \in I\}$ see [11]. Several authors have studied the theory of graded rings and modules see [5,14,16]. Primary ideal, graded ideal and graded primary ideal have been studied in [2,19,20]. In this paper we show that the amalgamation preserves the graduation of rings, modules, ideals and primary ideals. We investigate also some homological properties of the amalgamation over a graded ring. This paper is organized as follows:

In Sect. 2 we study the graded ideal, grade primary ideal amalgamated and we show that if I is a graded ideal of A, then $I \bowtie^f J$ is a graded ideal of $A \bowtie^f J$. We show also that I is a graded primary ideal of A if and only if $I \bowtie^f J$ is a graded primary ideal of $A \bowtie^f J$. In Sect. 3 we investigate also some homological properties of the amalgamation over a graded ring and we show that if (A, m) is a Noetherian local ring with dimension d, $f : A \longrightarrow B$ a ring homomorphism and $0 \neq J \subseteq J(B)$ an ideal such that J is a finitely

M. Thiaw—This paper is dedicated to my mother Sadio KEBE.

S. Melliani and O. Castillo (Eds.): ICPAMS 2021, LNNS 476, pp. 326–331, 2023.
https://doi.org/10.1007/978-3-031-12416-7_28

generated A-module, then $H^d_{m\bowtie^f J}(A \bowtie^f J)$ is $FP-gr$-injective if and only if $H^d_m(A)$ and $H^d_m(J)$ are $FP-gr$-injective. In this section we show also that if (A,m) is a graded Noetherian local ring with dimension d, $f: A \to B$ a ring homomorphism and $0 \neq J \subseteq J(B)$ an ideal such that J is a finitely generated A-module, then $H^d_{m\bowtie^f J}(A \bowtie^f J)$ is flat (resp. free, resp. projective) if and only if $H^d_m(A)$ and $H^d_m(J)$ are flat (resp. free, resp. projective).

2 Graded Ideal and Graded Primary Ideal in Amalgamated Algebra Along an Ideal

Definition 1. A ring A is called graded (or $\mathbb{Z}$-graded) if there exists a family of subgroups $\{A_n\}_{n\in\mathbb{Z}}$ of A such that:

1. $A = \oplus_{n\in\mathbb{Z}} A_n$ (as abelian groups)
2. $A_n \cdot A_m \subseteq A_{n+m}$ for all $n,m \in \mathbb{Z}$.

The elements of A_n are called homogeneous of degree n. If $x \in A$, then x can be written uniquely as $\sum_{n\in\mathbb{Z}} x_n$, where x_n is the component of x in A_n. We write $h(R) = \cup_{n\in\mathbb{Z}} A_n$.

Example 1. Consider $K[x] = \oplus_{n\in\mathbb{Z}} Kx^n$, where $Kx^n = 0$ if $n < 0$. Then $K[x] = \cdots 0 \oplus \cdots \oplus 0 \oplus k \oplus Kx \oplus \cdots$ is a graded ring.

Definition 2. Let A be a graded ring and M an A-module. We say that M is a graded A-module (or has an A-grading) if there exists a family of subgroups $\{M_n\}_{n\in\mathbb{Z}}$ of M such that:

1. $M = \oplus_{n\in\mathbb{Z}} M_n$ (as abelian groups)
2. $A_n \cdot M_m \subseteq M_{n+m}$ for all $n,m \in \mathbb{Z}$.

Definition 3. Let I be an ideal of A. Then I is a graded ideal of A if $I = \oplus_{n\in\mathbb{Z}}(I \cap A_n)$. clearly we have $\oplus_{n\in\mathbb{Z}}(I \cap A_n) \subseteq I$ and so I is a graded ideal of A if $I \subseteq \oplus_{n\in\mathbb{Z}}(I \cap A_n)$.

Definition 4. Let I be a graded ideal of A.

1. The graded radical of I ($Gr(I)$) is the set of all $x \in A$ such that for each $n \in \mathbb{Z}$ there exists $k > 0$ with $x_n^k \in I$. Note that, if r is a homogeneous element of A, then $r \in Gr(I)$ iff $r^n \in$ for some $n \in \mathbb{Z}$.
2. We say that I is a graded primary ideal of A if $I \neq A$, and whenever $a,b \in h(A)$, with $ab \in I$, then $a \in I$ or $b \in Gr(I)$ so.

Example 2. Let $A = \mathbb{Z}[i]$ (The Gaussian integers). Then A is a $\mathbb{Z}_2$-graded ring with $A_0 = \mathbb{Z}$, $A_1 = i\mathbb{Z}$. Let $I = 2A$ be a graded prime ideal. Then I is agraded primary ideal. But I is not a primary ideal because 2 is not irreducible element of $A = \mathbb{Z}[i]$.

Theorem 1. *1. Let A be a graded ring, I be an ideal of A. Then $A \bowtie I$ is a graded ring, where*

$$(A \bowtie I)_n = A_n \bowtie I = \{(a_n, i) | a_n \in A_n \text{ and } i \in I\}.$$

2. *Let $f : A \to B$ be a ring homomorphism and J be an ideal of B. If A is a graded ring, then $A \bowtie^f J$.*

Proof. • It is not difficult to show that $(A \bowtie I)_n$ is an additive subgroup for all $n \in \mathbb{Z}$.

- Show that $(A_n \bowtie I) \cdot (A_m \bowtie I) \subset (A_{n+m} \bowtie I)$, $\forall\, n, m \in \mathbb{Z}$.
  Let $(a_n, a_n + i) \in A_n \bowtie I$ and $(a_m, a_m + j) \in A_m \bowtie I$.
  We have $(a_n, a_n + i) \cdot (a_m, a_m + j) = (a_n a_m, a_n a_m + a_n j + i a_m + ij)$,
  since $A_n \cdot A_m \subset A_{n+m}$, then $a_n a_m \in A_{n+m}$ and since $a_n j + i a_m + ij \in I$,
  so $(a_n, a_n + i) \cdot (a_m, a_m + j) \in A_{n+m} \bowtie I$.
  Hence $(A_n \bowtie I) \cdot (A_m \bowtie I) \subset (A_{n+m} \bowtie I)$.
- Show that $A \bowtie I = \oplus_{n \in \mathbb{Z}} (A_n \bowtie I)$.
  Let $(a, a + i) \in A \bowtie I \Rightarrow a \in A = \oplus_{n \in \mathbb{Z}} A_n \Rightarrow a = \sum_{n \in \mathbb{Z}} a_n$ and $i = \sum_{n \in \mathbb{Z}} i_n$, where $a_n \in A_n$, $i_n \in I$.
  So $(a, a + i) = (\sum_{n \in \mathbb{Z}} a_n, \sum_{n \in \mathbb{Z}} (a_n + i_n)) = \sum_{n \in \mathbb{Z}} (a_n, a_n + i_n) \in \oplus_{n \in \mathbb{Z}} (A_n \bowtie I)$.
  So $A \bowtie I \subset \oplus_{n \in \mathbb{Z}} (A_n \bowtie I)$.
  Let $x \in \oplus_{n \in \mathbb{Z}} (A_n \bowtie I) \Rightarrow \exists\; a_n \in A_n$ and $i_n \in I$ such that $x = \sum_{n \in \mathbb{Z}} (a_n, a_n + i) = (\sum_{n \in \mathbb{Z}} a_n, \sum_{n \in \mathbb{Z}} a_n + i) \in A \bowtie I$.
  So $\oplus_{n \in \mathbb{Z}} (A_n \bowtie I) \subset A \bowtie I$ and therefore $A \bowtie I = \oplus_{n \in \mathbb{Z}} (A_n \bowtie I)$.
- Show that $(A_n \bowtie I) \cap (A_m \bowtie I) = \{(0,0)\}$.
  Let $x \in (A_n \bowtie I) \cap (A_m \bowtie I) \Rightarrow \begin{cases} x \in A_n \bowtie I \Rightarrow x = (a_n, a_n + i_n) \\ x \in A_m \bowtie I \Rightarrow x = (a_m, a_m + i_m) \end{cases} \Rightarrow$
  $\begin{cases} a_n = a_m \in A_n \cap A_m \Rightarrow a_n = a_m = 0 \\ i_n = i_m \Rightarrow i_n = i_m = 0, \text{ since } n \neq m. \end{cases}$
  Therefore $A \bowtie I$ is a graded ring.
- The second result is shown in the same way as 1.

Proposition 1. *Let I be a graded ideal of A. Then $I \bowtie^f J$ is a graded ideal of $A \bowtie^f J$.*

Proof. Let $(a, f(a) + j) \in I \bowtie^f J \Rightarrow a \in I$.
Since I is graded we have $I \subseteq \oplus_{n \in \mathbb{Z}} (I \cap A_n)$, so $a \in \oplus_{n \in \mathbb{Z}} (I \cap A_n) \Rightarrow a = \sum_{n \in \mathbb{Z}} x_n$, where $x_n \in I \cap A_n$.
We have $(a, f(a) + j) = (\sum_{n \in \mathbb{Z}} x_n, \sum_{n \in \mathbb{Z}} f(x_n) + j) = \sum_{n \in \mathbb{Z}} (x_n, f(x_n) + j) \in \oplus_{n \in \mathbb{Z}} (I \cap A_n \bowtie^f J)$.
Therefore $I \bowtie^f J$ is a graded ideal of $A \bowtie^f J$.

Proposition 2. *Let I an ideal of A, H an ideal of $f(A) + J$ such that $f(I)J \subseteq H \subseteq J$. If $I \bowtie^f H := \{(i, f(i) + h) | i \in I, h \in H\}$ is a graded ideal of $A \bowtie^f J$, then I is a graded ideal of A.*

Proof. After [17] $I \bowtie^f H$ is an ideal of $A \bowtie^f J$.
Assume that $I \bowtie^f H$ is a graded ideal of $A \bowtie^f J$.
Let $i \in I$,

$$\begin{aligned} \text{so } (i, f(i)) \in I \bowtie^f H &= \oplus_{n \in \mathbb{Z}} ((I \bowtie^f H) \cap (A \bowtie^f H)_n) \\ &= \oplus_{n \in \mathbb{Z}} ((I \bowtie^f H) \cap (A_n \bowtie^f H)) \\ &= \oplus_{n \in \mathbb{Z}} ((I \cap A_n) \bowtie^f H). \end{aligned}$$

$$\begin{aligned} \text{So } (i, f(i)) &= \textstyle\sum_{n \in \mathbb{Z}} (x_n, f(x_n) + h_n), \text{ where } x_n \in I \cap A_n \text{ and } h \in H \\ &= (\textstyle\sum_{n \in \mathbb{Z}} x_n, \sum_{n \in \mathbb{Z}} (f(x_n) + h_n)). \end{aligned}$$

So $i = \sum_{n \in \mathbb{Z}} x_n$, where $x_n \in I \cap A_n$, so $i \in \oplus_{n \in \mathbb{Z}} (I \cap A_n)$, so $I \subseteq \oplus_{n \in \mathbb{Z}} (I \cap A_n)$.
Therefore I is a graded ideal of A.

Theorem 2. *I is a graded primary ideal of A if and only if $I \bowtie^f J$ is a graded primary ideal of $A \bowtie^f J$.*

Proof. • Assume that I is a graded primary ideal of A.
Let $(a, f(a)+j_1)$, $(b, f(b)+j_2) \in h(A \bowtie^f J) = h(A) \bowtie^f J$
such that $(a, f(a)+j_1)(b, f(b)+j_2) = (ab, f(ab)+j_2 f(a)+j_1 f(b)+j_1 j_2) \in I \bowtie^f J$
$\Rightarrow ab \in I \Rightarrow a \in I$ or $b \in Gr(I)$ since I is a graded primary ideal.
So $(a, f(a)+j_1) \in I \bowtie^f J$ or $(b, f(b)+j_2) \in Gr(I) \bowtie^f J = Gr(I \bowtie^f J)$.
Hence $I \bowtie^f J$ is a graded primary ideal of $A \bowtie^f J$.

• Let $a, b \in h(A)$ such that $ab \in I$.
We have $(a, f(a))(b, f(b)) = (ab, f(ab)) \in I \bowtie^f J$ and since $I \bowtie^f J$ is a graded primary ideal, then we have:
$(a, f(a)) \in I \bowtie^f J$ or $(b, f(b)) \in Gr(I) \bowtie^f J \Leftrightarrow a \in I$ or $b \in Gr(I)$, so I is a graded primary ideal.

3 Homological Properties of Amalgamation over a Graded Module

Definition 5. A graded A-module M is called $FP-gr$-injective if $Ext_A^1(N, M) = 0$ for any finitely presented graded A-module N.

Theorem 3. *Let (A, m) be a Noetherian local ring with dimension d, $f : A \to B$ be ring homomorphism and $0 \neq J \subseteq J(B)$ an ideal such that J is a finitely generated A-module. Then $H^d_{m \bowtie^f J}(A \bowtie^f J)$ is $FP-gr$-injective if and only if $H^d_m(A)$ and $H^d_m(J)$ are $FP-gr$-injective.*

Proof. • We have $dimA = dim(A \bowtie^f J) = d$. Then we have the following isomorphisms

$$H^d_{m \bowtie^f J}(A \bowtie^f J) \cong H^d_m(A \bowtie^f J) \cong H^d_m(A \oplus J) \cong H^d_m(A) \oplus H^d_m(J).$$

Since $H^d_{m \bowtie^f J}(A \bowtie^f J)$ is a $FP-gr$-injective A-module, for any finite presented A-module M we have,

$$0 = Ext_A^1(M, H^d_{m \bowtie^f J}(A \bowtie^f J)) \cong Ext_A^1(M, H^d_m(A)) \oplus Ext_A^1(M, H^d_m(J)).$$

So $Ext_A^1(M, H^d_m(A)) = Ext_A^1(M, H^d_m(J)) = 0$, hence $H^d_m(A)$ and $H^d_m(J)$ are $FP-gr$-injective.

• The converse is showed in the same way.

Corollary 1. *Let (A, m) be a Noetherian local ring of dimension d and let $0 \neq I$ be an ideal of A. Then $H^d_{m \bowtie I}(A \bowtie I)$ is $FP-gr$-injective if and only if $H^d_m(A)$ and $H^d_m(I)$ are $FP-gr$-injective.*

Proof. If (A, m) is a Noetherian local ring of dimension d, then $A \bowtie I$ is a Noetherian local ring with maximal ideal $m \bowtie I$ of dimension d.

Proposition 3. *Let (A,m) be a graded Noetherian local ring with dimension d, $f : A \to B$ be ring homomorphism and $0 \neq J \subseteq J(B)$ an ideal such that J is a finitely generated A-module. Then $H^d_{m\bowtie^f J}(A \bowtie^f J)$ is flat (resp. free, resp. projective) if and only if $H^d_m(A)$ and $H^d_m(J)$ are flat (resp. free, resp. projective).*

Proof. Since $H^d_{m\bowtie^f J}(A \bowtie^f J)$ is flat (resp. free, resp. projective) we have

$$0 = Tor_1^A(H^d_{m\bowtie^f J}(A \bowtie^f J), A/m) \cong Tor_1^A(H^d_m(A), A/m) \oplus Tor_1^A(H^d_m(J), A/m).$$

So $Tor_1^A(H^d_m(A), A/m) = Tor_1^A(H^d_m(J), A/m) = 0$, hence $H^d_m(A)$ and $H^d_m(J)$ are flat (resp. free, resp. projective).

Acknowledgements. I thank the organizers of the International Conference on PDE and Applications, Modeling and Simulation which is an opportunity for us to share and receive new ideas.

References

1. Alaoui Ismaili, K., Mahdou, N.: Coherence in amalgamated algebra along an ideal. Linear Algebra Appl. Bull. Iranian Math. Soc. **41**, 625–632 (2015)
2. Al-Zoubi, K.: On graded uniformily primary submodule. Int. J. Pure Appl. Math. **93**(1), 23–29 (2014)
3. Anderson, D.D., Smith, E.: Weakly prime ideals. Houston J. Math. **29**(4), 831–840 (2003)
4. Anderson, D.D., Batanieh, M.: Generalizations of prime ideals. Commun. Algebra **36**, 686–696 (2008)
5. Anderson, D.D.: Group graded rings and modules. Commun. Algebra **7**(1), 79–106 (1979)
6. Atiyah, M.F., MacDonald, I.G.: Introduction to Commutative Algebra. Addison-Wesley, Reading (1969)
7. Atani, S.E., Farzalipour, F.: On weakly primary ideals. Georgian Math. J. **12**(3), 423–429 (2005)
8. Badawi, A.: On weakly semiprime ideals of commutative rings. Beiträge zur Algebra und Geometrie/Contrib. Algebra Geom. **57**(3), 589–597 (2016). https://doi.org/10.1007/s13366-016-0283-9
9. Corner, A.L.S.: On endomorphism rings of primary abelian groups. Q. J. Math. Oxford **20**, 277–296 (1969)
10. Chhiti, M., Jarrar, M., Kabbaj, S., Mahdou, N.: Prufer conditions in an amalgamated duplication of a ring along an ideal. Commun. Algebra **43**(1), 249–261 (2015)
11. D'Anna, M., Fontana, M.: An amalgamated duplication of a ring along an ideal: the basic properties. J. Algebra Appl. **6**(3), 241–252 (2007)
12. D'Anna, M., Finocchiaro, C.A., Fontana, M.: Amalgamated Algebras Along an Ideal. Commutative Algebra and Its Applications, pp. 241–252. Walter de Gruyter, Berlin (2009)
13. D'Anna, M., Finocchiaro, C.A., Fontana, M.: Properties of chains of prime ideals in an amalgamated algebra along an ideal. J. Pure Appl. Algebra **214**, 1633–1641 (2010)
14. Dade, E.C.: Graded Krull domains. Math. Zeitschrift **174**(3), 241–262 (1980)
15. Darani, A.Y.: Generalization of primary ideals in commutative rings. Novi Sad J. Math. **42**(1), 27–35 (2012)
16. Goto, S., Watanabe, K.: On graded rings, I. J. Math. Soc. Jpn **30**, 172–213 (1978)
17. Mahdou, N., et al.: Weakly prime ideals issued from an amalgamated algebra. Hacet. J. Math. Stat. **49**(3), 1159–1167 (2020)

18. Marley, T.: Finitely graded local cohomology modules and the depths of graded algebras. Proc. Amer. Math. Soc. **123**, 3601–3607 (1995)
19. Oral, K.H., Tekir, U., Agargun, A.G.: On graded prime and primary submodules. Turk. J. Math. **35**, 159–167 (2011)
20. Thiaw, M.: Weakly primary ideal and ϕ-primary ideal in amalgamated algebra along an ideal. JP J. Algebra Number Theory Appl. **48**(2), 167–175 (2020)

Check for updates

Generalized Derivations of Tensor Products of Algebras

Moussa Thiaw(✉)

African Institute for Mathematical Sciences (AIMS-SENEGAL), Gaston Berger University, Km 2 route de Joal centre IRD, Saint-Louis, Senegal
moussa.thiaw@aims-senegal.org, thiawskr@gmail.com

Abstract. Given B and C two associative unital algebras over A not necessarily commutative which are finitely generated, we show that if f is a generalized derivation of $B\otimes C$ and $(\beta_i)_{i\in I}$ a basis of C, then for all $i\in I$, there exists a generalized derivation f_i of B such that for all $x\in B$ we have $f(x\otimes 1)=\sum_{i\in I} f_i(x)\otimes\beta_i$. By denoting by $Der^G(B)$ the set of generalized derivations of all B and $Der(B)$ the set of derivations of all B and considering the left multiplier $L_b: B\rightarrow B$ defined by $L_b(x)=bx$, $\forall\, b,x\in B$, the right multiplier $R_b: B\rightarrow B$ defined by $R_b(x)=xb$, $\forall\, b,x\in B$ and $adb=L_b-R_b$, $\forall\, b\in B$.

We show also in this paper that every generalized derivation f of $B\otimes C$ can be written in the form, $f=adu+\sum_{j=1}^{p} L_{z_j}\otimes d_j+\sum_{i=1}^{q} g_i\otimes L_{w_i}$, where $u\in B\otimes C$, $z_j\in Z(B), w_i\in Z(C), d_j\in Der(C), g_i\in Der^G(B)$ and $adu: B\otimes C\rightarrow B\otimes C$ defined by $adu(x)=ux-xb$, $\forall\, x\in B\otimes C$.

Keywords: Derivation · Generalized derivation · Algebra · Tensor product

1 Introduction

The notion of generalized derivation was introduced by Brešar in 1991. Given an A-algebra B, a linear mapping $f: B\rightarrow B$ is a generalized derivation if there exists a derivation d of B such that the generalized rule $f(xy)=f(x)y+xd(y)$ is satisfied for all $x,y\in B$. We denote by $Der^G(B)$ the set of generalized derivations of all B and $Der(B)$ the set of derivations of all B.

Matez Brešar in his article entitled "Derivation of Tensor Product of Nonassociative Algebra" has shown that any derivation d of $B\otimes C$ can be written in the form $d=adu+\sum_{j=1}^{p} L_{z_j}\otimes f_j+\sum_{i=1}^{q} g_i\otimes L_{w_i}$, where $u\in N(B)\otimes N(C), z_j\in Z(B), w_i\in Z(C), f_i\in Der(C)$ and $g_i\in Der(B)$ (see [2]). By taking inspiration from this result and considering the left multiplier $L_b: B\rightarrow B$ defined by $L_b(x)=bx$, $\forall\, b,x\in B$, the right multiplier $R_b: B\rightarrow B$ defined by $R_b(x)=xb$, $\forall\, b,x\in B$ and $adb=L_b-R_b$, $\forall\, b\in B$, we show that every generalized derivation f of $B\otimes C$ can be written in the form, $f=adu+\sum_{j=1}^{p} L_{z_j}\otimes d_j+\sum_{i=1}^{q} g_i\otimes L_{w_i}$, where $u\in B\otimes C$, $z_j\in Z(B), w_i\in Z(C)$, $d_j\in Der(C)$ and $g_i\in Der^G(B)$.

It should be noted that this result does not exist to our knowledge in the literature, but we use the same demonstration techniques as in [2]. This paper is structured in two

This paper is dedicated to my mother Sadio KEBE.

S. Melliani and O. Castillo (Eds.): ICPAMS 2021, LNNS 476, pp. 332–342, 2023.
https://doi.org/10.1007/978-3-031-12416-7_29

parts, in the first part entitled "preliminary results" we recall some basic results of the generalized derivation. We also recall in this part the results in [2] to use it to establish our results. In the second part of this paper we present our main results.

2 Preliminaries Results

Definition 1. Let B an A-algebra. A linear mapping $f : B \longrightarrow B$ is a generalized derivation, if there exists a derivation d of B such that the generalized rule

$$f(xy) = f(x)y + xd(y)$$

is satisfied for all $x, y \in B$.

Notation: We denote by $Der^G(B)$ the set of derivations of all B.

Example 1. The maps of the form $f(x) = bx + d(x)$, where b is a fixed element of B and $d \in Der(B)$, are generalized derivations.

Definition 2. We call left multiplier any map $L : B \longrightarrow B$ defined by

$$L(xy) = L(x)y, \forall x, y \in B.$$

Definition 3. We call right multiplier any map $R : B \longrightarrow B$ defined by

$$R(xy) = yR(x), \forall x, y \in B.$$

Example 2. 1. The map $L_b : B \longrightarrow B$ defined by $L_b(x) = bx$, $\forall\, b, x \in B$ is a left multiplier.
2. The map $R_b : B \longrightarrow B$ defined by $R_b(x) = xb$, $\forall\, b, x \in B$ is a right multiplier.

We can easily see that $adu = L_u - R_u \in Der(B)$.

Proposition 1. *Let d a derivation of C. Then the map*

$$\begin{aligned} d_z :\ & B \otimes C \longrightarrow B \otimes C \\ & x \otimes y \longmapsto zx \otimes d(y), \forall z \in Z(B). \end{aligned}$$

is a derivation of $B \otimes C$.

Proof. See [2].

Proposition 2. *Let d a derivation of B. Then the map*

$$\begin{aligned} d_z :\ & B \otimes C \longrightarrow B \otimes C \\ & x \otimes y \longmapsto d(x) \otimes zy, \forall z \in Z(C). \end{aligned}$$

is a derivation of $B \otimes C$.

Proof. See [2].

Lemma 1

Proof. See [2].

Lemma 2. *Let $e_1,\cdots,e_n \in U$ be linearly independent. If $v_1,\cdots,v_n \in V$ such that*

$$e_1 \otimes v_1 + \cdots + e_n \otimes v_n = 0,$$

then $v_i = 0$, $\forall\, 1 \leq i \leq n$.

Proof. Let $f_i : U \longrightarrow V$ the linear map defined by $f_i(e_j) = \delta_{ij}$, $\forall\, j = 1,\cdots,n$. Consider the map $g : U \times V \longrightarrow V$

$$(x,y) \longmapsto f_i(x)y.$$

It is not difficult to see that g is bilinear, so according to the universal property of the tensor product it exists $h_i : U \otimes V \longrightarrow V$ such that

$$h_i(x \otimes y) = f_i(x)y.$$

So on the one hand we have

$$\sum_{j=1}^{n} h_i(e_j \otimes v_j) = \sum_{j=1}^{n} f_i(e_j)v_j = \sum_{j=1}^{n} \delta_{ij}v_j = v_i,$$

on the other hand we have

$$\sum_{j=1}^{n} h_i(e_j \otimes v_j) = h_i(\sum_{j=1}^{n} e_j \otimes v_j) = h_i(0) = 0.$$

Hence $v_i = 0$.

Lemma 3. *Let $e_1,\cdots,e_n \in U$ be linearly independent. If $v_1,\cdots,v_n \in V$ such that*

$$e_1 \otimes v_1 + \cdots + e_n \otimes v_n = \varepsilon_1 \otimes w_1 + \cdots + \varepsilon_m \otimes w_m,$$

for some $\varepsilon_1,\cdots,\varepsilon_m \in U$ and $w_1,\cdots,w_m \in V$. Then each v_i is a linear combination of $w_1,\cdots,w_m$.

Proof. Since the elements $e_1,\cdots,e_n \in U$ are linearly independent, then we can complete it in one basis of $vect\{e_1,\cdots,e_n,\varepsilon_1,\cdots,\varepsilon_m\}$.

Let $B = \{e_1,\cdots,e_n,e_{n+1},\cdots,e_p\}$ such basis.

So we have $\varepsilon_j = \sum_{i=1}^{p} \alpha_i^j e_i$.

Using the bilinearity of the tensor product we have

$$\begin{aligned} e_1 \otimes v_1 + \cdots + e_n \otimes v_n &= \sum_{i=1}^{p} \alpha_i^1 e_i \otimes w_1 + \cdots + \sum_{i=1}^{p} \alpha_i^m e_i \otimes w_m \\ &= \sum_{i=1}^{p} e_i \otimes \alpha_i^1 w_1 + \cdots + \sum_{i=1}^{p} e_i \otimes \alpha_i^m w_m \\ &= e_1 \otimes \sum_{j=1}^{m} \alpha_1^j w_j + \cdots + e_p \otimes \sum_{j=1}^{m} \alpha_p^j w_j \end{aligned}$$

So we have
$e_1 \otimes v_1 - e_1 \otimes \sum_{j=1}^{m} \alpha_1^j w_j + \cdots + e_n v_n - e_n \otimes \sum_{j=1}^{m} \alpha_n^j w_j - e_{n+1} \otimes \sum_{j=1}^{m} \alpha_{n+1}^j w_j - \cdots - e_p \otimes \sum_{j=1}^{m} \alpha_p^j w_j = 0$
$e_1 \otimes (v_1 - \sum_{j=1}^{m} \alpha_1^j w_j) + \cdots + e_n \otimes (v_n - \sum_{j=1}^{m} \alpha_n^j w_j) - e_{n+1} \otimes \sum_{j=1}^{m} \alpha_{n+1}^j w_j - \cdots - e_p \otimes \sum_{j=1}^{m} \alpha_p^j w_j = 0.$

So after the Lemma 2 we have $v_i = \sum_{j=1}^{m} \alpha_i^j w_j$, hence the result.

3 Results: Generalized Derivation and Tensor Products of Algebras

3.1 Generalized Derivation and Tensor Products

Theorem 1. *Let f a generalized derivation of C. Then the map*

$$\begin{aligned} f_z :\ & B \otimes C \longrightarrow B \otimes C \\ & x \otimes y \longmapsto zx \otimes f(y), \forall z \in Z(B). \end{aligned}$$

is a generalized derivation of $B \otimes C$.

Proof. Let $x, x' \in B$ and $y, y' \in C$.
Since $f \in Der^G(C)$, then it exists $d \in Der(C)$ such that $f(yy') = f(y)y' + yd(y')$.
We have

$$\begin{aligned} f_z((x \otimes y)(x' \otimes y')) &= f_z(xx' \otimes yy') = z(xx') \otimes f(yy') = z(xx') \otimes (f(y)y' + yd(y')) \\ &= z(xx') \otimes f(y)y' + z(xx') \otimes yd(y') \\ &= (zx)x' \otimes f(y)y' + x(zx') \otimes yd(y'), \text{ because } z \in Z(B) \\ &= ((zx) \otimes f(y))(x' \otimes y') + (x \otimes y)((zx') \otimes d(y')) \\ &= f_z(x \otimes y)(x' \otimes y') + (x \otimes y)d_z(x' \otimes y'). \end{aligned}$$

Therefore $f_z \in Der^G(B \otimes C)$.

Theorem 2. *Let g a generalized derivation of B. Then the map*

$$\begin{aligned} g_z :\ & B \otimes C \longrightarrow B \otimes C \\ & x \otimes y \longmapsto g(x) \otimes zy, \forall z \in Z(C). \end{aligned}$$

is a generalized derivation of $B \otimes C$.

Proof. The proof is similar to that of the previous theorem.

Proposition 3. *Let $g \in Der^G(B)$ and $z \in Z(C)$. Then $g \otimes L_z \in Der^G(B \otimes C)$.*

Proof. Let $x, x' \in B$ and $y, y' \in C$.
Since $g \in Der^G(B)$, then it exists $d \in Der(B)$ such that $g(xx') = g(x)x' + xd(x')$.
We have

$$\begin{aligned}
(g \otimes L_z)((x \otimes y)(x' \otimes y')) &= (g \otimes L_z)(xx' \otimes yy') = g(xx') \otimes L_z(yy') \\
&= (g(x)x' + xd(x')) \otimes (z(yy')) \\
&= (g(x)x') \otimes (z(yy')) + (xd(x')) \otimes (z(yy')) \\
&= (g(x)x') \otimes (z(yy')) + (x \otimes y)(d(x') \otimes (zy')), z \in Z(C) \\
&= (g(x) \otimes L_z(y))(x' \otimes y') + (x \otimes y)d_z(x' \otimes y') \\
&= ((g \otimes L_z)(x \otimes y))(x' \otimes y') + (x \otimes y)d_z(x' \otimes y')
\end{aligned}$$

Therefore $g \otimes L_z \in Der^G(B \otimes C)$.

Proposition 4. *Let $f \in Der^G(C)$ and $z \in Z(B)$. Then $L_z \otimes f \in Der^G(B \otimes C)$.*

Proof. The proof is similar to that of the previous proposition.

4 Results: Generalized Derivation and Tensor Products of Algebras

4.1 Generalized Derivation and Tensor Products

Theorem 3. *Let f a generalized derivation of C. Then the map*

$$\begin{aligned}
f_z : \ & B \otimes C \longrightarrow B \otimes C \\
& x \otimes y \longmapsto zx \otimes f(y), \forall z \in Z(B).
\end{aligned}$$

is a generalized derivation of $B \otimes C$.

Proof. Let $x, x' \in B$ and $y, y' \in C$.
Since $f \in Der^G(C)$, then it exists $d \in Der(C)$ such that $f(yy') = f(y)y' + yd(y')$.
We have

$$\begin{aligned}
f_z((x \otimes y)(x' \otimes y')) &= f_z(xx' \otimes yy') = z(xx') \otimes f(yy') = z(xx') \otimes (f(y)y' + yd(y')) \\
&= z(xx') \otimes f(y)y' + z(xx') \otimes yd(y') \\
&= (zx)x' \otimes f(y)y' + x(zx') \otimes yd(y'), \text{ because } z \in Z(B) \\
&= ((zx) \otimes f(y))(x' \otimes y') + (x \otimes y)((zx') \otimes d(y')) \\
&= f_z(x \otimes y)(x' \otimes y') + (x \otimes y)d_z(x' \otimes y').
\end{aligned}$$

Therfore $f_z \in Der^G(B \otimes C)$.

Theorem 4. *Let g a generalized derivation of B. Then the map*

$$\begin{aligned} g_z : \; & B\otimes C \longrightarrow B\otimes C \\ & x\otimes y \longmapsto g(x)\otimes zy, \forall z\in Z(C). \end{aligned}$$

is a generalized derivation of $B\otimes C$.

Proof. The proof is similar to that of the previous theorem.

Proposition 5. *Let* $g\in Der^G(B)$ *and* $z\in Z(C)$. *Then* $g\otimes L_z\in Der^G(B\otimes C)$.

Proof. Let $x,x'\in B$ and $y,y'\in C$.
Since $g\in Der^G(B)$, then it exists $d\in Der(B)$ such that $g(xx')=g(x)x'+xd(x')$.
We have

$$\begin{aligned} (g\otimes L_z)((x\otimes y)(x'\otimes y')) &= (g\otimes L_z)(xx'\otimes yy') = g(xx')\otimes L_z(yy') \\ &= (g(x)x'+xd(x'))\otimes(z(yy')) \\ &= (g(x)x')\otimes(z(yy'))+(xd(x'))\otimes(z(yy')) \\ &= (g(x)x')\otimes(z(yy'))+(x\otimes y)(d(x')\otimes(zy')), z\in Z(C) \\ &= (g(x)\otimes L_z(y))(x'\otimes y')+(x\otimes y)d_z(x'\otimes y') \\ &= ((g\otimes L_z)(x\otimes y))(x'\otimes y')+(x\otimes y)d_z(x'\otimes y') \end{aligned}$$

Therefore $g\otimes L_z\in Der^G(B\otimes C)$.

Proposition 6. *Let* $f\in Der^G(C)$ *and* $z\in Z(B)$. *Then* $L_z\otimes f\in Der^G(B\otimes C)$.

Proof. The proof is similar to that of the previous proposition.

4.2 Main Results

The following lemma and the following theorem are the main results of this paper.

Lemma 4. *Let B and C two associative A-algebras not necessarily commutative, f a generalized derivation of* $B\otimes C$ *and* $(\beta_i)_{i\in I}$ *a basis of C. Then for all* $i\in I$, *there exists a generalized derivation* f_i *of B such that for all* $x\in B$ *we have*

$$f(x\otimes 1)=\sum_{i\in I} f_i(x)\otimes\beta_i. \tag{1}$$

Proof. As in Lemma ?? for every $x\in B$, there exists $f_i(x)\in B$ such that (1) holds.
The linearity of f_i comes from the linearity of f.
Let $x,y\in B$.
On the one hand we have:

$$\begin{aligned} f(xy\otimes 1) &= f((x\otimes 1)(y\otimes 1)) \\ &= f(x\otimes 1)(y\otimes 1)+(x\otimes 1)d(y\otimes 1), \text{ where } d\in Der(B\otimes C), \\ &= \left(\textstyle\sum_{i\in I} f_i(x)\otimes\beta_i\right)(y\otimes 1)+(x\otimes 1)\textstyle\sum_{i\in I} d_i(y)\otimes\beta_i, \text{ it's after (??) and (1)} \\ &= \textstyle\sum_{i\in I}[(f_i(x)\otimes\beta_i)(y\otimes 1)]+\textstyle\sum_{i\in I}[(x\otimes 1)d_i(y)\otimes\beta_i] \\ &= \textstyle\sum_{i\in I} f_i(x)y\otimes\beta_i+\textstyle\sum_{i\in I} xd_i(y)\otimes\beta_i=\textstyle\sum_{i\in I}(f_i(x)y+xd_i(y))\otimes\beta_i(*). \end{aligned}$$

On the other hand we have:

$$f(xy\otimes 1)=\sum_{i\in I} f_i(xy)\otimes\beta_i,\ \text{it's after (1)}(**).$$

So after $(*)$ and $(**)$ we have

$\sum_{i\in I} f_i(xy)\otimes\beta_i=\sum_{i\in I}(f_i(x)y+xd_i(y))\otimes\beta_i$

So $\sum_{i\in I}(f_i(xy)-f_i(x)y-xd_i(y))\otimes\beta_i=0.$

Since the vectors β_i are linearly independent, then we have

$$f_i(xy)-f_i(x)y-xd_i(y))=0.$$

So we have

$$f_i(xy)=f_i(x)y+xd_i(y)),\ \text{with}\ d_i\in Der(B),\forall i\in I.$$

Therefore $f_i\in Der^G(B)$.

Theorem 5. *Let B and C two associative A-algebras not necessarily commutative which are finitely generated. Then every generalized derivation f of $B\otimes C$ can be written in the form*

$$f=adu+\sum_{j=1}^{p} L_{z_j}\otimes d_j+\sum_{i=1}^{q} g_i\otimes L_{w_i},$$

where $u\in B\otimes C$, $z_j\in Z(B), w_i\in Z(C)$, $d_j\in Der(C)$ and $g_i\in Der^G(B)$.

Proof. Let $(w_i)_{1\le i\le q}$ a basis of $Z(C)$. So by completing this one we get a basis $(w_i)_{1\le i\le q}\cup(\beta_i)_{1\le i\le l}$ of C.

Let $f\in Der^G(B\otimes C)$, $x\in B$ and $y\in C$.

So it exists $d\in Der(B\otimes C)$ such that

$$f(x\otimes y)=f((x\otimes 1)(1\otimes y))=f(x\otimes 1)(1\otimes y)+(x\otimes 1)d(1\otimes y).$$

Since $f\in Der^G(B\otimes C)$, then after the Lemma 4 it exists $g_i,h_i\in Der^G(B)$ such that

$$f(x\otimes 1)=\sum_{i=1}^{q} g_i(x)\otimes w_i+\sum_{i=1}^{l} h_i(x)\otimes\beta_i. \tag{2}$$

Also since $d\in Der(B\otimes C)$, then after the Lemma ?? it exists $d_j,d'_j\in Der(C)$ such that

$$d(1\otimes y)=\sum_{j=1}^{p} z_j\otimes d_j(y)+\sum_{j=1}^{m} r_j\otimes d'_j(y), \tag{3}$$

where $z_j\in Z(B)$ and $r_j\in B$.

So from (2) and (3) we have:

$$\begin{aligned} f(x\otimes y) &=[\textstyle\sum_{i=1}^{q} g_i(x)\otimes w_i](1\otimes y)+[\sum_{i=1}^{l} h_i(x)\otimes\beta_i](1\otimes y)+(x\otimes 1)\sum_{j=1}^{p} z_j\otimes d_j(y)\\ &\quad+(x\otimes 1)\textstyle\sum_{j=1}^{m} r_j\otimes d'_j(y)\\ &=\textstyle\sum_{i=1}^{q} g_i(x)\otimes w_iy+\sum_{i=1}^{l} h_i(x)\otimes\beta_iy+\sum_{j=1}^{p} xz_j\otimes d_j(y)+\sum_{j=1}^{m} xr_j\otimes d'_j(y). \end{aligned}$$

So we have:

$$f(x\otimes y)=\sum_{i=1}^{q} g_i(x)\otimes L_{w_i}(y)+\sum_{i=1}^{l} h_i(x)\otimes L_{\beta_i}(y)+\sum_{j=1}^{p} R_{z_j}(x)\otimes d_j(y)+\sum_{j=1}^{m} R_{r_j}(x)\otimes d'_j(y).$$

Since $z_j \in Z(B)$, then we have $R_{z_j}=L_{z_j}$.

So we have

$$f=\sum_{i=1}^{q} g_i\otimes L_{w_i}+\sum_{i=1}^{l} h_i\otimes L_{\beta_i}+\sum_{j=1}^{p} L_{z_j}\otimes d_j+\sum_{j=1}^{m} R_{r_j}\otimes d'_j \tag{4}$$

After the Proposition 3, $g_i\otimes L_{w_i}\in Der^G(B\otimes C)$ and $L_{z_j}\otimes d_j \in Der(B\otimes C)$.

We have $\delta := f-\sum_{j=1}^{p} L_{z_j}\otimes d_j-\sum_{i=1}^{q} g_i\otimes L_{w_i}$ is a generalized derivation of $B\otimes C$.

From (4) we have

$$\delta=\sum_{i=1}^{l} h_i\otimes L_{\beta_i}+\sum_{j=1}^{m} R_{r_j}\otimes d'_j.$$

So it remains to show that $\delta = adu$, for some $u\in B\otimes C$ to show the theorem.

Suppose it exists $1\le i\le l$ such that $h_i\neq 0$.

Let $\{h_1,\cdots,h_s\}$ a maximal linearly independent subset of $\{h_1,\cdots,h_l\}$.

We have

$$\begin{aligned}\sum_{i=1}^{l} h_i\otimes L_{\beta_i} &=\sum_{i=1}^{s} h_i\otimes L_{\beta_i}+\sum_{i>s}^{l} h_i\otimes L_{\beta_i}=\sum_{i=1}^{s} h_i\otimes L_{\beta_i}+\sum_{i>s}^{l}\sum_{j=1}^{s} \alpha_j h_j\otimes L_{\beta_i}\\ &=\sum_{i=1}^{s} h_i\otimes L_{\beta_i}+\sum_{j=1}^{s} h_j\otimes L_{\sum_{i>s}^{l}\alpha_j\beta_i}.\end{aligned}$$

So $\sum_{i=1}^{l} h_i\otimes L_{\beta_i}=\sum_{i=1}^{s} h_i\otimes L_{n_i}$, where n_i are linearly independent elements in $vect\{\beta_1,\cdots,\beta_l\}$.

In the same way, we suppose that $\{d'_1,\cdots,d'_t\}$ is a maximal linearly independent subset of $\{d'_1,\cdots,d'_m\}$.

We have $\sum_{j=1}^{m} R_{r_j}\otimes d'_j=\sum_{j=1}^{t} R_{m_j}\otimes d'_j$, where m_j are linearly independent elements in $vect\{r_1,\cdots,r_m\}$.

Hence we have

$$\delta=\sum_{i=1}^{s} h_i\otimes L_{n_i}+\sum_{j=1}^{t} R_{m_j}\otimes d'_j, \tag{5}$$

where $h_1,\cdots,h_s\in Der^G(B)$ and are linearly independent, $d'_1,\cdots,d'_t\in Der(C)$ and are linearly independent, $n_1,\cdots,n_s$ are elements of C which are linearly independent and $m_1,\cdots,m_t$ are elements of B which are linearly independent.

furthermore we have

$$vect\{n_1,\cdots,n_s\}\cap Z(C)=0 \tag{6}$$

$$vect\{m_1,\cdots,m_t\}\cap Z(B)=0. \tag{7}$$

Since $\delta \in Der^G(B\otimes C)$, then $\delta \in Der(B\otimes C)$, so we have:

$$\begin{aligned}\delta((1\otimes y)(x\otimes 1)) &= (1\otimes y)\delta(x\otimes 1)+\delta(1\otimes y)(x\otimes 1)+\textstyle\sum_{j=1}^{t} R_{m_j}(1)\otimes d'_j(y)](x\otimes 1)\\ &= (1\otimes y)[\textstyle\sum_{i=1}^{s} h_i(x)\otimes L_{n_i}(1)]+[\textstyle\sum_{j=1}^{t} R_{m_j}(1)\otimes d'_j(y)](x\otimes 1)\\ &= \textstyle\sum_{i=1}^{s} h_i(x)\otimes yn_i+\textstyle\sum_{j=1}^{t} m_j x\otimes d'_j(y)(\textstyle\sum_{i=1}^{s} h_i\otimes R_{n_i})(x\otimes y)+(\textstyle\sum_{j=1}^{t} L_{m_j}\otimes d'_j)(x\otimes y).\end{aligned}$$

So we have

$$\delta=\sum_{i=1}^{s} h_i\otimes R_{n_i}+\sum_{j=1}^{t} L_{m_j}\otimes d'_j \tag{8}$$

From (5) and (8) we have

$$\textstyle\sum_{i=1}^{s} h_i\otimes L_{n_i}+\textstyle\sum_{j=1}^{t} R_{m_j}\otimes d'_j=\textstyle\sum_{i=1}^{s} h_i\otimes R_{n_i}+\textstyle\sum_{j=1}^{t} L_{m_j}\otimes d'_j$$

So we have

$$\sum_{i=1}^{s} h_i\otimes L_{n_i}-\sum_{i=1}^{s} h_i\otimes R_{n_i}=\sum_{j=1}^{t} L_{m_j}\otimes d'_j-\sum_{j=1}^{t} R_{m_j}\otimes d'_j$$

Hence we have

$$\sum_{i=1}^{s} h_i\otimes (R_{n_i}-L_{n_i})=\sum_{j=1}^{t} (R_{m_j}-L_{m_j})\otimes d'_j.$$

So we have

$$\sum_{i=1}^{s} h_i\otimes adn_i=\sum_{j=1}^{t} adm_j\otimes d'_j.$$

$*$ Let $\alpha_i \in A$ such that $\sum_i \alpha_i adm_i=0 \Rightarrow ad(\sum_i \alpha_i m_i)=0 \Rightarrow \sum_i \alpha_i m_i \in Z(B)$, so after (7) we have $\sum_i \alpha_i m_i=0 \Rightarrow \alpha_i=0$ because the n_i are linearly independent.

So $\{adm_1,\cdots,adm_t\}$ is a linearly independent set.

So after the Lemma 3 each d'_j is a linear combination of the vectors $adn_i \Rightarrow \exists\, \lambda_{ij}\in A$ such that

$$d'_j=\sum_{i=1}^{s} \lambda_{ij} ad_{n_i}.$$

Therefore we have

$$\begin{aligned}\textstyle\sum_{i=1}^{s} h_i\otimes adn_i &= \textstyle\sum_{j=1}^{t} adm_j\otimes \textstyle\sum_{i=1}^{s} \lambda_{ij} ad_{n_i} \Rightarrow \textstyle\sum_{i=1}^{s} h_i\otimes adn_i-(\textstyle\sum_{i=1}^{s}(\textstyle\sum_{j=1}^{t} \lambda_{ij} adm_j)\otimes adn_i)\\ &\Rightarrow (\textstyle\sum_{i=1}^{s} h_i-\textstyle\sum_{i=1}^{s}\textstyle\sum_{j=1}^{t} \lambda_{ij} adm_j)\otimes adn_i=0.\end{aligned}$$

So we have

$$\sum_{i=1}^{s}(h_i-\sum_{j=1}^{t} \lambda_{ij} adm_j)\otimes adn_i=0.$$

From (7) $\{adn_1,\cdots,adn_m\}$ is a linearly independent set, so we have

$$h_i=\sum_{j=1}^{t} \lambda_{ij} adm_j.$$

So from 8 we have:

$$\begin{aligned}
\delta &= \sum_{i=1}^{s}(\sum_{j=1}^{t}\lambda_{ij}adm_j)\otimes R_{n_i}+\sum_{j=1}^{t}L_{m_j}\otimes\sum_{i=1}^{s}\lambda_{ij}ad_{n_i}\\
&= \sum_{i=1}^{s}\sum_{j=1}^{t}(\lambda_{ij}adm_j)\otimes R_{n_i})+\sum_{i=1}^{s}\sum_{j=1}^{t}(L_{m_j}\otimes\lambda_{ij}ad_{n_i})\\
&= \sum_{i=1}^{s}\sum_{j=1}^{t}\lambda_{ij}(adm_j\otimes R_{n_i}+L_{m_j}\otimes ad_{n_i})\\
&= \sum_{i=1}^{s}\sum_{j=1}^{t}\lambda_{ij}[(L_{m_j}-R_{m_j})\otimes R_{n_i}+L_{m_j}\otimes(L_{n_i}-R_{n_i})]\\
&= \sum_{i=1}^{s}\sum_{j=1}^{t}\lambda_{ij}(L_{m_j}\otimes L_{n_i}-R_{m_j}\otimes R_{n_i})=\sum_{i=1}^{s}\sum_{j=1}^{t}\lambda_{ij}ad(m_j\otimes n_i)\\
&= ad(\sum_{i=1}^{s}\sum_{j=1}^{t}\lambda_{ij}m_j\otimes n_i).
\end{aligned}$$

Pose

$$u=\sum_{i=1}^{s}\sum_{j=1}^{t}\lambda_{ij}m_j\otimes n_i$$

From the above we have $\delta = f-\sum_{j=1}^{p}L_{z_j}\otimes d_j-\sum_{i=1}^{q}g_i\otimes L_{w_i}$, so we have

$$adu=f-\sum_{j=1}^{p}L_{z_j}\otimes d_j-\sum_{i=1}^{q}g_i\otimes L_{w_i}$$

Hence we have

$$f=adu+\sum_{j=1}^{p}L_{z_j}\otimes d_j+\sum_{i=1}^{q}g_i\otimes L_{w_i},$$

with $f\in Der^G(B\otimes C)$, $g_i\in Der^G(B)$ and $d_j\in Der(C)$.

Acknowledgments. I thank the organizers of the International Conference on PDE and Applications, Modeling and Simulation which is an opportunity for us to share and receive new ideas.

References

1. Benkart, G.M., Osborn, J.M.: Derivations and automorphisms of nonassociative matrix algebras. Trans. Am. Math. Soc. **263**, 411–430 (1981)
2. Brešar, M.: Derivation of tensor products of nonassociative algebras. Linear Algebra Appl. **530**, 244–252 (2017)
3. Brešar, M.: A note on derivation. Math. J. Okayama Univ. **32**, 83–88 (1990)
4. Brešar, M.: Introduction to Noncommutative Algebra. Universitext, Springer, Cham (2014). https://doi.org/10.1007/978-3-319-08693-4
5. Creedon, T.: Products of derivations. Proc. Edinburgh Math. Soc. **41**, 407–410 (1998)
6. De Filipis, V., Huang, S.: Generalized derivation on semiprime rings. Bull. Korean Math. Soc. **48**(6), 1253–1259 (2011)
7. De Filippis, V.: Generalized derivations in prime rings and noncommutative Banach algebras. Bull. Korean Math. Soc. **45**, 621–629 (2008)
8. Hvala, B.: Generalized derivation in prime rings. Comm. Algebra **26**(4), 1147–1166 (1998)
9. Jung, Y.S., Park, K.H.: On generalized (α, β)-derivations and commutativity in prime rings. Bull. Korean Math. Soc. **43**(1), 101–106 (2006)

10. Komatsu, H., Nakajima, A.: Generalized derivations with invertible values. Commun. Algebra **32**(5), 1937–1944 (2004)
11. Lee, T.K.: Generalized derivations of left faithful rings. Commun. Algebra **27**(8), 4057–4073 (1998)
12. Quadri, M.A., Khan, M.S., Rehman, N.: Generalized derivations and commutativity of prime rings. Ind. J. Pure Appl. Math. **34**(98), 1393–1396 (2003)
13. Zam, S.A.: Derivation of tensor product algebra. Commun. Algebra **36**, 905–927 (2008)

A Generalized Fixed Points for Multi-valued Mappings in G-Metric Spaces and Applications

Noreddine Makran[1](✉), Abdelhak El Haddouchi[2], and Brahim Marzouki[1]

[1] Department of Mathematical Sciences, Mohammed Premier University, Oujda, Morocco
makranmakran83@gmail.com

[2] Department of Mathematical Sciences, Faculty of Sciences and Technologie, Er-Rachidia, Morocco

Abstract. In this work we are interested to prove a general fixed point theorem of multi-valued mappings satisfying a new type relation in $G-$metric spaces. The results in this paper generalize the results obtained in [3]. An example and application integral equation are given to illustrate the usability of the main results.

Keywords: Metric space · $G-$metric space · Implicit relation · Fixed point · Multivalued maps

Subject Classifications: MSC2010: 54H25 · 47H10

1 Introduction and Preliminary

In analysis, the fixed point theorems turn out to be very useful tools in mathematics, especially in solving differential and functional equations. In 1922, Banach [2] proved that each contraction map in a complete metric space has a unique fixed point. Mustafa and Sims [16] introduced a new notion of generalized metric space called a G-metric space. Mustafa, Sims and others studied fixed point theorems for mappings satisfying different contractive conditions [11,14,16,18–21]. Several authors have introduced a new class of generalized metric space, obtained several results in fixed point theory, (see [1–21]).

As a consequence of our work we obtain some results known in the case of multi-valued mappings that we will point out, and we give an application for an integral equation.

Let X be a G-metric space. We shall denote $B(X)$ the set of nonempty closed bounded subsets of X. Let $H_G(.,.,.)$ be the Hausdorff G-distance on $B(X)$, in [4] Kaewcharoen and Kaewkhao defined Hausdorf G-metric as, for $A, B, C \in B(X)$ we have

$$H_G(A, B, C) = \max\{\sup_{x\in A} G(x, B, C), \sup_{x\in B} G(x, A, C), \sup_{x\in C} G(x, B, A)\}$$

S. Melliani and O. Castillo (Eds.): ICPAMS 2021, LNNS 476, pp. 343–355, 2023.
https://doi.org/10.1007/978-3-031-12416-7_30

where

$$G(x,B,C)=d_G(x,B)+d_G(B,C)+d_G(x,C),\quad d_G(x,B)=\inf\{d_G(x,y),y\in B\},$$
$$d_G(B,C)=\inf\{d_G(x,y),x\in B,y\in C\},\quad G(x,y,C)=\inf\{G(x,y,z),z\in C\}.$$

Note that (see [17]) $d_G(x,y)$ is given as $d_G(x,y)=G(x,y,y)+G(x,x,y)$ which defines a metric on X. A mapping $T: X \longrightarrow B(X)$ is called a multivalued mapping. A point $x\in X$ is called a fixed point of T if $x\in Tx$.

Definition 1 ([12]). Let X be a nonempty set. Suppose that G: $X\times X\times X\longrightarrow R^+$ satisfies:

(G1) $G(x,y,z)=0 \quad if \quad x=y=z$,
(G2) $0<G(x,x,y),\ for\ all\quad x,y\in X, with \quad x\neq y$,
(G3) $G(x,x,y)\leq G(x,y,z)$, for all $x,y,z\in X$, with $z\neq y$,
(G4) $G(x,y,z)=G(x,z,y)=G(y,x,z)=...$ (symmetry in all three variables), and
(G5) $G(x,y,z)\leq G(x,a,a)+G(a,y,z),\forall x,y,z,a\in X$ (triangle inequality).

Then G is called a generalized metric and the pair (X,G) is called a G-metric space.

Example 1. [12] Let (X,d) be a metric space, then (X,G) is a G-metric space with:

$$G(x,y,z)=max\left\{d(x,y),d(x,z),d(y,z)\right\},\qquad \forall x,y,z\in X.$$

Proposition 1 ([12]). *Let (X,G) be a G-metric space, then for each $x,y,z,a\in X$, we have:*

(1) If $G(x,y,z)=0$ then $x=y=z$,
(2) $G(x,y,z)\leq G(x,x,y)+G(x,x,z)$,
(3) $G(x,y,y)\leq 2G(y,x,x)$,
(4) $G(x,y,z)\leq G(x,a,z)+G(a,y,z)$,
(5) $G(x,y,z)\leq\frac{2}{3}(G(x,y,a)+G(x,a,z)+G(a,y,z))$,

Definition 2 ([12]). Let (X,G) be a G-metric space. A sequence (x_n) in X is said to be

(1) G-Cauchy sequence if, for each $\varepsilon>0$, there exists a positive integer n_0 such that, for all $m,n,l\geq n_0$, $G(x_n,x_m,x_l)<\varepsilon$.
(2) G-convergent to a point $x\in X$ if, for each $\varepsilon>0$, there exists a positive integer n_0 such that, for all $m,n\geq n_0$, $G(x_n,x_m,x)<\varepsilon$.

Proposition 2 ([12]). *Let (X,G) be a G-metric space. Then, the following are equivalent:*

(1) the sequence (x_n) is G-Cauchy;
(2) $\forall\varepsilon>0$, $\exists n_0\in N$ such that, $\forall\, m,n\geq n_0$, $G(x_n,x_m,x_m)<\varepsilon$.

Proposition 3 ([17]). *Let (X, G) be a G-metric space. Then, the following are equivalent:*

(1) (x_n) *is G-convergent to x.*
(2) $\lim_{n \to +\infty} G(x_n, x_n, x) = 0.$
(3) $\lim_{n \to +\infty} G(x_n, x, x) = 0.$
(4) $\lim_{n,m \to +\infty} G(x_n, x_m, x) = 0.$

Definition 3 ([12]). A G-metric space X is called G-complete if every G-Cauchy sequence is G-convergent in X.

Definition 4 ([12]). Let (X, G) and (X', G') be two G-metric spaces. Then a function $T : X \longrightarrow X'$ is G-continuous at a point $x \in X$ if and only if it is G-sequentially continuous at x, that is, whenever (x_n) is G-convergent to x, (Tx_n) is G'-convergent to Tx.

Mustafa and Sims proved that each G-metric function $G(x, y, z)$ is jointly continuous in all three of its variables (see, Proposition 8 [12]).

2 Main Results

Definition 5. Let $\mathscr{F}$ be the set of all functions $F(t_1, t_2, ..., t_{11}) : R_+^{11} \longrightarrow R$ such that:
$(\mathscr{F}_1)$: F is continuous in variables t_1, t_2, t_3, t_6, t_7, t_{10}, t_{11} and non increasing in variables t_3, t_4,..., t_{11}.
$(\mathscr{F}_2)$: $\exists h \in [0, 1[$, such that $\forall u, v \geq 0$:
$F(u, v, v, u, u, u + v, 0, u, u, 0, u + v) \leq 0$ or $F(u, u, 0, 0, 0, u, v, 0, 0, v, u) \leq 0 \Rightarrow u \leq hv$.
$(\mathscr{F}_3)$: $\exists\, \alpha, \beta \geq 0$ such that $\forall u, v, w \geq 0$:
$F(u, v, 0, u + w, u + w, u, w, u + w, u + w, w, u) \leq 0 \Rightarrow u \leq \alpha v + \beta w$.

Example 2. $F(t_1, t_2, ..., t_{11}) = t_1 - \lambda max\{t_3, t_4, t_5\}$ with $0 \leq \lambda < 1$.

Example 3. $F(t_1, t_2, ..., t_{11}) = t_1 - \lambda t_2$ with $\lambda \in [0, 1)$.

Example 4. $F(t_1, t_2, ..., t_{11}) = t_1 - \lambda max\{t_2, t_3, t_4, t_5, t_6, t_8, t_{10}\}$ with $0 \leq \lambda < \frac{1}{2}$.

Example 5. $F(t_1, t_2, ..., t_{11}) = t_1 - \lambda(t_6 + t_7)$ with $0 \leq \lambda < \frac{1}{2}$.

Example 6. $F(t_1, t_2, ..., t_{11}) = t_1 - \lambda max\{t_4 + t_6, t_7 + t_{10}\}$ with $0 \leq \lambda < \frac{1}{3}$.

Example 7. $F(t_1, t_2, ..., t_{11}) = t_1 - \lambda max\{t_6 + t_7, t_8 + t_9, t_{10} + t_{11}\}$ with $0 \leq \lambda < \frac{1}{2}$.

Example 8. $F(t_1, t_2, ..., t_{11}) = t_1 - \lambda max\{t_6 + t_9, t_7 + t_{10}, t_8 + t_{11}\}$ with $0 \leq \lambda < \frac{1}{3}$.

Theorem 1. *Let (X, G) be a G-complete G-metric space, $T : X \longrightarrow B(X)$ such that $\forall x, y, z \in X$:*

$$F\begin{pmatrix} H_G(T(x), T(y), T(z)), G(x, y, z), G(x, T(x), T(x)), \\ G(y, T(y), T(y)), G(z, T(z), T(z)), G(x, T(y), T(y)), \\ G(y, T(x), T(x)), G(y, T(z), T(z)), G(z, T(y), T(y)), \\ G(z, T(x), T(x)), G(x, T(z), T(z)) \end{pmatrix} \le 0, \quad (1)$$

with $F \in \mathscr{F}$. Then, T have a fixed point $x \in X$. Moreover, if x is absolutely fixed for T (which means that $T(x) = \{x\}$), then the fixed point is unique.

For the proof of this theorem we need two lemmas.

Lemma 1. *Let (X, G) be a G-metric space and $A, B \in B(X)$. Then for each $a \in A$, we have*

$$G(a, B, B) \le H_G(A, B, B).$$

Proof.

$$H_G(A, B, B) \ge \sup_{x \in A} G(x, B, B) \ge G(a, B, B).$$

Lemma 2. *Let (X, G) be a G-metric space. If $A, B \in B(X)$ and $x \in A$, then for each $\varepsilon > 0$, there exists $y \in B$ such that*

$$G(x, y, y) \le H_G(A, B, B) + \varepsilon.$$

Proof.
By Lemma 1 and the characterization of inf we have, for each $\varepsilon > 0$, there exists $y \in B$ such that

$$\begin{aligned} G(x, y, y) &\le \inf\{G(x, z, z), z \in B\} + \varepsilon \\ &\le 2 \inf_{z \in B} (G(x, z, z) + G(z, x, x)) + \varepsilon \\ &= 2d_G(x, B) + \varepsilon = G(x, B, B) + \varepsilon \\ &\le H_G(A, B, B) + \varepsilon. \end{aligned}$$

Proof of Theorem 1.
Existence.
For $x_0 \in X$, and $x_1 \in T(x_0)$. According to (1), with $x = x_0$ and $y = z = x_1$ we have

$$F\begin{pmatrix} H_G(T(x_0), T(x_1), T(x_1)), G(x_0, x_1, x_1), G(x_0, T(x_0), T(x_0)), \\ G(x_1, T(x_1), T(x_1)), G(x_1, T(x_1), T(x_1)), G(x_0, T(x_1), T(x_1)), \\ G(x_1, T(x_0), T(x_0)), G(x_1, T(x_1), T(x_1)), G(x_1, T(x_1), T(x_1)), \\ G(x_1, T(x_0), T(x_0)), G(x_0, T(x_1), T(x_1)) \end{pmatrix} \le 0. \quad (2)$$

According to (G_5) we have

$$G(x_0, T(x_0), T(x_0)) \leq G(x_0, x_1, x_1) + G(x_1, T(x_0), T(x_0)) \\ = G(x_0, x_1, x_1)$$

by Lemma 1 and (G_5) we have

$$G(x_0, T(x_1), T(x_1)) \leq G(x_0, x_1, x_1) + G(x_1, T(x_1), T(x_1)) \\ \leq G(x_0, x_1, x_1) + H_G(T(x_0), T(x_1), T(x_1)).$$

By Lemma 1, $(\mathscr{F}_1)$ and (2) becomes:

$$F\begin{pmatrix} H_G(T(x_0), T(x_1), T(x_1)), G(x_0, x_1, x_1), G(x_0, x_1, x_1), H_G(T(x_0), R(x_1), T(x_1)), \\ H_G(T(x_0), T(x_1), T(x_1)), G(x_0, x_1, x_1) + H_G(T(x_0), T(x_1), T(x_1)), \\ 0, H_G(T(x_0), T(x_1), T(x_1)), H_G(T(x_0), T(x_1), T(x_1)), \\ 0, G(x_0, x_1, x_1) + H_G(T(x_0), T(x_1), T(x_1)) \end{pmatrix} \leq 0.$$

According to $(\mathscr{F}_2)$ we have:

$$H_G(T(x_0), T(x_1), T(x_1)) \leq h \;\; G(x_0, x_1, x_1).$$

By Lemma 2 we have for $\varepsilon = \frac{1}{2}(1-h)G(x_0, x_1, x_1)$, there exists $x_2 \in T(x_1)$ such that:

$$\begin{aligned} G(x_1, x_2, x_2) &\leq H_G(T(x_0), T(x_1), T(x_1)) + \varepsilon \\ &\leq hG(x_0, x_1, x_1) + \frac{1}{2}(1-h)G(x_0, x_1, x_1) \\ &= \frac{1}{2}(1+h)G(x_0, x_1, x_1) \\ &= h_1 G(x_0, x_1, x_1) \qquad \textit{with } h_1 = \frac{1}{2}(1+h) < 1. \end{aligned}$$

According to (1), with $x = x_1$ and $y = z = x_2$ we have

$$F\begin{pmatrix} H_G(T(x_1), T(x_2), T(x_2)), G(x_1, x_2, x_2), G(x_1, T(x_1), T(x_1)), \\ G(x_2, T(x_2), T(x_2)), G(x_2, T(x_2), T(x_2)), G(x_1, T(x_2), T(x_2)), \\ G(x_2, T(x_1), T(x_1)), G(x_2, T(x_2), T(x_2)), G(x_2, T(x_2), T(x_2)), \\ G(x_2, T(x_1), T(x_1)), G(x_1, T(x_2), T(x_2)) \end{pmatrix} \leq 0. \quad (3)$$

by Lemma 1, (G_{b5}), $(\mathscr{F}_1)$ and (3) becomes:

$$F\begin{pmatrix} H_G(T(x_1), T(x_2), T(x_2)), G(x_1, x_2, x_2), G(x_1, x_2, x_2), H_G(T(x_1), T(x_2), T(x_2)), \\ H_G(T(x_1), T(x_2), T(x_2)), G(x_1, x_2, x_2) + H_G(T(x_1), T(x_2), T(x_2)), \\ 0, H_G(T(x_1), T(x_2), T(x_2)), H_G(T(x_1), T(x_2), T(x_2)), \\ 0, G(x_1, x_2, x_2) + H_G(T(x_1), T(x_2), T(x_2)) \end{pmatrix} \leq 0.$$

According to $(\mathscr{F}_2)$ we get:

$$H_G(T(x_1), T(x_2), T(x_2)) \leq h \;\; G(x_1, x_2, x_2)$$

By Lemma 2 we have for $\varepsilon = \frac{1}{2}(1-h)G(x_1, x_2, x_2)$, there exists $x_3 \in T(x_2)$ such that:

$$\begin{aligned}
G(x_2, x_3, x_3) &\leq H_G(T(x_1), T(x_2), T(x_2)) + \varepsilon \\
&\leq hG(x_1, x_2, x_2) + \frac{1}{2}(1-h)G(x_1, x_2, x_2) \\
&= \frac{1}{2}(1+h)G(x_1, x_2, x_2) \\
&= h_1 G(x_1, x_2, x_2) \qquad \textit{with } h_1 = \frac{1}{2}(1+h) < 1.
\end{aligned}$$

By recurrence, we construct a sequence (x_n) as $x_n \in Tx_{n-1}$, for every $n \in N^*$, we have

$$G(x_n, x_{n+1}, x_{n+1}) \leq h_1 \ G(x_{n-1}, x_n, x_n)$$

from where

$$G(x_n, x_{n+1}, x_{n+1}) \leq h_1^n \ G(x_0, x_1, x_1).$$

Now we have to show that (x_n) is a Cauchy sequence. Let $m, n \in N^*$, then

$$\begin{aligned}
G(x_n, x_{n+m}, x_{n+m}) \leq{}& G(x_n, x_{n+1}, x_{n+1}) + G(x_{n+1}, x_{n+2}, x_{n+2}) + G(x_{n+2}, x_{n+3}, x_{n+3}) + \\
& ... + G(x_{n+m-2}, x_{n+m-1}, x_{n+m-1}) + G(x_{n+m-1}, x_{n+m}, x_{n+m})
\end{aligned}$$

On the other hand, we have :

$$\begin{aligned}
G(x_n, x_{n+m}, x_{n+m}) &\leq h_1^n G(x_0, x_1, x_1) + h_1^{n+1} G(x_0, x_1, x_1) + h_1^{n+2} G(x_0, x_1, x_1) + \\
&\quad ... + h_1^{n+m-2} G(x_0, x_1, x_1) + h_1^{n+m-1} G(x_0, x_1, x_1) \\
&\leq h_1^n (1 + h_1 + h_1^2 + ... + h_1^{m-2} + h_1^{m-1}) G(x_0, x_1, x_1) \\
&= h_1^n \left(\frac{1-h_1^m}{1-h_1}\right) G(x_0, x_1, x_1) \\
&\leq \left(\frac{h_1^n}{1-h_1}\right) G(x_0, x_1, x_1).
\end{aligned}$$

from where $\lim_{n\to\infty} G(x_n, x_{n+m}, x_{n+m}) = 0$ for $m \in N^*$. By Proposition 2, then (x_n) is a Cauchy sequence. As the $G-$metric space (X, G) is complete, there exists $x \in X$ such that $\lim_{n\to\infty} G(x_n, x_n, x) = 0$. Next we show that $x \in Tx$, indeed, by (1) we have

$$F\begin{pmatrix}
H_G(T(x_{n-1}), T(x), T(x)), G(x_{n-1}, x, x), G(x_{n-1}, T(x_{n-1}), T(x_{n-1})), \\
G(x, T(x), T(x)), G(x, T(x), T(x)), G(x_{n-1}, T(x), T(x)), \\
G(x, T(x_{n-1}), T(x_{n-1})), G(x, T(x), T(x)), \\
G(x, T(x), T(x)), G(x, T(x_{n-1}), T(x_{n-1})), G(x_{n-1}, T(x), T(x))
\end{pmatrix} \leq 0. \quad (4)$$

According to (G_5) we have

$$G(x_{n-1}, Tx, Tx) \leq G(x_{n-1}, x, x) + G(x, Tx, Tx),$$

$$\begin{aligned}
G(x_{n-1}, T(x_{n-1}), T(x_{n-1})) &\leq G(x_{n-1}, x_n, x_n) + G(x_n, T(x_{n-1}), T(x_{n-1})) \\
&= G(x_{n-1}, x_n, x_n)
\end{aligned}$$

and

$$\begin{aligned} G(x,T(x_{n-1}),T(x_{n-1})) &\leq G(x,x_n,x_n) + G(x_n,T(x_{n-1}),T(x_{n-1})) \\ &= G(x,x_n,x_n). \end{aligned}$$

So by $(\mathscr{F}_1)$, (4) becomes:

$$F\begin{pmatrix} H_G(T(x_{n-1}),T(x),T(x)), G(x_{n-1},x,x), G(x_{n-1},x_n,x_n), \\ G(x,T(x),T(x)), G(x,T(x),T(x)), G(x_{n-1},x,x) + G(x,Tx,Tx), \\ G(x,x_n,x_n), G(x,T(x),T(x)), \\ G(x,T(x),T(x)), G(x,x_n,x_n), G(x_{n-1},x,x) + G(x,Tx,Tx) \end{pmatrix} \leq 0.$$

letting $n \to \infty$ we obtain

$$F\begin{pmatrix} \liminf_{n\to\infty} H_G(T(x_{n-1}),T(x),T(x)), 0, 0, G(x,T(x),T(x)), G(x,T(x),T(x)), G(x,Tx,Tx), \\ 0, G(x,T(x),T(x)), G(x,T(x),T(x)), 0, G(x,Tx,Tx)) \end{pmatrix} \leq 0.$$

because we have $H_G(T(x_{n-1}),T(x_n),T(x_n)) \leq h\,G(x_{n-1},x_n,x_n)$, $x_{n+1} \in T(x_n)$

$$\begin{aligned} H_G(T(x_{n-1}),T(x),T(x)) &\leq H_G(T(x_{n-1}),\{x_{n+1}\},\{x_{n+1}\}) + H_G(\{x_{n+1}\},T(x),T(x)) \\ &\leq H_G(T(x_{n-1}),T(x_n),T(x_n)) + H_G(\{x_{n+1}\},\{x\},\{x\}) \\ &\quad + H_G(\{x\},T(x),T(x)) \\ &\leq h\,G(x_{n-1},x_n,x_n) + G(x_{n+1},x,x) + H_G(\{x\},T(x),T(x)), \end{aligned}$$

so we deduce that the sequence $(H_G(T(x_{n-1}),T(x),T(x)))_n$ is bounded.

Now, by $(\mathscr{F}_2)$, $\exists h \in [0,1[$, such that:

$$\liminf_{n\to\infty} H_G(T(x_{n-1}),T(x),T(x)) \leq hG(x,Tx,Tx). \tag{5}$$

On the other hand we show that $G(x,Tx,Tx) = 0$. Suppose that $G(x,Tx,Tx) > 0$, then

$$\begin{aligned} G(x,Tx,Tx) &\leq G(x,x_n,x_n) + G(x_n,Tx,Tx) \\ &\leq G(x,x_n,x_n) + H_G(T(x_{n-1}),T(x),T(x)). \end{aligned}$$

By (5) we have

$$\begin{aligned} G(x,Tx,Tx) &\leq \liminf_{n\to\infty} (G(x,x_n,x_n) + H_G(T(x_{n-1}),T(x),T(x))) \\ &\leq hG(x,Tx,Tx) < G(x,Tx,Tx) \end{aligned}$$

which is a contradiction, then $G(x,Tx,Tx) = 0$ from where $x \in Tx$.

Unicity

Suppose that $Tx = \{x\}$ and $y \in X$ is an other fixed point of T then by (1) we have

$$F\begin{pmatrix} H_G(T(x),T(y),T(y)), G(x,y,y), G(x,T(x),T(x)), \\ G(y,T(y),T(y)), G(y,T(y),T(y)), G(x,T(y),T(y)), \\ G(y,T(x),T(x)), G(y,T(y),T(y)), G(y,T(y),T(y)), \\ G(y,T(x),T(x)), G(x,T(y),T(y)) \end{pmatrix} \leq 0,$$

implies

$$F\begin{pmatrix} G(x,y,y),G(x,y,y),0,0,0,G(x,y,y), \\ G(y,x,x),0,0,G(y,x,x),G(x,y,y) \end{pmatrix} \leq 0.$$

According to ($\mathscr{F}_2$) we have $G(x,y,y) \leq h\,G(y,x,x)$, in the same way by (1) we find $G(y,x,x) \leq h\,G(x,y,y)$. Then $G(x,y,y) \leq h^2\,G(x,y,y)$ then $G(x,y,y) = 0$ from where $x = y$.

Example 9. Let $X = [0,+\infty[$. Define a mapping $T : X \longrightarrow B(X)$ by $Tx = [0, \frac{x}{4}]$.
Define a G-metric on X by

$$G(x,y,z) = \mid x-y \mid + \mid y-z \mid + \mid x-z \mid .$$

We prove that T check

$$H_G(Tx,Ty,Tz) \leq \frac{1}{3}\ max\{G(x,Tx,Tx), G(y,Ty,Ty), G(z,Tz,Tz)\}$$

Indeed, we have $d_G(x,y) = G(x,y,y) + G(y,x,x) = 4|x-y|$ for all $x,y \in X$.
Let $x,y,z \in X$. If $x = y = z = 0$ then $H_G(Tx,Ty,Tz) = 0$. Thus we may assume that x,y and z are not all zero. Without loss of generality we assume that $x \leq y \leq z$. Then

$$H_G(Tx,Ty,Tz)) = H_G\left(\left[0,\frac{x}{4}\right],\left[0,\frac{y}{4}\right],\left[0,\frac{z}{4}\right]\right) = \max\left\{\begin{array}{l} \sup\limits_{0\leq a\leq \frac{x}{4}} G\left(a,\left[0,\frac{y}{4}\right],\left[0,\frac{z}{4}\right]\right), \\ \sup\limits_{0\leq b\leq \frac{y}{4}} G\left(b,\left[0,\frac{x}{4}\right],\left[0,\frac{z}{4}\right]\right), \\ \sup\limits_{0\leq c\leq \frac{z}{4}} G\left(c,\left[0,\frac{x}{4}\right],\left[0,\frac{y}{4}\right]\right) \end{array}\right\}.$$

Since $x \leq y \leq z$, then $\left[0,\frac{x}{4}\right] \subseteq \left[0,\frac{y}{4}\right] \subseteq \left[0,\frac{z}{4}\right]$ which implies that

$$d_G\left(\left[0,\frac{x}{4}\right],\left[0,\frac{y}{4}\right]\right) = d_G\left(\left[0,\frac{y}{4}\right],\left[0,\frac{z}{4}\right]\right) = d_G\left(\left[0,\frac{x}{4}\right],\left[0,\frac{z}{4}\right]\right) = 0.$$

For each $0 \leq a \leq \frac{x}{4}$ we have

$$G\left(a,\left[0,\frac{y}{4}\right],\left[0,\frac{z}{4}\right]\right) = d_G\left(a,\left[0,\frac{y}{4}\right]\right) + d_G\left(\left[0,\frac{y}{4}\right],\left[0,\frac{z}{4}\right]\right) + d_G\left(a,\left[0,\frac{z}{4}\right]\right) = 0.$$

Also for each $0 \leq b \leq \frac{y}{4}$ we have

$$\begin{aligned} G\left(b,\left[0,\frac{x}{4}\right],\left[0,\frac{z}{4}\right]\right) &= d_G\left(b,\left[0,\frac{x}{4}\right]\right) + d_G\left(\left[0,\frac{x}{4}\right],\left[0,\frac{z}{4}\right]\right) + d_G\left(b,\left[0,\frac{z}{4}\right]\right) \\ &= d_G\left(b,\left[0,\frac{x}{4}\right]\right) = \inf_{0\leq e\leq \frac{x}{4}} d_G(b,e) \\ &= \begin{cases} 0, & \text{if} \quad 0 \leq b \leq \frac{x}{4} \\ 4\left(b-\frac{x}{4}\right), & \text{if} \quad \frac{x}{4} \leq b \leq \frac{y}{4} \end{cases} \end{aligned}$$

Implies

$$\sup_{0\leq b\leq \frac{y}{4}} G\left(b,\left[0,\frac{x}{4}\right],\left[0,\frac{z}{4}\right]\right) = (y-x).$$

Moreover, for each $0 \le c \le \frac{z}{4}$ we have

$$\begin{aligned}
G\left(c,\left[0,\frac{x}{4}\right],\left[0,\frac{y}{4}\right]\right) &= d_G\left(c,\left[0,\frac{x}{4}\right]\right) + d_G\left(\left[0,\frac{x}{4}\right],\left[0,\frac{y}{4}\right]\right) + d_G\left(c,\left[0,\frac{y}{4}\right]\right)\\
&= d_G\left(c,\left[0,\frac{x}{4}\right]\right) + d_G\left(c,\left[0,\frac{y}{4}\right]\right)\\
&= \begin{cases} 0, & \text{if} \quad 0 \le c \le \frac{x}{4}\\ 4\left(c-\frac{x}{4}\right), & \text{if} \quad \frac{x}{4} \le c \le \frac{y}{4}\\ 4\left(c-\frac{x}{4}\right) + 4\left(c-\frac{y}{4}\right), & \text{if } \frac{y}{4} \le c \le \frac{z}{4}\end{cases}
\end{aligned}$$

Implies

$$\sup_{0\le c\le \frac{z}{4}} G\left(c,\left[0,\frac{x}{4}\right],\left[0,\frac{y}{4}\right]\right) = [(z-x)+(z-y)].$$

Thus we deduce that

$$\begin{aligned}
H_G(Tx,Ty,Tz) &= \max\{0,(y-x),[(z-x)+(z-y)]\}\\
&= [(z-x)+(z-y)]\\
&\le [(z-x)+(z-x)]\\
&= 2(z-x). \qquad (*)
\end{aligned}$$

On the other hand, we have

$$\begin{aligned}
G(x,Tx,Tx) &= d_G\left(x,\left[0,\frac{x}{4}\right]\right) + d_G\left(\left[0,\frac{x}{4}\right],\left[0,\frac{x}{4}\right]\right) + d_G\left(x,\left[0,\frac{x}{4}\right]\right)\\
&= 2d_G\left(x,\left[0,\frac{x}{4}\right]\right) = 2\inf_{0\le e\le \frac{x}{4}} d_G(x,e) = 8(x-\frac{x}{4}) = 6x.
\end{aligned}$$

From where

$$max\{G(x,Tx,Tx),G(y,Ty,Ty),G(z,Tz,Tz)\} = 6z. \qquad (**)$$

By $(*)$ and $(**)$ we have

$$\begin{aligned}
H_G(Tx,Ty,Tz) &= 2(z-x)\\
&\le 2z = \frac{1}{3}\times 6z\\
&= \frac{1}{3}max\{G(x,Tx,Tx),G(y,Ty,Ty),G(z,Tz,Tz)\}
\end{aligned}$$

All the conditions of Theorem 1 are satisfied with F as in Example 2, then 0 is the unique absolutely fixed point of T.

If T single mappings, the we obtain the following :

Corollary 1 (Theorem 27 [3]). *Let (X,G) be a complete G-metric space, $T: X \longrightarrow X$ such that $\forall x,y,z \in X$. :*

$$F\begin{pmatrix} G(T(x),T(y),T(z)),G(x,y,z),G(x,T(x),T(x)),\\ G(y,T(y),T(y)),G(z,T(z),T(z)),G(x,T(y),T(y)),\\ G(y,T(x),T(x)),G(y,T(z),T(z)),G(z,T(y),T(y)),\\ G(z,T(x),T(x)),G(x,T(z),T(z))\end{pmatrix} \le 0, \qquad (6)$$

with $F \in \mathscr{F}$. Then, T have a unique fixed point, and T is G-continuous at this point.

Proof.
The unique fixed point of T becomes from Theorem 1.

The Continuity
Let (y_n) is G-Convergent to x. For all $n \in N$, by (1) we have

$$F\begin{pmatrix} G(T(x),T(y_n),T(y_n)), G(x,y_n,y_n), G(x,T(x),T(x)), G(y_n,T(y_n),T(y_n)), \\ G(y_n,T(y_n),T(y_n)), G(x,T(y_n),T(y_n)), G(y_n,T(x),T(x)), G(y_n,T(y_n),T(y_n)), \\ ,G(y_n,T(y_n),T(y_n)), G(y_n,T(x),T(x)), G(x,T(y_n),T(y_n)) \end{pmatrix} \leq 0.$$

As $\quad G(y_n,T(y_n),T(y_n)) \leq G(y_n,x,x) + G(x,T(y_n),T(y_n))$. We have

$$F\begin{pmatrix} G(x,T(y_n),T(y_n)), G(x,y_n,y_n), 0, G(y_n,x,x) + G(x,T(y_n),T(y_n)), \\ G(y_n,x,x) + G(x,T(y_n),T(y_n)), G(x,T(y_n),T(y_n)), G(y_n,x,x), \\ G(y_n,x,x) + G(x,T(y_n),T(y_n)), G(y_n,x,x) + G(x,T(y_n),T(y_n)), \\ G(y_n,x,x), G(x,T(y_n),T(y_n)) \end{pmatrix} \leq 0.$$

According to $(\mathscr{F}_3)$ we have $G(x,T(y_n),T(y_n)) \leq \alpha G(x,y_n,y_n) + \beta G(y_n,x,x)$ from where
$\lim_{n\to\infty} G(x,T(y_n),T(y_n)) = 0$, then $(T(y_n))$ is G-convergent to $x = T(x)$. Then T is G-continuous at x.

3 Consequences of the Main Result

From Corollary 1 and Example 4 a we obtain [Theorem 2.1 [14]].

From Corollary 1 and Example 6 a we obtain [Theorem 31 [3]].

From Corollary 1 and Example 6, $y = z$ a we obtain [Theorem 2.6 [14]].

From Corollary 1 and Example 7 a we obtain [Theorem 2.4 [14]].

From Corollary 1 and Example 8 a we obtain [Theorem 2.8 [14]].

4 Applications

Let $X = C([a,b],R)$ be the set of real continuous functions defined on $[a,b]$. Take the G-metric G: $X \times X \times X \longrightarrow R^+$ given by

$$G(x,y,z) = \sup_{t\in[a,b]} \mid x(t) - y(t) \mid + \sup_{t\in[a,b]} \mid x(t) - z(t) \mid + \sup_{t\in[a,b]} \mid y(t) - z(t) \mid \quad (7)$$

For all $x,y,z \in X$. Then (X,G) is G-metric spaces with.

Consider the following integral equation

$$x(t) = P(t) + \int_a^b M(t,u) f(u,x(u)) du, \qquad t \in [a,b], \qquad (8)$$

where $f : [a,b] \times R \longrightarrow R$ and $P : [a,b] \longrightarrow R$ are two continuous functions and $M : [a,b] \times [a,b] \longrightarrow R^+$ is a function such that $M(t,.) \in L^1([a,b])$ for all $t \in [a,b]$.

Consider the operator T: $X \longrightarrow X$ defined by

$$Tx(t) = P(t) + \int_a^b M(t,u)f(u,x(u))du, \quad t \in [a,b]. \tag{9}$$

Theorem 2. *Suppose that the following conditions are satisfied:*
(H_1) *there exists* $\theta : X \times X \longrightarrow R^+$ *for all* $u \in [a,b]$

$$| f(u,x(u)) - f(u,y(u)) | \leq \theta(x,y) | x(u) - y(u) | \quad \forall x,y \in X,$$

(H_2) *there exists* $\lambda \in [0,1)$, *such that*

$$\sup_{t\in[a,b]} \int_a^b M(t,u)\theta(x,y)du \leq \lambda.$$

Then the integral Eq. (8) has a unique solution in X.

Proof.
It is clear that any fixed point of (9) is a solution of (8). By conditions (H_1) and (H_2), we have

$$\begin{aligned}
\sup_{t\in[a,b]} | Tx(t) - Ty(t) | &= \sup_{t\in[a,b]} | \int_a^b M(t,u)f(u,x(u))du - \int_a^b M(t,u)f(u,y(u))du | \\
&= \sup_{t\in[a,b]} | \int_a^b M(t,u)[f(u,x(u)) - f(u,y(u))]du | \\
&\leq \sup_{t\in[a,b]} \int_a^b M(t,u)\theta(x,y) | x(u) - y(u) | du \\
&\leq \sup_{t\in[a,b]} | x(t) - y(t) | \times \sup_{t\in[a,b]} \int_a^b M(t,u)\theta(x,y))du \\
&\leq \lambda \sup_{t\in[a,b]} | x(t) - y(t) | .
\end{aligned}$$

Similarly we find

$$\sup_{t\in[a,b]} |Tx(t) - Tz(t)| \leq \lambda \sup_{t\in[a,b]} |x(t) - z(t)| \quad \text{and} \quad \sup_{t\in[a,b]} |Ty(t) - Tz(t)| \leq \lambda \sup_{t\in[a,b]} |y(t) - z(t)|.$$

Therefore

$$G(Tx(t), Ty(t), Tz(t)) \leq \lambda G(x(t), y(t), z(t)).$$

Then all conditions of Theorem 1 are satisfied with F as in Example 3 . Thus the operator T has a unique fixed point, that is the integral has a unique solution in X.

Example 10. The following integral equation has a solution in $X = (C[1,e], R)$.

$$x(t) = cos(t) + \int_1^e \frac{\ln(ut)}{e^2} x(u)du, \quad t \in [1,e]. \tag{10}$$

Proof.
Let T: $X \longrightarrow X$ defined by

$$Tx(t) = cos(t) + \int_1^e \frac{\ln(ut)}{e^2} x(u)du, \quad t \in [1, e].$$

By specifying $M(t,u) = \frac{\ln(ut)}{e^2}$, $f(u,x) = x$ and $P(t) = cos(t)$ in Theorem 2, we get :
(1) For all $x(.), y(.) \in X$, it is clear that the condition (H_1) in Theorem 2 is satisfied with $\theta = 1$.
(2)

$$\begin{aligned}\sup_{t\in[1,e]} \int_1^e \frac{\ln(ut)}{e^2} du &= \frac{1}{e^2} \sup_{t\in[1,e]} \int_1^e (\ln(u) + \ln(t))du \\ &= \frac{1}{e^2} \sup_{t\in[1,e]} [u\ln(u) - u + u\ln(t)]_1^e \\ &= \frac{1}{e^2} \sup_{t\in[1,e]} (\ln(t)(e-1) + 1) \\ &= \frac{1}{e}\end{aligned}$$

Therefore, all conditions of Theorem 2 are satisfied, hence the mapping T has a fixed point in X, which is a solution to Eq. (10).

Acknowledgments. The authors would like to thank referee sincerely for very helpful comments improving the paper.

The authors declare that there is no conflict of interest regarding the publication of this paper.

References

1. Ali, M.U., Kamran, T., Postolache, M.: Solution of volterra integral inclusion in b-metric spaces via new fixed point theorem. Nonlinear Anal. Model. Control **22**(1), 17–30 (2017)
2. Banach, S.: Sur les operations dans les ensembles abstraits et leur application aux equations integrales. Fundam. Math. **3**(1), 133–181 (1922)
3. El Haddouchi, A., Marzouki, B.: A generalized fixed point theorem in G-metric space. J. Anal. Appl. **17**(N2), 89–105 (2019)
4. Kaewcharoen, A., Kaewkhao, A.: Common fixed points for single-valued and multi-valued mappings in G-metric spaces. Int. J. Math. Anal. **536**, 1775–1790 (2011)
5. Kamran, T., Postolache, M., Ali, M.U., Kiran, Q.: Feng and Liu type F-contraction in b-metric spaces with application to integral equations. J. Math. Anal. **7**(5), 18–27 (2016)
6. Makran, N., El Haddouchi, A., Marzouki, B.: A common fixed point of multi-valued maps in b-Metric space. U.P.B. Sci. Bull., Series A, **82**(1) (2020). ISSN 1223–7027
7. Makran, N., El Haddouchi, A., Marzouki, B.: A generalized common fixed points of multivalued mappings in G_b-Metric spaces with an application. U.P.B. Sci. Bull., Series A, **83**(1) (2021). ISSN 1223-7027

8. Marzouki, B., El Haddouchi, A.: Generalized altering distances and fixed point for occasionally hybrid mapping. Fasciculi Math. **56**, 111–120 (2016)
9. Marzouki, B., El Haddouchi, A.: Common fixed point of multivalued maps. Nonliear Anal. Differ. Eqn. **4**, 1–7 (2016)
10. Marzouki, B., El Haddouchi, A.: A common fixed point theorem in modular space. Nonliear Anal. Differ. Eqn. **4**, 219–223 (2016)
11. Mustafa, Z., Sims, B.: Some remarks concerning D-metric spaces. In: Proceedings of the International Conference on Fixed Point Theory and Applications, Yokohama, Japan (2004)
12. Mustafa, Z., Sims, B.: A new approach to generalized metric spaces. J. Nonlinear Convex Anal. **7**(2), 289–297 (2006)
13. Mustafa, Z., Obiedat, H., Awawdeh, F.: Some fixed point theorem for mapping on complete G-metric spaces. Fixed Point Theory Appl. **2008**(1), 1–12 (2008). https://doi.org/10.1155/2008/189870
14. Mustafa, Z., Sims, B.: Fixed point theorems for contractive mappings in complete G-metric spaces. Fixed Point Theory Appl. **2009**(1), 1–10 (2009). https://doi.org/10.1155/2009/917175
15. Mustafa, Z., Shatanawi, W., Bataineh, M.: Existence of fixed point results in $G-$metric spaces, Int. J. Math. Math. Sci. **2009**, 10, Article ID 283028 (2009)
16. Mustafa, Z., Sims, B.: Fixed point theorems for contractive mappings in complete G-metric spaces. Fixed Point Theory Appl. **2009**, 10, Article ID 917175 (2009)
17. Mustafa, Z., Arshad, M., Khan, S.U., Ahmad, J., Jaradat, M.M.M.: Common fixed point for multivalued in G-metric spaces with applications. J. Nonlinear Sci. Appl. **10**, 2550–2564 (2017)
18. Saadati, R., Vaezpour, S.M., Vetro, P., Rhoades, B.E.: Fixed point theorems in generalized partially ordered G-metric spaces. Math. Comput. Modell. **52**, 797–801 (2010)
19. Shatanawi, W.: Coupled fixed point theorems in generalized metric spaces. Hacet. J. Math. Stat. **40**(3), 441–447 (2011)
20. Shatanawi, W.: Fixed point theory for contractive mappings satisfying Φ-maps in G-metric spaces. Fixed Point Theory Appl. **2010**(1), 1–9 (2010). https://doi.org/10.1155/2010/181650
21. Rao, K.P.R., Lakshmi, K.B., Mustafa, Z., Raju, V.C.C.: Fixed and related fixed point theorems for three maps in $G-$metric spaces. J. Adv. Stud. Topology **3**(4), 12–19 (2012)

On Intuitionistic Fuzzy Sumudu Transform for Intuitionistic Fuzzy Differential Equations

Razika Ettoussi, Said Melliani(✉), and Lalla Saadia Chadli

LMACS, Sultan Moulay Slimane University, BP 523 - 23000 Beni Mellal, Morocco
razika.imi@gmail.com, s.melliani@usms.ma, sa.chadli@yahoo.fr

Abstract. In this paper, we introduce a result for the intuitionistic fuzzy Sumudu transform. The related theorems and properties are proved in detail and we propose a procedure for solving first-order intuitionistic fuzzy differential equations by using the intuitionistic fuzzy Sumudu transform method. Finally, we present an example to illustrate this work.

1 Introduction

In 1965, Zadeh [20] first introduced the fuzzy set theory. Later many researchers have applied this theory to the well known results in the classical set theory.

The idea of intuitionistic fuzzy set was first published by Atanassov [7,8] as a generalization of the notion of fuzzy set. The notions of differential and integral calculus for intuitionistic fuzzy-set-valued are given using Hukuhara difference in intuitionistic Fuzzy theory [9]. The authors of papers[14,15] are discussed differential and partial differential equations under intuitionistic fuzzy environment respectively. The existence and uniqueness of the solution of intuitionistic fuzzy differential equations by using successive approximations method have been discussed in [11], while in [9] the theorem of the existence and uniqueness of the solution for differential equations with intuitionistic fuzzy data are proved by using the theorem of fixed point in the complete metric space, also the explicit formula of the solution are given by using the α-cuts method.

In the literature there are numerous integral transforms and widely used in physics, astronomy as well as in engineering. The integral transform method is also an efficient method to solve the differential equations. Watugala introduced in 1993 a new transform and named as Sumudu transform who used it to solve engineering control problems [18,19]. Sumudu transform based solutions to convolution type integral equations and discrete dynamic systems were later obtained by Asiru [4–6]. This work has motivated the authors to study the classical Sumudu transform in the intuitionistic fuzzy set theory.

This paper is organized as follows: in Sect. 2 we give preliminaries which we will use throughout this work. In Sect. 3 a result for intuitionistic fuzzy Sumudu transform is presented. In Sect. 4 we construct a procedure for solving first-order intuitionistic fuzzy differential equations. In the last section, we present an example for illustrate this work.

S. Melliani and O. Castillo (Eds.): ICPAMS 2021, LNNS 476, pp. 356–367, 2023.
https://doi.org/10.1007/978-3-031-12416-7_31

2 Preliminaries

Let us $[a, b] \subset \mathbb{R}$ be a compact interval.

Definition 1. we denote by

$$\mathrm{IF}_1 = \mathrm{IF}(\mathbb{R}) = \left\{ \langle u, v \rangle \ : \ \mathbb{R} \to [0,1]^2 , |\forall\ x \in \mathbb{R}| 0 \leq u(x) + v(x) \leq 1 \right\}$$

An element $\langle u, v \rangle$ of IF_1 is said an intuitionistic fuzzy number if it satisfies the following conditions

(i) $\langle u, v \rangle$ is normal i.e. there exists x_0, $x_1 \in \mathbb{R}$ such that $u(x_0) = 1$ and $v(x_1) = 1$.
(ii) u is fuzzy convex and v is fuzzy concave.
(iii) u is upper semi-continuous and v is lower semi-continuous
(iv) *supp* $\langle u, v \rangle = \mathrm{cl}\{x \in \mathbb{R} \ : | \ v(x) < 1\}$ is bounded.

so we denote the collection of all intuitionistic fuzzy number by IF_1.

Definition 2. [12] An intuitionistic fuzzy number $\langle u, v \rangle$ in parametric form is a pair $\langle u, v \rangle = \left((\underline{\langle u, v \rangle}^+, \overline{\langle u, v \rangle}^+), (\underline{\langle u, v \rangle}^-, \overline{\langle u, v \rangle}^-) \right)$ of functions $\underline{\langle u, v \rangle}^-(\alpha)$, $\overline{\langle u, v \rangle}^-(\alpha)$, $\underline{\langle u, v \rangle}^+(\alpha)$ and $\overline{\langle u, v \rangle}^+(\alpha)$, which satisfies the following requirements:

1. $\underline{\langle u, v \rangle}^+(\alpha)$ is a bounded monotonic increasing continuous function,
2. $\overline{\langle u, v \rangle}^+(\alpha)$ is a bounded monotonic decreasing continuous function,
3. $\underline{\langle u, v \rangle}^-(\alpha)$ is a bounded monotonic increasing continuous function,
4. $\overline{\langle u, v \rangle}^-(\alpha)$ is a bounded monotonic decreasing continuous function,
5. $\underline{\langle u, v \rangle}^-(\alpha) \leq \overline{\langle u, v \rangle}^-(\alpha)$ and $\underline{\langle u, v \rangle}^+(\alpha) \leq \overline{\langle u, v \rangle}^+(\alpha)$, for all $0 \leq \alpha \leq 1$.

Example
A Triangular Intuitionistic Fuzzy Number (TIFN) $\langle u, v \rangle$ is an intuitionistic fuzzy set in $\mathbb{R}$ with the following membership function u and non-membership function v :

$$u(x) = \begin{cases} \dfrac{x - a_1}{a_2 - a_1} & \text{if } a_1 \leq x \leq a_2 \\ \dfrac{a_3 - x}{a_3 - a_2} & \text{if } a_2 \leq x \leq a_3, \\ 0 & \text{otherwise} \end{cases}$$

$$v(x) = \begin{cases} \dfrac{a_2 - x}{a_2 - a_1^{'}} & \text{if } a_1^{'} \leq x \leq a_2 \\ \dfrac{x - a_2}{a_3^{'} - a_2} & \text{if } a_2 \leq x \leq a_3^{'}, \\ 1 & \textit{otherwise.} \end{cases}$$

where $a_1^{'} \leq a_1 \leq a_2 \leq a_3 \leq a_3^{'}$
This TIFN is denoted by $\langle u, v\rangle = \left\langle a_1, a_2, a_3; a_1^{'}, a_2, a_3^{'} \right\rangle$.
Its parametric form is

$$\underline{\langle u, v\rangle}^{+}(\alpha) = a_1 + \alpha(a_2 - a_1), \ \overline{\langle u, v\rangle}^{+}(\alpha) = a_3 - \alpha(a_3 - a_2)$$
$$\underline{\langle u, v\rangle}^{-}(\alpha) = a_1' + \alpha(a_2 - a_1'), \ \overline{\langle u, v\rangle}^{-}(\alpha) = a_3' - \alpha(a_3' - a_2)$$

For $\alpha \in [0, 1]$ and $\langle u, v\rangle \in \mathrm{IF}_1$, the upper and lower α-cuts of $\langle u, v\rangle$ are defined by

$$[\langle u, v\rangle]^{\alpha} = \{x \in \mathbb{R} : v(x) \leq 1 - \alpha\}$$

and

$$[\langle u, v\rangle]_{\alpha} = \{x \in \mathbb{R} : u(x) \geq \alpha\}$$

Remark 1. If $\langle u, v\rangle \in \mathrm{IF}_1$, so we can see $[\langle u, v\rangle]_{\alpha}$ as $[u]^{\alpha}$ and $[\langle u, v\rangle]^{\alpha}$ as $[1 - v]^{\alpha}$ in the fuzzy case.

We define $0_{\langle 1,0\rangle} \in \mathrm{IF}_1$ as

$$0_{\langle 1,0\rangle}(t) = \begin{cases} \langle 1, 0\rangle & t = 0 \\ \langle 0, 1\rangle & t \neq 0 \end{cases}$$

For $\langle u, v\rangle$, $\langle z, w\rangle \in \mathrm{IF}_1$ and $\lambda \in \mathbb{R}$, the addition and scaler-multiplication are defined as follows

$$\Big[\langle u, v\rangle \oplus \langle z, w\rangle\Big]^{\alpha} = \Big[\langle u, v\rangle\Big]^{\alpha} + \Big[\langle z, w\rangle\Big]^{\alpha}, \quad \Big[\lambda \langle z, w\rangle\Big]^{\alpha} = \lambda\Big[\langle z, w\rangle\Big]^{\alpha}$$
$$\Big[\langle u, v\rangle \oplus \langle z, w\rangle\Big]_{\alpha} = \Big[\langle u, v\rangle\Big]_{\alpha} + \Big[\langle z, w\rangle\Big]_{\alpha}, \quad \Big[\lambda \langle z, w\rangle\Big]_{\alpha} = \lambda\Big[\langle z, w\rangle\Big]_{\alpha}$$

Definition 3. Let $\langle u, v\rangle$ an element of IF_1 and $\alpha \in [0, 1]$, we define the following sets:

$$\Big[\langle u, v\rangle\Big]_l^{+}(\alpha) = \inf\{x \in \mathbb{R} \mid u(x) \geq \alpha\}, \quad \Big[\langle u, v\rangle\Big]_r^{+}(\alpha) = \sup\{x \in \mathbb{R} \mid u(x) \geq \alpha\}$$
$$\Big[\langle u, v\rangle\Big]_l^{-}(\alpha) = \inf\{x \in \mathbb{R} \mid v(x) \leq 1 - \alpha\}, \quad \Big[\langle u, v\rangle\Big]_r^{-}(\alpha) = \sup\{x \in \mathbb{R} \mid v(x) \leq 1 - \alpha\}$$

Remark 2.

$$\Big[\langle u, v\rangle\Big]_{\alpha} = \Big[\Big[\langle u, v\rangle\Big]_l^{+}(\alpha), \Big[\langle u, v\rangle\Big]_r^{+}(\alpha)\Big]$$
$$\Big[\langle u, v\rangle\Big]^{\alpha} = \Big[\Big[\langle u, v\rangle\Big]_l^{-}(\alpha), \Big[\langle u, v\rangle\Big]_r^{-}(\alpha)\Big]$$

On the space IF_1 we will consider the following metric,

$$\begin{aligned} d_\infty\Big(\langle u,v\rangle , \langle z,w\rangle \Big) &= \frac{1}{4}\sup_{0<\alpha\le 1}\Big|\Big[\langle u,v\rangle\Big]_r^+(\alpha) - \Big[\langle z,w\rangle\Big]_r^+(\alpha)\Big| \\ &+ \frac{1}{4}\sup_{0<\alpha\le 1}\Big|\Big[\langle u,v\rangle\Big]_l^+(\alpha) - \Big[\langle z,w\rangle\Big]_l^+(\alpha)\Big| \\ &+ \frac{1}{4}\sup_{0<\alpha\le 1}\Big|\Big[\langle u,v\rangle\Big]_r^-(\alpha) - \Big[\langle z,w\rangle\Big]_r^-(\alpha)\Big| \\ &+ \frac{1}{4}\sup_{0<\alpha\le 1}\Big|\Big[\langle u,v\rangle\Big]_l^-(\alpha) - \Big[\langle z,w\rangle\Big]_l^-(\alpha)\Big| \end{aligned}$$

Theorem 1. ([13])
The metric space (IF_1, d_∞) is complete.

Proposition 1. *For all α, $\beta \in [0,1]$ and $\langle u,v\rangle \in IF_1$*

(i) $\Big[\langle u,v\rangle\Big]_\alpha \subset \Big[\langle u,v\rangle\Big]^\alpha$
(ii) $\Big[\langle u,v\rangle\Big]_\alpha$ *and* $\Big[\langle u,v\rangle\Big]^\alpha$ *are nonempty compact convex sets in* $\mathbb{R}$
(iii) if $\alpha \le \beta$ *then* $\Big[\langle u,v\rangle\Big]_\beta \subset \Big[\langle u,v\rangle\Big]_\alpha$ *and* $\Big[\langle u,v\rangle\Big]^\beta \subset \Big[\langle u,v\rangle\Big]^\alpha$
(iv) If $\alpha_n \nearrow \alpha$ *then* $\Big[\langle u,v\rangle\Big]_\alpha = \bigcap_n \Big[\langle u,v\rangle\Big]_{\alpha_n}$ *and* $\Big[\langle u,v\rangle\Big]^\alpha = \bigcap_n \Big[\langle u,v\rangle\Big]^{\alpha_n}$

Let M any set and $\alpha \in [0,1]$ we denote by

$$M_\alpha = \{x \in \mathbb{R} \ : \ u(x) \ge \alpha\} \quad \text{and} \quad M^\alpha = \{x \in \mathbb{R} \ : \ v(x) \le 1-\alpha\}$$

Lemma 1. [13] *let* $\Big\{M_\alpha,\ \alpha \in [0,1]\Big\}$ *and* $\Big\{M^\alpha,\ \alpha \in [0,1]\Big\}$ *two families of subsets of* $\mathbb{R}$ *satisfies (i)–(iv) in Proposition 1, if* u *and* v *define by*

$$u(x) = \begin{cases} 0 & \text{if } x \notin M_0 \\ \sup\{\alpha \in [0,1] \ : \ x \in M_\alpha\} & \text{if } x \in M_0 \end{cases}$$
$$v(x) = \begin{cases} 1 & \text{if } x \notin M^0 \\ 1 - \sup\{\alpha \in [0,1] \ : \ x \in M^\alpha\} & \text{if } x \in M^0 \end{cases}$$

Then $\langle u,v\rangle \in IF_1$

Definition 4. [9] Let $F : [a,b] \to \text{IF}_1$ be an intuitionistic fuzzy valued mapping and $t_0 \in [a,b]$. Then F is called intuitionistic fuzzy continuous in t_0 iff:

$$(\forall \varepsilon > 0)(\exists \delta > 0)\Big(\forall t \in T \text{such as} \mid t - t_0 \mid < \delta\Big) \Rightarrow d_\infty(F(t), F(t_0)) < \varepsilon.$$

Definition 5. [9] A mapping $F : [a,b] \to \mathrm{IF}_1$ is said to be differentiable at $t_0 \in (a,b)$ if there exist $F'(t_0) \in \mathrm{IF}_1$ such that limits:

$$\lim_{\Delta t \to 0+} \frac{F(t_0 + \Delta t) \ominus F(t_0)}{\Delta t} \quad \textit{and} \quad \lim_{\Delta t \to 0+} \frac{F(t_0) \ominus F(t_0 - \Delta t)}{\Delta t}$$

exist and they are equal to $F'(t_0)$.

If $F : [a,b] \to \mathrm{IF}_1$ is differentiable at $t_0 \in [a,b]$, then we say that $F'(t_0)$ is the intuitionistic fuzzy derivative of $F(t)$ at the point t_0. Here the limit is taken in the metric space $(\mathrm{IF}_1, d_\infty)$. At the end points of $[a,b]$ we consider only the one-sided derivatives.

Proposition 2. ([17], *p. 82, Lebesgue's Theorem*). *Let f be a bounded function on $[a,b]$. Then f is Riemann-integrable on $[a,b]$ if and only if f is continuous a.e. on $[a,b]$.*

Proposition 3. ([3], *p. 276, Theorem 10.33*). *Assume f is Riemann-integrable on $[a,b]$ for every $b \geq a$, and assume there is a positive constant M such that $\int_a^\infty |f(t)|dt \leq M$ for every $b \geq a$, Then both f and $|f|$ are improper Riemann integrable on $[a,\infty)$. Also, f is Lebesgue-integrable on $[a,\infty)$ and the Lebesgue integral of f is equal to the improper Riemann integral of f.*

Definition 6. Let $f(t)$ be an intuitionistic fuzzy-valued function on $[a,b]$. Suppose that $f_l^+(t,\alpha), f_r^+(t,\alpha)$,
$f_l^-(t,\alpha)$ and $f_r^-(t,\alpha)$ are Riemann-integrable on $[a,b]$ for all $\alpha \in [0,1]$.
Let

$$A_\alpha = \left[\int_a^b f_l^+(t,\alpha)dt, \int_a^b f_r^+(t,\alpha)dt\right]$$

and

$$A^\alpha = \left[\int_a^b f_l^-(t,\alpha)dt, \int_a^b f_r^-(t,\alpha)dt\right]$$

Then we say that $f(t)$ is intuitionistic fuzzy Riemann-integrable on $[a,b]$ denoted as $\mathbb{IFRI}$ on $[a,b]$ and the membership function and the nonmembership function of $\int_a^b f(t)dt$ are defined by,

$$u_{\int_a^b f(t)dt}(y) = \sup_{0 \leq \alpha \leq 1} \alpha.\chi_{A_\alpha}(y)$$

$$v_{\int_a^b f(t)dt}(y) = \inf_{0 \leq \alpha \leq 1} \alpha.\chi_{A^\alpha}(y)$$

for all $y \in A^0$.

Theorem 2. [10] *Let $f(t)$ be an intuitionistic fuzzy-valued function on $[a,\infty)$ represented by*
$\left(f_l^+(t,\alpha), f_r^+(t,\alpha), f_l^-(t,\alpha), f_r^-(t,\alpha)\right)$, *for any fixed $\alpha \in [0,1]$, assume that $f_l^+(t,\alpha), f_r^+(t,\alpha), f_l^-(t,\alpha)$ and $f_r^-(t,\alpha)$ are Riemann-integrable on $[a,b]$ for*

every $b \geq a$, and assume there are four positive constants $M_l^+(\alpha)$, $M_r^+(\alpha)$, $M_l^-(\alpha)$ and $M_r^-(\alpha)$ such that

$$\int_a^\infty |f_l^+(t,\alpha)|dt \leq M_l^+(\alpha), \int_a^\infty |f_r^+(t,\alpha)|dt \leq M_r^+(\alpha)$$
$$\int_a^\infty |f_l^-(t,\alpha)|dt \leq M_l^-(\alpha), \int_a^\infty |f_r^-(t,\alpha)|dt \leq M_r^-(\alpha)$$

Then $f(t)$ is an improper intuitionistic fuzzy Riemann integrable on $[a,\infty)$ and the improper intuitionistic fuzzy Riemann-integral is an intuitionistic fuzzy number.
Furthermore, we have:

$$\int_a^\infty f(t,\alpha)dt = \left(\int_a^\infty f_l^+(t,\alpha)dt, \int_a^\infty f_r^+(t,\alpha)dt, \int_a^\infty f_l^-(t,\alpha)dt, \int_a^\infty f_r^-(t,\alpha)dt\right)$$

3 Intuitionistic Fuzzy Sumudu Transform

In order to establish results, some definitions are needed. $G(u)$ and $\mathbf{S}[f(t)]$ will be used as the notation for intuitionistic fuzzy Sumudu transform throughout this paper.

Definition 7. Let $f(t)$ be continuous intuitionistic fuzzy-value function on $[0,\infty)$.

Suppose that $f(ut)e^{-t}$ is improper intuitionistic fuzzy Riemann-integrable on $[0,\infty)$, then $\int_0^\infty f(ut)e^{-t}dt$ is called intuitionistic fuzzy Sumudu transform and is denoted as

$$G(u) = \mathbf{S}[f(t)] = \int_0^\infty f(ut)e^{-t}dt, \quad u \in [-\tau_1, \tau_1]$$

where the variable u is used to factor the variable t in the argument of the intuitionistic fuzzy-valued function.
We have

$$\int_0^\infty f(ut,\alpha)e^{-t}dt = \left(\int_0^\infty f_l^+(ut,\alpha)e^{-t}dt, \int_0^\infty f_r^+(ut,\alpha)e^{-t}dt, \int_0^\infty f_l^-(ut,\alpha)e^{-t}dt, \int_0^\infty f_r^-(ut,\alpha)e^{-t}dt\right)$$

Also by using the definition of classical Sumudu transform

$$\mathcal{S}(f_l^+(ut,\alpha)) = \int_0^\infty f_l^+(ut,\alpha)e^{-t}dt, \quad \mathcal{S}(f_r^+(ut,\alpha)) = \int_0^\infty f_r^+(ut,\alpha)e^{-t}dt$$
$$\mathcal{S}(f_l^-(ut,\alpha)) = \int_0^\infty f_l^-(ut,\alpha)e^{-t}dt, \quad \mathcal{S}(f_r^-(ut,\alpha)) = \int_0^\infty f_r^-(ut,\alpha)e^{-t}dt$$

Then, we get

$$\mathbf{S}[f(ut,\alpha)] = \left(\mathcal{S}(f_l^+(ut,\alpha)), \mathcal{S}(f_r^+(ut,\alpha)), \mathcal{S}(f_l^-(ut,\alpha)), \mathcal{S}(f_r^-(ut,\alpha))\right)$$

3.1 Duality Properties of Intuitionistic Fuzzy Laplace Transform and Intuitionistic Fuzzy Sumudu Transform

IFLT has a close relationship with IFST. It is necessary for us to be able to link between the two transforms in order to prove theorems and properties of the IFST. The definition for IFLT is given as follows.

Definition 8. [10]
Let $f(t)$ be continuous intuitionistic fuzzy-value function on $[0, \infty)$.

Suppose that $f(t)e^{-st}$ is improper intuitionistic fuzzy Riemann-integrable on $[0, \infty)$, then $\int_0^\infty f(t)e^{-st}dt$ is called intuitionistic fuzzy Laplace transform and is denoted as

$$F(s) = \mathbf{L}[f(t)] = \int_0^\infty f(t)e^{-st}dt, s > 0 \tag{1}$$

Theorem 3. *Let $f(t)$ be continuous intuitionistic fuzzy-value function. If F is the intuitionistic fuzzy Laplace transform of $f(t)$ and G is the intuitionistic fuzzy Sumudu transform of $f(t)$, then*

$$G(u) = \frac{F(\frac{1}{u})}{u} \tag{2}$$

Proof. By the definition of the intuitionistic fuzzy Sumudu transform of $f(t)$ we have

$$G(u) = \mathbf{S}[f(t)] = \int_0^\infty f(ut)e^{-t}dt, \quad u \in [-\tau_1, \tau_1]$$

We put $w = ut$ or $t = \frac{w}{u}$, then we have

$$\begin{aligned} G(u) &= \int_0^\infty f(w)e^{-\frac{t}{u}}\frac{dw}{u}, \quad u \in [-\tau_1, \tau_1] \\ &= \frac{1}{u}\int_0^\infty f(w)e^{-\frac{t}{u}}dw, \quad u \in [-\tau_1, \tau_1]. \end{aligned}$$

Thus,

$$G(u) = \frac{F(\frac{1}{u})}{u}$$

In the following corollary, we show that the roles of F and G in Eq. (1) can be interchanged.

Corollary 1. *Let $f(t)$ be continuous intuitionistic fuzzy-value function, having F and G for intuitionistic fuzzy Laplace transform and intuitionistic fuzzy Sumudu transform, respectively. Then*

$$F(s) = \frac{G(\frac{1}{s})}{s} \tag{3}$$

Proof. The proof of Eq. (3) can be obtained by changing u to $\frac{1}{u}$ in Eq. (2).

3.2 Fundamental Theorems and Properties of Intuitionistic Fuzzy Sumudu Transform

Theorem 4. *Let $f, g : \mathbb{R} \to IF_1$ be two continuous intuitionistic fuzzy valued functions. Suppose that c_1 and c_2 are arbitrary constants, then*

$$\boldsymbol{S}\big[c_1 f(t) \oplus c_2 g(t)\big] = c_1 \boldsymbol{S}(f(t)) \oplus c_2 \boldsymbol{S}(g(t))$$

Proof. By the definition of the intuitionistic fuzzy Sumudu transform of $f(t)$ we have

$$\begin{aligned}\mathbf{S}\big[c_1 f(t) \oplus c_2 g(t)\big] &= \int_0^\infty (c_1 f(ut) \oplus c_2 g(ut))e^{-t}dt, u \in [-\tau_1, \tau_1] \\ &= \int_0^\infty (c_1 f(ut)e^{-t} \oplus c_2 g(ut)e^{-t})dt\end{aligned}$$

By the Theorem 4.3 in [11], we have the linearity of the integral that allows us to write

$$\begin{aligned}\int_0^\infty (c_1 f(ut) \oplus c_2 g(ut))e^{-t}dt &= \int_0^\infty c_1 f(ut)e^{-t}dt \oplus \int_0^\infty c_2 g(ut)e^{-t}dt, u \in [-\tau_1, \tau_1] \\ &= c_1 \int_0^\infty f(ut)e^{-t}dt \oplus c_2 \int_0^\infty g(ut)e^{-t}dt\end{aligned}$$

Then, we get

$$\int_0^\infty (c_1 f(ut) \oplus c_2 g(ut))e^{-t}dt = c_1 \mathbf{S}[f(t)] \oplus c_2 \mathbf{S}[g(t)] \tag{4}$$

Thus,

$$\mathbf{S}[c_1 f(t) \oplus c_2 g(t)] = c_1 \mathbf{S}[f(t)] \oplus c_2 \mathbf{S}[g(t)].$$

Theorem 5. *Let $f : \mathbb{R} \to IF_1$ be continuous intuitionistic fuzzy valued function, and f is the primitive of f' on $[0, \infty)$. Then*

$$\boldsymbol{S}(f'(t)) = \frac{G(u) \ominus f(0)}{u}$$

Proof. By the definition of the intuitionistic fuzzy Sumudu transform of $f'(t)$ we have

$$\mathbf{S}[f'(t)] = \int_0^\infty f'(ut)e^{-t}dt, \quad u \in [-\tau_1, \tau_1]$$

By using integration by parts formula, we have

$$\mathbf{S}[f'(t)] = \frac{1}{u}\int_0^\infty f(ut)e^{-t}dt \oplus \left[\frac{1}{u}f(ut)e^{-t}\right]_0^\infty, \quad u \in [-\tau_1, \tau_1]$$

$$\begin{aligned}&= \frac{1}{u}\left[\int_0^\infty f(ut)e^{-t}dt \ominus f(0)\right] \\ &= \frac{G(u) \ominus f(0)}{u}\end{aligned}$$

Thus,

$$\mathbf{S}(f'(t)) = \frac{G(u) \ominus f(0)}{u}$$

4 Procedure for Solving First Order Intuitionistic Fuzzy Differential Equations

In this section,we consider the following first order intuitionistic fuzzy initial value problem in general form

$$\begin{cases} y'(t) = f(t, y(t)) \\ y(t_0) = \Big(y_l^+(t_0,\alpha), y_r^+(t_0,\alpha), y_l^-(t_0,\alpha), y_r^-(t_0,\alpha)\Big) \end{cases} \tag{5}$$

where $f : [t_0, T] \times \mathrm{IF}_1 \longrightarrow \mathrm{IF}_1$ is a continuous intuitionistic fuzzy mapping. By using intuitionistic fuzzy Sumudo transform method we have:

$$\mathbf{S}[y'(t)] = \mathbf{S}[f(t, y(t))] \tag{6}$$

By using Theorem 11, Eq. (6) can be written as follows:

$$\frac{\mathbf{S}[y(t)] \ominus y(t_0)}{u} = \mathbf{S}[f(t, y(t))]$$

Thus,

$$\begin{cases} \frac{\mathcal{S}[y_l^+(t,\alpha)] - y_l^+(t_0,\alpha)}{u} = \mathcal{S}[f_l^+(t, y(t), \alpha)], \\ \frac{\mathcal{S}[y_r^+(t,\alpha)] - y_r^+(t_0,\alpha)}{u} = \mathcal{S}[f_r^+(t, y(t), \alpha)], \\ \frac{\mathcal{S}[y_l^-(t,\alpha)] - y_l^-(t_0,\alpha)}{u} = \mathcal{S}[f_l^-(t, y(t), \alpha)], \\ \frac{\mathcal{S}[y_r^-(t,\alpha)] - y_r^-(t_0,\alpha)}{u} = \mathcal{S}[f_r^-(t, y(t), \alpha)]. \end{cases} \tag{7}$$

To solve the linear system (7), for simplicity we assume that:

$$\begin{cases} \mathcal{S}[y_l^+(t,\alpha)] = H_1(s,\alpha), \\ \mathcal{S}[y_r^+(t,\alpha)] = H_2(s,\alpha), \\ \mathcal{S}[y_l^-(t,\alpha)] = K_1(s,\alpha), \\ \mathcal{S}[y_r^-(t,\alpha)] = K_2(s,\alpha). \end{cases} \tag{8}$$

where $H_1(s,\alpha), H_2(s,\alpha), K_1(s,\alpha)$ and $K_2(s,\alpha)$ are solutions of system (7).

By using inverse classical Sumudu transform, $y_l^+(t,\alpha), y_r^+(t,\alpha), y_l^-(t,\alpha)$ and $y_r^-(t,\alpha)$ are computed as follows

$$\begin{cases} y_l^+(t,\alpha) = \mathcal{S}^{-1}[H_1(s,\alpha)], \\ y_r^+(t,\alpha) = \mathcal{S}^{-1}[H_2(s,\alpha)], \\ y_l^-(t,\alpha) = \mathcal{S}^{-1}[K_1(s,\alpha)], \\ y_r^-(t,\alpha) = \mathcal{S}^{-1}[K_2(s,\alpha)]. \end{cases} \tag{9}$$

5 Example

We consider the following intuitionistic fuzzy initial value problem (IFIVP)

$$\begin{cases} y'(t) = 3y(t) \\ y(0) = \Big(3+\alpha, 5-\alpha, 2+2\alpha, 6-2\alpha\Big) \end{cases} \tag{10}$$

the parametric form of the IFIVP is given by the following form:

$$y'(t,\alpha) = 3y(t,\alpha)$$

i.e.

$$\begin{cases} y_l'^{+}(t,\alpha) = 3y_l^{+}(t,\alpha), \\ y_r'^{+}(t,\alpha) = 3y_r^{+}(t,\alpha), \\ y_l'^{-}(t,\alpha) = 3y_l^{-}(t,\alpha), \\ y_r'^{-}(t,\alpha) = 3y_r^{-}(t,\alpha), \end{cases} \tag{11}$$

Applying intuitionistic fuzzy Sumudu Transform

$$\mathbf{S}[y'(t)] = 3\mathbf{S}[y(t)]$$

i.e.,

$$\begin{cases} \mathcal{S}[y_l'^{+}(t,\alpha)] = 3\mathcal{S}[y_l^{+}(t,\alpha)], \\ \mathcal{S}[y_r'^{+}(t,\alpha)] = 3\mathcal{S}[y_r^{+}(t,\alpha)], \\ \mathcal{S}[y_l'^{-}(t,\alpha)] = 3\mathcal{S}[y_l^{-}(t,\alpha)], \\ \mathcal{S}[y_r'^{-}(t,\alpha)] = 3\mathcal{S}[y_r^{-}(t,\alpha)]. \end{cases} \tag{12}$$

Therefore,

$$\begin{cases} \frac{\mathcal{S}[y_l^{+}(t,\alpha)] - y_l^{+}(0,\alpha)}{u} = 3\mathcal{S}[y_l^{+}(t,\alpha)], \\ \frac{\mathcal{S}[y_r^{+}(t,\alpha)] - y_r^{+}(0,\alpha)}{u} = 3\mathcal{S}[y_r^{+}(t,\alpha)], \\ \frac{\mathcal{S}[y_l^{-}(t,\alpha)] - y_l^{-}(0,\alpha)}{u} = 3\mathcal{S}[y_l^{-}(t,\alpha)], \\ \frac{\mathcal{S}[y_r^{-}(t,\alpha)] - y_r^{-}(0,\alpha)}{u} = 3\mathcal{S}[y_r^{-}(t,\alpha)]. \end{cases} \tag{13}$$

Thus,

$$\begin{cases} \mathcal{S}[y_l^{+}(t,\alpha)] = \frac{1}{1-3u} y_l^{+}(0,\alpha), \\ \mathcal{S}[y_r^{+}(t,\alpha)] = \frac{1}{1-3u} y_r^{+}(0,\alpha), \\ \mathcal{S}[y_l^{-}(t,\alpha)] = \frac{1}{1-3u} y_l^{-}(0,\alpha), \\ \mathcal{S}[y_r^{-}(t,\alpha)] = \frac{1}{1-3u} y_r^{-}(0,\alpha). \end{cases} \tag{14}$$

After that we replace the value of parametric form of $y(0)$ in the system (14), we have

$$\begin{cases} \mathcal{S}[y_l^+(t,\alpha)] = \frac{1}{1-3u}(3+\alpha), \\ \mathcal{S}[y_r^+(t,\alpha)] = \frac{1}{1-3u}(5-\alpha), \\ \mathcal{S}[y_l^-(t,\alpha)] = \frac{1}{1-3u}(2+2\alpha), \\ \mathcal{S}[y_r^-(t,\alpha)] = \frac{1}{1-3u}(6-2\alpha). \end{cases} \tag{15}$$

Now, After performing partition of fractions and by using inverse classical Sumudu transform, $y_l^+(t,\alpha), y_r^+(t,\alpha), y_l^-(t,\alpha)$ and $y_r^-(t,\alpha)$ are computed as follows

$$\begin{cases} y_l^+(t,\alpha) = (\frac{2\alpha+7}{4})\mathcal{S}^{-1}[\frac{1}{1-3u}], \\ y_r^+(t,\alpha) = (\frac{-2\alpha+11}{4})\mathcal{S}^{-1}[\frac{1}{1-3u}], \\ y_l^-(t,\alpha) = (\frac{4\alpha+5}{4})\mathcal{S}^{-1}[\frac{1}{1-3u}], \\ y_r^-(t,\alpha) = (\frac{-4\alpha+13}{4})\mathcal{S}^{-1}[\frac{1}{1-3u}]. \end{cases} \tag{16}$$

Thus,

$$\begin{cases} y_l^+(t,\alpha) = (\frac{2\alpha+7}{4})e^{3t}, \\ y_r^+(t,\alpha) = (\frac{-2\alpha+11}{4})e^{3t}, \\ y_l^-(t,\alpha) = (\frac{4\alpha+5}{4})e^{3t}, \\ y_r^-(t,\alpha) = (\frac{-4\alpha+13}{4})e^{3t}. \end{cases} \tag{17}$$

We remark that $y_l^+(t,\alpha) \leq y_r^+(t,\alpha)$; $y_l^-(t,\alpha) \leq y_r^-(t,\alpha)$ also the functions $y_l^+(t,\alpha), y_l^-(t,\alpha)$ are increasing with respect to α and the functions $y_r^+(t,\alpha), y_r^-(t,\alpha)$ are decreasing with respect to α.

So, this we shown that $\left(y_l^+(t,\alpha), y_r^+(t,\alpha), y_l^-(t,\alpha), y_r^-(t,\alpha)\right)$ is the parametric form of the solution of the problem (11).

References

1. Ahmad, N., Mamat, M., Kumar, J.K., Amir Hamzah, N.S.: Solving fuzzy duffing equation by the laplace transform decomposition. Appl. Math. Sci. **6**(59), 2935–2944 (2012)
2. Allahviranloo, T., Barkhordari Ahmadi, M.: Fuzzy laplace transforms. Soft Comput. **14**(3), 235–243 (2010). https://doi.org/10.1007/s00500-008-0397-6
3. Apostol, T.M.: Mathematical Analysis, 2nd edn. Addison-Welesy, Reading, MA (1974)
4. Asiru, M.A.: Sumudu transform and the solution of integral equations of convolution type. Int. J. Math. Educ. Sci. Technol. **32**(6), 906–910 (2001)
5. Asiru, M.A.: Further properties of the sumudu transform and its applications. Int. J. Math. Educ. Sci. Technol. **33**(3), 441–449 (2002)

6. Asiru, M.A.: Application of the sumudu transform to discrete dynamical systems. Int. J. Math. Educ. Sci. Technol. **34**(6), 944–949 (2003)
7. Atanassov, K.: Intuitionistic fuzzy sets. Fuzzy Sets Syst. **20**, 87–96 (1986)
8. Atanassov, K.: Intuitionistic Fuzzy Sets: Theory and Applications. Physica-Verlag, Heidelberg (1999)
9. Ettoussi, R., Melliani, S., Chadli, L.S.: Differential equation with intuitionistic fuzzy parameters. Notes Intuitionistic Fuzzy Sets **23**(4), 46–61 (2017)
10. Ettoussi, R., Melliani, S., Chadli, L.S.: On intuitionistic fuzzy laplace transforms for second order intuitionistic fuzzy differential equations. In: Melliani, S., Castillo, O. (eds.) Recent Advances in Intuitionistic Fuzzy Logic Systems. SFSC, vol. 372, pp. 153–168. Springer, Cham (2019). https://doi.org/10.1007/978-3-030-02155-9_13
11. Ettoussi, R., Melliani, S., Elomari, M., Chadli, L.S.: Solution of intuitionistic fuzzy differential equations by successive approximations method. Notes Intuitionistic Fuzzy Sets **21**(2), 51–62 (2015)
12. Keyanpour, M., Akbarian, T.: Solving intuitionistic fuzzy nonlinear equations. J. Fuzzy Set Valued Anal. **2014**, 1–6 (2014)
13. Melliani, S., Elomari, M., Chadli, L.S., Ettoussi, R.: Intuitionistic fuzzy metric spaces. Notes Intuitionistic Fuzzy sets **21**(1), 43–53 (2015)
14. Melliani, S., Chadli, L.S.: Introduction to intuitionistic fuzzy partial differential equations. Notes Intuitionistic Fuzzy sets **7**(3), 39–42 (2001)
15. Melliani, S., Chadli, L.S.: Introduction to intuitionistic fuzzy differential equations. Notes Intuitionistic Fuzzy sets **6**(2), 31–41 (2000)
16. Ramazannia Tolouti, S.J., Barkhordary Ahmadi, M.: Fuzzy laplace transform on two order derivative and solving fuzzy two order differential equation. Int. J. Industrial Math. **2**(4), 279–293 (2010)
17. Royden, H.L.: Real Analysis, 2nd edn. Macmillan, New York (1968)
18. Watugala, G.K.: Sumudu transform: a new integral transform to solve differential equations and control engineering problems. Int. J. Math. Educ. Sci. Technol. **24**(1), 35–43 (1993)
19. Watugala, G.K.: Sumudu transform a new integral transform to solve differential equations and control engineering problems. Math. Eng. Ind. **6**(4), 319–329 (1998)
20. Zadeh, L.A.: Fuzzy set. Inf. Control **8**(3), 338–353 (1965)

Entropy Solution of Nonlinear Elliptic $p(u)$-Laplacian Problem

Adil Abbassi, Chakir Allalou(✉), and Said Ait Temghart

Laboratory LMACS, FST of Beni Mellal, Sultan Moulay Slimane University, Beni Mellal, Morocco
adil.abbassi@usms.ma, chakir.allalou@yahoo.fr

Abstract. In this paper, we prove the existence of an entropy solution for a class of nonlinear nonlocal elliptic problem associated to the following equation

$$\begin{cases} b(u) - \text{div} a(x,u,\nabla u) - \text{div}\phi(u) = f \text{ in } \Omega \\ u = 0 \qquad\qquad\qquad\qquad\qquad\quad \text{on } \partial\Omega, \end{cases}$$

where the operator $-\text{div} a(x,u,\nabla u)$ is called $p(u)$-Laplacian.

1 Introduction

Let Ω be a bounded domain of $\mathbb{R}^N, N \geq 2$, with Lipschitz boundary $\partial\Omega$. In the present paper, we consider the following nonlinear problem

$$\begin{cases} -\text{div} a(x,u,\nabla u) - \text{div}\phi(u) = f \text{ in } \Omega \\ \\ u = 0 \qquad\qquad\qquad\qquad\qquad\quad \text{on } \partial\Omega, \end{cases} \tag{1}$$

The operator $\text{div} a(x,u,\nabla u)$ is more complicated than the $p(x)$-Laplacian in the term of nonlinearity, it is called $p(u)$-Laplacian. A prototype of this operator is $\text{div}\left(|\nabla u|^{p(u)-2} \cdot \nabla u\right)$. The variable exponent p depend both on the space variable x and on the unknown solution u. We were inspired by the work of Ouaro and Sawadogo (see [1]), where they considered the following nonlinear Fourier boundary value problem

$$\begin{cases} b(u) - \text{div}\, a(x,u,\nabla u) = f \text{ in} \Omega \\ a(x,u,\nabla u) \cdot \eta + \lambda u = g \quad \text{on} \partial\Omega, \end{cases}$$

it is a generalization of the following nonlinear problem

$$\begin{cases} b(u) - \text{div}\, a(x,\nabla u) = f \text{in} \Omega \\ a(x,\nabla u) \cdot \eta + \lambda u = g \quad \text{on} \partial\Omega \end{cases}$$

studied by Nyanquini and Ouaro in [11], where the authors used an auxiliary result due to Le in [12] to prove the existence of the weak solution when $f \in L^\infty(\Omega), g \in L^\infty(\partial\Omega)$ and by approximation methods they obtained the entropy solution when $f \in L^1(\Omega), g \in$

S. Melliani and O. Castillo (Eds.): ICPAMS 2021, LNNS 476, pp. 368–392, 2023.
https://doi.org/10.1007/978-3-031-12416-7_32

$L^1(\partial\Omega)$. In recent years, the study of partial differential equations and variational problems with variable exponent involving $p(u)$-Laplacian has received considerable attention in many models coming from various branches of mathematical physics, such as electrorheological fluid dynamicsn elastic mechanics, computer vision and image processing (see [15–18]). The $p(.)$-Laplacian type models with the exponent of nonlinearity p depending on the solution u itself have been used in the image analysis and computer vision (see [13]). By approximation method and convergent sequences in terms of Young measure. Ouaro and Sawadogo in [14] studied the existence and uniqueness results of weak solution and the structural stability result of nonlinear $p(u)$-Laplacian problem with homogeneous Neumann boundary condition. The notion of entropy solutions was introduced by Ph. Bénilan, L. Boccardo, T. Gallouet, R. Gariepy, M. Pierre, J.L. Vazquez in [19], it was adapted by many authors to study some nonlinear elliptic and parabolic problems with a constant or variable exponent and with Dirichlet, Fourier or Neumann boundary conditions (see [20–24]).

This paper is organized us follow. In Sect. 2 we introduce the basic assumptions and we recall some definitions, basic properties of generalised Sobolev spaces, also we give a brief review on Young measures [9, 10] that we will used later. The Sect. 3 is devoted to show the existence results of the entropy solution when the data are in L^1.

2 Preliminaries

In order to discuss problem (1), we must work in Lebesgue and Sobolev spaces with variable exponent, that depend on x and on $u(x)$. We need the Sobolev spaces $W^{1,\pi(\cdot)}(\Omega)$ where $\pi(\cdot) = p(.,p(\cdot))$. Let Ω be a bounded open subset of $\mathbb{R}^N$ $(N \geq 3)$ with the Lipschitz boundary $\partial\Omega$. We define

$$p_- := ess \inf_{(x,z)\in\Omega\times\mathbb{R}} p(x,z) \text{ and } p_+ := ess \sup_{(x,z)\in\Omega\times\mathbb{R}} p(x,z).$$

Definition 1. $L^{\pi(\cdot)}(\Omega)$ is the space of all measurable function $f : \Omega \longrightarrow \mathbb{R}$ such that the modular

$$\rho_{\pi(.)}(f) := \int_\Omega |f|^{\pi(x)} dx < +\infty.$$

If p_+ is finite, this space is equipped with the Luxembourg norm

$$\|f\|_{\pi(.)} := \inf\left\{\lambda > 0;\quad \rho_{\pi(.)}\left(\frac{f}{\lambda}\right) \leq 1\right\}.$$

$W^{1,\pi(\cdot)}(\Omega)$ is the space of all functions $f \in L^{\pi(\cdot)}(\Omega)$ such that the gradient of f belongs to $L^{\pi(\cdot)}(\Omega)$. The space $W^{1,\pi(\cdot)}(\Omega)$ is equipped with the norm

$$\|u\|_{1,\pi(.)} := \|u\|_{\pi(.)} + \|\nabla u\|_{\pi(.)}.$$

When $1 < p_- \leq \pi(.) \leq p_+ < +\infty$, all the above spaces are separable and reflexive Banach spaces.

Proposition 1. *(See [2], Proposition 2.3)*
For all measurable function $\pi : \Omega \to [p_-, p_+]$, the following properties hold.

i) $L^{\pi(\cdot)}(\Omega)$ *and* $W^{1,\pi(\cdot)}(\Omega)$ *are separable and reflexive Banach spaces.*
ii) $L^{\pi'(\cdot)}(\Omega)$ *can be identified with the dual space of* $L^{\pi(\cdot)}(\Omega)$*, and the following Hlder type inequality holds:*

$$\forall f \in L^{\pi(\cdot)}(\Omega), g \in L^{\pi'(\cdot)}(\Omega), \quad \left| \int_\Omega f g dx \right| \leq 2\|f\|_{L^{\pi(\cdot)}(\Omega)} \|g\|_{L^{\pi'(\cdot)}(\Omega)}$$

iii) One has $\rho_{\pi(.)}(f) = 1$ *if and only if* $\|f\|_{L^{\pi(\cdot)}(\Omega)} = 1$*; further, if* $\rho_{\pi(.)}(f) \leq 1$*, then* $\|f\|^{p_+}_{L^{\pi(\cdot)}(\Omega)} \leq \rho_{\pi(.)}(f) \leq \|f\|^{p_-}_{L^{\pi(\cdot)}(\Omega)}$*; if* $\rho_{\pi(.)}(f) \geq 1$*, then* $\|f\|^{p_-}_{L^{\pi(\cdot)}(\Omega)} \leq \rho_{\pi(.)}(f) \leq \|f\|^{p_+}_{L^{\pi(\cdot)}(\Omega)}$*. In particular, if* $(f_n)_{n\in\mathbb{N}}$ *is a sequence in* $L^{\pi(\cdot)}(\Omega)$*, then* $\|f_n\|_{L^{\pi(\cdot)}(\Omega)}$ *tends to zero (resp., to infinity) if and only if* $\rho_{\pi(.)}(f_n)$ *tends to zero (resp., to infinity), as* $n \to +\infty$*.*

Proposition 2. *(See [3,4])*
If $f \in W^{1,\pi(\cdot)}(\Omega)$*, the following properties hold:*

i) $\|f\|_{W^{1,\pi(\cdot)}(\Omega)} > 1 \Rightarrow \|f\|^{p_-}_{W^{1,\pi(\cdot)}(\Omega)} < \rho_{1,\pi(.)}(f) < \|f\|^{p_+}_{W^{1,\pi(\cdot)}(\Omega)}$;
ii) $\|f\|_{W^{1,\pi(\cdot)}(\Omega)} < 1 \Rightarrow \|f\|^{p_+}_{W^{1,\pi(\cdot)}(\Omega)} < \rho_{1,\pi(.)}(f) < \|f\|^{p_-}_{W^{1,\pi(\cdot)}(\Omega)}$;
iii) $\|f\|_{W^{1,\pi(\cdot)}(\Omega)} < 1$ *(respectively* $= 1; > 1$*)* $\Leftrightarrow \rho_{1,\pi(.)}(f) < 1$ *(respectively* $= 1; > 1$*).*

We give now some embedding results.

Proposition 3. *(See [2])*
Assume that $\pi : \Omega \to [p_-, p_+]$ *satisfying the log-Hlder continuity assumption:*

$$\exists L > 0, \quad \forall x, y \in \bar{\Omega}, x \neq y, \quad -(\log|x - y|)|\pi(x) - \pi(y)| \leq L.$$

i) Then, $\mathscr{D}(\Omega)$ *is dense in* $W^{1,\pi(\cdot)}(\Omega)$*.*
ii) $W^{1,\pi(\cdot)}(\Omega)$ *is embedded into* $L^{\pi^*(\cdot)}(\Omega)$*, where* $\pi^*(.)$ *is the Sobolev embedding exponent defined below. If q is a measurable variable exponent such that ess* $\inf_{x\in\Omega}(\pi^*(.) - q(.)) > 0$*, then the embedding of* $W^{1,\pi(\cdot)}(\Omega)$ *into* $L^{q(\cdot)}(\Omega)$ *is compact. For a given* $\pi(.)$*, a function taking values in* $[p_-, p_+]$*,* $\pi^*(.)$ *denotes the optimal Sobolev embedding defined for any* $x \in \Omega$ *by*

$$\pi^*(x) = \begin{cases} \frac{N\pi(x)}{N-\pi(x)} & \text{if } \pi(x) < N \\ \text{any real value} & \text{if } \pi(x) = N \\ +\infty & \text{if } \pi(x) > N. \end{cases}$$

We will use the so-called truncation function

$$T_k(s) := \begin{cases} s & \text{if } |s| \leq k \\ k\,\text{sign}_0(s) & \text{if } |s| > k \end{cases}, \quad \text{where } \text{sign}_0(s) := \begin{cases} 1 & \text{if } s > 0 \\ 0 & \text{if } s = 0 \\ -1 & \text{if } s < 0. \end{cases}$$

The truncation function possesses the following properties.

$$T_k(-s) = -T_k(s), \ |T_k(s)| = \min\{|s|, k\}.$$

We also need to truncate vector valued-function with the help of the mapping

$$h_m : \mathbb{R}^N \longrightarrow \mathbb{R}^N, \ h_m(\lambda) = \begin{cases} \lambda, & \text{if} |\lambda| \leq m \\ m\frac{\lambda}{|\lambda|} & \text{if} |\lambda| > m, \end{cases} \quad \text{where } m > 0.$$

A Brief Review on Young Measures

As stated in the introduction, the tool we use to prove the needed result is the Young measure. For the reader not familiar with this concept we recall some basic notions and properties (see [5–7] and references therein).

Theorem 1.

(i) *Let $\Omega \subset \mathbb{R}^N, N \in \mathbb{N}$, and a sequence $(v_n)_{n\in\mathbb{N}}$ of $\mathbb{R}^d$ -valued functions, $d \in \mathbb{N}$ such that $(v_n)_{n\in\mathbb{N}}$ is equi-integrable on Ω. Then, there exists a subsequence $(n_k)_{k\in\mathbb{N}}$ and a parametrized family $(\nu_x)_{x\in\Omega}$ of probability measures on $\mathbb{R}^d (d \in \mathbb{N})$, weakly measurable in x with respect to the Lebesgue measure in Ω, such that for all Carathodory function $F : \Omega \times \mathbb{R}^d \longrightarrow \mathbb{R}^t, t \in \mathbb{N}$, we have*

$$\lim_{k\longrightarrow+\infty} \int_\Omega F\left(x, v_{n_k}\right) dx = \int_\Omega \int_{\mathbb{R}^d} F(x, \lambda) d\nu_x(\lambda) dx. \tag{2}$$

(ii) *If $|\Omega| < \infty$ and ν_x is the Young measure generated by the sequence $(v_n)_{n\in\mathbb{N}}$, then there holds*

$$v_n \longrightarrow v \text{ in measure} \iff \nu_x = \delta_{v(x)} \text{ for a.e. } x \in \Omega.$$

(iii) *If the sequence $(u_n)_{n\in\mathbb{N}}$ generates the Young measure $\delta_{u(x)}$, then $(u_n, v_n)_{n\in\mathbb{N}}$ generates the Young measure $\delta_{u(x)} \otimes \nu_x$.*
the Young measure $\left(\delta_{u(x)} \otimes \nu_x\right)_{x\in\Omega}$ on $\mathbb{R}^{d_1+d_2}$. Whenever a sequence $(v_n)_{n\in\mathbb{N}}$ generates a Young measure $(\nu_x)_{x\in\Omega}$, following the terminology of [7] we will say that $(v_n)_{n\in\mathbb{N}}$ nonlinear weak- converges, and $(\nu_x)_{x\in\Omega}$ is the nonlinear weak-* limit of the sequence $(v_n)_{n\in\mathbb{N}}$. In the case where $(v_n)_{n\in\mathbb{N}}$ possesses a nonlinear weak-* convergent subsequence, we will say that it is nonlinear weak-* compact. It means that any equi-integrable sequence of measurable functions is nonlinear weak-* compact on Ω.*

Basic Assumptions
In this paper, we consider the following basic assumptions on the data for the study of the problem (1).
(A_1) f is a function such that $f \in L^1(\Omega)$.
Problem (1) is adapted into a generalized Leray-Lions framework under the assumption that $a : \Omega \times \left(\mathbb{R} \times \mathbb{R}^N\right) \to \mathbb{R}^N$ is a Carathodory function with:
(A_2) b is nondecreasing surjective and continuous function defined on $\mathbb{R}$ such that $b(0) = 0$.
(A_3) $a(x,z,0) = 0$ for all $z \in \mathbb{R}$, and a.e. $x \in \Omega$.
(A_4) $(a(x,z,\xi) - a(x,z,\eta)) \cdot (\xi - \eta) > 0$ for all $\xi, \eta \in \mathbb{R}^N, \xi \neq \eta$.
(A_5) $|a(x,z,\xi)|^{p'(x,z)} \leq C_1 \left(|\xi|^{p(x,z)} + \mathcal{M}(x)\right)$.
(A_6) $a(x,z,\xi) \cdot \xi \geq \frac{1}{C_2} |\xi|^{p(x,z)}$.
Here, C_1 and C_2 are positive constants and $\mathcal{M}$ is a positive function such that $\mathcal{M} \in L^1(\Omega)$. $p : \Omega \times \mathbb{R} \to [p_-, p_+]$ is a Carathodory function, $1 < p_- \leq p_+ < +\infty$ and $p'(x,z) = \frac{p(x,z)}{p(x,z)-1}$ is the conjugate exponent of $p(x,z)$.

3 Main Results

In this section we formulate and prove the main result of the paper.
Now, we give a definition of entropy solutions for the elliptic problem (1).
We consider

$$\mathcal{T}^{1,\pi(\cdot)}(\Omega) = \left\{u : \Omega \to \mathbb{R}, \text{ meas. such that } T_k(u) \in W^{1,\pi(\cdot)}(\Omega), \text{for any } k > 0\right\}.$$

Definition 2. A measurable function u for $\pi(.) = p(., u(.))$ is said to be an entropy solution for the problem (1), if

$$u \in \mathcal{T}^{1,\pi(.)}(\Omega), \;\; b(u) \in L^1(\Omega)$$

and for all $k > 0$,

$$\int_\Omega b(u) T_k(u-v)dx + \int_\Omega a(x,u,\nabla u)\nabla T_k(u-v)dx + \int_\Omega \phi(u)\nabla T_k(u-v)dx \leq \int_\Omega f T_k(u-v)dx, \tag{3}$$

$$\forall v \in W^{1,\pi(.)}(\Omega) \cap L^\infty(\Omega).$$

Theorem 2. *Assume that $(H_1) - (H_3)$ hold. Then there exists at least one entropy solution in the sense of the Definition 2 of the problem (1).*

The proof of Theorem (2) is divided into three parts.

Part 1: The Approximate Problem
We consider the following approximate problem

$$(\mathcal{P}_n) \begin{cases} T_n(b(u_n)) - \text{div} a(x,u_n,\nabla u_n) - \text{div}\phi(T_n(u_n)) - \varepsilon \Delta_{p_+} + \varepsilon |u_n|^{p_+-2} u_n = f_n & \text{in } \Omega \\ u_n = 0 & \text{on } \partial\Omega, \end{cases}$$

where

$$-\varepsilon \Delta_{p_+} u_n := -\sum_{i=1}^{N} \frac{\partial}{\partial x_i}\left(\left|\frac{\partial u_n}{\partial x_i}\right|^{p_+-2}\frac{\partial u_n}{\partial x_i}\right),$$

and $f_n = T_n(f)$. Then, $f_n \in L^\infty(\Omega)$. Moreover, $(f_n)_{n\in\mathbb{N}}$ converges strongly to f in $L^1(\Omega)$ such that $\|f_n\| \leq \|f\|$.

Theorem 3. *There exists at least one weak solution u_n for the problem $\mathscr{P}_n$ in the sense that $u_n \in W^{1,p_+}(\Omega)$,*

$$\begin{aligned}&\int_\Omega T_n(b(u_n))v dx + \int_\Omega a(x,u_n,\nabla u_n)\nabla v dx + \int_\Omega \phi(T_n(u_n))\nabla v dx \\ &+ \varepsilon \int_\Omega \left(|\nabla u_n|^{p_+-2}\nabla u_n \nabla v + |u_n|^{p_+-2} u_n v\right) dx = \int_\Omega f_n v dx.\end{aligned} \tag{4}$$

We define the operator A_n by

$$A_n u = Au + Bu,$$

where

$$\langle Au, v\rangle = \int_\Omega a(x,u,\nabla u)\nabla v dx,$$

and

$$\langle B_n u, v\rangle = \int_\Omega T_n(b(u_n))v dx + \int_\Omega \phi(T_n(u))\nabla v dx + \varepsilon \int_\Omega \left(|\nabla u|^{p_+-2}\nabla u \nabla v + |u|^{p_+-2} uv\right) dx,$$

with $u, v \in W^{1,p^+}(\Omega)$.

Proof of the Theorem 3. The proof is organized in three Steps.

Step 1: A_n is bounded.
By using Hölder type inequality and (A_5) with constant exponent p_+, we deduce that A is bounded.

Let $u \in W^{1,p^+}(\Omega)$

$$\begin{aligned}\langle B_n u, v\rangle &= \int_\Omega T_n(b(u))v dx + \int_\Omega \phi(T_n(u))\nabla v dx + \varepsilon \int_\Omega \left(|\nabla u|^{p_+-2}\nabla u \nabla v + |u|^{p_+-2} uv\right) dx, \\ &\leq \int_\Omega |b(u)||v| dx + \left(\int_\Omega |\phi(T_n(u))|^{p'_+} dx\right)^{\frac{1}{p'_+}} \|\nabla v\|_{p_+} + \varepsilon \int_\Omega \left(|\nabla u|^{p_+-1}\nabla v + |u|^{p_+-1} v\right) dx, \\ &\leq C\|v\|_1 + \left(\int_\Omega \sup_{|s|\leq n} |\phi(s)|^{p'_+} dx\right)^{\frac{1}{p'_+}} \|\nabla v\|_{p_+} + \varepsilon\left(\|\nabla u\|_{p_+}^{\frac{p_+}{p'_+}} \|\nabla v\|_{p_+} + \|u\|_{p_+}^{\frac{p_+}{p'_+}} \|v\|_{p_+}\right), \\ &\leq C\|v\|_{1,p_+} + \left(\int_\Omega (\sup_{|s|\leq n} |\phi(s)|+1)^{p'_+} dx\right)^{\frac{1}{p'_+}} \|\nabla v\|_{p_+} + \varepsilon\left(\|\nabla u\|_{p_+}^{\frac{p_+}{p'_+}} \|\nabla v\|_{p_+} + \|u\|_{p_+}^{\frac{p_+}{p'_+}} \|v\|_{p_+}\right), \\ &\leq C\|v\|_{1,p_+} + (\sup_{|s|\leq n} |\phi(s)|+1)(meas(\Omega)+1)^{p'_+} \|\nabla v\|_{p_+} + \varepsilon\left(\|\nabla u\|_{p_+}^{\frac{p_+}{p'_+}} \|\nabla v\|_{p_+} + \|u\|_{p_+}^{\frac{p_+}{p'_+}} \|v\|_{p_+}\right), \\ &\leq C(n)\|v\|_{1,p_+}.\end{aligned}$$

We deduce that A_n is bounded.

Step 2: A_n is coercive.
We have

$$\begin{aligned}\langle Au, u\rangle &= \int_\Omega a(x,u,\nabla u)\nabla u dx,\\ &\geq \frac{1}{C_2}\int_\Omega |\nabla u|^{p_+}dx,\\ &\geq C_3\|u\|^{p_+}_{1,p_+}.\end{aligned}$$

Therefore,

$$\frac{\langle A_n u, u\rangle}{\|u\|_{1,p_+}} \longrightarrow +\infty \quad \text{as} \quad \|u\|_{1,p_+} \longrightarrow +\infty.$$

Step 3: A_n is pseudo-monotone.
Let $(u_k)_k$ be a sequence in $W^{1,p_+}(\Omega)$ such that

$$\begin{cases} u_k \rightharpoonup u \ \ in \ \ W^{1,p_+}(\Omega) \\ A_n u_k \rightharpoonup \chi \ \ in \ \ W^{-1,p'_+}(\Omega) \\ \limsup\limits_{k\longrightarrow+\infty}\langle Lu_k, u_k\rangle = \langle \chi, u\rangle. \end{cases}$$

We will prove that $\chi = A_n u$.

As $T_n(b(u_k))u_k \geq 0$, by Fatou's Lemma, we deduce that

$$\liminf_{k\longrightarrow\infty}\int_\Omega T_n(b(u_k))u_k dx \geq \int_\Omega T_n(b(u))udx.$$

Using the Lebesgue dominated convergence Theorem, we have

$$\begin{aligned}&\lim_{k\longrightarrow+\infty}\left(\int_\Omega T_n(b(u_k))vdx + \int_\Omega \phi(T_n(u_k))\nabla v dx + \varepsilon\int_\Omega\left(|\nabla u_k|^{p_+-2}\nabla u_k\nabla v + |u_k|^{p_+-2}u_k v\right)dx\right)\\ &\quad = \int_\Omega T_n(b(u))vdx + \int_\Omega \phi(T_n(u))\nabla v dx + \varepsilon\int_\Omega\left(|\nabla u|^{p_+-2}\nabla u\nabla v + |u|^{p_+-2}uv\right)dx.\end{aligned}$$

Therefore, for k large enough,

$$\begin{aligned}&T_n(b(u_k)) + \phi(T_n(u_k)) + \varepsilon\left(|\nabla u_k|^{p_+-2}\nabla u_k + |u_k|^{p_+-2}u_k\right)\\ &\rightharpoonup T_n(b(u)) + \phi(T_n(u)) + \varepsilon\left(|\nabla u|^{p_+-2}\nabla u + |u|^{p_+-2}u\right) \quad \text{in } L^{p'_+}(\Omega).\end{aligned}$$

Thus,

$$Au_k \rightharpoonup \chi - \left(T_n(b(u)) + \phi(T_n(u)) + \varepsilon\left(|\nabla u|^{p_+-2}\nabla u + |u|^{p_+-2}u\right)\right) \text{ in } L^{p'_+}(\Omega), \text{ as } k \rightarrow +\infty.$$

Let us Prove that A is of Type of Calculus of Variation
Let us set

$$a_1(u,v,w) = \int_\Omega a(x,u,\nabla u)\nabla w dx.$$

Then, $w \mapsto a_1(u,v,w)$ is continuous on $W^{1,p_+}(\Omega)$, Thus

$$a_1(u,v,w) = \langle A(u,v),w\rangle, \quad A(u,v) \in W^{-1,p'_+}(\Omega),$$

and verify,

$$A(u,u) = Au, \quad \text{where} \quad Au := -diva(x,u,\nabla u).$$

- As $A(u,.)$ is bounded, we prove that $v \mapsto A(u,v)$ is hemi-continuous from $W^{1,p_+}(\Omega) \to W^{-1,p'_+}(\Omega)$. Since $a(x,u,\nabla(v_1+tv_2)) \rightharpoonup a(x,u,\nabla v_1)$ in $L^{p'_+}(\Omega)$ as $t \to +\infty$ and $u,v_1,v_2 \in W^{1,p_+}(\Omega)$ then, $a_1(u,v_1+tv_2,w) \to a_1(u,v_1,w)$ as $t \to +\infty$. In the same manner we prove that $u \mapsto A(u,v)$ is hemi-continuous from $W^{1,p_+}(\Omega) \to W^{-1,p'_+}(\Omega)$.
  Moreover, for all $u,v \in W^{1,p_+}(\Omega)$, we have

$$\begin{aligned}\langle A(u,u)-A(u,v),u-v\rangle &= \langle A(u,u),u-v\rangle - \langle A(u,v),u-v\rangle \\ &= a_1(u,u,u-v) - a_1(u,v,u-v) \\ &= \int_\Omega a(x,u,\nabla u)\nabla(u-v)dx - \int_\Omega a(x,u,\nabla v)\nabla(u-v)dx \\ &= \int_\Omega (a(x,u,\nabla u) - a(x,u,\nabla v))\nabla(u-v)dx \geq 0.\end{aligned}$$

- Let us suppose that $u_k \rightharpoonup u$ in $W^{1,p_+}(\Omega)$ and $\langle A(u_k,u_k) - A(u_k,v),u_k - v\rangle \to 0$. We prove that

$$\forall v \in W^{1,p_+}(\Omega), \ A(u_k,v) \rightharpoonup A(u,v) \quad \text{in } W^{-1,p'_+}(\Omega).$$

  As $u_k \rightharpoonup u$, we have

$$a(x,u_k,\nabla v) \rightharpoonup a(x,u,\nabla v) \quad \text{in } L^{p'_+}(\Omega).$$

  Thus,

$$A(u_k,v) \rightharpoonup A(u,v) \quad \text{in } W^{-1,p'_+}(\Omega).$$

- We suppose that $u_k \rightharpoonup u$ in $W^{1,p_+}(\Omega)$ and $A(u_k,v) \rightharpoonup \Theta$ in $W^{-1,p'_+}(\Omega)$. We prove that

$$\langle A(u_k,v),u_k\rangle \to \langle \Theta,u\rangle.$$

  Then $a(x,u_k,\nabla v) \to a(x,u,\nabla v)$ in $L^{p'_+}(\Omega)$, and thus, $a_1(u_k,v,u_k) \to a_1(u,v,u)$. Therefore,

$$\langle A(u_k,v),u_k\rangle = a_1(u_k,v,u_k) \to \langle A(u,v),u\rangle \quad \text{and } \Theta = A(u,v).$$

  Hence, A is of type Calculus of variation. Therefore, A is of type (M).
  So, we have immediately

$$Au = \chi - \left(T_n(b(u)) + \phi(T_n(u)) + \varepsilon\left(|\nabla u|^{p_+-2}\nabla u + |u|^{p_+-2}u\right)\right).$$

  Therefore, we deduce that $A_n u = \chi$.

Thus, $T_n(f) \subset W^{-1,p'_+}(\Omega)$, there exist at least one solution $u_n \in W^{1,p_+}(\Omega)$ of the problem

$$\langle A_n u_n, v\rangle = \langle T_n(f), v\rangle \quad \forall v \in W^{1,p_+}(\Omega).$$

That's completes the proof of Theorem 3.

Part 2: A Priori Estimates and Convergence Results
This part is done in three steps, we make a priori estimates, some convergence results and other based on the Young measure and nonlinear weak-∗ convergence.

Step 1: A priori estimates.

Lemma 1. *Suppose that* $(A_3)-(A_6)$ *hold with variable exponent* $\pi_n(.)$ *and* $f_n \in L^\infty(\Omega)$. *Let* u_n *be a weak solution of* $(\mathscr{P}_n)$. *Then, for all* $k>0$,

$$\int_\Omega |\nabla T_k(u_n)|^{\pi_n(.)}dx \leq C(k+1), \tag{5}$$

$$\int_\Omega |T_n(b(u_n))|\,dx \leq C_3. \tag{6}$$

Proof of Lemma 1.
We take $v = T_k(u_n)$ as a test function in the weak formulation (4), we get

$$\begin{aligned}\int_\Omega T_n(b(u_n))\,T_k(u_n)dx + \int_\Omega a(x,u_n,\nabla u_n).\nabla T_k(u_n)dx + \int_\Omega \phi(T_n(u_n)).\nabla T_k(u_n)dx \\ +\varepsilon \int_\Omega \left(|\nabla u_n|^{p_+-2}\nabla u_n \nabla T_k(u_n) + |u_n|^{p_+-2}u_n T_k(u_n)\right)dx = \int_\Omega f_n T_k(u_n)dx.\end{aligned} \tag{7}$$

Since the first and the last term of the left hand side of (7) are nonnegative, we deduce that

$$\int_\Omega a(x,u_n,\nabla u_n).\nabla T_k(u_n)dx \leq \int_\Omega |\phi(T_n(u_n))|.|\nabla T_k(u_n)|dx + \int_\Omega f_n T_k(u_n)dx. \tag{8}$$

Thanks to Young's inequality, we obtain

$$\int_\Omega a(x,u_n,\nabla u_n).\nabla T_k(u_n)dx \leq C_1 \int_\Omega |\phi(T_n(u_n))|^{(\pi_n(.))'}dx + \frac{\alpha}{2}\int_\Omega |\nabla T_k(u_n)|^{\pi_n(.)}dx + k\|f\|_{L^1(\Omega)}. \tag{9}$$

By using (A_6), we get

$$\int_\Omega |\nabla T_k(u_n)|^{\pi_n(.)}dx \leq C_1 \left(\sup_{|s|\leq n}|\phi(s)|+1\right)^{p'_+} dx + k\|f\|_{L^1(\Omega)}. \tag{10}$$

Hence,

$$\int_\Omega |\nabla T_k(u_n)|^{\pi_n(.)}dx \leq C(k+1). \tag{11}$$

From (7), we obtain

$$\begin{aligned}\int_\Omega T_n(b(u_n))\,T_k(u_n)\,dx &\le \int_\Omega f_n T_k(u_n)\,dx + \int_\Omega |\phi(T_n(u_n))|.|\nabla T_k(u_n)|dx\\ &\le k\|f\|_{L^1(\Omega)} + C(k).\end{aligned}$$

Dividing by k and letting k goes to 0, we obtain

$$\int_\Omega T_n(b(u_n))\,\mathrm{sign}_0(u_n)\,dx \le C_3.$$

Hence,

$$\int_\Omega |T_n(b(u_n))|\,dx \le C_3.$$

Lemma 2. *Assume that* $(A_3)-(A_6)$ *hold. If* u_n *is a weak solution of the problem* $(\mathscr{P}_n)$ *and* $f_n \in L^\infty(\Omega)$. *Then, for all* $k > 0$,

$$\int_\Omega |\nabla T_k(u_n)|^{p_-}dx \le C'(k+1). \tag{12}$$

Proof of Lemma 2
We have

$$\int_\Omega |\nabla T_k(u_n)|^{\pi_n(.)}dx \le Ck + C'. \tag{13}$$

we know that

$$\begin{aligned}\int_\Omega |\nabla T_k(u_n)|^{p_-}dx &= \int_{\{|\nabla T_k(u_n)|>1\}} |\nabla T_k(u_n)|^{p_-}dx + \int_{\{|\nabla T_k(u_n)|\le 1\}} |\nabla T_k(u_n)|^{p_-}dx\\ &\le \int_{\{|\nabla T_k(u_n)|>1\}} |\nabla T_k(u_n)|^{p_-}dx + meas(\Omega)\\ &\le \int_\Omega |\nabla T_k(u_n)|^{\pi_n(.)}dx + meas(\Omega)\end{aligned}$$

By using (13), we get

$$\int_\Omega |\nabla T_k(u_n)|^{p_-}dx \le C'(k+1). \tag{14}$$

Lemma 3. *For any* $k > 0$, *we have*

$$\|T_k(u_n)\|_{1,\pi_n(.)} \le C(k+1), \tag{15}$$

and for all $k \ge 1$,

$$meas(\{|u_n > k|\}) \le C. \tag{16}$$

Proof of Lemma 3
By (5), we have

$$\int_\Omega |\nabla T_k(u_n)|^{\pi_n(.)}dx \le C(k+1). \tag{17}$$

We also have

$$\begin{aligned}\int_\Omega |T_k(u_n)|^{\pi_n(.)}dx &= \int_{\{|u_n|\le k\}} |T_k(u_n)|^{\pi_n(.)}dx + \int_{\{|u_n|>k\}} |T_k(u_n)|^{\pi_n(.)}dx\\ &\le \int_{\{|u_n|\le k\}} k^{\pi_n(.)}dx + \int_{\{|u_n|>k\}} k^{\pi_n(.)}dx\\ &\le (1+k^{p_+})meas(\Omega) + (1+k^{p_+})meas(\Omega)\\ &\le 2(1+k^{p_+})meas(\Omega).\end{aligned}$$

We deduce that

$$\rho_{1,\pi_n(.)}(T_k(u_n)) \le C(k+1) + 2(1+k^{p_+})meas(\Omega).$$

For $\|T_k(u_n)\|_{1,\pi_n(.)} \ge 1$, we have

$$\|T_k(u_n)\|^{p_-}_{1,\pi_n(.)} \le \rho_{1,\pi_n(.)}(T_k(u_n)) \le C(k+1) + 2(1+k^{p_+})meas(\Omega),$$

which implies that

$$\begin{aligned}\|T_k(u_n)\|_{1,\pi_n(.)} &\le [C(k+1) + 2(1+k^{p_+})meas(\Omega)]^{\frac{1}{p_-}}\\ &:= C(k,f,p_+,p_-,meas(\Omega)).\end{aligned}$$

Therefore

$$\|T_k(u_n)\|_{1,\pi_n(.)} \le 1 + C(k,f,p_+,p_-,meas(\Omega)).$$

From the Lemma 3, we deduce that the sequence $(T_k(u_n))_{n\in\mathbb{N}}$ is uniformly bounded in $W^{1,\pi_n(.)}(\Omega)$ and also in $W^{1,p_-}(\Omega)$. Then there exists a subsequence still denoted $T_k(u_n)$, we can assume that for any $k>0$, $T_k(u_n)$ weakly converges to s_k in $W^{1,p_-}(\Omega)$ and also $T_k(u_n)$ strongly conerges to s_k in $L^{p_-}(\Omega)$. **Step 2: The convergence results.**

The proof of the following proposition use the Lemma 3.

Proposition 4. *We suppose that $(A_3)-(A_6)$ hold and let u_n be a weak solution of the problem $(\mathscr{P}_n)$, then the sequence $(u_n)_{n\in\mathbb{N}}$ is Cauchy in measure. In particular, there exists a measurable function u and a subsequence still denoted u_n such that $u_n \to u$ in measure, as $n\to\infty$.*

As $(u_n)_n \in \mathbb{N}$ is a Cauchy sequence in measure, so (up to a subsequence) it converges almost everywhere to some measurable function u.
As for any $k>0$, T_k is continuous, then $T_k(u_n)\to T_k(u)$ a.e $x\in\Omega$, so $s_k = T_k(u)$.
Therefore,

$$T_k(u_n) \rightharpoonup T_k(u) \text{ in } W^{1,p_-}(\Omega).$$

Hence,

$$T_k(u_n) \to T_k(u) \text{ in } L^{p_-}(\Omega) \text{ and a.e in } \Omega.$$

Step 3: The convergence in term of Young measure.
The following assertions are based on the Young measure and nonlinear weak$-*$ convergence results.

Assertion 1.
The sequence $(\nabla T_k(u_n))_{n\in\mathbb{N}}$ converges to a Young measure $\nu_x^k(\lambda)$ on $\mathbb{R}^N$ in the sense of the nonlinear weak$-*$ convergence and

$$\nabla T_k(u) = \int_{\mathbb{R}^N} \lambda d\nu_x^k(\lambda). \tag{18}$$

Proof.
Using Lemma 2, $\nabla T_k(u_n)$ is uniformly bounded in $L^{p_-}(\Omega)$, so, equi-integrable on Ω. Moreover, $\nabla T_k(u_n)$ weakly converges to $\nabla T_k(u)$ in $L^{p_-}(\Omega)$. Therefore, using the representation of weakly convergence sequences in $L^1(\Omega)$ in terms of Young measures, we can write

$$\nabla T_k(u) = \int_{\mathbb{R}^N} \lambda d\nu_x^k(\lambda).$$

Assertion 2.
$|\lambda|^{\pi(.)}$ is integrable with respect to the measure $\nu_x^k(\lambda)dx$ on $\mathbb{R}^N \times \Omega$, moreover, $T_k(u) \in W^{1,\pi(.)}(\Omega)$.

Proof.
We know that $p(.,u_n(.)) \to p(.,u(.))$ in measure on Ω. Using Theorem 1 *(ii)*, *(iii)* $(p(.,u_n(.)), \nabla T_k(u_n))_{n\in\mathbb{N}}$ converges on $\mathbb{R}\times\mathbb{R}^N$ to Young measure $\mu_x^k = \delta_{\pi(x)} \otimes \nu_x^k$. Thus, we can apply the weak convergence properties (22) to the Carathodory function $F_m(x,\lambda_0,\lambda) \in \Omega \times (\mathbb{R}\times\mathbb{R}^N) \mapsto |h_m(\lambda)|^{\lambda_0}$ with $m \in \mathbb{N}$. Then, we obtain

$$\begin{aligned}
\int_{\Omega\times\mathbb{R}^N} |h_m(\lambda)|^{\pi(x)} d\nu_x^k(\lambda)dx &= \int_{\Omega\times(\mathbb{R}\times\mathbb{R}^N)} |h_m(\lambda)|^{\lambda_0} d\mu_x^k(\lambda_0,\lambda)\,dx \\
&= \int_{\Omega}\int_{\mathbb{R}\times\mathbb{R}^N} F_m(x,\lambda_0,\lambda)\, d\mu_x^k(\lambda_0,\lambda)\,dx \\
&= \lim_{n\to+\infty} \int_{\Omega} F_m(x, p(x,u_n(x)), \nabla T_k(u_n(x)))\,dx \\
&= \lim_{n\to+\infty} \int_{\Omega} |h_m(\nabla T_k(u_n))|^{p(.,u_n(\cdot))}\,dx \\
&\le \lim_{n\to+\infty} \int_{\Omega} |\nabla T_k(u_n)|^{p(,u_n(\cdot))}\,dx \\
&\le C(k+1).
\end{aligned}$$

$h_m(\lambda) \to \lambda$, as $m \to +\infty$ and $m \mapsto h_m(\lambda)$ is increasing. Then, using Lebesgue convergence Theorem, we deduce from last inequality that

$$\int_{\Omega\times\mathbb{R}^N} |\lambda|^{\pi(x)} d\nu_x^k(\lambda)dx \le C(k+1).$$

Hence, $|\lambda|^{\pi(\cdot)}$ is integrable with respect to the measure $\nu_x^k(\lambda)dx$ on $\mathbb{R}^N \times \Omega$. From (18), the last inequality and Jensen inequality, we get

$$\int_\Omega |\nabla T_k(u)|^{\pi(x)}\, dx = \int_\Omega \left| \int_{\mathbb{R}^N} \lambda d\nu_x^k(\lambda) \right|^{\pi(x)} dx \leq \int_{\Omega\times\mathbb{R}^N} |\lambda|^{\pi(x)} d\nu_x^k dx < \infty$$

Thus, $\nabla T_k(u) \in L^{\pi(\cdot)}(\Omega)$. Moreover, $\int_\Omega |T_k(u)|^{\pi(\cdot)}\, dx \leq \max(k^{p_+}, k^{p_-})\,\mathrm{meas}(\Omega)$. Hence, $T_k(u) \in L^{\pi(\cdot)}(\Omega)$ and we conclude that $T_k(u) \in W^{1,\pi(\cdot)}(\Omega)$.

Assertion 3.

i) The sequence $\left(\Phi_n^k\right)_{n\in\mathbb{N}}$ defined by $\Phi_n^k := a\left(x, u_n, \nabla T_k\left(u_n\right)\right)$ is equi-integrable on Ω.

ii) The sequence $\left(\Phi_n^k\right)_{n\in\mathbb{N}}$ weakly converges to Φ^k in $L^1(\Omega)$ and we have

$$\Phi^k(x) = \int_{\mathbb{R}^N} a(x,u,\lambda) d\nu_x^k(\lambda). \tag{19}$$

Proof.

i) Using the growth assumption (A_5) with variable exponent $p\left(., u_n(.)\right)$ and relation (3.4), we deduce that $\left(\Phi_n^k\right)$ is bounded in $L^{\pi_n'(\cdot)}(\Omega)$, so, $L^{\pi_n'(\cdot)}-$ equi-integrable on Ω. Moreover, as $\pi_n'(.) > 1$, we obtain

$$\left|a\left(x, u_n, \nabla T_k\left(u_n\right)\right)\right| \leq 1 + \left|a\left(x, u_n, \nabla T_k\left(u_n\right)\right)\right|^{\pi_n'(\cdot)}.$$

Thus, for all subset $E \subset \Omega$, we have

$$\int_\Omega \left|a\left(x, u_n, \nabla T_k\left(u_n\right)\right)\right| dx \leq meas(E) + \int_\Omega \left|a\left(x, u_n, \nabla T_k\left(u_n\right)\right)\right|^{\pi_n'(\cdot)} dx.$$

Therefore, for $meas(E)$ small enough, $\left(\Phi_n^k\right)$ is equi-integrable on Ω.

ii) We set $\tilde{\Phi}_n^k = a\left(x, u(x), \nabla v_n\right)$ with $\nabla v_n = \nabla T_k\left(u_n\right) \cdot \chi_{S_n}$ where $S_n = \left\{x \in \Omega, |\pi(x) - \pi_n(x)| < \frac{1}{2}\right\}$.
Applying (A_5) with variable exponent $\pi(.)$ on $a\left(x, u(x), \nabla v_n\right)$, we have for all subset $E \subset \Omega$

$$\begin{aligned}\int_E \left|a\left(x, u(x), \nabla v_n\right)\right| dx &\leq C \int_E \left(1 + \mathscr{M}(x) + |\nabla v_n|^{\pi(\cdot)-1}\right) dx \\ &\leq C \int_E (1 + \mathscr{M}(x)) dx + \int_{E\cap S_n} \left|\nabla T_k\left(u_n\right)\right|^{\pi(\cdot)-1} dx\end{aligned}$$

The first term of the right hand side of the last inequality is small for $meas(E)$ small enough. For $x \in S_n, \pi(x) < \pi_n(x) + \frac{1}{2}$, thus

$$\int_{E\cap S_n} \left|\nabla T_k\left(u_n\right)\right|^{\pi(\cdot)-1} dx \leq \int_{E\cap S_n} \left(1 + \left|\nabla T_k\left(u_n\right)\right|^{\pi_n(\cdot)-\frac{1}{2}}\right) dx$$

and

$$\int_{\Omega} |\nabla T_k(u_n)|^{\left(\pi_n(\cdot)-\frac{1}{2}\right)(2\pi_n(\cdot))'}\, dx = \int_{\Omega} |\nabla T_k(u_n)|^{\pi_n(\cdot)}\, dx < \infty,$$

which is equivalent to saying $|\nabla T_k(u_n)|^{\pi_n(\cdot)-\frac{1}{2}} \in L^{(2\pi_n(\cdot))'}(\Omega)$. Now, using Hölder type inequality,

$$\begin{aligned}\int_{E\cap S_n} |\nabla T_k(u_n)|^{\pi(\cdot)-1}\, dx &\le \int_E \left(1+|\nabla T_k(u_n)|^{\pi_n(\cdot)-\frac{1}{2}}\right) dx \\ &\le meas(E) + 2\|\nabla T_k(u_n)\|_{L^{\pi_n(\cdot)}(\Omega)} \|\chi_E\|_{L^{2\pi_n(\cdot)}(\Omega)}.\end{aligned} \tag{20}$$

From Proposition 1

$$\begin{aligned}\|\chi_E\|_{L^{2\pi_n(\cdot)}(\Omega)} &\le \max\left\{ \left(\rho_{2\pi_n(\cdot)}(\chi_E)\right)^{\frac{1}{2P-}}, \left(\rho_{2\pi_n(\cdot)}(\chi_E)\right)^{\frac{1}{2P+}} \right\} \\ &= \max\left\{ (\text{meas}(E))^{\frac{1}{2P-}}, (\text{meas}(E))^{\frac{1}{2P+}} \right\}\end{aligned}$$

The right-hand side of (20) is uniformly small for $meas(E)$ small, and the equi-integrability of $\tilde{\Phi}_n^k$ follows. Therefore, (up to a subsequence) $\tilde{\Phi}_n^k$ weakly converges in $L^1(\Omega)$ to $\tilde{\Phi}^k$, as $n \to +\infty$.

Now, we prove that $\tilde{\Phi}^k = \Phi^k$; more precisely, we show that $\tilde{\Phi}_n^k - \Phi_n^k$ strongly converges in $L^1(\Omega)$ to 0. Let $\beta > 0$, by (3.4), $\int_{\Omega} |\nabla T_k(u_n)|^{\pi_n(\cdot)}\, dx$ is uniformly bounded, which implies that $\int_{\Omega} |\nabla T_k(u_n)|\, dx$ is finite, since

$$\int_{\Omega} |\nabla T_k(u_n)|\, dx \le \int_{\Omega} \left(1+|\nabla T_k(u_n)|^{\pi_n(x)}\right) dx.$$

By Chebyschev Inequality, we have

$$\text{meas}(\{|\nabla T_k(u_n)| > L\}) \le \frac{\int_{\Omega} |\nabla T_k(u_n)|\, dx}{L}.$$

Therefore, $\sup_{n\in\mathbb{N}} \text{meas}(\{|\nabla T_k(u_n)| > L\})$ tends to 0 for L large enough. Since $\tilde{\Phi}_n^k - \Phi_n^k$ is equiintegrable, there exists $\delta = \delta(\beta)$ such that for all $A \subset \Omega, \text{meas}(A) < \delta$ and $\int_A |\tilde{\Phi}_n^k - \Phi_n^k|\, dx < \frac{\beta}{4}$. Therefore, if we choose L large enough, we get $\frac{\int_{\Omega}|\nabla T_k(u_n)|dx}{L} < \delta$, so meas $(\{|\nabla T_k(u_n)| > L\}) < \delta$. Hence,

$$\int_{\{|\nabla T_k(u_n)|>L\}} \left|\tilde{\Phi}_n^k - \Phi_n^k\right| dx < \frac{\beta}{4}$$

By Lemma 2.21, we also have

$$\text{meas}\left(\left\{x \in \Omega; \sup_{\lambda\in K} |a(x,u_n(x),\lambda) - a(x,u(x),\lambda)| \ge \sigma\right\}\right) \longrightarrow 0$$

as $n \to +\infty$. Thus, by the above equi-integrability, for all $\sigma > 0$, there exists $n_0 = n_0(\sigma, L) \in \mathbb{N}$ such that for all $n \geq n_0$,

$$\int_{\{x\in\Omega;\sup_{|\lambda|\leq L}|a(x,u_n(x),\lambda)-a(x,u(x),\lambda)|\geq\sigma\}} \left|\tilde{\Phi}_n^k - \Phi_n^k\right| dx < \frac{\beta}{4}$$

Using the definition of Φ_n^k and $\tilde{\Phi}_n^k$, we have

$$\Phi_n^k - \tilde{\Phi}_n^k = a\left(x, u_n(x), \nabla T_k\left(u_n\right)\right) - a\left(x, u(x), \nabla T_k\left(u_n\right)\right) \text{on} S_n$$

Now, we reason on

$$S_{n,L,\sigma} := \left\{x \in \Omega;\ \sup_{|\lambda|\leq L} \left|a\left(x, u_n(x), \lambda\right) - a(x, u(x), \lambda)\right| < \sigma, \left|\nabla T_k\left(u_n\right)\right| \leq L\right\}$$

$$\begin{aligned}\int_{S_{n,L,\sigma}} \left|\tilde{\Phi}_n^k - \Phi_n^k\right| dx &\leq \int_{S_{n,L,\sigma}} \sup_{|\lambda|\leq L}\sup_n \left|a\left(x, u_n(x), \lambda\right) - a(x, u(x), \lambda)\right| dx \\ &\leq \sigma meas(\Omega).\end{aligned}$$

We observe that

$$\int_{S_n} \left|\tilde{\Phi}_n^k - \Phi_n^k\right| dx = \int_{S_n\cap S_{n,L,\sigma}} \left|\tilde{\Phi}_n^k - \Phi_n^k\right| dx + \int_{S_n\backslash S_{n,L,\sigma}} \left|\tilde{\Phi}_n^k - \Phi_n^k\right| dx$$

and

$$S_n \backslash S_{n,L,\sigma} \subset \left\{x \in \Omega;\ \sup_{|\lambda|\leq L} \left|a\left(x, u_n(x), \lambda\right) - a(x, u(x), \lambda)\right| \geq \sigma\right\} \cup \left\{\left|\nabla T_k\left(u_n\right)\right| > L\right\}.$$

Consequently, by choosing $\sigma = \sigma(\beta) < \frac{\beta}{4\,\text{meas}(\Omega)}$, we get

$$\int_{S_n} \left|\tilde{\Phi}_n^k - \Phi_n\right| dx < \frac{\beta}{4} + \frac{\beta}{4} + \frac{\beta}{4} = \frac{3\beta}{4}$$

for all $n \geq n_0(\sigma, L)$. Since $meas\left(\left\{x \in \Omega, \left|\pi(x) - \pi_n(x)\right| \geq \frac{1}{2}\right\}\right) \to 0$ for n large enough; which means that $meas\left(\Omega \backslash S_n\right)$ converges to 0 for n large enough. Thus,

$$\int_{\Omega\backslash S_n} \left|\tilde{\Phi}_n^k - \Phi_n^k\right| dx = \int_{\Omega\backslash S_n} \left|\Phi_n^k\right| dx \leq \frac{\beta}{4}$$

Therefore, for all $\beta > 0$ there exists $n_0 = n_0(\beta)$ such that for all $n \geq n_0$, $\int_{\Omega} \left|\tilde{\Phi}_n^k - \Phi_n^k\right| dx \leq \beta$. Hence, $\tilde{\Phi}_n^k - \Phi_n^k$ strongly converges to 0 in $L^1(\Omega)$. We prove that

$$\Phi^k(x) = \int_{\mathbb{R}^N} a(x, u(x), \lambda) d\nu_x^k(\lambda) \quad \text{a.e. } x \in \Omega \text{ and } \Phi^k \in L^{\pi'(\cdot)}(\Omega).$$

Notice that

$$\lim_{n\to+\infty}\int_\Omega |\nabla T_k(u_n)|\,(1-\chi_{S_n})\,dx = \lim_{n\to+\infty}\int_{\Omega\backslash S_n} |\nabla T_k(u_n)|\,dx = 0$$

since $(\nabla T_k(u_n))_{n\in\mathbb{N}}$ is equi-integrable and meas$(\Omega\backslash S_n)$ converges to 0 for n large enough. Therefore, $(\nabla T_k(u_n))_{n\in\mathbb{N}}$ and $\nabla T_k(u_n)\chi_{S_n}$ converge to the same Young measure $\nu_x^k(\lambda)$. Moreover, by applying Theorem 1 i) to the Carathodory function $F(x,(\lambda_0,\lambda)) := a(x,\lambda_0,\lambda)$, we infer that

$$\tilde{\Phi}(x) = \Phi(x) = \int_{\mathbb{R}^N} a(x,u(x),\lambda)d\nu_x^k(\lambda)\text{a.e. } x\in\Omega.$$

Using (A_5), it follows that $|a(x,u(x),\lambda)|^{\pi'(.)} \le C\left(\mathscr{M}(x)+|\lambda|^{\pi(\cdot)}\right)$. Thus, with Jensen Inequality, it follows that

$$\begin{aligned}\int_\Omega \left|\Phi^k(x)\right|^{\pi'(.)} dx &= \int_\Omega \left|\int_{\mathbb{R}^N} a(x,u(x),\lambda)d\nu_x^k(\lambda)\right|^{\pi'(\cdot)} dx\\ &\le \int_{\Omega\times\mathbb{R}^N} |a(x,u(x),\lambda)|^{\pi'(\cdot)}d\nu_x^k(\lambda)dx\\ &\le C\int_{\Omega\times\mathbb{R}^N}\left(\mathscr{M}(x)+|\lambda|^{\pi(\cdot)}\right)d\nu_x^k(\lambda)dx < \infty\end{aligned}$$

Hence, $\Phi^k \in L^{\pi'(\cdot)}(\Omega)$.

Assertion 4.

(a) For all $k' > k > 0$, we have $\Phi^k = \Phi^{k'}\chi_{\{|u|<k\}}$.
(b) For all $k>0$,

$$\int_\Omega \Phi^k\cdot\nabla T_k(u)dx \ge \int_{\Omega\times\mathbb{R}^N} a(x,u(x),\lambda)\cdot\lambda d\nu_x^k(\lambda)dx. \tag{21}$$

(c) The div-curl inequality holds:

$$\int_{\Omega\times\mathbb{R}^N}\left(a(x,u(x),\lambda)-a(x,u(x),\nabla T_k(u(x)))\,(\lambda-\nabla T_k(u(x)))\,d\nu_x^k(\lambda)dx \le 0.\right. \tag{22}$$

(d) For all $k>0$,

$$\Phi^k = a(x,u(x),\nabla T_k(u))\text{ for a.e. } x\in\Omega$$

and $\nabla T_k(u_n)$ converges to $\nabla T_k(u)$ in measure on Ω, as $n\to+\infty$.

Proof.

(a) Let $k' > k > 0$ and $g_n^k := a(x, u_n, \nabla T_{k'}(u_n))\chi_{[|u|<k]}$. By Assertion 3-ii), $(g_n^k)_{n\in\mathbb{N}}$ weakly con- verges to $\Phi^{k'}\chi_{[|u|<k]}$ in $L^1(\Omega)$. If we prove that $(g_n^k)_{n\in\mathbb{N}}$ weakly converges to Φ^k in $L^1(\Omega)$, then the wished result will come of the uniqueness of the limit. Let us put

$$h_n^k := a(x, u_n, \nabla T_{k'}(u_n))\chi_{[|u_n|<k]}$$

As $\nabla T_k(u_n) \equiv \nabla T_{k'}(u_n)\chi_{[|u_n|<k]}$, for all $k' > k > 0$, then, we get

$$h_n^k := a(x, u_n, \nabla T_{k'}(u_n))\chi_{[|u_n|<k]} \equiv a(x, u_n, \nabla T_k(u_n))$$

so, $(h_n^k)_{n\in\mathbb{N}}$ weakly converges to Φ^k in $L^1(\Omega)$ by Assertion 3-ii). Set

$$d_n^k := g_n^k - h_n^k = a(x, u_n, \nabla T_{k'}(u_n))\left(\chi_{[|u|<k]} - \chi_{[|u_n|<k]}\right)$$

On the one hand, thanks to Assertion 3-i), $(d_n^k)_{n\in\mathbb{N}}$ is equi-integrable. On the other hand $d_n^k \to 0$ a.e. on Ω. Indeed, $\chi_{[|u_n|<k]} = \chi_{(-k,k)}(u_n)$ and if $|u_n| \neq k$ a.e. on $\Omega, \chi_{(-k,k)}(.)$ is continuous on $\mathbb{R}$. In other words $\chi_{(-k,k)}(.)$ is continuous on the image of Ω by u a.e. $k > 0$. Moreover, $u_n \to u$ a.e. on Ω, then $\chi_{[|u_n|<k]} \to \chi_{[|u|<k]}$ a.e. in Ω. Now, using Vitali's Theorem $(d_n^k)_{n\in\mathbb{N}}$ strongly converges to 0 in $L^1(\Omega)$, so it weakly converges in $L^1(\Omega)$. Hence, $(g_n^k)_{n\in\mathbb{N}}$ and $(h_n^k)_{n\in\mathbb{N}}$ weakly converge to the same limit Φ^k in $L^1(\Omega)$.

(b) Let $\mathscr{S}$ be a set of $W^{2,\infty}$ functions $S : \mathbb{R} \to \mathbb{R}$ such that $S'(.)$ has a compact support. We construct a sequence $(S_M)_{M\in\mathbb{N}} \subset \mathscr{S}$ such that
- S'_M and S''_M are uniformly bounded;
- for all $M \in \mathbb{N}, S'_M = 1$ on $[-M+1, M-1]$, $\operatorname{supp} S' \subset [-M, M]$;
- the sequence $(b(z)S'_M(z))_{M\in\mathbb{N}}$ is non-decreasing for all $z \in \mathbb{R}$.

For all $\varphi \in C^\infty(\bar{\Omega}), v = \varphi S'_M(u_n)$ is an admissible test function in the weak formulation (4). We have

$$\begin{aligned}
&\int_\Omega T_n(b_n(u_n))S'_M(u_n)\varphi dx + \int_\Omega S'_M(u_n)\, a(x, u_n, \nabla T_M(u_n)).\nabla\varphi dx \\
&+ \int_\Omega S''_M(u_n)\, a(x, u_n, \nabla T_M(u_n)) \cdot \nabla T_M(u_n)\varphi dx + \int_\Omega S'_M(u_n)\,\phi(T_M(u_n)).\nabla\varphi dx \\
&+ \int_\Omega S''_M(u_n)\,\phi(T_M(u_n)) \cdot \nabla T_M(u_n)\varphi dx \\
&+ \varepsilon\int_\Omega \left[|\nabla u_n|^{p+-2}\nabla u_n \nabla(\varphi S'_M(u_n)) + |u_n|^{p+-2}u_n S'_M(u_n)\varphi\right] dx \\
&= \int_\Omega f_n S'_M(u_n)\varphi dx.
\end{aligned} \tag{23}$$

Since u_n converges to u a.e. in Ω, by continuity of b, S'_M and the compacteness of $\operatorname{supp} S'_M$, we obtain

$$\int_\Omega T_n(b(u_n))\, S'_M(u_n)\varphi dx \to \int_\Omega b(u)S'_M(u)\varphi dx, \text{as } n \to +\infty. \tag{24}$$

Moreover, we have $|f_n S'_M(u_n)\varphi| \le \|S'_M\|_{L^\infty(\mathbb{R}}|f||\varphi| \in L^1(\Omega)$, $f_n S'_M(u_n)\varphi \to f S'_M(u)\varphi$ a.e. in Ω. Thus, by Lebesgue dominated convergence Theorem

$$\int_\Omega f_n S'_M(u_n)\varphi dx \to \int_\Omega f S'_M(u)\varphi dx, \text{ as } n \to +\infty \tag{25}$$

Let us prove now, that

$$\int_\Omega S'_M(u_n)\, a(x,u_n,\nabla T_M(u_n))\cdot\nabla\varphi dx \to \int_\Omega S'_M(u)\Phi^M\cdot\nabla\varphi dx, \text{as } n\to+\infty \tag{26}$$

For all $L>0$, we have

$$\begin{aligned}\int_\Omega S'_M(u_n)\, a(x,u_n,\nabla T_M(u_n))\cdot\nabla\varphi dx &= \int_{\{|\nabla\varphi|\le L\}} S'_M(u_n)\,\Phi_n^M\cdot\nabla\varphi dx\\ &+\int_{\{|\nabla\varphi|>L\}} S'_M(u_n)\,\Phi_n^M\cdot\nabla\varphi dx\end{aligned} \tag{27}$$

For the first term of the right-hand side of (27), we have

$$\int_{\{|\nabla\varphi|\le L\}} S'_M(u_n)\,\Phi_n^M\cdot\nabla\varphi dx \to \int_{\{|\nabla\varphi|\le L\}} S'_M(u)\Phi^M\cdot\nabla\varphi dx, \text{as } n\to+\infty \tag{28}$$

Thanks $\Phi_n^M \to \Phi^M$ in $L^1(\Omega)$ and $\nabla\varphi S'_M(u_n)\chi_{\{|\nabla\varphi|\le L\}} \to^* \nabla\varphi S'_M(u)\chi_{\{|\nabla\varphi|\le L\}}$ in $L^\infty(\Omega)$. Furthermore, the second term of the right hand-side of (27) converges to zero for L large enough, uniformly in n. Indeed, using Hlder type inequality and the fact that $L^{p_+}(\Omega) \hookrightarrow L^{\pi_n(\cdot)}(\Omega)$, we get

$$\begin{aligned}&\left|\int_{\{|\nabla\varphi|>L\}} \Phi_n^M\nabla\varphi S'_M(u_n)\,dx\right|\\ &\le C\left\|S'_M\right\|_{L^\infty(\mathbb{R})}\left\|\Phi_n^M\right\|_{L^{\pi'(\cdot)}(\Omega)}\left\|\nabla\varphi\chi_{\{|\nabla\varphi|>L\}}\right\|_{L^{\pi}n()(\Omega)}\\ &\le C\left(p_-,\left\|S'_M\right\|_{L^\infty(\mathbb{R})},\text{meas}(\Omega)\right)\left\|\Phi_n^M\right\|_{L^{\pi'_n(\cdot)}(\Omega)}\|\nabla\varphi\|_{L^{p_+}(\Omega)}\text{meas}(\{|\nabla\varphi|>L\})\end{aligned}$$

From (A_5), (5) we obtain

$$\|\Phi_M^n\|_{L^{\pi'_n(\cdot)}(\Omega)} < C.$$

Moreover, $\varphi \in C^\infty(\bar\Omega)$ and $C^\infty(\bar\Omega)$ is dense in the space $W^{1,p_+}(\Omega)$. Then, by the fact that $\lim_{L\to+\infty}\text{meas}(\{|\nabla\varphi|>L\})=0$, we get

$$\text{meas}(\{|\nabla\varphi|>L\})\left\|\Phi_n^M\right\|_{L^{\pi'_n}(\Omega)}\|\nabla\varphi\|_{L^{p_+}(\Omega)}\to 0, \quad \text{as } L\to+\infty.$$

Hence, the second term of the right hand-side of (27) converges to zero, as L tends to infinity. Thus, as $n\to+\infty$ and $L\to+\infty$ in (27), we deduce (26). Let us consider the third term of left hand-side of (23), we obtain

$$\begin{aligned}\int_\Omega |S''_M(u_n)|\, a(x,u_n,\nabla T_M(u_n))\cdot\nabla T_M(u_n)\varphi dx &\le C\int_{\{|u_n|<M\}} |S''_M(u_n)|\, a(x,u_n,\nabla T_M(u_n))\cdot\nabla T_M(u_n)\,dx\\ &\le C'\int_{\{M-1<|u_n|<M\}} a(x,u_n,\nabla T_M(u_n))\,\nabla T_M(u_n)\,dx\\ &+C\int_{\{|u_n|\le M-1\}}\underbrace{S''_M(u_n)}_{=0}|\, a(x,u_n,\nabla T_M(u_n))\cdot\nabla T_M(u_n)\,dx,\end{aligned} \tag{29}$$

where $C = C\left(\|\varphi\|_{L^\infty(\Omega)}\right)$, $C' = C\left(\|S''_M\|_{L^\infty(\mathbb{R})}, \|\varphi\|_{L^\infty(\Omega)}\right)$ and $a(x,u_n,\nabla T_M(u_n))\cdot\nabla T_M(u_n)$ is finite. Otherwise,

$$\int_{\{M-1<|u_n|<M\}} a(x,u_n,\nabla T_M(u_n))\nabla T_M(u_n)\,dx \to 0, \quad \text{as } M \to +\infty$$

Since, $\lim_{M\to+\infty} \operatorname{meas}(\{M-1<|u_n|<M\}) = 0$ and $a(x,u_n,\nabla T_M(u_n))\nabla T_M(u_n)$ is equi-integrable.

In the same way, we show that

$$\int_\Omega S'_M(u_n)\phi(T_M(u_n)).\nabla\varphi dx \to \int_\Omega S'_M(u)\phi(T_M(u))\cdot\nabla\varphi dx, \text{ as } n\to+\infty, \tag{30}$$

and

$$\int_\Omega S''_M(u_n)\phi(T_M(u_n))\cdot\nabla T_M(u_n)\varphi dx \to 0, \text{as } n\to+\infty \tag{31}$$

Finally, using (24), (25), (26), (29), (30), (31) and passing to the limit in (23), as n tends to infinty and as ε goes to 0, we obtain

$$\int_\Omega b(u)S'_M(u)\varphi dx + \int_\Omega S'_M(u)\Phi^M.\nabla\varphi dx + \int_\Omega S'_M(u)\phi(T_M(u))\cdot\nabla\varphi dx = \int_\Omega fS'_M(u)\varphi dx. \tag{32}$$

For $k>0$ fixed, $T_k(u)\in W^{1,\pi(\cdot)}(\Omega)$ and the exponent $\pi(.)$ verify the log-Hlder continuity assumption. Therefore, $C^\infty(\bar{\Omega})$ is dense in $W^{1,\pi(\cdot)}(\Omega)$, so, we replace φ by $T_k(u)$. Now, for $M>k$, thanks to (a), we replace $\Phi^M.\nabla T_k(u)$ by $\Phi^k\cdot\nabla T_k(u)$ in (32).

S'_M converges a.e. to 1 on $\mathbb{R}$, as $M\to+\infty$, then using the monotone convergence theorem in the first term of left hand-side of (32) and the dominated convergence theorem in the other terms of (32), we get

$$\int_\Omega \left[b(u)T_k(u)+\Phi^k\cdot\nabla T_k(u)\right]dx + \int_\Omega \phi(u)\cdot\nabla T_k(u)dx = \int_\Omega fT_k(u)dx \tag{33}$$

The relation (7) is equivalent to

$$\begin{aligned}&\int_\Omega T_n(b(u_n))T_k(u_n)dx + \int_\Omega a(x,u_n,\nabla u_n).\nabla T_k(u_n)dx + \int_\Omega \phi(T_n(u_n)).\nabla T_k(u_n)dx\\ &+\varepsilon\int_\Omega \left(|\nabla u_n|^{p_+-2}\nabla u_n\nabla T_k(u_n) + |u_n|^{p_+-2}u_nT_k(u_n)\right)dx = \int_\Omega f_nT_k(u_n)dx.\end{aligned} \tag{34}$$

By Fatou's Lemma, we have

$$\liminf_{n\to+\infty}\int_\Omega T_n(b(u_n))T_k(u_n)dx \geq \int_\Omega b(u)T_k(u)dx. \tag{35}$$

Now, we consider the right hand side of (34). We have $|f_nT_k(u_n)| \leq k|f| \in L^1(\Omega)$, $f_nT_k(u_n) \to fT_k(u)$ a.e. in Ω. Thus, by Lebesgue dominated convergence Theorem

$$\int_\Omega f_nT_k(u_n)dx \to \int_\Omega fT_k(u)dx, \text{ as } n\to+\infty. \tag{36}$$

Using (34), we get

$$\begin{aligned}
&\int_\Omega fT_k(u)dx - \left(\int_\Omega b(u)T_k(u)dx + \int_\Omega \phi(u)\cdot\nabla T_k(u)dx\right)\\
&\geq \liminf_{n\to+\infty}\int_\Omega a(x,u_n,\nabla T_k(u_n))\nabla T_k(u_n)\,dx\\
&+\varepsilon\int_\Omega\left[|u_n|^{p_+-2}u_nT_k(u_n)+|\nabla u_n|^{p_+-2}\nabla u_n\nabla T_k(u_n)\right]dx\\
&\geq \liminf_{n\to+\infty}\int_\Omega a(x,u_n,\nabla T_k(u_n))\nabla T_k(u_n)\,dx.
\end{aligned}$$

Thus, by using (33), we obtain

$$\int_\Omega \Phi^k\nabla T_k(u)dx \geq \liminf_{n\to+\infty}\int_\Omega a(x,u_n,\nabla T_k(u_n))\nabla T_k(u_n)\,dx \tag{37}$$

(c) We have $m\mapsto a(x,u_n,h_m(\nabla T_k(u_n)))\cdot h_m(\nabla T_k(u_n))$ is increasing and converges to $a(x,u_n,\nabla T_k(u_n)).\nabla T_k(u_n)$ for m large enough. Thus, we deduce that

$$a(x,u_n,h_m(\nabla T_k(u_n)))\cdot h_m(\nabla T_k(u_n)) \leq a(x,u_n,\nabla T_k(u_n))\cdot\nabla T_k(u_n) = \Phi_n^k\cdot\nabla T_k(u_n)$$

Therefore, using (b) and Theorem 1, we get

$$\begin{aligned}
\int_\Omega \Phi^k\cdot\nabla T_k(u)dx &\geq \liminf_{n\to+\infty}\int_\Omega \Phi_n^k\cdot\nabla T_k(u_n)\,dx\\
&\geq \lim_{n\to+\infty}\int_\Omega a(x,u_n,h_m(\nabla T_k(u_n)))\cdot h_m(\nabla T_k(u_n))\,dx\\
&= \int_{\Omega\times\mathbb{R}^N} a(x,u,h_m(\lambda))\cdot h_m(\lambda)d\nu_x^k(\lambda)dx
\end{aligned}$$

Using Lebesgue convergence Theorem, we get for m large enough

$$\int_\Omega \Phi^k\cdot\nabla T_k(u)dx \geq \int_{\Omega\times\mathbb{R}^N} a(x,u,\lambda)\cdot\lambda d\nu_x^k(\lambda)dx. \tag{38}$$

We have

$$\begin{aligned}
&\int_{\Omega\times\mathbb{R}^N}\left(a(x,u(x),\lambda)-a(x,u(x),\nabla T_k(u(x)))\,(\lambda-\nabla T_k(u(x)))\,d\nu_x^k(\lambda)dx\right.\\
&=\int_{\Omega\times\mathbb{R}^N} a(x,u(x),\lambda)\cdot\lambda d\nu_x^k(\lambda)dx-\int_{\Omega\times\mathbb{R}^N} a(x,u(x),\lambda)\cdot\nabla T_k(u(x))d\nu_x^k(\lambda)dx\\
&-\int_{\Omega\times\mathbb{R}^N} a(x,u(x),\nabla T_k(u(x)))\cdot\lambda d\nu_x^k(\lambda)dx+\int_{\Omega\times\mathbb{R}^N} a(x,u(x),\nabla T_k(u(x)))\cdot\nabla T_k(u(x))d\nu_x^k(\lambda)dx\\
&=\int_{\Omega\times\mathbb{R}^N} a(x,u(x),\lambda)\cdot\lambda d\nu_x^k(\lambda)dx-\int_\Omega\left(\int_{\mathbb{R}^N} a(x,u(x),\lambda)d\nu_x^k(\lambda)\right)\nabla T_k(u(x))dx\\
&-\int_\Omega a(x,u(x),\nabla T_k(u(x)))\cdot\left(\int_{\mathbb{R}^N}\lambda d\nu_x^k\right)dx+\int_\Omega a(x,u(x),\nabla T_k(u(x)))\cdot\nabla T_k(u(x))\left(\int_{\mathbb{R}^N}d\nu_x^k\right)dx\\
&=\int_{\Omega\times\mathbb{R}^N} a(x,u(x),\lambda)\cdot\lambda d\nu_x^k(\lambda)dx-\int_\Omega \Phi^k\cdot\nabla T_k(u(x))dx\leq 0.
\end{aligned}$$

We pass from the first equality to the second equality by using Fubini-Tonelli Theorem and from the second equality to the third one by using (18), (19) and the fact that ν_x is probability measures on $\mathbb{R}^N$. Finally (38) give us the desired inequality. (d) Using (22) and the strict monotonicity assumption (A_3), we deduce that

$$\left(a(x,u(x),\lambda)-a\left(x,u(x),\nabla T_k(u(x))\right)\left(\lambda-\nabla T_k(u(x))\right)\right)=0 \text{ a.e. } x\in\Omega,\quad \lambda\in\mathbb{R}^N.$$

Thus, $\lambda=\nabla T_k(u(x))$ a.e. $x\in\Omega$ with respect to the measure ν_x^k on $\mathbb{R}^N$. Therefore, the measure ν_x^k reduces to the Dirac measure $\delta_{\nabla T_k(u(x))}$. Using (3.20), we obtain

$$\Phi^k=\int_{\mathbb{R}^N} a(x,u(x),\lambda)d\nu_x^k(\lambda)=a\left(x,u(x),\nabla T_k(u(x))\right) \text{ a.e. } x\in\Omega$$

Now, by using Theorem 1 – (i) we deduce that $\nabla T_k(u_n)$ converges in measure to $\nabla T_k(u)$.

Part 3: Passage to the Limit

Lemma 4. *u is an entropy solution of* (1).

Proof.
Let u_n be a weak solution of the problem (3.2). Then, by Assertion 4 $-(d),(\nabla T_k(u_n))_{n\in\mathbb{N}}$ converges to $\nabla T_k(u)$ in measure, thus (up to a subsequence still denoted $(\nabla T_k(u_n))_{n\in\mathbb{N}}),(\nabla T_k(u_n))_{n\in\mathbb{N}}$ converges to $\nabla T_k(u)$ a.e. Ω. Moreover, we deduce from Lemma 3.4 that $\nabla T_k(u_n)$ is uniformly bounded in $L^{p_-}(\Omega)$, so, p_-– equi-integrable on Ω. Then, by using Vitali's Theorem $\nabla T_k(u_n)\to\nabla T_k(u)$ in $L^{p_-}(\Omega)$, which implies that $\nabla T_k(u_n)\to\nabla T_k(u)$ in $L^1(\Omega)$.

Furthermore, thanks to Assertion 2, $u\in\mathscr{T}^{1,\pi(.)}(\Omega)$.

Since $T_n(b(u_n))\to b(u)$ a.e. in Ω, it follows from Fatou's Lemma that

$$\int_\Omega |b(u)|\le\liminf_{n\to+\infty}\int_\Omega |T_n(b(u_n))|\,dx\le C_3.$$

Hence, $b(u)\in L^1(\Omega)$.

Let $\varphi\in\mathscr{C}^\infty(\bar{\Omega})$, then we can choose $T_k(u_n-\varphi)$ as a test function in (4) to get

$$\begin{aligned}\int_\Omega T_n(b(u_n))T_k(u_n-\varphi)\,dx&+\int_\Omega a(x,u_n,\nabla u_n).\nabla T_k(u_n-\varphi)\,dx+\int_\Omega \phi(T_n(u_n))\nabla T_k(u_n-\varphi)\,dx\\ &+\varepsilon\int_\Omega\left[|\nabla u_n|^{p_+-2}\nabla u_n\nabla T_k(u_n-\varphi)+|u_n|^{p_+-2}u_nT_k(u_n-\varphi)\right]dx\\ &=\int_\Omega f_nT_k(u_n-\varphi)\,dx.\end{aligned}\tag{39}$$

For the first term of the left hand side of (39), we have

$$\begin{aligned}\int_\Omega T_n(b(u_n))T_k(u_n-\varphi)\,dx=&\int_\Omega [T_n(b(u_n))-T_n(b(\varphi))]\,T_k(u_n-\varphi)\,dx\\ &+\int_\Omega T_n(b(\varphi))T_k(u_n-\varphi)\,dx.\end{aligned}$$

Using Fatou's Lemma, we obtain

$$\liminf_{n\to+\infty}\int_\Omega T_n(b(u_n))T_k(u_n-\varphi)\,dx\ge\int_\Omega b(u)T_k(u-\varphi)dx.\tag{40}$$

For the third term of the left hand side of (39), we prove that

$$\lim_{n\to+\infty} \varepsilon \int_\Omega \left[|\nabla u_n|^{p_+-2}\nabla u_n \nabla T_k(u_n-\varphi) + |u_n|^{p_+-2}u_n T_k(u_n-\varphi)\right] dx \geq 0 \text{as } \varepsilon \to 0. \tag{41}$$

Setting $l = k + \|\varphi\|_{L^\infty(\Omega)}$ we have,

$$\begin{aligned}
\varepsilon \int_\Omega |\nabla u_n|^{p_+-2}\nabla u_n \nabla T_k(u_n-\varphi)\, dx &= \varepsilon \int_{\{|u_n-\varphi|<k\}} |\nabla T_l(u_n)|^{p_+-2}\nabla T_l(u_n) \nabla (T_l(u_n)-\varphi)\, dx \\
&= \varepsilon \int_{\{|u_n-\varphi|<k\}} |\nabla T_l(u_n)|^{p_+}\, dx - \varepsilon \int_{\{|u_n-\varphi|<k\}} |\nabla T_l(u_n)|^{p_+-2}\nabla T_l(u_n) \nabla \varphi dx \\
&\geq -\varepsilon \int_{\{|u_n-\varphi|<k\}} |\nabla T_l(u_n)|^{p_+-2}\nabla T_l(u_n) \nabla \varphi dx.
\end{aligned} \tag{42}$$

The sequence $\varepsilon \nabla T_l(u_n)$ is uniformly bounded in $L^{p_+}(\Omega)$. From, Assertion 4 $-(d), \nabla T_l(u_n)$ converges a.e. in Ω (up to a subsequence) to $\nabla T_l(u)$. So, by Vitali's Theorem, $\varepsilon \nabla T_l(u_n)$ converges to $\varepsilon \nabla T_l(u)$ in $L^{p_+}(\Omega)$. Thus, $\varepsilon |\nabla T_l(u_n)|^{p_+-2}\nabla T_l(u_n) \chi_{\{|u_n-\varphi|<k\}}$ converges to $\varepsilon |\nabla T_l(u)|^{p_+-2}\nabla T_l(u) \chi_{\{|u-\varphi|<k\}}$ in $L^{p'_+}(\Omega)$. Using (42), we obtain

$$\lim_{n\to+\infty} \varepsilon \int_\Omega |\nabla u_n|^{p_+-2}\nabla u_n \nabla T_k(u_n-\varphi)\, dx \geq -\varepsilon \int_{\{|u-\varphi|<k\}} |\nabla T_l(u)|^{p_+-2}\nabla T_l(u) \nabla \varphi dx.$$

Therefore,

$$\lim_{n\to+\infty} \varepsilon \int_\Omega |\nabla u_n|^{p_+-2}\nabla u_n \nabla T_k(u_n-\varphi)\, dx \geq 0, \text{as } \varepsilon \to 0. \tag{43}$$

Now, we prove that

$$\lim_{n\to+\infty} \varepsilon \int_\Omega |u_n|^{p_+-2}u_n T_k(u_n-\varphi)\, dx \geq 0, \text{as } \varepsilon \to 0.$$

We have

$$\begin{aligned}
\int_\Omega |u_n|^{p_+-2}u_n T_k(u_n-\varphi)\, dx &= \int_\Omega \left(|u_n|^{p_+-2}u_n - |\varphi|^{p_+-2}\varphi\right) T_k(u_n-\varphi)\, dx \\
&\quad + \int_\Omega |\varphi|^{p_+-2}\varphi T_k(u_n-\varphi)\, dx \\
&\geq \int_\Omega |\varphi|^{p_+-2}\varphi T_k(u_n-\varphi)\, dx
\end{aligned} \tag{44}$$

since $\left(|u_n|^{p_+-2}u_n - |\varphi|^{p_+-2}\varphi\right) T_k(u_n-\varphi)$ is nonnegative. Furthermore, $T_k(u_n-\varphi)$ converges weakly* to $T_k(u-\varphi)$ in $L^\infty(\Omega)$ and $|\varphi|^{p_+-2}\varphi \in L^{p'_+}(\Omega)$, so

$$\lim_{n\to+\infty} \int_\Omega |\varphi|^{p_+-2}\varphi T_k(u_n-\varphi)\, dx = \int_\Omega |\varphi|^{p_+-2}\varphi T_k(u-\varphi) dx \tag{45}$$

Combining (44) and (45), we obtain

$$\lim_{n\to+\infty} \varepsilon \int_\Omega |u_n|^{p_+-2}u_n T_k(u_n-\varphi)\, dx \geq 0, \text{as } \varepsilon \to 0 \tag{46}$$

Combining (43) and (46), we get (41).

For the first term of the left hand side of (39), we recall that $l = k + \|\varphi\|_{L^\infty(\Omega)}$ and we get

$$\int_\Omega a(x,u_n,\nabla u_n).\nabla T_k(u_n-\varphi)\,dx = \int_\Omega a(x,u_n,\nabla T_l(u_n)).\nabla(T_l(u_n)-\varphi)\,\chi_{\{|u_n-\varphi|<k\}}dx$$
$$= \int_\Omega a(x,u_n,\nabla T_l(u_n))\cdot\nabla T_l(u_n)\,\chi_{\{|u_n-\varphi|<k\}}dx - \int_\Omega a(x,u_n,\nabla T_l(u_n))\cdot\nabla\varphi\chi_{\{|u_n-\varphi|<k\}}dx. \tag{47}$$

Moreover, $a(x,u_n,\nabla T_l(u_n))\cdot\nabla T_l(u_n)\,\chi_{\{|u_n-\varphi|<k\}}$ is nonnegative and converges a.e. in Ω to $a(x,u,\nabla T_l(u))\,\nabla T_l(u)\chi_{\{|u-\varphi|<k\}}$. Thanks to Fatou's Lemma, we get

$$\liminf_{n\to+\infty}\int_\Omega a(x,u_n,\nabla T_l(u_n))\cdot\nabla T_l(u_n)\,\chi_{\{|u_n-\varphi|<k\}}dx \geq \int_\Omega a(x,u,\nabla T_l(u))\cdot\nabla T_l(u)\chi_{\{|u-\varphi|<k\}}dx. \tag{48}$$

For the last term of (47), first we prove that $a(x,u_n,\nabla T_l(u_n)).\nabla\varphi\chi_{\{|u_n-\varphi|<k\}}$ is equi-integrable. Let E be a subset of Ω.

$$\begin{aligned}\int_E a(x,u_n,\nabla T_l(u_n))\cdot\nabla\varphi\chi_{\{|u_n-\varphi|<k\}}dx &\leq \int_E |a(x,u_n,\nabla T_l(u_n))|\,|\nabla\varphi|dx\\ &\leq \int_E \frac{1}{\pi_n'(.)}|a(x,u_n,\nabla T_l(u_n))|^{\pi_n'(\cdot)}dx + \int_E \frac{1}{\pi_n(.)}|\nabla\varphi|^{\pi_n(\cdot)}dx\\ &\leq C_1\int_E\left(\mathscr{M}(x)+|\nabla T_l(u_n)|^{\pi_n(\cdot)}\right)dx + \int_E|\nabla\varphi|^{\pi_n(\cdot)}dx\end{aligned}$$

Moreover,

$$\begin{aligned}\int_E|\nabla\varphi|^{\pi_n(\cdot)}dx &= \int_{E\cap\{|\nabla\varphi|\leq 1\}}|\nabla\varphi|^{\pi_n(\cdot)}dx + \int_{E\cap\{|\nabla\varphi|>1\}}|\nabla\varphi|^{\pi_n(\cdot)}dx\\ &\leq \text{meas}(E) + \int_E|\nabla\varphi|^{p_+}dx,\end{aligned}$$

since $|\nabla\varphi|^{p_+}, \mathscr{M} \in L^1(\Omega)$ and $|\nabla T_l(u_n)|^{\pi_n(\cdot)}$ is equi-integrable. Then, we obtain

$$\lim_{\text{meas}(E)\to 0}\int_E a(x,u_n,\nabla T_l(u_n))\,\nabla\varphi\chi_{\{|u_n-\varphi|<k\}}dx = 0.$$

Furthermore,

$$a(x,u_n,\nabla T_l(u_n))\,\nabla\varphi\chi_{\{|u_n-\varphi|<k\}} \to a(x,u,\nabla T_l(u))\cdot\nabla\varphi\chi_{\{|u-\varphi|<k\}} \quad \text{a.e. in } \Omega.$$

By applying Vitali's Theorem, we obtain

$$\lim_{n\to+\infty}\int_\Omega a(x,u_n,\nabla T_l(u_n))\,\nabla\varphi\chi_{\{|u_n-\varphi|<k\}}dx = \int_\Omega a(x,u,\nabla T_l(u)).\nabla\varphi\chi_{\{|u-\varphi|<k\}}dx. \tag{49}$$

From (47)–(49) we obtain

$$\begin{aligned}\liminf_{n\to+\infty}\int_\Omega a(x,u_n,\nabla u_n)\cdot\nabla T_k(u_n-\varphi)\,dx &\geq \int_\Omega a(x,u,\nabla T_l(u))\,\nabla(T_l(u)-\varphi)\,\chi_{\{|u-\varphi|<k\}}dx\\ &= \int_\Omega a(x,u,\nabla u)\nabla T_k(u-\varphi)dx.\end{aligned} \tag{50}$$

For the second term of the left hand side of (39), we have $\phi_n(u_n) = \phi(T_{k+\|\varphi\|_\infty}(u_n))$, then

$$\lim_{n\to+\infty} \int_\Omega \phi(T_n(u_n))\nabla T_k(u_n-\varphi)\,dx = \int_\Omega \phi(u)\nabla T_k(u-\varphi)\,dx. \tag{51}$$

For the right hand side of (39), since $f_n \to f$ in $L^1(\Omega)$ and $T_k(u_n-\varphi) \to^* T_k(u-\varphi)$ in $L^\infty(\Omega)$, we have

$$\lim_{n\to+\infty} \int_\Omega f_n T_k(u_n-\varphi)\,dx = \int_\Omega f T_k(u-\varphi)dx. \tag{52}$$

From (39), (40), (41), (50), (51), (52), we obtain

$$\begin{aligned}
&\int_\Omega b(u)T_k(u-\varphi)dx + \int_\Omega a(x,u,\nabla u)\cdot\nabla T_k(u-\varphi)dx + \int_\Omega \phi(u)\nabla T_k(u-\varphi)\,dx \\
&\le \liminf_{n\to+\infty}\Bigg(\int_\Omega T_n(b(u_n))\,T_k(u_n-\varphi)dx + \int_\Omega a(x,u_n,\nabla u_n)\cdot\nabla T_k(u_n-\varphi)\,dx \\
&+ \int_\Omega \phi(T_n(u_n))\nabla T_k(u_n-\varphi)\,dx \\
&+ \varepsilon\int_\Omega \left[|\nabla u_n|^{p_+-2}\nabla u_n \nabla T_k(u_n-\varphi) + |u_n|^{p_+-2}u_n T_k(u_n-\varphi)\right]dx\Bigg) \\
&= \int_\Omega f T_k(u-\varphi)dx, \text{as } \varepsilon\to 0,
\end{aligned}$$

then

$$\int_\Omega b(u)T_k(u-\varphi)dx + \int_\Omega a(x,u,\nabla u)\cdot\nabla T_k(u-\varphi)dx + \int_\Omega \phi(u)\nabla T_k(u-\varphi)\,dx \le \int_\Omega f T_k(u-\varphi)dx \tag{53}$$

for $\varphi \in C^\infty(\bar{\Omega})$.

As $\pi(.)$ verifies the log-Hlder condition, $\mathscr{C}^\infty(\bar{\Omega})$ is dense in the space $W^{1,\pi(\cdot)}(\Omega)$. Moreover, $W^{1,\pi(\cdot)}(\Omega) \hookrightarrow W^{1,p_-}(\Omega) \hookrightarrow L^\infty(\Omega)$. Therefore, the inequality (53) holds true for $\varphi \in W^{1,\pi(\cdot)}(\Omega)\cap L^\infty(\Omega)$. Hence, u is an entropy solution of (1).

References

1. Ouaro, S., Sawadogo, N.: Nonlinear elliptic $p(u)$-Laplacian problem with Fourier boundary condition. CUBO A Math. **22**(01), 85–124 (2020)
2. Andreianov, B., Bendahmane, M., Ouaro, S.: Structural stability for variable exponent elliptic problems. I. The $p(x)$-Laplacian kind problems. Nonlinear Anal. **73**, 2–24 (2010)
3. Wang, L., Fan, Y., Ge, W.: Existence and multiplicity of solutions for a Neumann problem involving the $p(x)$-Laplace operator. Nonlinear Anal. **71**, 4259–4270 (2009)
4. Yao, J.: Solutions for Neumann boundary value problems involving p(x)-Laplace operator. Nonlinear Anal: (TMA) **68**, 1271–1283 (2008)
5. Ball, J.M.: A version of the fundamental theorem for young measures. In: Rascle, M., Serre, D., Slemrod, M. (eds.) PDEs and Continuum Models of Phase Transitions. Lecture Notes in Physics, vol. 344, pp. 207–215. Springer, Heidelberg (1989). https://doi.org/10.1007/BFb0024945

6. Hungerbhler, N.: A refinement of Ball's theorem on young measures. New York J. Math. **3**, 48–53 (1997)
7. Pedregal, P.: Parametrized Measures and Variational Principles. Progress in Nonlinear Differential Equations and their Applications, vol. 30. Birkhauser, Basel (1997)
8. Eymard, R., Galloet, T., Herbin, R.: Finite volume methods. In: Handbook of Numerical Analysis, North-Holland, vol. VII, pp. 713–1020 (2000)
9. Dolzmann, G., Hungerbhler, N., Mller, S.: The p-harmonic system with measure-valued right hand side. Ann. Inst. H. Poincar. Anal. Non Linaire **14**(3), 353–364 (1997)
10. Hungerbhler, N.: Quasi-linear parabolic systems in divergence form with weak monotonicity. Duke Math. J. **107**(3), 497–520 (2001)
11. Nyanquini, I., Ouaro, S.: Entropy solution for nonlinear elliptic problem involving variable exponent and Fourier type boundary condition. Afr. Mat. **23**(2), 205–228 (2012)
12. Le, V.K.: On sub-supersolution method for variational inequalities with Leray-Lions operators in variable exponent spaces. Nonlinear Anal. **71**, 3305–3321 (2009)
13. Chipot, M., de Oliveira, H.B.: Some results on the $p(u)$-Laplacian problem. Math. Ann. **375**(2019), 307–313 (2019)
14. Ouaro, S., Sawadogo, N.: Structural stability for nonlinear Neumann boundary $p(u)$-Laplacian problem. Discuss. Math. Differ. Incl. Control Optim. **39**, 81–117 (2019)
15. Antontsev, S.N., Rodrigues, J.F.: On stationary thermo-rheological viscous flows. Ann. Univ. Ferrara Sez. VII Sci. Math. **52**(1), 19–36 (2006)
16. Blomgren, P., Chan, T., Mulet, P.,Wong, C.: Total variation image restoration: numerical methods and extensions. In: Proceedings of the IEEE International Conference on Image Processing, vol. 3, 384-387. IEEE Computer Society Press, Piscataway (1997)
17. Ružicka, M.: Electrorheological Fluids, Modeling and Mathematical Theory. Lectures Notes in Math, vol. 1748. Springer, Berlin (2000)
18. Türola, J.: Image denoising using directional adaptive variable exponents model. J. Math. Imag. Vis. **57**, 56–74 (2017)
19. Bénilan, Ph., Boccardo, L., Gallouet, T., Gariepy, R., Pierre, M., Vazquez, J.L.: An L1 theory of existence and uniqueness of solutions of nonlinear elliptic equations. Ann. Sc. Norm. Super. Pisa Cl. Sci. **22**, 241–273 (1995)
20. Abassi, A., El Hachimi, A., Jamea, A.: Entropy solutions to nonlinear Neumann problemswith L1-data. Int. J. Math. Stat. **2**(S08), 4–17 (2008)
21. Jamea, A., Alaoui, A., El Hachimi, A.: Existence of entropy solutions to nonlinear parabolic problems with variable exponent and L1-data. Ric. Math. **67**(2), 785–801 (2018)
22. Ouaro, S., Noufou, S.: Nonlinear elliptic p(u)-Laplacian problem with Fourier boundary condition. Cubo Temuco **22**(1), 85–124 (2020)
23. Sabri, A., Jamea, A., Talibi, H.A.: Existence of entropy solutions to nonlinear degenerate parabolic problems with variable exponent and L1-data. Commun. Math. **28**, 67–88 (2020)
24. Sanchón, M., Urbano, J.M.: Entropy solutions for the p(x)-Laplace equation. Trans. Am. Math. Soc. **361**, 6387–6405 (2009)

Valuing Option Under Double Heston Jump-Diffusion Model with Stochastic Interest Rate and Approximative Fractional Brownian Motion

Siham Bayad[1], Abdelmajid El Hajaji[2], and Khalid Hilal[1](✉)

[1] LAMSC Laboratoty, Sultan Moulay Slimane University, Beni Mellal, Morocco
hilalkhalid2005@yahoo.fr

[2] LERSEM Laboratory, ENCGJ, University Chouaib Doukkali, El jadida, Morocco

Abstract. In the light of the current research, we propose a more general and realistic model based on approximative fractional Brownian motion studies. This framework presents an option pricing model under the double Heston Jump-Diffusion model, including approximative fractional motion with stochastic interest rate and stochastic intensity. The stochastic interest rate is determined using a two-factor Vasicek model. The negative interest rate is allowed for this model. Therefore, we are constructing a multi-factor model with a stochastic interest rate structure. We derive a closed-form pricing formula with an analytical solution for European options. Finally, some numerical results are presented to illustrate the value of a European call option comparing to other classical models.

1 Introduction

In 1997 Black & Scholes [4] published a groundbreaking paper in which they proposed an elegant model focused on Brownian motion to explain the complexities of the underlying asset price and presented a closed-form formula for European options. According to Duan and Wei [7], the Black-Scholes model cannot explain the phenomena of the asymmetric leptokurtic and also the volatility smile that is observed in the real market. Since That point, academic researchers have created different models by joining in the Black-Scholes model the non-constant volatility . The Scott [19] model, Hull and White [13] model, the Stein and Stein [22] model and the Wiggins [26] model. However, the majority of these stochastic volatility models are unsuitable for use. In 1993 Heston [11] describe the variance (the square of volatility) by Cox-Ingersoll-Ross process [5] and deriving a closed-form formula for European options.

On the other side, a single factor model cannot describe the shapes of the volatility smile with precision. Multi-factor stochastic volatility models are useful for expressing return data in various ways, such as using a stylized effect or

S. Melliani and O. Castillo (Eds.): ICPAMS 2021, LNNS 476, pp. 393–403, 2023.
https://doi.org/10.1007/978-3-031-12416-7_33

fitting the implied surface. We choose to investigate option pricing under two-factor stochastic volatility in this study since it is more appropriate for practical applications.

Otherwise, the financial market owns long-range persistence and self-similarity traits, and fractional Brownian motion has these two essential properties. Moreover, fractional Brownian motion is not a Markov process or semi-martingale; the classical Ito calculus cannot be used in this case. Wick products have been created by Hu and Oksendal [12] for analyzing it. In addition, Xiao and Al [27] used the Wick products to define a fractional stochastic integral. Björk and Hult [3] demonstrated that the model lacks an economic interpretation. To solve this problem is appropriate to use the mixed fractional Brownian motion [8,17,23,28]. Approximation Fractional Brownian motion [24] can also be used instead of fractional Brownian motion. Thao [24] showed that Approximation Fractional Brownian motion is a semi-martingale. Furthermore, many researchers (see [6]) adopted Approximation Fractional Brownian motion in building stochastic volatility models.

Many authors have worked on a hybrid model in recent years by incorporating the stochastic interest rate into stochastic models [9,10,14,21]. In addition, empirical studies show that using stochastic interest rates into option pricing models will contribute to improved model results [18].

Roughly speaking, permitting for changes in volatility and interest rate and the presence of jumps and the jump intensity changing over time indicate realistic asset return dynamics. In a parallel development, incorporating jump into models for pricing option also proposes describing the discontinuous behavior of the underlying asset (see [1,2,15,16,20]).

The rest of the paper is organized as follows. We adopt the double-Heston jump-diffusion (DHJD) model with approximative fractional Brownian motion, stochastic intensity, and interest rate follow a two-factor model in Sect. 2. In Sect. 3, we derive analytical pricing formula for European call option. In Sect. 4, we present some numerical illustrations. Finally, we conclude in Sect. 5.

2 The Model

We present some basic information on approximative fractional Brownian motion. At the first, we present an analysis of fractional Brownian motion $(B_t^H)_{t\geq 0}$ with the Hurst index $H \in (0,1)$. It is a Gaussian process with zero mean and the following covariance:

$$\mathbb{E}[B_t^H B_s^H] = \frac{1}{2}\Big(|t|^{2H} + |s|^{2H} - |Tt - s|^{2H}\Big). \tag{1}$$

The decomposition of a fractional Brownian motion B is as follows:

$$B_t^H = \frac{1}{\Gamma(H+\frac{1}{2})}\left[Z_t + \int_0^t (t-s)^{H-\frac{1}{2}} dW_s\right] \tag{2}$$

where

$$Z_t = \int_{-\infty}^{0} \left((t-s)^{H-\frac{1}{2}} - (-s)^{H-\frac{1}{2}}\right) dW_s, \tag{3}$$

W_t indicates standard Brownian motion, and Γ indicates the gamma function. It is sufficient to focus exclusively on the term:

$$B_t = \int_0^t (t-s)^{H-\frac{1}{2}} \tag{4}$$

that has a long-range memory. Note that The approximation of B_t is $\tilde{B}_t^{\epsilon,H}$ which can be expressed as [26]

$$\tilde{B}_t^{\varepsilon,H} = \int_0^t (t-s+\varepsilon)^{H-\frac{1}{2}} dW_s \tag{5}$$

where H is a long-memory parameter, ε is non negative approximation factor. Thao [24] proved that for $\varepsilon \to 0$, $(B_t^{\varepsilon,t})_\varepsilon$ converges uniformly to a non-Markov process. In addition, if $\varepsilon > 0$ then $B_t^{\varepsilon,t}$ is a semi-martingale [24]

$$d\tilde{B}_t^{\varepsilon,H} = (H - \frac{1}{2})\psi_t dt + \varepsilon^{H-\frac{1}{2}} dW_t^v \tag{6}$$

ψ_t is a stochastic processes expressed as

$$\psi_t = \int_0^t (t-s+\varepsilon)^{H-\frac{3}{2}} dW_s^\psi, \tag{7}$$

where$(W_t^\psi)_{t\in[0,T]}$ and $(W_t^v)_{t\in[0,T]}$, are independent standard Brownian motions.

Let$(\Omega, \mathcal{F}, (\mathcal{F}_t)_{t\in[0,T]}, \mathbb{Q})$ be a complete probability space with a filtration and $\mathbb{Q}$ presents a risk-neutral measure.The stock price S_t is expressed by the following dynamic system:

$$\begin{cases} \frac{dS_t}{S_t} = (r_1 + r_2 - \lambda_t \mu_J)\mathrm{dt} + \sqrt{v_t} dW_t^s + \sqrt{\hat{v}} d\hat{W}_t^s + (J-1)dN_t \\ dv_t = k_v(\theta - v_t)dt + \sigma_v \sqrt{v_t} d\tilde{B}_t^{\epsilon,H} \\ d\hat{v}_t = \hat{k}(\hat{\theta} - \hat{v})dt + \sigma_{\hat{v}} \sqrt{\hat{v}} dW_t^{\hat{v}} \\ d\lambda_t = k_\lambda(\theta_\lambda - \lambda_1)dt + \sigma_\lambda dW_t^\lambda \\ dr_1 = \alpha_1(\beta_1 - r_1)dt + \sigma_1 dW_t^{r_1} \\ dr_2 = \alpha_2(\beta_2 - r_1)dt + \sigma_2 dW_t^{r_2} \end{cases} \tag{8}$$

where $W_1^s, \hat{W}_t^s, W_t^v, W_t^{r_1}, W_t^{r_2}$ and W_t^λ are the standard Brownian motions. We assume that W_t^s is correlated with W_t^v, $dW_t^s.dW_t^v = \rho_1 dt$,$\hat{W}_t^s$ correlated with $W_t^{\hat{v}}$, $d\hat{W}_t^s dW_t^{\hat{v}} = \rho_2 dt$ and $W_t^{r_1}$ correlated with $W_t^{r_2}$, $dW_t^{r_1}.dW_t^{r_2} = \rho_r dt$. Any other Brownian motions are pairwise independent.

$v_t, \hat{v}_t$ are variances, and λ_t is the jump intensity. $k, \hat{k}$ and k_λ are mean reversion rates, θ, $\hat{\theta}$ and θ_λ are mean reversion levels, σ_v, $\sigma_{\hat{v}}$ and σ_λ are the volatilities of the variances. and the short rate is follow two-factor Vasicek model where the short rate is given as a sum of two factors r_1 and r_2, where β_1, β_2 are their mean-reversion , α_1, α_2 are theire mean-reversion speed, σ_1, σ_2 are their volatilities, N_t represents Poisson process with intensity λ_t and J represents the jump size, and we suppose that lnJ has an asymmetric double exponential distribution with density function $pdf_u(z)$:

$$pdf_u(z) = p\eta_1 e^{\eta_1 z} 1_{\geq 0} + q\eta_2 e^{\eta_2 z} 1_{z<0}, \tag{9}$$

where $\eta_1 > 1, \eta_2 > 0, p, q > 0$, and $p + q = 1$, where q and p represent the probabilities for positive and negative jumps, respectively. As a result we can obtain that $\mu_J = \mathbb{E}^{\mathbb{Q}}(J-1) = (p\eta_1/\eta_1 - 1) + (q\eta_2/\eta_2 + 1) - 1$.

We set $\tau = T - t$, $X_t = lnS_t$, $Y = lnJ$, the interest rate r are determined by the sum of the two factors r_1 and r_2 $(r = r_1 + r_2)$ and $k = lnK$, where T is the maturity date, and K is the strike price. In the risk-neutral world, the price of a call option $C(S, V1, V2, r, \lambda, t)$ at time $t \in [0, T]$ with strike price K and maturity date T is given by

$$C(S, v, \hat{v}, r_1, r_2, \lambda, t) = \mathbb{E}^{\mathbb{Q}}\left(e^{-\int_0^t r_s ds} max(S_T - K, 0)|\mathcal{F}_t\right) \tag{10}$$

we convert measure $\mathbb{Q}$ to the measure $\mathbb{Q}^S$ and the T forward measure $\mathbb{Q}^T$. By applying Radon-Nikodym derivatives,

$$\frac{d\mathbb{Q}}{d\mathbb{Q}^S} = \frac{e^X}{e^{-\int_0^T r_s ds + X_T}} \tag{11}$$

$$\frac{d\mathbb{Q}}{d\mathbb{Q}^T} = \frac{P(t,T)}{e^{-\int_0^t r_s ds}} \tag{12}$$

where

$$S = e^X = \mathbb{E}^{\mathbb{Q}}\left(e^{-\int_t^T r_s ds + X_T}|\mathcal{F}_t\right), \tag{13}$$

$P(t,T) := \mathbb{E}^{\mathbb{Q}}\left(e^{-\int_t^T r_s ds +}|\mathcal{F}_t\right)$, is the price at time t of a zero-coupon bond which matures at time T (see appendix). Then, we can have the following expression:

$$C(S, v, \hat{v}, r_1, r_2, \lambda, t) = S\mathbb{E}^{\mathbb{Q}^S}(1_{\{X_T > k\}}|\mathcal{F}_t) - KP(t,T)\mathbb{E}^{\mathbb{Q}^T}(1_{\{X_T > k\}}|\mathcal{F}_t) \tag{14}$$

we define

$$\varphi_S(u) := \mathbb{E}^{\mathbb{Q}^S}(e^{iuX_T}|\mathcal{F}_t), \tag{15}$$

$$\varphi_T(u) := \mathbb{E}^{\mathbb{Q}^T}(e^{iuX_T}|\mathcal{F}_t), \tag{16}$$

$$\varphi(u) := \mathbb{E}^{\mathbb{Q}}(e^{\int_t^T r_s ds + iuX_T}|\mathcal{F}_t), \tag{17}$$

where $\varphi_S(u)$ denotes the characteristic function under $\mathbb{Q}^S$, $\varphi_T(u)$ denotes the characteristic function under $\mathbb{Q}^T$, and $\varphi(u)$ denotes the discounted characteristic function under $\mathbb{Q}$. Furthermore, by using Radon-Nikodym derivatives we can have the following expression:

$$C(S,v,\hat{v},r_1,r_2,\lambda,t) = S\Big(\frac{1}{2}+\frac{1}{\pi}\int_0^\infty R\Big(\frac{e^{-iuk}\varphi(u-i)}{iu\varphi(-i)}\Big)du\Big) - KP(t,T)\Big(\frac{1}{2}+\frac{1}{\pi}\int_0^\infty R\Big(\frac{e^{-iuk}\varphi(u)}{iuP(t,T)}\Big)du\Big) \tag{18}$$

all we need to do is to derive the formula of $\varphi(u)$ to have the pricing formula.

Theorem 1. *If the asset price is governed by the dynamic system (1), the discounted characteristic function $\varphi(u;X,v,\hat{v},r_1,r_2,\lambda,\tau)$ takes the following form:*

$$\varphi(u;X,v,\hat{v},r_1,r_2,\lambda,\tau) = e^{C(u,\tau)+D_v(u,\tau)v+D_{\hat{v}}(u,\tau)\hat{v}+E(u,\tau)r_1+F(u,\tau)r_2+G(u,\tau)\lambda+iuX} \tag{19}$$

where

$$C(u,\tau) = \frac{2k_v\theta_v}{\sigma_v^2\varepsilon^{2H-1}}\Big[\frac{(k_v - iu\rho_1\sigma_v\varepsilon^{H-\frac{1}{2}} - d)\tau}{2} + ln\frac{2d}{2d+(k_v - iu\rho_1\sigma_v\varepsilon^{H-\frac{1}{2}} - d)(1-e^{-d\tau})}\Big]$$

$$+ \frac{2\hat{k}\hat{\theta}}{\sigma_{\hat{v}}^2}\Big[\frac{(\hat{k} - iu\rho_2\sigma_{\hat{v}} - \hat{d})\tau}{2} + ln\frac{2\hat{d}}{2d+(\hat{k} - iu\rho_2\sigma_{\hat{v}} - \hat{d})(1-e^{-\hat{d}\tau})}\Big]$$

$$+ (iu-1)\Big((\frac{\theta_1}{k_1}(k_1t - 1 - e^{-k_1t}) + \frac{\theta_2}{k_2}(k_2t - 1 - e^{-k_2t})\Big)$$

$$- \frac{\sigma_1^2}{4k_1^3}(iu-1)^2\Big(e^{-2k_1t} - 4e^{-k_1t} - 2k_1t + 3\Big) - \frac{\sigma_2^2}{4k_2^3}(iu-1)^2\Big(e^{-2k_2t} - 4e^{-k_2t} - 2k_2t + 3\Big)$$

$$+ \rho_r\sigma_1\sigma_2(iu-1)^2\frac{1}{k_1k_2}\Big(t + \frac{1}{k_2}e^{-k_2t} + \frac{1}{k_1}e^{-k_1t} - \frac{1}{k_2+k_1}e^{-(k_2+k_1)t} - \frac{1}{k_1} - \frac{1}{k_2}$$
$$+ \frac{1}{k_1+k_2}\Big)$$

$$+ \frac{2k_\lambda\theta_\lambda}{\sigma_\lambda^2}\Big[\frac{(k_\lambda - iu\rho_2\sigma_\lambda - \varsigma)\tau}{2} + ln\frac{2\varsigma}{2\varsigma+(k_\lambda - iu\rho_2\sigma_\lambda - \varsigma)(1-e^{-\hat{d}\tau})}\Big]$$

$$D_v(u,\tau) = ((iu)^2 - iu)\frac{1-e^{-d\tau}}{2d+(k_v - iu\rho_1\sigma_v\varepsilon^{H-\frac{1}{2}} - d)(1-e^{-d\tau})}$$

$$D_{\hat{v}}(u,\tau) = ((iu)^2 - iu)\frac{1-e^{-\hat{d}\tau}}{2d_{\hat{v}}+(\hat{k} - iu\rho_2\sigma_{\hat{v}} - \hat{d})(1-e^{-\hat{d}\tau})}$$

$$G(u,\tau) = 2\omega(u)\frac{1-e^{-\varsigma\tau}}{2\varsigma + (k_\lambda - \varsigma)(1-e^{-\varsigma\tau})}$$

$$E(u,\tau) = \frac{1}{k_1}(iu-1)(1-e^{-k_1\tau})$$

$$F(u,\tau) = \frac{1}{k_2}(iu-1)(1-e^{-k_2\tau})$$

$$d = \sqrt{(k_v - iu\rho_1\sigma_v\varepsilon^{H-\frac{1}{2}})^2 - \sigma^2\varepsilon^{2H-1}((iu)^2 - iu)},$$

$$\hat{d} = \sqrt{(\hat{k} - iu\rho_2\sigma_{\hat{v}})^2 - \sigma_{\hat{v}}^2((iu)^2 - iu)}$$

$$M(u) = \frac{p\eta_1}{\eta_1 - iu} + \frac{q\eta_2}{\eta_2 + iu} - 1$$

$$\omega(u) = M(u) - iu\mu_J, \varsigma = \sqrt{k_\lambda^2 - 2\sigma_\lambda^2\omega(u)}$$

Proof. $\varphi(u; X, v, \hat{v}, r_1, r_2, \lambda, \tau)$ satisfies a PIDE by applying the Feynman-Kac theorem:

$$-\frac{\partial\varphi}{\partial\tau} + (r_1 + r_2 - \lambda\mu_J - \frac{1}{2}(v+\hat{v}))\frac{\partial\varphi}{\partial x} + \frac{1}{2}(v+\hat{v})\frac{\partial^2\varphi}{\partial x^2} + (k_v(\theta_v - v) + (H - \frac{1}{2})\sigma_v\sqrt{v})\frac{\partial\varphi}{\partial v}$$

$$+ \frac{1}{2}\sigma_v^2\varepsilon^{2H-1}v\frac{\partial^2\varphi}{\partial v^2} + \hat{k}(\hat{\theta} - \hat{v})\frac{\partial\varphi}{\partial\hat{v}} + \frac{1}{2}\sigma_{\hat{v}}^2\hat{v}\frac{\partial^2\varphi}{\partial\hat{v}^2} + \rho_1\sigma_v v\varepsilon^{H-\frac{1}{2}}\frac{\partial^2}{\partial x\partial v} + \rho_2\sigma_{\hat{v}}\hat{v}\frac{\partial^2\varphi}{\partial x\partial\hat{v}} + k_1(\theta_1 - r_1)\frac{\partial\varphi}{\partial r_1}$$

$$+ \frac{1}{2}\sigma_1^2\frac{\partial^2\varphi}{\partial r_1^2} + k_2(\theta_1 - r_2)\frac{\partial\varphi}{\partial r_2} + \frac{1}{2}\sigma_2^2\frac{\partial^2\varphi}{\partial r_2^2} + \sigma_1\sigma_2\rho_r\frac{\partial^2\varphi}{\partial r_1\partial r_2} + k_\lambda(\theta_\lambda - \lambda)\frac{\partial\varphi}{\partial\lambda} + \frac{1}{2}\sigma_\lambda^2\frac{\partial^2\varphi}{\partial\lambda^2}$$

$$+ \lambda\int_{-\infty}^{+\infty}(\varphi(x+y) - \varphi(x))f(y)dy - r\varphi = 0 \tag{20}$$

If we assume that $\varphi(u; X, v, \hat{v}, r_1, r_2, \lambda, \tau)$ takes the form of

$$\varphi(u; X, v, \hat{v}, r_1, r_2, \lambda, \tau) = e^{C(u,\tau) + D_v(u,\tau)v + D_{\hat{v}}(u,\tau)\hat{v} + E(u,\tau)r_1 + F(u,\tau)r_2 + G(u,\tau)\lambda + iuX} \tag{21}$$

and substitute into Eq. (20), we can obtain

$$\begin{cases} \frac{\partial C}{\partial \tau} = k_v\theta_v D_v + \hat{k}\hat{\theta} D_{\hat{v}} + k_1\theta_1 E + k_2\theta_2 F + \frac{1}{2}\sigma_1^2 E^2 + \frac{1}{2}\sigma_2^2 F^2 + \rho_r\sigma_1\sigma_2 EF + Gk_\lambda\theta_\lambda \\ \frac{\partial D_v}{\partial \tau} = \frac{1}{2}\sigma_v^2\epsilon^{2H-1}D_v^2 + (\rho_1\sigma_v\epsilon^{H-\frac{1}{2}}iu - k_v)D_v + \frac{1}{2}iu(iu-1) \\ \frac{\partial D_{\hat{v}}}{\partial \tau} = \frac{1}{2}\sigma_{\hat{v}}^2 D_{\hat{v}}^2 + (\rho_2\sigma_{\hat{v}}iu - k_{\hat{v}})D_{\hat{v}} + \frac{1}{2}iu(iu-1) \\ \frac{\partial G}{\partial \tau} = \frac{1}{2}\sigma_\lambda^2 G^2 - k_\lambda G + M(u) - \mu_J iu \\ \frac{\partial E}{\partial \tau} = -k_1 E + iu - 1 \\ \frac{\partial F}{\partial \tau} = -k_1 F + iu - 1 \end{cases} \tag{22}$$

with boundary conditions $C(u,0) = D_v(u,0) = D_{\hat{v}}(u,0) = E(u,0) = F(u,0) = G(u,0) = 0$. by applying some algebraic calculations, we will obtain the result.

3 Numerical Discussion

We'll analyze European option prices under DHJDF with two-factor stochastic interest rate model parameters in this section. The parameters we use are listed in Table 1.

Table 1. Values of parameters.

Parameter	Value	Parameter	value
k_v	9.9772k1	$\hat{k}$	2.3388
θ_v	0.0189	$\hat{\theta}$	0.001
σ_v	0.8379	$\sigma_{\hat{v}}$	0.9957
ρ_1	−0.9764	ρ_2	−0.8178
v	0.0002	$\hat{v}$	0.0633
ε	0.00005	ρ_r	1
α_1	0.3322	α_2	0.26594
β_1	0.1	β_2	0.1
σ_1	0.02	σ_2	0.02
r_1	0.001	r_2	0.012
k_λ	2	σ_λ	0.1
θ_λ	0.001	λ	0.001
k_{r_1}	0.02	k_{r_2}	0.02
η_1	1.0333	η_2	19.7482
S	100	K	100

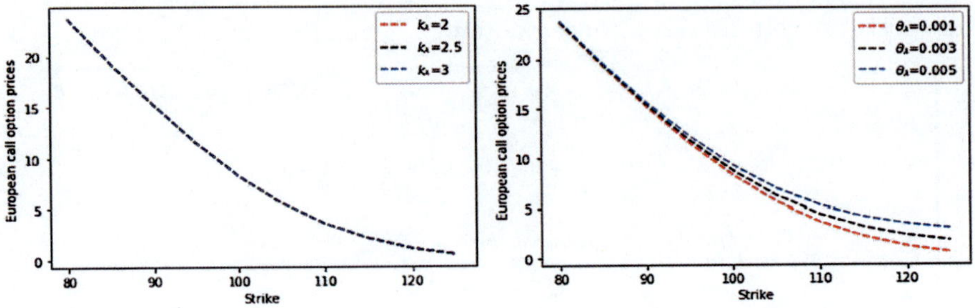

Fig. 1. The impact of k_λ and θ_λ on call option prices for $T = 1$.

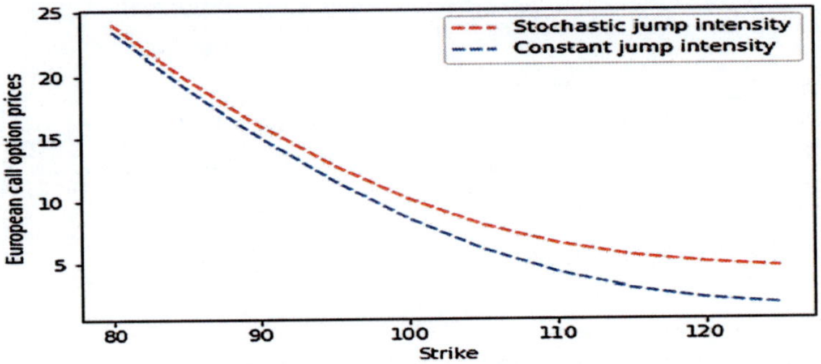

Fig. 2. The impact of the existence of the jump intensity process on call option prices for $T = 1$.

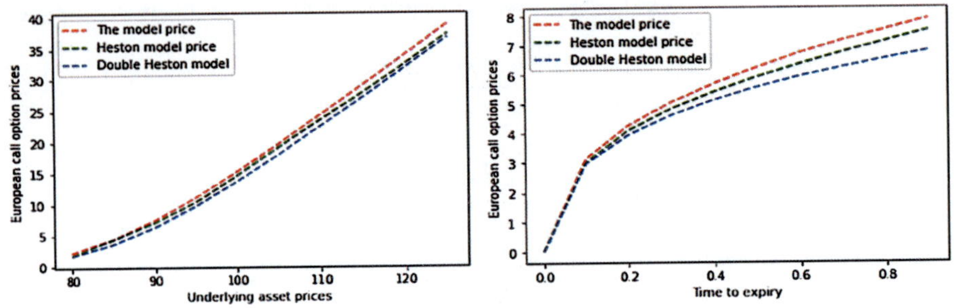

(a) (b)

Fig. 3. The model price, the Heston price and double Heston price with respect to the underlying asset price (a) and time to expiry (b).

Figure 1 shows that changes in the mean-reversion level θ_λ have a significant effect on call option prices, while changes in the mean-reversion rate k_λ have little effect on call option prices. The obtained results show that an increase in the value of θ_λ leads to an increase in the value of the call option price.

Figure 2 illustrate the effect of the presence of the jump intensity process on call option prices. It shows that the price of a call option with stochastic jump intensity is greater than the price of a call option with a constant jump intensity.

On the other hand. By using theoretical results of pricing formula, we can investigate the impact of incorporating a two-factor stochastic interest rate into DHJD model with approximative fractional Brownian motion and stochastic intensity under the chosen set of parameters. It can be distinctly observed that our price model's is high that the Heston's price. Specifically, depicted in Fig. 3 is the option prices with different time to expiry. Clearly, our price and the price of Heston are about the same when the time of expiry increases, the gap between our price and the Heston price increases. The reason that this phenomenon happens is increasing time to expiry implies a longer period of time for the interest rate changes which can thus definitely rate that can reflect the widened divide.

4 Conclusion

This paper introduces the European option under double Heston jump-diffusion hybrid model based on approximative fractional Brownian motion by adding interest rate follow two-factor Vasicek model and jump intensity follow a stochastic process. We derived a closed pricing formula for European option under this model by used the Radon-Nikodym derivative. The numerical results show that European call option prices under this model are higher than those under the double Heston model and Heston model.

Appendix

If the risk-free interest rate follows the Two-Vasicek model, then $P(r_1, r_2, t, T)$ should satisfy the following PDE problem:

$$\begin{cases} \frac{\partial P}{\partial t} + k_1(\theta_2 - r_1)\frac{\partial P}{\partial r_1} + k_2(\theta_2 - r_2)\frac{\partial P}{\partial r_1} + \frac{1}{2}\sigma_1^2\frac{\partial^2 P}{\partial r_1^2} + \frac{1}{2}\sigma_2^2\frac{\partial^2 P}{\partial r_2^2} + \rho_r\sigma_1\sigma_2\frac{\partial^2 P}{\partial r_1 \partial r_2} - (r_2 + r_2)P = 0 \\ P(r_1, r_2, T, T) = 1 \end{cases} \tag{23}$$

If we assume that $P(r_1, r_2, t, T)$ takes the form of

$$P(r_1, r_2, t, T) = e^{[A(\tau) - B_1(\tau)r_1 - B_1(\tau)_{r}2]} \tag{24}$$

and substitute it into PDE (23), we can obtain:

$$\begin{cases} \frac{\partial B_1}{\partial t} = 1 - k_1 B_1 \\ \frac{\partial B_2}{\partial t} = 1 - k_2 B_2 \\ \frac{\partial A}{\partial t} = -k_1\theta_1 B_1 - k_2\theta_2 B_2 + \frac{1}{2}\sigma_1^2 B_1^2 + \frac{1}{2}\sigma_2^2 B_2^2 + \rho_r\sigma_1\sigma_2 B_1 B_2 \end{cases} \tag{25}$$

with the terminal condition $B_1(0) = B_2(0) = A(0) = 0$ Then we have :

$$B_1(\tau) = \frac{1}{k_1}(1 - e^{k_1 \tau}) \tag{26}$$

$$B_2(\tau) = \frac{1}{k_1}(1 - e^{k_1 \tau}) \tag{27}$$

$$\begin{aligned} A(\tau) = & -\theta_1(\tau + \frac{1}{k_1}e^{-k_1\tau} - \frac{1}{k_1}) - \theta_2(\tau + \frac{2}{k_2}e^{-k_2\tau} - \frac{1}{k_2}) + \frac{\sigma_1^2}{k_1^2}(t + \frac{2}{k_1}e^{-k_1 t} - \frac{1}{2k_1}e^{-2k_1 t} - \frac{3}{2k_1}) \\ & + p_r\sigma_2\sigma_2\frac{1}{k_1k_2}(t + \frac{1}{k_1}e^{-k_1 t} + \frac{1}{k_2}e^{k_2 t} - \frac{1}{k_1+k_2}e^{-(k_1+k_2)t} + \frac{1}{k_1+k_2} - \frac{1}{k_1} - \frac{1}{k_2}) \\ & + \frac{\sigma_2^2}{k_2^2}(t + \frac{2}{k_2}e^{-k_2 t} - \frac{1}{2k_2}e^{-2k_2 t} - \frac{3}{2k_2}). \end{aligned} \tag{28}$$

References

1. Bakshi, G., Cao, C., Chen, Z.: Empirical performance of alternative option pricing models. J. Finance **52**(5), 2003–2049 (1997)
2. Bates, D.S.: Jumps and stochastic volatility: exchange rate processes implicit in deutsche mark options. Rev. Financ. Stud. **9**(1), 69–107 (1996)
3. Björk, T., Hult, H.: A note on Wick products and the fractional Black-Scholes model. Finance Stoch. **9**, 197–209 (2005)
4. Black, F., Scholes, M.: The pricing of options and corporate liabilities. J. Polit. Econ. **81**, 637–659 (1973)
5. Cox, J.C., Ingersoll, J.E., Ross, S.A.: A theory of the term structure of interest rates. Econometrica **53**, 385–407 (1985)
6. Chang, Y., Wang, Y., Zhang, S.: Option pricing under double Heston jump-diffusion model with approximative fractional stochastic volatility. Mathematics **9**, 126 (2021). https://doi.org/10.1155/2021/6634779
7. Duan, J., Wei, J.: Pricing foreign-currency and cross-currency options under GARCH. J. Deriv. **7**(1), 51–63 (1999)
8. El-Nouty, C.: The fractional mixed fractional Brownian motion. Statist. Probab. Lett. **65**(2), 111–120 (2003)
9. Goard, J.: Closed-Form formulae for European options under three-factor models. Commun. Math. Stat. **8**, 379–408 (2018)
10. Grzelak, L.A., Oosterlee, C.W., Van Weeren, S.: Extension of stochastic volatility equity models with the Hull-White interest rate process. Quant. Financ. **12**(1), 89–105 (2012)
11. Heston, S.L.: A closed-form solution for options with stochastic volatility with applications to bond and currency options. Rev. Financ. Stud. **6**(2), 327–343 (1993)
12. Hu, Y., Øksendal, B.: Fractional white noise calculus and application to finance. Infin. Dimens. Anal. Quantum Probab. Relat. Top. **6**(1), 1–32 (2003)
13. Hull, J., White, A.: The pricing of options on assets with stochastic volatility. J. Financ. **42**(2), 281–300 (1987)
14. Hull, J.C., White, A.D.: Using Hull-White interest rate trees. J. Deriv. Spring **3**(3), 26–36 (1996)
15. Jiang, G.J.: Testing option pricing models with stochastic volatility, random jumps and stochastic interest rates. Int. Rev. Financ. **3**(3–4), 233–272 (2002)

16. Kou, S.G.: A jump-diffusion model for option pricing. Manag. Sci. **48**(8), 1086–1101 (2002)
17. Mishura, Y.: Stochastic Calculus for Fractional Brownian Motions and Related Processes. Springer, Heidelberg (2008). https://doi.org/10.1007/978-3-540-75873-0
18. Rindell, K.: Pricing of index options when interest rates are stochastic: an empirical test. J. Bank. Financ. **19**(5), 785–802 (1995)
19. Scott, L.O.: Option pricing when the variance changes randomly: theory, estimation and an application. J. Financ. Quant. Anal. **22**(4), 419–438 (1987)
20. Scott, L.O.: Pricing stock options in a jump-diffusion model with stochastic volatility and interest rates: applications of Fourier inversion methods. Math. Financ. **7**(4), 413–426 (1997)
21. Schöbel, R., Zhu, J.: Stochastic volatility with an Ornstein-Uhlenbeck process: an extension. Eur. Rev. Financ. **3**(1), 23–46 (1999)
22. Stein, E.M., Stein, J.C.: Stock price distributions with stochastic volatility: an analytical approach. Rev. Financ. Stud. **4**(4), 727–752 (1991)
23. Sun, L.: Pricing currency options in the mixed fractional Brownian motion. Phys. A **392**(16), 3441–3458 (2013)
24. Thao, T.H.: An approximate approach to fractional analysis for finance. Nonlinear Anal. Real World Appl. **7**, 124–132 (2006)
25. Vasicek, O.: An equilibrium characterization of the term structure. J. Financ. Econ. **5**(2), 177–88 (1977)
26. Wiggins, J.B.: Option values under stochastic volatility: theory and empirical estimates. J. Financ. Econ. **19**(2), 351–372 (1987)
27. Xiao, W., Zhang, W., Xu, W.: Parameter estimation for fractional Ornstein-Uhlenbeck processes at discrete observation. Appl. Math. Model. **35**(9), 4196–4207 (2011)
28. Xiao, W., Zhang, W., Xu, W., Zhang, X.: The valuation of equity warrants in a fractional Brownian environment. Phys. A **391**(4), 1742–1752 (2012)

On a Nonlinear Equation $p(x)$-Elliptic Problem of Neumann Type by Topological Degree Method

Soukaina Yacini, Adil Abbassi, Chakir Allalou(✉), and Abderrazak Kassidi

Laboratory LMACS, FST of Beni-Mellal, Sultan Moulay slimane University, Beni Mellal, Morocco
chakir.allalou@yahoo.fr

Abstract. The aim of this work is devoted to study the existence of weak solutions for the nonlinear $p(x)$-elliptic problem,

$$-div\, a(x,u,\nabla u) = b(x)|u|^{p(x)-2}u + \lambda H(x,u,\nabla u) \quad \text{in} \quad \Omega,$$

in the Weighted Sobolev spaces Weighted-withe Exponent Variable.The existence is proved by using the topological degree, introduced by Berkovits.

1 Introduction

In this paper we discuss the existence of weak solutions for the following type Neumann problem given by

$$\begin{cases} -div\, a(x,u,\nabla u) = b(x)|u|^{p(x)-2}u + \lambda H(x,u,\nabla u) & \text{in} \quad \Omega, \\ a(x,u,\nabla u)\cdot \eta = 0 & \text{on} \quad \partial\Omega, \end{cases} \tag{1}$$

In the problem Ω is a bounded domain of $\mathbb{R}^N$, $N \geq 2$, λ is a real parameter and η the outer unit normal vector on $\partial\Omega$.

We assume also that $p(.)$ is log- Holder continuous function (in a sense to be precised in section 2 below) and $2 < p^- < p(x) < p^+ < \infty$, we mention that the $b \in L^\infty(\Omega)$, $b(x) > 0$ a.e. in Ω. The operator $-div\, a(x,u,\nabla u)$ is a Leray-Lions operator defined from $W^{1,p(x)}(\Omega,\omega)$ to its dual $(W^{1,p(x)}(\Omega,\omega))^*$, where $a : \Omega \times \mathbb{R} \times \mathbb{R}^N \to \mathbb{R}$, is a Carathéodory's function who satisfaid assumptions of growth, ellipticity and strict monotonicity (see assumption (7), (8) and (9) of section 3), who the nonlinear term $H : \Omega \times \mathbb{R} \times \mathbb{R}^N \to \mathbb{R}$ is a Carathéodory's functions has the growth condition (see assumption (10). When $a(x,u,\nabla u) = A(x,\nabla u)$ and $H(x,u,\nabla u) = f(x,\nabla u) + s(x)|u|^{p-1}u$, T. He et al. have proved in [12] for all large λ. Some important and interesting results can be found in [1,3,11,16]. In the case $a(x,u,\nabla u) = |\nabla u|^{p-2}u$ and $H(x,u,\nabla u) = f(x,\nabla u)$ with Dirichlet boundary conditions, has been established by S. Liu [15]. Furthermore in [2], A. Abbassi, C. Allalou and A.Kassidi solved problem (1) by topological degree methods in the case were p is constant (i.e. $p(.) = p$) within the framework of classical sobolev space $W^{1,p}(\Omega)$. In recent years, the research of differential aquation with variable exponent has been a particularly active field, with applications in electrorheological fluid (see [17,19]) and image processing (see [8,9].

S. Melliani and O. Castillo (Eds.): ICPAMS 2021, LNNS 476, pp. 404–417, 2023.
https://doi.org/10.1007/978-3-031-12416-7_34

Our goal in this article is to study the existence of weak solution of problem (1). The method used to solve the issue (1) is the topological degree, which is frequently utilized in the study of nonlinear equations, particularly elliptic equations. Brouwer created the first topological degree in 1912 for continuous mappings in finite dimensional Euclidean spaces [7]. then Leray and Schauder generalized it in 1934 for compact operators in Banach spaces of infinite dimension [14]. Later, the theory was constructed by Berkovits [5,6]. is organized as follows: in Sect. 2,we state some basic results for the weighted variable exponent LebesgueSobolev spaces and an outline of Berkovits degree theory. In Sect. 3, we give our basic assumption and some related lemmas to prepare for the proof of the main theorem. Finally, in the fourth section, we prove the existence of weak solutions of (1).

2 Mathematical Preliminaries

2.1 Classes of Mappings and Topological Degree

In this subsection, we recall some results and properties from the theory of topological degree, Let X be a real separable reflexive Banach space with dual X^* and with continuous dual pairing $\langle \cdot, \cdot \rangle$ between X^* and X in this order, and given a nonempty subset Ω of X, let $\overline{\Omega}$ and $\partial\Omega$ denote the closure and the boundary of Ω in X, respectively. The symbol $\to$ ($\rightharpoonup$) stands for strong (weak) convergence.

Definition 1. Let Y be another real Banach space. A operator $F : \Omega \subset X \to Y$ is said to be

1. bounded, if it takes any bounded set into a bounded set.
2. demicontinuous, if for any sequence $(u_n) \subset \Omega$, $u_n \to u$ implies $F(u_n) \rightharpoonup F(u)$.
3. compact, if it is continuous and the image of any bounded set is relatively compact.

Definition 2. A mapping $F : \Omega \subset X \to X^*$ is said to be

1. of type (S_+), if for any sequence $(u_n) \subset \Omega$ with $u_n \rightharpoonup u$ and $\limsup_{n\to\infty}\langle Fu_n, u_n - u\rangle \leq 0$, we have $u_n \to u$.
2. quasimonotone, if for any sequence $(u_n) \subset \Omega$ with $u_n \rightharpoonup u$, we have $\limsup_{n\to\infty}\langle Fu_n,\ u_n - u\rangle \geq 0$.

Definition 3. Let $T : \Omega_1 \subset X \to X^*$ be a bounded operator such that $\Omega \subset \Omega_1$. For any operator $F : \Omega \subset X \to X$, we say that

1. F satisfies condition $(S_+)_T$, if for any sequence $(u_n) \subset \Omega$ with $u_n \rightharpoonup u$, $y_n := Tu_n \rightharpoonup y$ and $\limsup_{n\to\infty}\langle Fu_n,\ y_n - y\rangle \leq 0$, we have $u_n \to u$.
2. F has the property $(QM)_T$, if for any sequence $(u_n) \subset \Omega$ with $u_n \rightharpoonup u$, $y_n := Tu_n \rightharpoonup y$, we have $\limsup_{n\to\infty}\langle Fu_n, y - y_n\rangle \geq 0$.

Let $\mathscr{O}$ be the collection of all bounded open set in X. For any $\Omega \subset X$, we consider the following classes of operators:

$$\mathscr{F}_1(\Omega) := \{F : \Omega \to X^* \backslash F \text{ is bounded, demicontinuous and satifies condition } (S_+)\},$$
$$\mathscr{F}_{T,B}(\Omega) := \{F : \Omega \to X \backslash F \text{ is bounded, demicontinuous and satifies condition } (S_+)_T\},$$
$$\mathscr{F}_T(\Omega) := \{F : \Omega \to X \backslash F \text{ is demicontinuous and satifies condition } (S_+)_T\},$$
$$\mathscr{F}_B(X) := \{F \in \mathscr{F}_{T,B}(\overline{E}) \backslash E \in \mathscr{O},\ \mathrm{T} \in \mathscr{F}_1(\overline{\mathrm{E}})\}.$$

Throughout the paper $T \in \mathscr{F}_1(\overline{E})$ is called an essential inner map to F.

Lemma 1 *([6], Lemmas 2.2 and 2.4). Lets $T \in \mathscr{F}_1(\overline{E})$ be continuous and $S : D_S \subset X^* \to X$ be demicontinuous such that $T(\overline{E}) \subset D_s$, where E is a bounded open set in a real reflexive Banach space X. Then the following statements are true:*

1. If S *is quasimonotone, then* $I + SoT \in \mathscr{F}_T(\overline{E})$, *where* I *denotes the identity operator.*
2. *If* S *is of class* (S_+), *then* $SoT \in \mathscr{F}_T(\overline{E})$.

Definition 4. Suppose that E is bounded open subset of a real reflexive Banach space X, $T \in \mathscr{F}_1(\overline{E})$ be continuous and let $F, S \in \mathscr{F}_T(\overline{E})$. The affine homotopy $\Lambda : [0,1] \times \overline{E} \to X$ defined by

$$\Lambda(t,\ u) := (1-t)Fu + tSu \qquad \text{for} \qquad (t,\ u) \in [0,1] \times \overline{E}$$

is called an admissible affine homotopy with the common continuous essential inner map T.

Remark 1 [6]. *The above affine homotopy satisfies condition* $(S_+)_T$.

Now, we introduce the Berkovits topological degree for the class $\mathscr{F}_B(X)$ for more details see [5,6].

Theorem 1. There exists a unique degree function

$$d : \{(F,E,h) \backslash E \in \mathscr{O},\ T \in \mathscr{F}_1(\overline{E}),\ F \in \mathscr{F}_{T,B}(\overline{E}),\ h \notin F(\partial E)\} \longrightarrow \mathbb{Z}$$

that satisfies the following properties:

1. *(*Normalization*)* For any $h \in E$, we have $d(I,E,h) = 1$.
2. *(*Additivity*)* Let $F \in \mathscr{F}_{T,B}(\overline{E})$. If E_1 and E_2 are two disjoint open subsets of E such that $h \notin F(\overline{E} \backslash (E_1 \cup E_2))$ then we have
$$d(F,E,h) = d(F,E_1,h) + d(F,E_2,h).$$
3. *(*Homotopy invariance*)* If $\Lambda : [0,1] \times \overline{E} \to X$ is a bounded admissible affine homotopy with a common continuous essential inner map and h: $[0,1] \to X$ is a continuous path in X such that $h(t) \notin \Lambda(t, \partial E)$ for all $t \in [0,1]$, then the value of $d(\Lambda(t,\cdot), E, h(t))$ is constant for all $t \in [0,1]$.
4. *(*Existence*)* if $d(F,E,h) \neq 0$, then the equation $Fu = h$ has a solution in E.

2.2 Weighted Variable Lebesgue and Sobolev Spaces

Now we give some definitions and elementary properties for the spaces of Lebesgue and Sobolev with Weight and to variable exponents $L^{p(x)}(\Omega,\omega)$, $W^{1,p(x)}(\Omega,\omega)$,

Let Ω be a bounded open subset of $\mathbb{R}^N (N \geq 2)$, we say that a real-valued continuous function $p(.)$ is log-Hölder continuous in Ω if

$$|p(x)-p(y)| \leq \frac{C}{|\log|x-y||} \tag{2}$$

$\forall x,y \in \overline{\Omega}$ such that $|x-y| < \frac{1}{2}$, with possible different constant C.

Denote

$$C_+(\overline{\Omega}) = \{p \in C(\overline{\Omega}) : \min_{x\in\overline{\Omega}} p(x) > 1\}.$$

We define

$$p^+ = \max_{x\in\overline{\Omega}} p(x), \quad p^- = \min_{x\in\overline{\Omega}} p(x).$$

For any $p \in C_+(\overline{\Omega})$, we introduce the variable exponent Lebesgue space by:

$$L^{p(x)}(\Omega) = \left\{u : \Omega \to \mathbb{R}; u \textit{ is measurable with } : \int_\Omega |u(x)|^{p(x)}dx < \infty\right\},$$

the space $L^{p(x)}(\Omega)$ under the norm "Luxembourg norm"

$$||u||_{L^{p(x)}(\Omega)} = ||u||_{p(x)} = \inf\left\{\lambda > 0 : \int_\Omega |\frac{u(x)}{\lambda}|^{p(x)}dx \leq 1\right\},$$

which is a separable and reflexive Banach space.

Definition 5. A function ω defined on Ω is called a weigh function if it is measurable and strictly positive a.e. in Ω.

We introduce for any $p \in C_+(\overline{\Omega})$, the weighted variable exponent Lebesgue space $L^{p(x)}(\Omega,\omega)$ that consists of all measurable real-valued functions u such that

$$L^{p(x)}(\Omega,\omega) = \{u : \Omega \to \mathbb{R}, \textit{measurable}, \int_\Omega |u(x)|^{p(x)}\omega(x)dx < \infty\}.$$

The dual space $L^{p(x)}(\Omega,\omega)$ denoted $L^{p'(x)}(\Omega,\omega^*)$ where $\omega^* = \omega^{1-p'(x)}$ with $\frac{1}{p(x)} + \frac{1}{p'(x)} = 1$ The Luxemburg norm of $L^{p(x)}(\Omega,\omega)$ is

$$||u||_{L^{p(x)}(\Omega,\omega)} = \inf\{\lambda > 0 : \int_\Omega |\frac{u(x)}{\lambda}|^{p(x)}\omega(x)dx \leq 1\}$$

becomes a normed space. When $\omega(x) \equiv 1$, we have $L^{p(x)}(\Omega,\omega) \equiv L^{p(x)}(\Omega)$.

The weighted variable exponent Sobolev space $W^{1,p(x)}(\Omega,\omega)$ is defined by

$$W^{1,p(x)}(\Omega,\omega) = \{u \in L^{p(x)}(\Omega); |\nabla u| \in L^{p(x)}(\Omega,\omega)\},$$

where the norm is

$$\|u\| = \|u\|_{W^{1,p(x)}(\Omega,\omega)} = \|u\|_{L^{p(x)}(\Omega)} + \|\nabla u\|_{L^{p(x)}(\Omega,\omega)}, \tag{3}$$

or, equivalently

$$\|u\|_{W^{1,p(x)}(\Omega,\omega)} = \inf\{\lambda > 0 : \int_\Omega |\frac{u(x)}{\lambda}|^{p(x)} + \omega(x)|\frac{\nabla u(x)}{\lambda}|^{p(x)} dx \leq 1\}$$

for all $u \in W^{1,p(x)}(\Omega,\omega)$.

Lemma 2 *(See* [10]*). (Generalised Hölder inequality)*

i) *For any functions $u \in L^{p(x)}(\Omega,\omega)$ and $v \in L^{p'(x)}(\Omega,\omega^*)$ we have* $|\int_\Omega uvdx| \leq (\frac{1}{p^-} + \frac{1}{p'^-})||u||_{p(x),\omega}||v||_{p'(x),\omega^*} \leq 2||u||_{p(x),\omega}||v||_{p'(x),\omega^*}$.

ii) *For all p, $q \in C_+(\bar{\Omega})$ such that $p(x) \leq q(x)$ a.e. in Ω, we have* $L^{q(x)}\Omega,\omega) \hookrightarrow L^{p(x)}\Omega,\omega)$ *and the embedding is continuous.*

Lemma 3 *(See* [10]*). Denote $\rho(u) = \int_\Omega \omega(x)|u(x)|^{p(x)}dx$ for all $u \in L^{p(x)}(\Omega,\omega)$. Then,*

$$|u|_{L^{p(x)}(\Omega,\omega)} < 1(=1;>1) \text{ if and only if } \rho(u) < 1(=1;>1), \tag{4}$$

$$\text{if } |u|_{L^{p(x)}(\Omega,\omega)} > 1 \text{ then } |u|^{p^-}_{L^{p(x)}(\Omega,\omega)} \leq \rho(u) \leq |u|^{p^+}_{L^{p(x)}(\Omega,\omega)}, \tag{5}$$

$$\text{if } |u|_{L^{p(x)}(\Omega,\omega)} < 1 \text{ then } |u|^{p^+}_{L^{p(x)}(\Omega,\omega)} \leq \rho(u) \leq |u|^{p^-}_{L^{p(x)}(\Omega,\omega)}. \tag{6}$$

Let introduce the integrability conditions used on the framework of these spaces
$(\mathbf{W}_1)$ $\omega \in L^1_{loc(\Omega)}$ and $\omega^{-\frac{1}{(p(x)-1)}} \in L^1_{loc}(\Omega)$;
$(\mathbf{W}_2)$ $\omega^{-s(x)} \in L^1(\Omega,\omega)$ with $s(x) \in (\frac{N}{p(x)},\infty) \cap [\frac{1}{p(x)-1},\infty)$.

Remark 2 [10].
(i) If ω is a positive measurable and finite function, then $L^{p(x)}(\Omega,\omega)$ is a reflexive Banach space.
(ii) Moreover, if (W_1) holds, then $W^{1,p(x)}(\Omega,\omega)$ is a reflexive Banach space.

Proposition 1 **([13], Proposition 2.3).** $W^{1,p(x)}(\Omega,\omega) \hookrightarrow\hookrightarrow L^{p(x)}(\Omega)$.

Proposition 2 *[10]. Let $p \in C_+(\overline{\Omega})$ satisfy the log–Hölder continuity condition. If (W_1) and (W_2) hold, then the estimate*

$$\|u\|_{L^{p(x)}(\Omega)} \leq C\|\nabla u\|_{L^{p(x)}(\Omega,\omega)}$$

holds, for every $u \in C_0^\infty(\Omega)$ with a positive constant C independent of u.

3 Basic Assumptions and Technical Lemmas

Throughout the paper, we assume that the following assumption hold true.

ASSUMPTION (H1)
Let Ω is a bounded open set of $\mathbb{R}^N (N \geq 2)$, $p \in C_+(\bar{\Omega})$.
$a : \Omega \times \mathbb{R} \times \mathbb{R}^N \to \mathbb{R}$ is a Caratheodory function, i.e., (measurable with respect to x in Ω for every (s, ξ) in $\mathbb{R} \times \mathbb{R}^N$ and continuous with respect to (s, ξ) in $\mathbb{R} \times \mathbb{R}^N$, for almost every x in Ω) which satisfies the following conditions there exist $k \in L^{p'(x)}(\Omega)$ and $\alpha > 0$, $\beta > 0$ such that for almost every $(x) \in \Omega$ all $(s, \xi) \in \mathbb{R} \times \mathbb{R}^N$

$$|a(x,s,\xi)| \leq \beta \omega^{1/p(x)}(x) \Big[k(x) + |s|^{p(x)-1} + \omega^{1/p'(x)}(x)|\xi|^{p(x)-1}\Big], \tag{7}$$

$$\Big[a(x,s,\xi) - a(x,s,\eta)\Big] \cdot (\xi - \eta) > 0 \;\forall\, \xi \neq \eta \in \mathbb{R}^N, \tag{8}$$

$$a(x,s,\xi) \cdot \xi \geq \alpha \omega |\xi|^{p(x)}. \tag{9}$$

where α, β are some positive constants and $k(x)$ is a positive function in $L^{p'(x)}(\Omega)$, ($p'(x)$ is the conjugate exponent of $p(x)$).

ASSUMPTION (H2)
Let $H : \Omega \times \mathbb{R} \times \mathbb{R}^N \to \mathbb{R}$ is a Carathéodory function such that H satisfies the gowth condition

$$|H(x,t,s,\xi)| \leq \rho\big(e(x) + |s|^{p(x)-1} + \omega^{1/p'(x)}(x)|\xi|^{p(x)-1}\big), \tag{10}$$

where ρ is a positive constant, $e(x)$ is a positive function in $L^{p(x)'}(\Omega)$.

Definition 6. We say that $u \in W^{1,p(x)}(\Omega, \omega)$ is is a weak solution of (1), if

$$\int_\Omega a(x,u,\nabla u)\nabla v dx = \int_\Omega b(x)|u|^{p(x)-2}uv dx + \lambda \int_\Omega H(x,u,\nabla u)v dx, \quad \forall v \in W^{1,p(x)}(\Omega, \omega).$$

Lemma 4 [4]. *Let $g \in L^{p(x)}(\Omega, \omega)$ and let $g_n \in L^{p(x}(\Omega, \omega)$, with $\|g_n\|_{L^{p(x)}(\Omega,\omega)} \leq c$, $1 < r(x) < \infty$. If $g_n(x) \to g(x)$ a.e. in Q, then $g_n \rightharpoonup g$ in $L^{p(x)}(\Omega, \omega)$, where ω is a weight function on Ω.*

Let us consider the nonlinear operator T from $W^{1,p(x)}(\Omega, \omega)$ into its dual of the form

$$Tu = -\operatorname{div} a(x,u,\nabla u),$$

then

$$\langle Tu, v\rangle = \int_\Omega a(x,u,\nabla u)\nabla v dx,$$

for all v in $W^{1,p(x)}(\Omega, \omega)$.

Lemma 5 *Assume that (7), (8) and (9) hold. Then*

- *T is bounded, coercive and continuous.*
- *T is a mapping of type (S_+).*

Proof:

- The operator T is bounded. Inded,
  Let $u,v \in W^{1,p(x)}(\Omega,\omega)$, by using the Hölder's inequality and (7) we have

$$\begin{aligned}
\left|\langle Tu,v\rangle\right| &= \left|\int_\Omega a(x,u,\nabla u)\nabla v\,dx\right| = \left|\int_\Omega a(x,u,\nabla u)\omega(x)^{\frac{1}{p(x)}}\nabla v\omega(x)^{\frac{-1}{p(x)}}\,dx\right| \\
&\leq \left(\frac{1}{p^-}+\frac{1}{p'^-}\right)\|a(x,u,\nabla u)\omega^{\frac{-1}{p(x)}}\|_{p'(x)}\|\nabla v\omega^{\frac{1}{p(x)}}\|_{p(x)} \\
&\leq \left(\frac{1}{p^-}+\frac{1}{p'^-}\right)\|a(x,u,\nabla u)\omega^{\frac{-1}{p(x)}}\|_{p'(x)}\|\nabla v\|_{p(x),\Omega} \\
&\leq 2\Big(\int_\Omega |a(x,u,\nabla u)\omega(x)^{\frac{-1}{p(x)}}|^{p'(x)}dx\Big)^{1/\gamma}\|v\| \\
&\leq 2\Big(\int_\Omega \Big|\beta\omega(x)^{1/p(x)}(k(x)+|u|^{p(x)-1}+\omega(x)^{1/p'(x)}|\nabla u|^{p(x)-1}\Big)\omega^{\frac{-1}{p(x)}}|^{p'(x)}dx\Big)^{1/\gamma}\|v\| \\
&\leq C_1.2^{2(p'^+-1)}\Big(\int_\Omega |(k(x)|^{p'(x)}+|u|^{p(x)}+\omega(x)|\nabla u|^{p(x)}dx\Big)^{1/\gamma}\|v\| \\
&\leq C_2\Big(\int_\Omega |(k(x)|^{p'(x)}dx+\int_\Omega |u|^{p(x)}+\omega(x)|\nabla u|^{p(x)}dx\Big)^{1/\gamma}\|v\| \\
&\leq C_2.(C_3+\|u\|)\|v\| \\
&\leq C_T.\|v\|,
\end{aligned}$$

$$\gamma = \begin{cases} p^- \text{ if } & \|a(x,u,\nabla u)\omega^{\frac{-1}{p(x)}}\|_{p'(x)} \leq 1 \\ p^+ \text{ if } & Verta(x,u,\nabla u)\omega^{\frac{-1}{p(x)}}\|_{p'(x)} \geq 1 \end{cases}$$

  Hence the operator T is bounded.
- The operator T is coercive.Inded for all $v \in W^{1,p(x)}(\Omega)$, we get from 9

$$\begin{aligned}
\frac{\langle Tv,v\rangle}{\|v\|} &= \frac{\int_\Omega a(x,v,\nabla v)\nabla v dx}{\|v\|} \\
&\geq \alpha\frac{\int_\Omega \omega(x)|\nabla v|^{p(x)}dx}{\|v\|} \\
&\geq \alpha\frac{\|v\|^\delta}{\|v\|} \\
&\geq \alpha\|v\|^{\delta-1} \to \infty \quad \text{ if } \|v\| \to \infty
\end{aligned}$$

  were

$$\delta = \begin{cases} p^+ \text{ if } & \|v\| \leq 1 \\ p^- \text{ if } & \|v\| \geq 1, \end{cases}$$

  Hence T is coercive.

- Now, we show that T is continuous, let $u_n \to u$ in $W^{1,p(x)}(\Omega,\omega)$. Then $\nabla u_n \to \nabla u$ in $\left(L^{p(x)}(\Omega,\omega)\right)^N$. Hence there exist a subsequence (u_k) of (u_n) and measurable functions h in $L^{p(x)}(\Omega,\omega)$ and g in $(L^{p(x)}(\Omega,\omega))^N$ such that

$$u_k(x) \to u(x) \quad \text{and} \quad \nabla u_k(x) \to \nabla u(x),$$

$$|u_k(x)| \leq h(x) \quad \text{and} \quad |\nabla u_k(x)| \leq |g(x)|,$$

for $a.e.x \in \Omega$ and all $k \in \mathbb{N}$. Since a satisfies the Carathéodory condition, we obtain

$$a(x,u_k,\nabla u_k) \to a(x,u,\nabla u) \qquad \text{a.e. } x \in \Omega. \tag{11}$$

According to (7), we have

$$|a(x,u_k,\nabla u_k)| \leq \beta \omega^{\frac{1}{p(x)}} \left(k(x) + |h(x)|^{p(x)-1} + \omega^{\frac{1}{p'(x)}} |g(x)|^{p(x)-1}\right),$$

for a.e. $x \in \Omega$. Since

$$\int_\Omega \beta \omega^{\frac{1}{p(x)}} |k(x) + |h(x)|^{p(x)-1} + \omega^{\frac{1}{p'(x)}} |g(x)|^{p(x)-1}|^{p'(x)} \omega^* dx \leq C \int_\Omega |k(x)|^{p'(x)} + |h(x)|^{p(x)} + \omega |g(x)|^{p(x)} dx < \infty$$

(because $K \in L^{p'(x)}(\Omega)$, $h \in L^{p(x)}(\Omega,\omega)$ and g in $(L^{p(x)}(\Omega,\omega))^N$),then

$$\beta \omega^{\frac{1}{p(x)}} \left(k(x) + |h(x)|^{p(x)-1} + \omega^{\frac{1}{p'(x)}} |g(x)|^{p(x)-1}\right) \in L^{p'(x)}(\Omega,\omega^*),$$

and by using (11) we have

$$\int_\Omega |a(x,u_k,\nabla u_k) - a(x,u,\nabla u)|^{p'(x)} \omega^* dx \longrightarrow 0.$$

The dominated convergence theorem imply that

$$a(x,u_k,\nabla u_k) \to a(x,u,\nabla u) \qquad \text{in} \quad \left(L^{p'(x)}(\Omega,\omega^*)\right)^N.$$

Therefore, for all $v \in W^{1,p(x)}(\Omega,\omega)$ we have $\langle Tu_n, v\rangle \to \langle Tu, v\rangle$, which implies that the operator T is continuous.
- It remains to prove that the operator T is of type (S_+).
Let $(u_n)_n$ be a sequence in $W^{1,p(x)}(\Omega,\omega)$ such that

$$\begin{cases} u_n \rightharpoonup u & \text{in } W^{1,p(x)}(\Omega,\omega) \\ \limsup\limits_{n\to\infty} \langle Tu_n, u_n - u\rangle \leq 0. \end{cases} \tag{12}$$

We will show that $u_n \to u$ in $W^{1,p(x)}(\Omega,\omega)$.

$$\lim_{n\to\infty} \langle Tu_n,\ u_n - u\rangle = \lim_{n\to\infty} \langle Tu_n - Tu,\ u_n - u\rangle = 0. \tag{13}$$

Hence

$$D_n \to 0 \quad \text{in } L^1(\Omega) \text{ as } n \to \infty,$$

where $D_n = \left[a(x,u_n,\nabla u_n) - a(x,u_n,\nabla u)\right]\left(\nabla u_n - \nabla u\right)$. thanks to (8), we have D_n is a positive function. Since $u_n \rightharpoonup u$ in $W^{1,p(x)}(\Omega,\omega)$ then, $u_n \to u$ a.e. in Ω and since $D_n \to 0$ a.e. in Ω, there exists a subset B in Ω with measure zero such that for all $x \in \Omega \backslash B$,

$$|u(x)| < \infty, \quad |\nabla u| < \infty, \quad K(x) < \infty, \quad u_n \to u, \quad D_n \to 0.$$

Taking $\xi_n = \nabla u_n$ and $\xi = \nabla u$, we have

$$\begin{aligned}
D_n(x) &= \Big[a(x,u_n,\xi_n) - a(x,u_n,\xi)\Big]\cdot(\xi_n - \xi)\\
&= a(x,u_n,\xi_n)\xi_n + a(x,u_n,\xi)\xi - a(x,u_n,\xi_n)\xi - a(x,u_n,\xi)\xi_n\\
&\geq \alpha\omega(x)|\xi_n|^{p(x)} + \alpha\omega(x)|\xi|^{p(x)}\\
&\quad - \beta\omega^{1/p(x)}(x)\Big(k(x) + |u_n|^{p(x)-1} + \omega^{1/p'(x)}(x)|\xi_n|^{p(x)-1}\Big)|\xi|\\
&\quad - \beta\omega^{1/p(x)}(x)\Big(k(x) + |u_n|^{p(x)-1} + \omega^{1/p'(x)}(x)|\xi|^{p(x)-1}\Big)|\xi_n|\\
&\geq \alpha\omega(x)|\xi_n|^{p(x)} - C_x[1 + \omega(x)^{1/p'(x)}|\xi_n|^{p(x)-1} + \omega(x)^{1/p(x)}|\xi_n|],
\end{aligned}$$

where C_x depending on x, without dependence on n. (since $u_n(x) \to u(x)$, then $(u_n)_n$ is bounded), we obtain

$$D_n(x) \geq |\xi_n|^{p(x)}\left(\alpha\omega(x) - \frac{C_x}{|\xi_n|^{p(x)}} - \frac{C_x\omega^{\frac{1}{p'(x)}}}{|\xi_n|} - \frac{C_x\omega^{\frac{1}{p(x)}}}{|\xi_n|^{p(x)-1}}\right)$$

by the standard argument $(\xi_n)_n$ is bounded almost everywhere in Ω. Indeed, if $|\xi_n| \to \infty$ in a measurable subset $E \in \Omega$, then

$$\lim_{n\to\infty}\int_\Omega D_n(x)dx \geq \lim_{n\to\infty}\int_E |\xi_n|^{p(x)}\left(\alpha\omega(x) - \frac{C_x}{|\xi_n|^{p(x)}} - \frac{C_x\omega^{\frac{1}{p'(x)}}}{|\xi_n|} - \frac{C_x\omega^{\frac{1}{p(x)}}}{|\xi_n|^{p(x)-1}}\right) = \infty$$

which is absurd since $D_n(x) \to 0$ in $L^1(\Omega)$. Let ξ^* an accumulation point of $(\xi_n)_n$, we have $|\xi^*| < \infty$ and by continuity of a, we obtain

$$\Big[a(x,u(x),\xi_n) - a(x,u(x),\xi)\Big]\cdot(\xi_n - \xi) = 0,$$

thanks to (8), we have $\xi^* = \xi$, the uniqueness of the accumulation point implies that $\nabla u_n(x) \to \nabla u(x)$ a.e. in Ω. Since the sequence $a(x,u,\nabla u_n)$ is bounded in $(L^{p'(x)}(\Omega,\omega^*))^N$ and $a(x,u,\nabla u_n) \to a(x,u,\nabla u)$ a.e. in Ω, Lemma 4 implies

$$a(x,u_n,\nabla u_n) \rightharpoonup a(x,u,\nabla u) \quad \text{in } (L^{p'(x)}(\Omega,\omega^*))^N.$$

Let us taking $\bar{y}_n = a(x,u,\nabla u_n)\nabla u_n$ and $\bar{y} = a(x,u,\nabla u)\nabla u$, then $\bar{y}_n \to \bar{y}$ in $L^1(\Omega)$, according to $A3$ the condition (9), we have

$$\alpha w(x)|\nabla u_n|^{p(x)} \leq a(x,u_n,\nabla u_n)\nabla u_n.$$

Let $z_n = |\nabla u_n|^{p(x)}\omega$, $z = |\nabla u|^{p(x)}\omega$ and $y_n = \frac{\bar{y}_n}{\alpha}$, $y = \frac{\bar{y}}{\alpha}$. Then, by Fatou's Lemma, we obtain

$$\int_\Omega 2y dx \leq \liminf_{n\to\infty} \int_\Omega (y_n + y - |z_n - z|) dx,$$

i.e., $0 \leq \limsup_{n\to\infty} \int_\Omega |z_n - z| dx$, hence

$$0 \leq \liminf_{n\to\infty} \int_\Omega |z_n - z| dx \leq \limsup_{n\to\infty} \int_\Omega |z_n - z| dx \leq 0,$$

this implies

$$\nabla u_n \to \nabla u \quad \text{in } (L^{p(x)}(\Omega,\omega))^N$$

we deduce that

$$u_n \to u \quad \text{in } W^{1,p(x)}(\Omega,\omega),$$

which completes our proof.

Lemma 6. *Suppose that the hypothese* (10) *holds. Then the operator* $S : W^{1,p(x)}(\Omega,\omega) \to \left(W^{1,p(x)}(\Omega,\omega)\right)^*$ *defined by*

$$\langle Su, v\rangle = -\int_\Omega \left(b(x)|u|^{p(x)-2}u + \lambda H(x,u,\nabla u)\right) v dx, \qquad \forall u, v \in W^{1,p(x)}(\Omega,\omega)$$

is compact.

The proof was divided into three steps.

Step 1
Let $\varphi : W^{1,p(x)}(\Omega,\omega) \to L^{p'(x)}(\Omega)$ be the operator setting by

$$\varphi u(x) := -b(x)|u(x)|^{p(x)-2}u(x) \quad \text{for} \quad u \in W^{1,p(x)}(\Omega,\omega) \quad \text{and} \quad x \in \Omega.$$

It is obvious that φ is continuous. Next we show that ϕ is bounded.
For all $u \in W^{1,p(x)}(\Omega,\omega)$, we have

$$\begin{aligned}
\|\varphi u\|^{\gamma}_{p'(x)} &= \int_\Omega |-b(x)|u|^{p(x)-2}u|^{p'(x)} dx \\
&\leq \|b^{p'^+}\|_\infty \int_\Omega |u|^{(p(x)-1)p'(x)} dx \\
&\leq \|b^{p'^+}\|_\infty \int_\Omega |u|^{p(x)} dx \\
&\leq C\|u\|^{\gamma'},
\end{aligned}$$

where

$$\gamma = \begin{cases} p^+ \text{ if } & \|\varphi u\|_{p'(x)} \leq 1 \\ p^- \text{ if } & \|\varphi u\|_{p'(x)} \geq 1. \end{cases}$$

and

$$\gamma' = \begin{cases} p^+ \text{ if } & \|u\| \leq 1 \\ p^- \text{ if } & \|u\| \geq 1. \end{cases}$$

This implies that φ is bounded on $W^{1,p(x)}(\Omega,\omega)$.

Step 2. We show that the operator ψ defined from $W^{1,p(x)}(\Omega,\omega)$ into $L^{p'(x)}(\Omega)$ by

$$\psi u(x) := -\lambda H(x,u,\nabla u) \quad \text{for} \quad u \in W^{1,p(x)}(\Omega,\omega) \quad \text{and} \quad x \in \Omega$$

is bounded and continuous.

Let $u \in W^{1,p(x)}(\Omega,\omega)$, by using the growth condition 10 we obtain,

$$\begin{aligned}
\|\psi u\|^{\theta}_{p'(x)} &\leq \int_{\Omega} |\lambda H(x,u,\nabla u)|^{p'(x)} dx \\
&\leq \int_{\Omega} \lambda^{p'(x)} \rho^{p'(x)} \left| e(x) + |u|^{p(x)-1} + \omega^{\frac{1}{p'(x)}} |\nabla u|^{p(x)-1} \right|^{p'(x)} dx \\
&\leq \int_{\Omega} (\rho\lambda)^{p'^{+}} 2^{p'^{+}-1} \left(|e(x)|^{p'(x)} + \left(|u|^{p(x)-1} + \omega^{\frac{1}{p'(x)}} |\nabla u|^{p(x)-1} \right)^{p'(x)} \right) dx \\
&\leq \int_{\Omega} (\rho\lambda)^{p'^{+}} 2^{p'^{+}-1} \left(|e(x)|^{p'(x)} + 2^{p'^{+}-1} \left(|u|^{(p(x)-1)p'(x)} + \omega |\nabla u|^{(p(x)-1)p'(x)} \right) dx \\
&\leq \int_{\Omega} (\rho\lambda)^{p'^{+}} 2^{2(p'^{+}-1)} \left(|e(x)|^{p'(x)} + |u|^{p(x)} + \omega |\nabla u|^{p(x)} \right) dx \\
&\leq C \int_{\Omega} |e(x)|^{p'(x)} dx + C \int_{\Omega} |u|^{p(x)} + \omega |\nabla u|^{p(x)} dx \\
&\leq C \|e\|^{\theta_1}_{p'(x)} + C \|u\|^{\theta_2} \\
&\leq C_{max} \left(\|u\|^{\theta_2} + 1 \right),
\end{aligned}$$

where $C_{max} = \max\left(C\|e\|^{\theta_1}_{p'(x)}, C \right)$, and

$$\theta = \begin{cases} p'^{+} & \text{if} \quad \|\psi u\|_{p'(x)} \leq 1 \\ p'^{-} & \text{if} \quad \|\psi u\|_{p'(x)} \geq 1. \end{cases}$$

$$\theta_1 = \begin{cases} p'^{+} & \text{if} \quad \|e\|_{p'(x)} \leq 1 \\ p'^{-} & \text{if} \quad \|e\|_{p'(x)} \geq 1. \end{cases} \quad and, \theta_2 = \begin{cases} p^{+} & \text{if} \quad \|u\| \leq 1 \\ p^{-} & \text{if} \quad \|u\| \geq 1. \end{cases}$$

Therefore ψ is bounded on $W^{1,p(x)}(\Omega,\omega)$. Next, we show that ψ is continuous, let $u_n \to u$ in $W^{1,p(x)}(\Omega,\omega)$,

then $u_n \to u$ in $L^{p(x)}(\Omega)$ and $\nabla u_n \to \nabla u$ in $\left(L^{p(x)}(\Omega,\omega)\right)^N$. Thus there exist a subsequence still denoted by (u_n) and measurable functions φ in $L^{p(x)}(\Omega)$ and σ in $\left(L^{p(x)}(\Omega,\omega)\right)^N$ such that

$$u_n(x) \to u(x) \quad \text{and} \quad \nabla u_n(x) \to \nabla u(x),$$

$$|u_n(x)| \leq \varphi(x) \quad \text{and} \quad |\nabla u_n(x)| \leq |\sigma(x)|,$$

for a.e. $x \in \Omega$ and all $n \in \mathbb{N}$. Since H satisfies the Carathéodory condition, we obtain

$$H(x,u_n(x),\nabla u_n(x)) \to H(x,u(x),\nabla u(x)) \quad \text{a.e. } x \in \Omega. \tag{14}$$

Thanks to (10) we obtain

$$|H(x,u_n(x),\nabla u_n(x))| \leq \rho \left(e(x) + |\varphi(x)|^{p(x)-1} + \omega^{\frac{1}{p'(x)}} \sigma(x)|^{p(x)-1} \right)$$

for a.e. $x \in \Omega$ and for all $k \in \mathbb{N}$. Since

$$e(x) + |\varphi(x)|^{p(x)-1} + |\omega^{\frac{1}{p'(x)}} \sigma(x)|^{p(x)-1} \in L^{p'(x)}(\Omega),$$

and from (14), we get

$$\int_{\Omega} |H(x, u_k(x), \nabla u_k(x)) - H(x, u(x), \nabla u(x))|^{p'(x)} dx \longrightarrow 0,$$

by using the dominated convergence theorem we have

$$\psi u_k \to \psi u \quad \text{in} \quad L^{p'(x)}(\Omega).$$

Thus the entire sequence (ψu_n) converges to ψu in $L^{p'(x)}(\Omega)$ and then ψ is continuous.

Step 3
Since the embedding $I : W^{1,p(x)}(\Omega, \omega) \to L^{p(x)}(\Omega)$ is compact, it is known that the adjoint operator $I^* : L^{p'(x)}(\Omega) \to \left(W^{1,p(x)}(\Omega, \omega)\right)^*$ is also compact. Therefore, the compositions $I^* o\varphi$ and $I^* o\psi$ from $W^{1,p(x)}(\Omega)$ into $\left(W^{1,p(x)}(\Omega, \omega)\right)^*$ are compact.

Finally, the composition

$$S = I^* o\phi + I^* o\varphi$$

is compact. which completes the present proof.

4 Main Results

Theorem 2. *Assume that the assumptions (7)–(10) hold. Then, the problem (1) has a weak solution u in* $W^{1,p(x)}(\Omega, \omega)$.

Proof. Let T, S be two operators from $W^{1,p(x)}(\Omega, \omega)$ into its dual as defined in Lemmas 5 and 6 respectively. Then $u \in W^{1,p(x)}(\Omega, \omega)$ is a weak solution of the problem (1) if and only if

$$Tu = -Su \tag{15}$$

According to Lemma 6 the operator S is bounded, continuous and quasimonotone. On the other hand, in light of the properties of the operator T given in Lemma 5 and since the operator T is strictly monotone. Moreover, note by using the Minty-Browder Theorem (see [18], Theorem 26 A), the inverse operator $G := T^{-1}: \left(W^{1,p(x)}(\Omega, \omega)\right)^* \to W^{1,p(x)}(\Omega, \omega)$ is bounded, continuous and satisfies condition (S_+).

Hence, Eq. (15) is equivalent to the abstract Hammerstein equation

$$u = Gv \quad \text{and} \quad v + SoGv = 0 \tag{16}$$

To solve Eq. (16), we will using the Berkovits topological degree introduced in Sect. 2. For that, we first proove that the set

$$B := \left\{ v \in \left(W^{1,p(x)}(\Omega, \omega)\right)^* \setminus \quad v + tSoGv = 0 \quad \text{for some} \quad t \in [0,1] \right\}$$

is bounded. Indeed, let $v \in B$ and take $u := Gv$.

Thanks to (7), (10), the Young's inequality, we have

$$
\begin{aligned}
\|Gv\|^{\theta'} &\leq \int_\Omega |u|^{p(x)}\,dx + \int_\Omega \omega|\nabla u|^{p(x)}\,dx \\
&\leq \int_\Omega |u|^{p(x)}\,dx + \frac{1}{\alpha}\int_\Omega a(x,u,\nabla u)\nabla u dx \\
&\leq \int_\Omega |u|^{p(x)}\,dx + \frac{1}{\alpha}\langle Tu,u\rangle = \int_\Omega |u|^{p(x)}\,dx + \frac{1}{\alpha}\langle v, Gv\rangle \\
&\leq \int_\Omega |u|^{p(x)}\,dx + \frac{t}{\alpha}|\langle S\circ Gv, Gv\rangle| \\
&\leq \int_\Omega |u|^{p(x)}\,dx + \frac{t}{\alpha}\int_\Omega b(x)|u|^{p(x)}dx + \frac{t}{\alpha}\int_\Omega |\lambda H(x,u,\nabla u)|u dx \\
&\leq \int_\Omega |u|^{p(x)}\,dx + \frac{t\|b\|_\infty}{\alpha}\int_\Omega |u|^{p(x)}dx + \frac{t}{\alpha}C_{p'}\int_\Omega |H(x,u,\nabla u)|^{p'(x)}dx + \frac{t}{\alpha}C_p\int_\Omega |u|^{p(x)}dx \\
&\leq C_1\int_\Omega |u|^{p(x)}dx + C_2(\|u\|^{\theta_2}+1) \\
&\leq Cst\left(\|Gv\|^{\theta'} + \|Gv\|^{\theta_2}+1\right),
\end{aligned}
$$

where

$$
\theta' = \begin{cases} p^+ \text{ if } & \|Gv\| \leq 1 \\ p^- \text{ if } & \|Gv\| \geq 1. \end{cases}
$$

This implies that $\{Gv \setminus v \in B\}$ is bounded.

Since the operator S is bounded, it is obvious from (16) that the set B is bounded in $\left(W^{1,p(x)}(\Omega,\omega)\right)^*$. Therefore, we can choose a positive constant R such that

$$
\|v\|_{\left(W^{1,p(x)}(\Omega,\omega)\right)^*} < R \quad \text{for all} \quad v \in B.
$$

It follows that

$$
v + tSoGv \neq 0 \quad \text{for all} \quad v \in \partial B_R(0) \quad \text{and all} \quad t \in [0,1].
$$

By Lemma 1 we get

$$
I + SoG \in \mathscr{F}_T(\overline{B_R(0)}) \quad \text{and} \quad I = ToG \in \mathscr{F}_T(\overline{B_R(0)}).
$$

Since the operators I, S and G are bounded, $I+SoG$ is also bounded. We conclude that

$$
I + SoG \in \mathscr{F}_{T,B}(\overline{B_R(0)}) \quad \text{and} \quad I \in \mathscr{F}_{T,B}(\overline{B_R(0)}).
$$

Consider an affine homotopy $\Lambda : [0,1]\times\overline{B_R(0)} \to W^{-1,p'(x)}(\Omega,\omega)$ given by

$$
\Lambda(t,v) := v + tSoGv \quad \text{for} \quad (t,v) \in [0,1]\times\overline{B_R(0)}.
$$

Applying Theorem 1, we have

$$
d(I+SoG, B_R(0), 0) = d(I, B_R(0), 0) = 1
$$

then, there exists a point $v \in B_R(0)$ such that

$$
v + SoGv = 0.
$$

which says that $u = Gv$ is a weak solution of (1). This completes the proof.

References

1. Abbassi, A., Azroul, E., Barbara, A.: Degenerate $p(x)$-elliptic equation with second membre in L^1. Adv. Sci. Technol. Eng. Syst. J. **2**(5), 45–54 (2017)
2. Abbassi, A., Allalou, C., Kassidi, A.: Topological degree methods for a Neumann problem governed by nonlinear elliptic equation. Moroccan J. Pure Appl. Anal. **6**(2), 231–242 (2020)
3. Aizicovici, S., Papageorgiou, N.S., Staicu, V.: Nodal solutions for nonlinear nonhomogeneous Neumann equations. Topol. Methods Nonlinear Anal. **43**, 421–438 (2014)
4. Bendahmane, M., Wittbold, P., Zimmermann, A.: Renormalized solutions for a nonlinear parabolic equation with variable exponents and L^1-data. J. Differ. Eqn. **249**, 1483–1515 (2010)
5. Berkovits, J.: On the Degree Theory for Nonlineair Mapping of Monotone Type, vol. 58. Guomalaimen Tiedeakatemia, Helsinki (1986)
6. Berkovits, J.: Extension of the Leray-Schauder degree for abstract Hammerstein type mappings. J. Differ. Eqn. **234**(1), 289–310 (2007)
7. Brouwer, L.: Uber abbilding von mannigfaltigktein. Matimatische annalen **71**(1), 97–115 (1912)
8. Chen, Y., Levine, S., Rao, M.: Variable exponent, linear growth functionals in image restoration. SIAM J. Appl. Math. **66**(4), 1383–1406 (2006)
9. Azorero, J.P.G., Alonso, I.P.: Existence and nonuniqueness for the p-Laplacian. Commun. Partial Differ. Eqns. **12**(12), 126–202 (1987)
10. Fan, X.L., Zhao, D.: On the spaces $L^{p(x)}(\Omega)$ and $W^{m,p(x)}(\Omega)$. J. Math. Anal. Appl. **263**, 424–446 (2001)
11. Hu, S., Papageorgiou, N.S.: Nonlinear Neumann problems with indefinite potential and concave terms. Commun. Pure Appl. Anal. **14**, 2561–2616 (2015)
12. He, T., Yao, Z.A., Sun, Z.: Multiple and nodal solutions for parametric Neumann problems with nonhomogeneous differential operator and critical growth. J. Math. Anal. Appl. **449**(2), 1133–1151 (2017)
13. Lahmi, B., Azroul, E., El Haiti, K.: Nonlinear degenerated elliptic problems with dual data and nonstandard growth. Math. Rep. **20**(1), 81–91 (2018)
14. Leray, J., Schauder, J.: Topologie et équations fonctionnelles. Ann. Sci. Ec. Norm. **51**, 45–78 (1934)
15. Liu, S.: Existence of solutions to a superlinear p-Laplacian equation. Electron. J. Differ. Eqn. **2001**(66), 1–6 (2001)
16. El Manouni, S., Papageorgiou, N.S., Winkert, P.: Parametric nonlinear nonhomogeneous Neumann equations involving a nonhomogeneous differential operator. Monatshefte Mat. **177**(2), 203–233 (2014). https://doi.org/10.1007/s00605-014-0649-8
17. Ruzicka, M.: Electrorheological fluids: modeling and mathematical theory. Lect. Notes Math. **1748**, 16–38 (2000)
18. Zeider, E.: Nonlinear Functional Analysis and its Applications, II\B: Nonlinear Monotone Operators. Springer, New York (1990). https://doi.org/10.1007/978-1-4612-0981-2
19. Zhikov, V.-V.: On the density of smooth functions in Sobolev-Orlicz spaces. Zap. Nauchn. Sem. S.- Peterburg. Otdel. Mat. Inst. Steklov. (POMI) **310**, 67–81 (2004)

Check for updates

Nonlinear Elliptic Problems in Weighted Variable Exponent Sobolev Spaces with Nonlocal Boundary Conditions

Soumia El Omari, Said Melliani(✉), and Chakir Allalou

LMACS, Sultan Moulay Slimane University, BP 523 - 23000 Beni Mellal, Morocco
s.melliani@usms.ma, chakir.allalou@yahou.fr

Abstract. In this paper, we study the existence and uniqueness result of weak solution for nonlinear elliptic problems with non-local boundary conditions in the Weighted variable exponent Sobolev spaces $W^{1,p(.)}(\Omega, \omega)$.

1 Introduction

Let Ω be an open bounded domain in $\mathbb{R}^N (N \geq 2)$ with a Lipschitz boundary $\partial\Omega$ such that $\partial\Omega = \Gamma_D \cup \Gamma_{Ne}$ and $\Gamma_D \cap \Gamma_{Ne} = \emptyset$. We aim to study the following problem

$$P(\beta,\rho,f,d)\begin{cases} \beta(u) - \nabla \cdot a(x,\nabla u) = f & \text{in } \Omega, \\ u = 0 & \text{on } \Gamma_D, \\ \rho(u) + \int_{\Gamma_{Ne}} a(x,\nabla u)\cdot\eta = d & \text{on } \Gamma_{Ne}, \\ u \equiv \text{constant} & \text{on } \Gamma_{Ne}, \end{cases}$$

we denote by η the outward unit normal vector on $\partial\Omega, \beta$ and ρ two continuous non-decreasing functions on $\mathbb{R}$ satisfying $\rho(0) = \beta(0) = 0$, a is a Leray-Lions type operator, $d \in \mathbb{R}$ and $f \in L^\infty(\Omega)$.

Our understanding of real world phenomena and our technology now are largely based on boundary values problems involving PDEs which allow us to approach from a mathematical point of view phenomena observed for example in the fields of physics and chemistry. These equations model many physical phenomena such as non linear elasticity, image processing...

In some of these problems, non local boundary conditions are imposed, it is of great interest in several areas of application by now. In typical non local problem, the PDE for an unknown function u at any point in domain Ω involves the non local behavior of u elsewhere in D, add to the local behavior of u in a neighborhood of that point. In addition to the mathematical interest of non local condition, this type of boundary condition arise in petroleum engineering model for well modeling example in a stratified petroleum reservoir of a $3D$ arbitrary geometry (see [1] and [2]).

In this work, we consider variable exponent, we have many applications to problems involving variable exponents as elastic mechanics, electrorheological fluids or image restoration (see [8–11]). Indeed, in our main problem in this paper, non local boundary conditions act on the average of the flux on the boundary, in contrast to the standard

S. Melliani and O. Castillo (Eds.): ICPAMS 2021, LNNS 476, pp. 418–432, 2023.
https://doi.org/10.1007/978-3-031-12416-7_35

case where the boundary conditions is given on the local values of the flux to be more precise, u verify the Dirichlet boundary condition on Γ_D

$$u = 0 \text{ on } \Gamma_D, \tag{1}$$

and u should also satisfy this non-local condition

$$\rho(u) + \int_{\Gamma_{Ne}} a(x, \nabla u) \cdot \eta = d \text{ on } \Gamma_{Ne}. \tag{2}$$

It is well-known that under only conditions (1) and (2),our problem $P(\beta, \rho, f, d)$ is ill-posed (see [4,7]). In order to make this problem well-posed, the unknown function u should be constant on Γ_{Ne}.

This problem was treated by Ouaro and Soma in 2017 in a Sobolev space (see [6]), and they proved the existence and the uniqueness of the same problem but this time in a Sobolev space with variable exponent in 2018 (see [5]).

In this article, we study a nonlinear elliptic problem with non local boundary conditions. We prove an existence and uniqueness result of weak solution to this problem in weighted sobolev space with variable exponent.

We introduce the theorem, which will be essential to establish the existence weak solutions to this main problem.

2 Preliminaries

In this section, we state some elementary properties of the (weighted) variable exponent Lebesgue-Sobolev spaces $L^{p(x)}(\Omega, \omega)$ which will be used in the next sections. The basic properties of the variable exponent Lebesgue-Sobolev spaces $W^{1,p(x)}(\Omega, \omega)$, that is, when $w(x) \equiv 1$ can be found in [12]. Let Ω be a bounded open subset of $\mathbb{R}$, $(N \geq 1)$. Let

$$C_+(\overline{\Omega}) = \left\{ p \in C(\overline{\Omega}) : \min_{x \in \Omega} p(x) > 1 \right\}.$$

For any $p \in C_+(\overline{\Omega})$, we define

$$p^+ = \max_{x \in \Omega} p(x), \quad p^- = \min_{x \in \Omega} p(x).$$

For any $p \in C_+(\overline{\Omega})$, we introduce the weighted variable exponent Lebesgue-Sobolev space $L^{p(x)}(\Omega, \omega)$ with weight ω on Ω, that consists of all measurable real-valued functions u such that

$$L^{p(.)}(\Omega, \omega) = \left\{ u = u(x) : u\omega^{1/p(x)} \in L^{p(.)}(\Omega) \right\}.$$

In this space, we define the Luxemburg norm

$$||u||_{L^{p(.)}(\Omega,\omega)} = ||u||_{p(.),\Omega,\omega} = inf \left\{ \sigma > 0 : \int_\Omega \omega(x) \left| \frac{u(x)}{\sigma} \right|^{p(x)} dx \leq 1 \right\}.$$

We denote by $W^{1,p(\cdot)}(\Omega,\omega)$ the space of all real-valued functions $u \in L^{p(\cdot)}(\Omega,\omega_0)$ such that the derivatives in the sense of distributions satisfy

$$\frac{\partial u}{\partial x_i} \in L^{p(\cdot)}(\Omega, w_i) \text{ for all } i = 1,\dots,N$$

i.e.

$$W^{1,p(\cdot)}(\Omega,\omega) = \left\{ u \in L^{p(\cdot)}(\Omega,\omega_0) / \frac{\partial u}{\partial x_i} \in L^{p(\cdot)}(\Omega, w_i) \text{ for all } i = 1,\dots,N \right\}.$$

This set of functions forms a Banach space under the norm

$$\|u\|_{1,p(\cdot),\Omega,w} = \inf\left\{ \mu > 0 : \int_\Omega \left(w_0(x) \left| \frac{u(x)}{\mu} \right|^{p(x)} + \sum_{i=1}^N w_i(x) \left| \frac{\frac{\partial u(x)}{\partial x_i}}{\mu} \right|^{p(x)} \right) dx \le 1 \right\}$$

is a norm on $W^{1,p(\cdot)}(\Omega,w)$ equivalent to $\|\cdot\|_{W^{1,p(\cdot)}(\Omega,w)}$. The theory of such spaces was developed in [4–8]. When $p(x)$ is a constant function, some results were proved in [1,3]. If $w_0(x) = w_1(x) = \ldots = w_N(x) = 1$, we write $W^{1,p(\cdot)}(\Omega)$ instead of $W^{1,p(\cdot)}(\Omega,w)$ and $\|u\|_{W^{1,p(\cdot)}(\Omega)}$ instead of $\|u\|_{W^{1,p(\cdot)}(\Omega,w)}$.

Lemma 2.1. 1 *Let* $\rho(u) = \int_\Omega \gamma(x)|u|^{|p(x)}dx$ *for* $u \in L^{p(\cdot)}(\Omega,\gamma)$. *We have*

(i) $\|u\|_{L^{p()}(\Omega,\gamma)} < 1 (= 1, > 1)$ *if and only if* $\rho(u) < 1 (= 1, > 1)$,
(ii) if $\|u\|_{L^{p(\cdot)}(\Omega,\gamma)} \le 1$, *then* $\|u\|^{p^+}_{L^{p(\cdot)}(\Omega,\gamma)} \le \rho(u) \le \|u\|^{p^-}_{L^{p(\cdot)}(\Omega,\gamma)'}$
(iii) if $\|u\|_{L^{p(\cdot)}(\Omega,\gamma)} \ge 1$, *then* $\|u\|^{p^-}_{L^{p(\cdot)}(\Omega,\gamma)} \le \rho(u) \le \|u\|^{p^+}_{L^{p(\cdot)}(\Omega,\gamma)}$

Finally, we recall that for $k > 0$ and $s \in \mathbb{R}$, the truncation function $T_k(.)$ given by

$$T_k(s) = \begin{cases} -k, & \text{if } s < -k, \\ s, & \text{if } |s| \le k, \\ k, & \text{if } s > k. \end{cases}$$

For $r \in \mathbb{R}$: Let $r \to r^+ := max(r,0)$

$$r \to sign_0^+(r) := \begin{cases} 1 & if \quad r > 0 \\ 0 & if \quad r \le 0. \end{cases}$$

For $k > 0$ we define $H_k^+ : \mathbb{R} \to \mathbb{R}$ by

$$H_k^+ := \begin{cases} 0 & if \quad r < 0 \\ \frac{r}{k} & if \quad 0 \le r \le k \\ 1 & if \quad r > k. \end{cases}$$

3 Essential Assumption and Main Result

3.1 Assumptions

Where ever the paper, we suppose that the following assumption hold true.

Assumption (H1)
Further, we assume in all our considerations that for $0 \leq i \leq N$,

$$\omega_i \in L^1_{loc}(\Omega) \quad and \quad \omega^{\frac{-1}{p(x)-1}} \in L^1_{loc}(\Omega).$$

Assumption (H2)
We consider Ω a bounded open set of $\mathbb{R}^N (N \geq 2)$.

The function $a : \Omega \times \mathbb{R} \times \mathbb{R}^N \to \mathbb{R}$ is a Carathéodory function satisfying the following conditions

$$|a_i(x,\xi)| \leq \beta \omega_i^{1/p(x)}(x) \Big[k(x) + \sum_{j=1}^{N} \omega_j^{1/p'(x)}(x)|\xi_j|^{p(x)-1}\Big], \tag{1}$$

$$\Big[a(x,\xi) - a(x,\eta)\Big] \cdot (\xi - \eta) > 0 \ \forall \xi \neq \eta \in \mathbb{R}^N, \tag{2}$$

$$a(x,\xi) \cdot \xi \geq \alpha \omega |\xi|^{p(x)}, \tag{3}$$

for all $(\eta, \xi) \in \mathbb{R}^N \times \mathbb{R}^N$ and for almost every $x \in \Omega$, where $k(x)$ is a positive function in $L^{p'(\cdot)}(\Omega)$ and α, β are two positive constants.

3.2 Existence Results

In what follows, we consider the following spaces:

$$W_D^{1,p(\cdot)}(\Omega,\omega) = \Big\{\varphi \in W^{1,p(\cdot)}(\Omega,\omega) : \varphi = 0 \, \text{on} \, \Gamma_D\Big\}$$

and

$$W_{Ne}^{1,p(\cdot)}(\Omega,\omega) = \Big\{\varphi \in W_D^{1,p(\cdot)}(\Omega,\omega) : \varphi \equiv \text{constant on } \Gamma_{Ne}\Big\}.$$

For any $v \in W_{Ne}^{1,p(\cdot)}(\Omega,\omega)$, let $v_{Ne} := v|_{\Gamma_{Ne}}$.

The concept of a solution for main problem $P(\beta,\rho,f,d)$ is given as follow:

Definition 3.1. *A solution of $P(\beta,\rho,f,d)$ is a measurable function $u : \Omega \to \mathbb{R}$ satisfying*

$$\begin{cases} u \in W_{Ne}^{1,p(\cdot)}(\Omega,\omega), \beta(u) \in L^1(\Omega) \, \textit{and for every} \, \varphi \in W_{Ne}^{1,p(\cdot)}(\Omega,\omega) \cap L^\infty(\Omega) \\ \int_\Omega a(x,\nabla u).\nabla\varphi dx + \int_\Omega \beta(u)\varphi dx = \int_\Omega f\varphi dx + (d - \rho(u)_{Ne})\,\varphi_{Ne}. \end{cases} \tag{4}$$

The next theorem represents the main result in this work.

Theorem 3.2. *Assume that* $(H1),(H2)$ *hold and let* $f \in L^{\infty}(\Omega)$. *Then the problem* $P(\beta,\rho,f,d)$ *admits a unique solution* u.

Proof.
In order to prove Theorem 3.2, let us introduce an auxiliary problem, in order to deduce useful a priori estimates.

This paper is organized as follow. In Step 1, we study the auxiliary problem, then in Step 2, we show the existence and uniqueness of solutions to the main problem $P(\beta,\rho,f,d)$.

STEP 1: Approximate Problem:
We consider a Leray-Lions type operator $\tilde{a}(x,\xi):\tilde{\Omega}\times\mathbb{R}^N\to\mathbb{R}^N$ satisfying (1)–(4). Let us consider the problem

$$P(\tilde{\beta},\tilde{\rho},\tilde{f},\tilde{d})\begin{cases}\tilde{\beta}(x,u)-\nabla\cdot\tilde{a}(x,\nabla u)=\tilde{f} \text{ in } \tilde{\Omega}\\ u=0 & \text{on } \Gamma_D\\ \tilde{\rho}(u)+\tilde{a}(x,\nabla u).\eta=\tilde{d} & \text{on } \tilde{\Gamma}_{Ne}\end{cases}$$

where the functions $\tilde{\beta},\tilde{\rho},\tilde{f}$ and $\tilde{d}$ are given by the following expressions:

- $\tilde{\beta}(x,s)=\beta(s)\chi_{\Omega}(x)$ for all $(x,s)\in\tilde{\Omega}\times\mathbb{R}$.
- $\tilde{\rho}(s)=\frac{1}{|\tilde{\Gamma}_{Ne}|}\rho(s)$, $\forall s\in\mathbb{R}$, where $|\tilde{\Gamma}_{Ne}|$ denotes the Hausdorff measure of $\tilde{\Gamma}_{Ne}$.
- $\tilde{f}(x)=(f\chi_{\Omega})(x)$ for all $x\in\tilde{\Omega}$.
- $\tilde{d}$ is a function in $L^{\infty}\left(\tilde{\Gamma}_{Ne}\right)$ such that

$$\int_{\tilde{\Gamma}_{Ne}}\tilde{d}\mathrm{d}\sigma=d. \tag{5}$$

Obviously, we have $\tilde{f}\in L^{\infty}(\tilde{\Omega})$.
The notion of a solution for the problem $P(\tilde{\beta},\tilde{\rho},\tilde{f},\tilde{d})$ is given in the next definition.

Definition 3.3. *A measurable function* $u:\tilde{\Omega}\to\mathbb{R}$ *is a solution for* $P(\tilde{\beta},\tilde{\rho},\tilde{f},\tilde{d})$ *if*

$$\begin{cases}u\in W_D^{1,p(x)}(\tilde{\Omega},\omega),\beta(u)\in L^1(\Omega)\,\textit{and for every}\,\tilde{\varphi}\in W_D^{1,p(x)}(\tilde{\Omega},\omega)\cap L^{\infty}(\Omega)\\ \int_{\Omega}\tilde{a}(x,\nabla u)\cdot\nabla\tilde{\varphi}\mathrm{d}x+\int_{\Omega}\beta(u)\tilde{\varphi}\mathrm{d}x=\int_{\Omega}f\tilde{\varphi}\mathrm{d}x+\int_{\tilde{\Gamma}_{Ne}}(\tilde{d}-\tilde{\rho}(u))\tilde{\varphi}\mathrm{d}\sigma.\end{cases} \tag{6}$$

The existence result for the problem $P(\tilde{\beta},\tilde{\rho},\tilde{f},\tilde{d})$ are given by the following theorem.

Theorem 3.4. *If we assume that the functions* $\tilde{\beta},\tilde{\rho},\tilde{f}$ *and* $\tilde{d}$ *are given as previous. So, the problem* $P(\tilde{\beta},\tilde{\rho},\tilde{f},\tilde{d})$ *admits at least one solution in the sense of Definition 3.3.*

Before proving Theorem 3.4, we need to study first an existence result to the following problem. For any $k>0$ we consider

$$P_k(\tilde{\beta},\tilde{\rho},\tilde{f},\tilde{d})\begin{cases}T_k\left(\tilde{\beta}(x,u_k)\right)-\nabla\cdot\tilde{a}(x,\nabla u_k)=\tilde{f} \text{ in } \tilde{\Omega}\\ u_k=0 & \text{on } \Gamma_D\\ T_k(\tilde{\rho}(u_k))+\tilde{a}(x,\nabla u_k)\cdot\eta=\tilde{d} & \text{on } \tilde{\Gamma}_{Ne}.\end{cases}$$

Hence, we show the following theorem.

Theorem 3.5. *If we assume that* $\tilde{\beta}, \tilde{\rho}, \tilde{f}$ *and* $\tilde{d}$ *are given as previous. Then, for any* $k > 0$ *the problem* $P_k(\tilde{\beta}, \tilde{\rho}, \tilde{f}, \tilde{d})$ *admits at least one solution* u_k *in the sense*

$$\begin{cases} u_k \in W_D^{1,p(.)}(\tilde{\Omega}, \omega) \text{ and for all } \tilde{\varphi} \in W_D^{1,p(.)}(\tilde{\Omega}, \omega) \\ \int_{\tilde{\Omega}} \tilde{a}(x, \nabla u_k) \cdot \nabla \tilde{\varphi} \mathrm{d}x + \int_{\Omega} T_k(\beta(u_k)) \tilde{\varphi} \mathrm{d}x = \int_{\Omega} f \tilde{\varphi} \mathrm{d}x + \int_{\tilde{\Gamma}_{Ne}} (\tilde{d} - T_k(\tilde{\rho}(u_k))) \tilde{\varphi} \mathrm{d}\sigma. \end{cases} \tag{7}$$

Furthermore, for any k large enough, we have some estimates on the solution u_k *of the problem* $P_k(\tilde{\beta}, \tilde{\rho}, \tilde{f}, \tilde{d})$

$$\begin{array}{l} |\beta(u_k)| \leq \theta_1 := \max\left\{\|f\|_\infty, (\beta \circ \rho_0^{-1})\left(|\tilde{\Gamma}_{Ne}| \, \|\tilde{d}\|_\infty\right)\right\} \text{ a.e. in } \Omega \\ |\tilde{\rho}(u_k)| \leq \theta_2 := \max\left\{\|\tilde{d}\|_\infty, (\tilde{\rho} \circ \beta_0^{-1})(\|f\|_\infty)\right\} \text{ a.e. in } \tilde{\Gamma}_{Ne}. \end{array} \tag{8}$$

Proof. In (7) we set $\tilde{\varphi} = H_\varepsilon(u_k - M), \varepsilon > 0$, where $M > 0$ is to be fixed later. We have

$$\begin{array}{l} \int_{\tilde{\Omega}} \tilde{a}(x, \nabla u_k) \cdot \nabla H_\varepsilon(u_k - M) \mathrm{d}x + \int_{\Omega} T_k(\beta(u_k)) H_\varepsilon(u_k - M) \mathrm{d}x \\ \quad = \int_{\Omega} f H_\varepsilon(u_k - M) \mathrm{d}x + \int_{\tilde{\Gamma}_{Ne}} \left(\tilde{d} - T_k(\tilde{\rho}(u_k))\right) H_\varepsilon(u_k - M) \mathrm{d}\sigma. \end{array} \tag{9}$$

The first term in (9) is non-negative.

Indeed,

$$\int_{\tilde{\Omega}} \tilde{a}(x, \nabla u_k) . \nabla H_\varepsilon(u_k - M) \mathrm{d}x = \frac{1}{\varepsilon} \int_{\{0 \leq u_k - M \leq \varepsilon\}} \tilde{a}(x, \nabla u_k) \cdot \nabla u_k \, \mathrm{d}x \geq 0.$$

From (9), we have,

$$\int_{\Omega} T_k(\beta(u_k)) H_\varepsilon(u_k - M) \mathrm{d}x \leq \int_{\Omega} f H_\varepsilon(u_k - M) \mathrm{d}x + \int_{\tilde{\Gamma}_{Ne}} (\tilde{d} - T_k(\tilde{\rho}(u_k))) H_\varepsilon(u_k - M) \mathrm{d}\sigma.$$

Then, we get

$$\begin{array}{l} \int_{\Omega} (T_k(\beta(u_k)) - T_k(\beta(M))) H_\varepsilon(u_k - M) \mathrm{d}x + \int_{\tilde{\Gamma}_{Ne}} (T_k(\tilde{\rho}(u_k)) - T_k(\tilde{\rho}(M))) H_\varepsilon(u_k - M) \mathrm{d}\sigma \\ \leq \int_{\Omega} (f - T_k(\beta(M))) H_\varepsilon(u_k - M) \mathrm{d}x + \int_{\tilde{\Gamma}_{Ne}} (\tilde{d} - T_k(\tilde{\rho}(M))) H_\varepsilon(u_k - M) \mathrm{d}\sigma. \end{array}$$

When $\varepsilon \to 0$ in the previous inequality, we obtain

$$\begin{aligned} &\int_{\Omega} (T_k(\beta(u_k)) - T_k(\beta(M))) \operatorname{sign}_0^+(u_k - M) \mathrm{d}x \\ &\quad + \int_{\tilde{\Gamma}_{Ne}} (T_k(\tilde{\rho}(u_k)) - T_k(\tilde{\rho}(M))) \operatorname{sign}_0^+(u_k - M) \mathrm{d}\sigma \\ &\qquad \leq \int_{\Omega} (f - T_k(\beta(M))) \operatorname{sign}_0^+(u_k - M) \mathrm{d}x \\ &\qquad \quad + \int_{\tilde{\Gamma}_{Ne}} (\tilde{d} - T_k(\tilde{\rho}(M))) \operatorname{sign}_0^+(u_k - M) \mathrm{d}\sigma. \end{aligned}$$

Equivalently we obtain

$$\begin{array}{l} \int_{\Omega} (T_k(\beta(u_k)) - T_k(\beta(M)))^+ \mathrm{d}x + \int_{\tilde{\Gamma}_{Ne}} (T_k(\tilde{\rho}(u_k)) - T_k(\tilde{\rho}(M)))^+ \mathrm{d}\sigma \\ \quad \leq \int_{\Omega} (f - T_k(\beta(M))) \operatorname{sign}_0^+(u_k - M) \mathrm{d}x + \int_{\tilde{\Gamma}_{Ne}} (\tilde{d} - T_k(\tilde{\rho}(M))) \operatorname{sign}_0^+(u_k - M) \mathrm{d}\sigma \end{array}$$

Since $\mathrm{Im}(\beta)=\mathrm{Im}(\rho)=\mathbb{R}$, we can fix $M=M_0=\max\left\{\beta_0^{-1}\left(\|f\|_\infty\right),\rho_0^{-1}\left(|\tilde{\Gamma}_{Ne}|\,\|\tilde{d}\|_\infty\right)\right\}$
From the previous inequality we have

$$\begin{aligned}&\int_\Omega\left(T_k\left(\beta\left(u_k\right)\right)-T_k\left(\beta\left(M_0\right)\right)\right)^+\mathrm{d}x+\int_{\tilde{\Gamma}_{Ne}}\left(T_k\left(\tilde{\rho}\left(u_k\right)\right)-T_k\left(\tilde{\rho}\left(M_0\right)\right)\right)^+\mathrm{d}\sigma\\&\quad\le\int_\Omega\left(f-T_k\left(\|f\|_\infty\right)\right)\mathrm{sign}_0^+\left(u_k-M_0\right)\mathrm{d}x+\int_{\tilde{\Gamma}_{Ne}}\left(\tilde{d}-T_k\left(\|\tilde{d}\|_\infty\right)\right)\mathrm{sign}_0^+\left(u_k-M_0\right)\mathrm{d}\sigma\end{aligned}$$

For $k>k_0:=\max\left\{\|f\|_\infty,\|\tilde{d}\|_\infty\right\}$, it follows that

$$\int_\Omega\left(T_k\left(\beta\left(u_k\right)\right)-T_k\left(\beta\left(M_0\right)\right)\right)^+\mathrm{d}x+\int_{\tilde{\Gamma}_{Ne}}\left(T_k\left(\tilde{\rho}\left(u_k\right)\right)-T_k\left(\tilde{\rho}\left(M_0\right)\right)\right)^+\mathrm{d}\sigma\le 0. \quad (10)$$

Thus

$$\begin{aligned}\int_\Omega\left(T_k\left(\beta\left(u_k\right)\right)-T_k\left(\beta\left(M_0\right)\right)\right)^+\mathrm{d}x\le 0\\ \int_{\tilde{\Gamma}_{Ne}}\left(T_k\left(\tilde{\rho}\left(u_k\right)\right)-T_k\left(\tilde{\rho}\left(M_0\right)\right)\right)^+\mathrm{d}\sigma\le 0.\end{aligned}$$

This means that

$$\begin{aligned}\int_\Omega\left(T_k\left(\beta\left(u_k\right)\right)-T_k\left(\beta\left(M_0\right)\right)\right)^+\mathrm{d}x=0\\ \int_{\tilde{\Gamma}_{Ne}}\left(T_k\left(\tilde{\rho}\left(u_k\right)\right)-T_k\left(\tilde{\rho}\left(M_0\right)\right)\right)^+\mathrm{d}\sigma=0.\end{aligned} \quad (11)$$

From (11) we conclude that

$$\begin{aligned}\left(T_k\left(\beta\left(u_k\right)\right)-T_k\left(\beta\left(M_0\right)\right)\right)^+&=0\,\text{a.e. in }\Omega\\ \left(T_k\left(\tilde{\rho}\left(u_k\right)\right)-T_k\left(\tilde{\rho}\left(M_0\right)\right)\right)^+&=0\,\text{a.e. on }\tilde{\Gamma}_{Ne}.\end{aligned}$$

Then, for any $k>k_0:=\max\left\{\|f\|_\infty,\|\tilde{d}\|_\infty\right\}$ we get

$$\begin{aligned}T_k\left(\beta\left(u_k\right)\right)\le T_k\left(\beta\left(M_0\right)\right)\text{ a.e. in }\Omega\\ T_k\left(\tilde{\rho}\left(u_k\right)\right)\le T_k\left(\tilde{\rho}\left(M_0\right)\right)\text{ a.e. on }\tilde{\Gamma}_{Ne}.\end{aligned} \quad (12)$$

From (12) we deduce that for every $k>k_1:=\max\left\{\|f\|_\infty,\|\tilde{d}\|_\infty,\beta\left(M_0\right),\tilde{\rho}\left(M_0\right)\right\}$ we get

$$\beta\left(u_k\right)\le\beta\left(M_0\right)\text{ a.e. in }\Omega$$

$$\tilde{\rho}\left(u_k\right)\le\tilde{\rho}\left(M_0\right)\text{ a.e. on }\tilde{\Gamma}_{Ne}.$$

Note that with the choice of M_0 and the fact that $\mathscr{D}(\beta)=\mathscr{D}(\rho)=\mathbb{R}$, for every $k>k_1:=\max\left\{\|f\|_\infty,\|\tilde{d}\|_\infty,\beta\left(M_0\right),\tilde{\rho}\left(M_0\right)\right\}$ we get

$$\begin{aligned}\beta\left(u_k\right)\le\max\left\{\|f\|_\infty,\left(\beta\circ\rho_0^{-1}\right)\left(|\tilde{\Gamma}_{Ne}|\,\|\tilde{d}\|_\infty\right)\right\}\text{ a.e. in }\Omega\\ \tilde{\rho}\left(u_k\right)\le\max\left\{\|\tilde{d}\|_\infty,\left(\tilde{\rho}\circ\beta_0^{-1}\right)\left(\|f\|_\infty\right)\right\}\text{ a.e. on }\tilde{\Gamma}_{Ne}.\end{aligned} \quad (13)$$

We need to prove that for any k large enough

$$\begin{aligned}\beta\left(u_k\right)\ge-\max\left\{\|f\|_\infty,\left(\beta\circ\rho_0^{-1}\right)\left(|\tilde{\Gamma}_{Ne}|\,\|\tilde{d}\|_\infty\right)\right\}\text{ a.e. in }\Omega\\ \tilde{\rho}\left(u_k\right)\ge-\max\left\{\|\tilde{d}\|_\infty,\left(\tilde{\rho}\circ\beta_0^{-1}\right)\left(\|f\|_\infty\right)\right\}\text{ a.e. on }\tilde{\Gamma}_{Ne},\end{aligned} \quad (14)$$

we can easily see that if u_k is a solution of $P_k(\tilde{\beta},\tilde{\rho},\tilde{f},\tilde{d})$, then $(-u_k)$ is a solution of

$$P_k(\widehat{\beta},\hat{\rho},\hat{f},\hat{d})\begin{cases}T_k(\widehat{\beta}(x,u))-\nabla\cdot\widehat{a}(x,\nabla u)=\widehat{f}\text{ in }\tilde{\Omega}\\ u=0 & \text{on }\Gamma_D\\ T_k(\hat{\rho}(u))+\widehat{a}(x,\nabla u)\cdot\eta=\hat{d} & \text{on }\widetilde{\Gamma}_{Ne},\end{cases}$$

where

$$\hat{a}(x,\xi) = -\tilde{a}(x,-\xi), \hat{\beta}(x,s) = -\tilde{\beta}(x,-s), \hat{\rho}(s) = -\tilde{\rho}(-s), \hat{f} = -\tilde{f} \text{ and } \hat{d} = -\tilde{d}.$$

Therefore, for every $k > k_2 := \max\left\{\|f\|_\infty, \|\tilde{d}\|_\infty, -\beta(-M_0), -\tilde{\rho}(-M_0)\right\}$ we get

$$-\beta(u_k) \leq \max\left\{\|f\|_\infty, (\beta \circ \rho_0^{-1})\left(|\tilde{\Gamma}_{Ne}|\,\|\tilde{d}\|_\infty\right)\right\} \text{ a.e. in } \Omega$$
$$-\tilde{\rho}(u_k) \leq \max\left\{\|\tilde{d}\|_\infty, (\tilde{\rho} \circ \beta_0^{-1})\left(\|f\|_\infty\right)\right\} \text{ a.e. on } \tilde{\Gamma}_{Ne}.$$

Which implies (14).

Thanks to the relations (13) and (14) we deduce (8).

As u_k is a solution of $P_k(\tilde{\beta},\tilde{\rho},\tilde{f},\tilde{d})$, using (8) and the fact that Ω is bounded, we obtain $\beta(u_k) \in L^1(\Omega)$. For $k = 1 + \max\{\theta_1, \theta_2\}$ fixed, by (8) we can see that the problem $P(\tilde{\beta},\tilde{\rho},\tilde{f},\tilde{d})$ admits at least one solution u.

Remark 3.6. *From the relations* (8) *and the fact that the functions β and ρ are non-decreasing, one sees that for k large enough the solution u of the problem $P(\tilde{\beta},\tilde{\rho},\tilde{f},\tilde{d})$ belongs to $L^\infty(\Omega) \cap L^\infty\left(\tilde{\Gamma}_{Ne}\right)$ and*

$$|u| \leq C(\beta, \theta_1) \quad \text{a.e. in } \Omega$$

and

$$|u| \leq C(\rho, \theta_2) \quad \text{a.e. on } \tilde{\Gamma}_{Ne}.$$

Now, we consider

$$\tilde{a}(x,\xi) := a(x,\xi)\chi_\Omega(x) + \frac{1}{\varepsilon^{p(x)}}|\xi|^{p(x)-2}\xi\chi_{\tilde{\Omega}\setminus\Omega}(x) \text{ for all } (x,\xi) \in \tilde{\Omega} \times \mathbb{R}^N$$

and we consider the following problem

$$P_\varepsilon(\tilde{\beta},\tilde{\rho},\tilde{f},\tilde{d}) \begin{cases} \tilde{\beta}(x,u_\varepsilon) - \nabla\cdot\left(a(x,\nabla u_\varepsilon) + \frac{1}{\varepsilon^{p(x)}}|\nabla u_\varepsilon|^{p(x)-2}\nabla u_\varepsilon\chi_{\tilde{\Omega}\setminus\Omega}(x)\right) = \tilde{f} & \text{in } \tilde{\Omega} \\ u_\varepsilon = 0 & \text{on } \Gamma_D \\ \tilde{\rho}(u_\varepsilon) + (a(x,\nabla u_\varepsilon))\cdot\eta = \tilde{d} & \text{on } \tilde{\Gamma}_{Ne}. \end{cases}$$

Thanks to Theorem 3.4, $P_\varepsilon(\tilde{\beta},\tilde{\rho},\tilde{f},\tilde{d})$ has at least one solution. Hence, there exists at least one measurable function $u_\varepsilon : \Omega \longrightarrow \mathbb{R}$ such that

$$\begin{cases} u_\varepsilon \in W_D^{1,p(.)}(\tilde{\Omega},\omega), \beta(u_\varepsilon) \in L^1(\Omega) \text{ and for every } \tilde{\varphi} \in W_D^{1,p(.)}(\tilde{\Omega},\omega) \cap L^\infty(\Omega) \\ \int_\Omega a(x,\nabla u_e)\cdot\nabla\tilde{\varphi}dx + \int_{\tilde{\Omega}\setminus\Omega}\frac{1}{\varepsilon^{p(x)}}|\nabla u_\varepsilon|^{p(x)-2}\nabla u_\varepsilon\cdot\nabla\tilde{\varphi}dx + \int_\Omega\beta(u_\varepsilon)\tilde{\varphi}dx \\ \quad = \int_\Omega f\tilde{\varphi}dx + \int_{\tilde{\Gamma}_{Ne}}\left(\tilde{d} - \tilde{\rho}(u_\varepsilon)\right)\tilde{\varphi}d\sigma. \end{cases} \tag{15}$$

Using Remark 3.6 we get $u_\varepsilon \in L^\infty(\Omega) \cap L^\infty\left(\tilde{\Gamma}_{Ne}\right)$.

Remark 3.7. *If u_ε is a solution of $P_e(\tilde{\beta},\tilde{\rho},\tilde{f},\tilde{d})$, by using the following test functions $\widetilde{\varphi} \in W_{Ne}^{1,p(.)}(\Omega,\omega)\cap L^\infty(\Omega)$ such that $\tilde{\varphi} \equiv$ constant on $\widetilde{\Omega}\backslash\Omega$, on sees that in (15)*

$$\int_{\tilde{\Omega}\backslash\Omega} \frac{1}{\varepsilon^{p(x)}}|\nabla u_\varepsilon|^{p(x)-2}\nabla u_\varepsilon.\nabla\tilde{\varphi}dx = 0$$

and the last term is equal to $\left(d - \int_{\tilde{\Gamma}_{Ne}} \tilde{\rho}\,(u_\varepsilon)\,\mathrm{d}\sigma\right)\tilde{\varphi}_{Ne}$ which implies

$$\int_\Omega a(x,\nabla u_\varepsilon).\nabla\tilde{\varphi}\mathrm{d}x + \int_\Omega \beta\,(u_\varepsilon)\,\tilde{\varphi}\mathrm{d}x = \int_\Omega f\tilde{\varphi}\mathrm{d}x + \left(d - \int_{\tilde{\Gamma}_{Ne}} \tilde{\rho}\,(u_\varepsilon)\,\mathrm{d}\sigma\right)\tilde{\varphi}_{Ne}. \quad (16)$$

A priori estimates on the solution u_ε of the problem $P_\varepsilon(\widetilde{\beta},\tilde{\rho},\tilde{f},\tilde{d})$is given by the next result.

Proposition 3.8. *Let consider u_ε a solution of $P_\varepsilon(\widetilde{\beta},\tilde{\rho},\tilde{f},\tilde{d})$. Hence, the following statements hold true:*

(i) $\displaystyle\int_\Omega |\nabla u_\varepsilon|^{p(x)}\,\omega\,\mathrm{d}x + \int_{\tilde{\Omega}\backslash\Omega} \frac{1}{\varepsilon^{p(x)}}|\nabla u_\varepsilon|^{p(x)}\,\omega\,\mathrm{d}x \le C\times\left(\|\tilde{d}\|_{L^1(\tilde{\Gamma}_{Ne})} + \|f\|_{L^1(\Omega)}\right)$,
where C is a positive constant independent of ε

(ii) $\displaystyle\int_\Omega |\beta\,(u_\varepsilon)|\,\mathrm{d}x + \int_{\tilde{\Gamma}_{Ne}} |\tilde{\rho}\,(u_\varepsilon)|\,\mathrm{d}\sigma \le \|\tilde{d}\|_{L^1(\tilde{\Gamma}_{Ne})} + \|f\|_{L^1(\Omega)}$.

Proof. We set $\tilde{\varphi} = u_\varepsilon$ in (15) to get

$$\begin{aligned}\int_\Omega a(x,\nabla u_\varepsilon).\nabla u_\varepsilon\mathrm{d}x + \int_{\tilde{\Omega}\backslash\Omega} \frac{1}{\varepsilon^{p(x)}}|\nabla u_\varepsilon|^{p(x)-2}\nabla u_\varepsilon.\nabla u_\varepsilon\mathrm{d}x + \int_\Omega \beta\,(u_\varepsilon)\,u_\varepsilon\mathrm{d}x \\ = \int_\Omega f u_\varepsilon\mathrm{d}x + \int_{\tilde{\Gamma}_{Ne}} (\tilde{d} - \tilde{\rho}\,(u_\varepsilon)\,\mathrm{d}\sigma)\,\nabla u_\varepsilon d\sigma.\end{aligned} \quad (17)$$

(i) Clearly, one has

$$\int_\Omega \beta\,(u_\varepsilon)\,u_\varepsilon\mathrm{d}x \ge 0,\ \int_{\tilde{\Omega}\backslash\Omega} \frac{1}{\varepsilon^{p(x)}}|\nabla u_\varepsilon|^{p(x)-2}\nabla u_\varepsilon\cdot\nabla u_\varepsilon\mathrm{d}x \ge 0$$

and

$$\begin{aligned}\int_\Omega f u_\varepsilon\mathrm{d}x &\le \int_\Omega |f||u_\varepsilon|dx \\ &\le C(\beta,\theta_1)\,\|f\|_{L^1(\Omega)}.\end{aligned}$$

$$\begin{aligned}\int_{\tilde{\Gamma}_{Ne}} \tilde{d} - \tilde{\rho}\,(u_\varepsilon)\,\mathrm{d}\sigma &= \int_{\tilde{\Gamma}_{Ne}} \tilde{d}u_\varepsilon d\sigma - \int_{\tilde{\Gamma}_{Ne}} \tilde{\rho}\,(u_\varepsilon)\,u_\varepsilon\mathrm{d}\sigma \\ &\le \int_{\tilde{\Gamma}_{Ne}} \tilde{d}u_\varepsilon d\sigma \\ &\le \int_{\tilde{\Gamma}_{Ne}} |\tilde{d}||u_\varepsilon|d\sigma \\ &\le c(\rho,\theta_2)||\tilde{d}||_{L^1(\tilde{\Gamma}Ne)}.\end{aligned}$$

In (17) for the last term, knowing the relation $a(x,\xi)\cdot\xi \geq \alpha\omega|\xi|^{p(x)}$, we get

$$\int_{\Omega} a(x,\nabla u_{\varepsilon})\cdot\nabla u_{\varepsilon}\mathrm{d}x \geq \alpha\int_{\Omega}|\nabla u_{\varepsilon}|^{p(x)}\,\omega\,\mathrm{d}x,$$

we have

$$\begin{aligned}\int_{\Omega}|\nabla u_{\varepsilon}|^{p(x)}\,\omega\,\mathrm{d}x &\leq \frac{1}{\alpha}\int_{\Omega} a(x,\nabla u_{\varepsilon})\cdot\nabla u_{\varepsilon}\mathrm{d}x\\ &\leq \frac{1}{\alpha}\left(\int_{\Omega} f u_{\varepsilon}dx + \int_{\tilde{\Gamma}_{Ne}}(\tilde{d}-\rho(u_{\varepsilon}))\,u_{\varepsilon}d\sigma\right)\\ &\leq \frac{c(\beta,\theta_1)}{\alpha}||f||_{L^1(\Omega)} + \frac{c(\rho,\theta_2)}{\alpha}||\tilde{d}||_{L^1(\tilde{\Gamma}_{Ne})}\\ &\leq C\times\left(\|\tilde{d}\|_{L^1(\tilde{\Gamma}_{Ne})} + \|f\|_{L^1(\Omega)}\right)\end{aligned} \tag{18}$$

with $C \geq \max\left\{\frac{C(\beta,\theta_1)}{\alpha},\frac{C(\rho,\theta_2)}{\alpha}\right\}$.

The terms $\int_{\Omega} a(x,\nabla u_e).\nabla u_e\,\mathrm{d}x$ and $\int_{\Omega}\beta(u_e)\,u_e\,\mathrm{d}x$ in (17), are non-negative so that we have

$$\int_{\tilde{\Omega}\setminus\Omega}\frac{1}{\varepsilon^{p(x)}}|\nabla u_{\varepsilon}|^{p(x)}\,\omega\,\mathrm{d}x \leq C\times\left(\|\tilde{d}\|_{L^1(\tilde{\Gamma}_{Ne})} + \|f\|_{L^1(\Omega)}\right). \tag{19}$$

Adding (18) and (19), we have (i).

(ii) In (15), we set $\tilde{\varphi} = T_k(u_{\varepsilon}), k > 0$, to have

$$\begin{aligned}\int_{\Omega} a(x,\nabla u_{\varepsilon})\cdot\nabla T_k(u_{\varepsilon})dx + \int_{\tilde{\Omega}\setminus\Omega}\frac{1}{\varepsilon^{p(x)}}|\nabla u_{\varepsilon}|^{p(x)-2}\nabla u_{\varepsilon}\cdot\nabla T_k(u_{\varepsilon})dx + \int_{\Omega}\beta(u_{\varepsilon})T_k(u_{\varepsilon})dx\\ + \int_{\tilde{\Gamma}_{Ne}}\rho(u_{\varepsilon})T_k(u_{\varepsilon})d\sigma = \int_{\Omega} fT_k(u_{\varepsilon})dx + \int_{\Gamma_{Ne}} fT_k(u_{\varepsilon})d\sigma.\end{aligned} \tag{20}$$

The first two terms in (20) are non-negative.

For the terms on the right-hand side of (20), we get

$$\begin{aligned}\int_{\Omega} fT_k(u_{\varepsilon})\,\mathrm{d}x + \int_{\tilde{\Gamma}_{Ne}}\tilde{d}T_k(u_{\varepsilon})\,\mathrm{d}\sigma &\leq k\left(\int_{\Omega}|f|\mathrm{d}x + \int_{\tilde{\Gamma}_{Ne}}\tilde{d}\,\mathrm{d}\sigma\right)\\ &= k\left(\|\tilde{d}\|_{L^1(\tilde{\Gamma}_{Ne})} + \|f\|_{L^1(\Omega)}\right).\end{aligned}$$

So, thanks to (20), we have

$$\int_{\Omega}\beta(u_{\varepsilon})\,T_k(u_{\varepsilon})\,\mathrm{d}x + \int_{\tilde{\Gamma}_{Ne}}\tilde{\rho}(u_{\varepsilon})\,T_k(u_{\varepsilon})\,\mathrm{d}\sigma \leq k\left(\|\tilde{d}\|_{L^1(\tilde{\Gamma}_{Ne})} + \|f\|_{L^1(\Omega)}\right).$$

We divide the previous inequality by k, as $k \to 0$ we obtain

$$\begin{aligned}\int_{\Omega}\beta(u_{\varepsilon})\,\mathrm{sign}(u_{\varepsilon})\,\mathrm{d}x + \int_{\tilde{\Gamma}_{Ne}}\tilde{\rho}(u_{\varepsilon})\,\mathrm{sign}(u_{\varepsilon})\,\mathrm{d}\sigma &= \int_{\Omega}|\beta(u_{\varepsilon})|\,\mathrm{d}x + \int_{\hat{\mathrm{N}}_{Ne}}|\tilde{\rho}(u_{\varepsilon})|\,\mathrm{d}\sigma\\ &\leq \left(\|\tilde{d}\|_{L^1(\hat{\mathrm{N}}_{Ne})} + \|f\|_{L^1(\Omega)}\right).\end{aligned}$$

The following result presents useful convergence.

Proposition 3.9. *As $\varepsilon \to 0$, we get*

(i) $u_\varepsilon \to u$ *a.e. in* Ω *and a.e. on* $\tilde{\Gamma}_{Ne}$ *with* $u \in W_D^{1,p^-}(\tilde{\Omega},\omega)\bigcap W_D^{1,p(.)}(\tilde{\Omega},\omega)$.
(ii) $\beta(u_\varepsilon) \to \beta(u)$ *in* $L^1(\Omega)$.
(iii) $\nabla u_\varepsilon \rightharpoonup \nabla u$ *in* $\left(L^{p^-}(\tilde{\Omega}\backslash\Omega,\omega)\right)^N$ *and* $\nabla u = 0$ *in* $\tilde{\Omega}\backslash\Omega$.
(iv) $\tilde{\rho}(u_e) \to \tilde{\rho}(u)$ *in* $L^1\left(\tilde{\Gamma}_{Ne}\right)$.
(v) $a(x,\nabla u_e) \rightharpoonup a(x,\nabla u)$ *in* $\left(L^{p'(.)}(\Omega,\omega)\right)^N$.

Proof. (i) $\forall 0 < \varepsilon < 1$ we get

$$\begin{aligned}
\int_{\tilde{\Omega}} |\nabla u_\varepsilon|^{p^-}\,\omega\,\mathrm{d}x &= \int_\Omega |\nabla u_\varepsilon|^{p^-}\,\omega\,\mathrm{d}x + \int_{\tilde{\Omega}\backslash\Omega} |\nabla u_\varepsilon|^{p^-}\,\omega\,\mathrm{d}x \\
&\leq \int_\Omega |\nabla u_\varepsilon|^{p^-}\,\omega\,\mathrm{d}x + \int_{\tilde{\Omega}\backslash\Omega} \frac{1}{\varepsilon^{p^-}} |\nabla u_\varepsilon|^{p^-}\,\omega\,\mathrm{d}x. \\
&\leq \int_{\Omega\bigcap[|\nabla u_\varepsilon|>1]} |\nabla u_\varepsilon|^{p^-}\,\omega dx + \int_{\Omega\bigcap[|\nabla u_\varepsilon|\leq 1]} |\nabla u_\varepsilon|^{p^-}\,\omega dx \\
&\quad + \int_{(\tilde{\Omega}\ \Omega)\bigcap[|\nabla u_\varepsilon|>\varepsilon]} \frac{1}{\varepsilon^{p^-}} |\nabla u_\varepsilon|^{p^-}\,\omega dx + \int_{(\tilde{\Omega}\ \Omega)\bigcap[|\nabla u_\varepsilon|\leq\varepsilon]} \frac{1}{\varepsilon^{p^-}} |\nabla u_\varepsilon|^{p^-}\,\omega dx \\
&\leq \int_{\Omega\bigcap[|\nabla u_\varepsilon|>1]} |\nabla u|^{p(x)}\,\omega dx + c_1 + \int_{(\tilde{\Omega}\ \Omega)\bigcap[|\nabla u|>\varepsilon]} \frac{1}{\varepsilon^{p(x)}} |\nabla u|^{p(x)}\,\omega dx + c_2 \\
&\leq c\left(||\tilde{d}||_{L^1(\tilde{\Gamma}_{Ne})} + ||f||_{L^1(\Omega)}\right) + c_1 + c\left(||\tilde{d}||_{L^1(\tilde{\Gamma}_{Ne})} + ||f||_{L^1(\Omega)}\right) + c_2
\end{aligned}$$

then, using Proposition 3.8 (i), the sequence $(|\nabla u_\varepsilon|)_{\in>0}$ is bounded in $L^{p^-}(\tilde{\Omega},\omega)$. Since $u_\varepsilon \in W_D^{1,p^-}(\tilde{\Omega},\omega)$ the sequence $(u_e)_{e>0}$ is bounded in $W_D^{1,p^-}(\tilde{\Omega},\omega)$. Up to a subsequence, when $\varepsilon \to 0$ we get

$$u_\varepsilon \rightharpoonup u \,\text{in}\, W_D^{1,p^-}(\tilde{\Omega},\omega), u_\varepsilon \to u \,\text{a.e. in}\, \tilde{\Omega} \,\text{and a.e. on}\, \tilde{\Gamma}_{Ne}.$$

From Proposition, we can see that the sequence $(|\nabla u_\varepsilon|)_{\varepsilon>0}$ is bounded in $L^{p(.)}(\Omega,\omega)$ and then the sequence $u_{\varepsilon>0}$ is bounded in $W_D^{1,p(.)}(\Omega,\omega)$.

Then, as $\varepsilon \to 0$,

$$u_\varepsilon \rightharpoonup u \quad in \quad W_D^{1,p(.)}(\Omega,\omega).$$

Using the fact that $\Omega \subset \tilde{\Omega}$, we conclude (i).

(ii) As $u_\varepsilon \to u$ a.e. in Ω and β is continuous, we deduce that $\beta(u_\varepsilon) \to \beta(u)$ a.e. in Ω. By using Fatou's lemma we get

$$\int_\Omega \beta(u)dx = \int \liminf \beta(u_\varepsilon) \leq \liminf \int_\Omega \beta(u_\varepsilon),$$

and thanks to Proposition 3.8 (ii), we obtain

$$\int_\Omega \beta(u)dx \leq ||\tilde{d}||_{L^1(\tilde{\Gamma}_{Ne})} + ||f||_{L^1(\Omega)}$$

we have $\beta(u) \in L^1(\Omega)$.

Knowing that $|\beta(u_\varepsilon)| \leq \theta_1$ a.e. in Ω, and the fact that $\beta(u_\varepsilon) \to \beta(u)$ a.e in Ω, thanks to Lebesgue dominated convergence theorem (as Ω is bounded), one sees that $\beta(u_\varepsilon) \to \beta(u)$ in $L^1(\Omega)$.

(iii) since $u_\varepsilon \rightharpoonup u$ in $W_D^{1,p_-}(\tilde{\Omega},\omega)$ we have $\nabla u_\varepsilon \rightharpoonup \nabla u$ in $\left(L^{p_-}(\tilde{\Omega},\omega)\right)^N$ and then $\nabla u_\varepsilon \rightharpoonup \nabla u$ in $\left(L^{p_-}(\tilde{\Omega}\backslash\Omega,\omega)\right)^N$. From Proposition 3.8 (i) we assert that $\left(\frac{1}{\varepsilon}|\nabla u_\varepsilon|\right)_{\varepsilon>0}$ is bounded in $L^{p_-}\left(\tilde{\Omega}\backslash\Omega,\omega\right)^N$. Which means that, there exists $\Theta \in \left(L^{p_-}(\tilde{\Omega}\backslash\Omega),\omega\right)^N$ such that

$$\frac{1}{\varepsilon}\nabla u_\varepsilon \rightharpoonup \Theta \text{ in } \left(L^{p_-}(\tilde{\Omega}\backslash\Omega,\omega)\right)^N \text{ as } \varepsilon \to 0.$$

For any $v \in \left(L^{(p_-)'}(\tilde{\Omega}\backslash\Omega,\omega)\right)^N$ we have

$$\int_{\tilde{\Omega}\backslash\Omega} \nabla u_\varepsilon \cdot v \, dx = \int_{\tilde{\Omega}\backslash\Omega} \varepsilon\left(\frac{1}{\varepsilon}\nabla u_\varepsilon\right) \cdot v \, dx = \int_{\tilde{\Omega}\backslash\Omega} \left(\frac{1}{\varepsilon}\nabla u_\varepsilon - \Theta\right) \cdot (\varepsilon v) dx + \varepsilon \int_{\tilde{\Omega}\backslash\Omega} \Theta . v \, dx.$$

As $(\varepsilon v)_{\varepsilon>0}$ converges strongly to zero in $\left(L^{(p_-)'}(\tilde{\Omega}\backslash\Omega,\omega)\right)^N$, passing to the limit, as $\varepsilon \to 0$, in (4.24) it yields

$$\nabla u_\varepsilon \rightharpoonup 0 \text{ in } \left(L^{p(-)}(\tilde{\Omega}\backslash\Omega,\omega)\right)^N$$

therefore, we obtain $\nabla u_\varepsilon \rightharpoonup \nabla u = 0$ in $\left(L^{p_-}(\tilde{\Omega}\backslash\Omega,\omega)\right)^N$.

(iv) As $u_\varepsilon \to u$ a.e. on $\tilde{\Gamma}_{Ne}$ and $\tilde{\rho}$ is continuous, one has $\tilde{\rho}(u_\varepsilon) \to \tilde{\rho}(u)$ a.e. on $\tilde{\Gamma}_{Ne}$. By using Fatou's lemma and Proposition 3.8 (ii)

$$\int_{\tilde{\Gamma}_{Ne}} \tilde{\rho}(u) d\sigma \leq \liminf \int_{\tilde{\Gamma}_{Ne}} \tilde{\rho}(u_\varepsilon) d\sigma \leq ||\tilde{d}||_{L^1(\tilde{\Gamma}_{Ne})} + ||f||_{L^1(\Omega)}$$

we obtain $\tilde{\rho}(u) \in L^1\left(\tilde{\Gamma}_{Ne}\right)$. By the estimate $|\tilde{\rho}(u_\varepsilon)| \leq \theta_2$ a.e. in $\tilde{\Gamma}_{Ne}$ and the Lebesgue dominated convergence theorem, we get (iv).

(v) The sequence $(a(x,\nabla u_\varepsilon))_{\varepsilon>0}$ is bounded in $\left(L^{p'(.)}(\Omega,\omega)\right)^N$. According to (1.1) we can extract a subsequence such that $a(x,\nabla u_\varepsilon) \to \Phi$ in $\left(L^{p'}(\Omega,\omega)\right)^N$. We need now to show that $\Phi = a(x,\nabla u)$ a.e. in Ω. The proof is divided into two steps.

Step 1: We prove that

$$\limsup_{\varepsilon\to 0} \int_\Omega a(x,\nabla u_\varepsilon) . \nabla(u_\varepsilon - u) \, dx \leq 0. \tag{21}$$

Let us set $\tilde{\varphi} = u_\varepsilon - u$ as a test function in (15), we get

$$\begin{aligned}\int_\Omega a(x,\nabla u_\varepsilon) \cdot \nabla(u_\varepsilon - u) \, dx + \int_{\tilde{\Omega}\backslash\Omega} \frac{1}{\varepsilon^{p(x)}} |\nabla u_\varepsilon|^{p(x)-2} \nabla u_\varepsilon \cdot \nabla(u_\varepsilon - u) \, dx \\ + \int_\Omega \beta(u_\varepsilon)(u_\varepsilon - u) \, dx + \int_{\tilde{\Gamma}_{Ne}} \tilde{\rho}(u_\varepsilon)(u_\varepsilon - u) \, d\sigma \\ = \int_\Omega f(u_\varepsilon - u) \, dx + \int_{\tilde{\Gamma}_{Ne}} \tilde{d}(u_\varepsilon - u) \, d\sigma.\end{aligned} \tag{22}$$

We have in mind

$$\int_{\Omega}\beta\left(u_{\varepsilon}\right)\left(u_{\varepsilon}-u\right)\mathrm{d}x=\int_{\Omega}\left(\beta\left(u_{\varepsilon}\right)-\beta(u)\right)\left(u_{\varepsilon}-u\right)\mathrm{d}x+\int_{\Omega}\beta(u)\left(u_{\varepsilon}-u\right)\mathrm{d}x.$$

Since $\int_{\Omega}\left(\beta\left(u_{\varepsilon}\right)-\beta(u)\right)\left(u_{\varepsilon}-u\right)\mathrm{d}x\geq 0$, it yields

$$\int_{\Omega}\beta\left(u_{\varepsilon}\right)\left(u_{\varepsilon}-u\right)\mathrm{d}x\geq\int_{\Omega}\beta(u)\left(u_{\varepsilon}-u\right)\mathrm{d}x.$$

By using the Lebesgue dominated convergence theorem we deduce that

$$\lim_{\varepsilon\to 0}\int_{\Omega}\beta(u)\left(u_{e}-u\right)\mathrm{d}x=0,$$

which means that

$$\limsup_{\varepsilon\to 0}\int_{\Omega}\beta\left(u_{\varepsilon}\right)\left(u_{\varepsilon}-u\right)\mathrm{d}x\geq 0. \tag{23}$$

As $\nabla u=0$ in $\widetilde{\Omega}\backslash\Omega$, one has

$$\begin{aligned}&\limsup_{\varepsilon\to 0}\int_{\tilde{\Omega}\backslash\Omega}\frac{1}{\varepsilon^{p(x)}}\left|\nabla u_{\varepsilon}\right|^{p(x)-2}\nabla u_{\varepsilon}\cdot\nabla\left(u_{\varepsilon}-u\right)\mathrm{d}x\\&=\limsup_{\varepsilon\to 0}\int_{\tilde{\Omega}\backslash\Omega}\frac{1}{\varepsilon^{p(x)}}\left|\nabla u_{\varepsilon}\right|^{p(x)-2}\nabla u_{\varepsilon}\cdot\nabla u\,\mathrm{d}x\geq 0.\end{aligned} \tag{24}$$

We get also

$$\begin{aligned}\int_{\tilde{\Gamma}_{Ne}}\tilde{\rho}\left(u_{\varepsilon}\right)\left(u_{\varepsilon}-u\right)\mathrm{d}\sigma&=\int_{\tilde{\Gamma}_{Ne}}\left(\tilde{\rho}\left(u_{\varepsilon}\right)-\tilde{\rho}(u)\right)\left(u_{\varepsilon}-u\right)\mathrm{d}\sigma+\int_{\tilde{\Gamma}_{Ne}}\tilde{\rho}(u)\left(u_{\varepsilon}-u\right)\mathrm{d}\sigma\\&\geq\int_{\tilde{\Gamma}_{Ne}}\tilde{\rho}(u)\left(u_{\varepsilon}-u\right)\mathrm{d}\sigma.\end{aligned}$$

Since $u_{\varepsilon}\to u$ a.e. on $\tilde{\Gamma}_{Ne}$, we obtain by using the Lebesgue dominated convergence theorem that

$$\lim_{\varepsilon\to 0}\int_{\tilde{\Gamma}_{Ne}}\tilde{\rho}(u)\left(u_{\varepsilon}-u\right)\mathrm{d}\sigma=0.$$

So,

$$\limsup_{\varepsilon\to 0}\int_{\tilde{\Gamma}_{Ne}}\tilde{\rho}\left(u_{\varepsilon}\right)\left(u_{\varepsilon}-u\right)\mathrm{d}\sigma\geq 0. \tag{25}$$

Using once more the Lebesgue dominated convergence theorem, we have

$$\lim_{\varepsilon\to 0}\int_{\Omega}f\left(u_{\varepsilon}-u\right)\mathrm{d}x=0 \tag{26}$$

and

$$\lim_{\varepsilon\to 0}\int_{\tilde{\Gamma}_{Ne}}\tilde{d}\left(u_{\varepsilon}-u\right)\mathrm{d}\sigma=0. \tag{27}$$

We passe to the limit in (22) when ε goes to zero and from (23)–(27), we have (21).

Step 2 : By the standard monotonicity arguments we show that $\Phi = a(x,\nabla u)$ a.e. in Ω. We consider $\varphi \in \mathscr{D}(\Omega)$ and $\lambda \in \mathbb{R}^*$. From (21) and (2) we get

$$\lambda \lim_{\varepsilon\to 0}\int_\Omega a(x,\nabla u_\varepsilon)\cdot\nabla\varphi dx \geq \limsup_{\varepsilon\to 0}\int_\Omega a(x,\nabla u_\varepsilon)\cdot\nabla(u_\varepsilon - u+\lambda\varphi)dx$$
$$\geq \limsup_{\varepsilon\to 0}\int_\Omega a(x,\nabla(u-\lambda\varphi))\cdot\nabla(u_\varepsilon - u+\lambda\varphi)dx.$$

Hence,

$$\lambda \lim_{\varepsilon\to 0}\int_\Omega a(x,\nabla u_\varepsilon).\nabla\varphi dx \geq \lambda\int_\Omega a(x,\nabla(u-\lambda\varphi))\cdot\nabla\varphi dx. \tag{28}$$

Let us divide (28) by $\lambda > 0$ and by $\lambda < 0$ respectively, passing to the limit as $\lambda \to 0$ we obtain

$$\lim_{\varepsilon\to 0}\int_\Omega a(x,\nabla u_\varepsilon).\nabla\varphi dx = \int_\Omega a(x,\nabla u)\cdot\nabla\varphi dx.$$

Then $\int_\Omega \Phi.\nabla\varphi dx = \int_\Omega a(x,\nabla u)\cdot\nabla\varphi dx$ this means that $\operatorname{div}(\Phi) = \operatorname{div} a(x,\nabla u)$ in $\mathscr{D}'(\Omega)$

Therefore, $\Phi = a(x,\nabla u)$ a.e. in Ω and we get $a(x,\nabla u_\varepsilon) \rightharpoonup a(x,\nabla u)$ in $\left(L^{p'}(\Omega,\omega)\right)^N$ as $\varepsilon \to 0$.

STEP 2: Existence and uniqueness of solutions to $P(\beta,\rho,f,d)$
Now we can show Theorem 3.2.

Proof. We get $u \equiv$ constant on $\widetilde{\Omega}\backslash\Omega$ because $\nabla u = 0$ in $L^{p(.)}(\tilde{\Omega}\backslash\Omega,\omega)$, then $u \in W_{Ne}^{1,p(.)}(\Omega,\omega)$ moreover, we have already prove that $\beta(u) \in L^1(\Omega)$ in the proof of Proposition 3.9 (ii). Before proving that u is a solution of $P(\beta,\rho,f,d)$, it suffices to prove the equality in (4).

$\forall\varphi \in W_{Ne}^{1,p(.)}(\Omega,\omega)\cap L^\infty(\Omega)$, let us consider the function $\tilde{\varphi} \in W_D^{1,p(.)}(\tilde{\Omega},\omega)\cap L^\infty(\Omega)$ such that $\tilde{\varphi} = \varphi\chi_\Omega + \varphi_{Ne}\chi_{\tilde{\Omega}\backslash\Omega}$. Thus, $\tilde{\varphi} \equiv$ constant on $\widetilde{\Omega}\backslash\Omega$. Thanks to Remark 3.7, such function $\tilde{\varphi}$ in the equality in (15) gives us

$$\int_\Omega a(x,\nabla u_\varepsilon)\cdot\nabla\varphi dx + \int_\Omega \beta(u_\varepsilon)\varphi dx = \int_\Omega f\varphi dx + \left(d - \int_{\tilde{\Gamma}_{Ne}}\tilde{\rho}(u_\varepsilon)d\sigma\right)\varphi_{Ne}. \tag{29}$$

Letting ε go to zero in (29) and thanks to the convergence in Proposition 3.9, one has

$$\int_\Omega a(x,\nabla u)\cdot\nabla\varphi dx + \int_\Omega \beta(u)\varphi dx = \int_\Omega f\varphi dx + d\varphi_{Ne} - \left(\lim_{\varepsilon\to 0}\int_{\tilde{\Gamma}_{Ne}}\tilde{\rho}(u_\varepsilon)d\sigma\right)\varphi_{Ne}$$
$$= \int_\Omega f\varphi dx + d\varphi_{Ne} - \left(\int_{\tilde{\Gamma}_{Ne}}\tilde{\rho}(u)d\sigma\right)\varphi_{Ne}$$
$$= \int_\Omega f\varphi dx + d\varphi_{Ne} - \left(\int_{\tilde{\Gamma}_{Ne}}\tilde{\rho}(u)_{Ne}\, d\sigma\right)\varphi_{Ne}$$
$$= \int_\Omega f\varphi dx + (d-\rho(u)_{Ne})\varphi_{Ne}$$

this means that u is a solution of $P(\beta,\rho,f,d)$.

Now, we prove the uniqueness part of Theorem 3.2. This proof is a straightforward consequence of the following lemma.

Lemma 3.10. *If we suppose that* u_1 *and* u_2 *are two solutions for the problems* $P(\beta,\rho,f_1,d_1)$ *and* $P(\beta,\rho,f_2,d_2)$, *respectively. Then*

$$(\rho(u_1)_{Ne} - \rho(u_2)_{Ne})^+ + \int_\Omega (\beta(u_1) - \beta(u_2))^+ \, dx \leq \|f_1 - f_2\|_{L^1(\Omega)} + |d_1 - d_2|. \quad (30)$$

Proof. Thanks to Lemma 2 in [6] the uniqueness has proved.

References

1. Ding, Y., Ha-Duong, T., Giroire, J., Moumas, V.: Modelling of single-phase flow for horizontal wells in a stratified medium. Comput. Fluids **33**(5–6), 715–727 (2004)
2. Giroire, J., Ha-Duong, T., Moumas, V.: A non-linear and non-local boundary condition for a diffusion equation in petroleum engineering. Math. Methods Appl. Sci. **28**(13), 1527–1552 (2005)
3. Showalter, R.E.: Monotone Operators in Banach Space and Nonlinear Partial Differential Equations, Mathematical Surveys and Monographs, vol. 49. American Mathematical Society, Providence (1996)
4. Lions, J.-L.: Quelques methodes resolution des problemes aux limites non lineaire. Paris (1969)
5. Ouaro, S., Soma, S.: Nonlinear elliptic problem involving non-local boundary conditions and variable exponent. Complex Variables Elliptic Eqn. **63**(3), 437–461 (2018)
6. Ouaro, S., Soma, S.: Non-local boundary conditions for nonlinear elliptic problems with bounded data and general functions. J. Nonlinear Evol. Eqn. Appl. Issn **2017**(5), 65–80 (2017)
7. Andreu, F., Igbida, N., Maz, J.M.: Obstacle problems for degenerate elliptic equation with nonhomogeneous nonlinear boundary conditions. Math. Models Methods Appl. Sci. **18**(11), 1869–1893 (2008)
8. Ruzicka, M.: Electrorheological Fluids: Modelling and Mathematical Theory. Lecture Notes in Mathematics, vol. 1748. Springer, Berlin (2002)
9. Rajagopal, K.R., Ruzicka, M.: Mathematical modeling of electrorheological materials. Continum Mech. Therm. **13**, 59–78 (2001)
10. Diening, L.: Theoretical and numerical results for electrorheological fluids [Ph.D. thesis]. University of Freiburg, Germany (2002)
11. Mihailescu, M., Radulescu, V.: A multiplicity result for a nonlinear degenerate problem arising in the theory of electrorheological fluids. Proc. R. Soc. A. **462**, 2625–2641 (2006)
12. Diening, L., Harjulehto, P., Hästö, P., Ruzicka, M.: Lebesgue and Sobolev Spaces with Variable Exponents. Lecture Notes in Mathematics, vol. 2017. Springer, Heidelberg (2011). https://doi.org/10.1007/978-3-642-18363-8

Intuitionistic Fuzzy Algebraic Field Extensions

A. Talhaoui, S. Melliani(✉), M. Elomari, and I. Bakhadach

LMACS, Sultan Moulay Slimane University, BP 523 - 23000 Beni Mellal, Morocco
ml.talhaoui@gmail.com, {s.melliani,m.elomari}@usms.ma

Abstract. In this paper we introduce the concept of intuitionistic fuzzy field, the intuitionistic Fuzzy field extensions and the intuitionistic fuzzy algebraic field extensions. and we have investigated these notions and shown some results using the intuitionistic fuzzy points and the $(\alpha, \beta)-$cut.

Keywords: Intuitionistic fuzzy subfield · Intuitionistic fuzzy field extension and the intuitionistic fuzzy algebraic field extensions

Introduction

The intuitionistic fuzzy sets are being studied extensively and being used in different fields, like environment, medical science, economics, some branch of engineering, etc. And it's thanks to the Bulgarian engineer K.T. Atanassov that the IFS have been introduced in 1983 [3], like a generalisation of the of Zadeh's fuzzy sets [8]. and in 1989 the notion of fuzzy subgroups, anti-fuzzy subgroups, fuzzy fields and fuzzy linear spaces was introduced by Biswas.R [4,5]. In 2017 B.Anandh and R. Giri introduced the notion of intuitionistic fuzzy subfield of a field with respect to (T,S)-norm [1]. In this paper, we introduce the notion of the intuitionistic fuzzy subfield and intuitionistic fuzzy field extension and intuitionistic algebraic field extensions and gave some of its properties using the intuitionistic fuzzy point and the $(\alpha, \beta)-$cut.

1 Preliminaries

We denote X the universe First we give the concept of intuitionistic fuzzy subset defined by Atanassov as a generalization of the concept of fuzzy set given by Zadeh.

Definition 1. [2,3] The intuitionistic fuzzy subsets (in shorts IFSS) defined on X as objects having the form

$$A = \{< x, \mu(x), \nu(x) >: x \in X\}$$

where the functions $\mu : X \to [0, 1]$ and $\nu : X \to [0, 1]$ denote the degree of membership and the degree of non-membership of each element $x \in X$ to the set A respectively, and $0 \leq \mu(x) + \nu(x) \leq 1$ for all $x \in X$.

S. Melliani and O. Castillo (Eds.): ICPAMS 2021, LNNS 476, pp. 433–436, 2023.
https://doi.org/10.1007/978-3-031-12416-7_36

For the sake of simplicity, we shall use the symbol $< \mu, \nu >$ for the intuitionistic fuzzy subset $A = \{< x, \mu(x), \nu(x) >: x \in X\}$.

Definition 2. [3] Let $A =< \mu_A, \nu_A >$ and $B =< \mu_B, \nu_B >$ IFSS of X. Then
$A \subset B$ iff $\mu_A \leq \mu_B$ and $\nu_A \geq \nu_B$
$A = B$ iff $A \subset B$ and $B \subset A$
$A^c =< \nu_A, \mu_A >$
$A \cap B =< \mu_A \wedge \mu_B, \nu_A \vee \nu_B >$
$A \cup B =< \mu_A \vee \mu_B, \nu_A \wedge \nu_B >$
$\Box A =< \mu_A, 1 - \mu_A >$, $<> A =< 1 - \nu_A, \nu_A >$

Definition 3. [6] Let $\alpha, \beta \in [0, 1]$ with $\alpha + \beta \leq 1$. An intuitionistic fuzzy point, written as $x_{(\alpha,\beta)}$ is defined to be an intuitionistic fuzzy subset of F, given by

$$x_{(\alpha,\beta)}(y) = \begin{cases} (\alpha, \beta) & if \quad x = y \\ (0, 1) & if \quad x \neq y \end{cases}$$

An intuitionistic fuzzy point $x_{(\alpha,\beta)}$ is said to belong in **IFS** $< \mu, \nu >$ denoted by $x_{(\alpha,\beta)} \in< \mu, \nu >$ if $\mu(x) \leq \alpha$ and $\nu(x) \geq \beta$ and we have for every $x, y \in F$

$$x_{(t,s)} + y_{(\alpha,\beta)} = (x + y)_{(t\wedge\alpha, s\vee\beta)},$$

$$x_{(t,s)} y_{(\alpha,\beta)} = (xy)_{(t\wedge\alpha, s\vee\beta)}.$$

Definition 4. [7] Let A be Intuitionistic fuzzy set of a univers set X. Than (α, β)−cut of A is a crisp subset $A_{(\alpha,\beta)}$ of the **IFS** A is given by:

$$A_{(\alpha,\beta)} = \{x : x \in X \textit{ such that } \mu_A(x) \geq \alpha, \nu_A(x) \leq \beta\}$$

where $\alpha, \beta \in [0, 1]$ with $\alpha + \beta \leq 1$

Definition 5. Let F be a field. An intuitionitic fuzzy set A of F is called an intuitionistic fuzzy subfield of F if and only if, $\forall x, y \in F$

i) $\mu(x - y) \geq \mu(x) \wedge \mu(y)$,
ii) $\nu(x_y) \leq \nu(x) \vee \nu(y)$,
iii) $\mu(xy) \geq \mu(x) \wedge \mu(y)$,
iv) $\nu(xy) \leq \nu(x) \vee \nu(y)$,
v) $\mu(x^{-1}) \geq \mu(x), x \neq 0$,
vi) $\nu(x^{-1}) \leq \nu(x), x \neq 0$.

- If A and B are intuitionistic fuzzy fields of F such that $A \supseteq B$, then we write **A/B** and call **A/B** an *intuitionistic fuzzy field extension.*
- Let **A/B** be an intuitionistic fuzzy field extension and C an intuitionistic fuzzy field of F such that $A \supseteq C \supseteq B$. Then C is called an *intuitionistic fuzzy intermediate field* of **A/B**.

Definition 6. Suppose that $c_{(\alpha,\beta)} \subseteq A$. Then $c_{(\alpha,\beta)}$ is said to be *intuitionistic fuzzy algebraic* over B if and only if there exists $n \in N, k_i \in F$, and $\lambda_i \in [0,1]$ with $(k_i)_{(\alpha_i,\beta_i)} \subseteq B$ for $i = 1,\dots, n$ and $k_n \neq 0$ such that $k_{n(\alpha_n,\beta_n)}(c_{(\alpha,\beta)})^n + \cdots + k_{1(\alpha_1,\beta_1)}c_{(\alpha,\beta)} + k_{0(\alpha_0,\beta_0)} = 0_{(\alpha,\beta)}$.

A/B is said to be *Intuitionistic fuzzy algebraic* if and only if every $c_{(\alpha,\beta)} \subseteq A$ is intuitionistic fuzzy algebraic over B.

Example: Let $A =< \mu_A, \nu_A >$ and $B =< \mu_B, \nu_B >$ defined by:
and

$$\nu_B : \mathbb{R} \to [0,1]$$

$$x \to \begin{cases} 0 & if\ x \in \mathbb{Q} \\ \frac{1}{4} & if\ not \end{cases}$$

$$\mu_B : \mathbb{R} \to [0,1]$$

$$x \to \begin{cases} 1 & si\ x \in \mathbb{Q} \\ \frac{1}{3} & if\ not \end{cases}$$

$$\mu_A : \mathbb{C} \to [0,1]$$

$$x \to \begin{cases} 1 & si\ x \in \mathbb{R} \\ \frac{1}{2} & if\ not \end{cases}$$

$$\nu_A : \mathbb{C} \to [0,1]$$

$$x \to \begin{cases} 0 & if\ x \in \mathbb{R} \\ \frac{1}{8} & if\ not \end{cases};$$

A/B is an intuitionistic fuzzy field extension. $i_{(\frac{2}{3};\frac{1}{3})}$ is an intuitionistic fuzzy algebraic over B (because $i^2_{(\frac{2}{3};\frac{1}{3})} + 1_{(1;0)} = 0_{(\frac{2}{3};\frac{1}{3})}$ et $X^2 + 1_1 \in B[x]$).

Proposition 1. *Suppose that $c_{(\alpha,\beta)} \subseteq A$. Then $c_{(\alpha,\beta)}$ is intuitionistic fuzzy algebraic over B if and only if c is algebraic over $B_{(\alpha,\beta)}$.*

Proof Now
$k_{n(\alpha_n,\beta_n)}(c_{(\alpha,\beta)})^n + \cdots + k_{1(\alpha_1,\beta_1)}c_{(\alpha,\beta)} + k_{0(\alpha_0,\beta_0)} = 0_{(\alpha,\beta)}$
if and only if
$(k_n c^n + \cdots + k_1 c + k_0)_{((\alpha_0\wedge\dots\wedge\alpha_n\wedge\alpha),(\beta_0\vee\dots\vee\beta_n\vee\beta))} = 0_{(\alpha,\beta)}$
and the latter condition is equivalent to $k_n c^n + \cdots + k_1 c + k_0 = 0$ and $\alpha_0 \wedge \dots \wedge \alpha_n \wedge \alpha = \alpha$ and $\beta_0 \vee \dots \vee \beta_n \vee \beta = \beta$.

$\Rightarrow$) Suppose that $c_{(\alpha,\beta)}$ is Intuitionistic fuzzy algebraic over B. Then with the notation in the **Definition 1.6**, $\alpha \leq \alpha_i \leq \mu_B(k_i)$ and $\beta \geq \beta_i \geq \nu_B(k_i)$ so $k_i \in B_{(\alpha,\beta)}$ for $i = 0, 1, \dots, n$. Hence c is algebraic over $B_{(\alpha,\beta)}$.

$\Leftarrow$) Conversely suppose that c is algebraic over $B_{(\alpha,\beta)}$. Then there exists $k_i \in B_{(\alpha,\beta)}, i = 0, 1, \dots, n$, such that $k_n c^n + \cdots + k_1 c + k_0 = 0$. Thus $\mu_B(k_i) \geq \alpha$ and $\nu_B(k_i) \leq \beta$ and the desired result holds with $\alpha_i = \mu_B(k_i)$ and $\beta_i = \nu_B(k_i)$ for $i = 0, 1, \dots, n$.

Proposition 2. *Suppose that $c_{(\alpha,\beta)} \subseteq A$ and lhat $0 \le \lambda \le \alpha \le 1$ and $1 \ge \gamma \ge \beta \ge 0$. If $c_{(\alpha,\beta)}$ is an intuitionistic fuzzy algebraic over B, then $c_{(\lambda,\gamma)}$ is intuitionistic fuzzy algebraic over B.*

Proof.
$k_{n(\lambda_n,\gamma_n)}(c_{(\alpha,\beta)})^n + \cdots + k_{1(\lambda_1,\gamma_1)}c_{(\alpha,\beta)} + k_0 = 0_{(\alpha,\beta)}$
implies
$k_{n(\lambda_n,\gamma_n)}(c_{(\lambda,\gamma)})^n + \cdots + k_{1(\lambda_1,\gamma_1)}c_{(\lambda,\gamma)} + k_{0(\lambda_0,\gamma_0)} = 0_{(\lambda,\gamma)}$ since $\lambda \le \alpha \le \lambda_i$ and $\gamma \ge \beta \ge \gamma_i$ for $i = 0, 1, \dots, n$.

Theorem 1. *A/B is intuitionistic fuzzy algebraic if and only if $A_{(\alpha,\beta)}/B_{(\alpha,\beta)}$ is algebraic for all $\alpha \in Im(\mu)$ and $\beta \in Im(\nu)$*

Proof

$\Rightarrow$) Suppose that A/B is intuitionistic fuzzy algebraic. Let $c \in A_{(\alpha,\beta)}$. Then $c_{(\alpha,\beta)} \subseteq A$ and so $c_{(\alpha,\beta)}$ is intuitionistic fuzzy algebraic over B. Thus c is algebraic over $B_{(\alpha,\beta)}$ by **proposition 1.1**.

$\Leftarrow$) Suppose that $A_{(\alpha,\beta)}/B_{(\alpha,\beta)}$ is algebraic for all $\alpha \in Im(\mu)$ and $\beta \in Im(\nu)$. Let $c_{(\alpha,\beta)}$. Let $\alpha = \mu(c)$ and $\beta = \nu(c)$. Then $\alpha \in Im(\mu)$ and $\beta \in Im(\nu)$ and $c \in A_{(\alpha,\beta)}$. Hence c is algebraic over $B_{(\alpha,\beta)}$. Thus $c_{(\alpha,\beta)} \subseteq A$ is intuitionistic fuzzy algebraic over B by **proposition 1.1**. By **proposition 1.3**, $c_{(\alpha,\beta)}$ is intuitionistic fuzzy algebraic over B.

References

1. Anandh, B., Giri, R.: Intuitionistic fuzzy subfields of a field with respect to $(T, S)-$norm. Adv. Fuzzy Math. **12**(4), 1007–1015 (2017). ISSN 0973-533X
2. Atanassova, K., Stoeva, S.: Intuitionistic fuzzy sets. In: Proceedings Polish Symposium on Interval and Fuzzy Mathematics, Poznan, pp. 23–26 (1983)
3. Atanassov, K.: Intuitionistic fuzzy sets. Fuzzy Sets Syst. **20**, 87–96 (1986)
4. Biswas, R.: Fuzzy subgroups and anti-fuzzy subgroups. Fuzzy Sets Syst. **35**, 121–124 (1990)
5. Biswas, R.: Fuzzy fields and fuzzy linear spaces redefined. Fuzzy Sets Syst. **33**(2), 257–259 (1989). North Holland
6. Jun, Y.B., Song, S.: Intuitionistic fuzzy semi preopen sets and Intuitionistic fuzzy-semi precontinuous mappings. J. Appl. Math. Comput. **19**(1–2), 467–474 (2005)
7. Sharma., P.K.: $(\alpha, \beta)-$ Cut of Intuitionistic Fuzzy Groups. Int. Math. Forum, **6**(53), 2605–2614 (2011)
8. Zadeh, L.A.: Fuzzy sets. Inf. Control **8**, 338–353 (1965)

Nanoparticles Shape Effect on Heat Transfer by Natural Convection of Nanofluid in a Vertical Porous Cylindrical Enclosure Subjected to a Heat Flux

Y. Foukhari(✉), M. Sammouda, M. Driouich, and S. Belhouideg

Sultan Moulay Slimane University, Beni Mellal, Morocco
youness.foukharifpb@usms.ac.ma

Abstract. The heat transfer by natural convection in a vertical cylindrical enclosure filled with a porous medium and saturated by a nanofluid with different nanoparticles shape is studied in the present work. The enclosure is maintained at a constant heat flux along its axis, the side wall is maintained at a constant temperature while the base walls are impermeables and isolated. The scalar equations system that governing the problem are solved by the use the finite difference method. The results obtained, presented in the term of a stream lines, isotherms and heat transfer expressed by the Nusselt number, showed that the flow patern of the nanofluid as well as the rate of heat transfer on the active walls are influenced by the nanoparticles concentration and shape of nanoparticles added to the base fluid. With cylindrical nanoparticles shape, the heat transfer is improved compared to with spherical nanoparticles.

1 Introduction

Several studies have focused on improving heat transmission through convection. Although the ideas for improvement have mostly influenced the geometry of systems and the physicochemical character of convective media, the study has only touched on the macroscopic or occasionally microscopic order of the process in terms of chronology. However, with the advent and rapid development of nanosciences and nanotechnologies in the second half of the 20th century, convection has reaped a large share of this new wealth and taken on a new facet of improvement: recent research has focused on the nanometric level of the convective medium's matter. Nanofluids are one of the byproducts of such richness. Nanofluids have unique physicochemical qualities, such as high thermal conductivity, that make them superior to traditional coolants in terms of heat transfer coefficient. The research conducted in this new approach has produced a large, but diverse bibliography:

In 1995 **Choi** [1] coined the term nanofluid. Heat exchangers, microchannels, cooling of electronic systems, vehicle cooling, nuclear reactors, buildings,

S. Melliani and O. Castillo (Eds.): ICPAMS 2021, LNNS 476, pp. 437–445, 2023.
https://doi.org/10.1007/978-3-031-12416-7_37

grain storage, and other examples are used because of their high thermal performance. In recent years, a great number of investigations on mixed convection using nanofluids have been conducted. **Wen and Ding** [2] used a similar experimental approach in an evenly heated circular tube, but they used aluminum trioxide (Al2O3) nanoparticles with diameters ranging from 27 to 56 nm. In the laminar regime, the use of the nanofluid considerably boosted the heat transfer coeffici nt. **Behzadmehr et al.** [3] investigated turbulent forced convection in a circular tube with 1% copper (Cu) nanoparticles in water.They established that an increase in the Reynolds number led Nusselt's number to grow in value. **Khanafer et al.** [4]] investigated the heat transmission of nanofluids in an enclosure using numerical simulations. For varied Grashof numbers, their results demonstrate that the mean Nusselt number increased with the volume fraction of the nanoparticles. **Xuan et al.** [5] investigated convective heat transfer and nanofluid flow properties experimentally. The convective heat transfer coefficient of the nanofluid increases with the speed flow rate and the volume fraction of nanoparticles, and it is greater than that of pure water, according to their results. There are just a few studies on the effect of particle shape on nanofluid viscosity in the literature. The viscosity of nanofluids, on the other hand, is strongly influenced by the form of the nanoparticles [6,7]. **Timofeeva et al.** [7] report that elongate particles, such as platelets and cylinders, have increased viscosity at the same volume percentage. Spherical particles or spheroids with a lower aspect ratio should be employed for lower viscosities. According to **Jeong et al.** [8]], the viscosity and thermal conductivity of the nanofluid with almost rectangular shape particles were 7,7% and 5,9% higher than that of the nanofluid with sphere shape particles, respectively.

2 Problem Configuration

Figure 1 shematically shows the problem under investigation. A vertical cylindrical enclosure of height H, radius R filled with a porous medium and saturated by a nanofluid with different nanoparticle shape is studied in the present work. The enclosure is maintained at a constant heat flux along its axis, the side wall is maintained at a constant temperature while the base walls are impermeables and isolated.

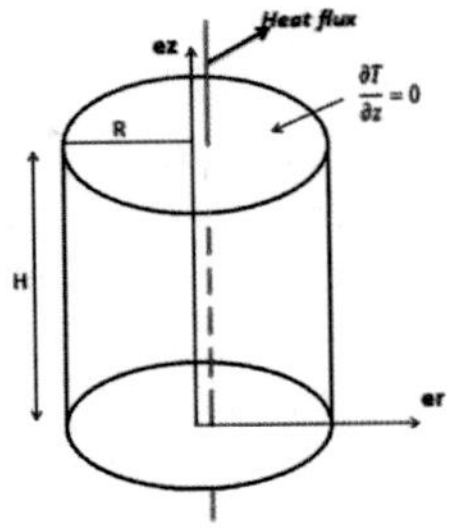

Fig. 1. Probleme shematic

3 Gouverning Equations and Boundary Conditions

A steady, laminar and incompressible flow is considered. Under these conditions the gouverning equations are expressed in cylindrical coordinate as :

The continuity equation:

$$\frac{1}{r}\frac{\partial (rU)}{\partial r} + \frac{\partial W}{\partial z} = 0 \tag{1}$$

The vorticity equation :

$$\begin{aligned}\frac{1}{\epsilon^2}\left(\frac{\partial (U\Omega)}{\partial r} + \frac{\partial (W\Omega)}{\partial z}\right) &= \left(-\overline{\lambda}\Sigma\frac{Pr_{fl}}{Da} - \overline{\lambda}\Lambda\Sigma\frac{Pr_{fl}}{r^2} - \frac{C_F}{\sqrt{Da}} \mid \vec{V} \mid\right)\Omega \\ &- \overline{\lambda}^2 \Gamma Ra_{th} Pr_{fl}\frac{\partial T}{\partial r} - \frac{C_F}{\sqrt{Da}}\left(U\frac{\partial \mid \vec{V} \mid}{\partial z} - W\frac{\partial \mid \vec{V} \mid}{\partial r}\right) \\ &+ \overline{\lambda}\Lambda\Sigma Pr_{fl}\left(\frac{\partial^2 \Omega}{\partial r^2} + \frac{\partial^2 \Omega}{\partial z^2} + \frac{1}{r}\frac{\partial \Omega}{\partial r}\right)\end{aligned} \tag{2}$$

The energy equation:

$$\frac{\partial T}{\partial t} + \frac{\partial (UT)}{\partial r} + \frac{\partial (WT)}{\partial z} + \frac{UT}{r} = \frac{\partial^2 T}{\partial r^2} + \frac{\partial^2 T}{\partial z^2} + \frac{1}{r}\frac{\partial T}{\partial r} \tag{3}$$

The vorticity function

$$\Omega = \frac{1}{r}\left(\frac{\partial^2 \psi}{\partial r^2} + \frac{\partial^2 \psi}{\partial z^2} - \frac{1}{r}\frac{\partial \psi}{\partial r}\right) \tag{4}$$

Velocity field:

$$U = \frac{1}{r}\frac{\partial \psi}{\partial z} \quad ; \quad W = -\frac{1}{r}\frac{\partial \psi}{\partial r} \tag{5}$$

where C_F, α^*_{nf}, λ^*_{nf}, Ω, U, W, $\overline{\lambda}$, Λ, Σ and Γ are respectively the inertial coefficient, the thermal diffusivity effective, the thermal conductivity effective, the vorticity function, the radial velocity, the axial velocity. Which are difined by:

$$C_F = \frac{1.75}{\sqrt{150\epsilon^3}} \quad ; \quad \overline{\lambda} = \frac{\lambda_{nf}}{\lambda^*_{nf}} \quad ; \quad \Lambda = \frac{\mu^*}{\mu_{nf}} \tag{6}$$

$$\lambda^*_{nf} = \epsilon\lambda_{nf} + (1-\epsilon)\lambda_s \tag{7}$$

$$\Gamma = \frac{\left((1-\phi) + \phi\frac{(\rho\beta_{Th})_{pr}}{(\rho\beta_{Th})_{fl}}\right)}{\left((1-\phi) + \phi\frac{\rho_{pr}}{\rho_{fl}}\right)\left(\frac{\frac{\lambda_{nf}}{\lambda_{fl}}}{(1-\phi)+\phi\frac{(\rho C)_{pr}}{(\rho C)_{fl}}}\right)^2} \tag{8}$$

$$\Sigma = \frac{(1-\phi)^{-2,5}\left((1-\phi) + \phi\frac{(\rho C)_{pr}}{(\rho C)_{fl}}\right)}{\frac{\lambda_{nf}}{\lambda_{fl}}\left((1-\phi) + \phi\frac{\rho_{pr}}{\rho_{fl}}\right)} \tag{9}$$

In the previous equations, the expressions of dimensionless numbers are given by:

$$Pr_{fl} = \frac{\mu_{fl}}{\rho_{fl}\alpha_{Th}} \quad ; \quad Ra_{Th} = \frac{\rho_{fl} g \beta_{Th} Q R_i^4}{\mu_{fl}\alpha_{Th}\lambda_{fl}} \quad ; \quad Da = \frac{K}{R_i^2} \tag{10}$$

Assuming a uniform dispersion of nanoparticles in the base fluid, the density of nanofluids are generally calculated using Pak and Cho's relation:

$$\rho_{nf} = (1-\phi)\,\rho_{fl} + \phi\rho_{pr} \tag{11}$$

The heat capacitance and the thermal expansion coefficient of the nanofluid are defined as:

$$(\rho\beta)_{nf} = (1-\phi)\,(\rho\beta)_{fl} + \phi\,(\rho\beta)_{pr} \tag{12}$$

$$(\rho C_p)_{nf} = (1-\phi)\,(\rho C_p)_{fl} + \phi\,(\rho C_p)_{pr} \tag{13}$$

Hamilton and Crosser (1962) offered the following relation to calculate the thermal conductivity coefficient of the solid-liquid mixture:

$$\frac{\lambda_{nf}}{\lambda_{fl}} = \frac{\lambda_{pr} + (n-1)\lambda_{fl} + (n-1)\,(\lambda_{pr} - \lambda_{fl})\,\phi}{\lambda_{pr} + (n-1)\lambda_{fl} - (\lambda_{pr} - \lambda_{fl})\,\phi} \tag{14}$$

Krieger et Daugherty (1956) offered the following relation to calculate the viscosity coefficient:

$$\mu_{nf} = \mu_{fl}\left(1 - \frac{\phi}{\phi_m}\right)^{-2} \quad whither \quad \phi_m = \frac{2}{0.321\delta + 3.02} \tag{15}$$

ϕ_m is the packing fraction and δ is the aspectration of nanoparticles. Where n is the empirical shape factor of nanoparticles defined by $n = \frac{3}{\chi}$ where χ is the nanoparticle sphericity and it is defined as the ration of the surface area of an equal volume sphere to the actual surface area of the particle.

Boundary Conditions:

For the Stream Lines

$$\psi = \frac{\partial\psi}{\partial r} = \frac{\partial\psi}{\partial r} = 0 \quad en \quad Z = 0, Z = H, r = R \tag{16}$$

$$\psi_{axe} = \frac{\partial\psi_{axe}}{\partial z} = 0 \quad en \quad r = 0 \tag{17}$$

For the Vorticity: At the Side Wall

$$\Omega \mid_{r=R} = \frac{2}{dr^2}\psi \mid_{r=R-dr} \tag{18}$$

For the Vorticity: At the Lower and Upper Wall

$$\Omega \mid_{z=0}= \frac{1}{r}\frac{2}{dz^2}\psi \mid_{z=dz} \tag{19}$$

$$\Omega \mid_{z=H}= \frac{1}{r}\frac{2}{dz^2}\psi \mid_{z=H-dz} \tag{20}$$

For the Vorticity: On the Axis

$$\Omega \mid_{r=0}= \frac{2}{dr^2}\psi \mid_{r=dr} \tag{21}$$

The local and global Nusselt numbers are given by

$$Nu_{local} = AR_r\frac{\lambda^*_{nf}}{\lambda_{fl}}\frac{1}{T} \mid_{heatflux} \quad ; \quad Nu_{global} = \frac{1}{\gamma}\int_0^{\gamma} Nu_{local}dz \tag{22}$$

4 Numerical Methods

The space discretization of the transport equations of the vorticity and energy will be done using the precise centered differences. Once discretized, the method of ADI (Alternation Direction Implicit) is used to solve the equations of energy and of vorticity. The equation of stream function is solved by the means of the method of S.O.R (Susccessive over Relaxation) [9], this method directly gives the value of stream function at the moment n+1 in the considered node. We consider that the convergence is reached if, with each step of time, the following test is checked.

$$\frac{\sum_i \sum_j \mid f_{i,j}^{n+1} - f_{i,j}^{n} \mid}{\sum_i \sum_j \mid f_{i,j}^{n} \mid} \leq 10^{-5} \tag{23}$$

f refers to vorticity Ω and Temperature T

5 Results and Discussion

The aim of this study is to investigate the effect of nanoparticles shape and nanoparticles concentration on the improvement of convective heat transfer.

For the cylindrical shape we notice that for a Rayleigh number ($Ra = 10^5$) Fig. 2, the flow intensifies ($\mid \psi \mid = 0.26$) and the isotherms are tighter on the upper side, and almost vertical. If we further increase the Rayleigh number to $Ra = 10^6$ Fig. 2, the flow becomes more intense ($\mid \psi \mid = 1.14$). Indeed, the increase in the Rayleigh number can be explained by an increase in the heating energy, the wire becomes hotter than the vertical walls, the fluid tends to move away from the hot zone to transfer the energy. from the hot side to the cold side. On the other hand for the spherical shape Fig. 3, we notice that the flow intensity becomes a little greater than the cylindrical shape.

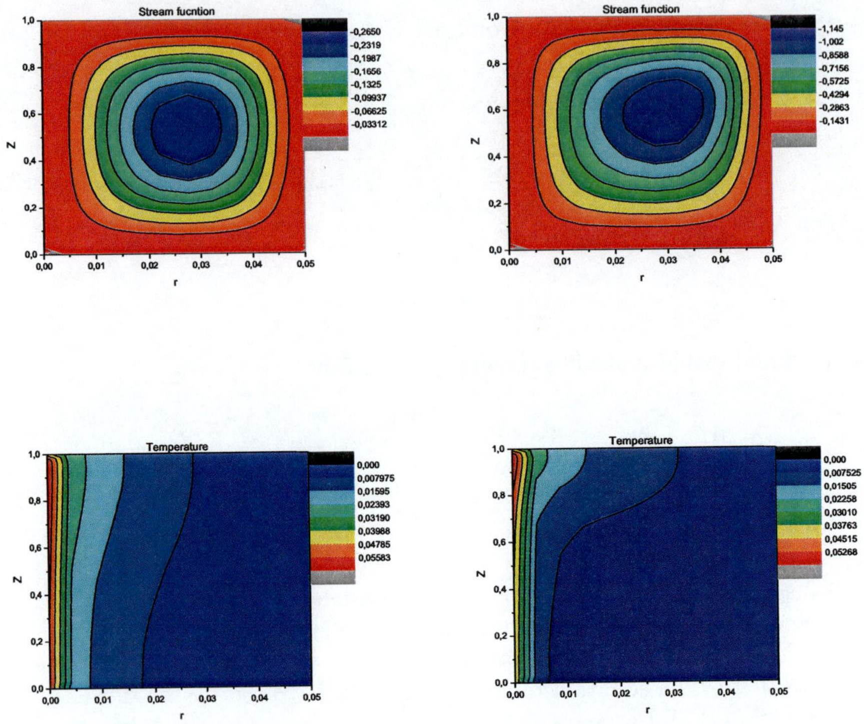

Fig. 2. Isotherms and stream function for a cylindrical shape and for a) $Ra = 10^5$, $\phi = 0.03$, b) $Ra = 10^6$, $\phi = 0.03$

The Fig. 4 shows a comparaison between a cylindrical and a spherical shape effect for the improvement of heat transfer, we remark that the hight values of Nusselt number come back to cylindrical shape, thus enhanced that the nanoparticles in a cylindrical shape has more influenced the heat transfer.

The Fig. 5 shows that the average Nusselt number increases for the cylindrical shape with the increase of Rayleigh number, also we remark that the low values of average Nusselt number for $\phi = 0.01$ and the hight ones for $\phi = 0.05$ that shows the heat transfer depends the nanoparticles concentration, thus the heat transfer is more important for the cylindrical shape.

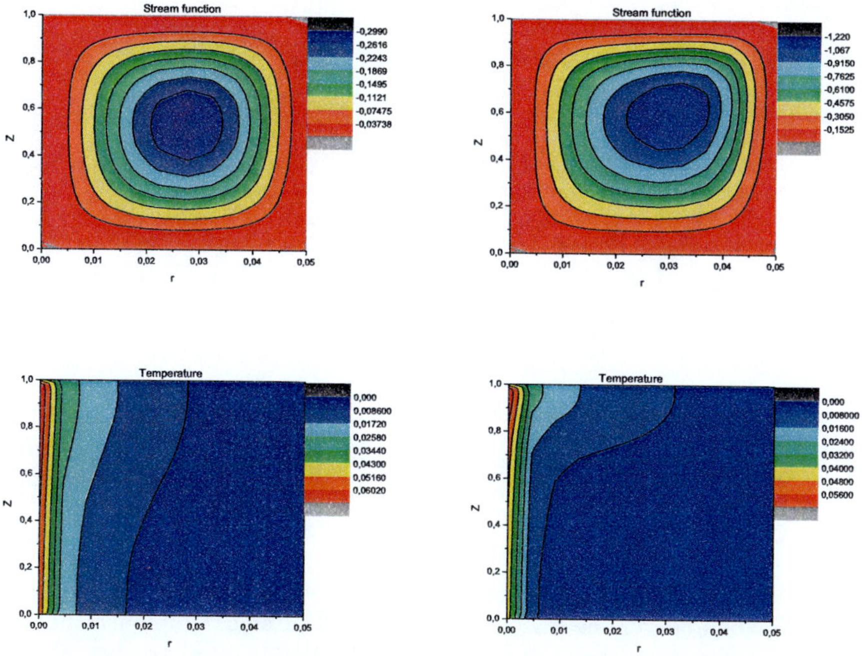

Fig. 3. Isotherms and stream function for a spherical shape and for a) $Ra = 10^5$, $\phi = 0.03$, b) $Ra = 10^6$, $\phi = 0.03$

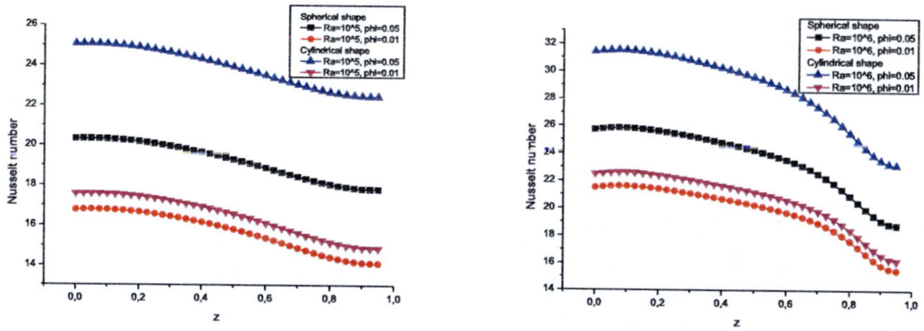

Fig. 4. Comparaison between the cylindrical and spherical shape of nanoparticles by Nusselt number for differents value of Rayleigh and nanoparticles concentration

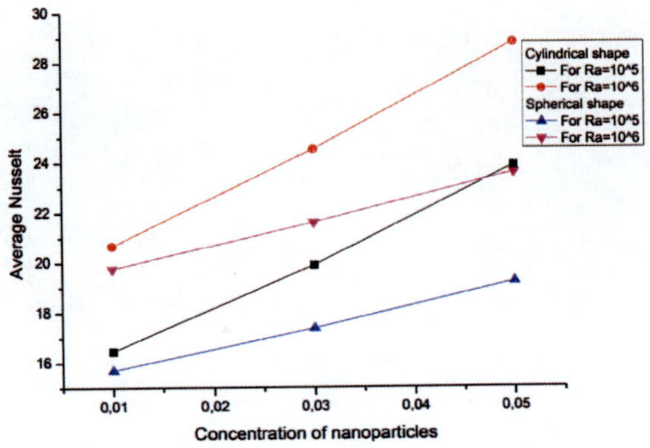

Fig. 5. The average Nusselt number for differents value of Rayleigh and concentration of nanoparticles

6 Conclusion

The numerical investigation was carried out using a numeric code to study the effect of shape and volume fraction on the flow field and the temperature field on natural convection in a vertical cylinder. The results obtained show that:

- Increasing the Rayleigh number and the concentration of nanoparticles leads to an increasement on the thermal conductivityof the nanofluid and that's means an improvement of heat transfer.
- The cylindrical shape of nanoparticles enhanced the heat transfer more than the spherical shape ones.

References

1. Choi, S.U., Eastman, J.A.: Enhancing thermal conductivity of fluids with nanoparticles. Argonne National Lab., IL (United States) (1995)
2. Wen, D., Ding, Y.: Experimental investigation into convective heat transfer of nanofluids at the entrance region under laminar flow conditions. Int. J. Heat Mass Transf. **47**(24), 5181–5188 (2004)
3. Behzadmehr, A., Saffar-Avval, M., Galanis, N.: Prediction of turbulent forced convection of a nanofluid in a tube with uniform heat flux using a two phase approach. Int. J. Heat Fluid Flow **28**(2), 211–219 (2007)
4. Khanafer, K., Vafai, K., Lightstone, M.: Buoyancy-driven heat transfer enhancement in a two-dimensional enclosure utilizing nanofluids. Int. J. Heat Mass Transf. **46**(19), 3639–3653 (2003)
5. Putra, N., Roetzel, W., Das, S.K.: Natural convection of nano-fluids. Heat Mass Transf. **39**(8), 775–784 (2003)
6. Timofeeva, E.V., Yu, W., France, D.M., et al.: Nanofluids for heat transfer: an engineering approach. Nanoscale Res. Lett. **6**(1), 1–7 (2011)

7. Timofeeva, E.V., Routbort, J.L., Singh, D.: Particle shape effects on thermophysical properties of alumina nanofluids. J. Appl. Phys. **106**(1), 014304 (2009)
8. Jeong, J., Li, C., Kwon, Y., et al.: Particle shape effect on the viscosity and thermal conductivity of ZnO nanofluids. Int. J. Refrig. **36**(8), 2233–2241 (2013)
9. Du Fort, E.C., Frankel, S.P.: Stability conditions in the numerical treatment of parabolic differential equations. Math.l Tables Other Aids Comput. **7**(43), 135–152 (1953)

Check for updates

Coupled System of Mixed Hybrid Fractional Differential Equations: Linear Perturbations of First and Second Type

Y. Allaoui, S. Melliani(✉), A. El Allaoui, and K. Hilal

LMACS, Sultan Moulay Slimane University, BP 523 - 23000 Beni Mellal, Morocco
s.melliani@usms.ma

Abstract. This paper studies the existence of solutions for a mixed system of hybrid fractional differential equations, it is a coupled hybrid fractional differential equations of first and second type. We make use of the standard tools of the fixed point theory to establish the main results. The existence and uniqueness result is elaborated with the aid of an example.

1 Introduction

Hybrid differential equations is a rich field of differential equations. It is quadratic perturbations of non linear differential equations. It has lately years been an object of increasing interest because of its vast applicability in several fields. For more details about hybrid differential equations, we refer to [1,3,5,6,8–11]

Motivated by [3,7]. The propose of this paper is to study the following coupled system of hybrid fractional differential equations with perturbations of first and second type.

$$\begin{cases} D_c^p\left[\dfrac{x(t)}{f_1(t,x(t),y(t))}\right] = h_1(t,x(t),y(t)), \quad 0<p<1, \quad t\in J=[0,a], \\ D_c^q\left[y(t)-f_2(t,x(t),y(t))\right] = h_2(t,x(t),y(t)), \quad 0<q<1, \quad t\in J, \\ x(0)=x_0, \\ y(0)=y_0, \end{cases} \tag{1}$$

where $f_1 \in \mathscr{C}(J\times\mathbb{R}\times\mathbb{R};\mathbb{R}\setminus\{0\})$ and $f_1, h_1, h_2 \in \mathscr{C}(J\times\mathbb{R}\times\mathbb{R};\mathbb{R})$.

We first start by recalling Lery Schauder alternative.

2 Preliminaries

In this section, we introduce some definitions and results which are used throughout this paper.

S. Melliani and O. Castillo (Eds.): ICPAMS 2021, LNNS 476, pp. 446–457, 2023.
https://doi.org/10.1007/978-3-031-12416-7_38

Definition 1. ([10]) Let $x \in C^n[0,\infty)$ and $n-1<\alpha<n$, where $n \in \mathbb{N}^*$, the Caputo's derivative of order α for function $x:[0,\infty) \longrightarrow \mathbb{R}$ can be written as

$$D^{\alpha}x(t) = \frac{1}{\Gamma(n-\alpha)} \int_0^t (t-s)^{n-\alpha-1} x^{(n)}(s)ds. \tag{2}$$

Definition 2. ([10]) The fractional integral of order α is defined as

$$I^{\alpha}x(t) = \frac{1}{\Gamma(\alpha)} \int_0^t (t-s)^{\alpha-1} x(s)ds, \tag{3}$$

which is called the Riemann-Liouville integral.

The following Lemmas are useful in what follows.

Lemma 1. *([3]) Assume that hypothesis* $x \longrightarrow \frac{x}{f_1(t,x)}$ *is increasing in* $\mathbb{R}$*, for each* $t \in J$*. Then for any* $h \in L^1(J,\mathbb{R}_+)$ *and* $0<p<1$*, the function* $x \in \mathscr{C}(J,\mathbb{R}_+)$ *is a solution of the fractional hybrid differential equation*

$$\begin{cases} D_c^p\left[\dfrac{x(t)}{f_1(t,x(t))}\right] = h_1(t), & t \in J, \\ x(0) = x_0, \end{cases}$$

if and only if x satisfies the hybrid integral equation

$$x(t) = f_1(t,x(t))\left[\frac{x_0}{f_1(0,x_0)} + \frac{1}{\Gamma(p)}\int_0^t (t-s)^{p-1}h_1(s)ds\right], \qquad t \in J.$$

Lemma 2. *([7]) Assume that hypothesis* $y \longrightarrow y - f_2(t,y)$ *is increasing in* $\mathbb{R}$ *for each* $t \in J$*. Then for any* $h: J \longrightarrow \mathbb{R}_+$ *and* $0<q<1$*, the function* $y \in \mathscr{C}(J,\mathbb{R}_+)$ *is a solution of the fractional hybrid Differential Equation*

$$\begin{cases} D_c^q[y(t) - f_2(t,y(t))] = h_2(t), & t \in J, \\ y(0) = y_0 \in \mathbb{R}. \end{cases}$$

if and only if y satisfies the hybrid integral equation

$$y(t) = y_0 - f_2(0,y_0) + f_2(t,y(t)) + \frac{1}{\Gamma(q)}\int_0^t (t-s)^{q-1}h_2(s)ds, \qquad t \in J.$$

Lemma 3. *(Lery Schauder alternative, [4])*
Let $\Lambda: X \longrightarrow X$ *be a completely continuous operator (i.e., map that is restricted to any bounded set X is compact). Let* $\mathscr{P}_\Lambda = \{x \in X : x = \delta\Lambda x \text{ for some } 0<\delta<1\}$*. Then either the set* $\mathscr{P}_\Lambda$ *is unbounded or* Λ *has at least one fixed point.*

3 Existence Result

In this section, we prove the existence and uniqueness result for (1) by using Banach fixed point theorem. Then we discuss the existence of solutions for this problem by means of Lery Schauder alternative.

To establish our results, we introduce the following assumptions:

(A₀) $(a)-$ The map $x \longrightarrow \dfrac{x}{f_1(t,x,y)}$ is increasing in $\mathbb{R}$ for each $t \in J,\, y \in \mathbb{R}$.

$(b)-$ The map $y \longrightarrow y - f_2(t,x,y)$ is increasing in $\mathbb{R}$ for each $t \in J,\, x \in \mathbb{R}$.

$(\mathbf{A}_1)$ There exists positive numbers μ_1, μ_2, ν such that

$$\nu \leq |f_1(t,x,y)| \leq \mu_1, \ |f_2(t,x,y)| \leq \mu_2,$$

for all $(t,x,y) \in J \times \mathbb{R} \times \mathbb{R}$.

$(\mathbf{A}_2)$ There exists positive numbers λ_1, λ_2, λ_3, λ_4 such that

$$|f_1(t,x,y) - f_1(t,x',y')| \leq \lambda_1|x-x'| + \lambda_2|y-y'|,$$

$$|f_2(t,x,y) - f_2(t,x',y')| \leq \lambda_3|x-x'| + \lambda_4|y-y'|,$$

for all $t \in J$ and $x,y,x',y' \in \mathbb{R}$.

$(\mathbf{A}_3)$ There exists a constants η_1, η_2, ξ_1, ξ_2 such that

$$|h_1(t,x,y) - h_1(t,x',y')| \leq \eta_1|x-x'| + \eta_2|y-y'|,$$

$$|h_2(t,x,y) - h_2(t,x',y')| \leq \xi_1|x-x'| + \xi_2|y-y'|,$$

for all $t \in J$ and x, y, x', $y' \in \mathbb{R}$.

Denote $E = \mathscr{C}(J,\mathbb{R}) \times \mathscr{C}(J,\mathbb{R})$, equipped with the norm

$$\|(x,y)\| = \|x\| + \|y\|,$$

where $\|x\| = \sup_{t \in J} |x(t)|$.

Notice that the space E with this norm is a Banach space. Seeing that as in lemma 1 and 2 we define on E the operator Λ by

$$\Lambda(x,y)(t) = \begin{pmatrix} \Lambda_1(x,y)(t) \\ \Lambda_2(x,y)(t) \end{pmatrix}$$

where

$$\Lambda_1(x,y)(t) = f_1(t,x(t),y(t)) \left[\frac{x_0}{f_1(0,x_0,y_0)} + \frac{1}{\Gamma(p)} \int_0^t (t-s)^{p-1} h_1(s),x(s),y(s))ds \right],$$

$$\Lambda_2(x,y)(t) = y_0 - f_2(0,x_0,y_0) + f_2(t,x(t),y(t)) + \frac{1}{\Gamma(q)} \int_0^t (t-s)^{q-1} h_2(s,x(t),y(t))ds.$$

As a results Λ is a solution to our problem under the assumption $(\mathbf{A}_0)$.
Let us set

$$\nu_1 = a^p \frac{\eta_1+\eta_2}{\Gamma(p+1)}, \ \nu_2 = a^q \frac{\xi_1+\xi_2}{\Gamma(q+1)}$$

$$\alpha_1 = \frac{a^p \mu_1 \eta_1}{\Gamma(p+1)} + \lambda_1 \left(\frac{|x_0|}{\nu} + \nu_1 r + \frac{a^p k_1}{\Gamma(p+1)} \right),$$

$$\beta_1 = \frac{a^p \mu_1 \eta_2}{\Gamma(p+1)} + \lambda_2 \left(\frac{|x_0|}{\nu} + \nu_1 r + \frac{a^p k_1}{\Gamma(p+1)} \right),$$

$$\alpha_2 = \lambda_3 + \frac{a^q \xi_1}{\Gamma(q+1)},$$

$$\beta_2 = \lambda_4 + \frac{a^q \xi_2}{\Gamma(q+1)},$$

$\gamma_1 = max(\alpha_1, \beta_1)$ and $\gamma_2 = max(\alpha_2, \beta_2)$, where $k_1 = \sup_{t\in J} |h_1(t,0,0)|$, $k_2 = \sup_{t\in J} |h_2(t,0,0)|$,
and

$$r \geq \frac{\mu_1 \frac{|x_0|}{\nu} + |y_0| + 2\mu_2 + \frac{a^p \mu_1 k_1}{\Gamma(p+1)} + \frac{a^q k_2}{\Gamma(q+1)}}{1-(\mu_1 \nu_1 + \nu_2)}, \tag{4}$$

Now we are in a position to state our first existence results. This result is based on Banach fixed point theorem.

Theorem 1. *Let assumptions $(\mathbf{A}_0)-(\mathbf{A}_3)$ be satisfied. Suppose, in addition that the following property is verified:*

$$\gamma_1 + \gamma_2 < 1$$

Then the problem (1) has a unique solution.

Proof. Proof Let us define the closed ball

$$B_r = \{(x,y) \in E : \ \|(x,y)\| \leq r\},$$

Then we shall check that $\Lambda B_r \subseteq B_r$.

For $(x,y) \in B_r$ and $t \in J$, we have

$$\begin{aligned}
&|\Lambda_1(x,y)(t)| \\
&= |f_1(t,x(t),y(t))| \left| \frac{x_0}{f_1(0,x_0,y_0)} + \frac{1}{\Gamma(p)} \int_0^t (t-s)^{p-1} h_1(s,x(s),y(s)) ds \right| \\
&\le \mu_1 \left(\frac{|x_0|}{|f_1(0,x_0,y_0)|} + \frac{1}{\Gamma(p)} \int_0^t (t-s)^{p-1} \left(|h_1(s,x(s),y(s)) - h_1(s,0,0)| + |h_1(s,0,0)| \right) ds \right) \\
&\le \mu_1 \left(\frac{|x_0|}{\nu} + \frac{1}{\Gamma(p)} \int_0^t (t-s)^{p-1} \left(\eta_1 |x(s)| + \eta_2 |y(s)| + k_1 \right) ds \right) \\
&\le \mu_1 \left(\frac{|x_0|}{\nu} + \frac{\eta_1 \|x\| + \eta_2 \|y\| + k_1}{\Gamma(p)} \int_0^t (t-s)^{p-1} ds \right) \\
&\le \mu_1 \left(\frac{|x_0|}{\nu} + a^p \frac{\eta_1 \|x\| + \eta_2 \|y\| + k_1}{\Gamma(p+1)} \right).
\end{aligned}$$

Hence

$$|\Lambda_1(x,y)(t)| \le \mu_1 \left(\frac{|x_0|}{\nu} + a^p \frac{\eta_1 r + \eta_2 r + k_1}{\Gamma(p+1)} \right) = \mu_1 \left(\frac{|x_0|}{\nu} + \nu_1 r + \frac{a^p k_1}{\Gamma(p+1)} \right). \quad (5)$$

On the other hand, we have

$$\begin{aligned}
&|\Lambda_2(x,y)(t)| \\
&= \left| y_0 - f_2(0,x_0,y_0) + f_2(t,x(t),y(t)) + \frac{1}{\Gamma(q)} \int_0^t (t-s)^{q-1} h_2(s,x(s),y(s)) ds \right| \\
&\le |y_0| + |f_2(0,x_0,y_0)| + |f_2(t,x(t),y(t))| + \frac{1}{\Gamma(q)} \int_0^t (t-s)^{q-1} |h_2(s,x(s),y(s))| ds \\
&\le |y_0| + 2\mu_2 + \frac{1}{\Gamma(q)} \int_0^t (t-s)^{q-1} \left(|h_2(s,x(s),y(s)) - h_2(s,0,0)| + |h_2(s,0,0)| \right) ds \\
&\le |y_0| + 2\mu_2 + \frac{1}{\Gamma(q)} \int_0^t (t-s)^{q-1} \left(\xi_1 |x(s)| + \xi_2 |y(s)| + k_2 \right) ds \\
&\le |y_0| + 2\mu_2 + \frac{\xi_1 \|x\| + \xi_2 \|y\| + k_2}{\Gamma(q)} \int_0^t (t-s)^{q-1} ds \\
&\le |y_0| + 2\mu_2 + a^q \frac{\xi_1 \|x\| + \xi_2 \|y\| + k_2}{\Gamma(q+1)}.
\end{aligned}$$

Hence

$$|\Lambda_2(x,y)(t)| \le |y_0| + 2\mu_2 + a^q \frac{\xi_1 r + \xi_2 r + k_2}{\Gamma(q+1)} = |y_0| + 2\mu_2 + \nu_2 r + \frac{a^q k_2}{\Gamma(q+1)}. \quad (6)$$

From (4), (5) and (6), we deduce that

$$\|\Lambda(x,y)\| \le r.$$

Thus $\Lambda B_r \subseteq B_r$.
For (x,y), $(x',y') \in B_r$ and $t \in J$, we have

$$\begin{aligned}
&|\Lambda_1(x,y)(t) - \Lambda_1(x',y')(t)| \\
&= \Bigg| f_1(t,x(t),y(t)) \left(\frac{x_0}{f_1(0,x_0,y_0)} + \frac{1}{\Gamma(p)} \int_0^t (t-s)^{p-1} h_1(s,x(s),y(s))ds \right) \\
&\quad - f_1(t,x(t),y(t)) \left(\frac{x_0}{f_1(0,x_0,y_0)} + \frac{1}{\Gamma(p)} \int_0^t (t-s)^{p-1} h_1(s,x'(s),y'(s))ds \right) \\
&\quad + f_1(t,x(t),y(t)) \left(\frac{x_0}{f_1(0,x_0,y_0)} + \frac{1}{\Gamma(p)} \int_0^t (t-s)^{p-1} h_1(s,x'(s),y'(s))ds \right) \\
&\quad - f_1(t,x'(t),y'(t)) \left(\frac{x_0}{f_1(0,x_0,y_0)} + \frac{1}{\Gamma(p)} \int_0^t (t-s)^{p-1} h_1(s,x'(s),y'(s))ds \right) \Bigg| \\
&\le \frac{|f_1(t,x(t),y(t))|}{\Gamma(p)} \int_0^t (t-s)^{p-1} |h_1(s,x(s),y(s)) - h_1(s,x'(s),y'(s))| ds \\
&\quad + \left| \frac{x_0}{f_1(0,x_0,y_0)} + \frac{1}{\Gamma(p)} \int_0^t (t-s)^{p-1} h_1(s,x'(s),y'(s))ds \right| |f_1(t,x(t),y(t)) - f_1(t,x'(t),y'(t))| \\
&\le \frac{\mu_1 (\eta_1 \|x-x'\| + \eta_2 \|y-y'\|)}{\Gamma(p)} \int_0^t (t-s)^{p-1} ds \\
&\quad + \left(\frac{|x_0|}{\nu} + \nu_1 r + \frac{a^p k_1}{\Gamma(p+1)} \right) (\lambda_1 \|x-x'\| + \lambda_2 \|y-y'\|) \\
&\le a^p \frac{\mu_1 (\eta_1 \|x-x'\| + \eta_2 \|y-y'\|)}{\Gamma(p+1)} + \left(\frac{|x_0|}{\nu} + \nu_1 r + \frac{a^p k_1}{\Gamma(p+1)} \right) (\lambda_1 \|x-x'\| + \lambda_2 \|y-y'\|) \\
&\le \left(\frac{a^p \mu_1 \eta_1}{\Gamma(p+1)} + \lambda_1 \left(\frac{|x_0|}{\nu} + \nu_1 r + \frac{a^p k_1}{\Gamma(p+1)} \right) \right) \|x-x'\| \\
&\quad + \left(\frac{a^p \mu_1 \eta_2}{\Gamma(p+1)} + \lambda_2 \left(\frac{|x_0|}{\nu} + \nu_1 r + \frac{a^p k_1}{\Gamma(p+1)} \right) \right) \|y-y'\| \\
&= \alpha_1 \|x-x'\| + \beta_1 \|y-y'\| \\
&\le \gamma_1 \left(\|x-x'\| + \|y-y'\| \right),
\end{aligned}$$

which yields

$$\|\Lambda_1(x,y) - \Lambda_1(x',y')\| \le \gamma_1 \left(\|x-x'\| + \|y-y'\| \right). \tag{7}$$

And we have

$$\begin{aligned}
&|\Lambda_2(x,y)(t) - \Lambda_2(x',y')(t)| \\
&= \Bigg| y_0 - f_2(0,x_0,y_0) + f_2(t,x(t),y(t)) + \frac{1}{\Gamma(q)} \int_0^t (t-s)^{q-1} h_2(s,x(s),y(s))ds \\
&\quad - y_0 + f_2(0,x_0,y_0) - f_2(t,x'(t),y'(t)) - \frac{1}{\Gamma(q)} \int_0^t (t-s)^{q-1} h_2(s,x'(s),y'(s))ds \Bigg| \\
&\le |f_2(t,x(t),y(t)) - f_2(t,x'(t),y'(t))| \\
&\quad + \frac{1}{\Gamma(q)} \int_0^t (t-s)^{q-1} |h_2(s,x(s),y(s)) - h_2(s,x'(s),y'(s))| ds \\
&\le \lambda_3 \|x-x'\| + \lambda_4 \|y-y'\| + \frac{\xi_1 \|x-x'\| + \xi_2 \|y-y'\|}{\Gamma(q)} \int_0^t (t-s)^{q-1} ds
\end{aligned}$$

$$
\begin{aligned}
&\le \lambda_3\|x-x'\|+\lambda_4\|y-y'\|+a^q\frac{\xi_1\|x-x'\|+\xi_2\|y-y'\|}{\Gamma(q+1)}\\
&\le \left(\lambda_3+\frac{a^q\xi_1}{\Gamma(q+1)}\right)\|x-x'\|+\left(\lambda_4+\frac{a^q\xi_2}{\Gamma(q+1)}\right)\|y-y'\|\\
&= \alpha_2\|x-x'\|+\beta_2\|y-y'\|\\
&\le \gamma_2\left(\|x-x'\|+\|y-y'\|\right).
\end{aligned}
$$

Then

$$\|\Lambda_2(x,y)-\Lambda_2(x',y')\|\le \gamma_2\left(\|x-x'\|+\|y-y'\|\right). \tag{8}$$

This implies by (7) and (8) that we have

$$
\begin{aligned}
\|\Lambda(x,y)-\Lambda(x',y')\| &= \|\Lambda_1(x,y)-\Lambda_1(x',y')\|+\|\Lambda_2(x,y)-\Lambda_2(x',y')\|\\
&\le (\gamma_1+\gamma_2)\left(\|x-x'\|+\|y-y'\|\right).
\end{aligned}
$$

Finally, we find that Λ is a contraction. This completes the proof.

Example 1. We give an example to illustrate our abstract results.

Consider the following coupled system

$$
\begin{cases}
D_c^{\frac{1}{2}}\left[\dfrac{x(t)}{\frac{1}{4}+\frac{1}{10}\cos|x(t)|+\frac{1}{20}\frac{|y(t)|}{1+|y(t)|}}\right]=\frac{1}{7}+\frac{1}{9}x(t)+\frac{1}{10}y(t), \quad t\in J=[0,1],\\
D_c^{\frac{1}{2}}\left[y(t)-\left(\frac{1}{4}+\frac{1}{10}\sin|y(t)|+\frac{1}{20}\frac{|x(t)|}{1+|x(t)|}\right)\right]=\frac{1}{7}+\frac{1}{9}y(t)+\frac{1}{10}x(t), \quad t\in J,\\
x(0)=0,\\
y(0)=0.
\end{cases} \tag{9}
$$

This problem can be abstracted into

$$
\begin{cases}
D_c^{\frac{1}{2}}\left[\dfrac{x(t)}{f_1(t,x(t),y(t))}\right]=h_1(t,x(t),y(t)), \quad t\in J=[0,1],\\
D_c^{\frac{1}{2}}\left[y(t)-f_2(t,x(t),y(t))\right]=h_2(t,x(t),y(t)), \quad t\in J,\\
x(0)=x_0,\\
y(0)=y_0,
\end{cases}
$$

where

$$f_1(t,x(t),y(t))=\frac{1}{4}+\frac{1}{10}\cos|x(t)|+\frac{1}{20}\frac{|y(t)|}{1+|y(t)|},\ h_1(t,x(t),y(t))=\frac{1}{7}+\frac{1}{9}x(t)+\frac{1}{10}y(t),$$

$$f_2(t,x(t),y(t))=\frac{1}{4}+\frac{1}{10}\sin|y(t)|+\frac{1}{20}\frac{|x(t)|}{1+|x(t)|},\ h_2(t,x(t),y(t))=\frac{1}{7}+\frac{1}{9}y(t)+\frac{1}{10}x(t),$$

and $x_0=0$, $y_0=0$.

It is easy to check that

$$\eta_1 = \frac{1}{9}, \eta_2 = \frac{1}{10}, \xi_1 = \frac{1}{10}, \xi_2 = \frac{1}{9}, k_1 = k_2 = \frac{1}{7}, \mu_1 = \mu_2 = \frac{2}{5}, \lambda_1 = \lambda_4 = \frac{1}{10}, \lambda_2 = \lambda_3 = \frac{1}{20}.$$

We have

$$\frac{\mu_1 \frac{|x_0|}{\nu} + |y_0| + 2\mu_2 + \dfrac{a^p \mu_1 k_1}{\Gamma(p+1)} + \dfrac{a^q k_2}{\Gamma(q+1)}}{1 - (\mu_1 \nu_1 + \nu_2)} \simeq 1.46.$$

Then, for $r = 1.5$, we have

$$\alpha_1 \simeq 0.303, \beta_1 \simeq 0.03, \alpha_2 \simeq 0.162, \beta_2 \simeq 0.225.$$

Consequently

$$\gamma = \gamma_1 + \gamma_2 \simeq 0.529 < 1.$$

Thus, all assumptions in Theorem 1 are satisfied and the problem (9) has a unique solution on J.

By using Lery Schauder alternative we obtain the second existence result.

Theorem 2. *Let assumptions* (A_0) *and* (A_1) *be satisfied. Suppose in addition that*

$$|h_1(t,x,y)| \leq \rho_0 + \rho_1 \|x\| + \rho_2 \|y\|, \; |h_2(t,x,y)| \leq \sigma_0 + \sigma_1 \|x\| + \sigma_2 \|y\|,$$

for each $(t,x,y) \in J \times \mathbb{R} \times \mathbb{R}$.
And

$$\frac{a^p \mu_1 \rho_1}{\Gamma(p+1)} + \frac{a^q \sigma_1}{\Gamma(q+1)} < 1, \; \frac{a^p \mu_1 \rho_2}{\Gamma(p+1)} + \frac{a^q \sigma_2}{\Gamma(q+1)} < 1.$$

Then the problem (1) has at least one solution.

Proof. Proof Let $\mathcal{M} \subseteq E$ be bounded.

Then we can find positive constants N_1, N_2 such that

$$|h_1(t,x(t),y(t))| \leq N_1, \; |h_2(t,x(t),y(t))| \leq N_2$$

for each $(x,y) \in \mathcal{M}$ and $t \in J$.
For $(x,y) \in \mathcal{M}, t \in J$, we have

$$\begin{aligned}
&|A_1(x,y)(t)| \\
&= |f_1(t,x(t),y(t))| \left| \frac{x_0}{f(0,x_0,y_0)} + \frac{1}{\Gamma(p)} \int_0^t (t-s)^{p-1} h_1(s,x(s),y(s)) ds \right| \\
&\leq \mu_1 \left(\frac{|x_0|}{\nu} + \frac{N_1}{\Gamma(p)} \int_0^t (t-s)^{p-1} ds \right) \\
&\leq \mu_1 \left(\frac{|x_0|}{\nu} + \frac{a^p N_1}{\Gamma(p+1)} \right).
\end{aligned}$$

Then

$$\|\Lambda_1(x,y)\| \leq \mu_1 \left(\frac{|x_0|}{v} + \frac{a^p N_1}{\Gamma(p+1)} \right). \tag{10}$$

In a similar manner, we obtain

$$\|\Lambda_2(x,y)\| \leq |y_0| + 2\mu_2 + \frac{a^q N_2}{\Gamma(q+1)}. \tag{11}$$

From (10) and (11), we deduce that Λ is uniformly bounded.

We will show that the operator Λ is equicontinuous.

Let $t_1, t_2 \in J$ with $t_1 < t_2$. For $(x,y) \in E$, we have

$$\begin{aligned}
&|\Lambda_1(x,y)(t_2) - \Lambda_1(x,y)(t_1)| \\
&= \left| f_1(t_2,x(t_2),y(t_2)) \left(\frac{x_0}{f(0,x_0,y_0)} + \frac{1}{\Gamma(p)} \int_0^{t_2} (t_2-s)^{p-1} h_1(s,x(s),y(s))ds \right) \right. \\
&\quad - f_1(t_2,x(t_2),y(t_2)) \left(\frac{x_0}{f(0,x_0,y_0)} + \frac{1}{\Gamma(p)} \int_0^{t_1} (t_1-s)^{p-1} h_1(s,x(s),y(s))ds \right) \\
&\quad + f_1(t_2,x(t_2),y(t_2)) \left(\frac{x_0}{f(0,x_0,y_0)} + \frac{1}{\Gamma(p)} \int_0^{t_1} (t_1-s)^{p-1} h_1(s,x(s),y(s))ds \right) \\
&\quad \left. - f_1(t_1,x(t_1),y(t_1)) \left(\frac{x_0}{f(0,x_0,y_0)} + \frac{1}{\Gamma(p)} \int_0^{t_1} (t_1-s)^{p-1} h_1(s,x(s),y(s))ds \right) \right| \\
&\leq \frac{|f_1(t_2,x(t_2),y(t_2))|}{\Gamma(p)} \int_0^{t_1} \left((t_1-s)^{p-1} - (t_2-s)^{p-1} \right) |h_1(s,x(s),y(s))|\, ds \\
&\quad + \frac{|f_1(t_2,x(t_2),y(t_2))|}{\Gamma(p)} \int_{t_1}^{t_2} (t_2-s)^{p-1} |h_1(s,x(s),y(s))|\, ds \\
&\quad + \left(\frac{|x_0|}{|f(0,x_0,y_0)|} + \frac{1}{\Gamma(p)} \int_0^{t_1} (t_1-s)^{p-1} |h_1(s,x(s),y(s))| ds \right) \\
&\quad \times |f_1(t_2,x(t_2),y(t_2)) - f_1(t_1,x(t_1),y(t_1))| \\
&\leq \frac{\mu_1 N_1}{\Gamma(p)} \int_0^{t_1} \left((t_1-s)^{p-1} - (t_2-s)^{p-1} \right) ds + \frac{\mu_1 N_1}{\Gamma(p)} \int_{t_1}^{t_2} (t_2-s)^{p-1} ds \\
&\quad + \left(\frac{|x_0|}{v} + \frac{a^p N_1}{\Gamma(p+1)} \right) |f_1(t_2,x(t_2),y(t_2)) - f_1(t_1,x(t_1),y(t_1))| \\
&\leq \frac{\mu_1 N_1}{\Gamma(p+1)} \left((t_2-t_1)^p + t_1^p - t_2^p \right) + \frac{\mu_1 N_1}{\Gamma(p+1)} (t_2-t_1)^p \\
&\quad + \left(\frac{|x_0|}{v} + \frac{a^p N_1}{\Gamma(p+1)} \right) |f_1(t_2,x(t_2),y(t_2)) - f_1(t_1,x(t_1),y(t_1))| \to 0
\end{aligned}$$

as $t_2 \to t_1$.

Similarly, we have

$$
\begin{aligned}
&|\Lambda_2(x,y)(t_2)-\Lambda_2(x,y)(t_1)| \\
&= \Big| y_0 - f_2(0,x_0,y_0) + f_2(t_2,x(t_2),y(t_2)) + \frac{1}{\Gamma(q)}\int_0^{t_2}(t_2-s)^{q-1}h_2(s,x(s),y(s))ds \\
&\quad - y_0 + f_2(0,x_0,y_0) - f_2(t_1,x(t_1),y(t_1)) - \frac{1}{\Gamma(q)}\int_0^{t_1}(t_1-s)^{q-1}h_2(s,x(s),y(s))ds \Big| \\
&\le |f_2(t_2,x(t_2),y(t_2)) - f_2(t_1,x(t_1),y(t_1))| \\
&\quad + \frac{1}{\Gamma(q)}\int_0^{t_1}\left((t_1-s)^{q-1}-(t_2-s)^{q-1}\right)|h_2(s,x(s),y(s))|ds \\
&\quad + \frac{1}{\Gamma(q)}\int_{t_1}^{t_2}(t_2-s)^{q-1}|h_2(s,x(s),y(s))|ds \\
&\le |f_2(t_2,x(t_2),y(t_2)) - f_2(t_1,x(t_1),y(t_1))| + \frac{N_2}{\Gamma(q+1)}\left((t_2-t_1)^q + t_1^q - t_2^q\right) \\
&\quad + \frac{N_2}{\Gamma(q+1)}(t_2-t_1)^q \to 0 \text{ as} t_2 \to t_1.
\end{aligned}
$$

Finally, we find that Λ is equicontinuous.

In the next step. Denote

$$\mathscr{P}_\Lambda = \{(x,y) \in E : (x,y) = \delta\Lambda(x,y),\ 0 \le \delta \le 1\}.$$

And

$$\rho = \min\left(1 - \frac{a^p\mu_1\rho_1}{\Gamma(p+1)} - \frac{a^q\sigma_1}{\Gamma(q+1)}, 1 - \frac{a^p\mu_1\rho_2}{\Gamma(p+1)} - \frac{a^q\sigma_2}{\Gamma(q+1)}\right).$$

$\mathscr{P}_r$ is bounded. Indeed

Let $(x,y) \in \mathscr{P}_\Lambda$. Thus for any $t \in J$

$$\begin{cases} x(t) = \delta\Lambda_1(x,y) \\ y(t) = \delta\Lambda_2(x,y). \end{cases}$$

Then

$$|x(t)| \le \mu_1\left(\frac{|x_0|}{\nu} + a^p\frac{\rho_0 + \rho_1\|x\| + \rho_2\|y\|)}{\Gamma(p+1)}\right).$$

Also

$$|y(t)| \le |y_0| + 2\mu_2 + a^q\frac{\sigma_0 + \sigma_1\|x\| + \sigma_2\|y\|}{\Gamma(q+1)}.$$

Which imply that

$$\|x\| \le \mu_1\left(\frac{|x_0|}{\nu} + a^p\frac{\rho_0 + \rho_1\|x\| + \rho_2\|y\|)}{\Gamma(p+1)}\right).$$

And

$$\|y\| \leq |y_0| + 2\mu_2 + a^q \frac{\sigma_0 + \sigma_1\|x\| + \sigma_2\|y\|}{\Gamma(q+1)}.$$

As a consequence, we have

$$\|x\| + \|y\| \leq \left(\frac{a^p\mu_1\rho_1}{\Gamma(p+1)} + \frac{a^q\sigma_1}{\Gamma(q+1)}\right)\|x\| + \left(\frac{a^p\mu_1\rho_2}{\Gamma(p+1)} + \frac{a^q\sigma_2}{\Gamma(q+1)}\right)\|y\|$$
$$+ \mu_1\left(\frac{|x_0|}{\nu} + \frac{a^p\rho_0}{\Gamma(p+1)}\right) + |y_0| + 2\mu_2 + \frac{a^q\sigma_0}{\Gamma(q+1)}.$$

Thus we obtain

$$\left(1 - \frac{a^p\mu_1\rho_1}{\Gamma(p+1)} - \frac{a^q\sigma_1}{\Gamma(q+1)}\right)\|x\| + \left(1 - \frac{a^p\mu_1\rho_2}{\Gamma(p+1)} - \frac{a^q\sigma_2}{\Gamma(q+1)}\right)\|y\|$$
$$\leq \mu_1\left(\frac{|x_0|}{\nu} + \frac{a^p\rho_0}{\Gamma(p+1)}\right) + |y_0| + 2\mu_2 + \frac{a^q\sigma_0}{\Gamma(q+1)}.$$

It follows that

$$\|(x,y)\| = \|x\| + \|y\| \leq \frac{\mu_1\left(\frac{|x_0|}{\nu} + \frac{a^p\rho_0}{\Gamma(p+1)}\right) + |y_0| + 2\mu_2 + \frac{a^q\sigma_0}{\Gamma(q+1)}}{\rho}$$

Thus, all assumptions of Lemma (3) are satisfied and this permits us to conclude that Λ has at least one fixed point. Which is a solution of the problem (1).

Conclusion

In this paper, extending the models given by [2] and [11],it has suggested a model which brings together the two situations. It is a coupled system of mixed Hybrid fractional differential equations: Linear perturbations of first and second type. we proved the existence of a solution for this problem using two fixed point theorems namely Banach fixed point theorem and Lery Schauder alternative. The main result is well illustrated by two examples.

References

1. Baleanu, D., Khan, H., Jafari, H., Khan, R.A., Alipour, M.: On existence results for solutions of a coupled system of hybrid boundary value problems with hybrid conditions. Adv. Differ. Equ. **2015**(1), 1–14 (2015). https://doi.org/10.1186/s13662-015-0651-z
2. Dhage, B.C., Jadhav, N.S.: Basic results in the theory of hybrid differential equations with linear perturbations of second type. Tamsui Oxf. J. Math. Sci. **44**(2), 171–186 (2013)
3. Dhage, C.B., Lakshmikantham, V.: Basic results on hybrid differential equations. Nonlinear Anal. Hybrid Syst. **4**, 414–424 (2010)

4. Granas, A., Dugundji, J.: Fixed Point Theory. Springer, New York, NY, USA (2003). https://doi.org/10.1007/978-0-387-21593-8
5. Hilal, K., Kajouni, A.: Boundary value problems for hybrid differential equations. Math. Theory Model. 2224–5804 (2015)
6. Kilbas, A.A., Srivastava, H.M., Trujillo, J.J.: Theory and applications of fractional differential equations. In: North-Holland Mathematics Studies, vol. 204. Elsevier, Amsterdam (2006)
7. Lu, H., Sun, S., Yang, D., Teng, H.: Theory of fractional hybrid differential equations with linear perturbations of second type. Bound. Value Probl. **2013**(1), 1–16 (2013). https://doi.org/10.1186/1687-2770-2013-23
8. Sitho, S., Ntouyas, S.K., Tariboon, J.: Existence results for hybrid fractional integro-differential equations. Bound. Value Probl. **2015**(1), 1–13 (2015). https://doi.org/10.1186/s13661-015-0376-7
9. Somjaiwang, D., Sa Ngiamsunthorn, P.: Existence and approximation of solutions to fractional order hybrid differential equations. Adv. Differ. Equ. **2016**(1), 1–11 (2016). https://doi.org/10.1186/s13662-016-0999-8
10. Zhou, Y., Jiao, F., Li, J.: Existence and uniqueness for fractional neutral differential equations with infinite delay. Nonlinear Anal. TMA **71**, 3249–3256 (2009)
11. Zhao, Y., Sun, S., Han, Z., Li, Q.: Theory of fractional hybrid differential equations. Comput. Math. Appl. **62**, 1312–1324 (2011)

On a Wilson Vector-Matrix Functional Equation

Youssef Aissi(✉) and Driss Zeglami

Department of Mathematics ENSAM, Moulay Ismail University,
B.P. 15290, Al Mansour, Meknes, Morocco
yaissi94@gmail.com

Abstract. The purpose of this paper is to investigate the vector-matrix functional equation

$$\frac{g(xy)+g(\sigma(y)x)}{2}=\Phi(y)g(x),\ \ x,y\in G,$$

where G is a group, g is a vector-valued function, Φ is a matrix-valued function, and $\sigma : G \to G$ is an involutive automorphism.

Keywords: Vector-matrix functional equation · d'Alembert · Wilson · Character · Quadratic equation · Morphism · Simultaneous triangularization · Linear algebra

2020 Mathematics Subject Classification: Primary 39B32 · 39B42 · 39B72

1 Set Up, Notation and Terminology

To formulate our results we work with the following set up, notation and terminology that will be used throughout the paper: G is a group and M is a monoid (a set with an associative composition rule with an identity element denoted by e). The map $\sigma : G \to G$ is an involutive automorphism of G. That σ is involutive means that $\sigma(\sigma(x)) = x$ for all $x \in G$. We denote by $M_n(\mathbb{C})$ the algebra of all complex $n \times n$ matrices and I_n its identity matrix, and we will view the elements of $\mathbb{C}^n$ as column vectors. By $\mathcal{M}(G)$ we denote the set of all characters $\chi : G \to \mathbb{C}^*$, by $\mathcal{A}(G)$ the vector space of all additive maps from G to $\mathbb{C}$ and we put $\mathcal{A}^{\pm}(G) := \{a \in \mathcal{A}(G) : a \circ \sigma = a\}$.

For any complex-valued function f on G we use the notation

$$f^e := \frac{f + f\circ\sigma}{2} \quad \text{and} \quad f^o := \frac{f - f\circ\sigma}{2}.$$

We say that f is σ-even (resp. σ-odd) if $f \circ \sigma = f$ (resp. $f \circ \sigma = -f$). The function f is central means $f(xy) = f(yx)$. Finally we denote by $\mathcal{F}$ the set of all vector-valued functions $g = \begin{pmatrix} g_1 \\ \vdots \\ g_n \end{pmatrix} : G \to \mathbb{C}^n$ with linearly independent components.

S. Melliani and O. Castillo (Eds.): ICPAMS 2021, LNNS 476, pp. 458–472, 2023.
https://doi.org/10.1007/978-3-031-12416-7_39

2 Introduction

The purpose of the present paper is to study the following functional equation

$$\frac{g(xy)+g(\sigma(y)x)}{2}=\Phi(y)g(x), \tag{2.1}$$

where the vector-valued function $g : G \to \mathbb{C}^2$ and the matrix-valued function $\Phi : G \to M_2(\mathbb{C})$ are the unknown functions to be determined.

The Eq. (2.1), that can be viewed as a 2-dimensional version of the scalar Wilson functional equation introduced in [12], has been treated, when σ is the group inversion, by Sinopoulos [7] under the assumption that G is a 2-divisible abelian group. It has been studied by Stetkaer [8] on an abelian group that need not be divisible by 2 and with a general involutive automorphism.

Recently, Aissi et al. [2] determined on groups the 2×2 matrix-valued solutions of the equation

$$\frac{\Phi(xy)+\Phi(\sigma(y)x)}{2}=\Phi(x)\Phi(y). \tag{2.2}$$

As continuation of [7,8], we will extend the setting of the solutions (g, Φ) of (2.1) from an abelian group to a group that need not be abelian.

Let $g : M \to \mathbb{C}^n$ and $\Phi : M \to M_n(\mathbb{C})$. We ask in Lemma 3.4 when Φ satisfies

$$\Phi(x)\Phi(y)=\Phi(x)\Phi(y) \text{ for all } x, y \in M. \tag{2.3}$$

This is a crucial key for solving (2.1).

Assume now that the pair $g : G \to \mathbb{C}^2$ and $\Phi : G \to M_2(\mathbb{C})$ with $g \in \mathcal{F}$ is a solution of (2.1). We will prove that Φ is σ-even, that both g^e and g^o satisfy (2.1) with the same Φ as for g and we derive the connection between (2.1) and (2.2) that if g is odd or in particular central (see Lemma 3.4) then Φ satisfies (2.2). With these results we express, in terms of characters of G and additive and bi-additive functions on G, all solutions of (2.1) under one of the two conditions: $\Phi : G \to M_2(\mathbb{C})$ satifies (2.3) or $g : G \to \mathbb{C}^2$ is central.

Note that results about the complex-valued functional equations

$$f(xy)+f(\sigma(y)x)=2f(x)f(y), \quad x, y \in G, \tag{2.4}$$
$$f(xy)+f(\sigma(y)x)=2f(x)g(y), \quad x, y \in G, \tag{2.5}$$
$$f(xy)+f(\sigma(y)x)=2f(x)+2f(y), \quad x, y \in G, \tag{2.6}$$
$$f(xy)+f(\sigma(y)x)=2f(x), \quad x, y \in G, \tag{2.7}$$

the sine addition law

$$f(xy)=f(x)g(y)+f(y)g(x), \quad x, y \in G, \tag{2.8}$$

and the symmetrized additive Cauchy equation

$$f(xy)+f(yx)=2f(x)+2f(y), \quad x, y \in G, \tag{2.9}$$

that were treated in [3–5,10], [9, Chap. 4], and [9, Chapter 2], respectively, are essential in finding the solutions of the functional Eq. (2.1). General results about similar scalar functional equations on groups are summarized in the monographs [1,9,11] that contains many references.

3 Basic Results

In this section we derive some properties of solutions of (2.1) on a monoid, which enable us to provide a connection between the Eqs. (2.1) and (2.2).

To deal with, let us recall some new tools, which will be used further on.

Lemma 3.1. *Let $\Phi : M \to M_2(\mathbb{C})$ be a function and let $\phi_1, \phi_2, \psi_1 : M \to \mathbb{C}$ be such that $\Phi = \begin{pmatrix} \phi_1 & \psi_1 \\ 0 & \phi_2 \end{pmatrix}$. Then, Φ is a solution of (2.2) if and only if*

$$\begin{cases} \phi_1(xy) + \phi_1(\sigma(y)x) = 2\phi_1(x)\phi_1(y) \\ \phi_2(xy) + \phi_2(\sigma(y)x) = 2\phi_2(x)\phi_2(y) \\ \psi_1(xy) + \psi_1(\sigma(y)x) = 2\phi_1(x)\psi_1(y) + 2\psi_1(x)\phi_2(y) \end{cases}, \quad x, y \in M. \tag{3.1}$$

Proof. Can be trivially shown. □

Lemma 3.2. *If $\Phi : M \to M_n(\mathbb{C})$, $g : M \to \mathbb{C}^n$ with $g \in \mathcal{F}$ is a solution of (2.1), then*

(1) $\Phi(e) = I_n$.
(2) $g^o(xy) = \Phi(y)g(x) - \Phi(x)g(\sigma(y))$, $x, y \in M$.
(3) $\begin{cases} \Phi^o(x)g^o(y) = -\Phi^o(y)g^o(x) \\ \Phi^e(x)g^e(y) = \Phi^e(y)g^e(x) \end{cases}, \quad x, y \in M.$
(4) If $g : M \to \mathbb{C}^n$ is central, then $\Phi : M \to M_n(\mathbb{C})$ satisfies

$$\Phi(x)\Phi(y) = \Phi(y)\Phi(x), \quad x, y \in M.$$

(5) If $g : M \to \mathbb{C}^n$ is odd, then $g : M \to \mathbb{C}^n$ is central.

Proof. Let $\Phi : M \to M_n(\mathbb{C})$, $g : M \to \mathbb{C}^n$ with $g \in \mathcal{F}$ be a solution of (2.1).

(1) Putting $y = e$ in (2.1) we get $(\Phi(e) - I_n)g = 0$. Since $g \in \mathcal{F}$ then $\Phi(e) = I_n$.
(2) Making the substitution $(\sigma(y), x)$ in (2.1), we obtain

$$g(\sigma(y)x) + g(\sigma(x)\sigma(y)) = 2\Phi(x)g(\sigma(y)), \quad x, y \in M. \tag{3.2}$$

If we subtract (2.1) from (3.2) we deduce easily the claimed result.
(3) We have

$$g^e(xy) + g^e(\sigma(y)x) = g(xy) + g(\sigma(y)x) - g^o(xy) - g^o(\sigma(y)x),$$

for all $x, y \in M$. So, from (2.1) and Lemma 3.2 (2) we get that

$$g^e(xy) + g^e(\sigma(y)x) = \Phi(y)g(x) + \Phi(\sigma(y))g(\sigma(x)), \quad x, y \in M.$$

This is equivalent to

$$\begin{aligned} g^e(xy) + g^e(\sigma(y)x) &= [\Phi^e(y) + \Phi^o(y)][g^e(x) + g^o(x)] \\ &\quad + [\Phi^e(y) - \Phi^o(y)][g^e(x) - g^o(x)], \quad x, y \in M. \end{aligned}$$

With simple computation we get

$$\frac{g^e(xy)+g^e(\sigma(y)x)}{2}=\Phi^e(y)g^e(x)+\Phi^o(y)g^o(x),\ x,y\in M. \tag{3.3}$$

Changing x and y by $\sigma(y)$ and x, respectively, in (3.3) we get

$$\frac{g^e(xy)+g^e(\sigma(y)x)}{2}=\Phi^e(x)g^e(y)-\Phi^o(x)g^o(y),\ x,y\in M. \tag{3.4}$$

From (3.3) and (3.4) we infer that

$$\Phi^e(y)g^e(x)+\Phi^o(y)g^o(x)=\Phi^e(x)g^e(y)-\Phi^o(x)g^o(y),\ x,y\in M.$$

This means that

$$\Phi^o(x)g^o(y)+\Phi^o(y)g^o(x)=\Phi^e(x)g^e(y)-\Phi^e(y)g^e(x),\ x,y\in M.$$

Hence,

$$\begin{cases}\Phi^o(x)g^o(y)=-\Phi^o(y)g^o(x)\\ \Phi^e(x)g^e(y)=\Phi^e(y)g^e(x)\end{cases},\ x,y\in M.$$

(4) Assume that $g:M\to\mathbb{C}^n$ is central. Replacing x by xz and $x\sigma(z)$ separately in (2.1), we get

$$g(xzy)+g(\sigma(y)xz)=2\Phi(y)g(xz),\ x,y\in M, \tag{3.5}$$

and

$$g(x\sigma(z)y)+g(\sigma(y)x\sigma(z))=2\Phi(y)g(x\sigma(z)),\ x,y\in M. \tag{3.6}$$

Since $g:M\to\mathbb{C}^n$ is central, then (3.5) and (3.6) are equivalent to

$$g(yxz)+g(\sigma(y)xz)=2\Phi(y)g(xz),\ x,y\in M, \tag{3.7}$$

and

$$g(\sigma(z)yx)+g(\sigma(z)\sigma(y)x)=2\Phi(y)g(\sigma(z)x),\ x,y\in M, \tag{3.8}$$

respectively. Adding the last two equations, we obtain

$$\begin{aligned}[g(yxz)+g(\sigma(z)yx)]+[g(\sigma(y)xz)+g(\sigma(z)\sigma(y)x)]\\ =2\Phi(y)[g(xz)+g(\sigma(z)x)],\end{aligned} \tag{3.9}$$

for all $x,y,z\in M$. So, by using (2.1) and the fact that g is central in (3.9) we obtain

$$\Phi(z)\Phi(y)g(x)=\Phi(y)\Phi(z)g(x).$$

As $g\in\mathcal{F}$ then $\Phi(x),\ x\in M$ commute pairwise.

(5) Suppose that $g:M\to\mathbb{C}^n$ is odd. From Lemma 3.2 (2) we get

$$g(xy)=\Phi(x)g(y)+\Phi(y)g(x),\ x,y\in M. \tag{3.10}$$

Hence, $g:M\to\mathbb{C}^n$ is central. This completes the proof.

□

We shall recall the definition of the triangularizability of a collection of linear transformations and prove an auxiliary lemma concerning a connection between Eqs. (2.1) and (2.2).

Definition 3.3. ([6]) A collection of linear transformations is triangularizable if there is a basis for the vector space such that all transformations in the collection have upper triangular matrix representations with respect to that basis.

Lemma 3.4. *Assume that $\Phi : M \to M_2(\mathbb{C})$ (i.e. the collection $\{\Phi(x) \mid x \in M\}$) is triangularizable. If the pair $g : M \to \mathbb{C}^2$ with $g \in \mathcal{F}$ and Φ is a solution of (2.1) then*

(1) Φ is σ-even.
(2) Both g^e and g^o satisfy (2.1) with the same Φ as for g.
(3) Φ is a solution of (2.2).

Proof. Let $\Phi : M \to M_2(\mathbb{C})$, $g =: \begin{pmatrix} g_1 \\ g_2 \end{pmatrix} : M \to \mathbb{C}^2$ with $g \in \mathcal{F}$ be a solution of (2.1). Since Φ is triangularizable then we may assume that $\Phi = \begin{pmatrix} \phi_1 & \psi_1 \\ 0 & \phi_2 \end{pmatrix}$.

(1) To prove (1) we distinguish between two cases:
Case 1: If $\phi_1 \neq \phi_2$, then we may assume that $\psi_1 = 0$. Since (g_1, ϕ_1) and (g_2, ϕ_2) are solutions of (2.5) and g_1, g_2 are non-zero, then from [3, Lemma 3.2] we derive that ϕ_1 and ϕ_2 are σ-even, which imply that Φ is σ-even.
Case 2: If $\phi_1 = \phi_2 =: \phi$, we have from [3, Lemma 3.2] that ϕ is σ-even, because the pair (g_2, ϕ) is a solution of (2.5) and $g_2 \neq 0$. From Lemma 3.2 (3) we see that $\psi_1^o(x)g_2^o(y) = -\psi_1^o(y)g_2^o(x)$ for all $x, y \in M$. This implies from [9, Exercise 1.1] that $\psi_1^o = 0$ or $g_2^o = 0$. If $\psi_1^o = 0$ then it is clear that Φ is σ-even. It remains to discuss the other case. Suppose that $g_2^o = 0$ i.e. g_2 is σ -even. From Lemma 3.2 (2) we get

$$g_1^o(xy) = \phi(y)g_1(x) + \psi_1(y)g_2(x) - \phi(x)g_1(\sigma(y)) - \psi_1(x)g_2(y), \ x, y \in M. \tag{3.11}$$

Since (g_2, ϕ) is a solution of (2.5) and g_2 is σ -even, then from [3, Lemma 3.2] we have $g_2 = g_2(e)\phi$ with $g_2(e) \neq 0$, because otherwise $g_2 \equiv 0$ which is in contradiction with $g \in \mathcal{F}$. Thus, (3.11) becomes

$$g_1^o(xy) = \phi(y)[g_1(x) - g_2(e)\psi_1(x)] + \phi(x)[g_2(e)\psi_1(y) - g_1(\sigma(y))], \ x, y \in M. \tag{3.12}$$

Putting $y = e$ in (3.12) we get by using Lemma 3.2 (1) that

$$g_1^o(x) = g_1(x) - g_2(e)\psi_1(x) - g_1(e)\phi(x), \ x \in M. \tag{3.13}$$

Replacing x by $\sigma(x)$ in (3.13) we get

$$-g_1^o(x) = g_1(\sigma(x)) - g_2(e)\psi_1(\sigma(x)) - g_1(e)\phi(x), \ x \in M. \tag{3.14}$$

If we subtract (3.13) from (3.14) we obtain

$$g_1^o(x) = g_1^o(x) - g_2(e)\psi_1^o(x),\ x \in M,$$

which implies that $\psi_1^o \equiv 0$, because $g_2(e) \neq 0$. Hence, Φ is σ-even.

(2) From Lemma 3.4 (1) and (3.3) we infer that

$$\frac{g^e(xy) + g^e(\sigma(y)x)}{2} = \Phi(y)g^e(x),\ x, y \in M. \tag{3.15}$$

Since

$$g^o(xy) + g^o(\sigma(y)x) = g(xy) + g(\sigma(y)x) - g^e(xy) - g^e(\sigma(y)x),\ x, y \in M.$$

Then, from (2.1) and (3.15) we get easily that

$$\frac{g^o(xy) + g^o(\sigma(y)x)}{2} = \Phi(y)g^o(x),\ x, y \in M,$$

which implies the claimed result.

(3) In order to show (3) we will distinguish between two cases:

Case 1: If $\phi_1 \neq \phi_2$, then we may assume that $\psi_1 = 0$, hence the Eq. (2.1) is equivalent to

$$\begin{cases} g_1(xy) + g_1(\sigma(y)x) = 2\phi_1(y)g_1(x) \\ g_2(xy) + g_2(\sigma(y)x) = 2\phi_2(y)g_2(x) \end{cases},\ x, y \in M. \tag{3.16}$$

Thus from [3, Proposition 3.4] we get that ϕ_1 and ϕ_2 are solutions of (2.4). By using Lemma 3.1 we deduce that Φ is a solution of (2.2).

Case 2: Here $\phi_1 = \phi_2 =: \phi$. Then, the Eq. (2.1) is equivalent to

$$\begin{cases} g_1(xy) + g_1(\sigma(y)x) = 2\phi(y)g_1(x) + 2\psi_1(y)g_2(x) \\ g_2(xy) + g_2(\sigma(y)x) = 2\phi(y)g_2(x) \end{cases},\ x, y \in M. \tag{3.17}$$

Since (g_2, ϕ) is a solution of (2.5), then from [3, Proposition 3.4] ϕ is a solution of (3.18). So, to conclude the claimed result it suffices, from Lemma 3.1, to show that the pair of functions (ψ_1, ϕ) satisfies the symmetrized sine addition formula, i.e.,

$$f(xy) + f(\sigma(y)x) = 2f(x)g(y) + 2f(y)g(x),\ x, y \in M, \tag{3.18}$$

where $f, g : M \to \mathbb{C}$ are complex valued functions. From Lemma 3.2 (3) and Lemma 3.4 (1) we have $\Phi(x)g^e(y) = \Phi(y)g^e(x)$ for all $x, y \in M$. This implies with simple computations that

$$\phi(x)g_1^e(y) + \psi_1(x)g_2^e(y) = \phi(y)g_1^e(x) + \psi_1(y)g_2^e(x),\ x, y \in M. \tag{3.19}$$

From Lemma 3.4 (2) we have (g^e, Φ) is a solution of (2.1). Since $g_2^e(e) = g_2(e) \neq 0$, then we deduce from [3, Lemma 3.2] that $g_2^e = g_2^e(e)\phi$. Replacing g_2^e by its expression in (3.19) we obtain

$$\phi(x)g_1^e(y) + g_2^e(e)\psi_1(x)\phi(y) = \phi(y)g_1^e(x) + g_2^e(e)\psi_1(y)\phi(x),\ x, y \in M. \tag{3.20}$$

This is equivalent to

$$\phi(x)[g_1^e(y) - g_2^e(e)\psi_1(y)] = \phi(y)[g_1^e(x) - g_2^e(e)\psi_1(x)], \; x, y \in M.$$

Since $\phi \neq 0$, then there exists $\lambda \in \mathbb{C}$ such that

$$g_2^e(e)\psi_1 = g_1^e - \lambda\phi.$$

So we have

$$\begin{aligned}
& g_2^e(e)[\psi_1(xy) + \psi_1(\sigma(y)x)] \\
&= g_1^e(xy) - \lambda\phi(xy) + g_1^e(\sigma(y)x) - \lambda\phi(\sigma(y)x) \\
&= 2\phi(y)g_1^e(x) + 2\psi_1(y)g_2^e(x) - 2\lambda\phi(x)\phi(y) \\
&= 2\phi(y)[g_2^e(e)\psi_1(x) + \lambda\phi(x)] + 2g_2^e(e)\phi(x)\psi_1(y) - 2\lambda\phi(x)\phi(y) \\
&= g_2^e(e)[2\phi(y)\psi_1(x) + 2\phi(x)\psi_1(y)], \; x, y \in M.
\end{aligned}$$

Since $g_2^e(e) = g_2(e) \neq 0$ then we get easily the claimed result.

□

Remark 3.5. Let $\Phi : M \to M_n(\mathbb{C})$ and $g : M \to \mathbb{C}^n$ be functions.

(1) If (g, Φ) is a solution of (2.1) such that $g \in \mathcal{F}$ and is central then Φ is σ-even.
(2) If Φ is a solution of (2.2), then the collection $\{\Phi(x) \mid x \in M\}$ is triangularizable.

4 Main Results

To determine the solutions (g, Φ) of (2.1) on a group with $g : G \to \mathbb{C}^2$ is a central function such that $g \in \mathcal{F}$, we need to know the solutions Φ of the functional Eq. (2.2) satisfying $\Phi(e) = I_2$. They are listed in [2, Theorem 5.1] that we cite here:

Theorem 4.1. *[2] The solutions $\Phi : G \to M_2(\mathbb{C})$ of the matrix functional Eq. (2.2) satisfying $\Phi(e) = I_2$ are the functions of the three forms below in which P ranges over $GL_2(\mathbb{C})$:*

(1)

$$\Phi = P \begin{pmatrix} \dfrac{\chi_1 + \chi_1 \circ \sigma}{2} & 0 \\ 0 & \dfrac{\chi_2 + \chi_2 \circ \sigma}{2} \end{pmatrix} P^{-1}, \tag{4.1}$$

where χ_1 and χ_2 are characters of G.

(2)

$$\Phi = P \begin{pmatrix} \dfrac{\chi + \chi \circ \sigma}{2} & \dfrac{\chi + \chi \circ \sigma}{2} a^+ + \dfrac{\chi - \chi \circ \sigma}{2} a^- \\ 0 & \dfrac{\chi + \chi \circ \sigma}{2} \end{pmatrix} P^{-1}, \tag{4.2}$$

where χ is a character of G such that $\chi \neq \chi \circ \sigma$ and $a^\pm \in \mathcal{A}^\pm(G)$.

(3)

$$\Phi = \chi P \begin{pmatrix} 1 & S+\psi \\ 0 & 1 \end{pmatrix} P^{-1}, \tag{4.3}$$

where χ is a character of G such that $\chi = \chi \circ \sigma$, ψ is a solution of the symmetrized additive Cauchy Eq. (2.9) such that $\psi \in \mathcal{N}(G,\sigma)$ and $S : G \to \mathbb{C}$ is a map of the form $S(x) = B(x,x)$, $x \in G$, where $B : G \times G \to \mathbb{C}$ is a bi-additive function of G such that $B(x,\sigma(y)) = -B(y,x)$.

The main result of the present paper reads as follows.

Theorem 4.2. *The functions $\Phi : G \to M_2(\mathbb{C})$ and $g : G \to \mathbb{C}^2$, with $g \in \mathcal{F}$ and is central, satisfy (2.1) if, and only if there exist $u, v \in \mathbb{C}^2$ and $P \in GL_2(\mathbb{C})$ such that*

$$\begin{cases} \Phi = P\dfrac{E+E\circ\sigma}{2}P^{-1} \\ g = P(Eu + E\circ\sigma v) \end{cases}, \tag{4.4}$$

where $E : G \to M_2(\mathbb{C})$ has one of the following six forms:

(1)

$$E = \begin{pmatrix} \chi_1 & 0 \\ 0 & \chi_2 \end{pmatrix},$$

where χ_1 and χ_2 are characters of G such that $\chi_1 \circ \sigma \neq \chi_1$ and $\chi_2 \circ \sigma \neq \chi_2$.

(2)

$$E = \begin{pmatrix} (1+a_1^-)\chi_1 & 0 \\ 0 & \chi_2 \end{pmatrix},$$

where χ_1 and χ_2 are characters of G such that $\chi_1 \circ \sigma = \chi_1$ and $\chi_2 \circ \sigma \neq \chi_2$, and $a_1^- \in \mathcal{A}^-(G)$.

(3)

$$E = \begin{pmatrix} (1+a_1^-)\chi_1 & 0 \\ 0 & (1+a_2^-)\chi_2 \end{pmatrix},$$

where χ_1 and χ_2 are characters of G such that $\chi_1 \circ \sigma = \chi_1$ and $\chi_2 \circ \sigma = \chi_2$, and $a_1^-, a_2^- \in \mathcal{A}^-(G)$.

(4)

$$E = \chi \begin{pmatrix} 1 & a^+ + a^- \\ 0 & 1 \end{pmatrix},$$

where $a^\pm \in \mathcal{A}^\pm(G)$, and χ is a character of G such that $\chi \circ \sigma \neq \chi$.

(5)

$$E = \chi \begin{pmatrix} 1 & S+\psi+a_3^- \\ 0 & 1 \end{pmatrix},$$

where χ is a character of G such that $\chi \circ \sigma = \chi$, $a_3^- \in \mathcal{A}^-(G)$, $\psi \in \mathcal{A}^+(G)$ and $S : G \to \mathbb{C}$ is a map of the form $S(x) = B(x,x)$, $x \in G$, where $B : G \times G \to \mathbb{C}$ is a bi-additive function of G such that $B(x,\sigma(y)) = -B(y,x)$

(6)

$$E=\chi\begin{pmatrix}1+a_2^- & \beta(a_2^-)^3+3\beta(a_2^-)^2+\psi+a_2^-\psi+a_4^-\\ 0 & 1+a_2^-\end{pmatrix},$$

where χ is a character of G such that $\chi\circ\sigma=\chi$, $\psi\in\mathcal{A}^+(G)$, $a_2^-,a_4^-\in\mathcal{A}^-(G)$ and $\beta\in\mathbb{C}$.

Proof. Let $\Phi: G\to M_2(\mathbb{C})$, $g: G\to\mathbb{C}^2$ be a solution of (2.1) such that g is a central function satisfying $g\in\mathcal{F}$. Since $g: G\to\mathbb{C}^2$ is central, then from Lemma 3.2 (4) and [6, Theorem 1.1.5] we infer that $\Phi: G\to M_2(\mathbb{C})$ is triangularizable, i.e., there exists an invertible matrix P such that

$$\Psi:=P\Phi P^{-1}=:\begin{pmatrix}\phi_1 & \psi_1\\ 0 & \phi_2\end{pmatrix},$$

where $\phi_1,\phi_2,\psi_1,\mathcal{G}_1,\mathcal{G}_2: G\to\mathbb{C}$. Using the notation $\mathcal{G}:=Pg=:\begin{pmatrix}\mathcal{G}_1\\ \mathcal{G}_2\end{pmatrix}$, (2.1) becomes

$$\frac{\mathcal{G}(xy)+\mathcal{G}(\sigma(y)x)}{2}=\Psi(y)\mathcal{G}(x),\ x,y\in G, \tag{4.5}$$

which is equivalent to

$$\begin{cases}\mathcal{G}_1(xy)+\mathcal{G}_1(\sigma(y)x)=2\phi_1(y)\mathcal{G}_1(x)+2\psi_1(y)\mathcal{G}_2(x)\\ \mathcal{G}_2(xy)+\mathcal{G}_2(\sigma(y)x)=2\phi_2(y)\mathcal{G}_2(x)\end{cases},\ x,y\in G. \tag{4.6}$$

From Lemma 3.2 (1) and Lemma 3.4 (3) we deduce that Ψ is a solution of (2.2) satisfying $\Psi(e)=I_2$. So, according to Theorem 4.1 we will break the job into three cases:

Case 1: If $\Psi=\begin{pmatrix}\frac{\chi_1+\chi_1\circ\sigma}{2} & 0\\ 0 & \frac{\chi_2+\chi_2\circ\sigma}{2}\end{pmatrix}$, where $\chi_1,\chi_2\in\mathcal{M}(G)$, then (4.6) becomes

$$\begin{cases}\mathcal{G}_1(xy)+\mathcal{G}_1(\sigma(y)x)=2\frac{\chi_1+\chi_1\circ\sigma}{2}(y)\mathcal{G}_1(x)\\ \mathcal{G}_2(xy)+\mathcal{G}_2(\sigma(y)x)=2\frac{\chi_2+\chi_2\circ\sigma}{2}(y)\mathcal{G}_2(x)\end{cases},\ x,y\in G. \tag{4.7}$$

Based on [3, Theorem 3.6] there exist $(\alpha_i,\beta_i)\in\mathbb{C}^2\setminus(0,0)$ and $a_i^-\in\mathcal{A}^-(G)$, $i\in\{1,2\}$ such that

$$\begin{cases}\mathcal{G}_i=\alpha_i\frac{\chi_i+\chi_i\circ\sigma}{2}+\beta_i\frac{\chi_i-\chi_i\circ\sigma}{2} & \text{if}\quad \chi_i\circ\sigma\neq\chi_i\\ \mathcal{G}_i=(\alpha_i+a_i^-)\chi_i & \text{if}\ \chi_i\circ\sigma=\chi_i\end{cases},\ i\in\{1,2\}.$$

Subcase 1.1: If $\chi_1\circ\sigma\neq\chi_1$ and $\chi_2\circ\sigma\neq\chi_2$, then we obtain

$$\begin{cases}\mathcal{G}_1=\alpha_1\frac{\chi_1+\chi_1\circ\sigma}{2}+\beta_1\frac{\chi_1-\chi_1\circ\sigma}{2}\\ \mathcal{G}_2=\alpha_2\frac{\chi_2+\chi_2\circ\sigma}{2}+\beta_2\frac{\chi_2-\chi_2\circ\sigma}{2}\end{cases}.$$

Hence, we have the desired result with

$$E = \begin{pmatrix} \chi_1 & 0 \\ 0 & \chi_2 \end{pmatrix}, \quad u = \frac{1}{2}\begin{pmatrix} \alpha_1 + \beta_1 \\ \alpha_2 + \beta_2 \end{pmatrix} \text{ and } v = \frac{1}{2}\begin{pmatrix} \alpha_1 - \beta_1 \\ \alpha_2 - \beta_2 \end{pmatrix}.$$

Subcase 1.2: If $\chi_1 \circ \sigma = \chi_1$ and $\chi_2 \circ \sigma \neq \chi_2$, then we have

$$\begin{cases} \mathcal{G}_1 = (\alpha_1 + a_1^-)\chi_1 \\ \mathcal{G}_2 = \alpha_2 \dfrac{\chi_2 + \chi_2 \circ \sigma}{2} + \beta_2 \dfrac{\chi_2 - \chi_2 \circ \sigma}{2} \end{cases}.$$

So, we get the claimed result with

$$E = \begin{pmatrix} (1 + a_1^-)\chi_1 & 0 \\ 0 & \chi_2 \end{pmatrix}, \quad u = \frac{1}{2}\begin{pmatrix} \alpha_1 + 1 \\ \alpha_2 + \beta_2 \end{pmatrix} \text{ and } v = \frac{1}{2}\begin{pmatrix} \alpha_1 - 1 \\ \alpha_2 - \beta_2 \end{pmatrix}.$$

Subcase 1.3: If $\chi_1 \circ \sigma \neq \chi_1$ and $\chi_2 \circ \sigma = \chi_2$, then as in Subcase 1.2 we can easily show that we have the desired result with

$$E = \begin{pmatrix} (1 + a_2^-)\chi_2 & 0 \\ 0 & \chi_1 \end{pmatrix}, \quad u = \frac{1}{2}\begin{pmatrix} \alpha_2 + 1 \\ \alpha_1 + \beta_1 \end{pmatrix} \text{ and } v = \frac{1}{2}\begin{pmatrix} \alpha_2 - 1 \\ \alpha_1 - \beta_1 \end{pmatrix}.$$

Subcase 1.4: If $\chi_1 \circ \sigma = \chi_1$ and $\chi_2 \circ \sigma = \chi_2$, then we have

$$\begin{cases} \mathcal{G}_1 = (\alpha_1 + a_1^-)\chi_1 \\ \mathcal{G}_2 = (\alpha_2 + a_2^-)\chi_2 \end{cases}.$$

So, we obtain the desired result with

$$E = \begin{pmatrix} (1 + a_1^-)\chi_1 & 0 \\ 0 & (1 + a_2^-)\chi_2 \end{pmatrix}, \quad u = \frac{1}{2}\begin{pmatrix} \alpha_1 + 1 \\ \alpha_2 + 1 \end{pmatrix} \text{ and } v = \frac{1}{2}\begin{pmatrix} \alpha_1 - 1 \\ \alpha_2 - 1 \end{pmatrix}.$$

Case 2: If $\Psi = \begin{pmatrix} \dfrac{\chi + \chi \circ \sigma}{2} & \dfrac{\chi + \chi \circ \sigma}{2} a^+ + \dfrac{\chi - \chi \circ \sigma}{2} a^- \\ 0 & \dfrac{\chi + \chi \circ \sigma}{2} \end{pmatrix}$, where χ is a character of G such that $\chi \neq \chi \circ \sigma$ and $a^\pm \in \mathcal{A}^\pm(G)$, then we obtain

$$\begin{aligned} \mathcal{G}_1(xy) + \mathcal{G}_1(\sigma(y)x) &= 2[\frac{\chi + \chi \circ \sigma}{2} a^+ + \frac{\chi - \chi \circ \sigma}{2} a^-](y)\mathcal{G}_2(x) \\ &\quad + 2\frac{\chi + \chi \circ \sigma}{2}(y)\mathcal{G}_1(x), \end{aligned} \tag{4.8}$$

and

$$\mathcal{G}_2(xy) + \mathcal{G}_2(\sigma(y)x) = 2\frac{\chi + \chi \circ \sigma}{2}(y)\mathcal{G}_2(x), \ x, y \in G.$$

Since $\chi \circ \sigma \neq \chi$ we deduce from [3, Theorem 3.6] that there exists $(\alpha_2, \beta_2) \in \mathbb{C}^2 \setminus (0, 0)$ such that

$$\mathcal{G}_2 = \alpha_2 \frac{\chi + \chi \circ \sigma}{2} + \beta_2 \frac{\chi - \chi \circ \sigma}{2}.$$

As a particular solution of (4.8) we have

$$H := \alpha_2 \frac{\chi + \chi\circ\sigma}{2} a^+ + \beta_2 \frac{\chi - \chi\circ\sigma}{2} a^+ + \alpha_2 \frac{\chi - \chi\circ\sigma}{2} a^- + \beta_2 \frac{\chi + \chi\circ\sigma}{2} a^-.$$

i.e.,

$$H(xy) + H(\sigma(y)x) = 2[\frac{\chi + \chi\circ\sigma}{2} a^+ + \frac{\chi - \chi\circ\sigma}{2} a^-](y)\mathcal{G}_2(x) \\ + 2\frac{\chi + \chi\circ\sigma}{2}(y)H(x). \tag{4.9}$$

Subtracting (4.8) from (4.9) then we get with $\Gamma := \mathcal{G}_1 - H$ that Γ is a solution of the equation

$$\Gamma(xy) + \Gamma(\sigma(y)x) = 2\frac{\chi + \chi\circ\sigma}{2}(y)\Gamma(x), \ x, y \in G. \tag{4.10}$$

Then, from [3, Theorem 3.6] there exists $(d_1, d_2) \in \mathbb{C}^2 \setminus (0,0)$ such that

$$\Gamma = d_1 \frac{\chi + \chi\circ\sigma}{2} + d_2 \frac{\chi - \chi\circ\sigma}{2}.$$

Hence,

$$\mathcal{G}_1 = \alpha_2 \frac{\chi + \chi\circ\sigma}{2} a^+ + \beta_2 \frac{\chi - \chi\circ\sigma}{2} a^+ + \alpha_2 \frac{\chi - \chi\circ\sigma}{2} a^- + \beta_2 \frac{\chi + \chi\circ\sigma}{2} a^- \\ + d_1 \frac{\chi + \chi\circ\sigma}{2} + d_2 \frac{\chi - \chi\circ\sigma}{2}.$$

So, in this case we get the claimed result with

$$E = \chi \begin{pmatrix} 1 & a^+ + a^- \\ 0 & 1 \end{pmatrix}, \ u = \frac{1}{2}\begin{pmatrix} d_1 + d_2 \\ \alpha_2 + \beta_2 \end{pmatrix} \text{ and } v = \frac{1}{2}\begin{pmatrix} d_1 - d_2 \\ \alpha_2 - \beta_2 \end{pmatrix}.$$

Case 3: If $\Psi = \chi \begin{pmatrix} 1 & S + \psi \\ 0 & 1 \end{pmatrix}$, where χ is a character of G such that $\chi = \chi\circ\sigma$, ψ is a solution of the symmetrized additive Cauchy Eq. (2.9) such that $\psi \in \mathcal{N}(G, \sigma)$ and $S : G \to \mathbb{C}$ is a map of the form $S(x) = B(x, x)$, $x \in G$, where $B : G \times G \to \mathbb{C}$ is a bi-additive function of G such that $B(x, \sigma(y)) = -B(y, x)$, then $\mathcal{G}_1$ and $\mathcal{G}_2$ satisfy the following equations:

$$\mathcal{G}_1(xy) + \mathcal{G}_1(\sigma(y)x) = 2\mathcal{G}_1(x)\chi(y) + 2\mathcal{G}_2(x)\chi(y)(S(y) + \psi(y)), \ x, y \in G, \tag{4.11}$$

and

$$\mathcal{G}_2(xy) + \mathcal{G}_2(\sigma(y)x) = 2\mathcal{G}_2(x)\chi(y), \ x, y \in G.$$

Since $(\mathcal{G}_2, \chi)$ is a solution of (2.5) and $\chi\circ\sigma = \chi$, then from [3, Theorem 3.6] there exist $\alpha_2 \in \mathbb{C}$ and $a_2^- \in \mathcal{A}^-(G)$ such that

$$\mathcal{G}_2 = (\alpha_2 + a_2^-)\chi.$$

Dividing the Eq. (4.11) by $\chi(x)\chi(y)$ we get

$$\mathcal{H}(xy)+\mathcal{H}(\sigma(y)x)=2\mathcal{H}(x)+2(\alpha_2+a_2^-(x))(S(y)+\psi(y)),\ x,y\in G, \quad (4.12)$$

where $\mathcal{H}:=\dfrac{\mathcal{G}_1}{\chi}$. Putting $x=e$ in (4.12) we get

$$\mathcal{H}^e(y)=\mathcal{H}(e)+\alpha_2(S(y)+\psi(y)),\ y\in G. \quad (4.13)$$

Since g is central, we deduce from (4.13) that $\psi: G\to\mathbb{C}$ is a σ-even additive function. Replacing x and y by $\sigma(x)$ and $\sigma(y)$ respectively in (4.12) we obtain

$$\mathcal{H}(\sigma(xy))+\mathcal{H}(\sigma(\sigma(y)x))=2\mathcal{H}(\sigma(x))+2(\alpha_2-a_2^-(x))(S(y)+\psi(y)), \quad (4.14)$$

for all $x,y\in G$. Subtracting (4.12) from (4.14) we get

$$\mathcal{H}^o(xy)+\mathcal{H}^o(\sigma(y)x)=2\mathcal{H}^o(x)+2a_2^-(x)(S(y)+\psi(y)),\ x,y\in G. \quad (4.15)$$

Replacing x and y by $\sigma(y)$ and x respectively in (4.15), we see that

$$\mathcal{H}^o(\sigma(y)x)-\mathcal{H}^o(xy)=-2\mathcal{H}^o(y)-2a_2^-(y)(S(x)+\psi(x)),\ x,y\in G. \quad (4.16)$$

Subtracting (4.15) from (4.16) we obtain that

$$\mathcal{H}^o(xy)=\mathcal{H}^o(x)+\mathcal{H}^o(y)+a_2^-(x)(S(y)+\psi(y))+a_2^-(y)(S(x)+\psi(x)), \quad (4.17)$$

for all $x,y\in G$.

Subcase 3.1: If $a_2^-\equiv 0$, then the equality (4.17) becomes

$$\mathcal{H}^o(xy)=\mathcal{H}^o(x)+\mathcal{H}^o(y),\ x,y\in G. \quad (4.18)$$

Thus,

$$\mathcal{H}=\mathcal{H}^e+\mathcal{H}^o=\mathcal{H}(e)+\alpha_2(S+\psi)+a_3^-, \quad (4.19)$$

where $a_3^-\in\mathcal{A}^-(G)$. Hence,

$$\mathcal{G}_1=[\mathcal{H}(e)+\alpha_2(S+\psi)+a_3^-]\chi. \quad (4.20)$$

Then we get with

$$E=\chi\begin{pmatrix}1 & S+\psi+a_3^-\\ 0 & 1\end{pmatrix},\ u=\frac{1}{2}\begin{pmatrix}\mathcal{H}(e)\\ \alpha_2+1\end{pmatrix}\ \text{and}\ v=\frac{1}{2}\begin{pmatrix}\mathcal{H}(e)\\ \alpha_2-1\end{pmatrix}.$$

that the solution of (2.1) has the form (4.4).

Subcase 3.2: If $a_2^-\not\equiv 0$, so, according to the associativity of the addition rule and (4.17) we have

$$\begin{aligned}
&\mathcal{H}^o((xy)z)\\
&=\mathcal{H}^o(xy)+\mathcal{H}^o(z)+a_2^-(xy)[S(z)+\psi(z)]+a_2^-(z)[S(xy)+\psi(xy)]\\
&=\mathcal{H}^o(x)+\mathcal{H}^o(y)+\mathcal{H}^o(z)+a_2^-(x)[S(y)+\psi(y)+S(z)+\psi(z)]\\
&+a_2^-(y)[S(x)+\psi(x)+S(z)+\psi(z)]+a_2^-(z)[S(x)+S(y)\\
&\qquad +B(x,y)+B(y,x)+\psi(xy)],\ x,y\in G,
\end{aligned}$$

and

$$\begin{aligned}
&\mathcal{H}^o(x(yz))\\
&= \mathcal{H}^o(x) + \mathcal{H}^o(yz) + a_2^-(x)[S(yz) + \psi(yz)] + a_2^-(yz)[S(x) + \psi(x)]\\
&= \mathcal{H}^o(x) + \mathcal{H}^o(y) + \mathcal{H}^o(z) + a_2^-(y)[S(z) + \psi(z) + S(x) + \psi(x)]\\
&+ a_2^-(z)[S(y) + \psi(y) + S(x) + \psi(x)] + a_2^-(x)[S(y) + S(z)\\
&\qquad + B(y,z) + B(z,y) + \psi(yz)], \ x, y \in G.
\end{aligned}$$

Hence,

$$\begin{aligned}
0 &= \mathcal{H}^o((xy)z) - \mathcal{H}^o(x(yz))\\
&= a_2^-(x)[\psi(y) + \psi(z)] + a_2^-(z)[B(x,y) + B(y,x) + \psi(xy)]\\
&- a_2^-(x)[B(y,z) + B(z,y) + \psi(yz)] - a_2^-(z)[\psi(y) + \psi(x)].
\end{aligned}$$

Since ψ is additive we infer that

$$a_2^-(z)[B(x,y) + B(y,x)] = a_2^-(x)[B(y,z) + B(z,y)], \ x, y, z \in G.$$

Choosing $z_0 \in G$ such that $a_2^-(z_0) \neq 0$ we get

$$B(x,y) + B(y,x) = a_2^-(x)\mathcal{Q}(y), \ x, y \in G, \tag{4.21}$$

where $\mathcal{Q}(y) = (a_2^-(z_0))^{-1}[B(y,z_0) + B(z_0,y)]$ for all $y \in G$. From (4.21) we deduce that $a_2^-(x)\mathcal{Q}(y) = a_2^-(y)\mathcal{Q}(x)$ for all $x, y \in G$, which yields that there exists $\beta \in \mathbb{C}$ such that $\mathcal{Q} = 6\beta a_2^-$. Substituting this into (4.21) we get $S = 3\beta(a_2^-)^2$. Replacing S by its expression in (4.12) we get

$$\mathcal{H}(xy)+\mathcal{H}(\sigma(y)x) = 2\mathcal{H}(x)+2(\alpha_2+a_2^-(x))(3\beta(a_2^-(y))^2+\psi(y)), \ x, y \in G. \tag{4.22}$$

Simple computations show that

$$F := a_2^-\psi + \beta(a_2^-)^3 + \alpha_2(\psi + 3\beta(a_2^-)^2)$$

is a particular solution of (4.22), which means that

$$F(xy)+F(\sigma(y)x) = 2F(x)+2(\alpha_2+a_2^-(x))(3\beta(a_2^-(y))^2+\psi(y)), \ x, y \in G. \tag{4.23}$$

Subtracting (4.22) from (4.23) and putting $Y := \mathcal{H} - F$, we see that Y is a solution of (2.7), then from [4, Theorem 3.2] there exist $a_4^- \in \mathcal{A}^-(G)$ and $\xi \in \mathbb{C}$ such that

$$\mathcal{H} = a_2^-\psi + \beta(a_2^-)^3 + \alpha_2(\psi + 3\beta(a_2^-)^2) + a_4^- + \xi.$$

Hence,

$$\mathcal{G}_1 = [a_2^-\psi + \beta(a_2^-)^3 + \alpha_2(\psi + 3\beta(a_2^-)^2) + a_4^- + \xi]\chi.$$

So the solution of (2.1) has the form (4.4) with

$$E = \chi\begin{pmatrix} 1 + a_2^- & \beta(a_2^-)^3 + 3\beta(a_2^-)^2 + \psi + a_2^-\psi + a_4^- \\ 0 & 1 + a_2^- \end{pmatrix},$$

$$u = \frac{1}{2}\begin{pmatrix} \xi \\ \alpha_2 + 1 \end{pmatrix} \text{ and } v = \frac{1}{2}\begin{pmatrix} \xi \\ \alpha_2 - 1 \end{pmatrix}.$$

Conversely, simple computations prove that the formulas above for (g, Φ) define solutions of (2.1) with $g : G \to \mathbb{C}^2$ is a central function such that $g \in \mathcal{F}$.

□

As consequences of Theorem 4.2 we have the following corollaries.

Corollary 4.3. *Let G be a compact group. The continuous solutions $\Phi : G \to M_2(\mathbb{C})$ and $g : G \to \mathbb{C}^2$ of (2.1) such that $g \in \mathcal{F}$ and is central are the functions of the following forms*

$$\begin{cases} \Phi = \frac{1}{2} P \begin{pmatrix} \chi_1 + \chi_1 \circ \sigma & 0 \\ 0 & \chi_2 + \chi_2 \circ \sigma \end{pmatrix} P^{-1} \\ g = P(\begin{pmatrix} \chi_1 & 0 \\ 0 & \chi_2 \end{pmatrix} u + \begin{pmatrix} \chi_1 \circ \sigma & 0 \\ 0 & \chi_2 \circ \sigma \end{pmatrix} v) \end{cases},$$

where $u, v \in \mathbb{C}^2$, $P \in GL_2(\mathbb{C})$ and χ_1, χ_2 are continuous characters of G.

Proof. The proof deduces easily from Lemmas 3.2 (1), 3.4 (3), [2, Corollary 6.3] and the proof of Theorem 4.1. □

Corollary 4.4. *The functions $\Phi : G \to M_2(\mathbb{C})$ and $g : G \to \mathbb{C}^2$ such that $g \in \mathcal{F}$ and is central satisfy the functional equation*

$$g(xy) = \Phi(y) g(x), \ x, y \in G, \tag{4.24}$$

if and only if, there exist $u \in \mathbb{C}^2$ and $P \in GL_2(\mathbb{C})$ such that

$$\begin{cases} \Phi = PEP^{-1} \\ g = PEu \end{cases},$$

where $E : G \to M_2(\mathbb{C})$ has, with χ, χ_1, χ_2 characters of G, and $\psi \in \mathcal{A}^+(G)$, one of the following two forms:

$$E = \begin{pmatrix} \chi_1 & 0 \\ 0 & \chi_2 \end{pmatrix} \ \text{or}\ E = \chi \begin{pmatrix} 1 & \psi \\ 0 & 1 \end{pmatrix}, \ x \in G.$$

Proof. The proof follows easily from Theorem 4.2 by taking $\sigma = id$. □

Remark 4.5. If the pair of functions $\Phi : M \to M_n(\mathbb{C})$ and $g : M \to \mathbb{C}^n$ with $g \in \mathcal{F}$ satisfies (4.24), then Φ is a solution of the equation

$$\Phi(xy) = \Phi(y)\Phi(x), \ x, y \in M. \tag{4.25}$$

Indeed, for all $x, y, z \in M$ we have $g(x(yz)) = \Phi(yz)g(x)$ and $g((xy)z) = \Phi(z)g(xy) = \Phi(z)\Phi(y)g(x)$. These give (4.25) because $g \in \mathcal{F}$.

References

1. Aczél, J., Dhombres, J.: Functional Equations in Several Variables. Cambridge University Press, New York (1989)
2. Aissi, Y., Zeglami, D., Ayoubi, M.: A variant of d'alembert's matrix functional equation. Annales Math. Sil. **35**, 21–43 (2021)
3. Fadli, B., Zeglami, D., Kabbaj, S.: A variant of Wilson's functional equation. Publ. Math. Debr. **87**, 415–427 (2015)
4. Fadli, B., Zeglami, D., Kabbaj, S.: A variant of Jensen's functional equation. Demonstratio Math. **49**(4), 413–420 (2016)
5. Fadli, B., Zeglami, D., Kabbaj, S.: A variant of the quadratic functional equation on semigroups. Proyecciones (Antofagasta) **37**, 45–55 (2018)
6. Radjavi, H., Rosenthal, P.: Simultaneous Triangularization. Springer, New York (2000). https://doi.org/10.1007/978-1-4612-1200-3
7. Sinopoulos, P.: Wilson's functional equation for vector and matrix functions. Proc. Am. Math. Soc. **125**, 1089–1094 (1997)
8. Stetkær, H.: D'Alembert's and Wilson's functional equations for vector and 2×2 matrix valued functions. Math. Scand. **87**, 115–132 (2000)
9. Stetkær, H.: Functional Equations on Groups. World Scientific Publishing Company, Singapore (2013)
10. Stetkær, H.: A variant of d'Alembert's functional equation. Aequationes Math. **89**, 657–662 (2015)
11. Szekélyhidi, L.: Convolution Type Functional Equations on Topological Abelian Groups. World Scientific Publishing, Singapore (1991)
12. Wilson, W.H.: On certain related functional equations. Bull. Amer. Math. Soc. **26**, 300–312 (1919)

Existence of Solution of Intuitionistic Fuzzy Transport Equation Based on Generalized Differentiability

Z. Belhallaj, S. Melliani(✉), and M. Elomari

LMACS, Sultan Moulay Slimane University, BP 523 - 23000 Beni Mellal, Morocco
{s.melliani,m.elomari}@usms.ma

Abstract. In this work, we study the existence of solution of transport equation with intuitionistic fuzzy data in cases of homogeneous and non-homogeneous approach, besides in the case of velocity parameter is fuzzy Intuitionistic, based on concept of generalized differentiability, and Zadeh's extension principle.

Keywords: Intuitionistic fuzzy differential equations · Generalized hukuhara difference · Intuitionistic fuzzy solution · Transport equation

1 Introduction

The concept of Intuitionistic Fuzzy Sets (IFS) are a significant part of the Fuzzy Set Theory [10], in 1983s Atanassov generalized the fuzzy set theory by presenting the basic elements of Intuitionistic Fuzzy Sets [1–3]. The Intuitionistic Fuzzy Differential Equations (IFDE) have drawn much attention from scientists and engineers at the end of its applicability in various fields, as artificial intelligence, robotics, fluid mechanics, heat and mass transfer, economics and social sciences for modeling and solving their respective problems. In [6] S. Melliani and L. S. Chadli introduce the concept of intuitionistic fuzzy differential equations, the sense of Hukuhara's difference in Intuitionistic Fuzzy Theory is giving in [9], the authors in [5] discussed the global existence of solutions to fuzzy hyperbolic function differential equation with generalized hukuhara derivatives. From this idea we investigate to study the existence of solution for particular case of hyperbolic equation such as Transport Equation under generalized differentiability in different cases.

The structure of this manuscript is organized as follows: In Sect. 2, a few basic results on intuitionistic fuzzy sets and the metric space, which have been examined in [7,8]. In Sect. 3 we study the transport equation with intuitionistic fuzzy data in homogeneous case. The transport equation in non-homogeneous case and with non-precise speed are studied in Sect. 4 and 5, respectively, and finally conclusion in Sect. 6.

2 Preliminaries

2.1 Intuitionistic Fuzzy Sets

An intuitionistic fuzzy set $A \in X$ is given by

$$A = \{(x, u_A(x), v_A(x)) | x \in X\}$$

S. Melliani and O. Castillo (Eds.): ICPAMS 2021, LNNS 476, pp. 473–484, 2023.
https://doi.org/10.1007/978-3-031-12416-7_40

where the function $u_A(x), v_A(x) : X \to [0,1]$ define respectively the degree of membership and degree of non-membership of the element $x \in X$ to the set A, which is a subset of X, and for every $x \in X$, $0 \leq u(x) + v(x) \leq 1$.

Obviously, every fuzzy set has the form

$$\{(x, u_A(x), u_{A^c}(x)) | x \in X\}$$

For each intuitionistic fuzzy set $A \in X$, we will call

$$\pi_A(x) = 1 - u_A(x) - v_A(x)$$

The intuitionistic fuzzy index of $x \in A$. It is obviously that $0 \leq \pi_A(x) \leq 1$.

2.2 Intuitionistic Fuzzy Numbers

An element $\langle u, v\rangle$ of $\mathbb{IF}_1$ is said an intuitionistic fuzzy number if it satisfies the following conditions:

(i) $\langle u, v\rangle$ is normal i.e. there exists $x_0, x_1 \in \mathbb{R}$ such that $u(x_0) = 1$ and $v(x_1) = 1$.
(ii) The membership function u is fuzzy convex i.e. $u(\lambda x_1 + (1-\lambda)x_2) \geq min(u(x_1), u(x_2))$.
(iii) The non-membership function v is fuzzy concave i.e. $v(\lambda x_1 + (1-\lambda)x_2) \leq max(v(x_1), v(x_2))$.
(iv) u is upper semi-continuous and v is lower semi-continuous.
(v) $Supp\langle u, v\rangle = cl\{x \in \mathbb{R} : |v(x) < 1\}$ is bounded.

We denote the collection of all intuitionistic fuzzy number by $\mathbb{IF}_1$.

For $\alpha \in [0,1]$ and $\langle u, v\rangle \in \mathbb{IF}^1$, the upper and lower α-cuts of $\langle u, v\rangle$ are defined by

$$[\langle u, v\rangle]^\alpha = \{x \in \mathbb{R} : v(x) \leq 1 - \alpha\}$$
and
$$[\langle u, v\rangle]_\alpha = \{x \in \mathbb{R} : u(x) \geq \alpha\}$$

Remark 1. If $\langle u, v\rangle \in \mathbb{IF}_1$, so we can see $[\langle u, v\rangle]_\alpha$ as $[u]^\alpha$ and $[\langle u, v\rangle]^\alpha$ as $[1-v]^\alpha$ in the fuzzy case.

We define $0_{(1,0)} \in \mathbb{IF}_1$ as

$$0_{(1,0)}(t) = \begin{cases} (1,0) & t = 0 \\ (0,1) & t \neq 0 \end{cases}$$

Let $\langle u, v\rangle$, $\langle u', v'\rangle \in \mathbb{IF}_1$ and $\lambda \in \mathbb{R}$, we define the following operations by:

$$(\langle u, v\rangle \oplus \langle u', v'\rangle)(z) = (sup_{z=x+y} min(u(x), u'(y)), inf_{z=x+y} max(v(x), v'(y)))$$

$$\lambda\langle u, v\rangle = \begin{cases} \langle \lambda u, \lambda v\rangle & if\ \lambda \neq 0 \\ 0_{(1,0)} & if\ \lambda = 0 \end{cases}$$

For $\langle u,v\rangle$, $\langle z,w\rangle \in \mathbb{IF}_1$ and $\lambda \in \mathbb{R}$, the addition and scalar-multiplication are defined as follows:

$$[\langle u,v\rangle \oplus \langle z,w\rangle]^{\alpha} = [\langle u,v\rangle]^{\alpha} + [\langle z,w\rangle]^{\alpha}, \quad [\lambda \langle z,w\rangle]^{\alpha} = \lambda[\langle z,w\rangle]^{\alpha}$$
$$[\langle u,v\rangle \oplus \langle z,w\rangle]_{\alpha} = [\langle u,v\rangle]_{\alpha} + [\langle z,w\rangle]_{\alpha}, \quad [\lambda \langle z,w\rangle]_{\alpha} = \lambda[\langle z,w\rangle]_{\alpha}$$

Definition 1. Let $\langle u,v\rangle$ an element of $\mathbb{IF}_1$ and $\alpha \in [0.1]$, we define the following sets:

$$[\langle u,v\rangle]_l^+(\alpha) = inf\{x \in \mathbb{R} | u(x) \geq \alpha\}, \quad [\langle u,v\rangle]_r^+(\alpha) = sup\{x \in \mathbb{R} | u(x) \geq \alpha\}$$
$$[\langle u,v\rangle]_l^-(\alpha) = inf\{x \in \mathbb{R} | v(x) \leq 1-\alpha\}, \quad [\langle u,v\rangle]_r^-(\alpha) = sup\{x \in \mathbb{R} | v(x) \leq 1-\alpha\}$$

Remark 2. $[\langle u,v\rangle]_{\alpha} = [[\langle u,v\rangle]_l^+(\alpha), [\langle u,v\rangle]_r^+(\alpha)]$, $[\langle u,v\rangle]^{\alpha} = [[\langle u,v\rangle]_l^-(\alpha), [\langle u,v\rangle]_r^-(\alpha)]$.

Proposition 1. *For all $\alpha, \beta \in [0,1]$ and $\langle u,v\rangle \in \mathbb{IF}_1$*

(i) $[\langle u,v\rangle]_{\alpha} \subset [\langle u,v\rangle]^{\alpha}$
(ii) $[\langle u,v\rangle]_{\alpha}$ *and* $[\langle u,v\rangle]^{\alpha}$ *are nonempty compact convex sets in* $\mathbb{R}$
(iii) If $\alpha \leq \beta$ *then* $[\langle u,v\rangle]_{\beta} \subset [\langle u,v\rangle]_{\alpha}$ *and* $[\langle u,v\rangle]^{\beta} \subset [\langle u,v\rangle]^{\alpha}$
(iv) If $\alpha_n \nearrow \alpha$ *then* $[\langle u,v\rangle]_{\alpha} = \cap_n[\langle u,v\rangle]_{\alpha_n}$ *and* $[\langle u,v\rangle]^{\alpha} = \cap_n[\langle u,v\rangle]^{\alpha_n}$

Let M any set and $\alpha \in [0,1]$ we denote by

$$M_{\alpha} = \{x \in \mathbb{R} : u(x) \geq \alpha\} \text{ and } M^{\alpha} = \{x \in \mathbb{R} : v(x) \leq 1-\alpha\}$$

Lemma 1. *Let $\{M_{\alpha}, \alpha \in [0,1]\}$ and $\{M^{\alpha}, \alpha \in [0,1]\}$ two families of $\mathbb{R}$ satisfies (i)–(iv) in Proposition 3.1, if u and v define by*

$$u(x) = \begin{cases} 0 & if\ x \notin M_0 \\ sup\{\alpha \in [0,1] : x \in M_{\alpha}\} & if\ x \in M_0 \end{cases}$$

$$v(x) = \begin{cases} 1 & if\ x \notin M_0 \\ 1 - sup\{\alpha \in [0,1] : x \in M^{\alpha}\} & if\ x \in M_0 \end{cases}$$

then $\langle u,v\rangle \in \mathbb{IF}_1$.

Lemma 2. *A mapping $d : \mathbb{IF}_1 \times \mathbb{F}_1 \rightarrow \mathbb{R}$ is said to be an intuitionistic fuzzy metric on $\mathbb{IF}_1$ if it satisfies the following conditions*

1. $d(\langle u_1,v_1\rangle, \langle u_2,v_2\rangle) \geq 0$, $\forall \langle u_1,v_1\rangle, \langle u_2,v_2\rangle \in \mathbb{F}_1$.
2. $d(\langle u_1,v_1\rangle, \langle u_2,v_2\rangle) = 0$ *iff* $\langle u_1,v_1\rangle = \langle u_2,v_2\rangle$.
3. $d(\langle u_1,v_1\rangle, \langle u_2,v_2\rangle) = d(\langle u_2,v_2\rangle, \langle u_1,v_1\rangle)$ $\forall \langle u_1,v_1\rangle, \langle u_2,v_2\rangle \in \mathbb{F}_1$.
4. $d(\langle u_1,v_1\rangle, \langle u_3,v_3\rangle) \leq d(\langle u_1,v_1\rangle, \langle u_2,v_2\rangle) + d(\langle u_2,v_2\rangle, \langle u_3,v_3\rangle)$. $\forall \langle u_1,v_1\rangle, \langle u_2,v_2\rangle, \langle u_3,v_3\rangle$.

On the space $\mathbb{IF}_1$ we will consider the following metric.

$$\begin{aligned} d_\infty(\langle u,v\rangle,\langle z,w\rangle) = & \frac{1}{4} sup_{0<\alpha\leq 1} \parallel \langle u,v\rangle_r^+(\alpha) - \langle z,w\rangle_r^+(\alpha) \parallel \\ & + \frac{1}{4} sup_{0<\alpha\leq 1} \parallel \langle u,v\rangle_l^+(\alpha) - \langle z,w\rangle_l^+(\alpha) \parallel \\ & + \frac{1}{4} sup_{0<\alpha\leq 1} \parallel \langle u,v\rangle_r^-(\alpha) - \langle z,w\rangle_r^-(\alpha) \parallel \\ & + \frac{1}{4} sup_{0<\alpha\leq 1} \parallel \langle u,v\rangle_l^-(\alpha) - \langle z,w\rangle_l^-(\alpha) \parallel \end{aligned}$$

where $\parallel . \parallel$ denotes the usual Euclidean norm in $\mathbb{R}^n$.

Proposition 2 [4]. *$(\mathbb{IF}_1, d_p)$ is a metric space.*

Definition 2. The generalized Hukuhara difference of two fuzzy numbers $\langle u,v\rangle$, $\langle u',v'\rangle \in \mathbb{IF}_1$ is defined as follows

$$\langle u,v\rangle \ominus_{gH} \langle u',v'\rangle = \langle w,z\rangle$$

$$\Leftrightarrow \begin{cases} \langle u,v\rangle = \langle u',v'\rangle + \langle w,z\rangle \\ or\ \langle u',v'\rangle = \langle u,v\rangle + (-1)\langle w,z\rangle \end{cases}$$

Definition 3. Let be $G:(a,b) \rightarrow W^1$ and $x_0 \in (a,b)$. It is said that F is strongly generalized differentiable on x_0, if $\exists\, G'^+(x_0)$, $\exists\, G'^-(x_0) \in E^1$, such that

(i) for all $h > 0$ sufficiently small, $\exists\, G^+(x_0+h) - G^+(x_0)$, $G^+(x_0) - G^+(x_0-h)$ and the limits (in the metric D)

$$\lim_{h\rightarrow 0} \frac{G^+(x_0+h) - G^+(x_0)}{h} = \lim_{h\rightarrow 0} \frac{G^+(x_0) - G^+(x_0-h)}{h} = G'^+(x_0)$$

Or

(ii) for all $h > 0$ sufficiently small, $\exists\, G^+(x_0) - G^+(x_0+h)$, $G^+(x_0-h) - G^+(x_0)$ and the limits

$$\lim_{h\rightarrow 0} \frac{G^+(x_0) - G^+(x_0+h)}{-h} = \lim_{h\rightarrow 0} \frac{G^+(x_0-h) - G^+(x_0)}{-h} = G'^+(x_0)$$

Or

(iii) for all $h > 0$ sufficiently small, $\exists\, G^+(x_0+h) - G^+(x_0)$, $G^+(x_0-h) - G^+(x_0)$ and the limits

$$\lim_{h\rightarrow 0} \frac{G^+(x_0) - G^+(x_0-h)}{h} = \lim_{h\rightarrow 0} \frac{G^+(x_0-h) - G^+(x_0)}{-h} = G'^+(x_0)$$

(iv) for all $h > 0$ sufficiently small, $\exists\, G^+(x_0) - G^+(x_0+h)$, $G^+(x_0) - G^+(x_0-h)$ and the limits

$$\lim_{h\rightarrow 0} \frac{G^+(x_0) - G^+(x_0+h)}{-h} = \frac{G^+(x_0) - G^+(x_0-h)}{-h} = G'^+(x_0)$$

Lemma 3. *Let $g : \mathbb{R} \times (0,\infty) \longrightarrow \mathbb{IF}^1$ be strongly generalized differentiable w.r.t x. Suppose there exists a continuous real-valued function $\Psi(x,t)$ such that $D(\frac{\partial g}{\partial} x(x,t), \chi_{\{0\}}) \leq \Psi(x,t)$ for $x \in \mathbb{R}$, $t \geq 0$. Then $G(x,t) = \int_0^t g(x,s)ds$ is strongly generalized differentiable w.r.t x and we have $\frac{\partial G}{\partial x}(x,t) = \int_0^t \frac{\partial g}{\partial x}(x,s)ds$.*

Proof. Let $g : \mathbb{R} \times (0,\infty) \longrightarrow \mathbb{IF}_1$ with $[g(x,t)]^\alpha = [g_l^-(\alpha)(x,t), g_r^-(\alpha)(x,t)]$ and $[g(x,t)]_\alpha$ = $[g_l^+(\alpha)(x,t), g_r^+(\alpha)(x,t)]$ be (i)-differentiable. We first observe that because of our hypothesis, we have uniformly the next derivatives w.r.t α.

$$\frac{\partial}{\partial x}\int_0^t g_l^-(\alpha)(x,s)ds = \int_0^t \frac{\partial g_l^-(\alpha)}{\partial x}(x,s)ds, \quad \frac{\partial}{\partial x}\int_0^t g_r^-(\alpha)(x,s)ds = \int_0^t \frac{\partial g_r^-(\alpha)}{\partial x}(x,s)ds$$
$$\frac{\partial}{\partial x}\int_0^t g_l^+(\alpha)(x,s)ds = \int_0^t \frac{\partial g_l^+(\alpha)}{\partial x}(x,s)ds, \quad \frac{\partial}{\partial x}\int_0^t g_r^+(\alpha)(x,s)ds = \int_0^t \frac{\partial g_r^+(\alpha)}{\partial x}(x,s)ds$$

where we have

$$\begin{aligned}
\lim_{h\longrightarrow 0} \frac{G(x+h,t) \ominus G(x,t)}{h} &= \lim_{h\longrightarrow 0} \frac{\int_0^t g(x+h,s)ds \ominus \int_0^t g(x,s)ds}{h} \\
&= \lim_{h\longrightarrow 0} \frac{\int_0^t (g(x+h,s) \ominus g(x,s))ds}{h} \\
&= \int_0^t \lim_{h\longrightarrow 0} \frac{g(x+h,s) \ominus g(x,s)}{h} ds \\
&= \int_0^t \frac{\partial g}{\partial x}(x,s)ds.
\end{aligned}$$

in the same sense we can show easily for $G(x,t) \ominus G(x-h,t)$, and we can demonstrate the assertion for the (ii)-differentiability.

Lemma 4. *Consider $g : \mathbb{R} \times (0,\infty) \longrightarrow \mathbb{IF}_1$ be a continuous intuitionistic fuzzy-valued function and $b \in (0,\infty)$. Then*

$$\lim_{h\longrightarrow 0} \frac{1}{h}\int_t^{t+h} \tilde{g}(x+(t-s)b+bh,s)ds = \tilde{g}(x,t)$$

Proof. Let $(x,t) \in \mathbb{R} \times (0,\infty)$, for given $\varepsilon > 0$ there exists $\rho_1 > 0$ such that for all $0 \leq \alpha \leq 1$ and $\mid x-y \mid < \rho_1$, $\mid t-s \mid < \rho_1$, we have $\mid \tilde{g}(\alpha)(y,t) - \tilde{g}(\alpha)(x,s) \mid < \varepsilon$.

Let $\rho = min\{\frac{\rho_1}{2b}, \rho_1\}$. For $\mid h \mid \leq \rho$, we have

$$\frac{1}{h}\int_t^{t+h} \mid \tilde{g}(\alpha)(x+(t-s)b+hb,s) - \tilde{g}(\alpha)(x,t) \mid ds \leq \varepsilon,$$

It illustrates that

$$\lim_{h\to 0} D(\frac{1}{h}\int_t^{t+h} \tilde{g}(x+(t-s)b+bh,s)ds, \tilde{g}(x,t)) = 0.$$

This completes the proof.

It is widely established that a intuitionistic function $g : (a,b) \to \mathbb{IF}_1$.

g is (i)-differentiable if and only if the functions $[g]^\alpha$, $[g]_\alpha$ are continuously differentiable with respect to x, uniformly with respect to $\alpha \in [0,1]$, as long as $[(g_l^-(\alpha))'(x), (g_r^-(\alpha))'(x)]$ and $[(g_l^+(\alpha))'(x), (g_r^+(\alpha))'(x)]$ defines a intuitionistic fuzzy number $g'(x) \in \mathbb{IF}_1$.

Likewise, g is (ii)-differentiable if and only if the functions $[g]^\alpha$, $[g]_\alpha$ are continuously differentiable with respect to x, uniformly with respect to $\alpha \in [0,1]$, as long as $[(g_r^-(\alpha))'(x), (g_l^-(\alpha))'(x)]$ and $[(g_r^+(\alpha))'(x), (g_l^+(\alpha))'(x)]$ defines a intuitionistic fuzzy number $g'(x) \in \mathbb{IF}_1$.

In the next sections, we submit a result only for existence of Hukuhara differences related to strongly generalized differentiability.

Theorem 1 [3]. *Assume that $g : (a,b) \to \mathbb{IF}_1$ be such that $[g]^\alpha = [g_l^-(\alpha), g_r^-(\alpha)]$ and $[g]_\alpha = [g_l^+(\alpha), g_r^+(\alpha)]$.*

Suppose that real valued functions $g_l^-(\alpha)$, $g_r^-(\alpha)$, $g_l^+(\alpha)$ and $g_r^+(\alpha)$ are differentiable w.r.t x.

1. *If the intervals $[(g_l^-(\alpha))'(x), (g_r^-(\alpha))'(x)]$ and $[(g_l^+(\alpha))'(x), (g_r^+(\alpha))'(x)]$ for all $\alpha \in [0,1]$ and $x \in (a,b)$, determine valid α-cuts of a fuzzy number, then the H-differences $g(x+h) \ominus g(x)$ and $g(x) \ominus g(x-h)$ exist for all $h > 0$ sufficiently small.*
2. *If the intervals $[(g_r^-(\alpha))'(x), (g_l^-(\alpha))'(x)]$ and $[(g_r^+(\alpha))'(x), (g_l^+(\alpha))'(x)]$ for all $\alpha \in [0,1]$ and $x \in (a,b)$, determine valid α-cuts of a fuzzy number, then the H-differences $g(x) \ominus g(x+h)$ and $g(x-h) \ominus g(x)$ exist for all $h > 0$ sufficiently small.*

3 Homogenous Transport Equation

In this chapter our focus is to investigate the transport equation of a mass with no sources with a intuitionistic fuzzy initial value and a positive/negative speed (it can be negative too):

$$\begin{cases} U_t = kU_x & in\ \mathbb{R} \times (0,\infty) \\ U(x,0) = P(x) \end{cases} \tag{1}$$

where P is intuitionistic fuzzy valued function, and k is a constant real number.

We say that U(x,t) is a solution of (1) on $\mathbb{R} \times (0,\infty)$ if U is strongly differentiable w.r to x, t and it verifies (1).

Our method is based on the constructive method in which we introduce the solution and examine that if checks the problem.

Theorem 2. *Suppose P is strongly generalized differentiable function on $\mathbb{R}$.*

Then $U(x,t) = P(x+kt)$ is a solution of (1), where be is a constant positive real number.

Proof. First, let P be (i)-differentiable, then $P(x+tk+th)\ominus P(x+tk)$ and $P(x+tk)\ominus P(x+tk-bh)$ for h sufficiently small exist and we have:

$$\begin{aligned} U_t &= \lim_{h\to 0}\frac{U(x,t+h)\ominus U(x,t)}{h} \\ &= \lim_{h\to 0}\frac{P(x+tk+kh)\ominus P(x+tk)}{h} \\ &= \lim_{h\to 0}k\frac{P(x+tk+kh)\ominus P(x+tk)}{kh} \\ &= \mathrm{kP}'(x+tk) \end{aligned}$$

As well as:

$$\begin{aligned} U_t &= \lim_{h\to 0}\frac{U(x,t)\ominus U(x,t-h)}{h} \\ &= \lim_{h\to 0}\frac{P(x+tk)\ominus P(x+tk-kh)}{h} \\ &= \lim_{h\to 0}k\frac{P(x+tk)\ominus P(x+tk-kh)}{kh} \\ &= \mathrm{kP}'(x+tk) \end{aligned}$$

In addition, we have:

$$\begin{aligned} U_x &= \lim_{h\to 0}\frac{U(x+h,t)\ominus U(x,t)}{h} \\ &= \lim_{h\to 0}\frac{P(x+tk+h)\ominus P(x+tk)}{h} \\ &= P'(x+tk) \end{aligned}$$

Likewise we give $\lim_{h\to 0}\frac{U(x,t)\ominus U(x-h,t)}{h}$.
These conclude the (i)-differentiability of U with respect to x and t.
It is clear that U verifies the equation and initial condition, when P is (ii)-differentiable in the same way, we can prove that it is a solution.

4 Nonhomogeneous Transport Equation

In this chapter we concentrate on the following non homogeneous intuitionistic fuzzy transport equation for which the second member and initial value may be intuitionistic fuzzy.

$$\begin{cases} U_t = kU_x + cU & \textit{in } \mathbb{R}\times(0,\infty) \\ U(x,0) = P(x) \end{cases} \tag{2}$$

where P is intuitionistic fuzzy valued function.

Theorem 3. *Let* $U:\mathbb{R}\times(0,\infty)\to\mathbb{IF}_1$ *be (i)-differentiable w.r.t x,t and let* $P:\mathbb{R}\to\mathbb{IF}_1$ *and let k and c are a positive real number.*

1. *If g is (i)-differentiable, then:* $U_1(x,t) = P(x+tk) + \int_0^t cU(x+(t-s)k,s)ds$ *is a (i)-differentiable solution of (2).*
2. *If g is (ii)-differentiable, then:* $U_2(x,t) = P(x+tk) \ominus (-1)\int_0^t cP(x+(t-s)k,s)ds$ *is a (ii)-differentiable solution of (2) provided the H-difference exists.*

Proof. • First we suppose g is (i)-differentiable. We show that $U_1(x,t)$ is (i)-differentiable and verifies the problem

$\frac{\partial U_1}{\partial x} = \lim_{h\to 0} \frac{U_1(x+h,t)\ominus U_1(x,t)}{h}$

$$= \lim_{h\to 0} \frac{(P(x+tk+h)+\int_0^t cU(x+h+(t-s)k,s)ds)}{h} \ominus \frac{(P(x+tk)+\int_0^t cU(x+(t-s)k,s)ds))}{h}$$

$$= \lim_{h\to 0} \frac{(P(x+tk+h)\ominus P(x+tk))}{h} + \lim_{h\to 0} \frac{\int_0^t c(U(x+h+(t-s)k,s)\ominus U(x+(t-s)k,s))ds}{h}$$

$$= P'(x+tk) + \lim_{h\to 0} \frac{c}{h}\int_0^t ((U(x+h+(t-s)k,s)\ominus U(x+(t-s)k,s))ds$$

$\frac{\partial U_1}{\partial x} = P'(x+tk) + \int_0^t c\frac{\partial}{\partial x}U(x+(t-s)k,s))ds.$

For derivative w.r.t t, from Lemma 4, we have

$\frac{\partial U_1}{\partial t} = \lim_{h\to 0} \frac{U_1(x,t+h)\ominus U_1(x,t)}{h}$

$$= \lim_{h\to 0} \frac{(P(x+(t+h)k)+\int_0^t cU(x+(t+h-s)k,s)ds)}{h} \ominus \frac{(P(x+tk)+\int_0^t cU(x+(t-s)k,s)ds))}{h}$$

$$= kP'(x+tb) + \lim_{h\to 0}\int_0^t \frac{c(U(x+(t-s)k+hk,s)\ominus U(x+(t-s)k,s))ds}{h} +$$
$$\lim_{h\to 0}\int_t^{t+h} \frac{cU(x+(t-s)k+hk,s)ds}{h}$$

$\frac{\partial U_1}{\partial t} = kP'(x+tk) + c.k\int_0^t \frac{\partial U}{\partial x}(x+(t-s)k,s)ds + cU(x,t).$

As a result, U_1 is a (i)-differentiable solution of (2).

• Now let P is (ii)-differentiable, from Lemma 3, U_2 is (ii)-differentiable w.r.t x and we have

$\frac{\partial U_2}{\partial x} = P'(x+tk) + (-1)\int_0^t -c\frac{\partial U}{\partial x}(x+(t-s)k,s)ds.$

We now calculate (ii)-derivative of U_2 w.r.t t, first we have

$$U_2(x,t)\ominus U_2(x,t+h)$$
$$= [P(x+tk)\ominus \int_0^t -cU(x+(t-s)k,s)ds]\ominus[P(x+tk+kh)\ominus \int_0^{t+h} -cU(x+(t-s)k+kh,s)ds]$$
$$= [P(x+tb)\ominus P(x+tk+kh)] + [\int_0^{t+h} -cU(x+(t-s)k+kh,s)ds \ominus \int_0^t -cU(x+(t-s)k,s)ds]$$
$$= [P(x+tk)\ominus P(x+tk+kh)] + \int_0^t -cU(x+(t-s)k+kh,s)ds \ominus \int_0^t -cU(x+(t-s)k,s)ds + \int_t^{t+h}$$
$$-cU(x+(t-s)k+kh,s)ds.$$

The multiplying with $-\frac{1}{h}$ and passing to limit with $h \to 0^+$, we obtain:

$$\frac{\partial U_2}{\partial t}(x,t) = \lim_{h\to 0} \frac{U_2(x,t)\ominus U_2(x,t+h)}{-h}$$
$$\frac{\partial U_2}{\partial t}(x,t) = kP'(x+tk) + c.k\int_0^t \frac{\partial U}{\partial x}(x+(t-s)k,s)ds + c.U(x,t).$$

5 Equation by Non-precise Speed

In this part, we investigate the following homogeneous intuitionistic fuzzy transport equation for which the speed and initial value can be intuitionistic fuzzy.

$$\begin{cases} U_t = kU_x & in\ \mathbb{R} \times (0,\infty) \\ U(x,0) = qP(x) \end{cases} \tag{3}$$

where k, q $\in \mathbb{IF}_1$ and P is a real valued function.

In the following theorem, using Zadeh's extension principle, we intuitionistic fuzzify the real function $P : \mathbb{R} \to \mathbb{R}$ to intuitionistic fuzzify function $\tilde{P} : \mathbb{IF}_1 \to \mathbb{IF}_1$. Since P is a continuous function. We have $[\tilde{P}(X)]^\alpha = P([X]^\alpha)$ and $[\tilde{P}(X)]_\alpha = P([X]_\alpha)$.

For instance, let $X = x + tk$ and P, k creasing function.

Then we have: $\forall \alpha \in [0,1]$

$$[\tilde{P}(x+tk)]^\alpha = P([x+tk]^\alpha), [\tilde{P}(x+tk)]_\alpha = P([x+tk]_\alpha).$$

So

$$[\tilde{P}(x+tk)]^\alpha = P([x+tk]^\alpha) = [P(x+tk_l^-(\alpha)), P(x+tk_r^-(\alpha))]$$

and

$$[\tilde{P}(x+tk)]_\alpha = P([x+tk]_\alpha) = [P(x+tk_l^+(\alpha)), P(x+tk_r^+(\alpha))]$$

Theorem 4. *Suppose $P \in C^2(\mathbb{R})$ is an integrable nonnegative monotone function and k, q $\in \mathbb{IF}_1$ let $\tilde{P} : \mathbb{IF}_1 \to \mathbb{IF}_1$ be the Zadeh's extension of P, consider:*

$$U : \mathbb{R} \times (0,\infty) \to \mathbb{IF}_1$$
$$U(x,t) = q\tilde{P}(x+tk)$$

1. *If k, q $\in \mathbb{IF}_1^+$ and P, P' are nondecreasing functions, then U is (i)-differentiable w.r.t both t, x and it is a solution of (3)*
2. *If k $\in \mathbb{IF}_1^-$, q $\in \mathbb{IF}_1^+$ and P nonincreasing and P' is nondecreasing function, then u is(i)-differentiable w.r.t t and (ii)-differentiable w.r.t x and satisfies (3).*
3. *If k $\in \mathbb{IF}_n^+$, q $\in \mathbb{IF}_n^-$ and P, P' are nondecreasing functions, then U is (i)-differentiable w.r.t both t, x and satisfies (3).*
4. *If k, q $\in \mathbb{IF}_n^-$ and P is nonincreasing and P' is decreasing function, then U is (i)-differentiable w.r.t t and (ii)-differentiable w.r.t x and satisfies (3).*

Proof. **Case (i):** It is easy to check that α-cuts:

$$[U]^\alpha = [q_l^-(\alpha)P(x+tk_l^-(\alpha)), q_r^-(\alpha)P(x+tk_r^-(\alpha))]$$
$$[U]_\alpha = [q_l^+(\alpha)P(x+tk_l^+(\alpha)), q_r^+(\alpha)P(x+tk_r^+(\alpha))]$$

Satisfy conditions is case (i) Theorem 1.

In fact, $[q_l^-(\alpha)k_l^-(\alpha)P'(x + tk_l^-(\alpha)), q_r^-(\alpha)k_r^-(\alpha)P'(x + tk_r^-(\alpha))]$ and $[q_l^+(\alpha)k_l^+(\alpha)$

$P'(x+tk_l^+(\alpha)), q_r^+ (\alpha)k_r^+(\alpha)P'(x+tk_r^+(\alpha))]$ are valid α-cuts of intuitionistic fuzzy number.

Then H-differences $U(x,t+h) \ominus U(x,t)$ and $U(x,t) \ominus U(x,t-h)$ exist.

In the same manner, since $[q_l^-(\alpha)P'(x+tk_l^-(\alpha)),\ q_r^-(\alpha)P'(x+tk_r^-(\alpha))]$ and $[q_l^+(\alpha)P'(x+tk_l^+(\alpha)), q_r^+(\alpha)\ P'(x+tk_r^+(\alpha))]$ forms a intuitionistic fuzzy number, H-differences $U(x+h,t) \ominus U(x,t)$ and $U(x,t) \ominus U(x-h,t)$ exist.

Now we illustrate that limits are uniformly with respect to $\alpha \in [0,1]$:

$$\lim_{h\to 0} \frac{q_l^-(\alpha)P(x+(t+h)k_l^-(\alpha)) - q_l^-(\alpha)P(x+tk_l^-(\alpha))}{h} = q_l^-(\alpha)k_l^-(\alpha)P'(x+tk_l^-(\alpha)) \tag{4}$$

$$\lim_{h\to 0} \frac{q_r^-(\alpha)P(x+(t+h)k_r^-(\alpha)) - q_r^-(\alpha)P(x+tk_r^-(\alpha))}{h} = q_r^-(\alpha)k_r^-(\alpha)P'(x+tk_r^-(\alpha)) \tag{5}$$

$$\lim_{h\to 0} \frac{q_l^+(\alpha)P(x+(t+h)k_l^+(\alpha)) - q_l^+(\alpha)P(x+tk_l^+(\alpha))}{h} = q_l^+(\alpha)k_l^+(\alpha)P'(x+tk_l^+(\alpha)) \tag{6}$$

$$\lim_{h\to 0} \frac{q_r^+(\alpha)P(x+(t+h)k_r^+(\alpha)) - q_r^+(\alpha)P(x+tk_r^+(\alpha))}{h} = q_r^+(\alpha)k_r^+(\alpha)P'(x+tk_r^+(\alpha)) \tag{7}$$

$$\lim_{h\to 0} \frac{q_l^-(\alpha)P(x+tk_l^-(\alpha)) - q_l^-(\alpha)P(x+(t-h)k_l^-(\alpha))}{h} = q_l^-(\alpha)k_l^-(\alpha)P'(x+tk_l^-(\alpha)) \tag{8}$$

$$\lim_{h\to 0} \frac{q_r^-(\alpha)P(x+tk_r^-(\alpha)) - q_r^-(\alpha)P(x+(t-h)k_r^-(\alpha))}{h} = q_r^-(\alpha)k_r^-(\alpha)P'(x+tk_r^-(\alpha)) \tag{9}$$

$$\lim_{h\to 0} \frac{q_l^+(\alpha)P(x+tk_l^+(\alpha)) - q_l^+(\alpha)P(x+(t-h)k_l^+(\alpha))}{h} = q_l^+(\alpha)k_l^+(\alpha)P'(x+tk_l^+(\alpha)) \tag{10}$$

$$\lim_{h\to 0} \frac{q_r^+(\alpha)P(x+tk_r^+(\alpha)) - q_r^+(\alpha)P(x+(t-h)k_r^+(\alpha))}{h} = q_r^+(\alpha)k_r^+(\alpha)P'(x+tk_r^+(\alpha)) \tag{11}$$

For (4) we have

$$\lim_{h\to 0} sup_{\alpha\in[0,1]} \mid \frac{q_l^-(\alpha)P(x+(t+h)k_l^-(\alpha)) - q_l^-(\alpha)P(x+tk_l^-(\alpha))}{h} - q_l^-(\alpha)k_l^-(\alpha)P'(x+tk_l^-(\alpha)) \mid$$
$$= \lim_{h\to 0} sup_{\alpha\in[0,1]} \mid q_l^-(\alpha)k_l^-(\alpha) \mid\mid \frac{P(x+(t+h)k_l^-(\alpha)) - P(x+tk_l^-(\alpha))}{hk_l^-(\alpha)} - P'(x+tk_l^-(\alpha)) \mid$$
$$= \lim_{h\to 0} sup_{\alpha\in[0,1]} \mid q_l^-(\alpha)k_l^-(\alpha) \mid\mid P'(x+tk_l^-(\alpha)+\zeta(h,\alpha)) - P'(x+tk_l^-(\alpha)) \mid$$
$$= \lim_{h\to 0} sup_{\alpha\in[0,1]} \mid q_l^-(\alpha)k_l^-(\alpha) \mid\mid P''(x+tk_l^-(\alpha)+\eta(h,\alpha))\zeta(h,\alpha) \mid$$
$$\leq \lim_{h\to 0} q_0 k_0^2 Q(x,t)h = 0,$$

where $\zeta(h,\alpha)$ is a point on the line segment between 0, $hk_l^-(\alpha)$ and $\eta(h,\alpha)$ is a point on the line segment between 0 and $\zeta(h,\alpha)$. An also, $\mid k_l^-(\alpha) \mid \leq max\{k_l^-(0), k_r^-(0)\} = k_0$ and $\mid q_l^-(\alpha) \mid \leq max\{q_l^-(0), q_r^-(0)\} = q_0$, for all $\alpha \in [0,1]$. Then $x+(t+h)k_0 \leq x+tk_l^-(\alpha)+\eta(h,\alpha)) \leq x+(t+h)k_0$, $\mid P''(x+tk_l^-(\alpha)+\eta(h,\alpha)) \mid \leq Q(x,t)$.

This concludes that

$$[U_t(x,t)]^\alpha = [q_l^-(\alpha)k_l^-(\alpha)P'(x+tk_l^-(\alpha)), q_r^-(\alpha)k_r^-(\alpha)P'(x+tk_r^-(\alpha))]$$

and

$$[U_t(x,t)]_\alpha = [q_l^+(\alpha)k_l^+(\alpha)P'(x+tk_l^+(\alpha)), q_r^+(\alpha)k_r^+(\alpha)P'(x+tk_r^+(\alpha))].$$

Case (ii): As per Zadeh's extension principle, α level sets U is defined as:

$$[U]^\alpha = [q_l^-(\alpha)P(x+tk_l^+(\alpha)), q_r^-(\alpha)P(x+tk_r^+(\alpha))]$$
$$[U]_\alpha = [q_l^+(\alpha)P(x+tk_l^-(\alpha)), q_r^+(\alpha)P(x+tk_r^-(\alpha))]$$

Also since g' is non positive function, then:

$$[q_l^-(\alpha)k_l^+(\alpha)P'(x+tk_l^+(\alpha)), q_r^-(\alpha)k_r^+(\alpha)P'(x+tk_r^+(\alpha))]$$

and

$$[q_l^+(\alpha)k_l^-(\alpha)P'(x+tk_l^-(\alpha)), q_r^+(\alpha)k_r^-(\alpha)P'(x+tk_r^-(\alpha))]$$

forms a intuitionistic fuzzy number, by Theorem 1, H-differences $U(x,t+h) \ominus U(x,t)$ and $U(x,t) \ominus U(x,t-h)$ exist. In a similar way, since $[q_l^+(\alpha)P'(x + tk_l^-(\alpha)), q_r^+(\alpha)P'(x + tk_r^-(\alpha))]$ and $[q_l^-(\alpha)P'(x + tb_l^+(\alpha)), q_r^-(\alpha)P'(x + tk_r^+(\alpha))]$ forms a intuitionistic fuzzy number, by theorem H-differences $U(x,t) \ominus U(x+h,t)$ and $U(x-h,t) \ominus U(x,t)$ exist.

Case (iii): As per to Zadeh's extension principle, α level sets U is defined as:

$$[U]^\alpha = [q_l^+(\alpha)P(x+tk_l^-(\alpha)), q_r^+(\alpha)P(x+tk_r^-(\alpha))]$$
$$[U]_\alpha = [q_l^-(\alpha)P(x+tb_l^+(\alpha)), q_r^-(\alpha)P(x+tk_r^+(\alpha))]$$

Also since P' is non negative function, then:

$$[q_l^+(\alpha)k_l^-(\alpha)P'(x+tk_l^-(\alpha)), q_r^+(\alpha)k_r^-(\alpha)P'(x+tk_r^-(\alpha))]$$

and

$$[q_l^-(\alpha)k_l^+(\alpha)P'(x+tk_l^+(\alpha)),q_r^-(\alpha)k_r^+(\alpha)P'(x+tk_r^+(\alpha))]$$

forms a intuitionistic fuzzy number, by Theorem 1, H-differences $U(x,t+h)\ominus U(x,t)$ and $U(x,t)\ominus U(x,t-h)$ exist. In a similar way, since $[q_l^-(\alpha)P'(x+tk_l^+(\alpha)),q_r^-(\alpha)P'(x+tk_r^+(\alpha))]$ and $[q_l^+(\alpha)P'(x+tk_l^-(\alpha)),q_r^+(\alpha)P'(x+tk_r^-(\alpha))]$ forms a intuitionistic fuzzy number, by theorem H-differences $U(x+h,t)\ominus U(x,t)$ and $U(x,t)\ominus U(x-h,t)$ exist.

Case (iv): As per to Zadeh's extension principle, α level sets U is defined as:

$$[U]^\alpha = [q_l^-(\alpha)P(x+tk_l^-(\alpha)),q_r^-(\alpha)P(x+tk_r^-(\alpha))]$$
$$[U]_\alpha = [q_l^+(\alpha)P(x+tk_l^+(\alpha)),q_r^+(\alpha)P(x+tk_r^+(\alpha))]$$

On the other hand, since P' is non-positive function, then $[q_l^-(\alpha)k_l^-(\alpha)P'(x+tk_l^-(\alpha)),q_r^-(\alpha)k_r^-(\alpha)P'(x+tk_r^-(\alpha))]$ and $[q_l^+(\alpha)k_l^+(\alpha)P'(x+tk_l^+(\alpha)), q_r^+(\alpha)k_r^+(\alpha)P'(x+tk_r^+(\alpha))]$ forms a fuzzy number. By Theorem 1, H-differences $U(x,t+h)\ominus U(x,t)$ and $U(x,t)\ominus U(x,t-h)$ exist. In similar way, since $[q_l^+(\alpha)P'(x+tb_l^+(\alpha)),q_r^-(\alpha)P'(x+tk_r^-(\alpha))]$ and $[q_l^-(\alpha)P'(x+tk_l^-(\alpha)),q_r^+(\alpha)P'(x+tk_r^+(\alpha))]$ forms a fuzzy number, H-differences $U(x,t)\ominus U(x+h,t)$ and $U(x-h,t)\ominus U(x,t)$ exist.

6 Conclusion

In this paper, we have studied the existence of a solution of transport equation in homogeneous and non-homogeneous cases with intuitionistic fuzzy data and we have present the solution in case when speed parameter is a intuitionistic fuzzy number.

References

1. Atanassov, K.T.: Intuitionistic Fuzzy Logics. Studies in Fuzziness and Soft Computing, vol. 351. Springer, Cham (2017). https://doi.org/10.1007/978-3-319-48953-7
2. Atanassov, K.T.: Intuitionistic Fuzzy Sets. VII ITKR Session, Sofia, 20–23 June 1983
3. Atanassov, K.: New operations defined over the intuitionistic fuzzy sets. Fuzzy Sets Syst. **61**(2), 137–142 (1994)
4. Bede, B., Stefanini, L.: Generalized differentiability of fuzzy-valued functions. Fuzzy Sets Syst. **230**, 119–141 (2013)
5. Long, H.V., Son, N.T.K., Tam, H.T.T.: Global existence of solutions to fuzzy partial hyperbolic functional differential equations with generalized Hukuhara derivatives. J. Intell. Fuzzy Syst. **29**(2), 939–954 (2015)
6. Melliani, S., Chadli, L.S.: Intuitionistic fuzzy differential equation. Notes Intuit. Fuzzy Sets **6**, 37–41 (2000)
7. Melliani, S., Elomari, M., Chadli, L.S., Ettoussi, R.: Intuitionistic fuzzy metric space. Notes Intuit. Fuzzy Sets **21**(1), 43–53 (2015)
8. Melliani, S., Elomari, M., Chadli, L.S.: Intuitioinistic fuzzy Hilbert space. In: Melliani, S., Castillo, O. (eds.) Recent Advances in Intuitionistic Fuzzy Logic Systems. SFSC, vol. 372, pp. 143–152. Springer, Cham (2019). https://doi.org/10.1007/978-3-030-02155-9_12
9. Melliani, S., Elomari, M., Chadli, L.S., Ettoussi, R.: Extension of Hukuhara difference in intuitionistic fuzzy set theory. Notes Intuit. Fuzzy Sets **21**(4), 34–47 (2015)
10. Zadeh, L.A.: Fuzzy sets. Inf. Control **8**, 338–353 (1965)

Author Index

S. Melliani and O. Castillo (Eds.): ICPAMS 2021, LNNS 476, pp. 485–486, 2023.
https://doi.org/10.1007/978-3-031-12416-7